D0164391

EXERCISE
PHYSIOLOGY

EXERCISE PHYSIOLOGY:
Human Bioenergetics and Its Applications

George A. Brooks

University of California, Berkeley

Thomas D. Fahey

DeAnza College

Macmillan Publishing Company
New York
Coller Macmillian Publishers
London

Copyright © 1985 Macmillan Publishing Co., a division of Macmillan, Inc.
Copyright © 1984, by John Wiley & Sons, Inc.

All rights reserved. No part of this book may be reproduced or transmitted
in any form or by any means, electronic or mechanical, including photo-
copying, recording, or any information storage and retrieval systems, without
permission in writing from the Publisher.

Macmillan Publishing Company
866 Third Avenue, New York, New York 10022
Collier Macmillan Canada, Inc.

Library of Congress Cataloging in Publication Data:

Brooks, George A.
   Exercise physiology

   Includes bibliographies and index.
   1. Exercise—Physiological aspects.   I. Fahey,
Thomas D.   II. Title.   [DNLM: 1. Exertion.   2. Metabolism.
3. Physiology. WE 105 B873e]
QP301.B885   1984   612'.76   83-14582
ISBN   0-02-315130-7

Printed in the United States of America

Printing   8 9 10 Year 0 1 2 3 4 5

To Daniel, Tommy, Timmy, and Mikey

PREFACE

Exercise physiology is the study of body function during exercise. We believe that no greater interest can be generated for the study of physiology than that involved in analyzing human performance during motor activities, particularly when the student's object of study is himself or herself. Because so many of us are engaged in life-styles that require high-energy outputs, it is important that we understand our physiological capacities for exercise, which largely determine our success and enjoyment in many areas. Studying exercise physiology not only allows students to understand the mechanisms governing their own performance but also enables them to be aware of performance in athletics, work, the performing arts, recreation, and preventive and rehabilitative medicine. Thus this book links the study of physiology to the reader's own self-interest.

Physical performance determines the outcome of so many areas of human activity. Our ability to fulfill a busy work schedule and to retain an energy reserve is essential for enjoyment in our lives. Physical capability no longer plays a determinant role in survival, but it contributes largely to the development and maintenance of physique and self-image. An individual's physical performance is the result of such factors as genetic endowment, state of maturation, nutrition, training, and environment—areas covered in this book.

In our contemporary society, degenerative diseases have replaced infectious diseases as major causes of debilitation and death—most significantly coronary heart disease (CHD). The causes of CHD are complex but lack of exercise and lack of proper active recreation are involved. Therefore, not only is proper physical exercise essential for physical development in the formative years, but it is also necessary for maintenance of physical capacity in younger and older adults. Today, exercise is used to diagnose coronary heart disease, to retard its development, and to treat it. Exercise conditioning and other forms of physical therapy are used to assist and improve recuperation from injury to, and surgery on, muscles and joints.

Many people have helped us write this book, including our teachers W.D. McArdle, J.A. Faulkner, R.E. Beyer, K.J. Hittelman, F.M. Henry, G.L. Rarick, and J.H. Wilmore. We have been inspired by many contemporary and past researchers, who are referenced at the end of our chapters. We also thank the authors of other texts, which we as students have used, including A.C. Guyton, P.-O. Åstrand, V.R. Edgerton, E. Fox, M. Kleiber, R.W. McGilvery, A.J. Vander, J.S. Sherman, and D.S. Luciano. We also thank reviewers of our manuscript, including J.B. Gale, R. Pate, K. Knight, K. Cureton, and R. Schroeder.

As authors we would appreciate hearing your views and opinions of our text. We encourage you to write us with your criticisms and suggestions, which will be seriously considered for incorporation into subsequent editions of this book.

George A. Brooks
Thomas D. Fahey

CONTENTS

6

CELLULAR OXIDATION OF PYRUVATE 97

7

LIPID METABOLISM 117

8

METABOLISM OF PROTEINS AND AMINO ACIDS   141

9

NEURAL–ENDOCRINE CONTROL OF METABOLISM   163

10

METABOLIC RESPONSE TO EXERCISE   189

13

VENTILATION AS A LIMITING FACTOR IN AEROBIC PERFORMANCE   271

14

THE HEART AND ITS CONTROL   279

15

THE CIRCULATION AND ITS CONTROL   295

16

CARDIOVASCULAR DYNAMICS DURING EXERCISE   313

17

EXCITABLE TISSUES   343

21

CONDITIONING FOR RHYTHMICAL ATHLETIC EVENTS 427

22

EXERCISE IN THE HEAT AND COLD 443

23

EXERCISE IN HIGH AND LOW PRESSURE ENVIRONMENTS   471

24

CORONARY HEART DISEASE   503

25

OBESITY AND BODY COMPOSITION   525

26

EXERCISE AND DISEASE   555

27

NUTRITION AND ATHLETIC PERFORMANCE   593

28

ERGOGENIC AIDS   611

29

SEX DIFFERENCES IN PHYSICAL PERFORMANCE   637

30

GROWTH AND DEVELOPMENT   661

31

AGING   683

32

FATIGUE DURING MUSCULAR EXERCISE   701

1

INTRODUCTION
The Limits of Human Performance

What are the limits of physical performance? Is it possible to run the 100-meter dash in 9 seconds (Figure 1-1)? Can the discus be hurled 300 feet (Figure 1-2)? Can the mile be run in less than 3 minutes (Figure 1-3), or is it possible to swim 100 m in less than 40 sec (Figure 1-4)? Can the routines of Olympic gymnasts become more complex and better executed than they are now (Figure 1-5)? Can leg muscles of paraplegics be used for locomotion (Figure 1-6)? Are these goals unrealistic, or do they underestimate the limits of human performance? These are important questions of interest not only to physical educators and athletics coaches, but also to a variety of others.

The answers are predictable. The shot-putter who achieves the distance of 80 ft will have to generate the necessary force to propel the implement that distance. An athlete with insufficient muscle mass and an inadequate metabolic capability will simply be incapable of this feat. Similarly, individuals with poor cardiovascular capacity, muscular strength, and coordination will lead very restricted lives. The study of exercise physiology can lead to a better understanding of the physical capacity and limitations of the human body, as well as its underlying mechanisms.

The limits of human performance have been of interest to a variety of scientists, including zoologists, physical anthropologists, human factors engineers, and exercise physiologists. Zoologists are interested in how the physical capabilities of various species relate to their survival and ecological impact.

FIGURE 1-1
Carl Lewis hits the tape in the 100-m dash, the premier sprint event in international competition.

(Bower, U.P.I.)

Physical anthropologists are concerned with the relationships among the physical capacities of early humans and their survival and development. Human factors engineers attempt to determine the physical demands in the work place to maximize efficiency and safety. Exercise physiologists are concerned with a wide range of questions related to the performance of single or repeated exercise bouts.

The researchers in each of these disciplines study the various aspects of exercise physiology by systematic analysis and manipulation of the various factors affecting performance. Coaches and physical educators are also exercise scientists. Although they seldom have expensive and sophisticated laboratory equipment to assist them, coaches and physical educators operate in the ultimate exercise physiology laboratory: the playing field. Like other scientists studying human performance, the coach's approach to a problem must be based on current knowledge of exercise physiology and the mechanisms dictating the probable physical responses to exercise.

FIGURE 1-2
Al Oerter has won four Olympic gold medals in the discus. Throwing the discus over 200 ft requires enormous power, coordination, and technique.

(Wayne Glusker.)

THE SCIENTIFIC BASIS OF EXERCISE PHYSIOLOGY: SOME DEFINITIONS

The scientific method involves the systematic solution of problems. The scientific approach to solving a problem involves the presentation of ideas (hypotheses), the collection of information (data) relevant to those hypotheses, and the acceptance or rejection of the hypotheses based on evaluation of the data (conclusions). Although the scientific method appears to be straightforward, the process of deriving appropriate hypotheses and systematically testing them can be complex. Nevertheless, it is evident that in our increasingly technological society, those who systematically analyze their problems and take appropriate steps to solve them are most likely to acquire satisfactory answers to their questions. Individuals who make the best use of the scientific method will be the most successful scientists, educators, coaches, and health professionals.

Physiology is a branch of biological science concerned with the function of organisms and their parts. The study of physiology is dependent on and intertwined with other disciplines such as anatomy, biochemistry, and biophysics. The reason for this interdependence is that the function of the human body must follow the natural laws of structure and function, which fall within the domain of these other disciplines.

FIGURE 1-3
Sebastian Coe sets a world's record in the mile run. Athletic feats such as this possibly represent the supreme example of speed, power, and endurance in human endeavor.

(J. Hollander, U.P.I.)

Exercise physiology is a branch of physiology that deals with the function of the body during exercise. As will be seen, there are definite physiological responses to exercise that are dependent on its intensity, duration, and frequency, the environmental circumstances, and the physiological status of the individual.

The question as to when a scientist is a physiologist or an exercise physiologist is one of more rhetorical than real significance. Let us say that a physiologist, being concerned chiefly with studies of biological function, may use exercise as one means to perturb an organism or system and observe its responses. An exercise physiologist is concerned with the process of exercise as well as its physiological implications. In this case, physiology is used to understand exercise. The boundaries of these classifications are often soft and ill defined. In actuality, from the standpoint of science, the categorization makes no difference so long as the approach to problem solving is the same.

SCIENCE IN PHYSICAL EDUCATION, ATHLETICS, AND ALLIED HEALTH PROFESSIONS

It is possible to ask the question: Are physical educators and coaches scientists? This depends on their approach to problem solving. Many teachers and coaches systematically evaluate their selection and training of individuals, so they can be considered scientists. The scientist-coach introduces an exercise stimulus and systematically gauges the response. The nonscientist-coach, in contrast, administers the training program according to whim and the dictates of tradition. This method relies on either mimicking the techniques of successful athletes or conforming tenaciously to traditional practices. Although nonscientist-coaches are sometimes successful, they are rarely innovative and seldom help an individual perform optimally. Systematic innovation is absolutely essential in any field in order for progress to occur.

FIGURE 1-4
Peter Rocca reaches for the wall and an Olympic medal in the 100-m backstroke. In internationally recognized swim competition there is no pure sprint event, but athletes rely on muscular endurance, aerobic endurance, and technique.

(Courtesy of Chris Georges and Swimming World Magazine.)

FIGURE 1-5
Nadia Cominec, the preeminent woman gymnast of modern times, completes a floor exercise routine. Coordination, strength, and balance are premier attributes of athletes such as Nadia.

(Wayne Glusker.)

Exercise and sport lend themselves to scientific analysis because the measures of success are easily quantified. If a football coach attempts to gain yardage with a particular play, the result is instantaneous; the play either succeeds or fails. Likewise, the distance a discus is thrown can be measured, the duration of a 100-m run can be timed, the number of baskets sunk in a game can be counted, and the distance an amputee can walk can be measured. The scientist-coach observes and quantifies the factors affecting these performances and systematically varies them to achieve success.

THE RELEVANCE OF PHYSIOLOGY FOR PHYSICAL EDUCATION, ATHLETICS, AND THE ALLIED HEALTH PROFESSIONS

Understanding the body during exercise is a primary responsibility of each physical educator and health science professional. Maximizing performance requires a knowledge of physiological processes. As competition becomes more intense, continued improvement will only be attained by careful consideration of the most efficient means of attaining biological adaptation.

Physical educators are increasingly working with a more heterogeneous population. New populations are interested in assuming an active life-style and even participating in competitive sports. Older adults are flocking to masters' sporting events, heart attack patients are resuming physical activity earlier than ever, and people with diseases such as asthma and diabetes are using exercise to improve their disabilities. These people need guidance by trained professionals who understand the responses to exercise in a variety of circumstances. Knowledge of exercise responses in these diverse populations requires a thorough understanding of both normal and abnormal physiology.

Degenerative diseases such as coronary heart disease and osteoarthritis have replaced infectious diseases as primary health problems. Many of these degenerative disorders are amenable to change through modification of life-style, such as ensuring regular exercise. The importance of exercise in a program of preventive medicine has reinforced the role of the physical educator as part of the interdisciplinary team concerned with health care and maintenance. The physical educator must speak the "language of science" to become a true professional and interact with the other professionals of the team.

THE BODY AS A MACHINE

In many ways the exercising human can be compared to a machine, such as an automobile. The machine converts one form of energy to another in the performance of work. Likewise, the athlete converts chemical energy to mechanical energy in the process of running, throwing, and jumping. The athlete, like the machine, can increase exercise intensity by increasing the rate at which energy is converted from one form to another. The athlete goes faster by increasing the metabolic rate and speeding the breakdown of fuels, which provides more energy for muscular work.

At their roots, motor activities are based on bioenergetics, which control and limit the performance of physical activities. In this sense, the body is a machine. The reasons for exercising can be quite varied, but when the exercise starts, the mechanisms of performance are determined by physical and chemical factors. Understanding how to select and prepare the biological apparatus for exercise, and understanding how the exercise affects the machine over both the short and the long term are important questions in exercise physiology and other fields.

This book emphasizes understanding the individual during exercise from the standpoint of the energetics that support the various activities. The discussion begins with energy and its importance to living organisms, emphasizing how we acquire, conserve, store, and release energy for everyday life. The functions of various physiological systems (ventilatory, circulatory, endocrine, etc.) are examined from the perspective of their function in support of physical performance and their place in the process of energy conversion. The immediate as well as the long-term effects of exercise are integrated into the text. Some background information on general principles of physiological response and the field of exercise physiology is offered next.

THE RATE-LIMITING FACTOR

In a complex biological machine such as an exercising human, many physiological processes are occurring simultaneously. For example, when a person runs a mile, the heart's contractility and beating frequency increase, hormones are mobilized, the metabolic rate is raised, and body temperature is elevated.

FIGURE 1-6
Nan Davis suffered three transections of her spinal cord in an automobile accident. She now pedals a cycle around the Wright State University campus by means of computer-controlled contractions of her paralyzed leg muscles.

(Courtesy of R. Glaser and J. Petrofsky, Wright State University.)

Despite the vast number of events occurring simultaneously, only a few usually control and limit the overall performance of the activity. Many scientists approach the understanding of physiological systems by studying the rate-limiting processes. Let us imagine an assembly line that manufactures a commodity such as an automobile. Although there are many steps in the manufacturing process, let us assume that one step, installing the engine, is the slowest. If we want to increase production, it will do us little good to increase the speed of the other steps, such as assembling the chassis. Rather, we should focus our attention on speeding up the process of installing the engine. For instance, we might hire some extra people to do that task, or we might remove some impediment to that process so that the workers can act more rapidly. As will be seen, the body is controlled by and adjusts to exercise in a similar fashion.

In athletics, successful coaches are those who can identify the rate-limiting factor, sometimes called a weakness, and improve the individual's capacity to perform that process. For instance, let us assume we are coaching a novice wrestler who has been a successful competitive weight lifter. It makes no sense to emphasize strength training. Rather, we should emphasize technique development and other aspects of fitness, such as endurance. We would strive only to maintain strength, while concentrating on the performance-limiting factors. Similarly, we would be ill advised to have a 400-m runner do 100 miles a week of road running, as this type of fitness is of minimal use to the athlete, and may even interfere with the required type of fitness.

MAXIMAL OXYGEN CONSUMPTION

The ability to supply energy for activities lasting more than 90 sec depends on the consumption and utilization of oxygen. Because most physical activities in daily life, in athletics, and in physical medicine require more than 90 sec, the consumption of O_2 provides the energetic basis of our existence. The consumption of O_2 (abbreviated $\dot{V}_{O_2}$) increases from rest, to easy, to difficult, and maximal work loads (Figure 1-7). The maximum rate at which an individual can consume oxygen ($\dot{V}_{O_2max}$) is an important determinant of the peak power output and of the maximal sustained power output or physical work capacity of which an individual is capable. Indeed, one definition of physical fitness is $\dot{V}_{O_2max}$.

As will be shown in later chapters, the capacity for $\dot{V}_{O_2max}$ depends on the capacity of the cardiovascular system. This realization that physical work capacity, $\dot{V}_{O_2max}$, and cardiovascular fitness are interrelated has resulted in a convergence of physical education (athletic performance) and medical (clinical) definitions of fitness. From the physical education–athletics perspective, cardiovascular function determines $\dot{V}_{O_2max}$, which in turn determines physical work capacity, or fitness. From the medicoclinical perspective, fitness involves free-

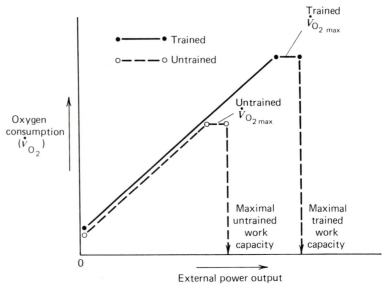

FIGURE 1-7

Relationship between oxygen consumption ($\dot{V}_{O_2}$) and external work rate (power output). In response to increments in power output, both trained and untrained individuals respond with an increase in $\dot{V}_{O_2}$. The greater ability of trained individuals to sustain a high power output is largely due to a greater maximal O_2 consumption ($\dot{V}_{O_2max}$).

dom from disease. Because cardiovascular disease represents the greatest threat to health of individuals in contemporary Western society, medical fitness is largely cardiovascular fitness. One of the major ways to determine cardiovascular fitness is to measure $\dot{V}_{O_2max}$. Therefore, $\dot{V}_{O_2}$ is not only an important parameter of metabolism, it is also a good measure of fitness for life in contemporary society.

FACTORS AFFECTING THE PERFORMANCE OF THE BIOLOGICAL MACHINE

Although the body can be compared to a machine, it would be simplistic, and indeed dehumanizing, to leave it at that. Unlike a machine, the body can adapt to physical stresses and improve its function. Likewise, in the absence of stress, functional capacity deteriorates. In addition, whereas particular types of machines have functions that are set at the time of manufacture, the performances of human machines are quite variable. Performance capabilities change continuously throughout life according to several time-honored principles that account for much of the observed individual differences. These principles are examined next.

Stress and Response

Physiological systems respond to appropriate stimuli. Sometimes the stimulus is called ''stress,'' and the response is called ''strain.'' Repeated stresses on physical systems frequently lead to adaptations resulting in an increase in functional capacity. Enlargement (hypertrophy) occurs in skeletal muscle as a result of the stress of weight training. However, not all stresses are appropriate to enhance the functioning of physiological systems. For instance, although cigarette smoking is a stress, it is not going to improve lung function. Smoking is an example of an inappropriate stimulus.

Physiologically, the purpose of any training session is to stress the body so that the response results in adaptation. Physical training is beneficial only as long as it forces the body to adapt to the stress of physical effort. If the stress is not sufficient to overload the body, then no adaptation occurs. If a stress is so great that it cannot be tolerated, then injury or overtraining result. The greatest improvements in performance will occur when the appropriate exercise stresses are introduced into the individual's training program.

Dr. Hans Selye has done much to make us aware of the phenomenon of stress–response–adaptation. He called this process the general adaptation syndrome (GAS). Selye described three stages involved in response to a stressor: alarm reaction, resistance development, and exhaustion. Each of these stages should be extremely familiar to every physical educator, athlete, coach, physical therapist, and other health professional who uses exercise to improve physical capacity.

Alarm reaction   The alarm reaction, the initial response to the stressor, involves the mobilization of systems and processes within the organism. During exercise, for example, the stress of running is combated by the strain of increased oxygen transport through an augmentation of cardiac output and a redistribution of blood flow to active muscle. The body has a limited capacity to adjust to various stressors; thus it must adapt its capacity so that the stressor is less of a threat to its homeostasis in the future.

Resistance development   The body improves its capacity or builds its reserves during the resistance stage of the GAS. This stage represents the goal of physical conditioning. Unfortunately, the attainment of optimal physiological resistance (a term known more commonly in athletics as physical fitness) does not occur in response to any random stressor. During physical training, for example, if the stress is below a critical threshold, then no training effect occurs. At the other extreme, if the stimulus cannot be tolerated, injury results.

The effectiveness of a stressor in creating an adaptive response is specific to an individual and relative to any given point in time and place. For example, running a 10-min mile may be exhausting to a sedentary 40-year-old man but would cause essentially no adapative response in a world-class distance runner.

Likewise, a training run that could easily be tolerated one day may be completely inappropriate following a prolonged illness. Environment can also introduce intraindividual variability in performance. An athlete will typically experience decreased performance in extreme heat and cold, at high altitude, and in polluted air.

Exhaustion   When the stress becomes intolerable, the organism enters the third stage of GAS, exhaustion (or distress). The stresses that result in exhaustion can be either acute or chronic. Examples of acute exhaustion include fractures, sprains, and strains. Chronic exhaustion (overtraining) is more subtle and includes stress fractures, emotional problems, and a variety of soft-tissue injuries.

The Overload Principle

Application of the appropriate stress is sometimes referred to as overloading the system. The "principle of overload" states that habitually overloading a system will cause it to respond and adapt. Overload is a positive stressor that can be quantified according to load, repetition, rest, and frequency.

Load refers to the intensity of the exercise stressor. In strength training, the load refers to the amount of resistance, whereas in running or swimming, load refers to speed. In general, the greater the load, the greater the fatigue and recovery time required.

Repetition refers to the number of times a load is administered. More favorable adaptation tends to occur (up to a point) when the load is administered more than once. In general, there is little agreement on the ideal number of repetitions in a given sport. The empirical maxims of sports training are in a constant state of flux as athletes become successful using overload combinations different from the norm. For example, in middle-distance running and swimming, interval training workouts have become extremely demanding as a result of the success of athletes who employed repetitions far in excess of those practiced only a few years ago.

Rest refers to the time interval between repetitions. Rest is vitally important for obtaining an adaptation and should be applied according to the nature of the desired physiological outcome. For example, a weight lifter who desires maximal strength would be more concerned with load and less concerned with rest interval than would a mountain climber or endurance runner. A short rest interval would impair the weight lifter's strength gain because inadequate recovery would make it impossible to exert maximal tension. Mountain climbers or runners, by contrast, are more concerned with muscular endurance than peak strength, so they would utilize short rest intervals to maximize this fitness characteristic. Rest also refers to the interval between training sessions. Because some responses to stress are prolonged, adaptation to stress requires adequate rest and recovery. Resting is a necessary part of training.

Frequency refers to the number of training sessions per week. In some sports, such as distance running, there has been a tendency toward more frequent training sessions. Unfortunately, this often leads to increases in overuse injuries due to overtraining. Although more severe training regimens have resulted in improved performance in many sports, these workouts must be tempered with proper recovery periods or injury will result.

Specificity

It has repeatedly been observed that stressing a particular system or body part does little to affect other systems or body parts. For example, doing repeated biceps curls with the right arm may cause the right biceps to hypertrophy, but the right triceps or left biceps will be little affected. Any training program should reflect the desired adaptation. The closer the training routine is to the requirements of competition, the better will be the outcome.

Reversibility

This is in a way a restatement of the principle of overload and emphasizes that whereas training may enhance performance, inactivity will lead to a performance decrement. For example, a college runner who built a robust circulatory system while in school should expect little or no residual capacity at 40 yr of age following 20 yr of subsequent inactivity.

Individuality

It is wise to note that we are all individuals, and that whereas physiological responses to particular stimuli are largely predictable, the precise responses and adaptations to those stimuli will vary among individuals. Therefore, the same training regimen may not equally benefit all those who participate.

THE DEVELOPMENT OF THE FIELD OF EXERCISE PHYSIOLOGY

Interest in exercise physiology can be traced to the beginnings of modern biology. Researchers such as Lavoisier, de la Place, Crawford, von Meyer, and Benedict provided basic observations that laid the groundwork for subsequent studies. Several late nineteenth- and early twentieth-century European physicians and biologists were outdoors enthusiasts who recorded their physiological responses to exercise and high altitude.

In England during the 1920s, A.V. Hill and associates were interested in studying the phenomenon of muscle contraction. Although a great deal of their work involved detailed studies on isolated muscles, numerous studies were performed on people during exercise. In many regards, A.V. Hill may be considered the "father of exercise physiology."

In Denmark during the early part of the twentieth century, August Krogh did extensive work studying the physiological responses to exercise and environment. Krogh and his students, such as Asmussen, advocated the importance of standardization and proper research design into exercise physiology experiments, which helped advance the study of exercise physiology from descriptive observations to experimental research. Today, in Copenhagen, the Krogh Institute serves as one of the important sites for studies on exercise physiology.

In the United States, during the early twentieth century, the Carnegie Foundation commissioned important studies on the energetics of muscular exercise. During the 1920s and through the mid-1940s, D.B. Dill and associates at the Harvard Fatigue Laboratory performed a great deal of work on physiological responses to exercise and environment.

Many other researchers, too numerous to mention here, contributed the groundwork to the study of exercise physiology. Today, knowledge in this field is obtained from a wide range of different types of studies. This book elaborates on these studies, but some of them deserve particular mention. The testing of physical work capacities of athletes, laborers, and a variety of people to determine their metabolic responses to exercise remains an important area of interest. Questions relating to the caloric cost of exercise, the efficiency of exercise, and the fuels used to support the exercise are addressed. In addition to traditional methods of respiratory determination, more recent techniques utilizing muscle sampling (biopsy), light and electron microscopy, enzymology, and radioactive tracers have contributed to our understanding of the metabolic responses to exercise.

Along these lines, research has been carried out and is under way to determine the optimal training techniques to use for particular activities. The training regimens to improve such qualities as muscular strength and running endurance are vastly different.

When exercise tests are performed in a clinical setting, the term "stress test" is often used. Under controlled exercise conditions cardiac and blood constituent responses are useful in determining the presence of underlying disease. Following an exercise stress test, an exercise prescription can be written to improve functional capacity.

Recent advances by J.S. Petrofsky and associates at Wright State University in Dayton, Ohio, in computerized control of paralyzed muscles now allow paraplegic and quadriplegic individuals to use their own muscle power to provide locomotion (Figures 1-6 and 1-8). As with nonparalyzed muscles, repeated overload of paralyzed muscles by computer-controlled electrical stimulation results in tremendous improvements in strength and endurance. Consequently, some muscles paralyzed for years can be trained and restored to normal strength.

Considerable interest has evolved in what might be termed "exercise biochemistry." Through the use of some new, but mostly established clinical and

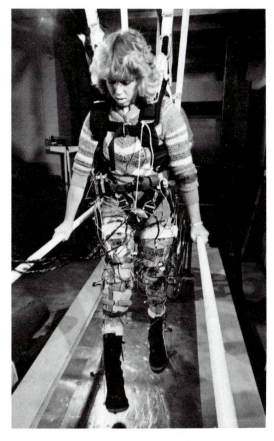

FIGURE 1-8
Nan Davis, a paraplegic, prepares to walk in the Wright State University National Center for Rehabilitation Engineering with the assistance of computer control of her muscles.

(Courtesy of J. Petrofsky.)

biochemical techniques, a variety of investigators have produced valuable and detailed information on the effects of exercise and training on muscle and on its chemistry and microanatomy.

The effects of environment on physical performance have long been a concern in exercise physiology. Environmental studies may have reached their peak of interest during preparations for the high-altitude Olympics held in Mexico City in 1968. At present, the physiological stress of work in the heat remains a particularly active area of research.

The effect of exercise on bones and connective tissue is another major area of interest in exercise physiology. How muscles increase in size and strength in response to heavy resistance training is a major question under study.

Summary

Although the study of exercise physiology is in its infancy compared to other sciences such as chemistry and physics, the area is bustling with activity. This area of research has a great deal of appeal because it concerns the limits of human potential. The results of these studies affect us all, whether we jog three times a week to improve our health, or whether we are concerned with the health of others.

Selected Readings

Asmussen, E. Muscle metabolism during exercise in man: A historical survey. In: Pernow, B. and B. Saltin (eds.), Muscle Metabolism During Exercise. New York: Plenum Press, 1971, p. 1–12.

Brooks, G.A. (ed.). Perspectives on the Academic Discipline of Physical Education. Champaign, Ill.: Human Kinetics, 1981.

Fahey, T. Getting into Olympic Form. New York: Butterick Publishing, 1980.

Fenn, W.O. History of the American Physiological Society: The Third Quarter Century 1937–1962. Washington, D.C.: The American Physiological Society, 1963.

Glaser, R.M., J.S. Petrofsky, J.A. Gruner, and B.A. Green. Isometric strength and endurance of electrically stimulated leg muscles of quadriplegics. Physiologist. 25: 253, 1982.

Handler, P. Basic research in the United States. Science. 204: 474–479, 1979.

Henry, F.M. Physical education: An academic discipline. J. Health Phys. Ed. Recreation. 35:32–33, 1964.

Hickson, R.C. Interference of strength development by simultaneously training for strength and endurance. Eur. J. Appl. Physiol. 45: 255–263, 1980.

Hill, A.V, C.N.H. Long, and H. Lupton. Muscular exercise, lactic acid, and the supply and utilization of oxygen (Pt. I–III). Proceedings of the Royal Society of London, Series B, 96:438–475, 1924.

Hill, A.V., C.N.H. Long, and H. Lupton. Muscular exercise, lactic acid, and the supply and utilization of oxygen (Pt. IV–VI). Proceedings of the Royal Society of London, Series B, 97: 84–138, 1924.

Hill, A.V., C.N.H. Long, and H. Lupton. Muscular exercise, lactic acid, and the supply and utilization of oxygen (Pt. VI–VIII). Proceedings of the Royal Society of London, Series B, 97: 155–176, 1924.

Lehninger, A.L. Biochemistry. New York: Worth, 1970.

Margaria, R., H.T. Edwards, and D.B. Dill. Possible mechanisms of contracting and paying oxygen debt and the role of lactic acid in muscular contraction. Am. J. Physiol. 106: 689–715, 1933.

Selye, H. The Stress of Life. New York: McGraw-Hill, 1976.

2

BIOENERGETICS

Physical education activities are energetic events. Understanding what energy is and how the body acquires, converts, stores, and utilizes it is the key to understanding how the body performs in sports, recreational, and occupational activities. The science that studies the principles limiting energetic events has been given two names: thermodynamics and energetics. With some limitations, the same principles that govern energetic events in the physical world (e.g., the explosion of dynamite) also govern events in the biological world, for example a sprinter's first step out of the starting blocks. The science that involves studies of energetic events in the biological world is call *bioenergetics*. In describing energy in the body there are two things to keep in mind. First, energy is not created; rather, it is acquired in one form and converted to another. Second, the conversion process is fairly inefficient, and much of the energy released will be in a nonusable form, heat. In this chapter, we discuss energy and begin to describe three energy systems that power muscular activities. They

The feat of Atlas holding up the world is a mythical representation of the physical, chemical, and energy forces which govern the universe. In the realm of human endeavor, the same laws of nature operate to control and limit muscular performance. (Robert Borneman/Photo Researchers.)

can function under different conditions, at different speeds, and for different durations. Together, these three energy systems determine our capacities for power events (e.g., putting the shot), speed events (e.g., 100-m sprint running), and endurance events (e.g., marathon running).

To beginning students of biology, the study of energetics seems abstract when it is learned that thermodynamics do not tell a student much about specific steps in a process or the time taken to complete a biological process.

However, if one understands that the ability to do work, or exercise, depends on the conversion of one form of energy to another, then the importance of studying energetics becomes apparent.

TERMINOLOGY

Before starting the formal presentation, the following definitions are necessary.

Energy   The capacity to do work.

Work   The product of a given force acting through a given distance.

System   An organized, functional unit. The boundaries of systems vary from situation to situation and depend on the process under consideration. Systems can vary from sub-light-microscopic cellular organelles such as mitochondria, to muscles, to intact individuals.

Surroundings   Exchanges of energy and matter frequently occur between systems and their environments. Once we have defined a system, all else comprises the surroundings.

Universe   Together, the system plus its surroundings make up the universe (Figure 2-1).

The above definition of work is a Newtonian or mechanical definition. In actuality, the performance of chemical and electrical work is more commonly

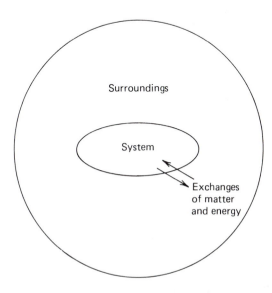

FIGURE 2-1
Universe system, and surroundings. The universe consists of the system(s) plus the surroundings. At any one instant, a reaction taking place in the system results in a decrease in the energy available to do work (free energy) of the system. Consequently, over the long term, an input of energy from the surroundings is required to maintain the system. The energy content of the universe is always constant.

done by cells than is mechanical work. There is, however, a common ground on which to equate these forms of energy, in that it is possible to exchange and to convert energy from one form to another. The physical science dealing with energy exchange is called *thermodynamics*.

The study of thermodynamics came out of the nineteenth century and the desire to predict the work output of machines, such as steam engines. Because heat was the most common form of energy utilized, the name thermodynamics came into use. However, as there are six primary forms of energy (thermal, chemical, mechanical, electrical, radiant, and atomic), a more appropriate term than thermodynamics is energetics. That branch of science that deals with energy exchanges in living things is called bioenergetics. Although a few limitations and unique properties involving heat and temperature are unique to biological systems, in general biological systems and other types of machines follow the same principles of energetics.

HEAT, TEMPERATURE, AND THE BIOLOGICAL APPARATUS

Steam and internal combustion engines are examples of machines that convert chemical energy (coal and gasoline) to heat. That heat energy is then converted to mechanical energy. Generally, the higher the heat the more the power produced. Biological engines differ from mechanical engines with respect to their ability to use heat. Biological engines *cannot* convert heat energy to other forms such as mechanical energy. In biological systems, heat is released as an essential but useless component of reactions in which other forms of work are accomplished. This is a fundamental difference between biological and mechanical engines.

Another characteristic of biological systems is their temperature dependence. Two factors merit consideration here. First, biological systems are very sensitive to small increments in temperature. Above 45°C, and certainly above 60°C, tissue proteins are denatured or degraded. Consequently, although a muscle might theoretically contract faster at 50 than at 35°C, increasing the temperature that much would literally cook the muscle.

Second, the rates of biological or enzymatic reactions are sensitive to temperature. Usually, the effect of temperature on reactions is studied by changing the temperature in multiples of 10°C. The resulting change is then called a "Q_{10} effect." Commonly it is found that increasing the temperature 10°C will double the rate of an enzymatic reaction. In this case the Q_{10} is 2. Figure 2-2 illustrates the Q_{10} effect. In that figure, note the large effect of temperature changes within the normal range of muscle temperature.

In describing these limitations and special cases involving effects of heat

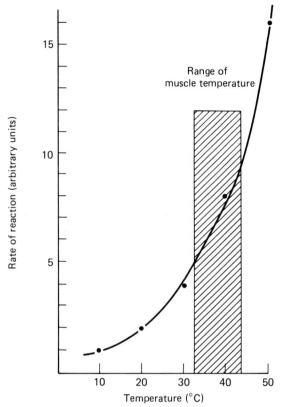

FIGURE 2-2
Illustration of a Q_{10} effect, where each 10°C increase in temperature doubles the rate of reaction. In the physiological range (shaded area) the curve is very steep.

and temperature on biological engines, it should be pointed out that changes in temperature can be advantageous as well as deleterious for an individual. For instance, even though a lizard cannot capture heat to do cellular work, it can sit in the sun to warm itself so that it can take advantage of the Q_{10} effect. Some rodents can make biochemical adaptations on exposure to extreme cold, allowing them to consume extra oxygen and generate heat without shivering. Thus they can go about their daily lives without shivering and shaking, which would endanger their survival. Another example would be the experience of many athletes that preliminary exercise ("warming up") makes for better performance. We shall see (Chapter 21) that preliminary exercise has several beneficial effects, one of which is to warm up muscle to take advantage of the local Q_{10} effect.

In particular instances, excessively high as well as low temperatures could be deleterious. Whereas warming up could increase the speed of particular enzymatic processes in muscle by a Q_{10} effect, temperatures of over 40°C have been observed to decrease the efficiency of oxygen use in muscle. This could negatively affect endurance. It has also been noted that cooling brain temperature by only a few degrees can affect thinking and cause disorientation. This could be especially serious for campers, swimmers, or divers, whose survival in a hostile environment often depends on their behavior.

LAWS OF THERMODYNAMICS

Repeated observations of events in the physical world reveal that two fundamental principles, or laws, always hold. The *first law of thermodynamics* states that energy can be neither created nor destroyed. Whenever there is an exchange of energy or matter between a system and its surroundings, the total energy content in the universe remains constant. The expression, "we get a burst of energy," is not quite correct. In actuality, mature adults consume food (parcels of chemical energy) and degrade the foodstuffs, converting some of the chemical energy to heat and cell work, and releasing the remaining chemical energy unused in body excrements.

Equation 2-1, which describes weight loss or gain in a healthy, mature adult, is not a statement of the first law but does conform to the first law.

$$\text{Energy in (food)} = \text{energy out (work)} + \text{energy out (heat)} \pm \qquad (2\text{-}1)$$
$$\text{energy stored (fat)}$$

Thus the only way obese people can lose fat is to eat less and exercise more. If the energy input (food) is less than that expended as the work and heat resulting from exercise, then the storage form (fat) will be reduced.

In addition to telling us that energy is not created or destroyed, but rather interconverted among forms, the first law tells us to be careful to account for all the energy.

The first law of thermodynamics implies that energy forms can be exchanged, but it does not tell us in what direction the exchanges will occur. The *second law of thermodynamics* tells us that processes always go in the direction of randomness, or disorder. This random, disordered form of energy is termed *entropy*. As the result of the second law, the entropy always increases. In biology the second law tells us that whenever energy is exchanged, the efficiency of exchange will be imperfect, and part of the energy will escape as entropy—usually in the form of heat. In looking at a system that is increasing in organization, such as a child growing, we are often tempted to ask: Is this not ordering of matter from nothing contrary to the second law? The answer is no. To allow a child to grow takes a tremendous input of energy from the environment. For a little growth, a lot of heat is released. In general, when one biological process moves some product toward a higher level of organization, it is driven by at least one "linked" or "coupled" entropic reaction.

Examples abound of concentrated units of energy being dispersed into random, entropic forms, such as the cooling of the sun. This being the case, we are frequently tempted to ask whether this is not completely bad for the biological and physical world. It is indeed bad in the sense that the ultimate destiny of the universe is randomness and disorder, which means that we are ultimately doomed. However, in the interim, the energetically downhill push toward the entropic condition allows us to capture energy in useful forms and to perform biological reactions that are uphill and require energy input. In this sense, for the present, the drive toward entropy is good.

EXERGONIC AND ENDERGONIC REACTIONS

An *exergonic* reaction is one that gives up energy. If heat is the form of energy given off, another term used is *exothermic*. In these terms, *erg* refers to work or energy, and *therm* refers to heat. *Spontaneous reaction* is still another synonymous term.

An example of a spontaneous reaction (A→B) is diagrammed in Figure 2-3. The energy content of the product B is less than that of the reactant A. The difference or change in energy is ΔE_1. In this example, ΔE is negative in sign because the energy content of B is less than that of A.

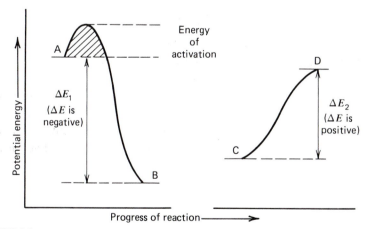

FIGURE 2-3
Examples of spontaneous (A→B) and nonspontaneous (C→D) reactions. Even though A→B is energetically downhill, the impediment imposed by the energy of activation may prohibit A→B unless energy is put in. Note that if C→D is driven by A→B, then the absolute value of ΔE_2 is less than ΔE_1.

Reaction (C→D) in Figure 2-3 is an example of an endergonic, nonspontaneous, uphill reaction. Here, the energy level of the product is greater than that of the reactant. This reaction will not occur unless there is an energy input. In this example ΔE_2 is positive in sign because the energy content of D is greater than that of C. Many important reactions and processes in the physical and biological world are endergonic and therefore require energy inputs. In the biological world, endergonic reactions such as C→D, are linked or coupled to, and driven by, exergonic reactions such as A→B. In the example given, the overall process is A→D. Note that the energy level of D, the final product, is less than that of A, the initial reactant.

Even though reaction A→B is downhill and "spontaneous," it is not likely to happen because there is an energy barrier, called *energy of activation,* that must first be overcome. In other words, even though the reaction A→B is classified as spontaneous, some energy has to be put in to activate the system and "prime the pump." As will be seen, several important biochemical pathways are begun with activating steps. It will also become evident that enzymes are important because they have the effect of lowering the energy of activation.

Although the energy of activation appears to be an impediment to things, the world as we know it depends on some processes having very high energies of activation. Oxides of nitrogen and other automobile exhaust emissions are currently very much in the news. Oxides of nitrogen are formed from nitrogen and oxygen according to the following reactions.

$$N_2 + 2O_2 \longrightarrow 2NO_2 \tag{2-2}$$

$$N + O_2 \longrightarrow NO_2 \tag{2-3}$$

These are spontaneous reactions with high energies of activation. In the automobile combustion chamber, extremes of temperature and pressure serve to activate this reaction and produce the noxious products. It is indeed fortunate that this reaction has a very high energy of activation, or else the atmosphere might catch fire.

Because initial reactants in a metabolic process must be activated to begin the process, and because enzymes can lower the energy of activation, there is enzymatic control over the process.

ENTHALPY

In the example of a spontaneous reaction just cited (A→B in Figure 2-3), we saw that energy was released (ΔE_1). This is because the energy content of the products was less than that of the reactants. More specifically,

$$\Sigma(EA) = \Sigma(EB) + \Delta E_1 \tag{2-4}$$

Here Σ means "content of." Accordingly, $\Sigma\,(E\mathrm{A})$ means content of energy in A.

In natural events, however, it is difficult to determine ΔE because some work is done on the atmosphere. Therefore, the term *enthalpy* (ΔH) is used because it takes into account pressure volume changes.

$$\Delta H = \Delta E + P\Delta V \qquad (2\text{-}5)$$

Fortunately, volume changes at constant pressures are extremely small, and for most biological reactions, $P\Delta V = 0$. Therefore,

$$\Delta E = \Delta H \qquad (2\text{-}6)$$

An example is seen in the important reaction of carbohydrate oxidation:

$$C_6H_{12}O_6 + 6O_2 \longrightarrow 6H_2O + 6CO_2 \qquad (2\text{-}7)$$

In the aqueous medium of a cell, the volume of O_2 consumed is equal to the volume of CO_2 formed, so the net volume change is zero.

The trend for biologists to use the terms energy and enthalpy interchangeably is somewhat confusing. Further, early theorists defined the terms enthalpy and entropy, which sound alike but are very different. Such terminology confusions must be carefully avoided.

FREE ENERGY

Although the second law of thermodynamics states that the entropy of the universe always increases, and it is the drive toward the entropic condition that allows physical and chemical reactions to occur, the entropic energy form is not directly useful in the performance of cell work. As implied by the first and second laws of thermodynamics, there exists another component to the total energy of a system. The term *free energy* is used because this energy is available, or free to do work. In biological reactions, such as those involved in muscle contraction, the free energy changes are of primary importance.

The following equations summarize what has been covered so far.

$$\text{Energy change (enthalpy)} = \text{change in energy available to} \qquad (2\text{-}8)$$
$$\text{do work} + \text{change in unavailable}$$
$$\text{energy}$$

For energy change let us substitute ΔH, and for unavailable energy let us substitute entropy. Therefore,

$$\Delta H = \text{Change in available energy} + \text{change in entropy} \qquad (2\text{-}9)$$

Recall that the energy available to do work was termed free energy. Therefore, we now have the following:

$$\Delta H = \text{Change in free energy} + \text{change in entropy} \qquad (2\text{-}10)$$

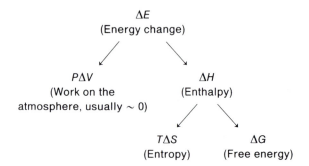

FIGURE 2-4
Enthalpy, free energy, and entropy. Only part of the energy change of a spontaneous reaction results in work. For a muscle this part may be only 25%.

Formally, free energy is noted as G, after the scientist Willard Gibbs, and entropy is noted as S.

Finally, with substitution, we obtain the following:

$$\Delta H = \Delta G + T\Delta S \qquad\qquad (2\text{-}11)$$

Where the Δ's refer to change, and T refers to absolute temperature. The temperature term is necessary to get entropy in appropriate units. Because of the second law, and by convention, as any real process proceeds the free energy carries a negative sign. Another way to remember that free energy (the useful component) carries a negative sign is to reason that a negative free energy term means that energy is given up by a reaction and is available to do work. Figure 2-4 summarizes the various components of chemical energy.

THE EQUILIBRIUM CONSTANT AND FREE ENERGY

The second law of thermodynamics states that there is always an increase in entropy of the universe. Entropy was defined as a random, disordered condition. When a reaction has reached the end of the road and entropy is at a maximum, the reaction has reached equilibrium. At chemical equilibrium there is no longer any net change from reactant to product. At equilibrium, then, there is no further potential to do work. In other words, the further from equilibrium the reactants of a process are, the more potential use the reaction has.

The importance of the equilibrium constant in determining the free energy change for a reaction is illustrated in Figure 2-5, where A is the reactant, B is the product, and each of the reactions has proceeded to equilibrium.

In a reaction such as (2-12), where very little or no product was formed, it

$$A \rightleftharpoons B \qquad K_{eq} = B/A, \text{ a small fraction} \qquad (2\text{-}12)$$

$$A \rightleftharpoons B \qquad\qquad\qquad\qquad\qquad\qquad (2\text{-}13)$$

$$A \rightleftharpoons B \qquad K_{eq} = B/A, \text{ a large integer} \qquad (2\text{-}14)$$

FIGURE 2-5
Reactant and product levels as related to K_{eq}. At time zero, when the reaction began, only the reactant A was present. Given enough time, (Equation 2-12) A will change to B and reach an equilibrium with B. At equilibrium (Equation 2-13), the net amount of A→B equals the net amount of B→A. Useful reactions, therefore, are those in which a large fraction of A changes to B and the K_{eq} is large (Equation 2-14).

is obvious that little happened, so there was little opportunity for energy to be exchanged. On the other end of the spectrum, with Reaction 2-14, almost all of the reactant has become product. In Reaction 2-14, much has happened chemically, and there has been opportunity to capture some of the energy given off.

The equilibrium constant K_{eq} is used to denote the concentrations of reactants to products at equilibrium.

$$K_{eq} = \frac{[B]}{[A]} = \frac{[products]}{[reactants]} \qquad (2\text{-}15)$$

Here the brackets indicate "concentration of." The K_{eq} of a reaction is an immutable constant at specified conditions of temperature and pressure.

In examining the K_{eq} and looking at the examples in Figure 2-5 (Equations 2-12 to 2-14), we can see that if the quotient is large, the reaction has potential use in driving a biological system.

Empirically, it has been determined that the free energy change of a reaction is simply related to its equilibrium constant:

$$\Delta G^{\circ\prime} = -RT \ln K'_{eq} \qquad (2\text{-}15)$$

where R is the gas constant (1.99 cal $\cdot$ mol deg^{-1}), T is absolute temperature, and $\ln K'_{eq}$ is the natural logarithm of the equilibrium constant determined at 25°C and pH 7.

The symbol $\Delta G^{\circ\prime}$ thus refers to the standard free energy change determined in the laboratory when a reaction takes place at 25°C, at 1 atmosphere pressure, and where the concentrations are maintained at 1 molal and in an aqueous medium at pH 7. In the notation the superscript ′ refers to pH 7, and ° refers to the other standard conditions, respectively. Although these "standard conditions" seem to be quite far removed from those in the body, categorizing reactions by this system allows us to compare free energy potentials of various reactions.

The relationship between K'_{eq} and $\Delta G^{\circ\prime}$ is further illustrated in Table 2-1.

TABLE 2-1

Relationship Between Equilibrium Constant K'_{eq} and Free Energy Change $\Delta G^{\circ\prime}$ Determined Under Standard Conditions (298 °K and pH 7)

K'_{eq}	$\Delta G^{\circ\prime}$ (kcal · mol^{-1})	
0.001	+4.09	Endergonic reactions
0.010	+2.73	Endergonic reactions
0.100	+1.36	Endergonic reactions
1.000	0.00	
10.000	−1.36	Exergonic reactions
100.000	−2.73	Exergonic reactions
1000.000	−4.09	Exergonic reactions

THE ACTUAL FREE ENERGY CHANGE

In order to consider a more realistic evaluation of free energy change, let us consider the reaction

$$r R + s S \longrightarrow p P + q Q \qquad (2\text{-}16)$$

By convention, for this reaction at equilibrium,

$$K'_{eq} = \frac{[P]^p \ [Q]^q}{[R]^r \ [S]^s} \qquad (2\text{-}17)$$

For the same reaction in a living cell, a determination of the actual, not equilibrium, concentrations gives the mass action ratio (MAR):

$$\text{MAR} = \frac{[P]^p \ [Q]^q}{[R]^r \ [S]^s} \qquad (2\text{-}18)$$

The formulas for the K'_{eq} and the MAR may look alike, but the values may be far apart. In fact, a secret of life is having MARs removed from the K'_{eq} in various reactions, or steps, of a process. It has been found that the actual (ΔG) and standard free energy changes ($\Delta G^{\circ\prime}$) are related by the equation

$$\Delta G = \Delta G^{\circ\prime} + RT \ \ln \text{MAR} \qquad (2\text{-}19)$$

Here we can clearly see what the effect of concentration change is on the actual free energy change for a reaction. Also, we can see that temperature has a direct effect on the free energy change. Though it is not directly shown, a change in pH would also affect the actual free energy. Some things to consider are the effects of exercise on muscle metabolite concentrations, pH, and temperature. These factors could affect exercise performance.

ENERGETICS AND ATHLETICS

Athletic activities can be classified into three groups: power, speed, and endurance events. Examples of these are the shot put, the 400-m sprint, and the marathon run, respectively. Success in each of these depends on energetics. Skeletal muscle has three energy systems, each of which is used in these three types of activities. In power events, where the activity lasts a few seconds or less, the muscle has several immediate energy sources (Figure 2-6). For rapid, forceful exercises lasting from a few seconds to approximately 1 min, muscle can depend on nonoxidative, or glycolytic energy sources, as well as immediate sources. For activities lasting 2 min or more, oxidative mechanisms become increasingly important. Before describing these three basic muscle energy sources, we need to describe the chemical–mechanical energy transduction of muscle contraction, as depicted in the following equation:

$$ATP + actin + myosin \underset{}{\overset{Ca^{2+}}{\rightleftharpoons}} actomyosin + P_i + ADP + energy \qquad (2\text{-}20)$$
$$\longrightarrow Contraction$$
$$\longleftarrow Relaxation \ and \ recovery$$

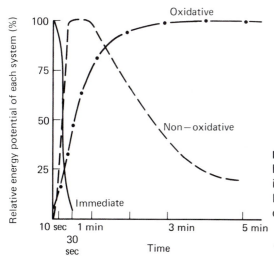

FIGURE 2-6
Energy sources for muscle as a function of activity duration. Schematic presentation showing how long each of the major energy systems can endure in supporting all-out work.

(SOURCE: From D.W. Edington and V.R. Edgerton, 1976. With permission.)

In the above reaction, actin and myosin are the two contractile proteins of muscle, and Ca^{2+} is calcium ion whose presence triggers the combination of actin and myosin. Inorganic phosphorus (P_i) is also produced by the reaction.

Each of the three muscle energy sources is mediated by specific enzymes or enzyme systems, as described in Table 2-2. In any muscle contraction, whether the activity is primarily one of power or endurance, the degradation of ATP supplies the immediate chemical energy to do work.

The immediate energy sources in muscle, as in most other cells, are three-fold. First there is ATP itself.

$$ATP + H_2O \longrightarrow ADP + P_i \qquad (2\text{-}21)$$

The second cellular source of immediate energy is creatine phosphate (CP). This high-energy phosphorylated compound exists in greater concentration in resting muscle than does ATP.

TABLE 2-2

Energy Sources of Muscular Work for Different types of Activities

	Power	Speed	Endurance
Duration of event	0 to 3 sec	4 to 50 sec	>2 min
Example of event	Shot put, discus, weight lifting	100 to 400-m run	≥1500-m run
Enzyme system	Single enzyme	One complex pathway	Several complex pathways
Enzyme location	Cytosol	Cytosol	Cytosol and mitochondria
Fuel storage site	Cytosol	Cytosol	Cytosol, blood, liver, adipose tissue
Rate of process	Immediate, very rapid	Rapid	Slower but prolonged
Storage form	ATP, creatine phosphate	Muscle glycogen and glucose	Muscle and liver glycogen, glucose; muscle, blood, and adipose tissue lipids; muscle, blood, and liver amino acids
Oxygen involved	No	No	Yes

$$CP + ADP \xrightarrow[\text{kinase}]{\text{creatine}} ATP + C \qquad (2\text{-}22)$$

This reaction is catalyzed by the enzyme creatine kinase.

The third immediate energy source involves an enzyme generally called adenylate kinase. The particular variety of this enzyme that occurs in muscle is termed myokinase. This enzyme has the ability to generate 1 ATP from 2 ADPs,

$$ADP + ADP \xrightarrow[\text{kinase}]{\text{adenylate}} ATP + AMP \qquad (2\text{-}23)$$

The nonoxidative energy source in muscle involves the breakdown of glucose (a sugar) and glycogen (stored muscle carbohydrate). These processes are specifically termed *glycolysis* and *glycogenolysis*. Muscle tissue is specialized in these processes and can break down glucose and glycogen rapidly. The nonoxidative energy source of muscle (glycolysis) can be summarized as follows:

$$\text{Glucose} \xrightarrow[\text{glycolysis}]{\text{rapid}} 2ATP + 2\,\text{lactate} \qquad (2\text{-}24)$$

Potential oxidative energy sources for muscle include sugars, carbohydrates, fats, and amino acids. Further, whereas sugars can be metabolized to an extent by glycolytic mechanisms (Equation 2-24), oxidative mechanisms allow far more energy to be liberated from a glucose molecule.

$$\text{Glucose} + O_2 \xrightarrow[\text{catabolism}]{\text{oxidative}} 36ATP + CO_2 + H_2O \qquad (2\text{-}25)$$

Fats can be catabolized only by oxidative mechanisms, but the energy yield is very large. For palmitate, an average-sized fatty acid,

$$\text{Palmitate} + O_2 \xrightarrow[\text{catabolism}]{\text{oxidative}} 130ATP + CO_2 + H_2O \qquad (2\text{-}26)$$

Amino acids, like fats, can be catabolized only by oxidative mechanisms. Prior to oxidation the nitrogen residue must be removed from amino acids. This is generally done by switching the nitrogen onto some other compound (transamination) or by a unique process of nitrogen removal in liver (oxidative deamination). Examples for alanine, a three-carbon amino acid, are given in the following:

$$\text{Alanine} + \alpha\text{-ketoglutarate} \xrightarrow[\substack{\text{pyruvate} \\ \text{transaminase}}]{\text{glutamate}} \text{pyruvate} + \text{glutamate} \qquad (2\text{-}27)$$

$$\text{Pyruvate} + O_2 \longrightarrow 15ATP + CO_2 + H_2O \qquad (2\text{-}28)$$

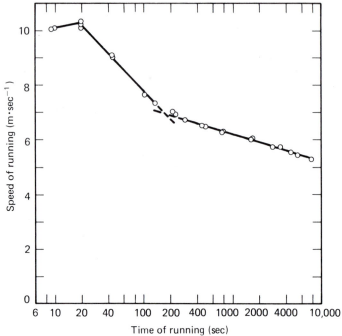

FIGURE 2-7
Logarithmic plot of average speed maintained versus time to event for men's world running records. Note the presence of three curve components that suggest the presence of three energy systems.

(SOURCE: Modified from R.W. McGilvery, 1975. With permission.)

Our interpretation that energy for the human engine comes from three sets of enzyme systems (Table 2-2) is supported by an analysis of the world's running records (Figure 2-7). A plot of running speed versus time reveals three distinct curve components.

Before describing the enzymatic controls of the three basic muscle energy sources, it is necessary first to discuss the essentials of metabolism in order that the biochemical pathways involved can be understood within the context of body function. This is the subject of the next chapter.

Summary

Performance in muscular activity depends on energetics. The utilization of energy by machines, both mechanical and biological, can be precisely defined. Basically two considerations govern physical and biological reactions involving energy exchanges: (1) Energy is not created but is acquired in one form and converted to another; and (2) the interconversion of energy among forms is inefficient, and a large fraction of the energy released will appear in an unusable form, usually heat. Muscles utilize three different systems of energy release during exercise, each of which differs in mechanism, capacity, and endurance. Consequently, muscular capacity is limited by these three systems of energy release.

Selected Readings

Bray, H.G. and K. White. Kinetics and Thermodynamics in Biochemistry. New York: Academic Press, 1966.

Edington, D.W. and V.R. Edgerton. The Biology of Physical Activity. Boston: Houghton Mifflin, 1976. p. 3–11.

Gibbs, C.L. and W.R. Gibson. Energy production of rat soleus muscle. Am. J. Physiol. 223:864–871, 1972.

Helmholtz, H. Über die Erhaltung der Kraft, Berlin (1847). Reprinted in Ostwald's Klassiker, No. 1, Leipzig, 1902.

Hill, T.L. Free Energy Transductions in Biology. New York: Academic Press, 1977.

Kleiber, M. The Fire of Life. New York: Wiley, 1961. p. 105–124.

Krebs, H.A. and H.L. Kornberg. Energy Transformations in Living Matter. Berlin: Springer-Verlag OHG, 1957.

Lehninger, A.L. Biochemistry. New York: Worth Publishing, 1970. p. 289–312.

Lehninger, A.L. Bioenergetics. Menlo Park, Cal.: W.A. Benjamin, Inc., 1973. p. 2–34.

McGilvery, R.W. Biochemical Concepts. Philadelphia: W.B. Saunders Co., 1975.

Merowitz, H.J., Entropy for Biologists. New York: Academic Press, 1970.

Mommaerts, W.F. H.M. Energetics at muscle contraction. Physiol. Rev. 49:427–508, 1969.

Wendt, I.R. and C.L. Gibbs. Energy production of rat extensor digitorum longus muscle. Am. J. Physiol. 224:1081–1086, 1973.

Wilkie, D.R. Heat work and phosphorylcreatine breakdown in muscle. J. Physiol. London 195:157–183, 1968.

Wilkie, D.R. The efficiency of muscular contraction. J. Mechanochem. Cell Motility 2:257–267, 1974.

3

BASICS OF METABOLISM

Metabolism can be defined as the sum total of processes occurring in a living organism. Because heat is produced by those processes, the metabolic rate is indicated by the rate of heat production. All processes of metabolism ultimately depend on biological oxidation, so measuring the rate of O_2 consumption will yield a good estimate of the rate of heat production, or *metabolic rate*. The maximum capability of an individual to consume oxygen ($\dot{V}_{O_2max}$) is highly related to that individual's ability to perform hard work over prolonged periods. A high capacity to consume and utilize O_2 indicates a high metabolic capacity.

ENERGY TRANSDUCTIONS IN THE BIOSPHERE

Our lives depend on conversion of chemical to other forms of energy. Those conversions, or transductions, of energy are limited by the two laws of thermodynamics, which apply to physical as well as biological energy transductions.

In the biological world (the biosphere) there are three major stages of energy transduction: photosynthesis, cell respiration, and cell work.

The *photosynthesis* of sugars is illustrated by the following equation:

$$6CO_2 + 6H_2O \xrightarrow{\text{sunlight}} C_6H_{12}O_6 + 6O_2 \qquad (3\text{-}1)$$

A photo from the classical study of human metabolic and cardio-ventilatory responses to exercise by H.M. Smith, 1922.

In photosynthesis the ΔG is positive in sign. Energy is put in.

Cell respiration can be illustrated by the following equation:

$$C_6H_{12}O_6 + 6O_2 \longrightarrow 6CO_2 + 6H_2O \qquad (3\text{-}2)$$

In cell respiration the ΔG is negative in sign, and the process is associated with the production of the important high-energy intermediate compound, ATP.

There are many types of cellular work, including mechanical, synthetic, chemical, osmotic, and electrical forms.

Muscle contraction (a chemical–mechanical energy transduction) can be illustrated by this equation:

$$\text{ATP} + \text{actin} + \text{myosin} \xrightarrow{\text{Ca}^{2+}} \text{actomyosin} + P_i + \text{ADP} + \text{heat} + \text{work} \qquad (3\text{-}3)$$

Here actin and myosin are the contractile proteins and the release of Ca^{2+} within the muscle cell triggers the reaction.

Although it may appear that our functioning depends on only two of these three major energy transductions (respiration and cell work), in reality we are ultimately dependent on photosynthesis. The products of photosynthesis give us the oxygen we breathe and the food that we eat. Cell respiration, as written, is a reversal of photosynthesis. Have you thanked the green plant today?

METABOLISM AND HEAT PRODUCTION IN ANIMALS

One characteristic of living animals is that they give off heat. This is because both the processes of cell respiration and cell work produce heat. As illustrated in Figure 3-1, for the body at rest the life processes result in heat production.

Scientists have developed two definitions of metabolism. A functional definition is that it is the sum of all transformations of energy and matter that occur within an organism. In other words, by this definition, metabolism is everything going on. It is not possible to measure that. Therefore, another operational definition has been developed, stating that metabolism is the rate of heat production. This definition takes advantage of the fact that all the cellular events result in heat (Figure 3-1). In this way, by determining the heat produced, one can obtain a measure of metabolism.

The basic unit of heat measurement is the calorie. Simply defined, a *calorie* is the heat required to raise the temperature of 1 gram of water 1 degree cen-

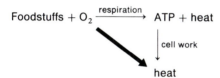

FIGURE 3-1
Metabolism and heat production. In the body at rest, all metabolic processes eventually result in heat production. Measuring heat production (calorimetry) gives the metabolic rate.

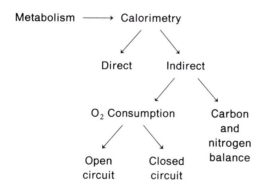

FIGURE 3-2
Relationship between metabolism and different methods of calorimetry. Because the processes of metabolism result in heat production, measuring heat production gives an estimate of the metabolic rate. Heat production can be measured directly (direct calorimetry), or it can be estimated from the O_2 consumption or from the carbon and nitrogen excreted (indirect calorimetry).

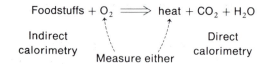

Foodstuffs + O_2 $\Longrightarrow$ heat + CO_2 + H_2O

Indirect
calorimetry

Measure either

Direct
calorimetry

FIGURE 3-3
Principle of indirect calorimetry (measuring O_2 consumption) as a basis of
estimating heat production. Instead of measuring the heat produced as
the result of biological reactions, we can measure the O_2 used to support
biological oxidations.

tigrade. Because the calorie is a very small quantity, frequently the term kilocalories (kcal) is used. A kilocalorie represents 1000 calories.

Because heat must be measured to determine metabolic rate, this procedure
is termed *calorimetry*. Several types of calorimetry are currently in use. These
are diagrammed in Figure 3-2.

As implied in Figure 3-2, not all methods of calorimetry involve the direct
measurement of heat, which is technically very difficult. Alternatively, it has
been determined that the measurement of oxygen consumption is a valid and
technically reliable procedure for measuring metabolic rate. The determination
of metabolic rate from O_2 consumption is termed *indirect calorimetry* because
it does not involve direct heat measurements. The principle of indirect calorimetry is illustrated in Figure 3-3. Still another method of indirect calorimetry
involves determination of the carbon and nitrogen content of excreted materials.

EARLY ATTEMPTS AT CALORIMETRY

In order to understand the relationship between heat production and O_2 consumption as alternative methods for determining metabolic rate, let us consider
some of the work of the eighteenth-century genius French chemist, Antoine
Lavoisier.

Because of his interest in studying living creatures, Lavoisier was able to
recognize certain characteristics of living animals: They gave off heat, and they
breathed. Dead animals did not give off heat and did not breathe. Lavoisier's
calorimeter, diagrammed in Figure 3-4, is simple but beautiful in its design.
By allowing the animal's warmth to melt ice, and then measuring the volume
of water produced, Lavoisier could calculate the heat produced by the animal
by knowing the quantity of heat required to melt a given quantity of ice. Such
a device is called a direct calorimeter because it determines metabolism by
measuring heat produced.

The respirometer of Lavoisier (Figure 3-5) was another device that was novel

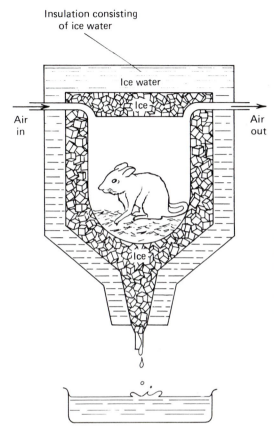

Insulation consisting
of ice water

Ice water

Ice

Air in

Air out

Ice

FIGURE 3-4
Lavoisier's calorimeter of 1780. The animal's body heat melts the ice. Measuring the amount of water formed allows estimation of the heat produced, knowing that 80 kcal of heat melts 1000 g of ice. The ice water surrounding the calorimeter provides a perfect (adiabatic) insulation because it is at the same temperature as the ice in the inner jacket around the animal's chamber. Therefore, the insulation will neither add heat to nor take the heat from the calorimeter.

(SOURCE: Based on original sources and M. Kleiber, 1961.)

for its time. With it Lavoisier could establish that something in the air (O_2) was consumed by the animal and that something else (CO_2) was produced in approximately equal amounts. Lavoisier also determined that matter gains weight when it burns. It has been thought previously that burning represented the loss of substance, sometimes called phlogiston.

With information obtained from his experiments, Lavoisier was able to interpret some earlier findings. For instance, Boyle had shown that air was necessary to have a flame, and Mayow had observed that a burning candle and an animal together in an airtight container expired at the same time. The fire of life and the fire of physical burning depended on the same substance in the air, which Lavoisier called *oxygène*.

The belief of Lavoisier and others that biological oxidation took place in the lungs has led to some confusion. Although it is true that breathing, or ventilation, takes place in the lungs and associated organs, respiration, or biological oxidation, takes place in most of the body's cells. Therefore, in this text we

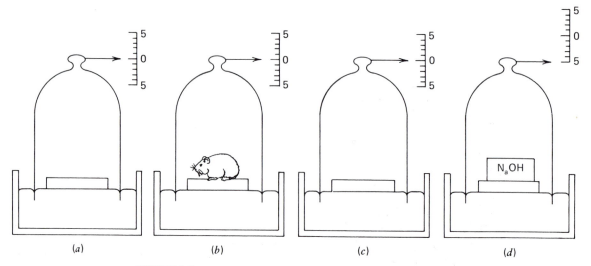

FIGURE 3-5

Lavosier's respirometer of 1784. (a) A glass bell jar rests on a bed of mercury. (b) An animal is placed in the jar from beneath the mercury seal and is left there for several hours. The apparent volume increases when the animal enters, but then the volume decreases very slowly. (c) The animal is removed, and the volume is observed to have decreased slightly. (d) Addition of NaOH (a CO_2 absorber) into the jar results in a decrease in the measured volume. From these volume changes, O_2 consumption (V_{O_2}) and carbon dioxide production (V_{CO_2}) can be measured: $V_{O_2} = V_a - V_d$; $V_{CO_2} = V_c - V_d$.

(SOURCE: Based on original sources and M. Kleiber, 1961.)

shall use the term *respiration* to denote cellular oxidations, and *ventilation* to denote pulmonary gas exchange.

Devices such as Lavoisier's respirometer are called indirect calorimeters because they estimate heat production by determining O_2 consumption or CO_2 production. Lavoisier's device is also referred to as a closed-circuit, indirect calorimeter, because the animal breathes gas within a sealed system.

Haldane's respirometer (Figure 3-6) is an example of an open-circuit, indirect calorimeter. This system is open to the atmosphere, and the animal breathes air. Today the type of calorimeters most frequently used are open-circuit, indirect designs.

The device of Atwater and Rosa (Figure 3-7) represents an important apparatus for the study of metabolism. Large enough to accommodate a person, it had the capability to determine heat production, O_2 consumption, and CO_2 production simultaneously. In this way the relationship between direct and indirect calorimetry was established. Thus it is now possible to predict metabolic

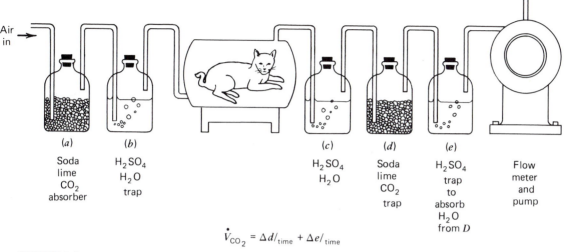

Air in →

(a)	(b)		(c)	(d)	(e)	
Soda lime CO_2 absorber	H_2SO_4 H_2O trap		H_2SO_4 H_2O	Soda lime CO_2 trap	H_2SO_4 trap to absorb H_2O from D	Flow meter and pump

$$\dot{V}_{CO_2} = \Delta d/_{time} + \Delta e/_{time}$$

FIGURE 3-6
Haldane's respirometer. This device is an open-circuit, indirect calorimeter, in which carbon dioxide and water vapor in air entering the system are removed by traps (a) and (b), respectively. Trap (c) removes the animal's expired H_2O vapor. Increase in weight of the soda lime CO_2 trap (d) gives the animal's CO_2 production.

(SOURCE: Based on original sources and M. Kleiber, 1961.)

rate (heat production) on the basis of determinations of O_2 consumption and CO_2 production in resting individuals.

The calorimeter illustrated in Figure 3-8 is called a bomb calorimeter. In this device foodstuffs are ignited and burned in O_2 under pressure. In this way the heats of combustion of particular foods can be determined.

Table 3-1 presents the relationships among caloric equivalents for combustions of various foodstuffs as determined by indirect and direct calorimetry as well as by bomb calorimetry. Perhaps the most interesting feature of this table is that, with a single exception, the caloric equivalents for the combustion of foodstuffs inside and outside the body are the same. Protein is the exception because nitrogen, an element unique to protein, is not oxidized within the body but is eliminated chiefly in the urine. Therefore, the caloric equivalent of protein metabolism is approximately 26% less than in a bomb calorimeter.

Table 3-1 also gives the caloric equivalents of foodstuffs in kilocalories per liter of O_2 consumed. Although fat, because of its relatively high carbon and hydrogen content, contains more potential chemical energy on a per-unit-weight basis, carbohydrates give more energy when combusted in a given volume of O_2.

Because of the established validity of indirect calorimetry, the metabolic

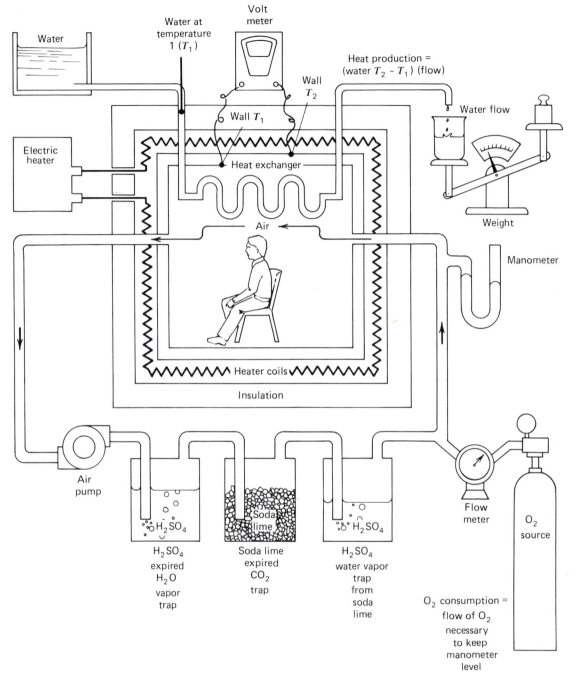

FIGURE 3-7

Atwater–Rosa calorimeter: A direct calorimeter suitable to accommodate a resting human and simultaneously determine that individual's O_2 consumption and CO_2 production. In this way direct and indirect calorimetry were correlated. O_2 Consumption is equal to the volume of O_2 added to keep the internal (manometer) pressure constant. In the calorimeter, heat loss through the walls is prevented by heating the middle wall (wall T_2) to the temperature of the inner wall (wall T_1). Metabolic heat production is then picked up in the water heat exchanger. (SOURCE: Based on original sources and M. Kleiber, 1961.)

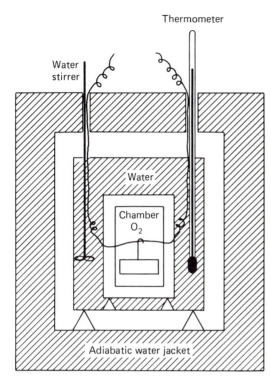

FIGURE 3-8
Bomb calorimeter. A food substance is attached to the ignition wires and placed in the chamber under several atmospheres of O_2 pressure. The sample is then ignited and burns explosively. The stirrer distributes the heat of combustion uniformly throughout the water surrounding the chamber. The thermometer detects the heat released.

(SOURCE: Based on M. Kleiber, 1961.)

heat production can be estimated by determining the O_2 consumption. In a similar way, determination of the ratio of CO_2 produced ($\dot{V}_{CO_2}$) to O_2 consumed ($\dot{V}_{O_2}$), gives an indication of the type of foodstuff being consumed. The $\dot{V}_{CO_2}/\dot{V}_{O_2}$ is usually referred to as the *respiratory quotient* (*RQ*) and reflects cellular processes. Equation 3-4 shows why the *RQ* of glucose, a sugar carbohydrate, is unity.

$$C_6H_{12}O_6 + 6O_2 \longrightarrow 6CO_2 + 6H_2O \qquad (3\text{-}4)$$

$$\frac{\dot{V}_{CO_2}}{\dot{V}_{O_2}} = \frac{6}{6} = 1$$

For the neutral fat trioleate, the *RQ* approximates 0.7.

$$C_{57}H_{104}O_6 + 80O_2 \longrightarrow 57CO_2 + 52H_2O \qquad (3\text{-}5)$$

$$RQ = \frac{57}{80} = 0.71$$

During hard exercise, an individual's *RQ* approaches 1.0, whereas during prolonged exercise, the *RQ* may be somewhat lower, 0.9 or less. Table 3-1 shows why it is an advantage for these changes in *RQ* to occur. During hard

TABLE 3-1

Caloric Equivalents of Foodstuffs Combusted Inside and Outside the Body

Food	kcal · liter O_2^{-1}	RQ ($\dot{V}_{CO_2}/\dot{V}_{O_2}$)	Inside Body (kcal · g^{-1})	Outside Body (kcal · g^{-1})
Carbohydrate	5.05	1.00	4.2	4.2
Fat	4.70	0.70	9.5	9.5
Protein	4.50	0.80	4.2	5.7 [a]
Mixed diet	4.82	0.82		
Starving individual	4.70	0.70		

[a] The amount of protein combusted outside the body is greater than that combusted inside the body (see text):

$$\frac{5.7 - 4.2}{5.7} = 26\% \text{ difference}$$

exercise, O_2 consumption can be limiting. Therefore, in oxidizing carbohydrate rather than fat the individual derives

$$\frac{5.0 - 4.7 \text{ kcal} \cdot \text{liter}^{-1} O_2}{4.7 \text{ kcal} \cdot \text{liter}^{-1} O_2}$$

or 6.4% more energy per unit O_2 consumed. During prolonged exercise, however, it makes sense that RQ decreases, indicating more fat is combusted. In prolonged work, glycogen supply rather than O_2 consumption can be limiting. Table 3-1 indicates that on a mass basis, fats provide about 9.5/4.2 kcal · g^{-1}, or 2.3 times as much energy as carbohydrate. Given this large difference, we can also see why endurance training improves the ability to use fat as a fuel.

The above kind of discussion is sometimes referred to as a "teleological argument," meaning that the purpose of something is assumed to explain its operation. In actuality, as will be shown, the reason that relatively more carbohydrate is used in hard exercise is related to the quantity and activity of glycolytic enzymes. Similarly, there are enzymatic explanations for the preponderance of fat used in prolonged exercise.

In order to obtain a precise estimate of metabolic rate and fuel used by means of indirect calorimetry, we must know a few other details besides the quantity of O_2 consumed and CO_2 produced. These additional parameters include the food ingested and the nitrogen excreted. To provide a relatively simple example of the utility of indirect calorimetry, let us consider a starving man, in whom there is no food input to account for and no large excretion of urinary nitrogen (Table 3-2).

In exercise physiology, current estimates of fuels combusted are usually simplified by assuming that there is no increase in the basal amino acid and protein degradation during exercise. The ventilatory exchange ratio R is then

TABLE 3-2
Calculation of Nitrogen-Free RQ on a Resting Starving Man

Given: (a) Protein is about 17% N by weight, or there is 1 g N/5.9 g protein
$(1/5.9 = 0.17)$.
(b) For protein $RQ = 4.9/5.9 = 0.83$, or 4.9 liters CO_2 are derived from
the catabolism of the protein associated with 1 g N, and 5.9 liters of O_2 are required to catabolize the protein.
The total O_2 consumption was **634** liters. The total CO_2 production was 461
liters, and urinary N losses were 14.7 g over 24 hours. We can use these
data to calculate the nitrogen-free RQ.

Calculations	Total CO_2 (liters)	Total O_2 (liters)
	461	634
In the urine there were 14.7 g N. The CO_2 produced by protein catabolism was $(14.7)(4.9) =$ 72.0 liters CO_2.	72	
The O_2 consumed associated with protein catabolism was $(14.7)(5.9) = 86.7$ liters O_2.		86.7
	389	547.3

Nonprotein $RQ = \dfrac{389}{547.3} = 0.71$

Heat production
From protein: $(14.7 \text{ g N}) (5.9 \text{ g protein/g N}) (4.2 \text{ kcal} \cdot \text{g}^{-1} \text{ protein}) = 364.3$ kcal
From fat: The nonprotein RQ was 0.71, so fat comprised the remaining
fuel. Therefore, $(547.3 \text{ liters } O_2) (4.7 \text{ kcal} \cdot \text{liter}^{-1} O_2) = 2572.3$ kcal.
Total heat production $= 364.3 + 2572.3$ kcal $= 2936.6$ kcal

used to represent the nonprotein RQ. As we shall see later (Chapter 8), this
assumption is not quite valid.

Although both RQ and R are given by the same formula ($\dot{V}_{CO_2}/\dot{V}_{O_2}$), over
any short period of measurement of gas exchange at the lungs, changes in CO_2
storage may cause R not to equal RQ. While RQ does not exceed 1.0, R can
reach 1.5 or higher. For the present, let us consider RQ to be the $\dot{V}_{CO_2}/\dot{V}_{O_2}$ in

FIGURE 3-9
Bicycle ergometers are convenient, stationary laboratory devices to control the external work rate (power output) while physiological responses are observed.

(Wayne Glusker.)

the cell where O_2 is consumed and CO_2 produced. Further, let us consider R to be the $\dot{V}_{CO_2}/\dot{V}_{O_2}$ measured at the mouth.

INDIRECT CALORIMETRY

For individuals at rest, indirect calorimetric determinations are of great use in studying metabolism. For instance, the effects of body size, growth, disease, gender, drugs, nutrition, age, and environment on metabolism can be determined. The resting metabolic rate per unit body mass is greater in males than in females, greater in children than in the aged, greater in small individuals than in large ones, and greater under extremes of heat and cold.

THE UTILITY OF INDIRECT CALORIMETRY DURING EXERCISE

Physical exercise represents a special metabolic situation. As Figure 3-1 indicates, for the body at rest, all the energy liberated within appears as heat. If

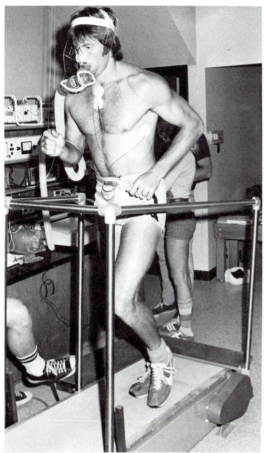

FIGURE 3-10
Treadmills are frequently used in the laboratory as a means to apply exercise stress and to record physiological responses on relatively stationary subjects during exercise. Compared to the bicycle ergometer (Figure 3-9), it is difficult to quantitate external work on the treadmill. However, the treadmill does allow subjects to walk or to run, which are perhaps more common modes of locomotion than is bicycling.

(Wayne Glusker.)

metabolism is constant, the quantity of heat produced within the body over a period of time will be the same as that leaving the body. However, during exercise some of the energy liberated within the body appears as physical work outside the body. Therefore, devices to measure external work performed, such as bicycle ergometers (Figure 3-9) and treadmills (Figure 3-10), are utilized.

During exercise, direct calorimeters such as the Atwater–Rosa calorimeter (Figure 3-7) are of little use, for several reasons. First, such devices are very expensive. Second, the heat generated by the ergometer, if it is electrically powered, may far exceed that of the subject. Third, body temperature increases during exercise because not all of the heat produced is liberated from the body. Therefore, the sensors in the walls of the calorimeter do not pick up all the heat produced. Finally, the body sweats during exercise, which also affects the calorimeter and changes the body mass. These changes in body mass and the unequal distribution of heat within the body make it very difficult to use direct calorimetry in exercise.

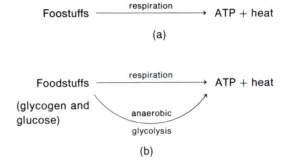

FIGURE 3-11
Respiration and ATP production. The validity of indirect calorimetric measurements depends on the O_2 consumption accurately representing the ATP formed. This is not always the case in exercise. In (a) the measurement would be valid. In (b) it would be invalid.

As with direct calorimetry during exercise, techniques of indirect calorimetry possess certain limitations. These are summarized in Figure 3-11. In order for determinations of $\dot{V}_{O_2}$ to reflect metabolism accurately, the situation in Figure 3-11a must hold. If there is another mechanism to supply energy, such as indicated in Figure 3-11b, then respiratory determinations do not completely reflect all metabolic processes. As will be seen, the body possesses the means to derive energy from the degradation of substances without the immediate use of O_2. These mechanisms include immediate sources and rapid glycogen (muscle carbohydrate) breakdown. Use of the $\dot{V}_{CO_2}/\dot{V}_{O_2}$ is also limited during exercise. Although over time the O_2 consumed by and CO_2 liberated at the lungs (the respiratory, or ventilatory exchange ratio) equals the respiratory quotient, the cellular events are not always represented in expired air. This is because the cells are fluid systems, and they are surrounded by other fluid systems on both the arterial and venous sides. When exercise starts, CO_2 is frequently stored in cells. When exercise is very difficult, the blood bicarbonate buffer system buffers lactic acid, and extra nonrespiratory CO_2 is produced (Chapter 11).

In the following example, lactic acid (HLA) is a strong acid whose level in muscle and blood increases during heavy work. It is known as a strong acid in physiological systems because it can readily dissociate a proton (H^+). To lessen the effect of protons generated from lactic acid during hard exercise, the body possesses a system of chemicals to lessen, or buffer, the effects of the acid. In the blood, the bicarbonate (HCO_3^-) –carbonic acid (H_2CO_3) system is the main way in which the effects of lactic acid are buffered. In equations 3-6 to 3-8, HCO_3^- neutralizes the H^+, but CO_2 is produced. This is eliminated at the lungs and appears in the breath. Consequently, during hard exercise, $R \neq RQ$. After exercise, metabolic CO_2 may be stored in cells, blood, and other body compartments to make up for that lost during exercise.

$$HLA \longrightarrow H^+ + LA^- \qquad (3\text{-}6)$$

$$H^+ + HCO_3^- \longrightarrow H_2CO_3 \qquad (3\text{-}7)$$

$$H_2CO_3 \longrightarrow H_2O + CO_2 \qquad (3\text{-}8)$$

Furthermore, during and immediately after exercise, urine production by the kidney is inhibited. Also, during exercise considerable nitrogen can be lost as urea in sweat. Therefore, it is difficult to determine the nitrogen excreted during exercise.

Determinations of indirect calorimetry are somewhat limited in their use by the fact that the respiratory gases give no specific information on the fuels used. If, for example, RQ is 1.0, then although we know that carbohydrate was the fuel catabolized, we do not know specifically which carbohydrate was involved. The possibilities could include, among others, glycogen, glucose, lactic acid, and pyruvic acid. However, radioactive and nonradioactive tracers to study metabolism at rest and during exercise have come into use in conjunction with indirect calorimetry to provide more detailed information on the specific fuels utilized.

Exercise is also a special situation in that the metabolic responses persist long after the exercise itself may have been completed. Consequently, physical activity results in an excess postexercise O_2 consumption (EPOC). This EPOC has sometimes been called the "O_2 debt" and has been used as a measure of anaerobic metabolism during exercise. A more detailed explanation of the O_2 debt is given later (Chapter 10); suffice it to say here that the mechanisms of the O_2 debt are complex and cannot be used to estimate anaerobic metabolism during exercise.

Whereas the body during exercise does present certain problems in the determination of metabolic rate, careful consideration of those various factors allows us to obtain important information about the metabolic responses to exercise. Estimations of $\dot{V}_{O_2}$ provide information on the cardioventilatory response to exercise. The caloric cost of various exercises can be estimated (Table 3-3), and information about the fuels used to support the exercise can be obtained.

Knowing that part of the energy liberated during exercise appears as external work is useful. By measuring the respiratory response to graded, submaximal exercise at specific external work rates, we can determine the fraction of the energy liberated within the human machine that appears as external work. This fraction is frequently reported as a percentage and is called efficiency.

An example of how the efficiency of the human body is calculated during

TABLE 3-3

Estimates of Caloric Expenditures of Sports Activities for a 70-kg Person[a]

Activity	Caloric Expenditure (kcal · min^{-1})	Activity	Caloric Expenditure (kcal · min^{-1})
Archery	4.6	Resting	1.2
Badminton	6.4	Running	
Basketball	9.8	8 min/mi	14.8
Canoeing	7.3	6 min/mi	17.9
Cycling	12.0	Squash	15.1
Field hockey	9.5	Swimming	
Fishing	4.4	Backstroke	12.0
Football	9.4	Crawl	11.1
Golf	6.0	Tennis	7.7
Gymnastics	4.7	Volleyball	3.6
Judo	13.8	Walking, easy	5.7

[a] For a more detailed listing, see Appendix II.

bicycle ergometer exercise is given below. In Figure 3-12 we see that the O_2 consumption of an individual increases in direct response to increments in work load while pedaling at constant speed. In this case efficiency can be calculated as here:

$$\text{Efficiency} = \frac{\text{change in work}}{\text{change in caloric equivalent of } O_2 \text{ consumption}} \quad (3\text{-}9)$$

The plateau steps in Figure 3-12a are referred to as "steady-rate" exercise. During the steady rate, the oxygen consumption ($\dot{V}_{O_2}$) is relatively constant and is directly proportional to the constant submaximal work load.

The calculation of body efficiency during bicycle exercise is given in Table 3-4. Here the calculated value of efficiency is 29.2%, which is close to a maximum value for bicycle ergometer work. Cycling at greater speeds and working at greater loads results in decrements in calculated efficiency. The efficiency of walking is slightly higher than that of cycling, but responds similarly to increments in speed and resistance. The reason it is usually easier to cycle from one place to another than to walk is that the rolling and wind resistance to cycling at a particular speed are far less than the work done in accelerating and decelerating the limbs during walking. Therefore, the reason it is usually easier to bicycle than to walk a given distance is that less work is done in cycling. Attempting to bicycle in soft sand will reveal that the work done to cover a given distance is far greater; yet measurements of the efficiency of movement would reveal no change or only a relatively small decrement.

In contrast to the bicycle ergometer, where the work done is the product of

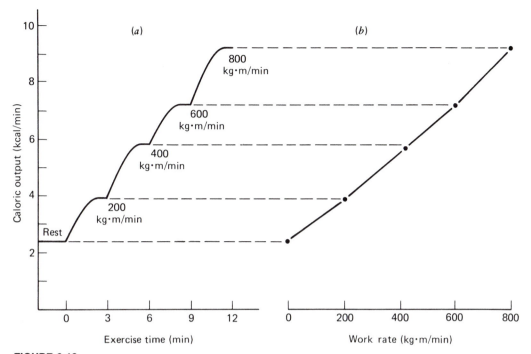

FIGURE 3-12
Respiratory response to graded submaximal bicycle ergometer work. Each 3 min, the work rate is increased 200 kg · m/min. The observed O_2 consumption ($\dot{V}_{O_2}$) is converted to kcal · min^{-1}. These values are then plotted as (a) a function of time and (b) function of the steady-rate work load. Note that a plot of the caloric cost of exercise against work rate (b) yields a straight line, or one that bends upward slightly.

the pedaling speed and the resistance, the calculation of work done in walking is more involved. This is because the body walking on a level treadmill does no external work. Estimates of the work done in walking, therefore, depend on applying an external work load that can be measured, or by estimating the work done internally in the body as a result of accelerating and decelerating the limbs.

The most common way to apply external work during walking is to have a subject climb an incline. In Figure 3-13, the vertical external work performed is in lifting the body mass the distance B–D. The work done is calculated according to either of two formulas.

$$\text{External work rate} = [\text{body weight (kg)}]\,[\text{speed (m} \cdot min^{-1})] \quad (3\text{-}10)$$
$$[\%\text{grade}/100]$$

$$\text{External work rate} = [\text{body weight (kg)}]\,[\text{speed (m} \cdot min^{-1})] \quad (3\text{-}11)$$
$$[\sin \Theta]$$

where $\sin \Theta$ is the angle $ACB = BD/CB$.

TABLE 3-4
Calculation of Body Efficiency During Cycling Exercise

Given:

$\dot{V}_{O_2}$ at 200 kg · m/min = 0.76 liter · min^{-1}

$\dot{V}_{O_2}$ at 400 kg · m/min = 1.08 liters · min^{-1}

$R = RQ = 1.0$

When $RQ = 1.0$, 1 liter $O_2 = 5$ kcal · min^{-1}

1 kg · m = 0.00234 kcal

$$\text{Efficiency} = \frac{\text{change in work output}}{\text{change in } \dot{V}_{O_2}}$$

$$\text{Efficiency} = \frac{400 - 200 \text{ kg} \cdot \text{m/min}}{1.08 - 0.76 \text{ liter} \cdot \text{min}^{-1}}$$

$$= \frac{200 \text{ kg} \cdot \text{m/min} \times 0.00234 \text{ kcal/kg} \cdot \text{m}}{0.32 \text{ liter} \cdot \text{min}^{-1} \times 5 \text{ kcal} \cdot \text{liter}^{-1} O_2}$$

$$= .292 = 29.2\%$$

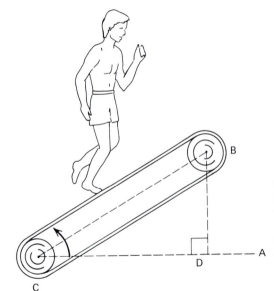

FIGURE 3-13
During horizontal treadmill walking, no external work is done. Therefore, it is impossible to calculate a value for body efficiency. However, a way to determine external work is to measure the work lifting the body up the hill. Refer to Equations 3-10 and 3-11 in the text for details of work rate calculation.

TABLE 3-5
Estimation of the Whole-Body Efficiency of Doing Vertical Work During
Steady-Rate Treadmill Walking at 3.0 km · hr^{-1}

Given: (a) Steady-rate caloric equivalent of $\dot{V}_{O_2}$ during horizontal, un-
graded walking (i.e., zero vertical work) at 3.0 km · hr^{-1} = 5
kcal · min^{-1}

(b) Steady-rate caloric equivalent of $\dot{V}_{O_2}$ while performing 375
kg · m/min of vertical work at 3.0 km · hr^{-1} = 7.9 kcal · min^{-1}

(c) 1.0 kg · m = 0.00234 kcal

$$\text{Efficiency} = \frac{\text{caloric equivalent of change in vertical work}}{\text{caloric equivalent of change in respiration}}$$

$$= \frac{(375 \text{ kg} \cdot \text{m/min} - 0 \text{ kg} \cdot \text{m/min}) \, (0.00234 \text{ kcal/kg} \cdot \text{m})}{7.92 - 5 \text{ kcal} \cdot \text{min}^{-1}}$$

$$= .30 \text{ or } 30\%$$

Recently, external work has been applied in studies of energetics by having
subjects walk against a horizontal impeding force (Figure 3-14). The work done
against the horizontal impeding force is calculated as follows:

$$\text{External work} = [\text{speed (m} \cdot \text{min}^{-1})] \, [\text{weight pulled (kg)}] \qquad \text{(3-12)}$$

An example of how to calculate the efficiency of performing external work
during incline walking is given in Table 3-5.

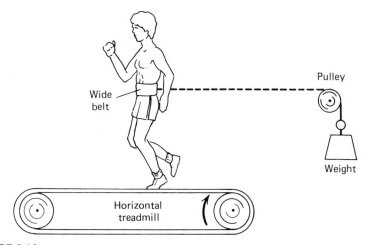

FIGURE 3-14
External work can be determined during horizontal treadmill walking by
having subjects pull a training weight. Refer to equation 3-12 in the text
for work rate calculation.

Another innovation for estimating the work involved in horizontal walking has been established by mechanical engineers at the University of California. They attached sensitive transducers to the joints so that their movements during walking could be recorded; these recordings, coupled with estimates of the masses of different body parts, made it possible to calculate the work done in moving the body parts and entire body on a computer. Because the various techniques of estimating work done in walking give similar results, it appears that the efficiencies with which the body does internal, horizontal, and lifting work during walking are similar.

Whereas the efficiency of the body during easy cycling and walking may be as high as 30%, it can only be surmised that the efficiency of running is somewhat lower. The lack of evidence on the efficiency of running exists because running is not a true steady-rate situation. During running, the metabolic rate is so high that both situations a and b in Figure 3-11 occur. Because $\dot{V}_{O_2}$ does not account for all the ATP supplied during running, a proper estimation of efficiency during running awaits development of the technical ability to estimate nonoxidative ATP supply during exercise.

Summary

Metabolism can be estimated in two ways: by direct determinations of heat production and by determinations of O_2 consumption. Determinations of metabolic rate provide valuable information about the status of an individual. In resting individuals, both methods provide similar results. During exercise, direct calorimetry is not feasible; therefore, indirect calorimetry must be used. During hard and prolonged exercise, indirect calorimetry may not provide a precise estimate of metabolic rate. Under these conditions, determinations of O_2 consumption still provide important information about the cardioventilatory systems.

Selected Readings

Asmussen, E. Aerobic recovery after anaerobiosis in rest and work. Acta Physiol. Scand. 11:197–210, 1946.

Atwater, W.O. and F.G. Benedict. Experiments on the metabolism of matter and energy in the human body. U.S. Dept. Agr. Off. Exp. Sta. Bull. 136:1–357, 1903.

Atwater, W.O. and E.B. Rosa. Description of new respiration calorimeter and experiments on the conservation of energy in the human body. U.S. Dept. Agr. Off. Exp. Sta. Bull., 63, 1899.

Benedict, F. and E.P. Cathcart. Muscular Work. Carnegie Institution of Washington, Publication 187, 1913.

Benedict, F.G. and H. Murchhauer. Energy Transformations during Horizontal Walking. Carnegie Institution of Washington, Publication 231, 1945.

Dickensen, S. The efficiency of bicycle pedaling as affected by speed and load. J. Physiol. London, 67:242–255, 1929.

Donovan, C.M. and G.A. Brooks. Muscular efficiency during steady-rate exercise II: Effects of walking speed on work rate. J. Appl. Physiol. 43:431–439, 1977.

Gaesser, G.A. and G.A. Brooks. Muscular efficiency during steady-rate exercise: effects of speed and work rate. J. Appl. Physiol. 38:1132–1139, 1975.

Haldane, J.S. A new form of apparatus for measuring the respiratory exchange of animals. J. Physiol. London 13:419–430, 1892.

Kleiber, M. Calorimetric measurements. In F. Über (ed.), *Biophysical Research Methods*. New York: Interscience, 1950.

Kleiber, M. The Fire of Life: An Introduction to Animal Energetics. New York: Wiley, 1961. p. 116–128, 291–311.

Krogh, A. and J. Lindhard. The relative value of fat and carbohydrate as sources of muscular energy. Biochem. J. 14:290, 1920.

Lavoisier, A.L. and R.S. de La Place. Mémoire sur la chaleur; Mémoires de l'Académie Royale (1789). Reprinted in Ostwald's Klassiker, No. 40, Leipzig, 1892.

Lloyd, B.B. and R.M. Zacks. The mechanical efficiency of treadmill running against a horizontal impeding force. J. Physiol. London 223: 355–363, 1972.

Ralston, H.J. Energy-speed relation and optimal speed during level walking. Intern. Z. Angew. Physiol. 17:277–283, 1958.

Smith, H.M. *Gaseous Exchange and Physiological Requirements for Level Walking.* Carnegie Institution of Washington, Publication 309, 1922.

Wilkie, D.R. The efficiency of muscular contraction. J. Mechanochem. Cell Motility. 2:257–267, 1974.

Zarrugh, M.Y., F.M. Todd, and H.J. Ralston. Optimization of energy expenditure during level walking. European J. Appl. Physiol. Occupational Physiol. 33:293–306, 1974.

4

ENERGY TRANSDUCTIONS IN CELLS

The mechanisms of energy conversion must be contained within each cell; therefore, cells require the presence of a substance that can receive energy input from energy-yielding reactions. Equally important, that substance must be able to yield energy to reactions requiring an energy input. In our cells, that substance is almost always adenosine triphosphate (ATP).

ATP—THE COMMON CHEMICAL INTERMEDIATE

We have discussed the processes of cell respiration and cell work. In the cell, respiration represents the conversion of the chemical energy of foodstuffs into a useful chemical form. Cell work represents the conversion of that useful form to other forms of energy. In order to function, this coupled system of energy-yielding and energy-utilizing reactions depends on having a substance that can act both as an energy receiver and as an energy resource, or donor. Most cells in most organisms utilize ATP as that form. Because of its central role in metabolism, ATP is frequently referred to as the "common chemical intermediate." If most species had not evolved to use ATP, they would have developed to use a related compound.

The history of the study of ATP is marked by several notable discoveries. In the 1920s, the work of Fletcher, Hopkins, and Hill led to the belief that the

Energy transduction in human muscle depends upon the enzymatic regulation of chemical energy release. Enzymatic control of chemical energy release in Jesse Owens' muscles allowed him to win four Olympic gold medals in 1936; the same enzymatic processes determine our abilities to perform routine as well as outstanding muscular activities. Jesse Owen pictured in a workout days prior to victories in the Berlin Olympics. (U.P.I.).

release of lactic acid in muscles was the stimulus for contraction. Subsequently, after Embden showed that rapid freezing of small isolated muscles after a contraction prevented the formation of lactic acid, it was concluded that lactate was not the cause of contraction. In the early 1930s, Lundsgaard showed that poisoning glucose metabolism in muscles with iodoacetic acid hastened fatigue and prevented lactate formation. Instead, there was a decrease in a phosphorylated compound (later shown to be creatine phosphate) and an increase in inorganic phosphate. In 1929 Fiske and Subbarow isolated and deduced the structure of ATP. Embden and Meyerhof independently found that ATP is formed by joining inorganic phosphate (P_i) to adenosine diphosphate (ADP) during the catabolism of glucose. Englehardt discovered that ATP was split to ADP by myosin, one of the two major contractile proteins of muscle. Szent Gyorgi found that injections of ATP into muscle fibers that had been soaked in glycerin to remove metabolites had two effects. An initial infusion of ATP caused the muscles to contract; a second injection frequently caused relaxation. Cori and

Cori observed that ATP had a biosynthetic function and could stimulate the aggregation of glucose (a simple sugar) into glycogen (a polymeric form of glucose).

In 1940, Lipman deduced from these various bits of information that ATP was the substance that linked energy-yielding and energy-using functions in the cell. This hypothesis was confirmed and generally accepted when Cain and Davies used the poison dinitroflurobenzene (DNFB) on small isolated muscles to show that contractions resulted in a utilization of ATP. In the normal muscle cell there is not much ATP present. This rather low concentration of ATP is probably more advantageous than not. As will be shown, the energy-yielding processes of metabolism are finely tuned to comparative levels of ATP, ADP, P_i, and AMP. By keeping the normal level of ATP rather low, any small utilization immediately markedly changes the relative level and stimulates the processes that generate ATP.

More cellular energy is stored in the form of creatine phosphate (CP) than in the form of ATP. The metabolisms of ATP and CP are linked by the reaction governed by the enzyme creatine kinase.

$$ADP + CP \xrightarrow[\text{kinase}]{\text{creatine}} ATP + C \qquad (4\text{-}1)$$

The poison DNFB blocks the action of creatine kinase.

$$ADP + CP \xrightarrow[\text{kinase}]{\text{creatine} \quad \text{DNFB}}$$
$$\text{(reaction blocked)} \qquad (4\text{-}2)$$

In most cells, such as muscle, there is enough CP and the action of creatine kinase is so rapid that it is difficult to determine an overall decrease in ATP level after a twitch or contraction. Therefore, while the muscle uses ATP as an immediate energy source for contraction, the ATP is replenished almost immediately by CP. By utilizing DNFB, Cain and Davies demonstrated that muscle contractions are powered directly by ATP.

Structure of ATP

The structure of ATP is given in Figure 4-1. One of the group of compounds called nucleotides, ATP contains the nitrogenous base (adenine), a five-carbon sugar (ribose), and three phosphates. Removal of the terminal phosphate results in adenosine diphosphate (ADP); and cleaving two phosphates gives adenosine monophosphate (AMP). With no phosphates the compound is the nucleotide, adenosine. In the cell, ATP is negatively charged, and the terminal phosphates of each ATP molecule usually associate with a magnesium ion (Mg^{2+}). The Mg^{2+} is usually required for enzymatic activity.

The reactions in which ATP is split to ADP to liberate energy involve water.

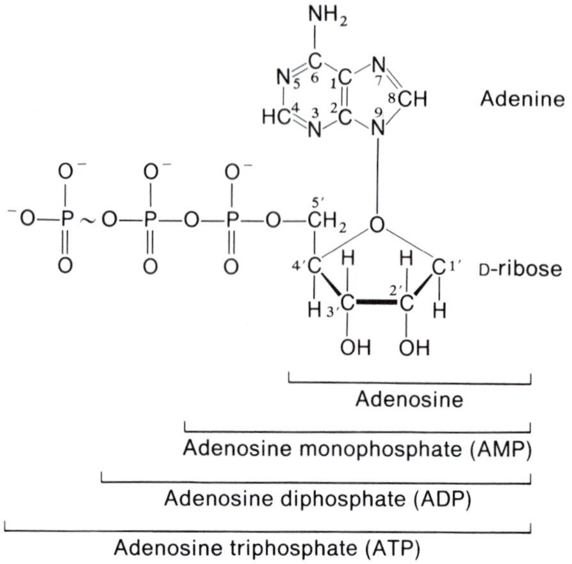

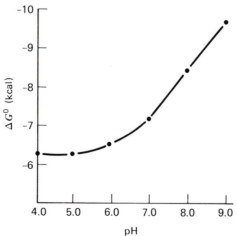

FIGURE 4-1
The structure of adenosine triphosphate (ATP). At pH 7 the molecule is ionized. Usually, the terminal phosphate (~bond) is hydrolyzed to provide energy.

Accordingly these reactions are called *hydrolyses,* meaning "split by water."

$$ATP + H_2O \xrightarrow[\text{enzyme}]{\text{ATPase}} ADP + P_i \qquad (4\text{-}3)$$

The standard free energy of ATP hydrolysis ($\Delta G^{\circ\prime}$) is -7.3 kcal $\cdot$ mol^{-1}. As pointed out earlier, the factors of metabolite concentration, temperature, and pH affect the actual free energy of hydrolysis (ΔG) in the cell. The effect of

FIGURE 4-2
Effect of pH on ΔG° of ATP at 25°C (298° K).

pH on the $\Delta G°$ of ATP is given in Figure 4-2. In the working muscle, ΔG for ATP is probably close to -11 kcal $\cdot$ mol^{-1}.

Three factors operate to give ATP a relatively high free energy of hydrolysis. First, the negative charges of the phosphates repel each other. Second, the products ADP and P_i form "resonance hybrids," which means that they can share electrons in ways to reduce the energy state. Third, ADP and ATP have the proper configurations to be accepted by enzymes that regulate energy-yielding and energy-requiring reactions.

The hydrolysis of ATP almost always involves splitting of the terminal phosphate group. Therefore, it is possible to write ATP as ADP$\sim$P, where the $\sim$P is the "high-energy" phosphate. Similarly, creatine phosphate can be abbreviated as C$\sim$P. Exergonic reactions of metabolism that hydrolyze the second phosphate group of ATP are infrequent; the myokinase reaction is a notable exception (Equation 2-23). There are probably no cellular reactions of energy supply that cleave the first phosphate from ATP.

ATP—the High-Energy Chemical Intermediate

In actuality the potential chemical energy of hydrolyzing the terminal phosphate of ATP is intermediate with regard to hydrolysis of other phosphorylated biological compounds (Figure 4-3). The term "high energy," when applied to

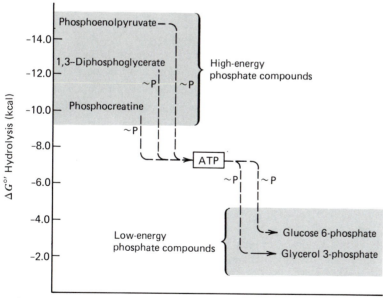

FIGURE 4-3
Standard free energies of hydrolysis of ATP as compared to those of related physiological compounds. Because of its intermediate position, ATP can transfer phosphate group energy to "low-energy" phosphate compounds, whereas ADP can accept phosphate group energy from "high energy" compounds.

ATP, more properly refers to the fact that the probability of hydrolysis transferring energy is high.

Figure 4-3 illustrates several aspects of the role of ATP in cellular energy transfer. First, the intermediate level of ATP in the scale of phosphate metabolites means that the phosphorylation of ADP to ATP can be driven by phosphorylated compounds of higher energy. These are energy-conserving reactions. The hydrolysis of ATP to ADP and P_i can transfer energy to other compounds or do cell work. These are energy-yielding reactions.

Second, it should be noted that many biological reactions of energy transfer involve phosphate exchange. The ATP $\leftrightarrows$ ADP + P_i system functions as a shuttle, or energy bridge, for exchanging phosphate energy.

ENZYMATIC REGULATION OF METABOLISM

The biochemical pathways that result in cellular energy transfer are discussed in the following chapters. Each of these pathways involves many steps, each of which is catalyzed by a specific enzyme. Chapter 2 noted some functions of enzymes. Although enzymes cannot change the equilibria of reactions, they can lower the energies of activation, thereby allowing spontaneous reactions to proceed. Additionally, by linking exergonic to endergonic reactions, enzymes facilitate endergonic processes.

The precise mechanisms by which enzymes operate are not completely understood, but some details of enzyme action are known. As illustrated in Figure 4-4a, enzymes are usually large molecules, in most cases with only a single site at which a reactant or *substrate* attaches. This site is called the *active site*. With a few exceptions, only the appropriate substrate interacts with an enzyme to induce a fit between the active site of the enzyme and the substrate. The combined substrate and enzyme are referred to as the *enzyme–substrate complex*. After the enzyme catalyzes the reaction, the products are released. Foreign substances that have the capability of competing for and occupying enzymatic sites are frequently poisons. An example is the hallucinogen LSD,

FIGURE 4-4

(a) Enzymes function, in part, by virtue of the substrate inducing a fit at the active site. By mechanisms still not completely understood, enzymes lower energies of activation thereby increasing the probability of the reaction occurring. (b) Some enzymes have additional (modulatory) sites at which factors other than substrates bind. When binding occurs at these modulator sites, the three-dimensional shape of the enzyme is adjusted such that the probability of substrate binding at the active site is affected. Modulators can stimulate or inhibit catalytic function. (See facing page.)

(SOURCE: Modified from A.J. Vander, J.H. Sherman, and D.S. Luciano, 1980. With permission.)

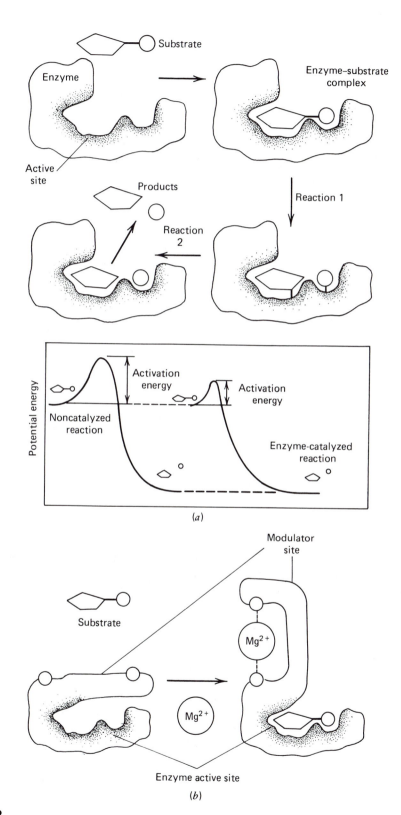

(a)

(b)

which competes with the chemical messenger serotonin in the brain. "Bad" and recurring "free" trips occur because LSD binds in such a way that serotonin has difficulty in displacing it.

In Figure 4-4b it should be noted that certain enzymes have binding sites other than the active site. When appropriate substances bind to these other sites, they affect the *configuration,* or shape, of the active site. Therefore, the binding of these other factors affects the interaction of the enzyme with its substrate, thereby changing the rate at which the enzyme can function. Consequently, these other factors that change catalytic rates of particular enzymes are termed *modulators.*

Modulators can be classified into two groups: Those that increase catalytic rates of enzymes are termed *stimulators,* and those that slow enzymatic function are *inhibitors.* In the control of energy metabolism, ATP is the classic example of an inhibitor, whereas ADP and P_i are usually stimulators.

In resting muscle, high levels of ATP inhibit carbohydrate, fat, and amino acid catabolism. However, when a muscle starts to twitch, ATP is degraded to ADP and P_i. When these are formed, in the locus of their presence they stimulate the mechanisms of ATP resupply.

The effect of various modulators on particular enzymes is called *allosterism,* referring to the fact that modulators change the spatial orientation, or shapes of parts of enzymes. When there are several modulators capable of affecting the catalytic rate of a particular enzyme, that enzyme is said to be *multivalent.* An example of a multivalent allosteric enzyme is phosphofructokinase (PFK), a key enzyme of carbohydrate catabolism (Chapter 5).

Different enzymes have different properties that have great effects on metabolism. Among these properties are the rate at which the enzyme functions. Maximum velocity (V_{max}) is an important descriptive parameter of enzymes. The Michaelis–Menten constant (K_M) describes the interaction of substrate and enzyme. The K_M is defined as the substrate concentration that gives half of V_{max}. V_{max} and K_M are illustrated in Figure 4-5. The enzymes glucokinase and hexokinase have the same catalytic function, to phosphorylate and activate sugars for further metabolism. These enzymes, however, have different catalytic activities, as measured by different V_{max} and K_M values.

Summary

Processes of food catabolism in cells are usually linked to the process of ATP restitution. In this way approximately 50% of the potential chemical energy released from the foodstuffs is captured in the common chemical intermediate, ATP. ATP, together with its storage form, creatine phosphate (CP), then serve as the immediate cellular energy sources on which endergonic processes depend. ATP and CP not only supply immediate cellular energy sources, but their relative levels also stimulate or inhibit processes of energy metabolism. At rest,

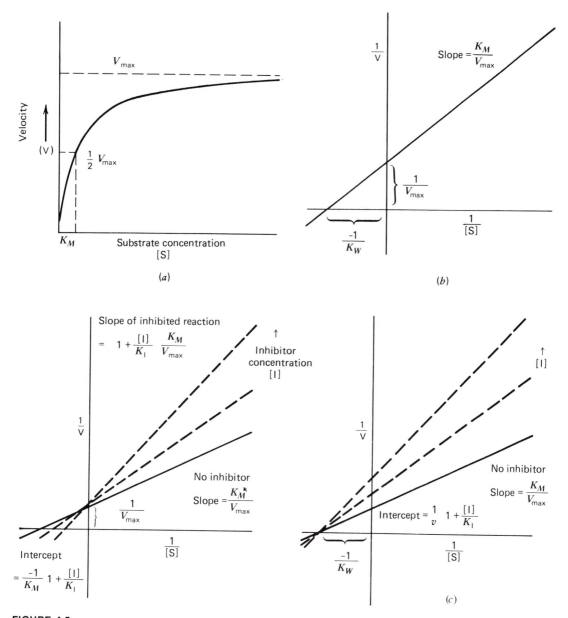

FIGURE 4-5

(a) Relationship between the maximum rate of catalysis for an enzyme (V_{max}) and its saturation by the substrate (Michaelis–Menten constant, K_M). The K_M is the substrate concentration that gives 50% of V_{max}. Enzymes can vary in both V_{max} and K_M. (b) The Lineweaver-Burk double-reciprocal plot is a means for accurately predicting V_{max} and K_M, as well as the effect of inhibitors. (c) Lineweaver-Burk plots of competitive (left) and noncompetitive inhibition (right). Different types of inhibitors have differential effects on K_M and V_{max}.

normal high levels of ATP and CP inhibit energy metabolism. When exercise starts, the utilization and decreased levels of ATP and CP, and the increased levels of ADP and P_i stimulate processes of energy metabolism.

Selected Readings

Cain, D.F. and R.E. Davies. Breakdown of adenosine triphosphate during a single contraction of working muscle. Biochem. Biophys. Res. Comm. 8:361–366, 1962.

Embden, G. and H. Lawaczeck. Über den zeitlichen Verlauf der Milchsäurebildung bei der Muskel Kontraktion. Z. Physiol. Chem. 170:311–315, 1928.

Fletcher, W.M. and F.G. Hopkins. Lactic acid in amphibian muscle. J. Physiol. 35:247–309, 1907.

Hill, A.V. The oxidative removal of lactic acid. J. Physiol. 48:x–xi, 1914.

Hill, A.V. and P. Kupalov. Anaerobic and aerobic activity in isolated muscle. Proc. Roy. Soc. (London) Ser. B. 105:313–322, 1929.

Lehninger, A.L. Biochemistry. New York: Worth Publishing, 1970. p. 313–327.

Lehninger, A.L. Bioenergetics Menlo Park, Cal.: W.A. Benjamin, Inc., 1978. p. 38–51.

Lipmann, F. Metabolic generation and utilization of phosphate bond energy. Advan. Enzymol. 1:99–162, 1941.

Lundsgaard, E. Phosphagen—and PyrophoPhatumsatz in Iodessigsäure-vegifteten Muskeln. Biochem. Z. 269:308–328, 1934.

Meyerhof, O. Die Energlewandlugen in Muskel I. Über die Bezie-hungen der Milchsäure Warmebildung und Arbeitstung des Muskels in der Anaerobiose. Arch. Gesamte Physiol. 182:232–282, 1920.

Meyerhof, O. Die Energie-wandlungen in Muskel. II Das Schicksal der Milschsäure in der Erholungs-periode des Muskels. Arch. Ges. Physiol. 182:284–317, 1920.

Mommarts, W.F.H.M. Energetics of muscle contraction. Physiol. Dev. 49:427–508, 1969.

Szent-Györgi, A. Myosin and Muscular Contraction. Basel: Karger Publ., 1942.

Szent-Györgi, A. Chemistry of Muscular Contraction. New York: Academic Press, 1951.

Wilkie, D.R. Thermodynamics and the interpretation of biological heat measurements. Prog. Biophys. Biophys Chem. 260–298, 1961.

5

GLYCOGENOLYSIS, GLYCOLYSIS AND GLUCONEOGENESIS: THE CELLULAR DEGRADATION AND SYNTHESIS OF SUGAR AND CARBOHYDRATE

Of the three main foodstuffs (carbohydrates, fats, and proteins), only carbohydrates can be degraded without the direct participation of oxygen. The main product of dietary sugar and starch digestion is glucose, which is released into the blood. The simple sugar glucose enters cells, including muscle and liver, and is either used directly or stored for later use. The first stage of cellular glucose catabolism is called glycolysis. Glucose molecules not undergoing glycolysis can be linked together to form the carbohydrate storage form called glycogen. Glycogen stored in muscle is broken down in a process called glycogenolysis. These processes occur in all cells but are specialized in some muscle cells and red blood cells. Glycogenolysis and glycolysis can provide the energy to sustain powerful muscular contractions for a period of a few seconds to about a minute. In prolonged exercise, glycogenolysis in muscle still provides a large part of the fuel for contractions. At the same time, glycogenolysis in liver provides glucose, which can circulate via the bloodstream to working muscle.

Lothar Thomas on the way to winning the 1 km race at the World's Track Cycling Championship in Brno, Czechoslovakia. Success in sprint activities, such as this, require great muscular power and the ability of muscles to operate upon chemical energy stores which do not immediately require O_2 for metabolism. Glycogen (stored carbohydrate) breakdown in muscle provides the majority of the non-oxidative energy. (U.P.I.).

THE DIETARY SOURCES OF GLUCOSE

Glucose, a six-carbon sugar, is the primary product of photosynthesis. Although parts of plants produce glucose, other parts utilize this sugar as a fuel much as we do. Still other parts of plants store glucose by linking the molecules together to form starch. Plants such as potatoes that store large amounts of starch are very useful to us as foodstuffs. Plants can also link glucose molecules together in a complex pattern to form cellulose for structural purposes.

The dietary sources of glucose are numerous. After their digestion and assimilation from the small intestine, most starches and dietary sugars reach the blood in the form of glucose. Large starch molecules are fairly rapidly split into disaccharides and glucose by the action of the pancreatic enzyme amylases. Sugars other than glucose are largely isomerized, that is, converted to glucose by the wall of the small intestine. Therefore, whether the dietary form is white or brown sugar, rice starch or glycogen, the vast majority of dietary sugars and starches reach the blood as glucose (Figure 5-1). Of course, it will take longer for a macrobiotic meal of brown rice, granola, and stone grit to be digested than a sugared drink, but the substance that reaches the blood will be

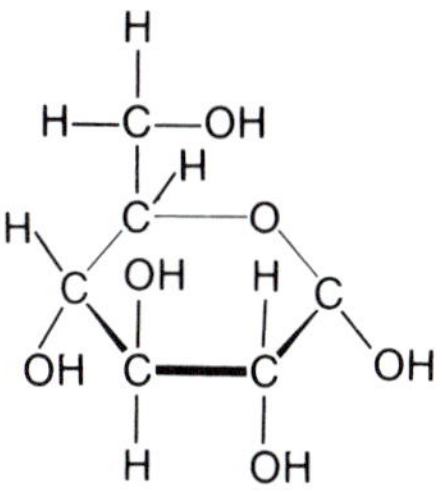

FIGURE 5-1
Structure of glucose, a simple sugar. Five carbons and an oxygen atom serve to create a hexagonal ring conformation. Shaded lines represent the three-dimensional platelike structure.

the same. Although cellulose is a polymer of glucose, humans do not have the enzymes to digest cellulose.

The uptake of glucose by cells depends on several factors, including the type of tissue, the levels of glucose in the blood and tissue, the presence of insulin, and the physiological status of the tissue. Most tissues, possibly with the exception of muscle during contraction, require insulin in order to take in glucose. The mechanism of the unique capacity by which muscle takes up glucose during exercise is under investigation. Nerve and brain tissues usually consume large amounts of glucose, and the liver can also take up large amounts. Traditionally, it has been thought that the liver serves to store and to release glucose, but does not utilize it as a fuel. Under appropriate circumstances, however, liver cells can use glucose as a fuel. Fat, or adipose cells, take up glucose when its level and that of the hormone insulin are high in the blood. In adipose cells, the presence of glucose stimulates fat synthesis (Figure 5-2).

GLYCOLYSIS

The metabolic pathway of glucose breakdown in mammalian cells is termed glycolysis, the dissolution of sugar. The process of glycolysis is frequently referred to as a metabolic pathway because the process proceeds by a specific route, in either 11 or 12 specific steps, each of which is catalyzed and regulated by a specific enzyme.

The study and appreciation of glycolysis predates recorded history. This is because yeast can carry glycolysis a few extra steps and produce ethyl alcohol and vinegar in a slightly longer process called *fermentation*. There is some evidence that the earliest communities in ancient Egypt were organized around structures that served as bakeries and breweries. It was probably discovered quite by accident that grain left to soak fermented to become beer, but it is

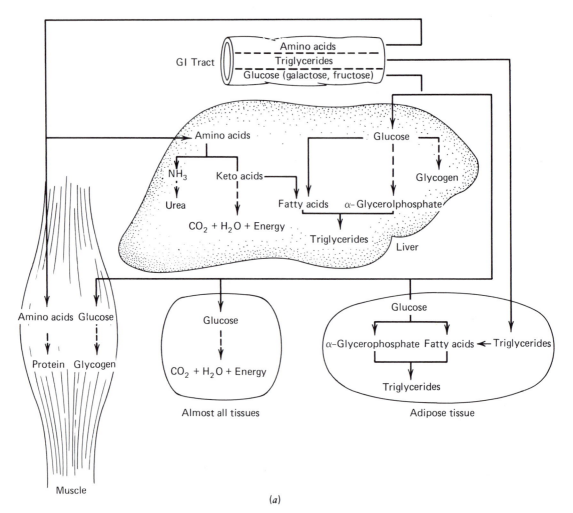

(a)

FIGURE 5-2
Flowchart of glucose and other metabolites in an individual (a) immediately after a meal and (b) during fasting.

(SOURCE: Modified from A.J. Vander, J.H. Sherman, and D.S. Luciano, 1980. With permission.)

quite ironic that the earliest Middle Eastern societies were organized like the contemporary college town—around the consumption of beer. Louis Pasteur was a famous man before he developed any vaccines to prevent disease. Pasteur discovered a method to keep ordinary table wine from fermenting to vinegar. Today, his process, called *pasteurization* is more frequently used to preserve milk.

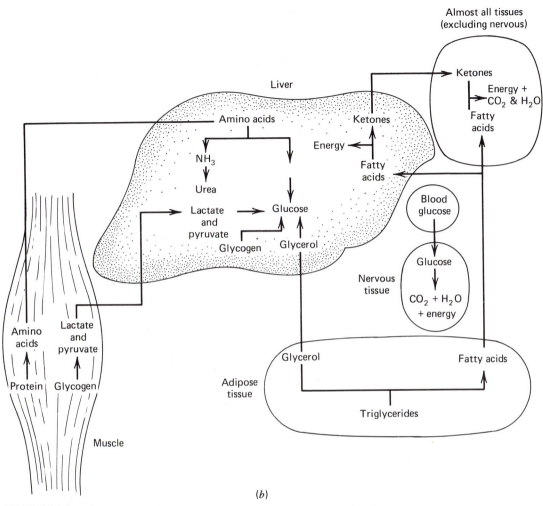

FIGURE 5-2 *(continued).*

GLYCOLYSIS IN MUSCLE

Glycolysis is a very active process in skeletal muscle, which is often termed a "glycolytic tissue." In particular, pale or white skeletal muscles contain large quantities of glycolytic enzymes. As will be discussed later, some highly oxidative red muscle fibers are capable of rapid glycolysis, but glycolysis is the main energy source for white muscle fibers during exercise (Figure 5-3). Because it is the predominant metabolic pathway of energy conservation in amphibians and reptiles, glycolysis is frequently considered a "primitive pathway." Because their hearts and lungs are poorly developed, amphibians and

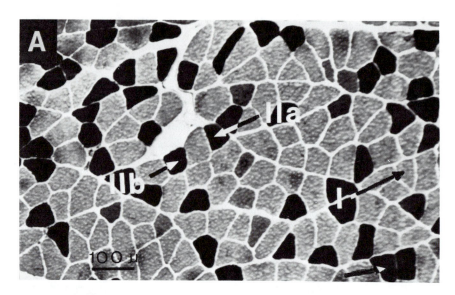

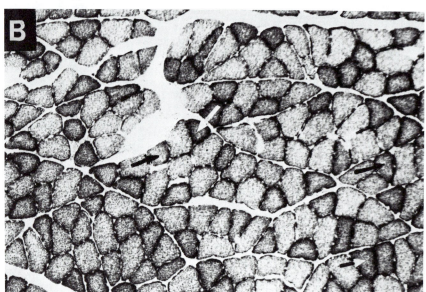

FIGURE 5-3

Serial transverse sections of rodent hindlimb skeletal muscle incubated as follows: (a) myofibrillar ATPase, pH 9.4; (b) succinate dehydrogenase (SDH); and (c) myofibrillar ATPase, pH 4.3. Fibers are classified as type I or type II based on low or high staining intensity, respectively, of the alkaline myofibrillar ATPase reaction (a). Type II fibers that demonstrate "acid reversal" of ATPase activity (a vs. c) are further classified as type IIa and IIb based on high or low activity, respectively, of SDH (b). Type II fibers that do not demonstrate "acid reversal" are classified IIc (c) and are found

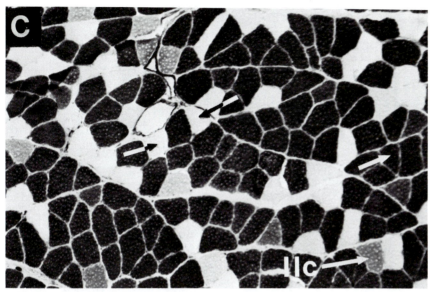

FIGURE 5-3 *(continued).*

rarely in normal muscle but frequently in fetal and regenerating muscle. It is likely that the undifferentiated type IIc fibers are transitory rather than being designated for permanent use. Type I fibers are sometimes also called "slow-oxidative" (SO) fibers; type IIa fibers are sometimes also called "fast-oxidative glycolytic" (FOG) fibers, and type IIb fibers are sometimes also called "fast glycolytic" (FG) fibers. Types I and IIa fibers are red in color, whereas type IIb is white. Refer to Figure 17-8.

(Micrographs courtesy of T.P. White.)

reptiles cannot utilize oxidative metabolism for bursts of activity and must therefore rely on glycolysis.

"AEROBIC," "ANAEROBIC," AND "NONROBIC" GLYCOLYSIS

There are two forms of glycolysis: aerobic ("with oxygen") and anaerobic ("without oxygen"). Historically, these terms were developed by scientists, such as Pasteur, who studied glycolysis in yeast and other unicellular organisms in test tubes where air was either present or was flushed out by gassing with substances such as nitrogen. Pasteur noted that when oxygen was present (aerobic), glycolysis is slow. We now know also that the product is pyruvic acid. When O_2 was removed (anaerobic), Pasteur found glycolysis to be rapid. Today we also know that anaerobic glycolysis proceeds an additional step beyond pyruvic acid to form lactic acid (Figure 5-4).

The early experimentation and terminology has led to some confusion in

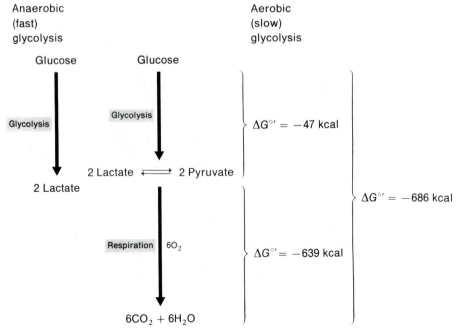

FIGURE 5-4

In anaerobic (fast) glycolysis, lactic acid is the product. In aerobic (slow) glycolysis, pyruvic acid is the main product. The terms aerobic (with O_2) and anaerobic (without O_2) refer to the test tube conditions used by early researchers to speed up or slow down glycolysis. In real life, pyruvate and lactate pools are in equilibrium, and the rapidity of glycolysis largely determines the product formed. Note the far greater energy released under aerobic conditions.

(SOURCE: Modified from A.L. Lehninger, 1973. With permission.)

contemporary physiology, because when early scientists observed increased levels of lactic acid in muscle and blood as the result of exercise, they assumed that the tissue was anaerobic (without oxygen) during exercise. As will be seen, there are several reasons for lactic acid formation; no oxygen or insufficient oxygen supply is only one of them. Further, a constant level of lactic acid in the blood does not mean that no lactic acid was formed, only that production and removal were equivalent. Also, there is good evidence that lactic acid is always produced, even at rest. Therefore, the terms aerobic and anaerobic as applied to mammalian systems are archaic. We can deal with these terms and use them so long as we are aware of their actual meaning. It is to be hoped that more descriptive terms such as "rapid" (for anaerobic) and "slow" (for aerobic) will come into use. Because much of glycolysis has little to do with the presence of O_2, the process itself is essentially "nonrobic"—that is, it does not involve O_2.

The glycolytic pathway is depicted in two ways in Figures 5-5 and 5-6. Figure 5-5 is a detailed presentation that will serve as a reference. Figure 5-6 is designed to contribute to an understanding of how the pathway is controlled. In glycolysis, one six-carbon sugar is split into two three-carbon carboxylic acids. Pyruvic and lactic acid each possess a carboxyl group (Figure 5-7). At physiological pH, these molecules dissociate a hydrogen ion (H^+) and therefore are acids. The terms pyruvate and lactate, properly meaning salts of the respective acids, are generally used interchangeably with pyruvic acid and lactic acid.

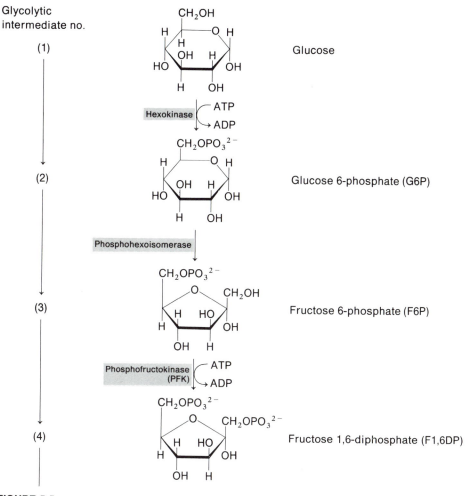

FIGURE 5-5

Detailed presentation of the glycolytic pathway. Glycolytic intermediates are identified by name to the right of each structure and by number to the left; catalyzing enzymes are noted on the left of each reaction step.

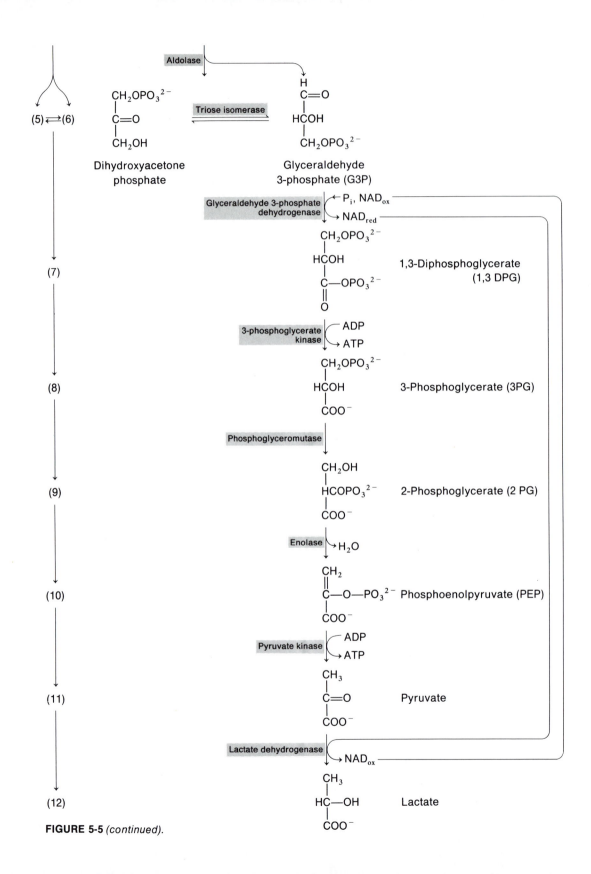

FIGURE 5-5 *(continued).*

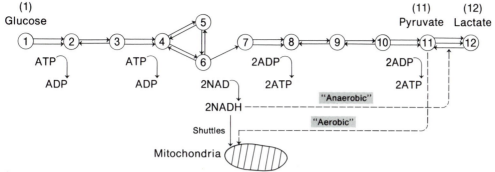

FIGURE 5-6
Simplified portrayal of glycolysis showing the beginning and ending substances, the sites of ATP utilization and production, and the sites of NADH formation and utilization.

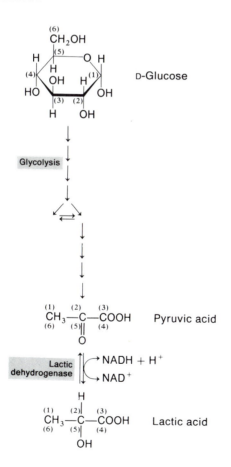

FIGURE 5-7
Chemical structures of glucose and of pyruvic and lactic acids. Small numbers in parentheses identify the carbon atoms in the original glucose structure. At physiological pH, lactic and pyruvic acids dissociate a hydrogen ion.

(SOURCE: Modified from A.L. Lehninger, 1973. With permission.)

The substance diagrammed in Figure 5-8*a* is called nicotinamide adenine dinucleotide (NAD^+). Because of its unique structure, NAD^+ can exist in two forms, NAD^+ (oxidized) and NADH (reduced, Figure 5-8*b*). We will encounter NAD^+ frequently, because it acts to transfer hydrogen ions and electrons within cells. Examination of its structure (Figure 5-8*a*) reveals that NAD^+ contains the nucleotides adenine and nicotinamide. We have already encountered aden-

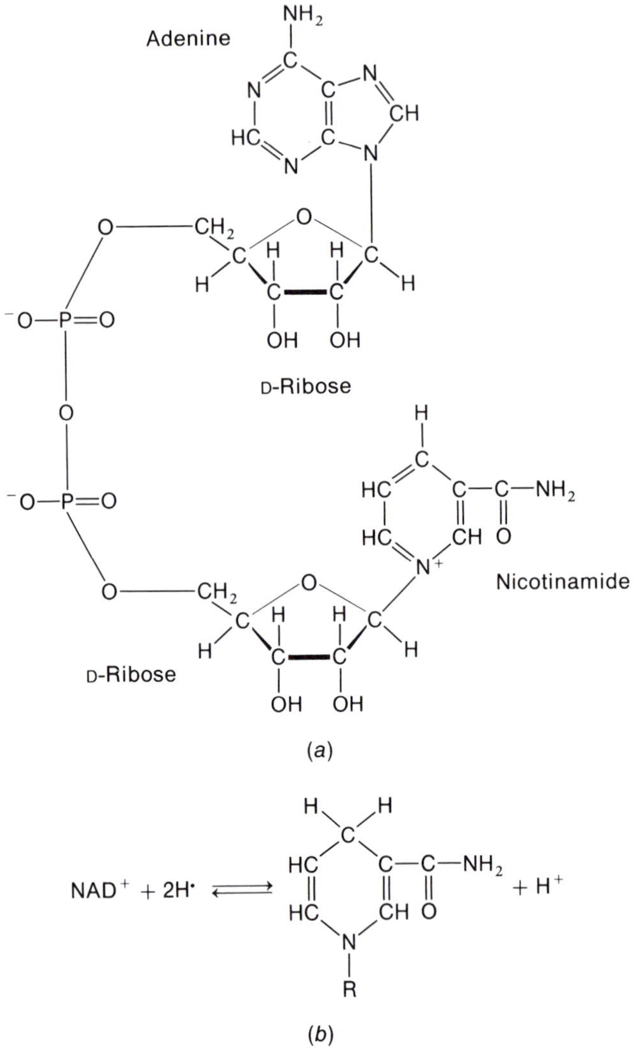

(a)

$$NAD^+ + 2H^\bullet \rightleftharpoons \quad + H^+$$

(b)

FIGURE 5-8

Structures of NAD^+ and NADH. NAD can exist in two forms: (a) without added hydrogen and electrons, that is, oxidized or (b) with added hydrogen and electrons, that is, reduced. NAD serves to transfer these high-energy species within cells:

$NAD^+ + 2\ H^\bullet \rightleftharpoons NADH + H^+$

ine in the structure of ATP (Figure 4-1). Nicotinamide is the product of a B vitamin. The dietary deficiency disease pellagra is typified by poor energy metabolism. Flavine adenine dinucleotide (FAD) is a similar compound to NAD^+. FAD can be reduced to $FADH_2$.

A simplified flowchart for glycolysis (Figure 5-9) reveals that step 6 involves the reduction (adding of hydrogen and electrons) of NAD^+ to NADH. Under resting (aerobic) conditions, (Figure 5-9a), NADH transports or "shuttles" the hydrogen and electrons to mitochondria (those cellular organelles where most oxygen is consumed). Under these circumstances, the end product of glycolysis, pyruvate, will also be consumed by the mitochrondria. However, if there is insufficient mitochrondrial activity, such as during maximal exercise (Figure 5-9b), then NADH will be oxidized and pyruvate reduced to form lactate in the cytoplasm as the result of rapid glycolysis.

$$\text{Pyruvate} + \text{NADH} + \text{H}^+ \xrightarrow{\text{lactic dehydrogenase}} \text{lactate} + \text{NAD}^+ \qquad (5\text{-}1)$$

The formation of lactate or pyruvate, then, depends on mitochondrial activity and not on the presence of oxygen.

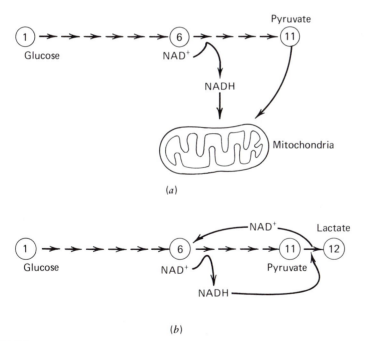

(a)

(b)

FIGURE 5-9
Illustration of the Pasteur effect. (a) When glycolysis is slow, mitochondria can accept the NADH and pyruvate formed. (b) When mitochondrial function is inadequate because of low enzymatic activity or insufficient O_2 supply, glycolysis is rapid.

CYTOPLASMIC—MITOCHONDRIAL SHUTTLE SYSTEMS

Several scientists has elucidated the pathways and controls of the two glycolytic–mitochrondrial shuttle systems. These are diagrammed in simple form in Figure 5-10. The malate–aspartate shuttle predominates in the heart, and the glycerol phosphate shuttle in skeletal muscle. In addition to differences in mechanism and tissue specificity, the shuttles differ in the energy level of the

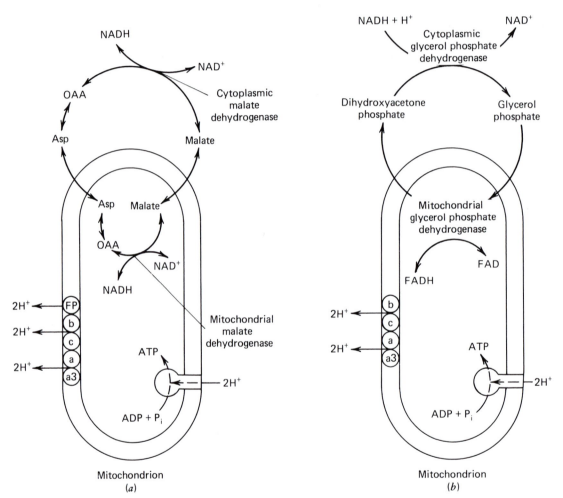

FIGURE 5-10

Shuttle systems for moving reducing equivalents generated from glycolysis in the cytoplasm to mitochondria. (a) The malate–aspartate shuttle is the main mechanism in the heart. (b) The glycerol phosphate shuttle is the main mechanism in skeletal muscle. Each proton pair pumped out of mitochondria results in a sufficient chemical and osmotic energy gradient to form an ATP molecule. NADH ≃ 3ATP; FADH ≃ 2ATP.

(SOURCE: Modified from A. L. Lehninger, 1971. With permission.)

product that becomes available within mitochondria. As will be seen, NADH will generate three ATP molecules within mitochrondria for each atom of oxygen consumed (ATP : O = ADP : O = P_i : O = 3), and FADH will generate two ATPs within mitochondria (P_i : O = 2).

ATP YIELD BY GLYCOLYSIS

At the beginning of glycolysis (see Figure 5-5) there are two activating steps where ATP is consumed. However, following the cleavage of the six-carbon molecule into two three-carbon molecules, there are two steps where an ATP molecule is formed. Because these steps occur twice each for each six-carbon glucose that enters the pathway, the gross ATP yield is $2 \times 2 = 4$. However, if we can then subtract the two ATP molecules used for activation, the net yield is 2.

Under "aerobic" conditions the formation of two ATPs per glucose is complemented by the formation for mitochrondrial consumption of two NADHs. Depending on the shuttle system used, these are equivalent to another four to six ATPs.

$$(2NADH) (3ATP/NADH) = 6ATP \qquad (5\text{-}2)$$

$$(2FADH) (2ATP/FADH) = 4ATP \qquad (5\text{-}3)$$

"Anaerobic" glycolysis is summarized in the following equation.

$$Glucose + 2P_i + 2ADP \xrightarrow[\text{glycolysis"}]{\text{"anaerobic}} 2lactate + 2ATP + H_2O \qquad (5\text{-}4)$$

"Aerobic" glycolysis is summarized in the following equation.

$$Glucose + 2P_i + 2ADP + 2NAD^+ \xrightarrow[\text{glycolysis"}]{\text{"aerobic}} 2pyruvate + 2ATP + 2NADH + 2H_2O \quad (5\text{-}5)$$

THE EFFICIENCY OF GLYCOLYSIS

The formation of only two ATP's from each glucose in glycolysis is seemingly a small number. Therefore, glycolysis is sometimes called an "inefficient" pathway. By the definition of number of ATP molecules formed from a glucose molecule, glycolysis is inefficient. Fortunately, skeletal muscle (white and red glycolytic fibers) can break down glucose rapidly and produce significant qualities of ATP for short periods of glycolysis.

In actuality, an appropriate energetic consideration reveals that the efficiency of glycolysis is excellent. The energy change (ΔH) from glucose to lactate is -47 kcal $\cdot$ mol^{-1}. If two ATPs are formed, and $\Delta G^{\circ\prime}$ for ATP $= -7.3$ kcal $\cdot$ mol^{-1}, then

$$\text{Efficiency} = \frac{2(-7.3)}{-47} = 31\% \tag{5-6}$$

If ΔG for ATP is -11 kcal $\cdot$ mol^{-1},

$$\text{Efficiency} = \frac{2(-11)}{-47} = 46.8\% \tag{5-7}$$

This latter figure of almost 50% compares favorably with that of oxidative enzymatic processes. Therefore, it is not correct to assume that glycolysis, or "anaerobic metabolism," is inefficient, because the enzymes have conserved a good deal of the energy released.

THE CONTROL OF GLYCOLYSIS

Phosphofructokinase

Several mechanisms serve to regulate glycolysis. Usually the dominant factor is the activity of the enzyme phosphofructokinase (PFK). PFK, which catalyzes the third step of glycolysis, is a multivalent, allosteric enzyme. Known modulators are given in Table 5-1. PFK is probably the rate-limiting enzyme when glycolysis is rapid during exercise, but factors such as ATP, C~P, and citrate affect the conformation of PFK to slow its activity during rest. When exercise starts, immediate changes in the relative concentrations of modulators of PFK serve to increase its activity. Hexokinase, pyruvate kinase, and lactic dehydrogenase are other important controlling enzymes whose activities are modulated.

The Cellular "Energy Charge"

Earlier it was shown that the contractile actin–myosin system splits ATP to ADP and P_i, and that adenyl kinase can act to maintain ATP level by forming ATP and AMP from two ADP's. The contents of the ATP–ADP–AMP system can therefore exist in one of three forms. The "energy charge" of the cell is defined as follows:

$$\text{Adenine nucleotide energy charge} = \left(\frac{[\text{ADP}] + 2\,[\text{ATP}]}{[\text{AMP}] + [\text{ADP}] + [\text{ATP}]} \right) 1/2 \tag{5-8}$$

TABLE 5-1
Control Enzymes of Glycolysis

Enzyme	Stimulators	Inhibitors
Phosphofructokinase	ADP, P_i, AMP, ↑pH, (NH$_4^+$),	ATP, C~P, citrate
Pyruvate kinase		ATP, C~P
Hexokinase		Glucose 6-phosphate
Lactic dehydrogenase		ATP

If all the adenine nucleotide is in the form of ATP, the energy charge is 1.0. If all the adenine nucleotide is in the form of AMP, the energy charge is 0.0. In the cell the energy charge is usually around 0.8. Even small decrements below that serve to activate ATP-yielding systems.

Thermodynamic Control

As indicated in Figure 5-6, there are several exergonic steps where there are significant free energy changes. These reactions are catalyzed by the enzymes hexokinase, phosphofructokinase, pyruvate kinase, and lactic dehydrogenase. These reactions serve to keep glycolysis going in the direction of product as the other steps have small free energy changes and are freely reversible. The reverse of glycolysis is not possible without energy input and the intervention of specific enzymes. This reversal process is called *gluconeogenesis,* meaning "making new glucose." Gluconeogenesis is a capability of muscle, but gluconeogenesis is presently not believed to be a primary function of muscle. Gluconeogenesis is a function of tissues such as liver and kidney.

Control by Lactic Dehydrogenase

The terminal enzyme of glycolysis, which results in the formation of lactic acid from pyruvic acid, is lactic dehydrogenase (LDH). As shown earlier (Figure 5-9a), LDH is in competition with mitochondria for pyruvate. Ordinarily, the uptake of pyruvate in mitochondria is such that the presence of LDH has little effect. However, LDH is an enzyme of significant content in muscle, especially white skeletal muscle. Furthermore, the K'_{eq} of LDH is large and the reaction proceeds actively to completion. Therefore, some lactate is always formed.

There are two basic types of LDH: muscle (M) and heart (H). These LDH types are found predominantly in white skeletal muscle and heart, respectively. The equilibria of the two types are identical, but they differ in their affinities for reactants (substrates) and products. The M type has a high affinity for the substrate pyruvate, and therefore has higher biological activity than the H type which has a lower affinity for pyruvate.

Each molecule of LDH has four subunits. Considering the two basic types of LDH, there are then five possible arrangements: M_4, M_3H_1, M_2H_2, M_1H_3, and H_4. The population of these *isozymes* of LDH varies among tissues, with M_4 being highest in white skeletal muscle and lowest in heart.

The biological activity of LDH, then, depends on its concentration and isozyme type. Active recruitment or use of white skeletal muscle will inevitably result in lactate production regardless of the presence of O_2.

Control by Substrate Availability

Glucose uptake   In order to perform glycolysis, the cell must have glucose or its cellular storage form, glycogen. As noted at the beginning of the chapter, the entry of glucose into most cells depends on the level of insulin present.

During prolonged exercise the circulating level of insulin falls, but active muscle continues to be able to take up glucose. By limiting glucose uptake by other tissues, the fall in insulin level may indirectly serve to enhance uptake by active muscle during prolonged exercise. However, if blood glucose level falls during prolonged exercise, muscle glycolysis will be impaired for lack of the initial substrate or fuel.

GLYCOGENOLYSIS

Skeletal muscle glycolysis is heavily dependent on the intramuscular storage form, glycogen. During heavy muscular exercise, glycogen may supply most of the immediate glucosyl residues. The structure of glycogen (Figure 5-11) consists mostly of end-to-end (C1–C6) linkages, with a few cross (C1–C4) linkages. The storage of glucose units as glycogen is dependent on activity of

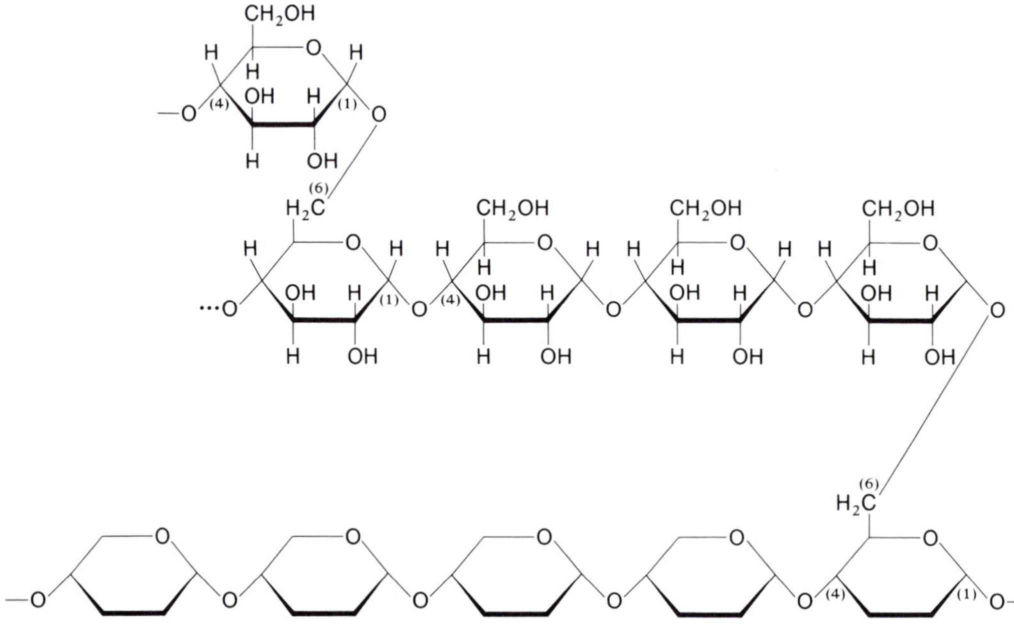

(a)

FIGURE 5-11
(a) The structure of glycogen is seen to be a polymer of glucose units. The linkages exist mainly end to end (carbon 1 to carbon 4 linkages), but there is also some cross-bonding (carbon 1 to carbon 6 linkages). (b) A pinwheel-like structure results from C1—C4 and C1—C6 linkages. The hexagons represent glucosyl units.

(SOURCE: Modified from R.W. McGilvery, Biochemical Concepts, 1975, With permission.)

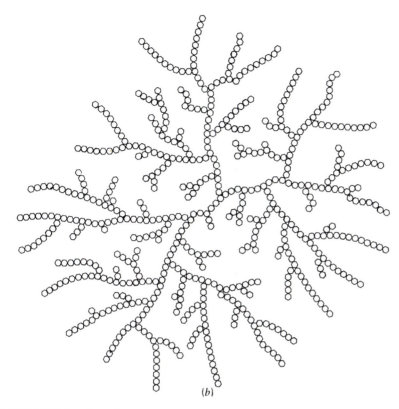

(b)

FIGURE 5-11 (continued).

the enzyme glycogen synthetase (Figure 5-12). Breakdown of glycogen is dependent on the enzyme phosphorylase. The activity of phosphorylase appears to be controlled by two mechanisms. One system is hormonally mediated and depends on the extracellular action of epinephrine and the intracellular action of cyclic AMP (cAMP), the "intracellular hormone" (Figure 5-13).

This mechanism is too slow to explain the rapid glycolysis during onset of heavy exercise. Therefore, a mechanism mediated by Ca^{2+} released from the sarcoplasmic reticulum may constitute a parallel control mechanism (Figure 5-14).

During rest, cellular uptake of glucose is usually sufficient to support glycogen synthesis and glycolysis. During maximal exercise, glycogenolysis is probably sufficient to support rapid glycolysis. During prolonged exercise, the depletion of intramuscular glycogen and liver glycogen can result in decreased capacity of substrate for glycolysis. Moreover, it is very likely also that the ability to glycolyze limits the ability to utilize fat. Therefore, glycogen depletion during exercise may be doubly important. The interaction between carbohydrate and fat metabolism is discussed more later on.

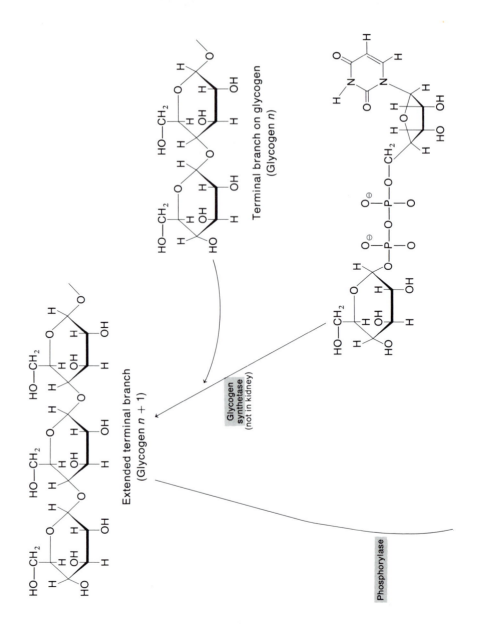

Terminal branch on glycogen
(Glycogen *n*)

Extended terminal branch
(Glycogen *n* + 1)

Glycogen
synthetase
(not in kidney)

Phosphorylase

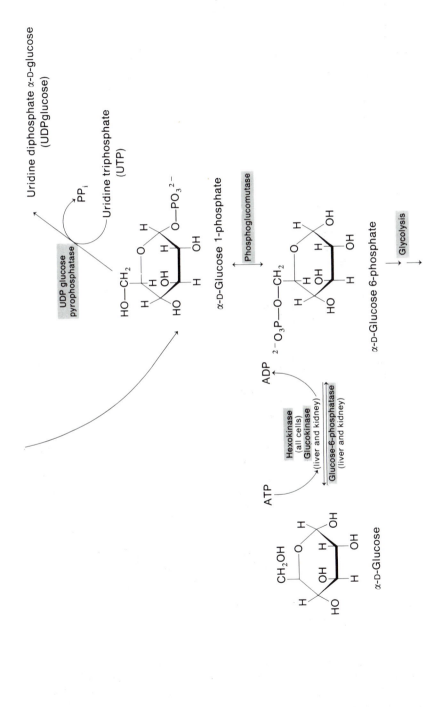

FIGURE 5-12

Relationship among hexokinase, glucose-6-phosphatase, glycogen synthetase, and phosphorylase and other enzymes for storing and utilizing glucose units.

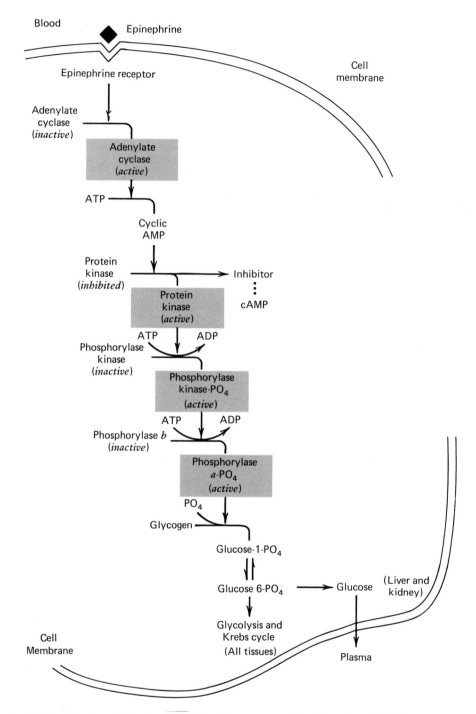

FIGURE 5-13

The breakdown of glycogen in muscle is controlled by the enzyme phosphorylase. (a) Phosphorylase b (the inactive form) can be converted to phosphorylase a (the active form) through a series of events initiated by the hormone epinephrine. This mechanism involves cyclic AMP (cAMP), the intracellular hormone. During sudden bursts of activity the cAMP mechanism is too slow to account for the observed glycogenolysis. Intracellular factors such as Ca^{2+} and AMP increase the catalytic activity of phosphorylase. (SOURCE: Modified from A.J. Vander, J.H. Sherman, and D.S. Luciano, 1980. With permission.)

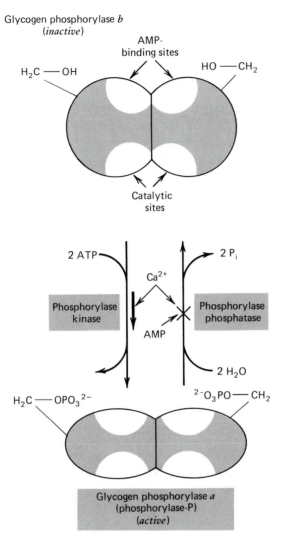

FIGURE 5-14

The conversion of phosphorylase *b* (the inactive form of the enzyme) to phosphorylase *a* (the active form) depends on the stimulation of phosphorylase kinase by Ca^{2+}. Calcium ion is released immediately when muscles contract, and this mechanism helps to link pathways of ATP supply with those of ATP utilization. During exercise the level of AMP increases; this helps to minimize the reconversion of phosphorylase *a* to *b* by inhibiting phosphorylase phosphatase.

(SOURCE: Modified from R.W. McGilvery, 1975. With permission.)

GLUCONEOGENESIS

Although skeletal muscle itself does not have the enzymatic apparatus to make glucose, muscle does participate in gluconeogenesis, the making of new glucose.

Cori and Cori were among the first to recognize that the lactate and pyruvate produced by skeletal muscle could circulate to the liver and be made into glucose. The glucose so produced could then recirculate to muscle. This cycle of carbon flow is called the Cori cycle (Figure 5-15). Because rapid glycolysis in

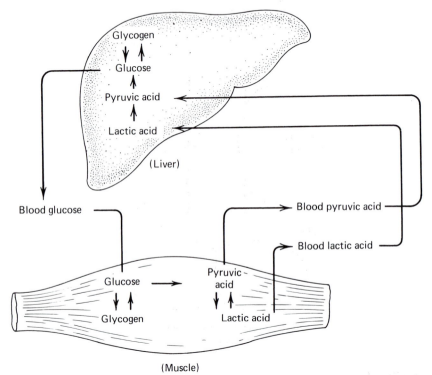

FIGURE 5-15
The Cori cycle, showing that pyruvate and lactate formed in muscle can circulate to liver and kidney. There carboxylic acids can be synthesized to glucose. The glucose thus formed can then reenter the circulation.

FIGURE 5-16
Gluconeogenesis is the process of making new glucose. This process occurs mainly in liver and to some extent in the kidney. Gluconeogenesis depends on the activities of specialized enzymes that can bypass the exergonic steps in glycolysis by catalyzing energy inputs from ATP and NADH into endergonic steps. In this scheme, note the central role played by glucose 6-phosphate (G6P). (See facing page.) For other abbreviations used, see Figure 5-5.

(SOURCE: Modified from A.L. Lehninger, 1971. With permission.)

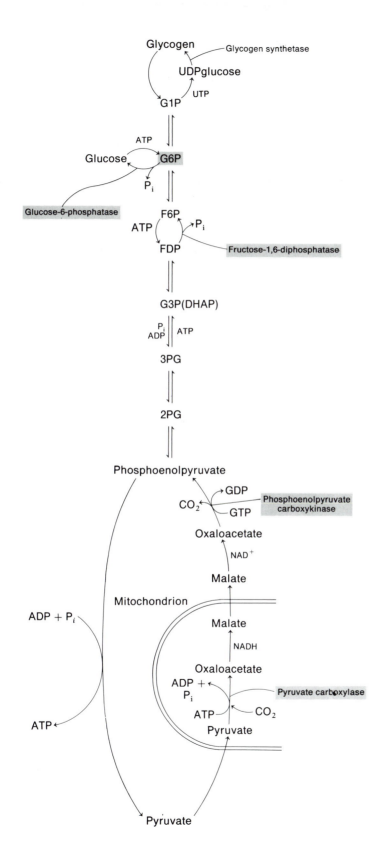

skeletal muscle inevitably results in lactate production because of the activity and K'_{eq} of LDH, gluconeogenesis is an efficient way to reutilize the products of glycolysis, thereby providing for the maintenance of blood glucose and prolonged muscle glycolysis. The Cori cycle is complemented by the glucose-alanine cycle, which is discussed later in this book (Chapter 8).

Because glycolysis is an exergonic process, gluconeogenesis must be endergonic. Gluconeogenic tissues such as liver and kidney therefore possess enzymes that bypass the controlling exergonic reactions of glycolysis (Figure 5-16). For each glucose molecule formed, 6 high-energy phosphate bonds and 2 NADH's—an equivalent of 12 ATPs—are required.

Gluconeogenesis may be considered to consist of three parts: the raising of pyruvate to glucose 6-phosphate (G6P), the formation of glucose, and the synthesis of glycogen. Skeletal muscle is incapable of quantitatively converting pyruvate to phosphoenolpyruvate (PEP) or of releasing free glucose. This is because muscle has low activities of the enzymes PEP carboxykinase, and glucose-6-phosphatase. However, although kidney is a gluconeogenic tissue, it lacks glycogen synthetase and does not store significant amounts of glycogen (Figure 5-12).

As suggested in Figure 5-16, the metabolism of G6P largely determines glycolytic carbon flow. The importance of G6P as a junction of carbon flow in the control of metabolism cannot be overestimated. What happens to G6P can influence glycogen synthesis, glucose release, and glycolysis.

Control at the Pathway Beginning

In regulating a process, control is often exercised at the beginning of the process. The entry of material into glycolysis via action of the enzymes hexokinase and phosphorylase, and the control exerted by PFK are good examples of the control of biological pathways at their beginning.

EFFECTS OF TRAINING ON GLYCOLYSIS

Studies on the effects of training on the glycolytic capability of skeletal muscle have mainly utilized the technique of catalysis. In this experimental approach, a muscle sample of the experimental individual (animal or human) is taken. Then, either an attempt is made to isolate the enzyme, or, more usually, the muscle sample is homogenized and the activity of the enzyme is studied. This can be done by observing the disappearance of a substrate or the appearance of a product.

Compared to its considerable effect on oxidative enzymes of muscle (Chapter 6), endurance training appears to have relatively insignificant effects on catalytic activities of glycolytic enzymes. This may be because skeletal muscle, especially fast-twitch muscle (Figure 5-3 and Chapter 17), has an intrinsically high glycolytic capability.

Endurance training apparently has little effect on most glycolytic enzymes. Reports of the effects of endurance training on PFK vary, so we may conclude for the present that endurance training has no great effect on PFK activity.

Hexokinase activity increases significantly in endurance-trained animals. This adaptation, which is thought to facilitate the entry of sugars from the blood into the glycolytic pathway of muscle, would be of benefit in prolonged exercise where the liver serves to supply muscle with glucose.

Endurance training has been observed to *decrease* LDH activity in fast gly-colytic muscle and to influence the LDH isozymes in muscle to include more of the heart type. This adaptation may serve to decrease lactate production per unit glycolytic carbon flow in endurance-trained muscle during exercise, and to allow some muscle fibers to take up and oxidize lactate produced in other fibers. Studies of the effects of speed and power training on activities of glyco-lytic enzymes are not as numerous as studies on the effects of endurance training. However, as with endurance training, the effects of speed and power training on the specific activities of glycolytic enzymes are not great. If the muscle hypertrophies as the result of speed and power training, then the total catalytic activity of the muscle will probably improve.

Limitations of the Techniques

Although studies of the catalytic activity of muscle glycolytic enzymes are use-ful in terms of predicting the glycolytic capacity of muscle, these types of studies actually say very little about whether, or how, the enzyme will operate in the body during exercise. The application of isotope tracer techniques to exercise studies should soon make it possible to confirm the above interpreta-tion of the effect of training on glycolysis during exercise, which is based on studies using catalytic techniques.

In the past, investigators attempted to determine glycolytic activity during exercise by assaying blood lactate concentration after exercise. However, it has since been determined that the kinetics and pathways of lactate removal during and after exercise are very rapid and diverse. Therefore, attempts at quantifying glycolysis from postexercise blood lactate level are pointless. The accumulation of lactate during exercise means solely that production was greater than re-moval. This topic is discussed in more detail in Chapter 10.

Summary

Glycolysis is a specialized function in muscle. Large quantities of glycolytic enzymes in muscle can support powerful contractions for brief periods. In these situations, muscle glycogenolysis provides most of the substrate for glycolysis.

During prolonged work, hepatic (liver) glycogenolysis can supply significant amounts of glucose, which can circulate to active muscle.

When glycolysis is very rapid and mitochondrial activity is inadequate, lactic acid will be formed. When mitochondrial activity is inadequate, the cellular energy charge falls, activating PFK, which accelerates glycolysis. When mitochondria cannot accept the pyruvate formed as the result of glycolysis, pyruvate will accumulate and be converted to lactic acid. This is because the equilibrium of LDH is in the direction of product.

Selected Readings

Baldwin, K.M., W.W. Winder, R.L. Terjung, and J.O. Holloszy. Glycolytic enzyme in different types of skeletal muscle: adaptation to exercise. Am. J. Physiol. 225:962–966, 1973.

Barnard, R.J., V.R. Edgerton, T. Furukawa, and J.B. Peter. Histochemical, biochemical and contractile properties of red, white, and intermediate fibers. Am. J. Physiol. 220:410–414, 1971.

Barnard, R.J. and J.B. Peter. Effect of training and exhaustion on hexokinase activity of skeletal muscle. J. Appl. Physiol. 27:691–695, 1969.

Cori, C.F. Mammalian carbohydrate metabolism. Physiol. Rev. 11:143–275, 1931.

Crabtree, B. and E.A. Newsholme. The activities of phosphorylase, hexokinase, phosphofructokinase, lactic dyhydrogenase and glycerol 3-phosphate dehydrogenases in muscle of vertebrates and invertebrates. Biochem. J. 126:49–58, 1972.

Davies, K.J.A., L. Packer, and G.A. Brooks. Exercise bioenergetics following spring training. Arch. Biochem. Biophys. 215:260–265, 1982.

Depocas, F., J. Minaire, and J. Chattonet. Rates of formation of lactic acid in dogs at rest and during moderate exercise. Can. J. Physiol. Pharmacol. 47:603–610, 1964.

Donovan, C.M. and G.A. Brooks. Endurance training affects lactate clearance, not lactate production. Am. J. Physiol. 244 (Endocrinol. Metab. 7): E83–E92, 1983.

Eldridge, F.L. Relationship between turnover rate and blood concentration of lactate in exercising dogs. J. Appl. Physiol. 39:231–234, 1975.

Embden, G., E. Lehnartz, and H. Hentschel. Der zeitliche Verlauf der Milschsäurebildung bei der Muskelkontraktion. Mitteilung. z. Physiol. Chem. 176:231–248, 1928.

Gollnick, P.D., R.B. Armstrong, W.L. Sembrowich, and B. Saltin. Glycogen depletion pattern in human skeletal muscle fibers after heavy exercise. J. Appl. Physiol. 34:615–618, 1973.

Gollnick, P.D., R. Armstrong, C. Saubert, W. Sembrowich, R. Shepherd, and B. Saltin. Glycogen depletion patterns in human skeletal muscle fibers during prolonged work. Pflügers Arch. 344:1–12, 1973.

Gollnick, P.D. and L. Hermansen. Biochemical adaptations to exercise: anaerobic metabolism. In J.H. Wilmore (Ed.), Exercise and Sport Sciences Reviews, vol. 1. New York: Academic Press, Inc., 1973. p. 1–43.

Gollnick, PD. and D.W. King. Energy release in the muscle cell. Med. Sci. Sports. 1:23–31, 1969.

Gollnick, P.D., K. Piehl, and B. Saltin. Selective glycogen depletion pattern in human muscle fibers after exercise of varying intensity and various pedaling rates. J. Physiol. London 241:45–57, 1974.

Gollnick, P.D., P.J. Struck, and T.P. Bogyo. Lactic dehydrogenase activities of rat heart and skeletal muscle after exercise and training. J. Appl. Physiol. 22:623–627, 1967.

Hermansen, L. Anaerobic energy release. Med Sci. Sports. 1:32–38, 1969.

Hickson, R.C., W.W. Heusner, and W.D. Van Huss. Skeletal muscle enzyme alterations after sprint and endurance training. J. Appl. Physiol. 40:868–872, 1976.

Issekutz, B., W.A.S. Shaw, and A.C. Issekutz. Lactate metabolism in resting and exercising dogs. J. Appl. Physiol. 40:312–319, 1976.

Jost, J.P. and H.V. Rickenberg. Cyclic AMP. Ann. Rev. Biochem. 40:741, 1971.

Katz, J., J. Rostami, and A. Dunn. Evaluation of glucose turnover, body mass and recycling with reversible and irreversible tracers. Biochem. J. 194:513–524, 1981.

Krebs, H.A. and M. Woodford. Fructose 1, 6-diphosphatase in striated muscle. Biochem. J. 94:436–445, 1965.

Koshland, D.E. and K.E. Neet. The catalytic and regulatory properties of enzymes. Annual Reviews of Biochemistry. 37:359, 1968.

Lamb, D.R., J.B. Peter, R.N. Jeffress, and H.A. Wallace. Glycogen, hexokinase, and glycogen synthetase adaptations to exercise. Am. J. Physiol. 217:1628–1632, 1969.

Lehninger, A.L., Biochemistry. New York: Worth Publishing, 1971. p. 313–335.

Lehninger, A.L. Bioenergetics. New York: W.A. Benjamin, 1973. p. 53–72.

Mansour, Phosphofructokinase Curr. Topics Cell Regul. 5:1, 1972.

McGilvery, R.W. Biochemical Concepts. Philadelphia: W.B. Saunders Co., 1975. p. 230–266.

Meyerhof, O. Die Energie-wandlungen in Muskel III: Kohlenhydrat und Milschsäureumsatz in Froschmuskel. Archiv. Fuer Die Gesante Physiologie. 185:11–25, 1920.

Newsholme, E.A. The regulation of phosphofructokinase in muscle. Cardiology. 56:22, 1971.

Opie, L.H. and E.A. Newsholme. The activities of fructose 1, 6-diphosphatase, phosphofructokinase and phosphenolpyruvate carboxykinase in white and red muscle. Biochem. J. 103:391–399, 1967.

Scrutton, M.C. and M.F. Utter. The regulation of glycolysis and gluconeogenesis in animal tissues. Ann. Rev. Biochem. 37:269–302, 1968.

Taylor, A.W., J. Stothart, R. Thayer, M. Booth, and S. Rao. Human skeletal muscle debranching enzyme activities with exercise and training. Europ. J. Appl. Physiol. 33:327–330, 1974.

Taylor, A.W., R. Thayor, and S. Rao. Human skeletal muscle glycogen synthetase activities with exercise and training. Can. J. Physiol. Pharmacol. 50:411–415, 1972.

Vander, A.J., J.H. Sherman, and D.S. Luciano. Human Physiology, 3rd Ed. New York: McGraw-Hill, 1980.

York, J., L.B. Oscai, and D.G. Penny. Alterations in skeletal muscle lactic dehydrogenase isozymes following exercise training. Biochem. Biophys. Res. Commun. 61:1387–1393, 1974.

6

CELLULAR OXIDATION OF PYRUVATE

Physical activities lasting a minute or more absolutely require the presence and use of oxygen in active muscle. Far more energy can be realized from a substrate through oxidation than glycolytic processes. Within muscle cells are specialized structures called mitochondria, which link the breakdown of foodstuffs, the consumption of oxygen, and the production of ATP. As opposed to glycolysis, which involves carbohydrate materials exclusively, cellular oxidative mechanisms allow for the continued metabolism of carbohydrates as well as for the breakdown of derivatives of fats and proteins. Even though the processes of cellular oxidation are far removed from the anatomical sites of breathing and the pumping of blood, it is the processes of cellular oxidation that breathing and beating of the heart serve. Seen in their proper perspective, the two most familiar physiological processes (breathing and the heartbeat) play a key role in energy transduction.

MITOCHONDRIAL STRUCTURE

Cellular oxidation takes place in cellular organelles called mitochondria (mitochondrion, singular). Pyruvate (a product of glycolysis) and also products of lipid and amino acid metabolism are metabolized within mitochondria. These organelles (mitochondria) have been called ''the powerhouses of the cell.'' It

Steve Ovett and Sebastian Coe battle for the Moscow Olympic 800 m run title. This day Ovett prevailed, but Coe was victorious in the 1500 m run a few days later. Performances such as these require the highest sustained metabolic power output humanly possible. Such performances ultimately depend upon the ability to consume and utilize oxygen. (U.P.I.).

is within mitochondria that almost all oxygen is consumed and ADP phosphorylated to ATP. Consequently, those activities that last more than a minute are powered by mitochondria. As suggested in Figure 6-1, mitochondria appear in two places in skeletal muscle. A significant population of mitochondria is immediately beneath the cell membrane (sarcolemma), and these are called subsarcolemmal mitochondria. These mitochondria are in a position to receive O_2 provided by the arterial circulation. Subsarcolemmal mitochondria are believed to provide the energy to maintain the integrity of the sarcolemma. Energy-requiring exchanges of ions and metabolites across the sarcolemma are most probably supported by subsarcolemmal mitochondrial activity. Deeper within muscle cells are the intermyofibrillar mitochondria. As the name implies, these mitochondria exist among the contractile elements of the muscle. These mitochondria probably have a higher activity per unit mass (specific activity) than do subsarcolemmal mitochondria, and the intermyofibrillar mitochondria probably provide most of the ATP for energy transduction during contraction.

Because of their appearance in cross-sectional electron micrographs and in micrographs of isolated mitochondria, it has long been believed that the mitochondria exist as discrete capsule-shaped organelles. However, recent work on diaphragm muscle by the Russian scientist Skulachev indicates that mitochondria are interconnected in a network (the mitochondrial reticulum), much like

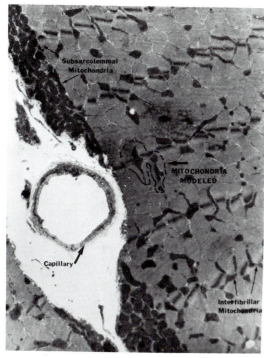

FIGURE 6-1

Mitochondria are found to be distributed within two areas of a muscle cell. Subsarcolemmal mitochondria are found immediately beneath the cell membrane (i.e., the sarcolemma). Intermyofibrillar mitochondria are found among the muscle cell's contractile elements.

(Micrograph courtesy of E.A. Munn, C. Greenwood, S.P. Kirkwood, L. Packer, and G.A. Brooks.)

the sarcoplasmic reticulum. Instead of thousands of mitochondria, a muscle cell may contain relatively few subsarcolemmal and intermyofibrillar mitochondria, but each mitochondrion has thousands of branches.

Figure 6-2 is the result of collaborative efforts of researchers in Berkeley, California and Babraham, England. From serial cross sections through rat soleus muscle (Figure 6-2*a*), a particular "mitochondrion" can be identified, traced, and a model constructed knowing the magnification and section thickness. In subsequent slices through the muscle (i.e., serial sections), the same "mitochondrion" can be modeled. These models of individual sections can then be stacked up to reveal the appearance of a mitochondrial reticulum (Figure 6-2*b*).

Red pigmented muscle fibers obtain their color, in part, because of their content of mitochondria, which are red. By comparison, pale muscle fibers contain few mitochondria (Figure 5-3).

Figure 6-1 is an electron micrograph of a muscle mitochondrion, whereas Figure 6-3 provides a schematic representation of a mitochondrion. Area 1, the outer membrane, functions as a barrier to maintain important internal constituents (e.g., NADH) and to exclude exterior factors. The outer membrane does contain specific transport mechanisms to regulate the influx and efflux of various materials. Area 2, the intermembrane space, also contains enzymes for

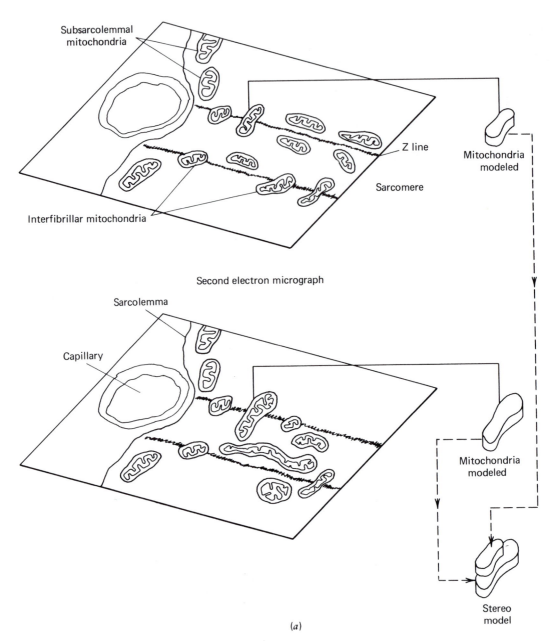

First electron micrograph

Subsarcolemmal mitochondria

Mitochondria modeled

Z line

Sarcomere

Interfibrillar mitochondria

Second electron micrograph

Sarcolemma

Capillary

Mitochondria modeled

Stereo model

(a)

FIGURE 6-2
In all probability, "mitochondria" do not exist as separate, individual enti-
ties within muscle cells but rather as parts of a network, or "reticulum."
Evidence for the mitochondrial reticulum in limb skeletal muscles was obtained by modeling of mitochondria seen in electron micrographs of serial
cross sections through rat soleus muscles. The process of modeling is illustrated in (a), and the resulting model is pictured in (b).

(Model (part b) courtesy of E.A. Munn and C. Greenwood, Babraham, England.)

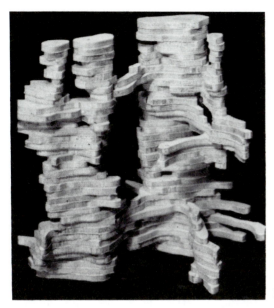

FIGURE 6-2 (*b*).

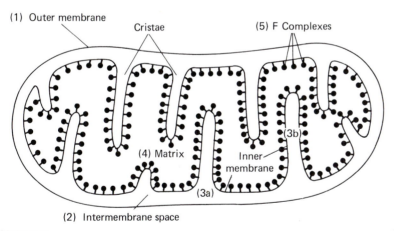

(1) Outer membrane

Cristae

(5) F Complexes

(4) Matrix

Inner membrane

(3b)

(3a)

(2) Intermembrane space

FIGURE 6-3
Schematic representation of a mitochondrion (reticulum fragment). Areas indicated are (1) the outer membrane, (2) the intermembrane space, (3) the inner membrane, and (4) the matrix. The inner membrane is diagrammed as existing in either of two orientations, either juxtaposed to the outer membrane (3a) or protruding into the matrix (3b). F complexes (5) protrude from the inner membrane, which is a mobile phospholipid structure. Inward folds of the inner membrane form the cristae. See text for description of the functions of each area.

exchange and transport. Area 3, the inner membrane, can be divided into two surfaces. Area 3a functions, in part, along with the outer and inner membrane constituents in a transport capacity. For example, the enzyme carnitine transferase is found on the inner wall of fragmented mitochondria. Carnitine transferase is involved in moving lipids into mitochondria. Area 3b of the inner membrane is referred to as the cristal membrane, so called because it is made up of many folds, or cristae (crista, singular). The cristal membrane is the main mitochondrial site where oxidative phosphorylation takes place. Because area 3a is confluent with area 3b, and the chemical compositions are similar, the a part of the inner membrane that is juxtaposed to the outer membrane (i.e., 3a) can phosphorylate as well.

In thinking about these mitochondria it must be kept in mind that the membrane constituents are not rigidly locked into a position, like a leg is attached to the trunk. Rather, the inner membranes change in conformation (shape) as mitochondria function, and the membrane constituents themselves display a high degree of mobility such that various components can move along and within the membranes.

The actual mitochondrial site of phosphorylation has been found to be the F complex, which appears as a ball on a stalk on the mitochondrial inner membrane. The F complex is alternatively termed the "elementary particle," and is made up of two subunits, the stalk (F_0 complex) and ball (F_1 complex).

Area 4, the matrix, is not simply a space but contains nearly half the mitochondrial protein. In the matrix are located the Krebs cycle enzymes. We describe those in detail next.

THE KREBS CYCLE

Pyruvate gains entry to the mitochondrial matrix via a carrier mechanism located in the outer membrane. The sequence of events of pyruvate metabolism, catalyzed first by the enzyme complex pyruvate dehyrogenase and then by the Krebs cycle enzymes, is illustrated in Figures 6-4 and 6-5.

The series of enzymes depicted in Figures 6-4 and 6-5 are called the Krebs cycle after Sir Hans Krebs, who did much of the work elaborating the pathway. This series of enzymes is also often referred to as the "citric acid cycle" (the first constituent is citric acid) and the "tricarboxylic acid cycle" (the initial constituents have three carboxyl groups, the TCA cycle).

Although the TCA cycle is generally referred to as a cycle, it is important to realize that it is an imperfect cycle. Various substances can leave the cycle, and others can gain entry to the cycle. This will be discussed later in detail. Figure 6-4 is intended as a reference. Figure 6-5 reveals the purpose of pyruvate dehydrogenase and the TCA cycle. These are decarboxylation (CO_2 for-

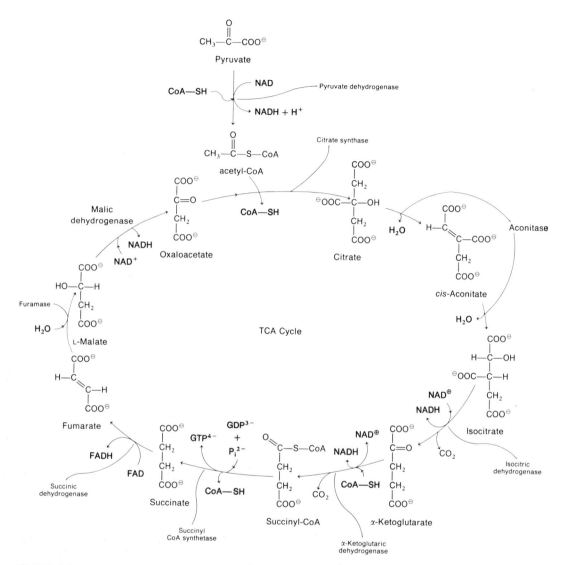

FIGURE 6-4
Detailed diagram of the Krebs cycle intermediates and catalyzing enzymes.
As the result of one acetyl-CoA entering and traversing the cycle, one GTP
(guanosine triphosphate, energetically equivalent to an ATP) and several
high-energy reducing equivalent compounds (NADH and FADH) are
formed. Those substances result in significant mitochondrial electron
transport and ATP production.

(SOURCE: Modified from R.W. McGilvery, 1975.)

mation), ATP production, and, most importantly, NADH production. It can be seen from Figure 6-5 that there are four places where NAD^+ is reduced to NADH, and one place each where FADH and ATP are formed. Recalling that each NADH is equivalent to three ATP's, and that each FADH is equivalent to two ATP's, the purpose of the TCA cycle is revealed. This is to continue the metabolism of pyruvate from glucose (as well as the intermediate products of

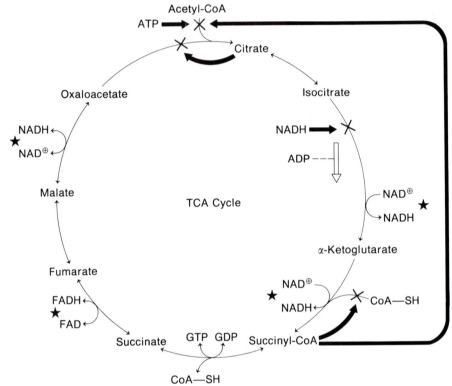

FIGURE 6-5

"Regulation of the citric acid cycle. The starred reactions require oxidized coenzymes; the ratio of oxidized to reduced coenzymes is governed by the availability of ADP and P_i for oxidative phosphorylation, thereby regulating the cycle to demand. However, the nearly irreversible isocitrate and α-keto-glutarate dehydrogenase reactions can still proceed at low NAD levels. They are controlled through inhibition by NADH and succinyl-co-enzyme A, respectively. Also, isocitrate dehydrogenase is activated by ADP. Finally, when the high-energy phosphate supply is high and GTP accumulates, the accumulating succinyl-CoA inhibits the initial citrate synthase reaction in competition with oxaloacetate. This effect also tends to balance the initial part of the cycle that consumes oxaloacetate against the final part that produces oxaloacetate. With the exception of succinic dehydrogenase, which bond to the inner membrane, TCA cycle enzymes exist in the mitochordrial matrix."

(SOURCE: From R.W. McGilvery, 1975.)

lipid and protein catabolism) and to trap part of the energy released in the forms of ATP and the high-energy reduced compounds NADH and FADH.

The enzyme pyruvate dehydrogenase is a complex that probably has two parts and functions in two steps. As implied by the name, an NADH molecule is formed. NADH is termed the "universal hydrogen carrier." Additionally, pyruvate functions in concert with coenzyme A (CoA), which is referred to as a coenzyme or cofactor. A cofactor is a low molecular weight substance whose presence is required to allow an enzyme to work. CoA is sometimes referred to as coenzyme 1, as it was the first such factor discovered; it is sometimes also referred to as the universal acetate carrier. In Figure 6-6 the terminal sulfur bond of CoA is shown to be the site at which acetate binds. As the result of pyruvate dehydrogenase, pyruvate is decarboxylated, releasing a CO_2 molecule, and the remaining two-carbon unit (acetate) is combined with CoA to

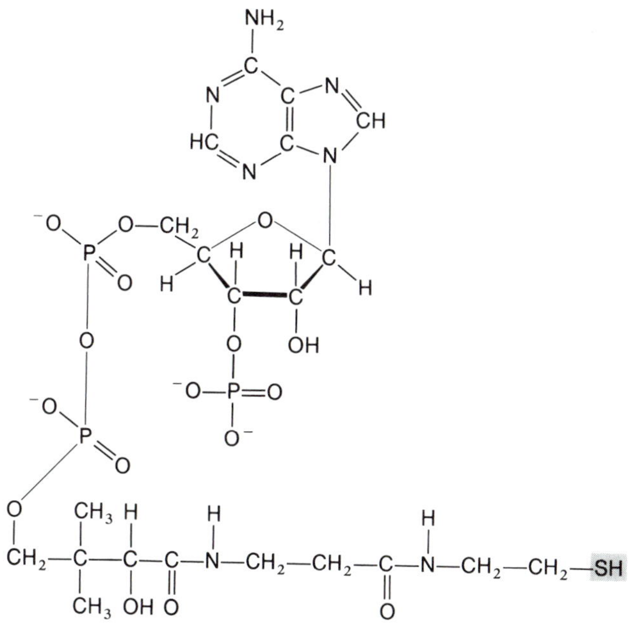

(a)

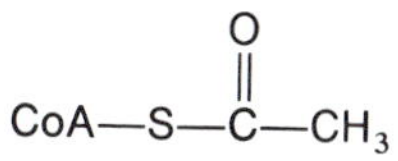

(b)

FIGURE 6-6

(a) Structure of coenzyme A (CoA) showing the binding site (-SH) for acetate. (b) Acetyl-CoA is formed from the union of CoA and acetate.

give acetyl-CoA. Pyruvate dehydrogenase is an important enzyme, not only because of its complex function, but also because it is a rate-limiting enzyme.

Acetyl-CoA is the entry substance into the TCA cycle. Acetyl-CoA can be formed from pyruvate as well as from fatty and amino acids. Under the influence of the enzyme citrate synthetase, acetyl-CoA and oxaloacetic acid (OAA) condense to give citric acid. As we shall see, the presence of OAA may be a regulating factor controlling the rate of the TCA cycle.

Several steps into the TCA cycle is the enzyme isocitrate dehydrogenase (IDH). This is the rate-limiting enzyme of the TCA, much like PFK in glycolysis. Like PFK, IDH is an allosteric enzyme that is stimulated by ADP. IDH, together with the other dehydrogenases of the TCA cycle, as well as pyruvate dehydrogenase, are sensitive to the redox (reduction–oxidation) potential in the cell. Simply defined, the redox potential is the NADH/NAD$^+$. The dehydrogenases are inhibited by a high redox potential and are stimulated by a decline in redox potential.

THE ELECTRON-TRANSPORT CHAIN

The electron-transport chain (ETC) is located on the mitochondrial inner membrane, probably in both areas 3a and 3b. The ETC constituents are arranged in the sequence indicated in Figure 6-7. We owe much of our present understanding of how mitochondria function to Peter Mitchell and his chemiosmotic theory of oxidative phosphorylation. The term oxidative phosphorylation refers to two separate processes that usually function together. Oxidation is a spontaneous process that is linked or coupled to the phosphorylation, the union of P_i with ADP to make ATP. Phosphorylation is an endergonic process that is driven by oxidation. It is important to note that although the linkage in oxidative phosphorylation is tight, certain situations can cause the linkage to be loosened or uncoupled, for example, heat buildup in muscles from prolonged work.

The function of the ETC is simply described as follows. Reducing equivalents containing a high-energy hydrogen and electron gain entry to the beginning of the chain. The hydrogen and electron then move from areas of electronegativity (NAD$^+$) toward electropositivity (atomic oxygen). Along the ETC the electron is stripped from the hydrogen, which continues along the chain; the resulting proton (H$^+$) is pumped outside the mitochondrion (Figure 6-7). For each NADH entering, three pairs or a total of six protons are pumped. For each FADH entering, two pairs of protons are pumped. Outside the mitochondria a region of decreased pH and positive charge is created. This chemical and osmotic potential ultimately supplies the energy to phosphorylate ADP. It is theorized that portals exist through the outer wall of the mitochondria, which apparently can channel the potential energy of the external proton field to the ATPase on the inner membrane (Figure 6-7).

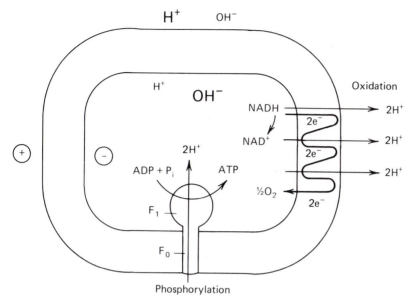

FIGURE 6-7
Orientation of components of the electron-transport chain (ETC) within the mitochondrial inner membrane. Entry of hydride ions (H·) from high-energy reduced compounds (e.g., NADH) into the ETC results in the oxidation of those compounds, electron transport, and the expulsion of protons (H^+). This creates a chemical and electrical gradient across the membrane. Entry of protons through specific portals into the F_0–F_1 complex (elementary particle or stalk and ball) provides the energy for the phosphorylation of ADP to ATP.

According to Mitchell, pairs of protons enter the mitochondria and are directed to the stalk or F_0 part of the elementary particle. These protons attack oxygen of phosphate groups previously bound to the ball (or F_1) part of the elementary particle. The combination of the phosphate and atomic oxygen results in the formation of water and an energized phosphate ion. This phosphate ion is then in position to unite with ADP, which has also previously been bound to F_1. This forms ATP. Phosphorylation may thus be written as two partial reactions.

$$2H^+ + PO_4^{3-} \longrightarrow H_2O + PO_3^- \tag{6-1}$$

$$PO_3^- + ADP^{3-} \longrightarrow ATP^{4-} \tag{6-2}$$

The formation of ATP according to the chemiosmotic theory of Mitchell represents a reversal of the hydrolysis of ATP. Recalling Equation 4-3, the hydrolysis of ATP can be written as:

$$ATP + HOH \underset{\text{phosphorylation}}{\overset{\underset{\text{hydrolysis}}{\longrightarrow}}{\underset{\text{ATPase}}{\longleftarrow}}} ADP + P_i$$

In the mitochondrion the phosphorylation of ADP is made possible by linking ATP production to the formation of water. With energy input, the mitochondrial ATPase is driven against its equilibrium, toward ATP production.

Function of the ETC

The electron-transport chain can function because of three basic factors. First, each of the constituents can exist in reduced (higher energy, with electron) and oxidized (lower energy, without electron) forms. Second, the ETC constituents are sequentially arranged in close proximity on the inner membrane forming a "respiratory assembly." Distance, therefore, does not impede electron movement. Third, and most important, the ETC constituents are arranged such that the redox potential of each constituent is greater (i.e., more positive) than that of the previous constituent. Electrons can then move from NADH (electronegative redox potential) (Figure 6-7) to atomic oxygen (electropositive). Oxygen is termed "the final electron acceptor."

Control of the ETC

We have seen that the adenine nucleotide charge serves to regulate both glycolysis and the Krebs cycle. It should therefore not be surprising that ADP and ATP, respectively, serve to stimulate and to inhibit the ETC. The control of muscle metabolism is therefore elegantly simple. As soon as muscle contracts, ATP is split and ADP is formed. The change in the relative amounts of these substances then sets in motion biochemical events to re-form the spent ATP. When exercise stops, the cellular respiratory mechanisms soon reestablish normal levels of ATP, ADP, and AMP. Consequently, whole-body O_2 consumption rapidly declines toward resting levels after exercise.

The Number of ATP from a Glucose Molecule

In our view, the breakdown of glucose and glucose 6-phosphate in muscle is extremely important because it can occur very rapidly and therefore can supply energy rapidly. Further, glycolysis can occur in presence or absence of oxygen. Energy requirements in muscle can, therefore, be supplied by glycolysis for finite periods of time under anaerobic conditions.

Under aerobic conditions, for each glucosyl unit a net of two ATP are formed in the cytoplasm and the energy equivalents of two cytoplasmic NADH can be shuttled into the mitochondria. Recalling that the reducing equivalent energy of cytoplasmic NADH can give rise to either NADH or FADH within mitochondria, and that the P:O of NADH is 3 and for FADH the P:O is 2, then under aerobic conditions glycolysis yields four to six ATP in mitochondria in addition to the two cytoplasmic ATP. Together with the ATP formed in the TCA cycle, and the ATP produced as the result of reducing equivalents donated from the TCA cycle to the ETC, the ATP yield per glucose is 36 to 38 molecules of ATP per molecule of glucose.

As stated earlier, when glycolysis is "anaerobic," a net of two ATP are formed per glucose.

When the substrate for glycolysis originates from glycogen, that is, when glucose 6-phosphate is supplied by glycogenolysis, it is more difficult to estimate the ATP yield per glucosyl unit cleaved from glycogen. When a molecule of glucose entered the glycolytic pathway through the action of the enzyme hexokinase, a molecule of ATP was required to energize glucose to the level of glucose 6-phosphate. In glycogenolysis, this energizing step is bypassed, and theoretically one might consider that the energy of one phorylation is saved. Thus the ATP yield for a glucosyl unit derived from glycogen under anaerobic conditions may be 3. However, it may be incorrect to assume that glycogen yields relatively more ATP than does glucose. It takes energy to synthesize glycogen from glucose or glucose 6-phosphate. Although glycogen synthesis may have preceded glycogenolysis in time by hours or days, the energy requirement for the process should be taken into account. Further ATP was required to activate the phosphorylase, the enzyme catalyzing glycogenolysis. In starting glycolysis from glucose, each molecule has to be phosphorylated. In glycogenolysis the enzyme, rather than the substrate, is phosphorylated.

The activity of phosphorylase is much higher in muscle than is hexokinase. Consequently entry of glucosyl units into glycolysis during exercise is more rapid from glycogen than from glucose. Therefore, it follows that the ATP yield for anaerobic glycolysis during heavy exercise reflects the dominant role of glycogenolysis and is closer to 3 than to 2.

EFFECTS OF TRAINING ON SKELETAL MUSCLE MITOCHONDRIA

A paper by John Holloszy in 1967 ushered in a new era of study that has been called "exercise biochemistry." In his initial paper and in subsequent reports with his associates, Holloszy has identified a number of specific effects of endurance training on skeletal muscle mitochondria and respiratory capacity. Researchers in other laboratories have also obtained similar results. Table 6-1 summarizes some of the results of these studies. In response to the endurance-training procedure utilized, several enzymes of the TCA cycle and constituents of the ETC have been observed to double in activity.

A major question in exercise biochemistry has been: How is the increase in muscle mitochondrial capacity (Table 6-1) accomplished? Do the mitochondria adapt in specific ways, such as by an increase in the number or density of enzymes on the mitochondrial cristae, or are there simply more mitochondria (a larger mitochondrial reticulum)? The answer to this question has been investigated by scientists at the University of California. Their results clearly indicate that mitochondria do not increase in specific activity, that is, enzymatic activity per unit of mitochondrial protein (Table 6-2, top). Rather, there are

TABLE 6-1

Effect of Endurance Training on Respiratory Capacity of Rat Skeletal Muscle Mitochondria[a]

Muscle	Group	Citrate Synthase ($\mu mol \cdot min^{-1} \cdot g^{-1}$) Muscle	Carnitine Palmityl-transferase	Cytochrome c (nmol $\cdot$ g^{-1})
Fast-twitch white	Exercised	18 ± 1	0.20 ± 0.02	6.3 ± 0.7
	Sedentary	10 ± 1	0.11 ± 0.01	3.2 ± 0.3
Fast-twitch red	Exercised	70 ± 4	1.20 ± 0.09	28.4 ± 2.1
	Sedentary	36 ± 3	0.72 ± 0.06	16.5 ± 1.6
Slow-twitch red	Exercised	41 ± 3	1.20 ± 0.05	—
	Sedentary	24 ± 2	0.63 ± 0.07	—
Heart	Exercised	160 ± 4	—	46.6 ± 0.9
	Sedentary	158 ± 1	—	47.1 ± 1.0

[a] In response to endurance training mitochondrial components in different types of rat skeletal muscles double in concentration. The heart does not change.

SOURCE: Data from Baldwin et al., Oscai et al., and Holloszy (see references at the end of the chapter).

more mitochondria or there is a more elaborate reticulum (Table 6-2, bottom). With a few minor exceptions such as the labile, nonstructural enzyme α-glycerol phosphate dehydrogenase, most mitochondrial constituents increase in direct proportion to the amount of mitochondrial material. Further, the Berkeley group showed no change in protein:lipid ratio with training. Training affects the mechanisms of mitochondrial replication and destruction. As the result of endurance training, skeletal muscle contains more mitochondrial material whose activity is the same as that of untrained individuals.

Examinations of muscle from untrained and trained animals under the electron microscope (Figure 6-8) provide results consistent with the biochemical analyses (Tables 6-1 and 6-3). Mitochondria in tissues of trained rats may appear to be more numerous. Recalling the concept of the mitochondrial reticulum, it may be said that the reticulum is more elaborate in response to training.

As dramatic as the effects of training on respiratory capacity of muscle are, at least two problems need to be resolved. First, what specific aspects of endurance training result in these changes? Why does progressive resistance (weight) training not improve mitochondrial capacity? Obviously, endurance training presents some stimulus to the muscle that weight training does not. The identities of these training stimuli that cause mitochondrial protein to in-

TABLE 6-2

Specific Activity (Top) and Mitochondrial Content (Bottom) in Skeletal Muscle from Endurance-Trained Rats and Sedentary Controls[a-c]

Parameter	Mitochondrial Protein Specific Activity $(nmol \cdot mg^{-1})$	
	Control Group ($n = 10$)	Endurance Group ($n = 10$)
Cytochrome a	0.37 ± 0.02	0.35 ± 0.08
Cytochrome b	0.32 ± 0.02	0.36 ± 0.02
Cytochrome c ($+c_1$)	0.91 ± 0.05	0.82 ± 0.03
Flavoprotein	1.13 ± 0.08	1.04 ± 0.05
b/a	0.86	1.03
c ($+c_1$)$/a$	2.46	2.34
Flavoprotein$/a$	3.05	2.97

Parameter	Mitochondrial Content of Muscle $(mg \cdot g^{-1})$		
	Control Group ($n = 9$)	Endurance Group ($n = 9$)[b]	Percentage Difference of Control from Endurance Trained
Pyruvate-malate oxidase	15.5 ± 0.7	25.1 ± 1.2	$+62$
Succinate oxidase	20.1 ± 1.0	43.6 ± 2.1	$+117$
Palmitoyl carnitine oxidase	21.1 ± 2.6	50.4 ± 4.7	$+138$
Cytochrome oxidase	19.2 ± 0.8	38.3 ± 1.8	$+99$
Succinate dehydrogenase	14.9 ± 0.8	30.9 ± 1.8	$+108$
NADH dehydrogenase	17.5 ± 0.8	27.9 ± 1.7	$+59$
Choline dehydrogenase	25.9 ± 1.7	55.7 ± 3.8	$+115$
Cytochrome c ($+c_1$)	16.2 ± 0.6	31.6 ± 1.2	$+95$
Cytochrome a	18.2 ± 0.7	36.5 ± 1.6	$+101$
Average	18.7 ± 1.2	37.8 ± 3.7	$+99$

[a] Values are mg mitochrondrial protein $\cdot$ g wet muscle^{-1} (means $\pm$ SE). Mitochondrial content of muscle was calculated as muscle activity/mitochondrial specific activity, for the relevant oxidases, dehydrogenases, and cytochromes.

[b] All endurance-trained values were significantly higher than controls, $P < 0.01$ (one-tailed t test).

[c] In terms of specific activity (units of activity or component content per unit of mitochondrial protein), no difference is seen between endurance-trained and sedentary control groups (top table), whereas the muscle mitochondrial content doubles (bottom table).

SOURCE: Data from Davies, et al. (see references at the end of the chapter).

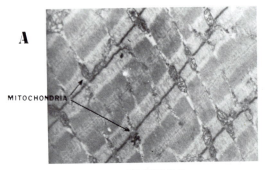

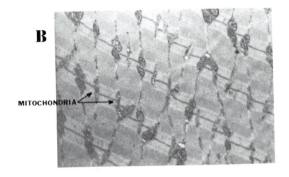

FIGURE 6-8
Electron micrographs of mitochondria in tissues of (a) untrained and (b) trained rats. Mitochondria in trained animals appear to be more numerous, probably because there is a more elaborate mitochondrial reticulum.

(Micrographs courtesy of P.D. Gollnick.)

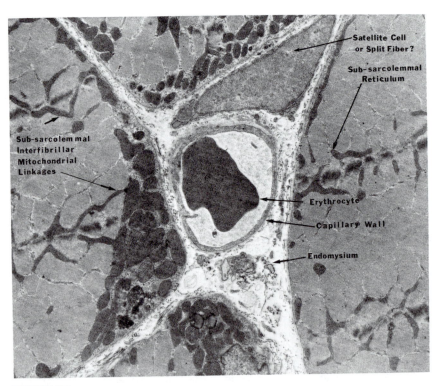

FIGURE 6-9
Electron micrograph of rat deep (red) vastus thigh muscle showing linkages between subsarcolemmal and interfibrillar mitochondria.

(Micrograph courtesy of S.P. Kirkwood, E.A. Munn, L. Packer, and G.A. Brooks.)

crease in response to endurance training and contractile protein to increase in response to heavy resistance training are under investigation.

A second major question concerns the significance of the doubling of mitochondrial activity. Table 6-1 indicates large increments in respiratory capacities of mitochondria in all three types of skeletal muscle fibers. In humans and in laboratory animals, training may increase severalfold the length of time a submaximal work load can be endured. However, in both humans and laboratory animals, maximal O_2 consumption capacity increases only 10 to 15% in response to endurance training. In results obtained by the U.C. group, muscle mitochondrial cytochrome oxidase activity correlated .92 with running endurance but only .70 with $\dot{V}_{O_2max}$. (Table 6-3). Consequently, it appears that mitochondrial capacity may be better related to endurance than $\dot{V}_{O_2max}$. The reason for this needs to be elaborated. The increase in mitochondrial mass due to endurance training may also be the means of increasing fat utilization as a fuel during exercise (Chapter 7). It may be that training has a greater effect on subsarcolemmal mitochondria, thereby improving the ability to maintain the integrity of the cell membrane and thereby improve endurance during heavy exercise. Training may have a lesser effect on interfibrillar mitochondrial, those mitochondria supportive of muscular contraction. Alternatively, it is likely that $\dot{V}_{O_2max}$ is limited by blood flow (cardiac output) but that endurance is dependent on mitochondrial function. From this perspective, it may be that exercise exhausts the functional capability of mitochondria and that having a large, elaborate reticulum, or many mitochondria, retards the eventual fatigue point. This question is being investigated.

TABLE 6-3

Correlation Matrix for Muscle Oxidases, $\dot{V}_{O_2max}$, and Maximal Endurance in Rats on a Treadmill [a,b]

	Pyruvate-malate Oxidase	Palmitoyl Carnitine Oxidase	$\dot{V}_{O_2max}$ (weight normalized)	Maximal Endurance
Cytochrome oxidase	0.95	0.93	0.74	0.92
Pyruvate-malate oxidase	—	0.89	0.68	0.89
Palmitoyl carnitine oxidase	—	—	0.71	0.91
$\dot{V}_{O_2max}$ (weight normalized)	—	—	—	0.70

[a] All correlations reported were statistically significant ($P < 0.01$).

[b] Running endurance of rats as measured on a standardized treadmill test correlates significantly better with skeletal muscle mitochondrial activity (i.e., cytochrome oxidase activity) than with maximal O_2 consumption ($\dot{V}_{O_2max}$).

SOURCE: Data from Davies, et al. (see references at the end of the chapter).

Doubling the mitochondrial content in muscle due to training (as determined by biochemical procedures) is consistent with an elaboration of the mitochondrial reticulum due to training. Elaboration of the mitochondrial reticulum due to training, however, has yet to be demonstrated. Similarly, it has not yet been established whether subsarcolemmal mitochondria are arranged in a reticulum, or whether the subsarcolemmal reticulum (if it exists) is continuous with the intrafibrillar reticulum. On the basis of electronmicroscopic information (Figure 6-9), it appears that the mitochondrial reticulum is continuous from subsarcolemmal to deep regions within a muscle cell.

Summary

Oxygen supplied to active muscle by the lungs, heart, and blood serves to support cellular production of ATP. Through the process of respiration, derivatives of carbohydrate as well as fats and proteins can yield substantial sources of potential energy for phosphorylating ADP to ATP. Endurance training has the unique effect of improving respiratory capacity of skeletal muscle. This is accomplished by increasing the amount of mitochondrial material in muscle.

Selected Readings

Bakeeva, L.E., Y.S. Chentsov, and V.P. Skulachev. Mitochondrial framework (Reticulum Mitochondriale) in rat diaphragm muscle. Biochim. Biophys. Acta. 501:349–369, 1978.

Baldwin, K., G. Klinkerfuss, R. Terjung, P.A. Molé, and J.O. Holloszy. Respiratory capacity of white, red, and intermediate muscle: adaptive response to exercise. Am. J. Physiol. 22:373–378, 1972.

Barnard, R., V.R., Edgerton, T. Furukawa, and J.B. Peter. Histochemical, biochemical and contractile properties of red, white, and intermediate fibers. Am. J. Physiol. 220:410–414, 1971.

Barnard, R.J., V.R. Edgerton, and J.B. Peter. Effect of exercise on skeletal muscle I. Biochemical and histochemical properties I. Appl. Physiol. 28:762–766, 1970.

Barnard, R.J., and J.B. Peter. Effect of exercise on skeletal muscle III: Cytochrome changes. Am. J. Physiol. 31:904–908, 1971.

Burke, F., F. Cerny, D. Costill, and D. Fink. Characteristics of skeletal muscle in competitive cyclists. Med. Sci. Sports. 9:109–112, 1977.

Chance, B. and C.R. Williams. The respiratory chain and oxidative phosphorylation. In Advances in Enzymology, vol. 17. New York: Interscience, 1956. p. 65–134.

Costill, D.L., J. Daniels, W. Evans, W. Fink, G. Krahenbuhl, and B. Saltin. Skeletal muscle enzymes and fiber composition in male and female track athletes. J. Appl. Physiol. 40:149–154, 1976.

Costill, D.L., W.J. Fink, and M.L. Pollock. Med. Sci. Sports. 8:96–100, 1976.

Davies, K.J.A., J.J. Maguire, G.A. Brooks, P.R. Dallman, and L. Packer. Muscle mitochondrial bioenergetics, oxygen supply, and work capacity during dietary iron deficiency and repletion. Am. J. Physiol. (Eendocrinol. Metab. 5): E418–E427, 1982.

Davies, K.J.A., L. Packer, and G.A. Brooks. Biochemical adaptation of mitochondria, muscle, and whole-animal respiration to endurance training. Arch. Biochem. Biophys. 209:538–553, 1981.

Davies, K.J.A., L. Packer, and G.A. Brooks. Exercise Bioenergetics following sprint training. Arch. Biochem. Biophys, 215:260–265, 1982.

Dohm, G.L. R.L. Huston, E.W. Askew, and H.L. Fleshood. Effect of exercise, training and diet on muscle citric acid cycle enzyme activity. Can. J. Biochem. 51:849–854, 1973.

Ernster L. and Z. Drahota (eds.). Mitochondria Structure and Function. New York: American Press, 1969.

Gale, J.B. Skeletal muscle mitochondrial swelling with exhaustive exercise. Med. Sci. Sports. 6:182–187, 1974.

Gollnick, P.D. and D.W. King. Effect of exercise and training on mitochondria of rat skeletal muscle. Am. J. Physiol. 216:1502–1509, 1969.

Holloszy, J.O. Effects of exercise on mitochondrial oxygen uptake and respiratory enzyme activity in skeletal muscle. J. Biol. Chem. 242:2278–2282, 1967.

Holloszy, J.O. Adaptation of skeletal muscle to endurance exercise. Med. Sci. Sports. 7:155–164, 1975.

Holloszy, J.O. and F.W. Booth. Biochemical adaptation to endurance exercise in muscle. Ann. Rev. Physiol. 38:273–291, 1976.

Holloszy, J.O., L.B. Oscai, P.A. Molé, and I.J. Don. Biochemical adaptations to endurance exercise in skeletal muscle. In: Pernow, B. and B. Saltin (eds.) Muscle Metabolism During Exercise. New York: Plenum Press, 1971.

Keilin, D. The History of Cell Respiration and Cytochromes. New York: Cambridge University Press, 1966.

King, D.W. and P. Gollnick. Ultrastructure of rat heart and liver after exhaustive exercise. Am. J. Physiol. 218:1150–1155, 1970.

Krebs, H.A. and H.L. Kornberg. Energy Transformation in Living Matter, Berlin: Springer-Verlag OHG, 1957.

Lehninger, A.L. Biochemistry. New York: Worth Publishing, 1970. pp. 337–416.

Lehninger, A.L. Bioenergetics. Menlo Park, Cal.: W.A. Benjamin, Inc., 1973. pp. 75–98.

Lowenstein, J.M. The tricarboxylic acid cycle. In: Greenberg, D.M. (ed.), Metabolic Pathways, vol. 1. New York: Academic Press, Inc., 1967. p. 146–270.

McGilvery, R.W. Biochemical Concepts. Philadelphia: W.B. Saunders Co., 1975. p. 157–249.

Mitchell, P. Chemiosmotic coupling in oxidative and photosynthetic phosphorylation. Biol. Rev. 41:455, 1965.

Mitchell, P. and J. Moyle. Estimation of membrane potential and pH difference across the cristae membrane of rat liver mitochondria. Eur. J. Biochem. 7:471–484, 1969.

Molé, P., K. Baldwin, R. Terjung, and J.O. Holloszy. Enzymatic pathways of pyruvate metabolism in skeletal muscle: adaptations to exercise. Am. J. Physiol. 224:50–54, 1973.

Munn, E.A. The Structure of Mitochondria. London: Academic Press, Inc., 1974.

Oscai, L.B., P.A. Molé, B. Brei, and J.O. Holloszy. Cardiac growth and respiratory enzyme levels in male rat subjected to a running program. Am. J. Physiol. 220:1238–2141, 1971.

Oscai, L.B., P.A. Molé, and J.O. Holloszy. Effects of exercise on cardiac weight and mitochondria in male and female rats. Am. J. Physicol. 220:1944–1948, 1971.

Terjung, R.L. Muscle fiber involvement during training of various intensities. Am. J. Physiol. 230:946–950, 1976.

Terjung, R.L., K.M. Baldwin, W.W. Winder, and J.O. Holloszy. Glycogen repletion in different types of muscle and in liver after exhaustive exercise. Am. J. Physiol. 226:1387–1391, 1974.

Terjung, R.L., G.H. Kinkerfuss, K.M. Baldwin, W.W. Wider, and J.O. Holloszy. Effect of exhausting exercise on rat heart mitochondria. Am. J. Physiol. 225:300–305, 1973.

Tzagoloff, A. Mitochondria. New York: Plenum Press, 1982.

Vander, A.J., J.H. Sherman, and D.S. Luciano. Human Physiology, 3rd ed. New York: McGraw-Hill, 1980.

7

LIPID METABOLISM

Dietary sources contain a variety of different kinds of lipids (fats), but triglycerides are quantitatively the most important. The high energy content of triglycerides and their vast, efficient storage reserve in adipose tissue, liver, and skeletal muscle provide an almost inexhaustible fuel supply for muscular exercise. However, despite the large quantity of lipid available as a fuel, the processes of lipid utilization are slow to be activated and proceed at rates significantly slower than those processes controlling sugar and carbohydrate catabolism. Despite these limitations, fats constitute an important segment of the fuel used during prolonged work, and significant increases in the ability to use fats as fuel during exercise can effectively slow sugar and carbohydrate metabolism. The sparing of glucose and glycogen metabolism by fats during prolonged exercise in highly trained individuals slows the depletion of these essential nutrients. Thus, increased fat oxidation capability due to training and genetic endowment greatly enhances endurance.

LIPID DEFINED

A lipid can be defined very simply on the basis of its physical solubility characteristics: It is a substance that is soluble in organic solvents but not in water. The solubility characteristics of lipids are determined by their nonpolar physical

Frank Shorter finishes and wins the Munich Olympic marathon. Success in endurance activities, such as the marathon, depend critically upon the ability to oxidize fats as an energy source. In humans, the ability to utilize fats is only indirectly related to the amount of stored body fat. (U.P.I.).

characteristics. Figure 7-1 gives structural formulas for several common lipids of physiological importance. As can be imagined from the very general definition of lipids, many different substances with different structures and functions are included in the broad definition. Fatty acids such as stearate (Figure 7-1*a*), are long-chain carboxylic acids. Fatty acids can combine with glycerol (Figure 7-1*b*) at its hydroxyl groups to form mono-, di-, or triglycerides. In Figure 7-1*c,* fatty acids have been substituted at each of the three hydroxyl groups of glycerol to yield a triglyceride. Triglycerides constitute the vast majority of the lipid, and calories, taken into and stored within the body.

Naturally occurring fatty acids vary in size, but they commonly contain an even number (14 to 24) of carbon atoms in a straight chain. Because fatty acids are synthesized from acetyl-CoA (two-carbon units) and are catabolized in two-carbon units, the even number of carbon atoms in the fatty acid chain is no accident.

Stearate (Figure 7-1*a*) is one of the most abundant saturated fatty acids in the diet. Unsaturation means that not all the valence bonds in the carbon chain are filled (saturated) with hydrogen; this also means that a carbon-to-carbon double bond exists.

Figure 7-1*d* illustrates linoleate, a polyunsaturated fatty acid. Like stearate, linoleate contains 18 carbon atoms, but linoleate contains two double bonds.

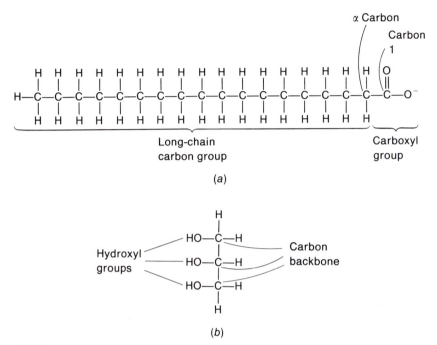

FIGURE 7-1
Typical lipids and important metabolites in lipid metabolism. (a) A saturated fatty acid, stearate, (b) Glycerol, (c) A triglyceride, stearyl-palmitoyl-linoleal-glycerol, (d) A polyunsaturated fatty acid, linoleate, (e) A common phospholipid, lecithin, (f) Cholesterol, (g) α-Glycerolphosphate or glycerol-3-phosphate, (h) Acetoacetate, a ketone.

The bonds exist at three-carbon intervals (e.g., between carbons 9 and 10, and between 12 and 13), which is a frequent arrangement of double bonds in fatty acids. Because mammals have lost the ability to synthesize unsaturated fatty acids with double bonds beyond the ninth carbon, fatty acids such as linoleate and oleate, which has a single unsaturation between C9 and C10, are essential and must be in the diet.

The presence of polyunsaturated fatty acids (two or more double bonds) in the diet has, in the past, been thought to lower levels of cholesterol in the blood. Elevated blood cholesterol has been implicated as a risk factor predisposing people to coronary heart disease. Consequently, substitution of polyunsaturated dietary fats for saturated fats has been recommended as a dietary manipulation to protect against coronary artery disease. This is discussed more in Chapter 24.

Other common lipids include phospholipids such as lecithin (Figure 7-1e), and cholesterol (Figure 7-1f). Phospholipids are crucial constituents of membranes, whereas cholesterol is necessary for the formation of steroid hormones and the bile salts used in lipid digestion.

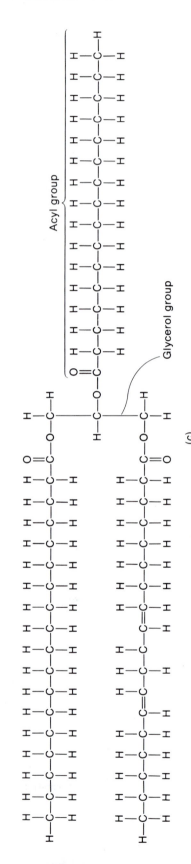

FIGURE 7-1 (continued).

(c)

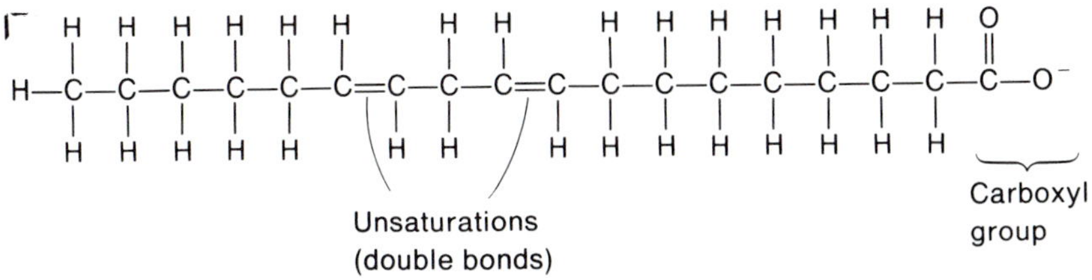

Unsaturations (double bonds)

Carboxyl group

(d)

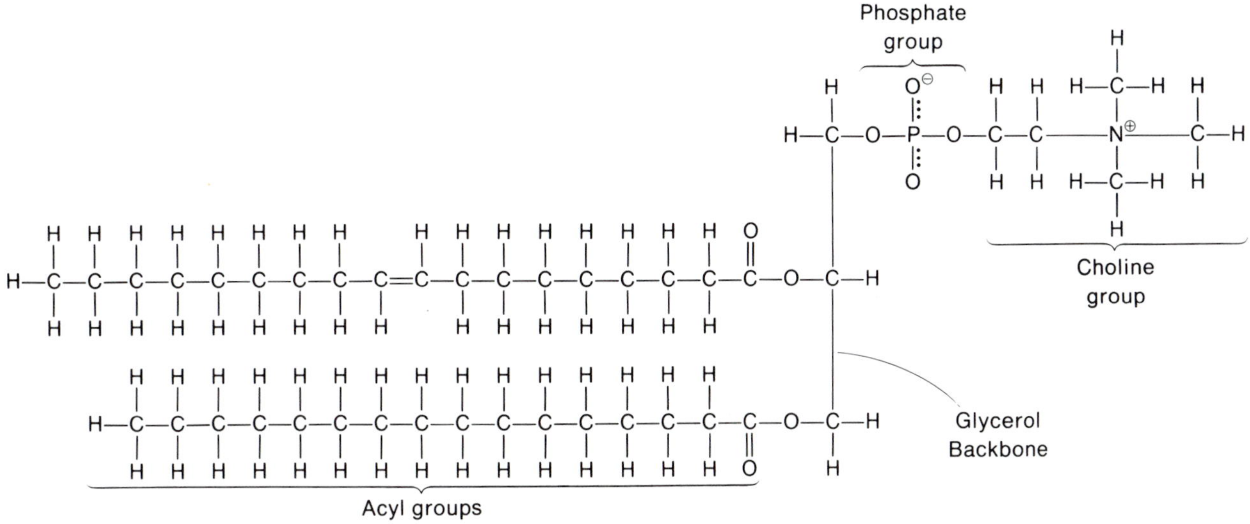

Phosphate group

Choline group

Acyl groups

Glycerol Backbone

(e)

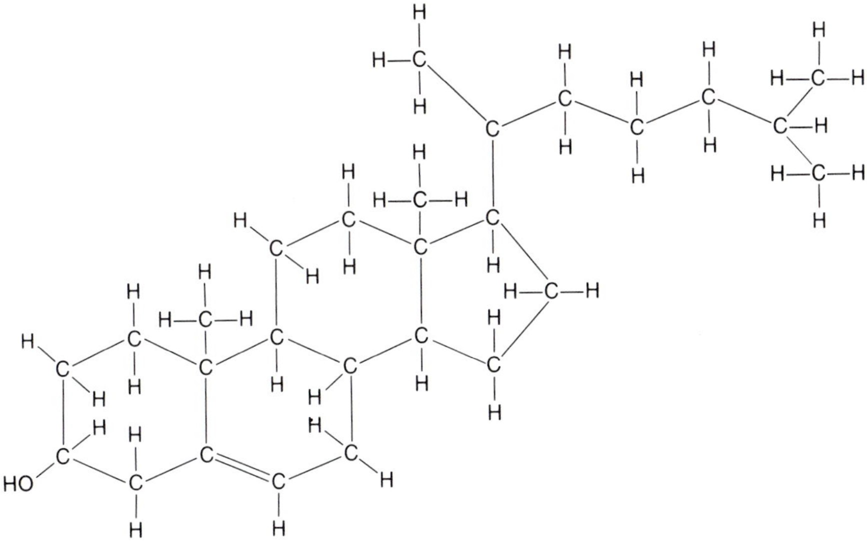

(f)

FIGURE 7-1 (continued).

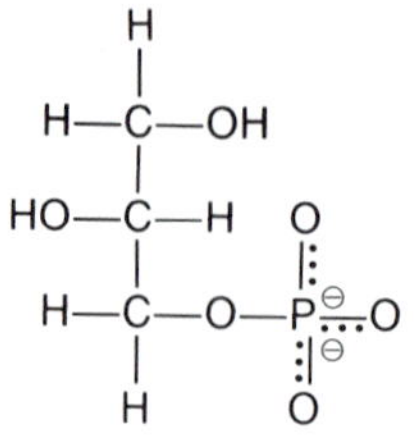

$$H-\overset{\displaystyle H}{\underset{\displaystyle H}{\overset{\displaystyle |}{\underset{\displaystyle |}{C}}}}-OH$$

(structure g)

(g)

(structure h with Keto group label)

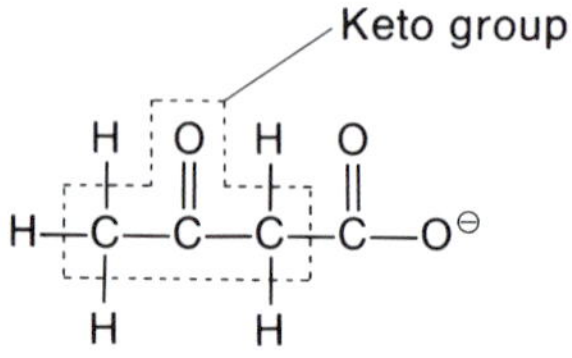

(h)

FIGURE 7-1 *(continued).*

ESTERIFICATION AND HYDROLYSIS

The making of triglycerides is called esterification because it involves attachment of a fatty acid to glycerol by means of an oxygen atom. *Lipolysis* is the process of triglyceride dissolution. Esterification and lipolysis are essentially reversals of each other. Triglycerides synthesized in plants or animals are consumed by humans. Digestion involves the lipolysis of triglycerides, whereas storage of consumed lipid involves the reesterification of fatty acids delivered to adipose tissue. The mobilization of triglycerides stored in adipose for utilization involves another lipolysis. Synthesis, digestion, storage, and mobilization can be simply diagrammed as a series of reversal reactions.

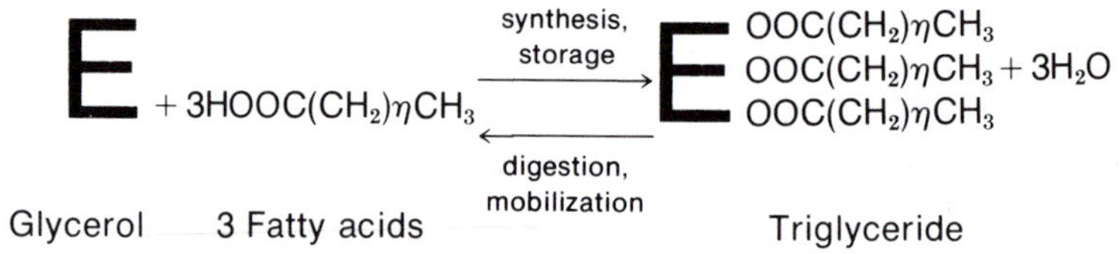

$$E + 3HOOC(CH_2)\eta CH_3 \underset{\substack{\text{digestion,}\\\text{mobilization}}}{\overset{\substack{\text{synthesis,}\\\text{storage}}}{\rightleftharpoons}} E\begin{array}{l} OOC(CH_2)\eta CH_3\\ OOC(CH_2)\eta CH_3 + 3H_2O\\ OOC(CH_2)\eta CH_3 \end{array} \qquad (7.1)$$

Glycerol   3 Fatty acids   Triglyceride

FATS IN THE DIET

Triglycerides constitute the vast majority of lipid mass consumed in the diet. Dietary triglycerides, along with ingested cholesterol and phospholipids, undergo no significant digestion in the stomach. Rather, fat digestion begins in the small intestine by the emulsification (disperion into a water medium) of fats by the action of bile salts from the liver via the gallbladder. Bile salts act by reducing the surface (interfacial) tension of the fat globules so that the agitation action of the small intestine can disperse them. In this emulsified form, fat globules are then hydrolyzed into monoglycerides, free fatty acids (FFA), and glycerol by the action of pancreatic lipase, a digestive enzyme. Following lipolysis, digestion of fats is again furthered by the action of bile salts, which allow the products to aggregate into very small fatty particles called *micelles*. When micelles come into contact with epithelial cells lining the small intestine, they allow the diffusion of lipid digestion products into the wall of the intestine.

After entry into the epithelial cell, monoglycerides are further hydrolyzed by an intracellular lipase. Then, the fatty acids are reconstituted with α-glycerol phosphate (Figure 7-1g), an activated form of glycerol, into triglycerides by the epithelial cell endoplasmic reticulum. These new triglycerides, along with absorbed cholesterol and phospholipids, are encased into globules with a protein coat by the endoplasmic reticulum. These particles, termed *chylomicrons*, are then extruded from the epithelial cells and released into the lymph. From there chylomicrons eventually enter the great veins in the neck through the thoracic duct. Thus, the route through the lymphatics constitutes the major route by which dietary lipids gain entry into the central circulation. A small amount of fatty acids, particularly short-chain fatty acids, however, are absorbed directly into the portal circulation and reach the liver directly.

Chylomicrons and Lipoproteins

Dietary fats appearing in the blood after a meal (cf. Figure 5-2a) are largely cleared (removed) from the plasma within several hours. Two major mechanisms exist for this.

One mechanism for the removal of chylomicron lipid from the blood is by transport of the chylomicron into liver cells. The liver is highly adapted to utilize lipid as a fuel and to convert lipids from chylomicrons into a variety of lipid-containing products including lipoproteins. Between meals, and during fasting and starvation, the liver is particularly active in producing lipoproteins to maintain circulating levels of lipid. Lipoproteins from the liver are mixtures of triglycerides, phospholipids, cholesterol, and proteins, like chylomicrons from the small intestine. Also, as with chylomicrons, protein encapsulates the lipid of lipoproteins to facilitate their solubilization and transport in the blood.

Lipoproteins may be classified by means of their density. Increased lipid content reduces the density of lipoproteins. The constituents of lipoproteins and chylomicrons are given in Table 7-1. In general, the low-density (LDL) and very-low-density lipoproteins (VLDL) function to transport triglycerides from liver to adipose tissue. High-density lipoproteins (HDL) are thought to be in-

TABLE 7-1
Major Constitutents of the Lipoproteins (%)

Type of Lipoprotein	Triglycerides	Phospholipids	Cholesterol	Protein
Chylomicrons	85	7	7	1
Very-low-density lipoproteins	50	18	23	9
Low-density lipoproteins	10	22	47	21
High-density lipoproteins	8	29	30	33

volved in cholesterol catabolism, and a high circulating level of HDL in relation to LDL (HDL/LDL) is thought to be beneficial in terms of protection from coronary heart disease (Chapter 24).

Lipoprotein lipase is an enzyme located in capillary walls of most tissues in the body. Particularly in adipose tissue and in heart and skeletal muscle, relatively large quantities of the enzyme are found. The major portion of chylomicrons, as well as LDL and VLDL, are cleared from the blood by the action of lipoprotein lipase. After a meal, when blood glucose and insulin levels are elevated, adipose lipoprotein lipase is activated. The rate of hydrolysis of triglyceride in lipoproteins arriving at adipose is then accelerated. Free fatty acids are thereby made available for storage in adipose.

It should be noted that control of muscle lipoprotein lipase is essentially opposite that in adipose. During exercise when insulin falls and glucagon rises (i.e., when the glucagon/insulin rises), muscle lipoprotein lipase is activated and that in adipose is inhibited.

An elevated insulin level following a meal promotes fat storage in other ways. Insulin allows the entry of glucose into cells, including adipose cells (adipocytes). Additionally, a by-product of glucose metabolism, α-glycerol phosphate, provides the glycerol backbone for the synthesis of adipose triglyceride.

THE UTILIZATION OF LIPIDS DURING EXERCISE

As noted earlier, lipids provide an important energy source for prolonged exercise. As will become obvious in the following presentation, lipid utilization is a complicated process that usually begins at one site (adipose tissue) and is completed in another site (skeletal muscle mitochondria). The metabolism at each of these and at intervening sites is well controlled, so that the ultimate process, lipid oxidation, represents a highly integrated process.

The processes of lipid metabolism during exercise can be summarized as follows:

1. *Mobilization*   The breakdown of adipose triglyceride.
2. *Circulation*   The transport of free fatty acids (FFA) from adipose to muscle.
3. *Uptake*   The entry of FFA into muscles from blood.
4. *Activation*   Raising the energy level of fatty acids preparatory to catabolism.
5. *Translocation*   The entry of activated fatty acids into mitochondria.
6. *β Oxidation*   The catabolism to acetyl-CoA of activated fatty acids and the production of reducing equivalents (NADH and FADH).
7. *Mitochondrial Oxidation*   Krebs cycle and electron-transport chain activity.

Mobilization

Figure 7-2 illustrates the processes of mobilization, circulation, and uptake. In addition to the lipoprotein lipase in adipose tissue, adipocytes contain a second lipase enzyme called "hormone-sensitive lipase." The control and action of these two lipases are essentially reversed. Whereas the lipoprotein lipase is stimulated by insulin and glucose and promotes fat storage, the hormone-sensitive lipase stimulates fat breakdown, is inhibited by insulin, and is stimulated by other hormones including the catecholamines (epinephrine and norepinephrine) and growth hormone.

The activity of hormone-sensitive lipase is directly controlled by the presence of cyclic AMP, which, in turn, is regulated by the adenylate cyclase system. Thus, the activation of fat breakdown is much like the initiation of glucose breakdown (Figure 5-5). Two activators of the hormone-sensitive lipase system (epinephrine and growth hormone) reach adipose tissue via the circulation, whereas norepinephrine is released locally by sympathetic nerve endings within adipose tissue. Compared to the release of growth hormone, which is slow, the release of the catecholamines is relatively rapid. The catecholamines are therefore responsible for the initiation of lipolysis at exercise onset. Blood levels of growth hormone take 10 to 15 min to increase during exercise, and the activation of adipocyte lipase activity by growth hormone requires protein synthesis. Growth hormone helps to maintain lipolysis during prolonged exercise.

Glycerol released into the circulation from adipose as the result of lipolysis is soluble in the blood. Glycerol, however, has a minor role as a substrate in muscle during exercise. Rather, glycerol represents a significant gluconeogenic precursor during exercise. Because the uptake by muscle of FFA released from

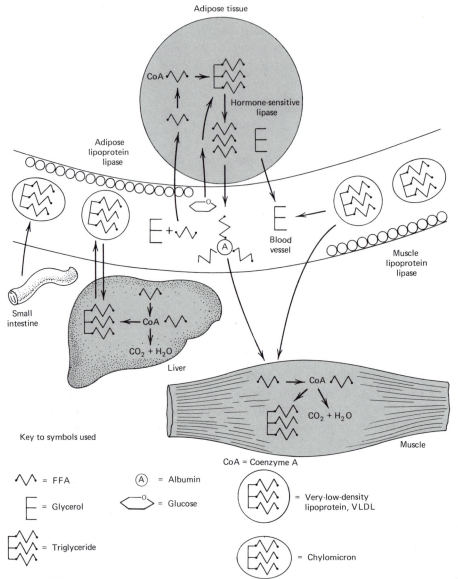

Key to symbols used

CoA = Coenzyme A

$\wedge\wedge$ = FFA

(A) = Albumin

$\underset{|}{\mathsf{E}}$ = Glycerol

$\langle$-o-$\rangle$ = Glucose

$\underset{|}{\mathsf{E}}\wedge\wedge$ = Triglyceride

= Very-low-density lipoprotein, VLDL

= Chylomicron

FIGURE 7-2
Mobilization and circulation. Most of the lipid oxidized in muscle during exercise is delivered by the circulating blood. The blood contains free fatty acids (FFA) that are released from adipose tissue after mobilization of triglyceride stores by activation of the hormone sensitive lipase. The blood also contains lipoproteins (both chylomicrons from the GI tract and VLDL from the liver), that deliver a much smaller quantity of lipid to muscle during exercise. Thus the ability to use fat as a fuel during exercise depends on the arterial blood level (a function of mobilization) and muscle blood flow (a function of the circulation).

(SOURCE: Modified from R.J. Havel. With permission.)

TABLE 7-2
Blood Lipid Levels in Normal Resting
Adults

Lipid Type	Blood Level (mg · dl^{-1})
Free fatty acids	8–20
Triglycerides	50–150
Cholesterol	150–250
Phospholipid	150–250

adipose is greater than the hepatic uptake of glycerol during exercise, glycerol levels rise sooner and relatively more than do FFA. Therefore, the elevation of blood glycerol during exercise provides a rough index of the rate of adipose lipolysis.

Because they are insoluble in aqueous media, fatty acids must be carried in the blood. The blood protein albumin serves to carry almost all the FFA and therefore a majority of the total lipids transported in the blood. Even though the quantity of lipid existing as FFA in the blood at any one time is a small portion of the total blood lipid content (Table 7-2), the turnover (entry and exit) of blood FFA is very rapid. Therefore, the contribution of FFA to the fuel supply during rest and exercise far exceeds the contribution of other blood lipids such as triglycerides.

Circulation and Uptake

By means of detailed radiotracer studies it has been demonstrated that half of the arterial FFA are removed in the muscle capillary bed during each circulation of blood through muscle. Therefore, several physiological as well as biochemical factors affect lipid metabolism during exercise.

Because the uptake of FFA in lipid delivery depends to a large extent on the arterial fatty acid content, the rate of adipose lipolysis will directly affect the uptake of FFA by muscle. Because the extraction of FFA by muscle remains about 50%, regardless of FFA level or rate of muscle blood flow, a key to lipid oxidation during exercise is the arterial level of FFA. As will be noted in Chapter 28 on ergogenic aids, any factor (e.g., caffeine) that stimulates adipose lipolysis and raises blood FFA levels could promote exercise endurance.

The rate of blood flow through muscle is a major determinant of FFA uptake and utilization during exercise. Although this factor is usually overlooked in a discussion of the control of fat metabolism, the enhancement of cardiac output and muscle blood flow by endurance training is a major factor determining the ability to oxidize lipids during exercise. The greater the muscle blood flow, the greater the delivery, uptake, and utilization of FFA by muscle during exercise.

The uptake into muscle of fatty acids from blood albumin is accomplished via a specific receptor site on the sarcolemma. As a result of the number of receptor sites, the activity, and the high solubility of fatty acids within the membrane (a lipid within a lipid), the entry of fatty acids from blood into muscle cells is rapid, and the muscle membrane fatty acid receptor sites are able to accommodate differing quantities of fatty acids provided by variations in blood fatty acid concentration and muscle blood flow.

Activation and Translocation

The first step in intracellular metabolism of lipids resembles the first step in glycolysis; this process, the activation of the fatty acid, is diagrammed in Figure 7-3. The activation process of fatty acids raises them to a higher energy level and involves ATP. However, the process differs from the activation in glycolysis in that the fatty acid is attached to coenzyme A (CoA), with the formation of a CoA derivative, termed fatty acyl-CoA.

The site of fatty acyl-CoA formation is the cytoplasm. However, the site of fatty acid oxidation is the mitochondrion. Because the outer membrane of the mitochondrion is either impermeable or selectively permeable to a wide variety of substances, usually materials gain entry into or exit from mitochondria by a transport mechanism. For fatty acids the mechanism involves a carrier, carnitine, and an enzyme, carnitine translocase. This mechanism (Figure 7-3) involves the stripping off of CoA and its return to the cytosol, and the acceptance of the fatty acid by carnitine with the formation of fatty acyl carnitine. The fatty acyl carnitine is free to move across the mitochondrial membrane, where on the inner side, carnitine translocase catalyzes the reverse reaction, leaving carnitine within the membrane, and releasing fatty acyl-CoA into the mitochondrial matrix.

In general, and particularly during exercise, the catalytic activity of carnitine translocase is an extremely important step in the overall control of fat oxidation. Translocation of fatty acid derivatives into the mitochondrial organelles (where the fats can be oxidized) is directly affected by the translocation process. Although no oxygen is directly involved, translocation is a very "aerobic" process, for it depends on the number of mitochondria present. Genetic endowment with red skeletal muscle rich in mitochondria, or endurance training to increase the mitochondrial mass, increases the amount of carnitine translocase that is present in constant amounts within mitochondria. Therefore, one of the reasons the lean, endurance-trained athlete is more successful than an obese, sedentary counterpart in using fats as a fuel is the number of mitochondria, which control the amount of activated fatty acids to be oxidized.

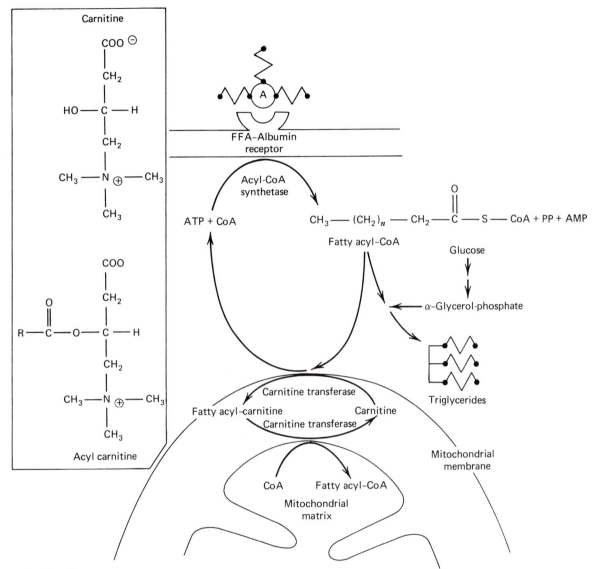

FIGURE 7-3

Free fatty acid (FFA) activation and carnitine translocation. Prior to either oxidation or storage as triglyceride in muscle, fatty acids delivered by the circulation must be activated; this involves ATP and coenzyme A (CoA). In order to gain entry into the mitochondrial matrix for oxidation, the activated fatty acid must combine with a carrier substance, carnitine, which exists in the mitochondrial membrane. An enzyme, carnitine transferase, facilitates both the formation of the fatty acyl–carnitine complex, and dissociation of the complex and release of the activated fatty acid into the matrix. The translocation of activated fatty acids from the cytosol into mitochondria is a rate-limiting step in lipid metabolism.

β Oxidation

The β-oxidation cycle, located in the mitochondrial matrix (Figure 7-4), serves several purposes. First it degrades the fatty acyl-CoA to acetyl-CoA by cleaving off the carbon atoms two at a time. Starting from the carboxyl end of a fatty acid (Figure 7-1a), the first carbon in is the α carbon and the second is the β carbon. In the β-oxidation pathway, cleavage occurs between the α and β carbons. The acetyl-CoA formed as the result of β oxidation can then enter the tricarboxylic acid (TCA) cycle, each acetyl-CoA resulting in the formation of 12 ATPs. For each cycle of the β-oxidation pathway, two carbons are lost from fatty acyl-COA. For palmitate, containing 16 carbons, there will be 8 acetyl-CoA molecules formed, but only 7, $n/2 - 1$, cycles of the β oxidation as the last unit formed will be acetyl-CoA itself, which will enter the TCA cycle rather than traversing the β-oxidation pathway.

The rate-limiting step in β oxidation is probably the last step, that catalyzed

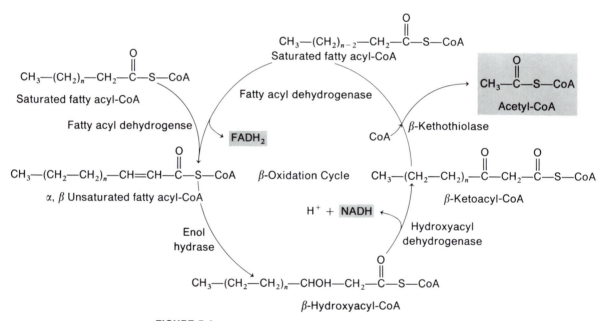

FIGURE 7-4

In the β-oxidation cycle, activated long-chain saturated fatty acids that have been translocated into the mitochondrial matrix are processed with the formation of three products (usually) for each cycle of the metabolic apparatus: acetyl-CoA, FADH, and NADH. These products will result in significant mitochondrial ATP production. The β-oxidation cycle is controlled by product inhibition. A reduced state of mitochondrial cofactors (i.e., NADH and FADH) will inhibit the dehydrogenases, and acetyl-CoA will inhibit the thiolase. Conversely, oxidized cofactors (NAD^+ and FAD) and low levels of acetyl CoA will stimulate the β-oxidation cycle.

by the enzyme β-ketothiolase. This enzyme is inhibited by its product, acetyl-CoA. Thus when the acetyl-CoA level is elevated, such as after a meal rich in carbohydrates, fat catabolism is slowed. When the acetyl-CoA level is depleted, such as during exercise, fat utilization is promoted.

As seen in Figure 7-4, a second function of the β-oxidation pathway is to produce the high-energy reducing equivalents NADH and FADH. For each cycle of the β-oxidation pathway, one each of NADH and FADH will be formed. These are collectively worth $3 + 2 = 5$ ATP.

In calculating the number of reducing equivalents formed from a fatty acid in β oxidation, remember that the number of cycles is $(n/2 - 1)$. For palmitate (16 carbons) there are 7 cycles, and 7 NADH and 7 FADH formed. Therefore, from a single palmitate, there are 129 ATP's formed. It may be helpful to make this calculation, remembering the ATP used in activation which produces AMP.

Mitochondrial Oxidation

After conversion to acetyl-CoA, metabolism of residues of fatty acids is the same as residues from sugar and carbohydrate metabolism. The formation of citrate in the Krebs cycle represents a common entry point for metabolism of acetyl-CoA derived from the various fuel sources (Figure 6-4).

INTRAMUSCULAR TRIGLYCERIDES AND LIPOPROTEINS AS FUEL SOURCES

Radiotracer studies have demonstrated that approximately half the lipid used by muscle is delivered by the circulation. Triglyceride stores within the muscle itself provide the other half. Lipid droplets can be seen in the electron microscope (Figure 7-5) to exist near mitochondria in red skeletal muscle. Control of lipolysis within muscle is apparently much like that in adipose tissue and is hormonally mediated.

As noted earlier, skeletal muscle does possess some lipoprotein lipase activity. Therefore, triglycerides in circulating lipoproteins are also consumed by muscle during prolonged exercise. These triglycerides join together with intramuscular triglycerides and free fatty acids delivered by the circulation to form a pool (reservoir) from which the lipids to be metabolized during exercise are selected.

TISSUE SPECIFICITY IN LIPID UTILIZATION

Various tissues in the body such as heart and liver are highly adapted for lipid utilization. Other cells such as brain and red blood cells rely almost exclusively on glycolysis for an energy supply. Among skeletal muscle cells, the ability to

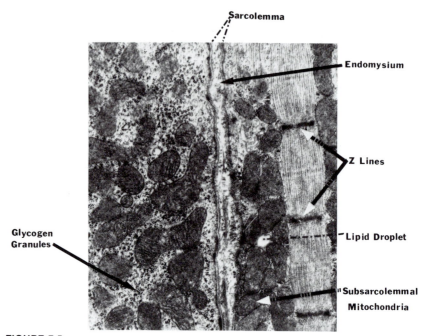

FIGURE 7-5
Electron micrograph of a red skeletal muscle fiber showing mitochondria, lipid droplets near the mitochondria, and glycogen granules.

(SOURCE: Micrograph courtesy of J.B. Gale.)

utilize fats as an energy source varies greatly. White, fast-contracting fibers, with less than optimal blood supply and low mitochondrial density, are limited in the ability to utilize fat. These white, fast-contracting fibers (sometimes called fast glycolytic fibers or type IIb fibers; see Figure 5-3) depend on glycogenolysis and glycolysis for energy supply. In contrast, red skeletal muscle fibers, with rich blood supply, many capillaries, and high myoglobin as well as mitochondrial contents, are well adapted to utilize fats as a fuel. The ability of muscle tissue to sustain prolonged activity (muscle tissue being made up of a heterogeneous mixture of muscle cells) by fat metabolism varies greatly. Endurance training has been demonstrated to double a cell's mitochondrial capacity to utilize fat. However, red skeletal muscles, on a unit weight basis, can possess 10 times the ability to utilize fats as white muscle cells (Table 6-1). Therefore, both genetics and training affect the capacity for fat utilization during exercise.

KETONES AS FUELS

The pathways described above constitute the major way in which mammals utilize fats. Another somewhat indirect mechanism exists whereby fats can be

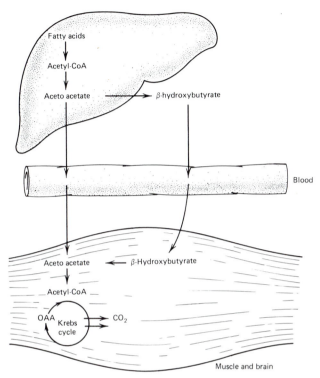

FIGURE 7-6
Acetoacetate and β-hydroxybutyrate are formed in liver and kidney from
fatty acid catabolism. These "ketones" are released into the circulation
and are taken up by brain (e.g., during starvation), and red skeletal muscle
(e.g., during prolonged exercise).

(SOURCE: Modified from R.W. McGilvery, 1975.)

used as fuels. Under conditions of carbohydrate starvation, such as during fasting, prolonged exercise, and diabetes (see below), ketones can be used as fuels.
During this process, diagrammed in Figure 7-6, the liver catabolizes fat to
acetyl-CoA, and then converts acetyl-CoA to ketones. The ketones then circulate and can enter cells such as muscle, nerves, and brain. In particular, during
dietary starvation the ability to sustain life depends on the ability to form and
utilize ketones. Within a day of fasting, the liver's glycogen supply is exhausted in the attempt to maintain blood glucose level. The brain, and to some
extent the kidney, depend heavily on glucose as a fuel. Entry of fatty acids
into the brain is very limited, thus in the first days of starvation, skeletal muscle
is catabolized rapidly to provide material for glucose synthesis. This rapid loss
of lean tissue cannot be maintained very long without severely affecting the
individual. However, after several days of starvation, ketone levels in the blood
elevate to the point where they can provide an alternative energy source for the

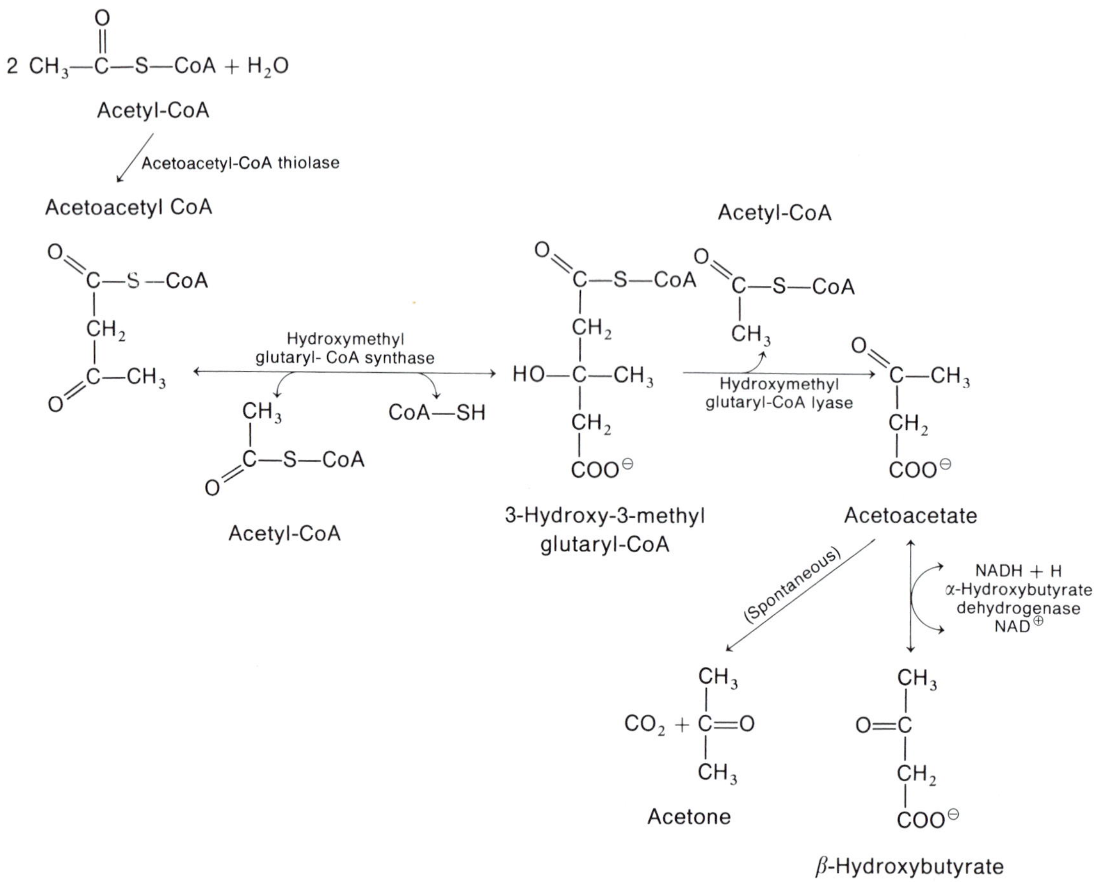

FIGURE 7-7

In the liver and kidney, acetoacetyl coenzyme A is formed from acetyl-CoA produced in fatty acid catabolism. Acetoacetyl-CoA is then converted to acetoacetate and β-hydroxybutyrate. A small amount of acetone is also formed from the spontaneous decarboxylation of acetoacetate.

brain and other nervous tissues. During untreated diabetes, blood glucose levels can be extremely high, but the cells can be starving for glucose because it cannot enter without insulin, which is lacking. Very high levels of ketones result, which frequently constitutes another problem as ketones are acidic and can affect an individual's physiology and biochemistry.

A ketone is a compound with the essential structure given in Figure 7-1h. Starting from acetyl-CoA, the basic chemistry of ketone formation is given in Figure 7-7. From acetyl-CoA, there are two main products (acetoacetate and β-hydroxybutyrate) and one minor product (acetone). Acetone is the most obvious product as it is extremely volatile and can be smelled on someone's breath. Acetoacetate can be formed in both the liver and kidneys. Much of the acetoacetate formed in the liver is converted to β-hydroxybutyrate. β-Hydrox-

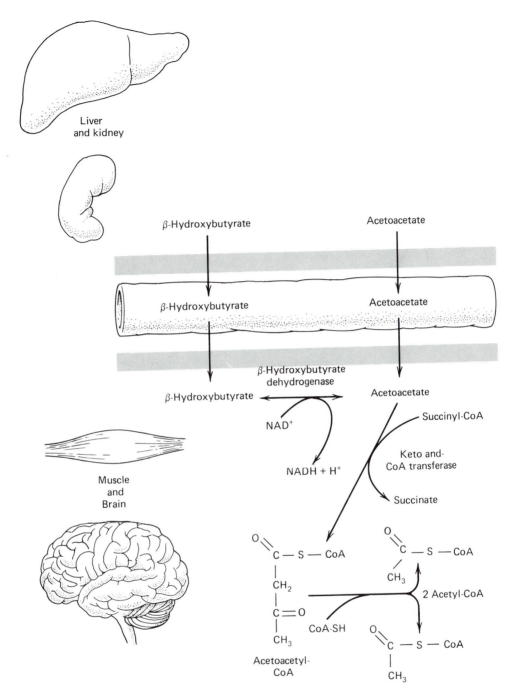

FIGURE 7-8

Ketones released from liver and kidney can circulate to and enter muscle and brain. In these tissues, β-hydroxybutyrate is converted to acetoacetate, which is converted to two molecules of acetyl-CoA. Each acetyl-CoA is equivalent to 12ATP after metabolism in the Krebs cycle and mitochondrial oxidative phosphorylation.

ybutyrate formed in the liver is a more reduced compound than acetoacetate. Therefore, when β-hydroxybutyrate reaches the muscles and brain, it can cause the formation of an extra NADH. As diagrammed in Figure 7-8, both aceto-acetate and β-hydroxybutyrate result in acetyl-CoA formation in muscle and brain.

It has been demonstrated that endurance training increases the muscular content of enzymes of ketone utilization. In prolonged exercise, an elevation of blood ketones is a usual result. However, when trained individuals exercise, the increase in blood ketones is not as dramatic as in untrained individuals because of the uptake and utilization of ketones by working skeletal muscle.

LIPID METABOLISM DURING VARIOUS ACTIVITIES

Lipid is a primary fuel sustaining prolonged aerobic activities. In those activities lasting only a few seconds or minutes, immediate energy sources and glycolysis meet the preponderance of energy requirements. It is not until activities are sustained for more than a few minutes that fat metabolism becomes significant. Through endurance training and dietary manipulation, muscle and liver glycogen reserves can be boosted to sustain continuous exercise at an intensity of 70 to 80% of $\dot{V}_{O_2}$max over the duration of an hour or more. For more prolonged activities (e.g., the marathon) the ability to utilize fat becomes critical. In both human and animal experiments, glycogen depletion has been well correlated with fatigue. Although muscle glycogen depletion is not the only cause of fatigue, it is at present clear that in prolonged exercise the exhaustion of glycogen depots compromises continued performance. An explanation of this phenomenon is offered in Chapter 32. However, to the extent that lipids can substitute for glucose and glycogen as fuel sources, lipids can spare the consumption of glucose and glycogen, thereby prolonging the activity.

There are several reasons for carbohydrates being preferable as fuels for muscular exercise. As discussed earlier (Table 3-1), carbohydrates yield more energy per unit of O_2 consumed. Second and more critical, however, the pathways of carbohydrate catabolism are more direct and function much more rapidly than those mechanisms of lipid catabolism. Even though the mechanisms of fat metabolism may proceed at their limit during maximal exercise, the contribution to the total energy flux may not contribute in a quantitatively significant way. Adequate levels of acetyl-CoA from carbohydrate catabolism will inhibit β oxidation and cause a backlog of lipid carbon flow. Furthermore, there is good evidence that high levels of lactic acid resulting from rapid glycogenolysis and glycolysis inhibit mobilization of fatty acids from adipose.

Mitochondrial Adaptation to Enhance Fat Oxidation

Several groups of researchers have estimated that skeletal muscle in an untrained individual possesses a greater oxidative capacity than can be supplied by O_2 delivery through the circulation. Why, then, does muscle mitochondrial

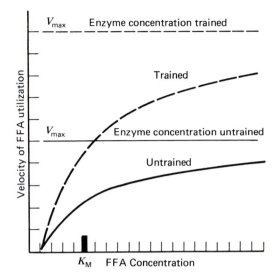

FIGURE 7-9
Effect of doubling enzyme concentrations on the utilization of high K_M substrates, such as fatty acids. At any given fatty acid concentration the rate of utilization is greatly increased. The K_M in trained individuals is equal to the V_{max} in untrained.

(SOURCE: Modified from P.D. Gollnick and B. Saltin, 1983. Refer to Figure 4–5.)

mass increase 100% after training but maximal cardiac output only increase 10 to 20%? The answer may be related to an effect on the muscle's ability to utilize FFA's as fuels. Although the mitochondrial enzymes specific to fatty acid utilization (e.g., carnitine translocase) constitute only a small fraction of the mitochondrial mass, the fact that mitochondria are assembled as complete units requires that the whole organelle be synthesized to enhance the capacity for fat utilization.

In muscles of trained individuals during exercise an increased ability to generate ATP and citrate from β oxidation would inhibit phosphofructokinase (PFK) and pyruvate dehydrogenase (PDH), thereby slowing the rate of glycolysis and the catabolism of glucose and glycogen.

The effect of mitochondrial proliferation due to training has been described by Gollnick and Saltin as having the effect of raising the apparent V_{max} of fat oxidation (Figure 7-9). Thus, at any given FFA concentration, the absolute utilization of fat is greater. Therefore, for reasons related to the kinetics of enzymes of fat utilization, the effect of endurance training on muscle mitochondrial mass may be explained.

Summary

During muscular activities of prolonged duration, it is essential to provide fuels at a rate sufficient to generate the ATP for muscular work aerobically and also to maintain an adequate level of blood glucose. Unfortunately, blood glucose and liver glycogen stores are inadequate to supply the fuels of exercise as well as to maintain blood glucose and muscle glycogen levels over protracted periods. Therefore, the ability to use fats as fuels is essential. The abundant supply of fats and their high energy content per unit weight provides, even in healthy lean people, a vast potential energy reserve. Moreover, the utilization of fats as fuels spares glucose and glycogen, which have essential roles in sustaining high rates of energy metabolism. In comparison to the use of carbohydrates and sugars as fuels, fats cannot be used except by oxidative processes, and thus the rate of fat catabolism is less than the rate of sugar and glycogen utilization. The capacity for fat utilization is specialized in red (oxidative) tissues such as liver and heart. Red skeletal muscle is likewise rich in mitochondrial content and can utilize fats at significant rates. Endurance training greatly enhances the ability to utilize fats by improving the circulation and mitochondrial content in skeletal muscle.

Selected Readings

Ahlborg, G., L. Hagenfeldt, and J. Wahren. Influence of lactate infusion on glucose during prolonged exercise in man. J. Clin. Invest. 53:1080–1090, 1974.

Ahlborg, G., L. Hagenfeldt, and H. Wahren. Influence of lactate infusion on glucose and FFA metabolism in man. Scand. J. Clin. Lab Invest. 36:193–201, 1976.

Carlson, L., L. Ekelund, and S. Froberg. Concentration of triglycerides, phospholipids, and glycogen in skeletal muscle and FFA and beta hydroxybutyric acid in blood in man in response to exercise. Eur. J. Clin. Invest. 1:248–254, 1976.

Costill, D., E. Coyle, G. Dalsky, W. Evans, W. Fink, and D. Hoopes. Effect of elevated FFA and insulin on muscle glycogen usage during exercise. J. Appl. Physiol. 43:695–699, 1977.

Dagenais, G., R. Tancredi, and K. Zierler. Free fatty acid oxidation by forearm muscle at rest, and evidence for an intramuscular lipid pool in human forearm. J. Clin. Invest. 58:421–431, 1976.

Davies, K.J.A., L. Packer, and G.A. Brooks. Biochemical adaptation of mitochondria, muscle, and whole-animal respiration to endurance training. Arch. Biochem. Biophys. 209:538–553, 1981.

Fitts, R.H., F.W. Booth, W.W. Winder, and J.O. Holloszy. Skeletal muscle respiratory capacity, endurance, and glycogen utilization. Am. J. Physiol. 228:1029–1033, 1975.

Frieberg, S., R. Klein, D. Trout, M. Bodgonoff, and E. Estes. The characteristics of peripheral transport of ^{14}C-labelled palmitic acid. J. Clin. Invest. 9:1511–1515, 1960.

Gollnick, P.D. Free fatty acid turnover and the availability of substrates as a limiting factor in prolonged exercise. Ann. N.Y. Acad. Sci. 301:64–71, 1977.

Gollnick, P.D. and B. Saltin. Hypothesis: Significance of skeletal muscle oxidative enzyme enhancement with endurance training. Clin. Physiol. 2:1–12, 1983.

Guyton, A.C. Textbook of Medical Physiology. Philadelphia: W.B. Saunders Co., 1981. p. 787, 825.

Havel, R.J. Lipid as an energy sources. In: Briskey, E.J. (ed.), Physiology and Biochemistry of Muscle as a Food. Madison: University of Wisconsin Press, 1970. p. 109–622.

Havel, R., G. Carlson, L. Ekelund, and A. Holmgren. Turnover rate and oxidation of different free fatty acids in man during exercise. J. Appl. Physiol. 19:613–618, 1964.

Havel, R.J., A. Naimark, and C.F. Borchgrerink. Turnover rate and oxidation of free fatty acids of blood plasma in man during exercise. J. Clin. Invest. 42:1054–1063, 1963.

Havel, R.J., B. Pernow, and N. Jones. Uptake and release of free fatty acids and other metabolites in legs of exercising men. J. Appl. Physiol. 236:90–99, 1967.

Hickson, R., M. Rennie, R. Conlee, W. Winder, and J. Holloszy. Effects of increased plasma FFA on glycogen utilization and endurance. J. Appl. Physiol. 43:829–833, 1977.

Issekutz, B., H.I. Miller, P. Paul, and K. Rodahl. Source of fat oxidation in exercising dogs. J. Appl. Physiol. 207:583–589, 1964.

Issekutz, B., H. Miller, P. Paul, and K. Rodahl. Aerobic work capacity and plasma FFA turnover. J. Appl. Physiol. 20:293–296, 1965.

Issekutz, B., W. Shaw, T. Issekutz. Effect of lactate on FFA and glycerol turnover in resting and exercising dogs. J. Appl. Physiol. 39:349–353, 1975.

Jones, N., G. Heigenhauser, A. Kuksis, C. Matsos, J. Sutton, and C. Toews. Fat metabolism in heavy exercise. Clin. Sci. 59:469–478, 1980.

Kozlowski, S., L. Budohoski, E. Pohoska, and K. Nazar. Lipoprotein lipase activity in the skeletal muscle during physical exercise in dogs. Pflugers Arch. 322:105–107, 1979.

Lehninger, A.L. Biochemistry. New York: Worth Publishing,Inc., 1970. p. 417–432.

Lehninger, A.L. Bioenergetics. Menlo Park, Cal.: W.A. Benjamin, Inc., 1973. p. 75–98.

Masoro, E.J. Physiological Chemistry of Lipids in Mammals. Philadelphia: W. B. Saunders Co., 1968.

McGilvery, R. W. Biochemical Concepts. Philadelphia: W.B. Saunders, Co., 1975. p. 345–358, 463–476.

Molé, P., L. Oscai, and J. Holloszy. Adaptations in muscle to exercise. J. Clin. Invest. 50:2323–2330, 1971.

Newsholme, E. Carbohydrate metabolism in vivo: Regulation of the blood glucose level. Clin. Endocrin. Metab. 5:543–578, 1976.

Newsholme, E. The control of fuel utilization by muscle during exercise and starvation. Diabetes 28: Suppl. 1–7, 1979.

Paul, P. FFA metabolism in normal dogs during steady-state exercise at different work loads. J. Appl. Physiol. 28:127–132, 1970.

Pruett, E. FFA mobilization during and after prolonged severe muscular work in men. J. Appl. Physiol. 29:809–815, 1970.

Rodbell, M. Modulation of lipolysis in adipose tissue by fatty acid concentration in the fat cell. Ann. N.Y. Acad. Sci. 131:302–314, 1965.

Terblanche, S.E., R.D. Fell, A.C. Juhlin-Dannfelt, B.W. Craig, and J.O. Holloszy. Effects of glycerol feeding before and after exercise. J. Appl. Physiol.: Respirat. Environ. Exercise Physiol. 50:94–101, 1981.

Winder, W., K. Baldwin, and J. Holloszy. Exercise-induced increase in the capacity of rat skeletal muscle to oxidize ketones. Can. J. Physiol. Pharmacol. 53:86–91, 1974.

Zierler, K. Fatty acids as substrates for heart and skeletal muscle. Circl. Res. 38:459–463, 1976.

8

METABOLISM OF PROTEINS AND AMINO ACIDS

Of the three major categories of foodstuffs, only proteins have the characteristics necessary to form body structures and enzymes. Proteins are assemblages of individual amino acid units. Amino acids contain peptide (—NH_2) groups that can be chemically linked to the carboxyl groups of other amino acids. These linkages are called peptide bonds and are the basis of protein structures. Because of the difficulties of studying amino acid and protein metabolism, and because of their essential structural roles, for a long time it was thought that amino acids do not serve as fuels to sustain muscular work. However, it is now apparent that particular amino acids are integrated into the flow of substrates powering the body during prolonged exercise. Furthermore, amino acids may be important not only because they are fuel sources, but also because they give rise to glucose (through the process of gluconeogenesis) as well as support the utilization of other fuels, such as fats.

STRUCTURE OF AMINO ACIDS AND PROTEINS

There are over 20 amino acids in the body. A generalized structure of an amino acid is given in Figure 8-1. Some of the body's amino acids can be synthesized internally from existing amino acids and other substances; amino acids that the body can synthesize are termed *nonessential*. Ten other amino acids cannot be

Bruce Jenner as he sets a world record in winning the Montreal Olympic Decathlon. Performances such as these depend upon having a sizable muscle mass as well as the ability to draw upon body reserves of carbohydrates, fats, and amino acids and proteins as fuel sources. (Wide World.)

synthesized in the body and are *essential* constituents in the diet. Generally, these essential amino acids have structures that are quite different from structures in intermediary metabolism. Some essential amino acids contain ring structures, branched end chains, or sulfur.

By convention, the structures of amino acids and other organic acids are described by considering the carboxyl carbon as carbon 1 in the structure. By convention also, the carboxyl end of an acid is usually placed at the right when drawing it (see Figure 8-1). The carbon next to carbon 1 (i.e., carbon 2) is termed the α carbon. When a carboxylic acid has an amino group ($-NH_3^+$) on the α carbon, it can be classified as an amino acid. Amino acids are sometimes referred to as α-amino carboxylic acids.

Amino groups on amino acids provide the structure necessary for attachment to carboxyl groups of other amino acids. When a bond is formed between the α amine of one amino acid and the carboxyl group of another, a peptide bond is formed; this is diagrammed in Figure 8-2. Given the more than 20 body amino acids, it is possible to assemble these in an almost infinite number of ways. Hence, it is possible for the body to synthesize a large number of very

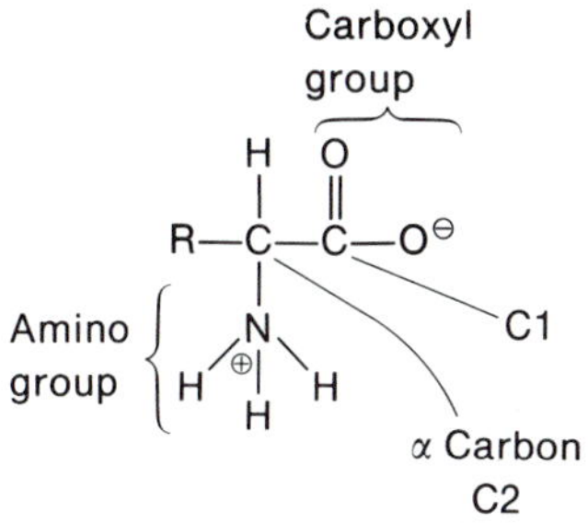

FIGURE 8-1
General structure of an amino acid, an α-amino carboxylic acid. At physiological pH, most amino acids carry a negative ($-$) charge, and usually a positive ($+$) charge as well. Note: R groups are side chains.

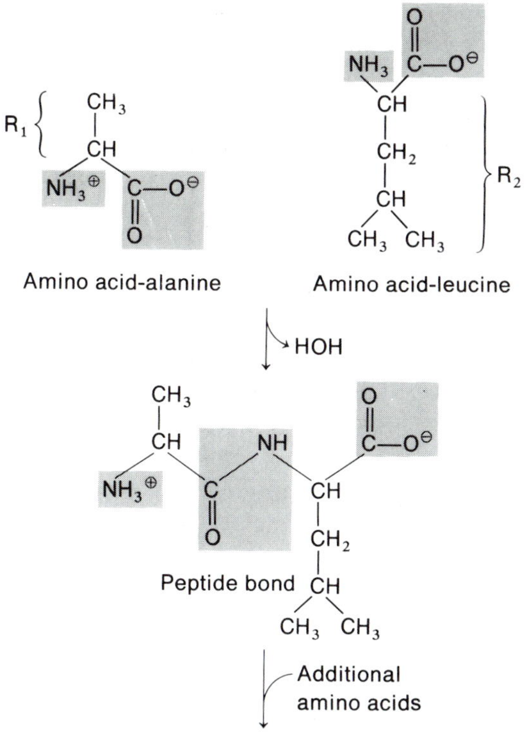

FIGURE 8-2
Proteins are formed by the linkage of the carboxyl group of one amino acid to the amino group of another. Such linkages are called peptide bonds, and are enzymatically formed and broken.

(SOURCE: Modified from A.J. Vander, J.H. Sherman and D.S. Luciano, 1980. With permission.)

different protein structures as well as smaller amino acid structures and enzymes.

PROTEINS IN THE DIET

Meats and vegetables contribute the bulk of dietary proteins. Digestion of proteins begins in the stomach under the influence of the digestive enzyme pepsin, which is very active in acidic media; thus the stomach also secretes HCl. There is almost no absorption of amino acids or proteins in the stomach. Digestion continues in the small intestine.

When contents of a protein meal leave the stomach, only about 15% leave as amino acids. In the small intestine, the larger by-products are attacked by the pancreatic enzymes trypsin, chymotrypsin, and carboxypolypeptidase. Together these proteolytic enzymes reduce most of the proteins in the meal contents to amino acids and slightly larger units, polypeptides. Epithelial cells of the small intestine contain additional amino polypeptidase and dipeptidase enzymes, which are responsible for hydrolyzing remaining peptide bonds.

Absorption of amino acids occurs through the mucosal cells lining the small intestine. The absorption process is relatively more rapid than the digestive processes in the stomach and small intestine. Consequently, during the 2 to 3 hr it takes to digest and absorb protein, only low levels of amino acids exist free in the digestive organs.

Amino acids are absorbed by specific carrier mechanisms. There are four such carrier systems, one each for neutral, basic, and acidic amino acids as well as one for proline and hydroxyproline. The transport of amino acids is an active process (requiring ATP) and is linked to the transport of sodium ions. Apparently, the carriers for amino acid transport, like those for glucose, exist on the brush border of intestinal villi and have sites for binding of both sodium and the amino acid. The carrier mechanism is thought to operate because of the sodium gradient across the brush border; sodium movement pulls the carrier and attached amino acid into the mucosal cells. From there the amino acids diffuse into the portal circulation.

THE AMINO ACID POOL

When amino acids from the diet enter the circulation (Figure 5-2a), they are entering into one of the important compartments comprising the body's amino acid pool (Figure 8-3). Liver and skeletal muscle are other major compartments comprising the total pool. Amino acids in these compartments are in direct equilibrium with those in the blood. Therefore, amino acid metabolism in one

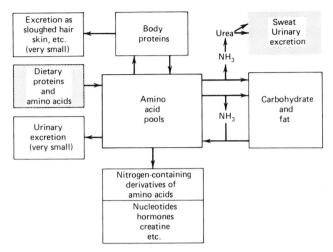

FIGURE 8-3

Amino acids and proteins enter the body through one route (foods) and leave through one major route (urea). The body's proteins and amino acids are not bound up in static compounds and structures, but the amino acids exist as part of various pools, some of which turn over slowly and some of which turn over more rapidly. When the total amount of dietary protein nitrogen entering a person equals that leaving, the person is said to be in nitrogen balance.

(SOURCE: Modified from A.J. Vander, J.H. Sherman, and D.S. Luciano, 1980. With permission.)

compartment affects amino acids not only in that compartment, but also in other compartments. There are relatively few amino acids in the blood compartment compared to the quantity of amino acids and proteins in the other compartments. However, movement (flux) through the blood is very rapid, and the blood provides communication and exchange with the other compartments.

The continuous elimination of nitrogenous products from the body (Figure 8-3) is due to the ongoing catabolism of proteins and amino acids in the body. Different proteins turn over at different rates, but the half-life (time for removal of half the protein existing at any one time) ranges from a few days to a few months. Therefore, constant input of new amino acids into the central pool is required to compensate for loss and a constant rearrangement of material within the body. The fact that a central pool of amino acids exists that is in balance with other compartments is of great advantage to the organism. In this way, the amino acid requirements in any one compartment can be met for a period of time even though the diet does not contain the proper amount or combination of amino acids required by the compartment. Ultimately, however, the central blood pool and its allied compartments will be depleted if dietary protein input is inadequate.

Nitrogen Balance

When the dietary input (in terms of grams of nitrogen per day) is equal to that excreted, the individual is in nitrogen balance. Most young, healthy adults on an adequate, balanced diet will be in nitrogen balance. Children given an adequate diet will be in *positive* nitrogen balance, indicating the storage of amino acids and the synthesis of lean tissue. Adults who engage in weight lifting, eat large amounts of protein, and/or take anabolic steroid drugs can also enter into positive nitrogen balance. This is discussed further in Chapter 28. People who are sick or injured are frequently in *negative* nitrogen balance. In mature adults, nitrogen balance can usually be maintained on 0.57 g of good quality dietary protein per kilogram of body weight per day. Allowing an almost 100% error for individual differences and fluctuations in daily requirements, having 1.0 g of protein per kilogram of body weight per day is a usual daily recommended protein consumption. The normal, balanced North American diet usually contains at least this quantity. In the belief that they require extra amounts of dietary protein, athletes frequently consume several times the recommended quantity (i.e., $1.0 \text{ g} \cdot \text{kg}^{-1} \cdot \text{day}^{-1}$). Protein requirements of athletes are discussed in Chapter 27.

The Removal of Nitrogen from Amino Acids— The Role of Glutamate

Before amino acids can be used as fuels, the nitrogen-containing amine group (or groups) must be removed. Removal of nitrogen is especially important for carnivores and for humans on high-protein diets. Consequently, use of such diets to sustain endurance activities, where large supplies of sugars and carbohydrates are utilized, is both expensive (in terms of the dollar cost of the protein foods) and inefficient (in terms of the extra metabolism involved).

Nitrogen is removed from amino acids by two major mechanisms: (1) oxidative deamination and (2) transamination. Although there are diverse routes by which amino acids can be deaminated or transaminated, a basic strategy in

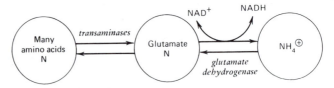

FIGURE 8-4
The most common means of removing nitrogen from amino acids is to transaminate the amino acid with α-ketoglutarate to form glutamate. The glutamate is then oxidized by NAD^+ through the action of glutamic dehydrogenase with the formation of ammonium ion ($NH_4^{(+)}$). This is a specialized enzymatic process in the liver.

(SOURCE: Modified from R.W. McGilvery, 1975. With permission.)

nitrogen removal is to convert amino acids to glutamate. The pathways of glutamate metabolism are well developed, and the funneling of amino acids into glutamate provides a common route for dealing with amino acid nitrogen (Figure 8-4).

Oxidative deamination   Oxidative deamination is a process that occurs in the mitochondrial matrix and involves NAD^+ as the oxidizing agent. The oxidative deamination of glutamate is given below.

$$H_2O + NH_3^+ - \overset{\displaystyle COO^-}{\underset{\displaystyle \underset{\textstyle COO^-}{\overset{\textstyle |}{CH_2}}}{\overset{\textstyle |}{\underset{\textstyle \overset{\textstyle |}{CH_2}}{\overset{\textstyle |}{C} - H}}}} + NAD^+ \underset{\text{Dehydrogenase}}{\overset{\text{Glutamate}}{\rightleftharpoons}} \overset{\displaystyle COO^-}{\underset{\displaystyle \underset{\textstyle COO^-}{\overset{\textstyle |}{CH_2}}}{\overset{\textstyle |}{\underset{\textstyle \overset{\textstyle |}{CH_2}}{\overset{\textstyle |}{C} = O}}}} + NH_4^+ + NADH \qquad (8\text{-}1)$$

Glutamate $\qquad\qquad\qquad\qquad$ α-Ketoglutarate

This reaction is catalyzed by glutamate dehydrogenase, an enzyme that is freely reversible and can function in either direction depending on the conditions. When sufficient quantities of substrates are available to provide material for the tricarboxylic acid (TCA) cycle and to reduce NAD^+ to NADH, the reaction can function to form glutamate. When there is a shortage of substrates, then glutamate can be broken down to two useful products for energy metabolism: α-ketoglutarate (α-KG) and NADH. α-Ketoglutarate is a TCA cycle intermediate, and NADH can yield several ATP molecules.

Transamination   Transamination is by far the most common route for exchange of amino acid nitrogen. As with deamination, major transaminases involve the amino acid glutamate. As implied in the name, transamination involves the transfer of an amine group. More specifically, transaminations involve the transfer of an amine from an amino acid to a keto acid. The resulting products are a new amino acid and a keto analog of the original amino acid. Like amino acid dehydrogenases, transaminases are enzymes that can function in either direction depending on the circumstances.

A very common and important transaminase is glutamate-pyruvate transaminase (GPT). This is a major route by which alanine is utilized.

$$\text{Alanine} + \alpha\text{-ketoglutarate} \underset{}{\overset{\substack{\text{Glutamate-pyruvate} \\ \text{Transaminase}}}{\rightleftharpoons}} \text{pyruvate} + \text{glutamate} \qquad (8\text{-}2)$$

A second common transaminase is glutamate-oxaloacetate transaminase (GOT).

$$\text{Oxaloacetate} + \text{glutamate} \xrightleftharpoons[]{\underset{\text{Transaminase}}{\overset{\text{Glutamate-oxaloacetate}}{}}} \text{aspartate} + \alpha\text{-ketoglutarate} \quad (8\text{-}3)$$

This reaction yields the TCA cycle intermediate (α-ketoglutarate) as well as the amino acid aspartate. Passage through aspartate is the major route by which most nitrogen is excreted from the body.

The Excretion of Nitrogenous Wastes

Small quantities of nitrogen are excreted as ammonia and other compounds (Figure 8-3), but by far the most nitrogen excreted is in the form of urea. The urea cycle (Figure 8-5) is a process centralized in the liver. Sometimes this process of urea synthesis is termed the "other Krebs cycle," for it was elaborated in good measure through the efforts of Sir Hans Krebs. As in the TCA cycle, in the urea cycle there is a union of two compounds, one entering the cycle (in this case carbamoyl phosphate) and the other being the last constituent in the cycle (in this case ornithine).

Formation of carbamoyl phosphate from ammonia and CO_2 is an energy-requiring and irreversible process. Normally, carbamoyl phosphate synthetase activity in liver is sufficient to "cleanse" liver and blood of almost all ammonium ions. In a subsequent step of the urea cycle, aspartate, which is derived from transamination of glutamate, enters the cycle so that a net of two nitrogens enters the cycle. Consequently, the product of the cycle (urea) contains two nitrogens. The urea synthesized in the liver is released into the blood. It is the function of the kidneys to remove the circulating urea and to secrete it into the urine.

Also of note in the urea cycle (Figure 8-5) is the formation of fumarate; this TCA cycle intermediate is a useful compound for gluconeogenesis.

Sites of Amino Acid and Protein Degradation

Skeletal muscle proteins provide the major store of amino acids in the body. Other significant though lesser amino acid sources are proteins in liver, blood, and intestinal wall. Skeletal muscle's role in the destruction of amino acids is a limited one, but it is capable of catabolizing considerable amounts of its protein content into amino acid. Skeletal muscle also contains large quantities of transaminase enzymes and can exchange amine groups among amino acids and keto acids. With the exception of the branched-chain amino acids, skeletal muscle has limited capability for net degradation of amino acids. This is because transaminases can change the type of amino acids present but not the molal quantity of amino acids. Oxidative deamination and the urea cycle are processes of the liver; thus the liver is the major site of amino acid degradation.

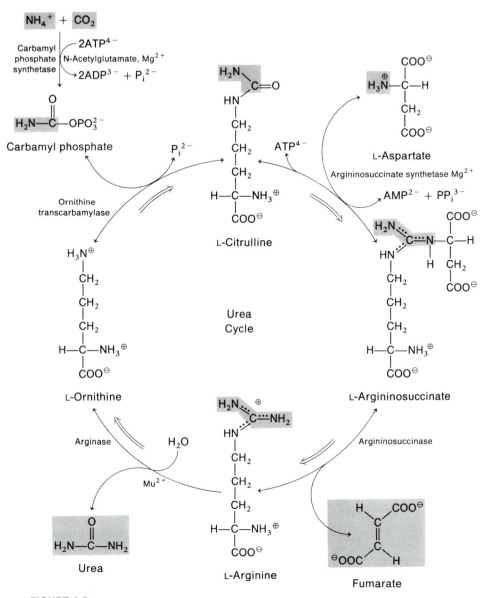

FIGURE 8-5
Urea is formed in the liver; the carbon and the nitrogen of urea come from
$NH_4^{(+)}$ (top left), and the other nitrogen comes from aspartate (top right).
Urea is released into the blood from the liver and is removed by the kidney
and sweat glands. Three ATP molecules are utilized and one fumarate
(TCA cycle intermediate) is formed as the result of each urea molecule
synthesized.

(SOURCE: From R.W. McGilvery, 1975. With permission.)

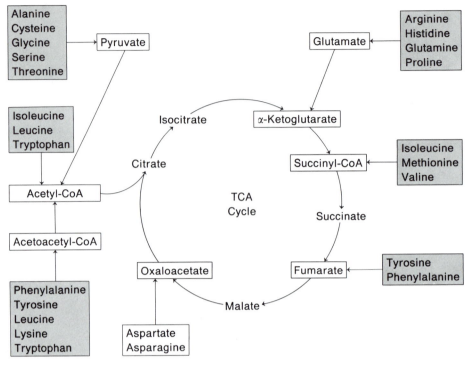

FIGURE 8-6
Carbon skeletons of the various amino acids gain access to the TCA cycle by various pathways. Leucine carbon enters only as acetyl-CoA.

(SOURCE: From A.L. Lehninger, 1970. With permission.)

The Fate of Amino Acid Carbon Skeletons

Whereas urea is the major end product of nitrogen from degraded amino acids, there are two general routes by which carbon skeletons of amino acids are degraded: (1) by converting the carbon atoms to glucose and (2) by converting the carbon to the ketone acetoacetate or to acetyl-CoA. The former process is said to be glucogenic; the latter process is ketogenic (Figure 8-6). After removal of their nitrogen groups, most amino acid residues appear as pyruvate or as TCA cycle intermediates.

GLUCONEOGENIC AMINO ACIDS

Amino acids can sometimes supply the major portion of carbon used in gluconeogenesis. During fasting and starvation, catabolism of proteins to amino acids and the conversion of amino acids are very important processes in maintaining the levels of glucose essential for brain and kidney function. Additionally, gluconeogenesis from amino acids occurs early each morning. Even after a good supper, the liver cannot store sufficient glycogen to maintain blood glucose levels until the nutrients in the breakfast become available. In the early morning, then, the liver increases the rate of gluconeogenesis as the potential for glycogenolysis falls.

In the last several years it has been realized that prolonged exercise creates in the body a situation analogous to fasting. In both, caloric demands are in excess of supply, and there is a requirement for glucose. It will become evident that, during fasting and prolonged exercise, some of the same mechanisms come into play to maintain substrate supply.

Amino acids that give rise to pyruvate, malate, and oxaloacetate can become precursors to phosphoenolpyruvate (PEP), which can be converted to glucose. As noted earlier, this is an expensive process for several reasons. Amino acids are frequently the most costly component in the diet. If the amino acids are derived from body proteins, then their cost is even more dear. Additionally, from the standpoint of energetics, gluconeogenesis from amino acids through PEP is expensive because it requires significant energy input. Under conditions when glucose supply is adequate, pyruvate, malate, and oxaloacetate can be converted to fatty acids.

Pathways of Phosphoenolpyruvate Formation

Formation of PEP can occur in several ways. The process usually begins in mitochondria, and mitochondrial processes are often complemented by cytoplasmic processes.

Pyruvate can be converted to PEP in a two-step process. Pyruvate is first converted to oxaloacetate by the enzyme pyruvate carboxylase, which is stimulated by the presence of acetyl-coenzyme A (acetyl-CoA).

$$
\begin{array}{l}
\text{CH}_3 \\
\mid \\
\text{C}{=}\text{O} + \text{CO}_2 + \text{ATP} \xrightarrow[\text{Carboxylase}]{\text{Pyruvate}} \quad \text{CH}_2 + \text{ADP} + \text{P}_i \\
\mid \qquad\qquad\qquad\qquad\qquad\qquad \mid \\
\text{COO}^- \qquad\qquad\qquad\qquad\qquad\quad \text{C}{=}\text{O} \\
\qquad\qquad\qquad\qquad\qquad\qquad\qquad\;\; \mid \\
\text{Pyruvate} \qquad\qquad\qquad\qquad\qquad\;\; \text{COO}^- \\
\end{array}
\tag{8-4}
$$

Oxaloacetate

Oxaloacetate is then converted to phosphoenolpyruvate.

$$
\begin{array}{l}
\text{COO}^- \\
\mid \\
\text{C}{=}\text{O} + \text{GTP} \xrightarrow[\text{Carboxykinase}]{\text{Phosphopyruvate}} \begin{array}{l}\text{COO}^- \\ \mid\; \text{C}{-}\text{OPO}_3^- + \text{CO}_2 + \text{GDP} \end{array} \\
\mid \qquad\qquad\qquad\qquad\qquad\qquad\quad \text{OPO}^{-3} \\
\text{CH}_2 \\
\mid \qquad\qquad\qquad\qquad\qquad\qquad\qquad\;\; \mid \\
\text{COO}^- \qquad\qquad\qquad\qquad\qquad\qquad \text{CH}_3 \\
\end{array}
\tag{8-5}
$$

Oxaloacetate   Phosphoenolpyruvate

Malate can probably be converted directly to PEP by malic enzyme.

(8-6)

$$\underset{\text{Malate}}{\begin{array}{c} COO^- \\ | \\ HO-C-H \\ | \\ CH_2 \\ | \\ COO^- \end{array}} + ATP \underset{\text{Enzyme}}{\overset{\text{Malic}}{\rightleftarrows}} \underset{\text{PEP}}{\begin{array}{c} COO^- \\ | \\ C-OPO_3 \\ | \\ CH_3 \end{array}} + ADP + CO_2$$

BRANCHED-CHAIN AMINO ACIDS

The branched-chain amino acids (leucine, isoleucine, and valine) are essential amino acids. They are unusual in that they are catabolized mainly in skeletal muscle—where carbon skeletons provide an oxidizable source of substrate and where their nitrogen residues participate in alanine formation. The metabolism of branched-chain amino acids (Figure 8-7) begins with a transamination and formation of glutamate and a keto acid. The glutamate so formed can then donate a nitrogen to pyruvate and form alanine (Equation 8-2). The second step in leucine catabolism is the dehydrogenase step (Figure 8-7), which is also a decarboxylase. The remaining carbon atoms of leucine are then converted either to acetyl-CoA or to acetoacetate. The fate of these products is oxidation. Leucine is therefore purely ketogenic. The catabolism of the other branched-chain amino acids proceeds somewhat differently. Isoleucine forms both acetyl-CoA and succinyl-CoA and is therefore both ketogenic and glucogenic. Valine produces succinyl-CoA and is glucogenic.

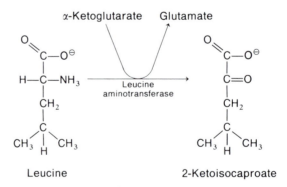

FIGURE 8-7
The catabolism of leucine begins with a transamination followed by a dehydrogenation that is also a decarboxylation. The transamination product glutamate can react to form alanine. The carbon skeleton of leucine is ketogenic.

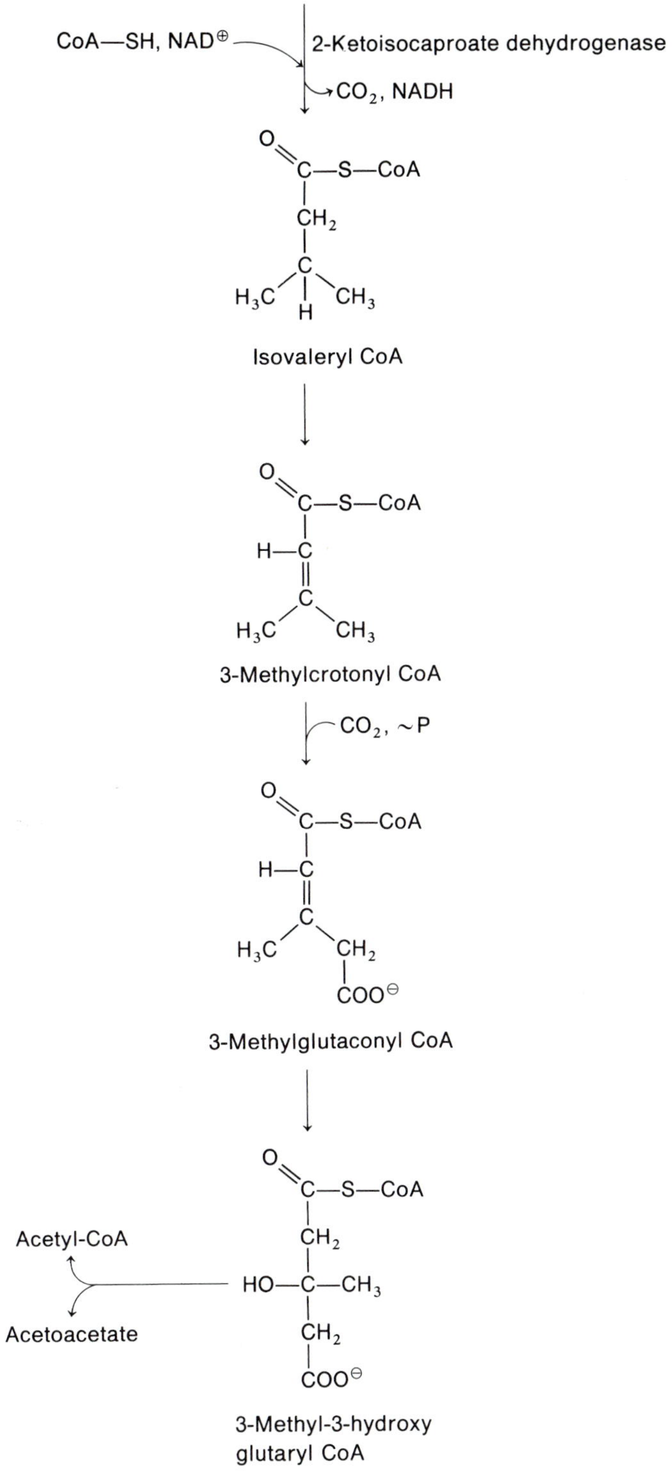

CoA—SH, NAD$^{\oplus}$ 2-Ketoisocaproate dehydrogenase

CO$_2$, NADH

Isovaleryl CoA

3-Methylcrotonyl CoA

CO$_2$, ~P

3-Methylglutaconyl CoA

Acetyl-CoA

Acetoacetate

3-Methyl-3-hydroxy glutaryl CoA

FIGURE 8-7 *(continued)*.

THE GLUCOSE—ALANINE CYCLE

For some time it has been known that the body's reserves of amino acids provide precursors for gluconeogenesis during fasting and starvation. The mechanism by which blood glucose homeostasis is maintained is the glucose–alanine cycle (Figure 8-8). During fasting, proteins are degraded rapidly to provide 100 g (400 kcal) of glucose per day, which is used almost exclusively by the brain, nerves, and kidney. Amino acids from degraded muscle proteins circulate to the liver, where deamination, transamination, and gluconeogenesis take place. Of the amino acids reaching the liver, alanine is by far the most important; half or more of the amino acids taken up by liver are in the form of alanine. During

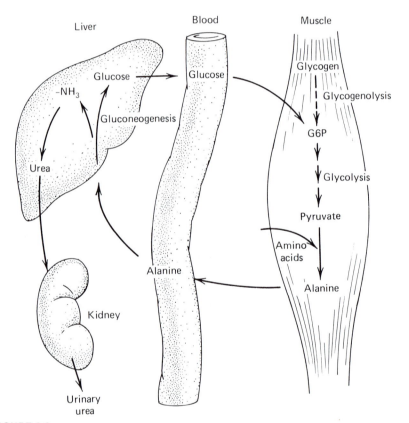

FIGURE 8-8

The glucose–alanine cycle as proposed by Cahill complements the Cori cycle and provides a means for the transportation of carbon atoms from skeletal muscle to the liver for gluconeogenesis. According to Odessey and Goldberg, the branched-chain amino acids (mainly leucine) provide the nitrogen for amino acid formation in muscle. According to them, the carbon source for alanine formation is glycolytic.

fasting and starvation, the arterial level of alanine largely determines the rate of gluconeogenesis.

The alanine formed in muscle and released into the circulation most likely does not represent the catabolism of a protein-rich in alanine content (i.e., a polyalanine). Rather, the alanine is newly (de novo) synthesized in muscle. At present considerable controversy exists over the sources of both the carbon and nitrogen in alanine synthesis. Goldberg and Odessey have provided results to indicate that branched-chain amino acids released from liver and muscle protein may provide the nitrogen precursor. The carbon skeletons provided in such a process would yield a source of oxidizable substrate in muscle. Other scientists have provided data to suggest that other amino acids also provide nitrogen for alanine synthesis.

The source of the carbon atoms for alanine synthesis has provided even greater scientific controversy. Goldberg and associates have obtained results to indicate that the carbon source is glycolytic. Other scientists, including groups

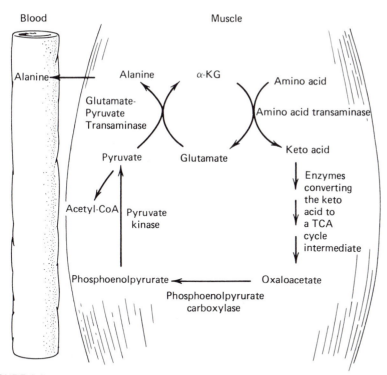

FIGURE 8-9
The scheme of Goldstein and Newsholme whereby muscle amino acids could provide *both* nitrogen and carbon precursors to alanine in situations such as starvation and prolonged exercise when blood glucose is low and glycogen is depleted.

(SOURCE: L. Goldstein, and E.A. Newsholme, 1976.)

lead by Newsholm and Kipnis, believe that the carbon skeletons for alanine synthesis are derived from other amino acids. The mechanism of Newsholm and Goldstein is displayed in Figure 8-9.

Which scientific group is correct with respect to the precursors of alanine synthesis remains to be determined. Perhaps both mechanisms operate with one or the other predominating at different times. Whereas the model displayed in Figure 8-8 may operate when fasting starts, the quantity of glucose and glycogen present in muscle and liver would be inadequate to sustain the cycle for very long unless other substances donated carbon atoms. Therefore, in prolonged starvation or exercise, the mechanism in Figure 8-9 may predominate.

Amino Acid Metabolism During Exercise

Because the glucose–alanine cycle contributes to glucose homeostasis during fasting and starvation, it was logical to hypothesize that the cycle also operates during prolonged exhaustive exercise. This hypothesis was put forth by Felig and associates, including Wahren and Ahlborg. Their results, based on quantitative measurements of muscle alanine release, liver uptake, and liver glucose release, indicate greatly elevated glucose–alanine cycle activity during exercise as compared to rest. According to their data, the glucose–alanine cycle may provide 5% of the total fuels used during exercise.

Additionally, White and Brooks, utilizing injection of [14C]leucine and [14C] alanine into animals during rest and exercise, demonstrated that the oxidations of the substances increases during exercise in proportion to the rate of oxygen consumption. Because several amino acids contribute to maintaining glucose homeostasis and substrate supply during exercise, perhaps the total contribution by all amino acids to fuel supply is closer to 10 than 5%. This percentage may not seem significant, but when it is recalled that record-breaking performances in athletics never supersede the existing standard by 2%, the 5 to 10% of the fuel supply contributed by amino acids may be critical. Additionally, amino acids, like glycolytic substances, may have anaplerotic and cataplerotic roles in sustaining fatty acid metabolism during prolonged work.

Muscle Proteolysis During Exercise

White and Brooks demonstrated that circulating levels of alanine and leucine increase in animals during prolonged hard exercise. Because alanine can be synthesized in muscle de novo, the significance of the result is difficult to interpret. However, leucine is an essential amino acid, and the results indicating an increase in its level must indicate elevated proteolysis in liver and muscle. During similar types of experiments, Dohm and associates have demonstrated that activities of particular enzymes that catabolize sarcoplasmic proteins are elevated after prolonged exhausting exercise. Further, Lemon and Mullin demonstrated that humans excrete increased amounts of urea in sweat during prolonged exercise. It therefore appears that proteins and amino acids are catabolized to support physical exercise.

GLUTAMATE AND GLUTAMINE AS AMMONIA SCAVENGERS

It has long been known that muscle liberates ammonia (NH_4^+) when it contracts. If creatine phosphate stores are maintained by intermediary metabolism in contracting muscle, then the ammonia must be provided by a source other than creatine phosphate. That source is thought to be the purine nucleotide cycle described by Lowenstein (Figure 8-10). The purine nucleotide cycle is very active in contracting muscle; the cycle produces IMP (inosine monophosphate) from AMP, a TCA cycle intermediate fumarate, and ammonia. The AMP is potentially useful in regulating metabolism, and fumarate provides material that will eventually form oxaloacetate and combine with acetyl-CoA in operation of the TCA cycle. Ammonia is one of the stimulators of a key glycolytic enzyme (phosphofructokinase), but accumulation of ammonia can be toxic to the tissue. The accumulation of ammonia is minimized by formation of glutamate and glutamine (Equation 8-7). Studies by Meyer and Terjung indicate that PNC activity is particularly high in fast skeletal muscle during exercise.

$$\text{(8-7)}$$

Rudderman has observed that under some conditions, contracting muscle releases glutamine in amounts comparable with those of alanine. Brooks and co-workers have produced radiochromatograms from blood and other tissues of exercised animals showing incorporation of isotopic label from glycolytic metabolites into glutamine and alanine (cf. Figure 10-6). Glutamate is apparently not quantitatively released from muscle. Glutamine is particularly well adapted for preventing ammonia toxicity during exercise, as it carries two amine groups to liver and kidney for disposal. During recovery, accumulated glutamine and glutamate function in the synthesis of amino acids and proteins.

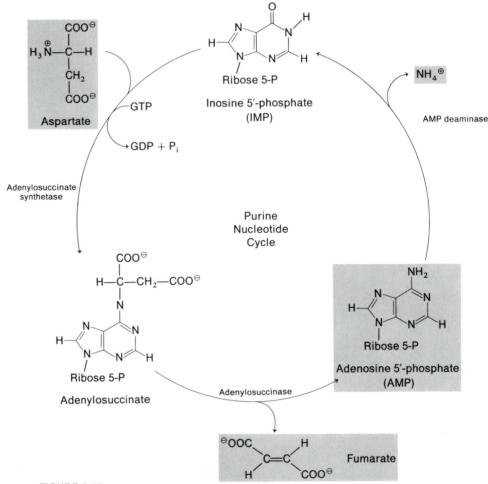

FIGURE 8-10
The purine nucleotide cycle (PNC) of Lowenstein, whereby the deamination of AMP (lower right) to IMP results in the deamination of aspartate to fumarate and the formation of $NH_4^{(+)}$. During muscular contraction, the PNC may result in purine nucleotide (AMP) consumption but the anaplerotic formation of fumarate. The starting point of the cycle is at AMP (lower right).

(SOURCE: J.M. Lowenstein, 1972.)

EFFECTS OF ENDURANCE TRAINING ON AMINO ACID METABOLISM

It has long been known that muscle contains high levels of transaminase enzymes. Additionally, Molé and associates have demonstrated that endurance training can double the levels of important transaminases such as GPT. Utilizing radioactive tracers, Henderson, Black, and Brooks demonstrated that endurance-trained animals have an increased ability to utilize the branched-chain amino acid leucine as a fuel during exercise. They also demonstrated that while not exercising, trained animals utilize amino acids more efficiently than do untrained ones. Therefore, although trained animals utilize relatively greater amounts of amino acids during exercise, their nutritional protein needs are similar to those of sedentary animals.

Summary

Although it has long been believed that amino acids and proteins play no significant role in supplying fuels for muscular exercise, results of recent studies reveal that amino acids and proteins are integrally involved in the metabolic adjustments to exercise. Amino acids and proteins appear to be involved in at least three important ways. During prolonged exercise, the exchanges of energy and matter are quite complex. In particular, the Krebs (TCA) cycle supports a number of functions, some of which deplete metabolic intermediates from the cycle. For TCA cycle activity to continue, it must be filled back up. Amino acids can contribute to this important (anaplerotic) function.

During prolonged exercise, the amino acid alanine represents an important gluconeogenic precursor. The glucose–alanine cycle, whereby alanine formed in skeletal muscle is released into the circulation and reaches the liver where conversion to glucose takes place, represents an important adjunct to the Cori cycle for maintaining blood glucose homeostasis during exercise.

Recently it has been discovered that amino acids, including even the essential amino acid leucine, are oxidized as fuels to support muscular exercise. The contribution of amino acids to the substrate supply is relatively small (5–10%), but because of their large tissue masses, amino acids and proteins represent a potentially significant fuel supply to support prolonged exercise bouts.

When amino acid- and protein-containing foods are consumed, nitrogenous products are excreted. Similarly, when body proteins are catabolized, nitrogenous wastes are excreted. Between the consumption of amino acids and proteins and the excretion of urea, important metabolic processes occur. These processes occur during exercise as well as during rest.

Selected Readings

Adibi, S.A., E.L. Morse, and P.M. Amin. Amino acid levels in plasma, liver, and skeletal muscle during protein deprivation. Am. J. Physiol. 225:408–414, 1973.

Ahlborg, G., P. Felig, L. Hagenfeldt, R. Hendler, and J. Wahren. Substrate turnover during prolonged exercise in man: Splanchnic and leg metabolism of glucose free fatty acids, and amino acids. J. Clin. Invest. 53:1080–1974.

Aragon, J.J. and J.M. Lowenstein. Purine-nucleotide cycle: Comparison of the levels of citric acid cycle intermediates with operation of the purine nucleotide cycle in rat skeletal muscle during exercise and recovery from exercise. Eur. J. Biochem. 110:371–377, 1980.

Buse, M.G., J. Biggers, C. Drier, and J. Buse. The effect of epinepherine, glucagon, and the nutritional state on the oxidation of branched chain aminol acids and pyruvate by isolated hearts and diphragms of the rat. J. Biol. Chem. 248:697–706, 1973.

Calloway, D.H., A.C.F. Odell, and S. Margen. Sweat and miscellaneous nitrogen losses in human balance studies. J. Nutrition. 101:775–786, 1971.

Calloway, D.H. and H. Spector. Nitrogen balance as related to caloric and protein intake in active young men. Am. J. Clin. Nutr. 2:405–411, 1954.

Cathcart, E.P. The Physiology of Protein Metabolism. London: Longmans, Green and Co., 1921. p. 129–132.

Cathcart, E.P. The influence of muscle work on protein metabolism. Physiol. Rev. 5:225–243, 1925.

Celejowa, I. and M. Homa. Food intake, nitrogen and energy balance in Polish weight lifters during training camp. Nutr. and Metab. 12:259–274, 1970.

Consolazio, C.F., H.L. Johnson, R.A. Nelson, J.G. Dramize, and J.H. Skala. Nitrogen metabolism during intensive physical training in the young adult. Am. J. Clin. Nutr. 28:29–35, 1975.

Cuthbertson, D.P., J.L. McGirr, and H.N. Munro. A study of the effect of overfeeding on protein metabolism in man. IV. The effect of muscular work at different levels of energy intake with particular reference to the timing of the work in relation to the taking of food. Biochem. J. 31:2293–2305, 1937.

Dohm, G.L., A.L. Hecker, W.E. Brown, G.J. Klain, F.R. Puente, E.W. Askew, and G.R. Beecher. Adaptation of protein metabolism to endurance training: Increased amino acid oxidation in response to training. Biochem. J. 164:705–708, 1977.

Dohm, G.L., G.J. Kasperek, E.B. Tapscott, and G.R. Beecher. Effect of exercise on synthesis and degradation of muscle protein. Biochem. J. 188:255–262, 1980.

Dohm, G.L., F.R. Puente, C.P. Smith, and A. Edge. Changes in tissue protein levels as a result of endurance exercise. Life Sci. 23:845–850, 1978.

Dohm, G.L., R.T. Williams, G.J. Kasperek, and A.M. van RiJ. Increased excretion of area and N^+-methylhistidine in rats and humans after a bout of exercise. J. Appl. Physiol. 52:458–466, 1982.

Felig, P. and J. Wahren. Amino acid metabolism in exercising man. J. Clin. Invest. 50:2703–2714, 1971.

Goldberg, A.L. and T.W. Chang. Regulation and significance of amino acid metabolism in skeletal muscle. Fed. Proc. 37:2301–2307, 1978.

Goldberg, A.L. and R. Odessey. Oxidation of amino acids by diaphragms. Am. J. Physiol. 223:1384–1391, 1972.

Goldstein, L. and E.A. Newsholme. The formation of alanine from amino acids in diaphragm muscle of the rat. Biochem. J. 154:555–558, 1976.

Gontzea, I., P. Sutzescu, and S. Dumitriache. The influence of muscular activity on nitrogen balance and on the need of man for proteins. Nutr. Reports Inter. 10:35–43, 1974.

Gontzea, I., P. Sutzescu, and S. Dumitriache. The influence of adaptation to physical effort on nitrogen balance in man. Nutr. Reports Inter. 11:231–233, 1975.

Guyton, A.C. Textbook of Medical Physiology. Philadelphia: W.B. Saunders Co., 1981. p. 816–825, 861–867.

Hagg, S.A., E.L. Morse, and S.A. Adibi. Effect of exercise on rates of oxidation, turnover, and plasma clearance of leucine in human subjects. Am. J. Physiol. 242:E407–E410, 1982.

Henderson, S.A., A.L. Black, and G.A. Brooks. Leucine turnover and oxidation on trained and untrained rats during rest and exercise. Med. Sci. Sports Exercise 15:98, 1983.

Iyengar, A. and B.S. Narasinga Roa. Effect of varying energy and protein intake on nitrogen balance in adults engaged in heavy manual labor. Br. J. Nutr. 41:19–25, 1979.

Lehninger, A.L. Biochemistry. New York: Worth Publishing, Inc., 1970. p. 433–454.

Lemon, P.W. and J.P. Mullin. Effect of initial muscle glycogen levels on protein catabolism during exercise. J. Appl. Physiol: Respirat. Environ. Exercise Physiol. 48:624–629, 1980.

Lowenstein, J.M. Ammonia production in muscle and other tissues. The purine nucleotide cycle. Physiol. Rev. 52:382–414, 1972.

Manchester, K.L. Oxidation of amino acids by isolated rat diaphragm and the influence of insulin. Biochim. Biophys. Acta. 100:295–298, 1965.

Manchester, K.L. Control by insulin of amino acid accumulation in muscle. Biochem. J. 117:457–466, 1970.

McGilvery, R.W. Biochemical Concepts. Philadelphia: W.B. Saunders, Co., 1975. p. 359–384.

Molé, R.A. and R.E. Johnson. Disclosure by dietary modification of an exercise induced protein catabolism in man. J. Appl. Physiol. 31:185–190, 1971.

Odessey, R. and A.L. Goldberg. Oxidation of leucine by rat skeletal muscle. Am. J. Physiol. 223:1376–1383, 1972.

Rennie, M.J., R.H.T. Edwards, D. Halliday, C.T.M. Davies, D.E. Mathews, and D.J. Millward. Protein metabolism during exercise. In: Warterlow, J.C. and J.M.L. Stephensen (eds.), Nitrogen Metabolism in Man. London: Applied Science Publishers, 1981. p. 509–523.

Ruderman, N.B. Amino acid metabolism and gluconeogenesis. Ann. Rev. of Med. 26:245–258, 1975.

Ruderman, N.B. and M. Berger. The formation of glutamine and alanine in skeletal muscle. J. Biol. Chem. 249:5500–5506, 1974.

Tischler, M.E. and A.F. Goldberg. Amino acid degradation and effect of leucine on pyruvate oxidation in rat atrial muscle. Am. J. Physiol. (Endocrinol. Metab). 238:E480–E486, 1980.

Vander, A.J., J. H. Sherman, and D.S. Luciano. Human Physiology, New York: McGraw-Hill, 1980. p. 402–478.

Wahren, J., P. Felig, R. Hendler, and G. Ahlborg. Glucose and alanine metabolism during recovery from exercise. J. Appl. Physiol. 34:838–845, 1973.

White, T.P. and G.A. Brooks. [U-^{14}C] glucose, -alanine, and -leucine oxidation in rats at rest and two intensities of running. Am. J. Physiol. 240 (Endocrinol. Metab. 3): E155–165, 1981.

Wolfe, R.R., R.D. Goodenough, M.H. Wolfe, G.T. Royle, and E.R. Nadel. Isotopic analysis of leucine and area metabolism in exercising humans. J. Appl. Physiol. 52:458–466, 1982.

NEURAL—ENDOCRINE CONTROL OF METABOLISM

A primary consideration during the stress of exercise is the maintenance of nearly "normal resting" levels of blood and cellular metabolites. In particular, the maintenance of blood glucose levels at about 4 mM (90 mg · dl^{-1}) is critical. Exercise causes increased glucose uptake from the blood. During exercise, blood glucose level can be maintained by increased release into blood of glucose as well as fuels that may serve as alternatives. The coordinated physiological response to maintain blood glucose homeostasis during exercise is governed by two related body systems: the autonomic nervous system (ANS) and the endocrine (hormonal) system. Chemical mediators are released by both the ANS and hormonal system to have the desired effects of maintaining an adequate blood glucose level. When blood glucose falls during prolonged hard exercise, fatigue is imminent. *Hormones* are chemical substances that are secreted into body fluids, usually by endocrine glands. The target tissues of hormones can be anatomically close to or quite far removed from the glands of secretion, and the targets can be one or several tissues. At the targets of their action, hormones have powerful effects on metabolism. In the resting person, metabolism is largely controlled by hormones. In general, hormonal secretion is controlled by a negative-feedback system in which rising levels of the substance released as a result of hormone action inhibit secretion of the hormone. For example, following a substantial carbohydrate meal, the blood glucose level rises. This increase in blood glucose stimulates the secretion of insulin, which

Competitors in the Tour de France climb through the Pyrénées mountains. In this event all the body's energy reserves are called upon. The effects of hormones in regulating metabolism are always important, especially in human endeavors such as the Tour de France. (Wide World.)

increases glucose uptake from blood. When blood glucose level falls during exercise, so does insulin secretion. The falling insulin level slows the rate of glucose removal from the blood and helps to maintain a "normal" blood glucose level.

The autonomic (automatic, unconscious) nervous system is composed of two subcomponents: the sympathetic and parasympathetic nervous systems. The *sympathetic* nervous system governs the "fight-or-flight reflex" mechanism; sympathetic activity increases contractility and frequency of heart contraction, and mobilizes fatty acids, glycogen, and glucose for fuels. The *parasympathetic* nervous system governs resting functions. Parasympathetic activity slows heart rate and stimulates fuel storage. A wide range of tissues are innervated by the ANS. Through increased or decreased sympathetic activity, and decreased or increased parasympathetic activity (they work in opposite directions, but in coordination), a broad range of tissues are either geared up for exercise or geared down for resting. The ANS has local effects by releasing hormones from its nerve endings at specific target tissues. However, like the effects of circulating hormones, the ANS has diffuse effects as well because its nerve endings reach most tissues and because some of the chemical messenger secreted (e.g., norepinephrine) reaches the circulation. The ANS and the endocrine system allow coordinated responses of various tissues. For instance, muscle work requires glucose as a fuel; the glucose is supplied by gut, liver, and kidney. The release of hormones and ANS activity represent a strain response to the stress of exercise. Exercise training tends to diminish the release of hormones and ANS activity in response to given exercise tasks.

GLUCOSE HOMEOSTASIS

The maintenance of nearly "normal resting" blood glucose levels is always of primary importance, including during prolonged hard exercise. Glucose from the blood and its storage form (glycogen) in muscle are necessary for continued muscular activity. Glucose and glycogen are important fuel sources, and they have anaplerotic effects in allowing fat utilization during activity. Because glucose is usually the only fuel acceptable to the brain and other central nervous system (CNS) tissues, the maintenance of a reasonable blood glucose level during prolonged work to supply fuel for the brain may be even more important than supplying substrate to muscle, which has recourse to alternative fuels. The actions of several hormones are to maintain blood glucose levels; these hormones are said to be *glucoregulatory*.

The period of exercise represents a time when the draw on blood glucose reserves is accelerated. This draw (or uptake) can be compensated for in two general ways. One way is via the increased release into the blood of glucose from the gut, liver, and kidney. In addition to release from the gut of digested contents from a previous meal, the liver can release glucosyl units previously stored as glycogen, and the liver and kidney can attempt to make new glucose from precursor molecules (i.e., gluconeogenesis). Those precursors to glucose (a six-carbon molecule) are mainly three-carbon molecules (lactate, pyruvate, glycerol, alanine). In a recently fed individual, digestion products of the meal raise the blood glucose level (Figure 5-2a). The elevated glucose concentration causes insulin to be secreted. Insulin promotes uptake and utilization of glucose by most tissues, glycogen synthesis in muscle and liver, and triglyceride synthesis in adipose tissue. When a meal has not been eaten for several hours or exercise has intervened, glucose is taken up from the blood, lowering its level there. A lowered blood glucose level inhibits insulin secretion, which in turn decreases glucose uptake by nonactive tissues, leaving or sparing the existing supply for muscle, brain, and nerve.

Several hormones are said to be insulin *antagonists;* that is, their actions oppose those of insulin. Whereas insulin stimulates glycogen storage, epinephrine and glucagon stimulate glycogen breakdown (glycogenolysis in muscle and liver, respectively). Similarly, whereas insulin stimulates protein synthesis, cortisol promotes protein catabolism and release of amino acids from muscle into venous blood. Several of the amino acids in muscle are glycogenic when metabolized in liver and kidney.

By raising the circulating levels of alternative substances to glucose, and by delivering these to active tissues, the body has a second approach to the problem of maintaining blood glucose during exercise. When those substances (fatty acids, triglycerides, and the amino acid leucine) are utilized by contracting muscle, the utilization of glucose is spared. Insulin antagonists such as epinephrine and growth hormone mobilize fatty acids and triglycerides during exercise. Use of these fuels slows glucose and glycogen utilization, thereby postponing the time when the fall in blood glucose level becomes critical.

CHARACTERISTICS OF HORMONES

A hormone is a chemical messenger that is produced and stored in glandular tissues. Because endocrine glands are ductless, hormones are released (secreted) into body fluids such as blood or lymph. In this way hormones circulate throughout the body and affect a variety of particular (target) tissues. Most hormones are generalized in their action and circulate widely. Other hormones, such as acetylcholine, are local hormones and are released by parasympathetic and skeletal muscle nerve endings. Local hormones have specific local effects, and they are usually metabolized within a limited area. At least one hormone, norepinephrine, is both local and general. Norepinephrine released by sympathetic nerve endings has local effects; a significant amount of the sympathetic release of norepinephrine also reaches the circulation. This circulating norepinephrine, along with norepinephrine released into the circulation from the adrenal medulla, has general effects as well.

Chemically there are two basic types of hormones: (1) steroids and (2) large polypeptides or small proteins. *Steroids* are produced from cholesterol by the adrenal cortex and gonads. *Polypeptide hormones* are derived from amino acids in the other endocrine glands.

Negative Feedback and the Control of Hormonal Secretion

Hormones have powerful effects on metabolism. They exert their influence in concentrations of only nanograms per milliliter ($ng \cdot ml^{-1}$), where "nano" means 10^{-9}. Therefore, the precise regulation of hormonal secretion is essential for normal functioning as well as for adjusting to a variety of stressful situations such as exercise or high altitude (or even exercise at high altitude). Hormonal secretion (production in and release from endocrine glands) is regulated by the negative-feedback mechanism, by which the secretion of a hormone is inhibited (turned off) if a particular end result of the hormonal action is achieved. In other words, a positive result has a negative effect on the hormonal secretion. Should hormonal secretion not have the desired effect, then by negative feedback the hormonal secretion will be stimulated.

MECHANISMS OF HORMONAL ACTION

The wide variety of hormonal actions appears to be accomplished by only a relatively few basic mechanisms. The specific effects of hormones on target tissues are accomplished by the binding of the hormones to stereospecific binding sites on membranes of target tissues. Polypeptide hormones interact with receptors on the cell's surface, whereas steroid hormones have mobility through the cell membrane and interact with the nucleus. Binding of the hormone may then (1) affect the permeability of the target cell membrane to a metabolite or ion, (2) activate an enzyme or enzyme system, or (3) activate the genetic apparatus to manufacture intracellular proteins or other substances.

For instance, the binding of insulin to most types of cells increases permeability of those cells to glucose; the binding of epinephrine to sarcolemma causes glycogenolysis in muscle by activation of the cAMP cascade (cf. Figure 5-13), and growth hormone stimulates protein synthesis in most cells.

CYCLIC AMP—THE "INTRACELLULAR HORMONE"

The binding of many hormones to cell membranes of target tissues causes the formation of cyclic 3', 5'-adenosine monophosphate (cyclic AMP, or cAMP) in the cell (Figure 9-1). Depending on the hormone and target cell, the cAMP is formed can then activate a variety of enzyme systems. Previously (Chapter 5) it was seen that epinephrine causes glycogenolysis in muscle, and glucagon causes glycogenolysis in liver. Both of these hormones act through cAMP. Additionally, epinephrine activates lipolysis in adipose tissue through a cAMP mechanism.

INSULIN AND GLUCAGON—THE IMMEDIATE CONTROL OF BLOOD GLUCOSE LEVEL

Insulin is secreted in the β cells of the pancreatic islets of Langerhans (Figure 9-2). Insulin is a small protein hormone that stimulates glucose uptake by many cells, of which muscle and adipose tissue are quantitatively most important. Brain cells and erythrocytes depend on glucose for fuel but not on insulin for glucose uptake. Increased glucose uptake usually stimulates glycogen synthesis

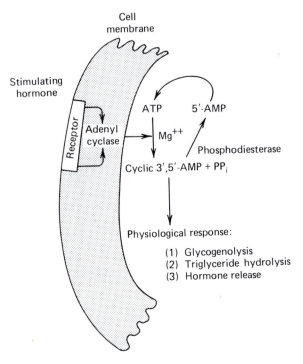

FIGURE 9-1

Many hormones exert their cellular action through a cyclic AMP (cAMP) mechanism. Through this mechanism, the circulating hormone binds to a specific receptor on a target cell's surface. As a result, the cell's level of cAMP, the intracellular hormone, is elevated, depending on the target cell, specific physiological responses occur.

(SOURCE: Modified from A.C. Guyton, Textbook of Medical Physiology. Copyright W.B. Saunders, Co., Philadelphia, 1976, p. 918. With permission.)

in muscle and fat synthesis in adipose tissue. The effect of this glucose uptake from blood is lowered blood glucose levels.

Facilitated Transport—The Mechanism of Insulin Action

The mechanism of insulin action is mainly through facilitating the transport of glucose through the cell membrane (Figure 9-3). Insulin promotes the combination of extracellular glucose with a glucose carrier within the membrane. The glucose carrier then migrates across the membrane toward the inside, where the rapid phosphorylation of glucose to glucose 6-phosphate (G6P) ensures a low intracellular concentration of free glucose. By promoting cellular G6P formation (which in turn is metabolized and then affects the concentration of other metabolites), insulin then stimulates glycogen synthesis and the preferential cellular metabolism of glucose. In adipose tissue, insulin stimulates lipoprotein

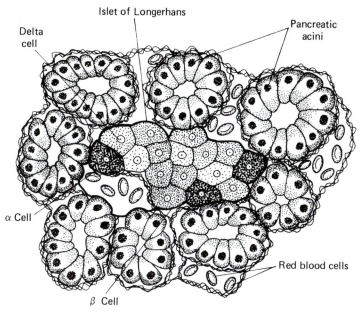

FIGURE 9-2
The pancreatic acini secrete digestive juices into the small intestine, whereas the islets of Langerhans secrete insulin (β cells), glucagon (α cells), and somatostatin (δ cells).

(**SOURCE**: Modified from A.C. Guyton, Textbook of Medical Physiology. Copyright W.B. Saunders Co., Philadelphia, 1976, p. 959. With permission.)

lipase and fatty acid uptake (Figure 7-2). In fat cells, insulin-facilitated glucose uptake stimulates triglyceride synthesis by providing α-glycerol phosphate (cf. Figure 7-1g).

Role of the Liver in Stabilizing Blood Glucose Level

The actions of the liver are crucial to the overall process of glucose homeostasis. When blood glucose levels are high, the liver stores glucose as glycogen for later periods when blood glucose is low. When blood glucose levels fall, the liver releases glucose into the blood. Both glycogenolysis and gluconeogenesis assist in this process.

Although the liver cell membrane is highly permeable in either direction to glucose, this permeability is not thought to be regulated by insulin. Instead, insulin causes liver to control blood glucose in another way. When the blood glucose level is high, insulin is secreted. The elevation in insulin levels stimulates the formation of glucokinase. This enzyme (Chapter 5) phosphorylates glucose to G6P and causes the uptake of glucose by liver. The phosphate group of G6P prevents efflux from the liver while G6P serves as a substrate for glycogen synthesis.

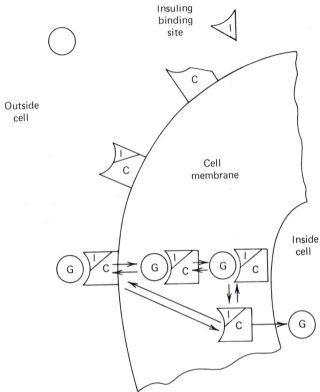

FIGURE 9-3
Insulin (I) binds to a cell surface receptor and has the effect of increasing the movement of the glucose–carrier (G–C) complex through the cell membrane.

The activity of glucokinase (sometimes called high-K_M hexokinase; Chapter 5) is very sensitive to the circulating level of glucose. When the blood glucose level falls, the activity of glucokinase is decreased, whereas the activity of the bypass enzyme for glucokinase (glucose-6-phosphatase) is activated. Glucose-6-phosphatase forms glucose from G6P; the hepatic glucose so formed will then follow the concentration gradient and move into the circulation, thereby replenishing depleted blood glucose (Figure 5-12).

Insulin and Hepatic Fat Metabolism

High levels of circulating glucose and insulin promote glycogen synthesis in liver. However, the liver can store at most 5 to 6% of its wet weight (5 g · 100 g^{-1}) as glycogen. Thereafter the excess G6P would stimulate glycolysis and lead to acetyl-CoA and fatty acid formation. These, together with α-glycerol phosphate, would promote triglyceride synthesis in liver. Triglycerides produced in liver are combined into very-low-density lipoproteins (VLDL) and circulate with chylomicrons from the gastrointestinal (GI) tract to adipose tissue, where most fat is stored. Conversely, when insulin levels are low, triglycerides are hydrolyzed in adipose tissue (Chapter 7), and free fatty acids and glycerol are released into the blood. These then circulate to supply fuels alternative to glucose (Figure 7-2).

Low levels of glucose and insulin in the blood cause fatty acids to become greatly elevated. Through the process called the *glucose–fatty acid cycle,* very low insulin and high fatty acid levels greatly reduce sugar and carbohydrate catabolism. The K_M of fatty acids is higher than the normal physiological levels. Consequently an elevation in circulating level of free fatty acids (FFA) greatly promotes the catabolism of FFA to acetyl-CoA. Acetyl-CoA from FFA inhibits pyruvate dehydrogenase, slowing entry of pyruvate into the TCA cycle. High acetyl-CoA levels also elevate mitochrondrial and cytoplasmic citrate and ATP levels. These inhibit phosphofructokinase (the rate-limiting reaction in glycolysis). This limitation in turn causes G6P to accumulate, which further slows cellular glucose uptake and utilization.

Excess acetyl-CoA formation from FFA can be condensed into acetoacetic acid (Figure 7-7). This ketone can be converted into β-hydroxybutyrate and acetone, two other ketones. Therefore, insulin lack can lead to ketone formation. During prolonged exercise, blood insulin levels fall and blood ketones inevitably rise. Fortunately, the rise is counterbalanced by the fact that muscle and brain can utilize the ketones as fuels. Endurance training promotes ketone oxidation in muscle. However, in starving or diabetic individuals, elevated keto acids can result in an acidotic state that can cause severe discomfort, coma, and death.

The Insulin Response to Exercise

The requirements for glucose in muscle during even moderate-intensity exercise tend to cause a decline in blood glucose. This decline can for a time be compensated for by the release of glucose mainly from liver, but also from kidney to some extent. For a time during exercise, blood glucose level may actually rise as a result of this accelerated release. Eventually, however, glucose uptake exceeds release and blood glucose level falls, with concomitant decline in insulin levels (Figure 9-4). This decline in blood glucose and insulin levels helps to minimize glucose uptake by nonactive tissues, therefore sparing blood glucose for active muscle and brain. Falling glucose and insulin levels help to spare blood glucose and muscle glycogen by enhancing lipolysis and making FFA available in the circulation for active and nonactive tissues alike.

Training and Insulin Release in Exercise

The general effect that training has on hormonal secretion is to reduce the hormonal response during exercise. Glucoregulatory hormones that are released

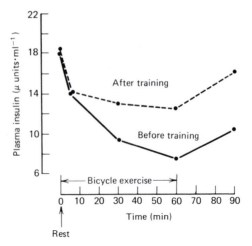

FIGURE 9-4
During prolonged exercise, as blood glucose level falls, so does the level of insulin. After training, the decrease in insulin is not as pronounced during exercise.

(SOURCE: Modified from F. Gyntelberg et al., 1977.)

during exercise (e.g., glucagon and catecholamines) are released to a lesser extent in trained individuals. In trained individuals during exercise, insulin does not fall as far as in the untrained (Figure 9-4). This lesser decrement in circulating insulin may be associated with more normal (higher) blood glucose levels in the trained during exercise. In trained individuals, increased FFA utilization and gluconeogenesis result in better control of blood glucose levels.

Glucagon—The Insulin Antagonist

The α cells of the pancreas (Figure 9-2) secrete the protein structure hormone glucagon. Whereas insulin is secreted when blood glucose levels are high, promoting removal of glucose from the blood, glucagon is secreted when blood glucose levels are low, acting to raise blood glucose levels. Glucagon has two effects on hepatic metabolism: (1) enhanced glycogenolysis and (2) increased gluconeogenesis.

Glucagon activates the adenylate cyclase cascade mechanism (Figure 5-13) in liver. Glucagon has a much smaller role in muscle glycogenolysis, which is activated by epinephrine.

Glucagon level in blood not only responds to glucose but also follows the blood alanine level. When alanine and other amino acids are released from muscle as a result of the actions of the catabolic steroid hormone cortisol and proteolytic enzymes, glucagon promotes hepatic amino acid uptake and gluconeogenesis from amino acids (Chapter 8; Figure 5-16). During prolonged exercise (Figure 9-5) the blood glucagon level rises as glucose and insulin levels fall. In this way both insulin and glucagon responses help maintain blood glu-

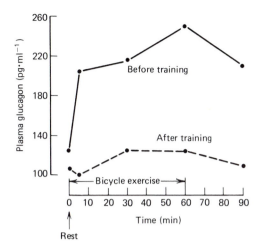

FIGURE 9-5

As part of the neuroendocrine response to exercise, blood glucagon rises to help maintain blood glucose levels. In trained individuals, the rise in glucagon is much less pronounced.

(SOURCE: Modified from F. Gyntelberg et al., 1977.)

cose homeostasis. Gluconeogenesis is accelerated not only during exercise but also in fasting. In the average fasting adult, about 100 g (400 kcal) of glucose can be produced each day. This production approximates the obligatory needs of the CNS for glucose.

As with blood insulin level (Figure 9-4), after training the glucagon response to exercise is dampened (Figure 9-5).

THE AUTONOMIC NERVOUS SYSTEM AND THE ADRENAL MEDULLA

As mentioned earlier, the autonomic nervous system (ANS) is composed of the sympathetic and parasympathetic nervous systems. The parasympathetic nervous system (Figure 9-6) has the effect of controlling resting functions. Parasympathetic function is dominated by the vagus or tenth cranial nerve (X in the figure). The nerve processes of the parasympathetic nervous system are composed of two neurons, each of which releases acetylcholine (ACH). Therefore, it is the ACH released from the ending of the second parasympathetic neuron that affects the target tissue. In keeping with its general function of controlling resting metabolism, parasympathetic activity has effects such as slowing the heart rate and stimulating digestion.

The sympathetic nervous system (Figure 9-7) controls fight-or-flight responses. Like the parasympathetic system, the sympathetic nerve processes are composed of two neurons. The first releases ACH (as in the parasympathetics), but the second usually releases norepinephrine (also known as noradrenaline).

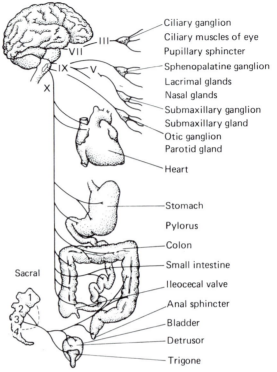

Ciliary ganglion
Ciliary muscles of eye
Pupillary sphincter
Sphenopalatine ganglion
Lacrimal glands
Nasal glands
Submaxillary ganglion
Submaxillary gland
Otic ganglion
Parotid gland
Heart
Stomach
Pylorus
Colon
Small intestine
Ileocecal valve
Anal sphincter
Bladder
Detrusor
Trigone

Sacral

FIGURE 9-6
The parasympathetic nervous system.

(SOURCE: Modified from A.C. Guyton, Textbook of Medical Physiology. Copyright W.B. Saunders, Co., Philadelphia, 1976, p. 711. With permission.)

Occasionally, sympathetic terminal nerve endings release ACH; thus they are called *cholinergic* sympathetics.

As with the endocrine system, chemical mediators released from the autonomic nerve endings bind to receptors in the cell membranes of target tissues. The chemical mediators frequently have effects by changing postsynaptic target tissue membrane permeability to ions. For instance, in the heart ACH tends to promote entry of Cl^-, whereas norepinephrine tends to increase entries of Na^+ and Ca^{2+}; ACH slows the heart by lowering the resting membrane potential, whereas norepinephrine speeds the heart by stimulating cation influx.

As indicated in Figure 9-7, sympathetic activity stimulates secretion from the adrenal medulla of norepinephrine and also epinephrine. Together, epinephrine and norepinephrine are called the catecholamines; they function together to effect powerful physiological responses. These responses are brought about not only by changes in ion permeability (norepinephrine and epinephrine), but

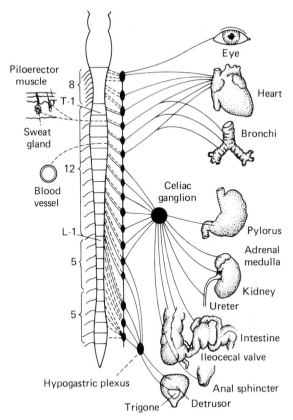

FIGURE 9-7
The adrenal medulla is innervated by the sympathetic nervous system.

(SOURCE: A.C. Guyton, Texbook of Medical Physiology. Copyright W.B. Saunders, Co., Philadelphia, 1976, p. 710. With permission.)

also by activating enzymes such as adenylate cyclase (epinephrine). The ratio of epinephrine to norepinephrine in adrenal secretions is 4:1. However, the circulating level of norepinephrine exceeds that of epinephrine fivefold. Therefore, much of the circulating norepinephrine originates from sympathetic release.

The catecholamines interact with two receptors, referred to as α and β receptors. Norepinephrine affects mainly the α receptors, whereas epinephrine affects both α and β receptors. The β receptors can be further subdivided into two groups (β_1 and β_2), depending on the actions of sympathomimetic drugs (Appendix III). Actions of these receptors are outlined in Table 9-1. It may be noted here that the term *adrenergic* means "activated or transmitted by epinephrine (adrenaline.)"

TABLE 9-1
Adrenergic Receptors and Their Functions

Effect	Receptor
Vasoconstriction	α
Vasodilatation	β_2
Cardiac acceleration	β_1
Increased myocardial contractility	β_1
Bronchodilatation	β_2
Calorigenesis	β_2
Glycogenolysis	β_2
Lipolysis	β_1
Intestinal relaxation	α
Pilomotor erection	α
Bladder sphincter contraction	α

Effects of Exercise Intensity and Training on Catecholamine Responses

Unless it is very prolonged and results in a fall in blood glucose levels, moderate exercise has no effect or a minimal effect on circulating catecholamine levels. However, as the level of exercise increases to an intensity of 50 to 70% of $\dot{V}_{O_2max}$, the blood catecholamine levels increase dramatically (Figure 9-8). Because during hard exercise blood levels of catecholamines rise before blood levels of glucose fall, and because norepinephrine levels continue to exceed epinephrine levels by approximately four- to fivefold, this catecholamine release is mediated by the sympathetic nervous system. Where a person is to perform in a major athletic competition, sympathetic activity may cause a slight rise in catecholamine levels prior to exercise.

As in the case with insulin (Figure 9-4) and glucagon (Figure 9-5), the release of catecholamines is minimized by endurance training. After training, exercise bouts of given absolute or relative intensities present less of a stress, and this is reflected in the lesser catecholamine response to exercise (Figure 9-9).

Epinephrine and Blood Glucose Homeostasis

Epinephrine secretion has powerful effects on blood glucose and carbohydrate metabolism during exercise. In muscle, through β-receptor action, epinephrine activates the adenylate cyclase mechanism (Figure 5-13). That mechanism, along with changes in intramuscular free Ca^{2+}, serves to stimulate glycogenolysis. Whereas a fall in blood glucose level can stimulate catecholamine release in the resting individual, immediately prior to exercise or at the beginning of

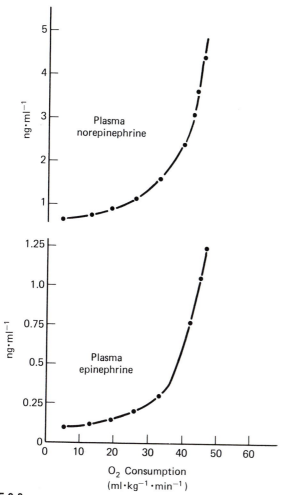

FIGURE 9-8

Circulating levels of catecholamines (epinephrine and norepinephrine) depend on the relative intensity of exercise. Moderate exercise intensities result in almost no increase in catecholamine levels. Beyond 50 to 70% of $\dot{V}_{O_2max}$, catecholamine levels rise disproportionately.

(SOURCE: Modified from V.J. Keul, et al., 1981.)

exercise, the sympathetic fight-or-flight mechanism results in catecholamine release.

Catecholamine secretion also has the effect of stimulating glycogenolysis in liver. Compared to the sensitivity of the mechanisms of liver glycogenolysis to glucagon, the sensitivity to epinephrine is much less. However, the rapid release of large amounts of catecholamines during exercise of high intensity (Figure 9-8) is sufficient to elevate blood glucose significantly by stimulating hepatic glycogenolysis.

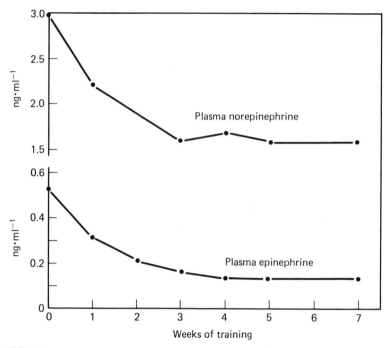

FIGURE 9-9

Blood catecholamine levels decrease during exercise bouts of given intensity as the result of endurance training.

(SOURCE: Modified from W.W. Winder, et al., 1978.)

In addition to affecting glucose and glycogen metabolism directly, epinephrine has an equally important indirect effect on blood glucose homeostasis during exercise. This is through the effect of a fatty acid mobilization from adipose tissue (Figure 7-2). By stimulating the hormone-sensitive lipase, epinephrine acts at exercise onset to raise arterial fatty acid levels. This rapid effect of epinephrine in stimulating lipolysis is followed by a slower but more prolonged effect of growth hormone in maintaining lipolysis.

GROWTH HORMONE RESPONSE TO CONTINUOUS AND INTERMITTENT EXERCISE

The integration of the neural and endocrine systems is no better illustrated than in describing the release of growth hormone from the anterior pituitary gland. The neural–endocrine integration is illustrated in Figure 9-10. The hypothalamus receives neural inputs and is sensitive to blood metabolite levels (e.g., glucose). In response to neural and blood-borne input, the hypothalamus syn-

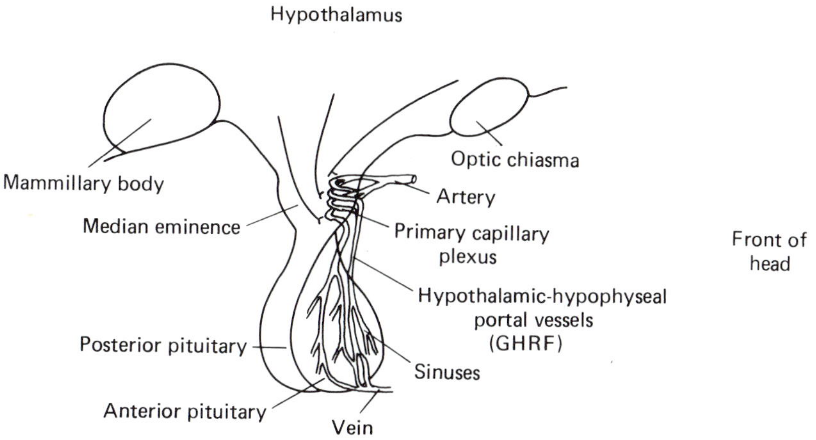

Hypothalamus

FIGURE 9-10

The hypothalamus of the brain and the pituitary hypophysis are connected by a series of neurons and a blood (portal) system.

(SOURCE: Modified from A.C. Guyton, Textbook of Medical Physiology, Copyright W.B. Saunders, Co., Philadelphia, 1976, p. 921. With permission.)

thesizes chemical factors that either inhibit or stimulate the synthesis and release of anterior pituitary hormones. One of the anterior pituitary hormones is growth hormone (GH).

Growth hormone, a polypeptide molecule released from the anterior pituitary, stimulates protein synthesis, especially in the young. In young as well as older individuals, GH is one of the major lipolytic hormones. Therefore, GH directly stimulates fat metabolism and indirectly suppresses carbohydrate metabolism. The action of GH in stimulating lipolysis in adipose tissue is through synthesis of enzymatic or protein factors that stimulate lipase activity.

In the fasting individual, low blood glucose levels stimulate the release of growth hormone-releasing factor (GHRF) from the hypothalamus. The factors responsible for GH release during exercise are less well understood. At various times, low blood glucose, high blood lactate, low blood pH, and elevated body temperature have been thought to be responsible for GH secretion during exercise. Through various experimental manipulations, scientists have gradually eliminated these factors in the direct regulation of GH release during exercise.

In Figure 9-11 the GH response to continuous exercise at about 50% of $\dot{V}_{O_2max}$ is contrasted with the GH response to intermittent exercise (1 min exercise, 1 min rest), where the exercise power output was twice as great as in continuous exercise. In Figure 9-11a a lag of approximately 15 min is indicated between exercise onset and accelerated GH release. Additionally, because the activity of GH on adipose lipolysis is indirect, GH at best has a delayed effect on FFA release from adipose tissue during exercise. In Figure 9-11b, blood

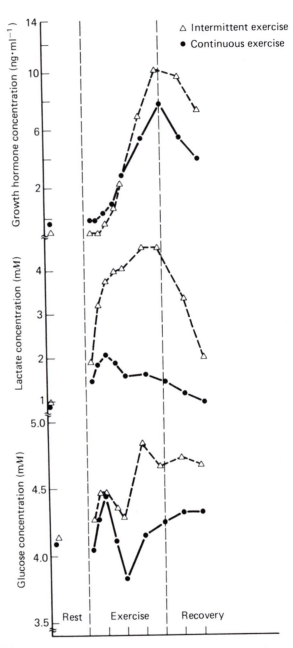

FIGURE 9-11

After about a 15-min lag, human growth hormone is released in response to both continuous and intermittent exercise bouts (a). In these experiments, growth hormone levels do not correlate with levels of blood lactate (b) or blood glucose (c).

(SOURCE: Modified from A. Karagiorgos et al., 1979.)

lactate responses to continuous and intermittent exercise bouts are graphed. Although blood lactate was several times higher in intermittent exercise, GH levels were not significantly different. Additionally, although intermittent exercise resulted in great differences in blood alanine and pyruvate levels, these could not also have significant effects on GH release. Blood glucose level was higher during intermittent (Figure 9-11c) than in continuous exercise, so that glucose could not have been an important regulatory factor for GH release during exercise. Furthermore, body temperature did not correlate with GH level. Therefore, for the present, neural factors are implicated as exerting primary control over GH secretion during exercise.

CORTISOL AND THE PITUITARY—ADRENAL AXIS

The steroid hormone cortisol assists in maintaining blood glucose homeostasis by stimulating amino acid release from muscle, by stimulating hepatic gluconeogenesis from amino acids, and by helping mobilize FFA from adipose tissue.

The mechanism of cortisol secretion is illustrated in Figure 9-12. Stress (either physical or emotional) or declining blood glucose levels stimulate the hypothalamus to secrete corticotropin-releasing factor (CRF). In turn, CRF stimulates the anterior pituitary to release adrenocorticotropin (ACTH), which causes the adrenal cortex to release cortisol into the circulation. Cortisol has a number of effects, including providing negative feedback on its own secretion (Figure 9-12).

Of the four corticosteroids released by the adrenal gland, two (cortisol and cortisone) stimulate glucose formation and are termed *glucocorticoids*. Two other corticosteroids (aldosterone and deoxycorticosterone) are important in electrolyte metabolism and are termed *mineralocorticoids*. Aldosterone is of primary importance for sodium resorption in the kidney and ultimately for fluid and electrolyte balance. This will be discussed in Chapter 22. Cortisol and aldosterone are the major corticosteroids; cortisone and deoxycorticosterone are released in lesser amounts.

The action of ACTH is to stimulate adenylate cyclase activity in adrenocortical cells. The resulting formation of cAMP stimulates secretion of cortisol as well as other corticosteroids. The particular steroid secreted appears to depend on the particular action of the adrenocortical cell stimulated.

Role of Cortisol in Prolonged Exercise and During Recovery from Exhausting Exercise

During starvation or during prolonged, hard exercise, ACTH will be secreted in response to the level of stress and to falling blood glucose levels. In turn,

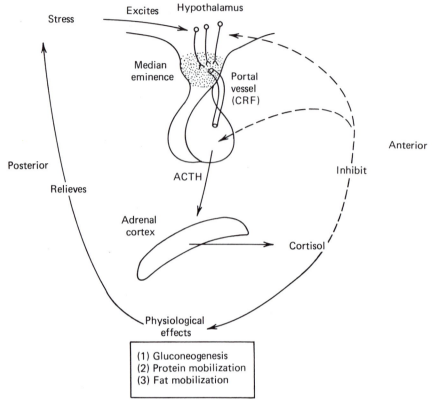

FIGURE 9-12

In response to stress, the secretion of cortisol is stimulated by the adrenal cortex by a series of events that are initiated in the hypothalamus.

(SOURCE: Modified from A.C. Guyton, Textbook of Medical Physiology. Copyright W.B. Saunders, Co., Philadelphia, 1976, p. 952. With permission.)

ACTH will stimulate cortisol release, which will stimulate proteolysis in muscle (Figure 9-13). Some of the resulting amino acids formed in muscle will be released directly into venous blood, and some will be channeled through alanine for release into venous blood. During the period of recovery from exhausting exercise, the return to normal blood glucose levels is a primary physiological concern. Gaesser and Brooks, and Fell and co-workers have observed that there can be normalization of blood glucose levels and restitution of cardiac glycogen in exercise-exhausted animals that were subsequently starved. The source of the glucosyl units is thought to be muscle protein. Cortisol (through the mobilization of muscle proteins) and glucagon (through the stimulation of gluconeogenesis from amino acids) are thought to affect glucose and glycogen replenishment in recovery.

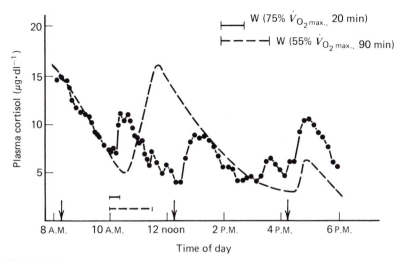

FIGURE 9-13

Blood cortisol level follows a daily pattern that is affected by circadian rhythms, eating, and exercise. Cortisol levels are high during the early morning but fall after breakfast (first arrow). After a lag when exercise starts, cortisol levels increase at a rate proportional to the exercise intensity, but reach a final level dependent on the duration of exercise. Later in the day, when cortisol levels are again low, they rise again after lunch and supper (second and third arrows, respectively.) W, Work.

(SOURCE: Modified from G. Brandenberger and M. Follenius, 1975.)

THE PERMISSIVE ACTION OF THYROID HORMONE

The thyroid gland (in the neck) secretes amino acid–iodine bound hormones thyroxine (T_4) and triiodothyronine (T_3). Of these, thyroxine is released in the greatest amount, whereas triiodothyronine has the greatest activity. Under the influence of thyroid-stimulating hormone (TSH) released from the pituitary, the thyroid secretes and releases thyroxine. Most of this is bound to plasma proteins; the bound T_4 is in equilibrium with free T_4, which is available to interact with target tissues.

Most cells in the body are targets for T_4. Target cells apparently metabolize T_4 to T_3, which is the active form of the hormone. The thyroid itself releases some T_3, but the circulating levels of T_3 cannot be accounted for by thyroid secretion. Thyroxine and triiodothyronine generally stimulate metabolism and increase the rate of such processes as oxygen consumption, protein synthesis, glycogenolysis, and lipolysis. Thyroid hormones appear to act indirectly in promoting the general increase in metabolic rate by potentiating the effects of other hormones. This potentiating or enhancement of other hormonal actions is termed "permissiveness." Thyroid hormones have the effect of raising cellular cAMP levels, and therefore their action (via T_3) may be to stimulate adenylate cyclase. In this manner, the effects of those hormones that act through cAMP would be amplified.

Compared to the glycoregulatory hormones (insulin, glucagon, epinephrine, and cortisol), the levels of thyroid hormones and ACTH do not change much during exercise or as the result of training. This is because the secretion of these hormones is linked to their utilization during exercise (Figure 9-14). In the case of the thyroid hormones, measurements of circulating levels do not give a clear indication of the increased utilization in active individuals. Winder and associates have clearly shown the turnover of thyroxine to be increased as a result of physical activity.

POSTERIOR PITUITARY AND ADH SECRETION

Antidiuretic hormone (ADH) functions together with aldosterone to maintain fluid and electrolyte balance. ADH is an amino acid protein hormone released from the posterior pituitary (Figure 9-10). Two types of stimuli appear to stimulate ADH secretion: osmolality and arterial pressure.

Particular hypothalamic nuclei are sensitive to osmolality (electrolyte concentration) in arterial blood. When severe sweating causes an increase in blood osmolality (sweat is more dilute than plasma, so sweating results in a concentration of blood constitutents), the hypothalamic supraoptic nuclei transmit action potentials to the posterior pituitary (neural hypophysis). These signals result in ADH secretion. In response to a fluid overload and hemodilution, the supraoptic nuclei decrease in their activity, and ADH release is suppressed. In the kidney, ADH allows the reabsorption of H_2O after glomerular filtration. Consequently, sweating stimulates ADH secretion, which stimulates renal H_2O retention and a decrease in urine volume. In response to a fluid overload, ADH secretion is suppressed and urine volume increases.

The second mechanism of ADH secretion involves pressure receptors in the left atrium and other vascular baroreceptors. When dehydration results in a fall in blood pressure, the vascular baroreceptors develop action potentials that are transmitted to the hypothalamus via afferent neurons. The end result is increased stimulation of the posterior pituitary and ADH secretion. Thus, a fall in blood pressure results in a decrease in urinary output, which helps to minimize further dehydration.

During exercise, where significant sweating is involved and hemoconcentration is a usual result, levels of ADH and aldosterone increase considerably. In the case of exercise, increased plasma osmolality is probably the major factor stimulating ADH secretion.

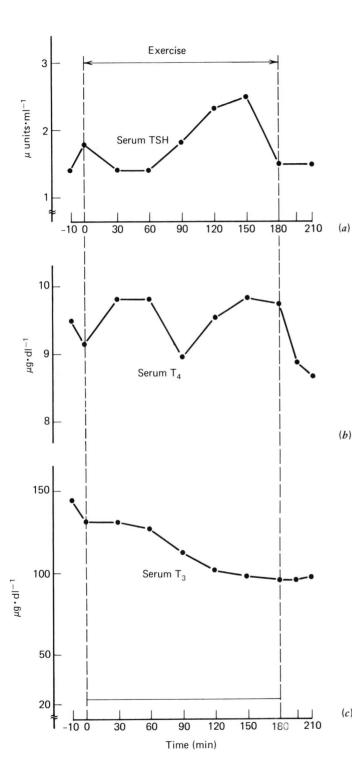

FIGURE 9-14
During prolonged submaximal exercise, changes in circulating levels of pituitary–thyroid hormones are small in comparison to the changes in glucoregulatory hormones (e.g., insulin and glucagon). The level of TSH tends to increase, whereas T_4 remains constant but then falls during recovery. The level of T_3 falls continuously during exercise. Increased T_4 and T_3 utilization during exercise is matched by thyroid hormone release during exercise.

(SOURCE: Modified from P. Berchtold et al., 1978.)

TABLE 9-2

Summary of Metabolic Effects of Hormones

Metabolic Effect	Hormone(s)
Cellular glucose uptake	Insulin
Glycolysis and glycogen synthesis	Insulin
Triglyceride synthesis	Insulin
Decrease in blood glucose level	Insulin
Liver glycogenolysis	Epinephrine (nonspecific)
	Glucagon (specific)
	Norepinephrine (?)[a]
Liver gluconeogenesis	Glucagon
Muscle glycogenolysis	Epinephrine
	Norepinephrine (?)
Lipolysis	Cortisol
	Epinephrine
	Growth hormone
Protein synthesis	Growth hormone
	Insulin
Protein catabolism	Cortisol
Increase in blood glucose level	Epinephrine
direct effect	Glucagon
Indirect effect	Cortisol
	Epinephrine
	Glucagon
Increased metabolic rate	Epinephrine
	Norepinephrine
	Thyroxine

[a] Question mark indicates that specificity of effect is unknown.

Summary

The maintenance of "normal" blood glucose levels and provision of a supply of alternative fuels to glucose depends on the integrated functioning of several systems. This integration is affected by transmitter substances released from the sympathetic nervous system and endocrine hormonal system. Hormones have powerful effects on metabolism even if present in only minute amounts. Table 9-2 summarizes the effects of the various hormones.

During exercise, increased demand for glucose by contracting muscle tends to cause a fall in blood glucose level. This fall is compensated for by a decrease in circulating insulin level that slows the uptake of glucose by most tissues, sparing the remaining glucose for active muscle and other tissues with an essential requirement for glucose (e.g., brain).

Glucose homeostasis is also maintained by the stimulation of liver glycogen breakdown and glucose release, and by the stimulation of the synthesis of glucose from other substances (gluconeogenesis). The hormone glucagon (which is synthesized in the pancreas, as is insulin) is important in the control of hepatic glycogenolysis and gluconeogenesis. The gluconeogenic function of glucagon is assisted by the steroid hormone cortisol, which stimulates proteolysis in muscle and the release from muscle of amino acids (e.g., alanine) that serve as gluconeogenic precursors.

By stimulating the release of fatty acids into the circulation from adipose tissue, epinephrine and growth hormone provide alternative fuels to blood glucose and muscle glycogen. Because glucose and glycogen are necessary for continued functioning during prolonged, hard exercise, the "sparing" of glucose and glycogen use by increasing FFA use postpone the time when glucose and glycogen stores become critical. Epinephrine has a rapid effect after exercise onset of stimulating lipolysis in adipose tissue, whereas growth hormone becomes increasingly more important for maintaining lipid mobilization as exercise duration progresses beyond 15 min.

Just as the hormonal response is geared to maintain a "normal" level of blood glucose during exercise, training is also geared to preserve glucose homeostasis during exercise. Given exercise bouts of absolute or relative intensity, trained individuals experience less metabolic stress and have dampened endocrine responses. In trained individuals, blood insulin level does not fall as much, and catecholamines, growth hormone, and glucagon do not rise as much as in untrained individuals.

Selected Readings

Berchtold, P., M. Berger, H.J. Cüppers, J. Herrmann, E. Nieschlag, K. Rudorff, H. Zimmerman, and H.L. Krüskemper. Non-gluco regulatory hormones (T_4, T_3, vT_3, TSH, and testosternone) during physical exercise in juvenile type diabetics. Horm. Metab. Res. 10:269–273, 1978.

Brandenberger, G. and M. Follenius. Influence of timing and intensity of muscular exercise on temporal patterns of plasma cortisol levels. J. Clin. Endocrinol. Metab. 40:845–849, 1975.

Buckler, J.M.H. The relationship between exercise body temperature and plasma growth hormone levels in a human subject. J. Physiol. London 214:25–26P, 1971.

Felig, P., J. Wahren, R. Hendler, and G. Ahlborg. Plasma glucagon levels in exercising man. N. Engl. J. Med. 287:184–185, 1972.

Galbo, H., J.J. Holst, N.J. Christensin, and J. Hilsted. Glucagon and plasma catecholamines during beta-receptor blockade in exercising man. J. Appl. Physiol. 40:855–863, 1976.

Gyntelberg, F., M.J. Rennie, R.C. Hickson, and J.O. Holloszy. Effect of training on the response of plasma glucagon to exercise. J. Appl. Physiol.: Respirat. Environ. Exercise Physiol. 43:302–305, 1977.

Hagberg, J.M., R.C. Hickson, J.A. McLane, A.A. Ehsani, and W.W. Winder. Disappearance of norepinephrine from the circulation following strenous exercise. J. Appl. Physiol.: Respirat. Environ. Exercise Physiol. 47:1311–1314, 1979.

Häggendal, J., L.H. Hartley, and B. Saltin. Arterial nonadrenaline concentration during exercise in relation to relative work levels. Scand. J. Clin. Lab. Invest. 26:337–342, 1970.

Hartley, L.J., J.W. Mason, R.P. Hogan, L.G. Jones, T.A. Kotchen, E.H. Mougey, R.E. Wherry, L.L. Pernington, and P.T. Ricketts. Multiple hormonal responses to prolonged exercise in relation to physical training. J. Appl. Physiol. 33:607–610, 1972.

Karagiorgos, A., J.F. Garcia, and G.A. Brooks. Growth hormone response to continuous and intermittent exercise. Med. Sci. Sports. 11:302–307, 1979.

Keul, V.J., M. Lehmann, and K. Wybitul. Zur wiung von burnitrolol auf hertzfrequenz, metabolishe Grössen bei Korperarbeit und Leistungsverhalten. Arzneim.-Gorsch/Drug Res. 31:1–16, 1981.

Lassarre, C., F. Girard, J. Durand, and J. Raynaund. Kinetics of human growth hormone during sub-maximal exercise. J. Appl. Physiol. 37:826–830, 1974.

Mason, J.W., L.H. Hartley, T. Kotchen, F.E. Wherry, L.L. Pernington, and L.G. Jones. Plasma thyroid-stimulating hormone response in anticipated of muscular exercise in the human. J. Clin. Endocrinol. Metab. 37:403–406, 1973.

Meyer, R.A. and R. Terjung. Ammonia and IMP in different skeletal muscle fibers after exhaustive exercise in rats. J. Appl. Physiol.: Respirat. Environ. Exercise Physiol. 49:1037–1041, 1980.

Péronnet, F., J. Cléroux, H. Perrault, D. Cousineau, J. de Champlain, and R. Nadeau. Plasma norepinephrine response to exercise before and after training in humans. J. Appl. Physiol.: Respirat. Environ. Exercise Physiol. 51:812–815, 1981.

Pruett, E.D.R. Plasma insulin concentrations during prolonged work near maximal oxygen uptake. J. Appl. Physiol. 29:155–158, 1970.

Sutton, J.R. Hormonal and metabolic responses to exercise in subjects of high and low work capacities. Med. Sci. Sports. 10:1–6, 1978.

Sutton, J.R. and L. Lazarus. Growth hormone in exercise: comparison of physiological and pharmacological stimuli. J. Appl. Physiol. 41:523–527, 1976.

Sutton, J.R., N.L. Jones, and C.J. Toews. Growth hormone secretion in acid-base alteration at rest and during exercise. Clin. Sci. Mol. Med. 50:241–247, 1976.

Terjung, R.L. and W.W. Winder. Exercise and thyroid function. Med. Sci. Sports Exercise. 7:20–26, 1975.

Terjung, R.L. and C.M. Tipton. Plasma thyroxine and thyroid-stimulating hormone levels during sub-maximal exercise in humans. Am. J. Physiol. 220:1840–1845, 1971.

Tharp, G.D. The role of glucocorticoids in exercise. Med. Sci. Sports. 7:6–11, 1975.

Winder, W.W. Time course of the T_3– and T_4– induced increase in rat soleus muscle mitochondria. Am. J. Physiol: Cell Physiol. 5:C132–C138, 1979.

Winder, W.W., M.A. Beattie, and R.T. Holman. Endurance training attenuates stress hormone responses to exercise in fasted rates. Am. J. Physiol. 243 (Regulatory Integrative Comp. Physiol. 12): R179–R184, 1982.

Winder, W.W., J.M. Hagberg, R.C. Hickson, A.A. Ehsani, and J.A. McLane. Time course of sympathoadrenal adaptation to endurance exercise training in man. J. Appl. Physiol. 45:370–374, 1978.

Winder, W.W., R.C. Hickson, J.M. Hagberg, A.A. Ehsani, and J.A. McLane. Training-induced changes in hormonal and metabolic responses to sub-maximal exercise. J. Appl. Physiol.:Respirat. Environ. Exercise Physiol. 46:766–771, 1979.

Winder, W.W. and R.W. Heinger. Effect of exercise on degradation of thyroxine in the rat. Am. J. Physiol. 224:572–575, 1973.

10

METABOLIC RESPONSE TO EXERCISE

It has long been known that respiratory (V_{O_2}) responses to exercise are at times inadequate to provide the ATP necessary to sustain work. Although aerobic metabolism has been well understood, the role of glycolytic (anaerobic) metabolism has not. Therefore, in this chapter it is also necessary to discuss three errant concepts concerning "anaerobic" metabolism and exercise. These are (1) that anaerobic metabolism during exercise results in an "O_2 debt" to be repaid after exercise, (2) that lactic acid is a "dead-end metabolite" that is only formed and not removed during exercise, and (3) that the elevation of blood lactic acid level during exercise represents anaerobiosis (O_2 insufficiency) in muscle. Because of the pandemic scope of these three misconceptions, it is necessary to discuss them along with contemporary explanations of lactic acid metabolism during exercise and recovery.

WHY MEASURE THE METABOLIC RESPONSE TO EXERCISE?

Knowing what the metabolic response to exercise is often constitutes the most important means of evaluating what the immediate as well as long-term effects of exercise on the body will be. The exercise itself is often described in terms of the metabolic response it elicits. Comparison of the metabolic response during a particular exercise bout with the resting metabolic rate, or with the max-

The effects of exercise are immediate and prolonged. (A) Some athletes in an indoor mile relay. (B) An athlete tossing the caber.

imal metabolic response, describes the absolute as well as relative intensity of the exercise bout. Exercise bouts are therefore described in terms of the absolute $\dot{V}_{O_2}$, multiples of the resting metabolic rate (METS), or as a percentage of $\dot{V}_{O_2max}$. Determination of metabolic response to exercise also allows estimation of the energy (calorie) cost of the exercise. Knowledge of the caloric cost of exercise can be important if the nutritional requirements of the exercise bout need to be provided for or if the efficiency of the body during the performance of the exercise is to be calculated.

VALIDITY OF INDIRECT CALORIMETRY IN MEASURING EXERCISE RESPONSE

Metabolic responses to exercise are almost exclusively studied using indirect calorimetry (i.e., measurements of O_2 consumption). In the main, attempts to validate indirect calorimetry as a measure of the metabolic rate have been performed on resting individuals. These and the limited attempts to validate indirect calorimetry on exercising individuals yield excellent results. However, one must be sure to ask the following question when utilizing O_2 consumption measures to estimate metabolic rate during exercise: Is indirect calorimetry valid for these circumstances? For exercise bouts that are not overly intense or too long in duration, the $\dot{V}_{O_2}$ can be counted on to provide an excellent measure of metabolic rate.

In order for indirect calorimetry to be valid, all the ATP used must come from respiration (O_2 consumption). Figure 10-1 is similar to Figure 3-12, except that Figure 10-1*b* includes the blood lactate level, which is seen to increase nonlinearly after a work intensity of approximately 60% of $\dot{V}_{O_2}$. For this reason, efficiency measures are frequently made at metabolic rates that require less than 60% of $\dot{V}_{O_2 max}$, before blood lactate accumulates.

In addition to confidence that there is no anaerobic glycolysis, several other assumptions have to be made in utilizing indirect calorimetry to estimate metabolic rate during exercise. One is that ATP and CP stores are maintained. This, like the assumption of insignificant anaerobic glycolysis, addresses the question of whether all the ATP utilized for exercise is provided by respiration.

The remaining assumption is that of insignificant amino acid or protein catabolism during exercise. This assumption is most likely invalid (cf. Chapter 8), but it is a compromise for convenience and necessity. The assumption allows use of the respiratory exchange ratio (R) as the nonprotein RQ (convenience); the assumption also eliminates the need to collect nitrogenous excretions and to match N_2 excretion through sweat and urine in time with exercise (necessity). The error incurred by this assumption is usually small.

As seen in Chapter 3, and as illustrated in Figure 10-1*a*, the steady-rate O_2 consumption responses can be used to represent the caloric, or O_2 cost of exercise—provided, of course, that all the ATP required for exercise is supplied by respiration.

THE EXCESS POSTEXERCISE OXYGEN CONSUMPTION (EPOC), OR THE "O₂ DEBT"

Estimating the metabolic cost of hard or maximal exercise has long fascinated physiologists. This problem is far more complex than understanding steady-rate exercise. Because the $\dot{V}_{O_2}$ obviously does not represent the metabolic rate during extreme exercise, investiagors have tried to take advantage of the extra O_2 consumed by the body during recovery to estimate the total "O_2 cost" of exercise. For a long time it was thought that if the O_2 consumption during exercise was inadequate to meet energy demands—that is, if there is a "deficit" in the O_2 consumption (Figure 10-2)—then the body borrows on its energy reserves (or credits). After exercise, then, the body was thought to pay back those credits, plus some interest. The extra O_2 consumed, above a resting baseline during recovery, was referred to as the O_2 debt and was used as a measure

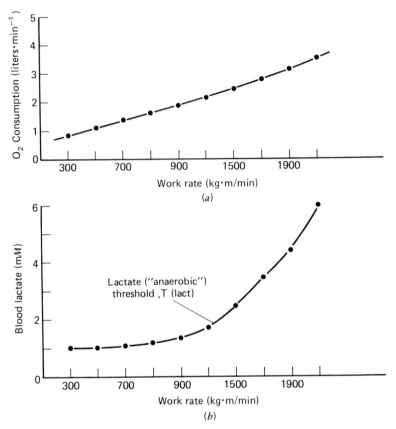

FIGURE 10-1

O_2 Consumption (a) and blood lactate (b) responses to a continuous, progressive cycle ergometer test in healthy young subjects. $\dot{V}_{O_2}$ Responds in almost a linear fashion (a), whereas lactate level does not change at first but then begins to rise nonlinearly (b). The lactate inflection point, or lactate threshold, T(lact), represents the point at which lactate entry into the blood exceeds its removal.

(Modified from E.H. Hughes, S.C. Turner, and G.A. Brooks, J. Appl. Physiol.: Respirat. Environ. Exercise Physiol. 52:1598–1607, 1982.)

of the anaerobic metabolism during exercise. As may be surmised, this explanation of the relationships between O_2 deficit and O_2 debt is too simplistic. Measurement of O_2 deficit or debt may have a purpose in contemporary research, but the O_2 debt after exercise is an inadequate measure of the anaerobic metabolism during exercise.

After exercise, O_2 consumption does not return to resting levels immediately but rather in a curvilinear fashion (Figure 10-3a). Plotting the post exercise O_2 consumption on semilogarithmic coordinates reveals that the curve is composed

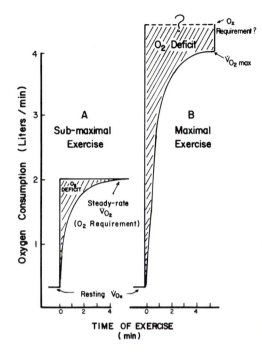

A
Sub-maximal
Exercise

B
Maximal
Exercise

FIGURE 10-2
Oxygen consumption before and during continuous bouts of steady-rate submaximal and maximal exercise. In submaximal exercise (a), the O_2 deficit can be estimated as the difference between the steady-rate $\dot{V}_{O_2}$ and the actual $\dot{V}_{O_2}$ prior to attainment of the steady-rate $\dot{V}_{O_2}$. In maximal exercise (b), the O_2 deficit cannot be estimated with certainty for lack of a precise value of the O_2 required. For this reason, scientists have attempted to utilize the "O_2 debt" (Figure 10-3) as a measure of anaerobic metabolism during exercise.

[From G.A. Brooks (ed.), Perspectives on the Academic Discipline of Physical Education, Human Kinetics Publishers, Champaign, Ill., 1981. pp. 97–120.]

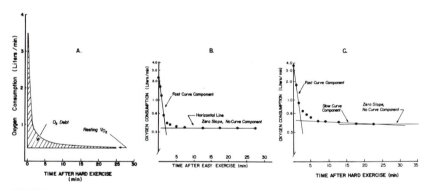

FIGURE 10-3
O_2 Consumption after a period of exercise requiring a constant work output. (a) Plotted on linear coordinates, the O_2 debt is the area under the postexercise $\dot{V}_{O_2}$ curve above the resting $\dot{V}_{O_2}$ baseline. (b) and (c) $\dot{V}_{O_2}$ after easy and hard exercise bouts, plotted on semilogarithmic coordinates. Note one postexercise $\dot{V}_{O_2}$ curve component after easy exercise (b), and two components after hard exercise (c).

[From G.A. Brooks (ed.), Perspectives on the Academic Discipline of Physical Education, Human Kinetics Publishers, Champaign, Ill., 1981. pp. 97–120.]

of one exponential component after mild exercise (Figure 10-3b), or two exponential components after moderate to maximal exercise (Figure 10-3c). The extra O_2 consumed during recovery above a resting baseline has been referred to by Brooks and associates as the excess postexercise oxygen consumption (EPOC). This phenomenon of extra O_2 consumption during recovery has long been observed and was termed the O_2 debt by A.V. Hill and co-workers. Because so much confusion exists at present about the mechanism of the postexercise O_2 consumption, it is necessary at present to address the topic in some detail.

CLASSICAL O_2 DEBT THEORY: THE EARLY TWENTIETH CENTURY

The exponential decline in O_2 consumption after exercise was first reported by August Krogh in Denmark. That report was apparently noted by A.V. Hill in England, who was at that time performing detailed studies on the energetics of small muscles, isolated from frogs. From the earlier work of Fletcher and Hopkins in England, it was known that muscles stimulated to contract produced lactic acid and heat, whether O_2 was present or not. It was also noted that in order for lactic acid to be removed, O_2 must be present. Hill observed that the muscle recovering from a contraction displayed a second burst of heat release (a latent heat), which approximated in magnitude the initial heat associated with contraction. However, the latent, or recovery heat appeared only if O_2 was present. According to Hill's calculations, the recovery heat represented the heat that would be released if one-fifth of the lactic acid formed during the contraction was combusted. According to Hill, then, the combustion coefficient of lactic acid (i.e., the fraction of lactate oxidized in recovery) was $\frac{1}{5}$.

In the second decade of the twentieth century, Otto Meyerhof in Germany discovered that the precursor of lactic acid was glycogen. When frog muscle preparations were made to contract, glycogen broke down (disappeared), and similar amounts of lactic acid appeared. During recovery, lactic acid disappeared, and glycogen was re-formed in only slightly smaller amounts. According to Meyerhof, the combustion coefficient of lactate was $\frac{1}{3}$.

Shortly thereafter, A.V. Hill and his associates in England experimented on humans recovering from exercise. Those experiments represented a brilliant attempt to unify the biochemical and energetic results obtained on isolated muscles with an understanding of human physiology. They knew that during contraction, frog muscles broke down glycogen and formed lactic acid. They also knew that during recovery those isolated muscles combusted a small portion ($\frac{1}{5}$ to $\frac{1}{3}$) of the lactate formed and that a majority of the lactate ($\frac{2}{3}$ to $\frac{4}{5}$) was reconverted to glycogen. It was logical to assume that a similar phenomenon oc-

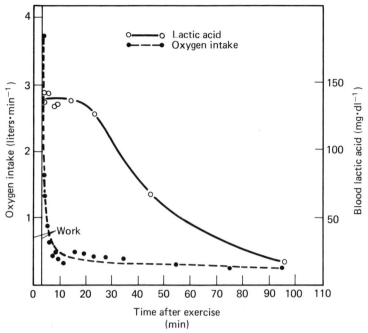

FIGURE 10-4
Oxygen consumption and blood lactate levels in a subject after a hard 3-min treadmill run. Initially blood lactate did not appear to change, whereas $\dot{V}_{O_2}$ declined rapidly. Later, $\dot{V}_{O_2}$ declined slowly and blood lactate declined. On this basis Margaria, Edwards, and Dill separated the "O₂ debt" into fast ("alactacid") and slow ("lactacid") components. Contemporary research reveals these terms to be inappropriate. See text for details.

(SOURCE: Modified from R. Margaria et al., 1933.)

curred in humans during exercise. The Hill–Meyerhof theory of the O₂ debt (as it came to be known) was that during recovery, $\frac{1}{5}$ of the lactate was oxidized and $\frac{4}{5}$ was reconverted to glycogen. In 1922 Hill and Meyerhof shared a Nobel prize awarded, in part, for the work described here, in which detailed biochemical and biophysical measurements on isolated muscles were linked to the intact functioning human.

Subsequent to Hill and associates, Margaria, Edwards, and Dill at the Harvard Fatigue Laboratory directed their efforts to studying recovery metabolism. Following some of the exercise protocols utilized by these investigators (approximately 3 to 8 min of difficult exercise), blood lactate did not decline immediately; rather, there was a delay in the decline of blood lactate (Figure 10-4). Therefore, Margaria and associates surmised that the first fast phase of the post exercise O₂ consumption curve, which was not temporally associated with a change in blood lactate, had nothing to do with lactate metabolism. This phase (Figure 10-4) they termed ''alactacid,'' meaning not associated with lactate metabolism. They proposed that the slow postexercise O_2 consumption curve, which temporarily coincided with the decline in blood lactate, was due to the reconversion of lactate to glycogen. They termed the slow phase the ''lactacid'' phase. Therefore, Margaria and co-workers departed from the work of Hill and co-workers and interpreted the O_2 debt as being due to two factors. It will become evident that Margaria and co-workers were incorrect in their interpretation, because they could not know that lactate was rapidly entering and leaving the blood immediately after exercise. The world, however, came to know and accept the hypothesis of ''lactic'' and ''alactic O_2 debts.''

Since its formulation in the 1920s, and subsequent modification in the 1930s, the O_2 debt theory has been the target of several serious challenges. Even though the challenges have gone unanswered in the scientific literature, O_2 debt theory has persisted in both textbooks and popular literature. This is an anomaly in science.

One of the first challenges to O_2 debt theory was by Ole Bang, who showed by using exercises of varied intensities and durations that the results of Margaria and colleagues were fortuitous results of the duration of their experiments. With prolonged exercise, Bang showed that blood lactate level reaches a maximum after about 10 min of exercise and then declines whether exercise ceases or not (Figure 10-5b). In some cases, basal lactate levels can be achieved during exercise itself. After exercise, however, there is always an ''O_2 debt,'' with predictable kinetics to be ''paid.'' By contrast, using exercise bouts lasting only a few minutes, Bang found that the concentration of lactic acid in the blood reached a maximum after exercise had ended, and depending on the intensity of exercise, remained elevated long after the oxygen intake had returned to preexercise levels (Figure 10-5a). These results cannot be reconciled with the idea that lactic acid determines oxygen consumption after exercise.

THE METABOLIC FATE OF LACTIC ACID AFTER EXERCISE

What happens to lactic acid after exercise (i.e., the metabolic fate of lactic acid) has been studied using radioisotopes in animals by Brooks, Brauner, and Cassens (1973), and by Brooks and Gaesser (1980). In the over 60-yr history of interest in this subject, there have been remarkably few attempts to determine directly what happens to lactic acid after exercise in intact mammals. In their study, Brooks and Gaesser injected [U-^{14}C]lactate into rats at the point of fatigue from hard exercise. At various times during recovery, blood, liver, kidney, heart, and muscle tissues were sampled. Metabolites were separated and quantified using two-dimensional radiochromatography (illustrated in Figure 10-6). From these radiochromatograms, the pathways of lactate metabolism

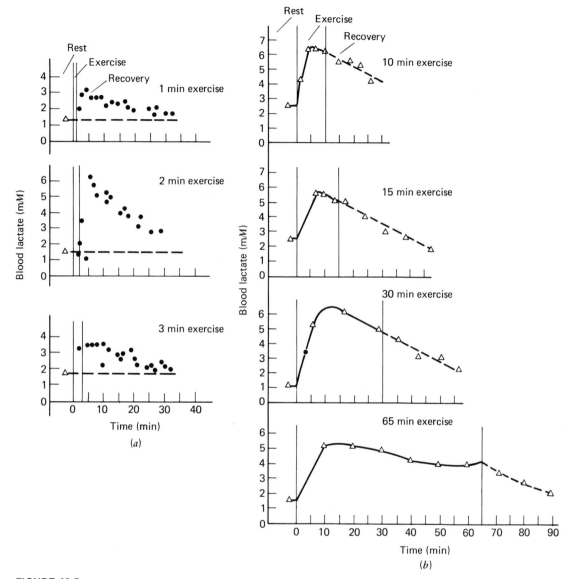

FIGURE 10-5
Blood lactate levels in subjects after bicycle ergometer exercise of the
same intensity but different durations. In series (a), involving short exercise
bouts of from 1 to 3 min, blood lactate clearly increases even though exercise has stopped. In series (b), involving longer periods of exercise, lactate
is clearly decreasing even though exercise continues at a steady rate. Although blood lactate varies greatly, the kinetics of the postexercise O_2 consumption curve (O_2 debt) are unaffected.

(SOURCE: Modified from O. Bang, 1936.)

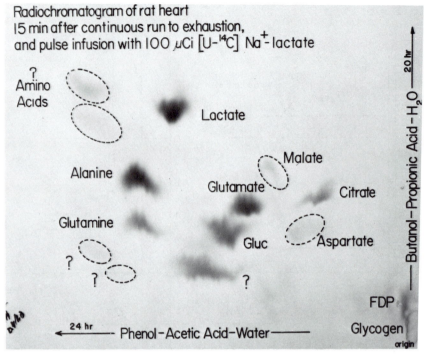

FIGURE 10-6

Two-dimensional radiochromatogram from the heart of a rat, 15 min after exercise to exhaustion and injection with [U-¹⁴C]lactate. Note incorporation of carbon label from tracer lactate into a wide variety of compounds.

(From Brooks and Gaesser, 1980, reprinted with permission of the American Journal of Physiology.)

were traced. The metabolic endpoints reached are displayed in Figure 10-7. Results of these experiments reveal that the metabolic pathways taken by lactate during recovery are diverse. Lactate is mainly oxidized after exercise, but it does participate in a number of other processes.

Lactate as a Carbon Reservoir During Recovery

As indicated in Figure 10-6, lactate can traverse a number of different metabolic pathways. Lactate, in effect, sits close to a metabolic crossroads. Because of the proximity of lactate to the TCA cycle, its entry into that cycle and subsequent oxidation constitute a major pathway of metabolism. In addition to serving as an oxidizable substrate, lactate can also serve as a gluconeogenic precursor or be incorporated into amino acids and proteins. The pathways of lactic acid metabolism after exercise appear to depend to some extent on the internal metabolic conditions when exercise stops. High levels of lactate and near-normal concentrations of other substrates such as liver glycogen and blood

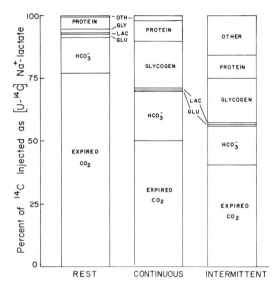

FIGURE 10-7

Histograms representing the quantitative recovery of tracer carbon injected as [U-^{14}C]lactate 4 hr after injection under three conditions. Note that oxidation is the major pathway of removal, and that after exercise, reconversion to glycogen represents less than 20% of the total.

(From Brooks and Gaesser, 1980, reprinted with permission of the American Journal of Physiology.)

glucose appear to favor lactate oxidation. However, the effects of prolonged exercise leading to exhaustion—such as glycogen depletion and hypoglycemia (low blood sugar)—may favor lesser oxidation and greater conversion of lactate to glucose (gluconeogenesis).

Lactic Acid Does Not Cause the O_2 Debt

Even though a majority of the lactate present when recovery begins may be oxidized, lactic acid cannot be said to cause the O_2 debt. Because lactate merely serves to supply fuel to power the recovery processes, the combustion of lactic acid does not result in extra O_2 consumption. Lactate is converted to pyruvate, and this pyruvate in effect substitutes for the pyruvate that would have been supplied by glucose or glycogen. Levels of these compounds may be low after prolonged or difficult exercise.

EXERCISE-RELATED DISTURBANCES TO MITOCHONDRIAL FUNCTION

If lactic acid accumulation does not by itself cause the extra O_2 consumption after exercise, what does? The answer may well lie in the mitochondria and the cellular sites at which O_2 is consumed.

Temperature

The major metabolic waste during exercise is that of heat. Exercise heat production elevates temperature in active muscle as well as other tissues. Using mitochondria isolated from muscle and liver, Brooks and co-workers (1971, 1972) demonstrated that elevated tissue temperatures produced by exercise cause an increase in the rate of mitochondrial O_2 consumption and an eventual decline in energy-trapping efficiency. A mitochondrion and a mechanical model for oxidative phosphorylation are diagrammed in Figure 10-8. It must be emphasized that the process of oxidation (O_2 consumption) and phosphorylation (ATP production) are separate (Chance and Williams, 1956). Ordinarily, the two processes are linked in that oxidation provides the energy for phosphorylation. In

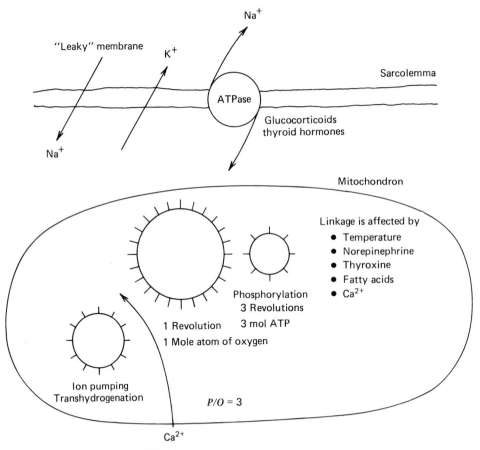

FIGURE 10-8

Diagram of a mitochondrion with the coupling mechanism between oxidation and phosphorylation depicted as a mechanical linkage. Factors suspected of affecting the linkage between oxidation and phosphorylation during the postexercise period are indicated. In addition, ion pumping and ATP utilization by the sarcolemma are promoted by muscle contraction and by thyroid hormones and glucocorticoids.

(SOURCE: Modified from Brooks, 1981.)

Figure 108b the linkage is simplistically diagrammed as a mechanical gear linkage, a situation analogous to the automobile. The engine can run full speed, consume oxygen, and utilize fuel, but unless the transmission and differential are in good operating order, nothing meaningful will happen in terms of rotation of the driving wheels. Elevated temperatures of the magnitude experienced in heavy exercise have the effect of loosening the coupling or linkage between oxidation and phosphorylation. Since the beginning of the twentieth century, when Dubois (1921) thoroughly studied basal metabolic rate (BMR), it has been known that the BMR is increased in humans during fever about 13% per degree Celsius increase in body temperature. Elevated tissue temperatures in humans after exercise have been demonstrated to be associated with EPOC volume (Claremont, Nagle, Reddan, & Brooks, 1975).

Fatty Acids and Ions

One very interesting physiological phenomenon, not generally familiar to students interested in exercise, is the phenomenon of nonshivering thermogenesis in certain animals. This well-studied phenomenon provides an interesting model for comparison to the individual recovering from exercise. When placed in cold environments most mammals will shiver to develop adequate body heat. After a week or 10 days, some species, such as the rat, will stop shivering yet continue to generate adequate heat to maintain internal temperature. The animal's basal metabolic rate will have increased two- to threefold. There are several theories about how the rat accomplishes this. One theory is that oxidation is uncoupled from phosphorylation (Smith & Hoijer, 1962). The sites of this uncoupling may be skeletal muscle as well as brown fat deposits in the body core. Brown fat deposits in hibernating animals have a similar capability, and in this instance, the brown fat functions to maintain an adequate temperature around the hibernating animal's vital organs. Thus the specific agent that uncouples oxidation from phosphorylation may be a fatty acid (Hittelman, Lindberg, & Cannon, 1969). It is possible that the lipolysis and release of fatty acids that occur during exercise have a similar effect on O_2 consumption after exercise.

Another theory to explain the mechanism of nonshivering thermogenesis is that the hormone norepinephrine directly or indirectly causes cell membranes to become more permeable to sodium ion (Na^+) and potassium ion (K^+) (Horowitz, 1978). In order to compensate for the ion leaks, the cell membrane Na^+–K^+ pump activity is increased. This pump requires an energy input in the form of ATP. Increased ATP demands for the pump are met by mitochondrial oxidative phosphorylation. Consequently, O_2 consumption is increased. It has also been discovered that thyroid hormone and glucocorticoids can contribute to Na^+–K^+ pump activity (Horowitz, 1979) (Figure 10-8). Because exercise affects both Na^+ and K^+ balance, and hormone levels, these factors probably serve to elevate O_2 consumption after exercise.

Calcium Ion

Carafoli and Lehninger (1971) have shown that Ca^{2+} has similar effects on mitochondria isolated from all tissues on all animals studied. The effect of calcium ion is to increase oxygen consumption in two ways. First, mitochondria have a great affinity for calcium ions and, given the chance, will sequester Ca^{2+} at a high rate. This process requires energy, which is reflected in an increased rate of O_2 consumption that is not associated with ATP generation. Second, increased amounts of Ca^{2+} within mitochondria ultimately affect the mechanism linking oxidation and phosphorylation. Again, the result is an increased rate of O_2 consumption.

During each muscle contraction of exercise, Ca^{2+} is liberated from the sarcoplasmic reticulum. In heart as well as skeletal muscle, some of this Ca^{2+} is probably taken up by mitochondria. The mechanism by which mitochondria, during recovery, dispose of the Ca^{2+} accumulated during exercise is at present not understood. However, it is likely that Ca^{2+} accumulated by mitochondria during exercise affects mitochondrial respiration during exercise as well as recovery.

DELETION OF THE TERM, "O_2 DEBT"

Throughout this discussion, the descriptive term elevated (or excess) postexercise O_2 consumption (EPOC) has been used, as has the classical term, O_2 debt. Because it is evident that there exists at present no complete explanation of postexercise metabolic phenomena, it may be advantageous to substitute another term for the O_2 debt. Other investigators agree that a name change is due, but there exists some disagreement as to the term that should be selected. An alternative suggested by Harris (1969) was "phlogiston debt." This whimsical suggestion, which projects O_2 debt theory back to the days of alchemy, was obviously meant to draw attention to the problem of terminology. An alternative suggested by Stainsby and Barclay was the term "recovery O_2." A disadvantage of this term is that it does not completely escape the implication of a mechanism or explanation. It is now considered probable that two factors responsible for significant portions of the elevated postexercise O_2 consumption (i.e., the calorigenic effects of catecholamines and temperature) occur during recovery from exercise but are not obligatory and necessary for recovery. Therefore, the purely descriptive term, "elevated postexercise O_2 consumption" (EPOC), is probably most appropriate.

LACTIC ACID TURNOVER DURING EXERCISE (PRODUCTION VERSUS REMOVAL)

Detailed studies utilizing both radioactive and nonradioactive tracers have shown that lactic acid is a dynamic metabolite both at rest and during exercise. At rest and during easy exercise, lactic acid is produced and removed at equal rates. This balance of production and removal is called turnover. Even though a metabolite such as lactic acid turns over very rapidly, its concentration in the blood may not change so long as the removal (from the blood) keeps pace with the production (entry into the blood). Tracer studies have clearly shown that for a given blood lactate level, the turnover of lactic acid during exercise is several times greater than at rest. Therefore, if the situation is that blood lactic acid level remains at resting levels during exercise (Figure 1b, left-hand portion of curve), then it is erroneous to conclude that there is no lactic acid production occurring.

At heavier exercise intensities, blood lactate level increases as compared to rest (Figures 10-1b and 10-5). If the exercise is maintained long enough, the lactate level can remain elevated, but by a constant amount. If the lactate level is constant, then lactate production and removal rates are equal. Tracer studies indicate that the metabolic clearance rate (i.e., turnover rate divided by the blood lactate level) is increased in heavy exercise. This means that lactate removal is concentration dependent; lactic acid needs to be raised to a higher level to force its removal.

In other exercise situations, blood lactate falls toward resting levels after an initial rise (Figure 10-5b), or it continues to increase throughout the exercise (Figure 10-1b). In the former case of declining blood lactate, removal exceeds production once circulation supplies O_2 necessary for muscle respiration and lactate in the blood is delivered to sites of removal. In the latter case of continuously rising lactate levels, production exceeds removal.

PRINCIPLES OF TRACER METHODOLOGY

A tracer is a molecule in which a normally occurring atom or atoms provided by nature is substituted for by a less frequently occurring isotope of that same elemental atom or atoms. Isotopes of an element have the same atomic number but different weight. They can be either radioactive or nonradioactive. The radioactive isotopes are usually far easier to detect and follow in the body or expired air, but because of radioactive hazard, these isotopes are today infrequently used in human experimentation. Carbon, molecular weight 12 (^{12}C), is the most abundant isotope of carbon; it is not radioactive. Carbon-14 (^{14}C) is radioactive and occurs naturally but infrequently in nature. Carbon-14 has a half-life of over 5000 yr. This stability allows it to be produced in a cyclotron and to be chemically incorporated into a variety of compounds. These compounds can be stored for a time and then used to follow metabolic pathways in organisms. Carbon-11 (^{11}C) is another radioactive isotope of carbon with a half-life of about 20 min. This isotope is being used more and more in studies on humans because it breaks down so rapidly, dissipating its potential for radiation exposure. To utilize this isotope, scientists must have a cyclotron close at hand to produce the material. Carbon-13 (^{13}C) is a nonradioactive isotope of carbon that is naturally abundant in nature. About 1% of all carbon is the heavier isotope, ^{13}C. This isotope is more difficult to utilize and trace in metabolism because it is not radioactive, but ^{13}C is becoming the isotope of preference for human experimentation for the same reason. The presence of carbon-13 is usually detected on the basis of its mass.

In addition to carbon, isotopes of hydrogen are frequently used to study metabolism. Hydrogen-2 (deuterium, ^{2}H) is not radioactive, whereas hydrogen-3 (tritium, ^{3}H) is. Simultaneous use of isotopes of carbon and hydrogen (e.g., ^{14}C and ^{3}H) has been of great use in determining how metabolites are split apart and joined together in metabolism.

There are two basic strategies for utilizing isotopes to study metabolism: (1) pulse injection and (2) continuous infusion. In the first case, all the isotope is injected into a rapidly mixing pool (e.g., the blood) in a single rapid (bolus) injection. In the second case, the isotope is given at a continuous, set rate. A third approach is the primed continuous infusion technique, a combination of the other two techniques.

The process of using tracers can be illustrated by the analogy of a rain barrel (Figure 10-9a). Observing the unchanging water level in the barrel, one might conclude that nothing is happening to the H$_2$O. Hidden from view, however, are the facts of entry into and removal from the barrel. Suppose that the ob-

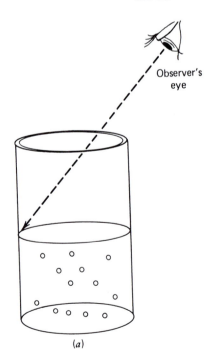

Observer's eye

(a)

FIGURE 10-9
Principle of the pulse (bolus) injection tracer technique. (a) Noting no change in water level, one is tempted to conclude that there is no water movement through the barrel. (b) The observer decides to perform an experiment and adds a dye that colors the H$_2$O. This is time zero (t_0) of the experiment. At various times after t_0 (i.e., $t_1, t_2 . . . . t_i$) the observer records the color in the barrel (c). The more rapid the H$_2$O flux through the barrel, the more rapidly the color changes. In real-life experiments, the dye represents an isotopic tracer and the barrel represents the vascular compartment of the body.

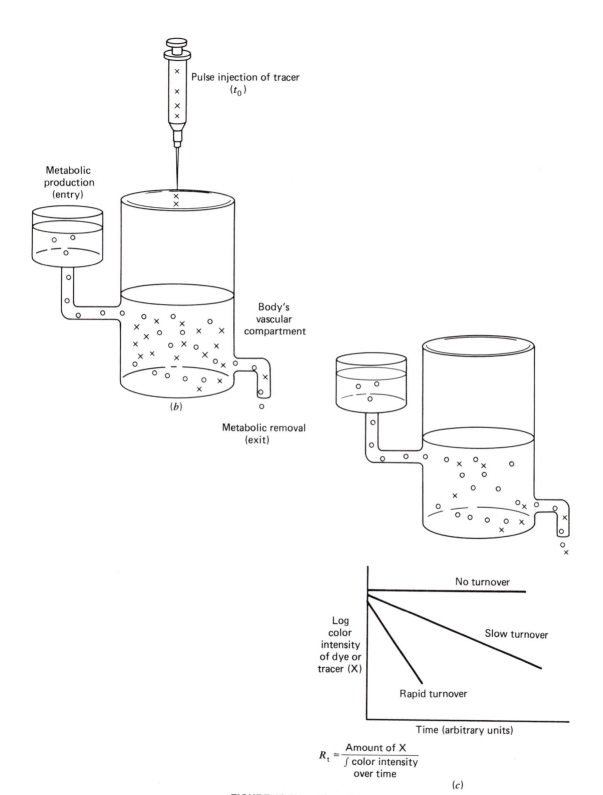

Pulse injection of tracer
(t_0)

Metabolic production (entry)

Body's vascular compartment

(b)

Metabolic removal (exit)

Log color intensity of dye or tracer (X)

No turnover

Slow turnover

Rapid turnover

Time (arbitrary units)

$$R_t = \frac{\text{Amount of X}}{\int \text{color intensity over time}}$$

(c)

FIGURE 10-9 *(continued).*

server decides to perform an experiment by placing a dye in the water. The intensity of color developed in the water will depend on the amount of dye used and the volume of H_2O; the greater the concentration of dye, the darker the color (Figure 10-9b). If there is no movement (flux) of water through the barrel, the intensity of color developed will remain constant. If, however, new clear water enters and previously contained colored water is removed, then the color of water in the barrel will fade. The speed with which the color fades indicates the speed of movement (flux) through the barrel (Figure 10-9c). This is analogous to the pulse injection technique, where the isotopic tracer is represented by the dye.

A second way of using dye (representing isotopic tracer in our analogy) to study movement of water through the rain barrel would be to add dye to the

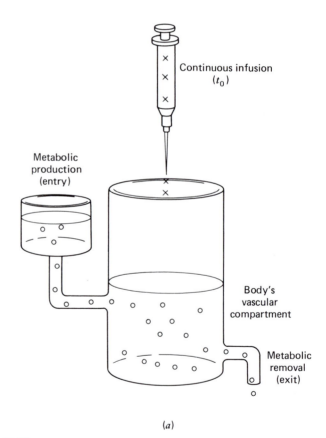

(a)

FIGURE 10-10
Principle of continuous infusion tracer technique, extending the analogy of Figure 10-9. In an alternative experiment, the observer decides to add the dye at a continuous rate. The color intensity at t_0 (a) increases (b) until it reaches an equilibrium (c). The more rapid the H_2O flux through the barrel, the less color intensity is developed.

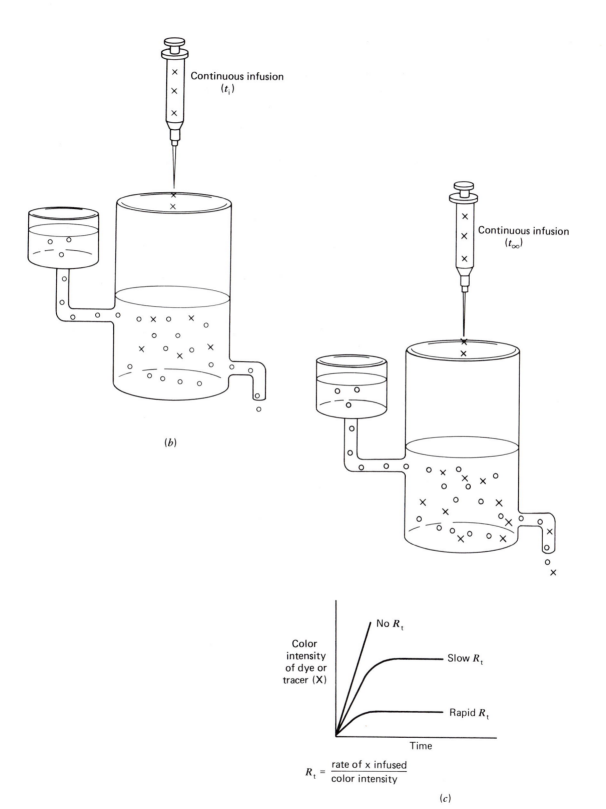

Continuous infusion (t_i)

(b)

Continuous infusion (t_∞)

Color intensity of dye or tracer (X)

No R_t

Slow R_t

Rapid R_t

Time

$$R_t = \frac{\text{rate of x infused}}{\text{color intensity}}$$

(c)

FIGURE 10-10 (continued).

water at a continuous, constant rate by placing the dye in a syringe and pushing it into the barrel water at a constant rate. After the infusion of dye started (Figure 10-10a), the clear water would gradually become darker (Figure 10-10b) until an equilibrium point was reached (Figure 10-10c). At that point, the entry and removal of dye would be proportional to the entry into and removal of water from the barrel. The darker the water becomes, the slower the movement of water through the barrel; the clearer the water remains, the more rapid the movement of water. This is analogous to the continuous infusion technique by studying metabolite flux in organisms.

In the third type of tracer approach, the primed continuous infusion method, tracer is added in a bolus, after which a continuous infusion is begun. This approach essentially relies on the same calculations as the continuous infusion method, but the initial bolus is intended to decrease the time to equilibrium.

EFFECT OF ENDURANCE TRAINING ON LACTATE METABOLISM DURING EXERCISE

At the University of California, Donovan and Brooks studied the effects of endurance training on lactate turnover during rest and exercise using the primed continuous infusion technique and both ^{14}C and ^{3}H tracers. In those studies, trained animals had lower lactate levels during both easy and hard exercises than did untrained animals. However, lactate turnover rates in trained animals during exercise were the same as in untrained animals. Unchanged blood lactate levels in trained animals concealed the fact that lactate production was the same in trained as in untrained animals. The difference due to training was the greater lactate clearance rate from the blood in trained animals (Figure 10-11).

THE "ANAEROBIC THRESHOLD": A MISNOMER

As exercise intensity increases, $\dot{V}_{O_2}$ increases linearly, but blood lactate level does not change until about 60% of $\dot{V}_{O_2max}$ has been reached. Thereafter, blood lactate level increases nonlinearly (Figure 10-1b). The inflection point in the blood lactate curve has mistakenly been termed the "anaerobic threshold." However, the blood lactate inflection point, per se, gives no information about anaerobic metabolism; rather, it reflects balance between lactate entry into and removal from the blood.

As described by Jones and Ehrsam (1982), the anaerobic threshold model is a longstanding paradigm that has recently by advanced by Wasserman and co-workers (Figure 10-12). The model, which proposed causal linkages between

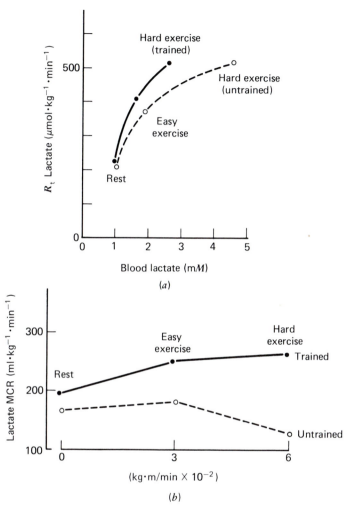

FIGURE 10-11
The effects of endurance training on lactate metabolism during steady-rate, submaximal exercise. (a) The relationship between lactate turnover (as measured with [U-¹⁴C]lactate) and blood lactate concentration. For a given lactate production (turnover, R_t), lactate concentration is much lower in trained rats. (b) The relationship between the metabolic clearance rate of lactate (MCR = R_t/concentration) and external work rate in trained and untrained rats.

(SOURCE: Modified from Donovan and Brooks, 1983.)

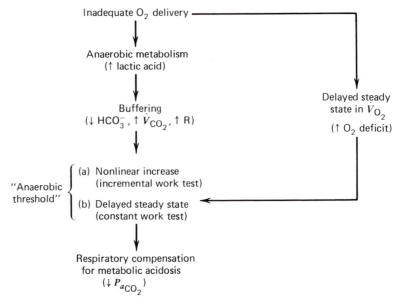

FIGURE 10-12

The model of the "anaerobic threshold" of Wasserman and co-workers is a permutation of O_2 deficit–O_2 debt theory. According to this model, tissue anaerobiosis results in lactate production and entry into the venous blood. This entry of fixed (nonvolatile) acid displaces the volatile acid CO_2 from the blood through the bicarbonate buffer system (Chapter 11). The change in CO_2 flux to the lungs is sensed and ventilation is stimulated. Although this model usually provides a method of determining the point at which blood lactate level starts to increase, under some circumstances it fails, which is understandable on theoretical grounds (see text).

(SOURCE: Modified from Wasserman et al., 1973. With permission of the Journal of Applied Physiology.)

muscle O_2 insufficiency (anaerobiosis), lactate production, and changes in pulmonary ventilation, was particularly attractive because it offered the possibility of detecting muscle anaerobiosis by changes in breathing. Unfortunately, it was found that various factors affect the anaerobic threshold determination, including nutritional status, body mass, mode of exercise, and speed of movement. Perhaps most revealing of the fallacy of the anaerobic threshold model were studies on McArdle's syndrome patients by Hagberg and others at Washington University in St. Louis. McArdle's syndrome is a genetically linked disease in which victims lack the enzyme phosphorylase (cf. Figure 5-13), which renders them incapable of catabolizing glycogen and forming lactic acid. Nevertheless, McArdle's syndrome patients demonstrate ventilatory, or "anaerobic," thresholds at the usual place on the lactate–work curve even though there is no change in blood lactate (Figure 10-13).

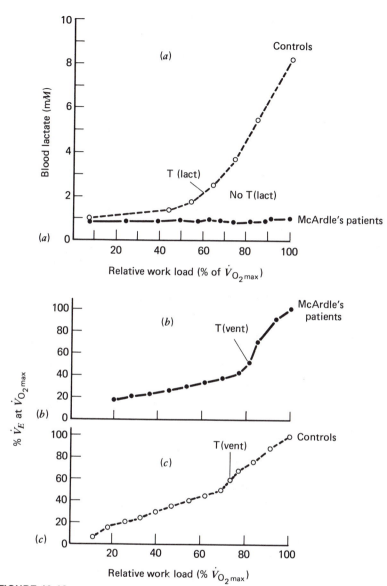

FIGURE 10-13
Blood lactate and ventilatory ($\dot{V}_E$) responses in McArdle's syndrome patients and normal controls during continuous, progressive exercise tests. (a) Controls reach a lactate threshold, T(lact), at about 70% of $\dot{V}_{O_2max}$, whereas McArdle's patients show no lactate inflection. (b) McArdle's patients display a ventilatory threshold, T(vent), despite the fact that there is no T(lact). (c) The usual ventilatory change is observed in controls. These results show that T(vent) is not solely due to T(lact).

(SOURCE: Modified from Hagberg et al., 1982.)

CAUSES OF THE LACTATE INFLECTION POINT

If the inflection point in blood lactic acid concentration is not necessarily due to muscle anoxia, then (1) why is there a lactate inflection point, (2) why is there a ventilatory inflection point, and (3) why do the two inflection points sometimes coincide?

As exercise intensity increases, several factors may operate to change the blood lactate level. Given that blood lactate concentration depends on the entry into and exit from the blood, any factor that affects entry and/or removal changes the blood concentration. Fast-contracting white skeletal muscle fibers produce lactic acid when they are recruited to contract, whether O_2 is present or not. This is a result of the high activity of muscle- (M-)type lactic dehydrogenase (LDHase) (Chapter 5). Moritani has provided electromyographic evidence that more fast white lactate-producing fibers are recruited as exercise intensity increases. This change in recruitment pattern appears to coincide with the change in blood lactate level.

There may be a neuroendocrine component to the explanation of lactate inflection point. As exercise difficulty increases, afferent (nervous) signals to the CNS increase in their frequency and amplitude. This activates the "fight-or-flight" response mechanism, and results in sympathetic autonomic nervous output to the various circulatory beds (including those in muscles, liver, kidney, and adipose tissue) and to the adrenal gland. The neurotransmitter released by the sympathetic nerve endings, norepinephrine, causes local vasoconstriction in arterioles, which increases resistance to blood flow in those areas. In contracting muscle itself, local factors override the constrictor effect and cause arterioles and precapillary sphincters to dilate, allowing relatively more of the cardiac output to flow to active areas.

Sympathetic, autonomic stimulation of the adrenal gland, specifically its inner segment called the adrenal medulla, results in the secretion of epinephrine (adrenaline) into the circulation. Sympathetic stimulation of the α cells in the pancreas causes release of glucagon. Epinephrine and glucagon cause glycogenolysis in muscle and liver, respectively (cf. Chapter 5). In muscle, glycogen is broken down rapidly to glucose 6-phosphate (G6P) by activated phosphorylase. Glucose released from liver and picked up in muscle is converted to G6P by hexokinase. Elevated levels of G6P increase glycolytic rate and formation of pyruvate. Because in this case the rate of glycolysis is not necessarily matched to the activity of the TCA cycle, the "excess" pyruvate produced will be converted to lactate. In other words, difficult exercise results in the secretion of hormones that cause glycolysis to be greatly accelerated, resulting in pyruvate production in excess of TCA cycle needs. The excess of pyruvate is converted to lactate by LDHase. This lactate formed in muscle appears in the blood.

The autonomic effect of vasoconstriction has another effect on lactate level.

Blood flow is shunted toward active muscle but away from other tissues, such as liver and kidney (areas of increased resistance). These organs usually remove lactate and produce glucose. As blood flow to these gluconeogenic tissues decreases, they can compensate somewhat by extracting a greater fraction of the lactate passing through them in the systemic circulation. This ability to increase the fractional extraction, however, is limited, and blood lactate level increases because of limited capacity of gluconeogenic tissues to remove the lactate.

The control of breathing during exercise is described in detail in Chapter 12. Various mechanisms ensure that ventilation is matched to the metabolic rate. The motor cortex affects the breathing centers in the brain to adjust the frequency and depth of breathing. These neural mechanisms operate rapidly and account for most of the control of breathing during exercise. The fine-tuning of breathing to match exercise intensity depends on neural signals, or metabolite levels in blood reaching the brain. The carotid and aortic bodies (Figure 12-16) are sensitive to the partial pressure of O_2 and to the concentration of hydrogen ion (H^+). Carbon dioxide tension in blood affects the cerebrospinal fluid, which in turn affects respiratory centers in the brain. Receptors in joints and muscles are sensitive to movement and transmit signals to the brain.

Of the "peripheral" controls of breathing, two factors appears to be most important: (1) the pH of arterial blood, which affects the arterial chemoreceptors and (2) (to a lesser extent) the pH of the cerebrospinal fluid, which directly affects the respiratory center. The arterial pH is influenced by the blood lactate level and the venous delivery (flux) of CO_2 to the lungs. When CO_2 is dissolved in water, carbonic acid is formed. This acid dissociates a proton, as does lactic acid.

Activity level, metabolism, CO_2 production, and lactate production are usually closely matched. Therefore, CO_2 flux to the lungs is highly correlated with pulmonary minute ventilation. This effect, however, does not amount to a cause. If there is a disturbance in blood pH, such as reduced glycolysis and lactate production due to glycogen depletion or McArdle's syndrome, or if acidifying or alkalinizing agents are ingested, CO_2 flux to the lungs and arterial pH will be affected. The usual relationships among pulmonary ventilation, CO_2 flux to the lungs, and arterial pH will be disturbed. The ventilation will continue to be highly related to the metabolic rate because the neural controls of ventilation are of primary importance, whereas the peripheral factors are of secondary importance.

Summary

Aerobic responses dominate the metabolic adjustments to exercise that lasts more than a minute. So important are the aerobic responses to exercise that the anaerobic responses have often been classified in terms of oxygen equivalency.

This tendency has resulted in the evolution of three terms: O_2 debt, O_2 deficit and anaerobic threshold. These terms originally provided theoretical bases for observed phenomena, but these theories have been shown to be incorrect. Because these terms have been used so widely, not only in exercise physiology and sports, but also in many other situations, it is necessary for the reader to understand their uses and limitations.

After exercise, O_2 consumption remains elevated for a while. This extra O_2 consumption has in the past been called the "O_2 debt," to represent a repayment for the nonoxidative (anaerobic) metabolism performed during exercise. The extra O_2 consumed after exercise was thought to support the oxidation of a minor fraction (1/5) of the lactate formed during exercise, thereby providing the energy to convert the remainder (4/5) to glycogen. In reality, however, the cause of the excess postexercise O_2 consumption (EPOC) is the general disturbance to homeostasis brought on by exercise. Exercise causes elevation in tissue temperatures, changes in intra- and extracellular ion concentrations, and changes in metabolite and hormone levels. Because these physiological changes persist into recovery, their effects serve to elevate O_2 consumption immediately after exercise.

Immediately after exercise, lactic acid provides a readily metabolizable reservoir of substrate. Depending on the conditions of recovery, this lactate reservoir may be mostly oxidized, or it may also contribute to the reestablishment of normal blood glucose levels. Lactic acid oxidation does not in itself result in elevated O_2 consumption, because the lactate substitutes for other substrates. Rather, the energy necessary to return the organism to the preexercise condition results in EPOC, with lactate merely supplying the fuel.

Lactic acid is produced constantly, not just during hard exercise. In fact, lactic acid may be the most dynamic metabolite during exercise; lactate turnover exceeds that of any other metabolite yet studied. The constancy of the blood lactic acid level, during rest or exercise, means that the entry into and removal of lactate from the blood are in balance. Increasing blood lactate levels indicate that entry exceeds removal; declining levels indicate the opposite. Radiotracer studies have shown the turnover of lactic acid during exercise to be several times greater for a given blood lactate level during exercise than during rest. Similarly, for a given blood lactate level, blood lactate removal is several times greater in trained than in untrained animals.

During exercise tests of increasing intensity, at about 60% of $\dot{V}_{O_2max}$, blood lactate levels begin increasing dramatically. This inflection point in blood lactic acid level has mistakenly been called the "anaerobic threshold," meaning the point at which muscle becomes O_2 deficient. This is an archaic concept akin to the beliefs that muscle develops an O_2 debt during exercise and that lactic acid is a "dead-end metabolite" during exercise. Several factors appear to be responsible for the lactate inflection point during graded exercise: hormonally mediated accelerations in glycogenolysis and glycolysis, recruitment of fast glycolytic muscle fibers, and a redistribution of blood flow from lactate-removing, gluconeogenic tissues to lactate-producing, glycolytic tissues.

Selected Readings

Abramson, H.A., M.G. Eggleton, and P. Eggleton. The utilization of intravenous sodium r-lactate. III. Glycogen synthesis by the liver. Glood sugar. Oxygen consumption. J. Biol. Chem. 75:763–778, 1927.

Ahlborg, G. and P. Felig. Lactate and glucose exchange across the forearm, legs, and splanchnic bed during and after prolonged leg exercise. J. Clin. Invest. 69:45–54, 1982.

Alpert, N.R. and W.S. Root. Relationship between excess respiratory metabolism and utilization of intravenously infused sodium racemic lactate and sodium L-(+)-lactate. Am. J. Physiol. 177:455–562, 1954.

Alpert, N.R. Lactate production and removal and the regulation of metabolism. Ann. N.Y. Acad. Sci. 119:995–1012, 1965.

Bang, O. The lactate content of the blood during and after muscular exercise in man. Skand. Arch. Physiol. 74 (Suppl 10): 49–82, 1936.

Barnard, R.J. and M.L. Foss. Oxygen debt: effect of beta-adrenergic blockade on the actacid and alactacid components. J. Appl. Physiol. 27:813–816, 1969.

Barnard, R.J., M.L. Foss, and C.M. Tipton. Oxygen debt: Involvement of the Cori cycle. Int. Z. angew. Physiol. 28:105–119, 1970.

Bendall, J.R. and A.A. Taylor. The Meyerhof quotient and the synthesis of glycogen from lactate in frog and rabbit muscle. Biochem J. 118:887–893, 1970.

Benedict, F.G. and E.P. Cathcart. Muscular Work. Washington, D.C.,: Carnegie Institution of Washington, Publ. 187, 1913.

Bennett, A.F. and P. Licht. Relative contributions of anaerobic and aerobic energy production during activity in Amphibia. J. Comp. Physiol. 87:351–360, 1973.

Brooks, G.A. and C.M. Donovan. Effect of endurance training on glucose kinetics during exercise. Am. J. Physiol. (Endocrinol. Metab. 7). In Press.

Brooks, G.A., K.E. Brauner, and R.G. Cassens. Glycogen synthesis and metabolism of lactic acid after exercise. Am. J. Physiol. 224:1162–1166, 1973.

Brooks, G.A. and G.A. Gaesser. End points of lactate and glucose metabolism after exhausting exercise. J. Appl. Physiol.: Respirat. Environ. Exercise Physiol. 49:1057–1069, 1980.

Brooks, G.A., K.J. Hittelman, J.A. Faulkner, and R.E. Beyer. Temperature, skeletal muscle mitochondrial functions, and oxygen debt. Am. J. Physiol. 220:1053–1059, 1971.

Brooks, G.A., K.J. Hittelman, J.A. Faulkner, and R.E. Beyer. Temperature, liver mitochondrial respiratory functions, and oxygen debt. Med. Sci. Sports. 2:72–74, 1971.

Brooks, G.A., K.J. Hittelman, J.A. Faulkner, and R.F. Beyer. Tissue temperatures and whole-animal oxygen consumption after exercise. Am. J. Physiol. 221:427–431, 1971.

Cain, S.M. Exercise O_2 debts of dogs at ground level and at altitude with and without β-block. J. Appl. Physiol. 30:838–843, 1971.

Cain, S.M. and C.K. Chapler. Effects of norepinephrine and α-block on O_2 uptake and blood flow in dog hindlimb. J. Appl. Physiol.: Respirat. Environ. Exercise Physiol. 51:1245–1250, 1981.

Carafoli, E. and A.L. Lehninger. A survey of the interaction of calcium ions with mitochondria from different tissues and species. Biochem. J. 122:681–690, 1971.

Chance, B. and C.R. Williams. The respiratory chain and oxidative phosphorylation. In Advances in Enzymology, vol. 17. New York: Interscience, 1956. p. 65–134.

Chapler, C.K., W.M. Stainsby, and L.B. Gladden. Effect of changes in blood flow, norepinephrine, and pH on oxygen uptake by resting skeletal muscle. Can. J. Physiol. Pharmacol. 58:93–96, 1980.

Claremont, A.D., F. Nagle, W.D. Reddan, and G.A. Brooks. Comparison of metabolic, temperature, heart rate and ventilatory responses to exercise at extreme ambient temperatures (0°C and 35°C). Med. Sci. Sports. 7:150–154, 1975.

Cori, C.F. Mammalian carbohydrate metabolism. Physiol. Rev. 11:143–275, 1931.

Depocas, F., Y. Minaire, and J. Chatonnet. Rate of formation and oxidation of lactic acid in dogs at rest and during moderate exercise. Can. J. Physiol. Pharmacol. 47:603–610, 1969.

Donovan, C.M. and G.A. Brooks. Endurance training affects lactate clearance, not lactate production. Am. J. Physiol. (Endocrinol. Metab. 7):E83–E92, 1983.

Drury, D.R. and A.N. Wick. Metabolism of lactic acid in the intact rabbit. Am. J. Physiol. 184:304–308, 1956.

Drury, D.R. and A.N. Wick. Chemistry and metabolism of L(+) and D(−) lactic acids. Ann. N.Y. Acad. Sci. 119:1061–1069, 1965.

Dubois, E.F. The basal metabolism in fever. J.A.M.A. 77:352–355, 1921.

Eggleton, M.G. and C.L. Evans. The lactic acid content of the blood after muscular contraction under experimental conditions. J. Physiol. (London) 70:269–293, 1930.

Eldridge, F.L., L. T'so, and H. Chang. Relationship between turnover rate and blood concentration of lactate in normal dogs. J. Appl. Physiol. 37:316–320, 1974.

Eldridge, F.L. Relationship between turnover rate and blood concentration of lactate in exercising dogs. J. Appl. Physiol. 39:231–234, 1975.

Fell, R., J. McLane, W. Winder, and J. Holloszy. Preferential resynthesis of muscle glycogen in fasting rats after exhausting exercise. Am. J. Physiol. 238 (Regulatory Integrative Comp. Physiol. 7): R328–R332, 1980.

Fletcher, W.M. and F.G. Hopkins. Lactic acid in amphibian muscle. J. Physiol. 35:247–309, 1907.

Freminet, A., E. Bursaux, and C.F. Poyart. Effect of elevated lactataemia on the rates of lactate turnover and oxidation in rats. Pflügers Arch. 346:75–86, 1974.

Gaesser, G.A. and G.A. Brooks. Glycogen repletion following continuous and intermittent exercise to exhaustion. J. Appl. Physiol: Respirat. Environ. Exercise Physiol. 49:722–728, 1980.

Gladden, L., W. Stainsby, and B. MacIntosh. Norepinephrine increases canine, skeletal muscle $\dot{V}O_2$ during recovery. Med. Sci. Sports Exerc. 14:471–476, 1982.

Gleeson, T.T. Metabolic recovery from exhaustive activity by a large lizard. J. Appl. Physiol.: Respirat. Environ. Exercise Physiol. 48:689–694, 1980.

Hagberg, J.M., J.P. Mullin, and F.J. Nagle. Effect of work intensity and duration on recovery O_2. J. Appl. Physiol.: Respirat. Environ. Exercise Physiol. 48:540–544, 1980.

Hagberg, J.M., E.F. Coyle, J.E. Carroll, J.M. Miller, W.H. Martin, and M.H. Brooke. Exercise hyperventilation in patients with McArdle's disease. J. Appl. Physiol.: Respirat. Environ. Exercise Physiol. 52:991–994, 1982.

Harris, P. Lactic acid and the phlogiston debt. Cardiov. Res. 3:381–390, 1969.

Harris, P., M. Bateman, T.J. Bayley, K.W. Donald, J. Gloster, and T. Whitehead.

Observations on the course of the metabolic events accompanying mild exercise. Quart. J. Exp. Physiol. 53:43–64, 1968.

Harris, R., R. Edwards, E. Hultman, L.-O. Nordesjö, B. Nyland, and K. Sahlin. The time course of phosphorylcreatine resynthesis during recovery of the quadriceps muscle in man. Pflügers Arch. 367:137–142, 1976.

Hartree, W. and A.V. Hill. The recovery heat-production in muscle. J. Physiol. 56:367–381, 1922.

Henry, F.M. and J. DeMoor. Lactic and alactic oxygen consumption in moderate exercise of graded intensity. J. Appl. Physiol. 8:608–614, 1956.

Hermansen, L. and I. Stensvold. Production and removal of lactate during exercise in man. Acta Physiol. Scand. 86:191–201, 1972.

Hill, A.V. The energy degraded in the recovery processes of stimulated muscles. J. Physiol. 46:28–80, 1913.

Hill, A.V. The oxidative removal of lactic acid. J. Physiol. 48: Proc. Physiol. Soc. x–xi, 1914.

Hill, A.V. and H. Lupton. Muscular exercise, Lactic acid, and the supply and utilisation of oxygen. Quart. J. Med. 16:135–171, 1923.

Hill, A.V., C.N.H. Long, and H. Lupton. Muscular exercise, lactic acid and the supply and utilisation of oxygen. Pt. I–III. Proc. Roy. Soc. B., 96:438–475, 1924.

Hill, A.V., C.N.H. Long, and H. Lupton. Muscular exercise, lactic acid and the supply and utilisation of oxygen. Pt. IV–VI. Proc. Roy. Soc. B. 97:84–138, 1924.

Hill, A.V., C.N.H. Long, and H. Lupton. Muscular exercise, lactic acid and the supply and utilisation of oxygen. Pt. VII–VIII. Proc. Soc. B. 97:155–176, 1924.

Hittelman, K.J., O. Lindberg, and B. Cannon. Oxidative phosphorylation and compartmentation of fatty acid metabolism in brown fat mitochondria. Europ. J. Biochem. 11:183–192, 1969.

Horwitz, B.A. Metabolic aspects of thermogenesis: neuronal and hormonal control. Fed. Proc. 38:2147–2149, 1979.

Hubbard, J.L. The effect of exercise on lactate metabolism. J. Physiol. 231:1–18, 1973.

Issekutz, B., W.A.S. Shaw, and A.C. Issekutz. Lactate metabolism in resting and exercising dogs. J. Appl. Physiol. 40:312–319, 1976.

Ivy, J.L., D.L. Costill, P.J. Van Handel, D.A. Essig, and R.W. Lower. Alteration in the lactate threshold with changes in substrate availability. Int. J. Sports Med. 2:139–142, 1981.

Jöbsis, F.F. and W.N. Stainsby. Oxidation of NADH during contractions of circulated mammalian skeletal muscle. Resp. Physiol. 4:292–300, 1968.

Jones, N. and R. Ehrsam. The anaerobic threshold. In: Terjung, R. (ed.), Exercise and Sports Sciences Reviews. 10:49–83, 1982.

Jorfeldt, L. Metabolism of L(+)-lactate in human skeletal muscle during exercise. Acta. Physiol. Scand. Suppl. 338, 1970.

Jorfeldt, L., A. Juhlin-Dannfeldt, and J. Karlsson. Lactate release in relation to tissue lactate in human skeletal muscle during exercise. J. Appl. Physiol.: Respirat. Environ. Exercise Physiol. 44:350–352, 1978.

Kayne, H.L. and N.R. Alpert. Oxygen consumption following exercise in the anesthetized dog. Am. J. Physiol. 206:51–56, 1964.

Keul, J., E. Doll, and D. Keppler. The substrate supply of the human skeltal muscle at rest, during, and after work. Experientia. 23:974–979, 1967.

Knuttgen, H. Oxygen debt, lactate, pyruvate, and excess lactate after muscular work. J. Appl. Physiol. 17:639–644, 1962.

Knuttgen, H.G. Oxygen debt after submaximal exercise. J. Appl. Physiol. 29:651–657, 1970.

Knuttgen, H. Lactate and oxygen debt: An introduction. In: Pernow, B. and B. Saltin (eds.), Muscle Metabolism During Exercise. New York: Plenum Press, 1971. p. 361–369.

Krebs, H.A., R. Hems, M.J. Weidmann, and R.N. Speake. The fate of isotopic carbon in kidney cortex synthesizing glucose and lactate. Biochem. J. 101:242–249, 1966.

Krebs, H.A. and M. Woodford. Fructose 1,6-diphosphatase in striated muscle. Biochem. J. 94:436–445, 1965.

Lee, S.-H. and E.J. Davis. Carboxylation and decarboxylation reactions. J. Biol. Chem. 254:420–430, 1979.

Lundsgaard, E. Untersuchungen über Muskel-kontraktionen ohne Milchsaürebildung. Biochem. Z. 217:162–177, 1930.

MacDougall, J., G. Ward, D. Sale, and J. Sutton. Muscle glycogen repletion after high intensity intermittent exercise. J. Appl. Physiol.: Respirat. Environ. Exercise Physiol. 42:129–132, 1977.

Maehlum, S., P. Felig, and J. Wahren. Splanchnic glucose and muscle glycogen metabolism after glucose feeding during postexercise recovery. Am. J. Physiol. 235 (Endocrinol. Metab. Gastrointest. Physiol.4): E255–E260, 1978.

Maehlum, S. and L. Hermansen. Muscle glycogen concentration during recovery after prolonged severe exercise in fasting subjects. Scand. J. Clin. Lab. Invest. 38:557–560, 1978.

Margaria, R., H.T. Edwards, and D.B. Dill. The possible mechanisms of contracting and paying the oxygen debt and the role of lactic acid in muscular contraction. Am. J. Physiol. 106:689–715, 1933.

Mazzeo, R.S., G.A. Brooks, D.A. Schoeller, and T.F. Budinger. Pulse injection ^{13}C-tracer studies of lactate metabolism in humans during rest and two levels of exercise. Biomed. Mass. Spect. 9:310–314, 1982.

McLane, J.A. and J.O. Holloszy. Glycogen synthesis from lactate in the three types of skeletal muscle. J. Biol. Chem. 254:6548–6553, 1979.

Meyerhof, O. Die Energieumwandlungen im Muskel. I. Über die Beziehungen der Milchsaüre zur Wärmebildung und Arbeitsleistung des Muskels in der Anaerobiose. Pflügers Arch. ges. Physiol. 182:232–283, 1920.

Meyerhof, O. Über die Energieumwandlungen im Muskel. II. Das Schicksal der Milchsaüre in der Erholungsperiode des Muskels. Pflügers Arch. ges. Physiol. 182:284–317, 1920.

Meyerhof, O. Die Engergieumwandlungen im Muskel. III. Kohlenhydrat- und Milchsaureumsatz im Froschmuskel. Pflügers Arch. ges. Physiol. 185:11–32, 1920.

Moorthy, K.A. and M.K. Gould. Synthesis of glycogen from glucose and lactate in isolated rat soleus muscle. Arch. Biochem. Biophys. 130:399–407, 1969.

Newsholme, E. Substrate cycles: their metabolic, energetic and thermic consequences in man. Biochem. Soc. Symp. 43:183–205, 1978.

Opie, L.H. and E.A. Newsholme. The activities of fructose 1,6- diphosphatase, phosphofructokinase and phosphoenolypyruvate carboxykinase in white muscle and red muscle. Biochem. J. 103:391–399, 1967.

Piiper, J. and P. Spiller. Repayment of O_2 debt and resynthesis of high energy phosphates in gastrocnemius muscle of the dog. J. Appl. Physiol. 28:657–662, 1970.

Richter, E., N. Ruderman, H. Gavras, E. Belur, and H. Galbo. Muscle glycogenolysis during exercise: dual control by epinephrine and contractions. Am. J. Physiol. 242 (Endocrinol. Metab. 5): E25–E32, 1982.

Ross, R.D., R. Hems, and H.A. Krebs. The rate of gluconeogenesis from various precursors in the perfused rat liver. Biochem. J. 102:942–951, 1967.

Rowell, L.B., K.K. Kraning II, T.O. Evans, J.W. Kennedy, J.R. Blackmon, and F. Kusumi. Splanchic removal of lactate and pyruvate during prolonged exercise in man. J. Appl. Physiol. 21:1773–1783, 1966.

Ryan, W., J. Sutton, C. Toews, and N. Jones. Metabolism of infused L(+)-lactate during exercise. Clin. Sci. 56:139–146, 1979.

Sacks, J. and W. Sacks. Carbohydrate changes during recovery from muscular contraction. Am. J. Physiol. 112:565–572, 1935.

Saltin, B., K. Nazar, D.L. Costill, E. Stein, E. Jansson, B. Essen, and P.D. Gollnick. The nature of the training response: peripheral and central adaptations to one-legged exercise. Acta Physiol. Scand. 96:289–305, 1976.

Scheen, A., J. Juchmes, and A. Cession-Fossion. Critical analysis of the "Anaerobic Threshold" during exercise at constant workloads. Eur. J. Appl. Physiol. 46:367–377, 1981.

Scrutton, M.C. and M.F. Utter. The regulation of glycolysis and gluconeogenesis in animal tissues. Ann. Rev. Biochem. 37:250–302, 1968.

Segal, S.S. and G.A. Brooks. Effects of glycogen depletion and workload on postexercise O_2 consumption and blood lactate. J. Appl. Physiol: Respirat. Environ. Exercise Physiol. 47:514–521, 1979.

Stainsby, W.N. and J.K. Barclay. Exercise metabolism: O_2 deficit, steady level O_2 uptake and O_2 uptake for recovery. Med. Sci. Sports. 2:177–186, 1970.

Warnock, L.G., R.E. Keoppe, N.F. Inciardi, and W.E. Wilson. L(+) and D(−) lactate as precursors of muscle glycogen. Ann. N.Y. Acad. Sci. 119:1048–1060, 1965.

Wasserman, K., B.J. Whipp, S.N. Koyal, and W.L. Beaver. Anaerobic threshold and respiratory gas exchange during exercise. J. Appl. Physiol. 35:236–243, 1973.

Welch, H.G. and W.N. Stainsby. Oxygen debt in contracting dog skeletal muscle *in situ*. Resp. Physiol. 3:229–242, 1967.

Welch, H.G., J.A. Faulkner, J.K. Barclay, and G.A. Brooks. Ventilatory response during recovery from muscular work and is relation with O_2 debt. Med. Sci. Sports. 2:15–19, 1970.

11

THE WHY OF PULMONARY VENTILATION

The movement of air in and out of the pulmonary system is called breathing or ventilation. Four specific purposes are accomplished by breathing: (1) the exchange of O_2, (2) the exchange of CO_2, (3) the control of blood acidity, and (4) oral communication. In general, breathing is essential to the cellular bioenergetic processes of life. Gases such as O_2 and CO_2 move from areas of high concentration (or pressure) to areas of lower concentration or pressure. The pressure exerted by a particular type of gas is termed *partial pressure*.

Because it is in the alveoli (air sacs) of the lung that O_2 and CO_2 are exchanged between the atmosphere and the blood, the process of ventilation results in a relatively higher partial pressure of O_2 in the lungs' alveoli than in the metabolizing tissues or the venous blood draining those tissues. Therefore, by keeping the alveolar partial pressure of O_2 at about 105 mmHg, ventilation results in a positive pressure gradient whereby O_2 moves from alveoli into the blood, which then circulates around the body to deliver O_2 to metabolizing tissues.

Through the action of ventilation, the partial pressure of the metabolic waste CO_2 is kept relatively low in the alveoli. This creates a negative pressure gradient for CO_2 to move from tissues, through blood and alveoli, to the atmosphere. The rapidity and depth of breathing affects the amount of O_2 and CO_2 exchanged between body and atmosphere. Metabolic CO_2 dissolved in body fluid forms carbonic acid, which then dissociates a proton or H^+ ($CO_2 + H_2O$

221

At the start of this Montreal Olympic 400 m individual medley final it is obvious that O_2 consumption capacity and breathing will affect the order of finish. Compared to other athletes who depend upon a supply of O_2 from the air, swimmers are faced with the problem of regulating breathing in accord with stroke mechanics; this is especially true for medley performers who must master four strokes and the complementary breathing patterns. (Alain Dejean/Sygma.)

$\rightleftarrows H_2CO_3 \rightleftarrows H^+ + HCO_3^-$). Because ventilation affects CO_2 exchange and storage, and because CO_2 affects H^+ concentration (or pH), then ventilation affects blood acid–base balance (i.e., the balance of acids and bases in the blood.)

BREATHING, VENTILATION, AND RESPIRATION

Respiration refers to the cellular utilization of O_2. During the eighteenth century the great French scientist Lavoisier and others believed that biological oxidation occurred in the lungs. For this reason, ventilation or breathing came to be known as respiration. By convention in the study of physiology the three terms ventilation, breathing, and respiration are used synonymously. Properly speaking, however, ventilation is the breathing of air in and out of the pulmonary system (nose, mouth, trachea, lungs), whereas respiration is the cellular utilization of O_2. Ventilation is but one step leading to respiration. Respiration is the major cellular mechanism of energy conversion.

Rhythmicity in Ventilation

We breath continually by means of a complex neural control mechanism. Sometimes, such as during heavy exercise, the vigor of ventilation makes breathing noticeable; usually, however, breathing goes on unnoticed. Breathing is usually controlled at an unconscious level, but sometimes we can volitionally modify the pattern of breathing. The trumpeter and the opera singer must breath continually, but they best coordinate ventilation with the musical requirement. Similarly, the swimmer must coordinate breathing into complex stroke mechanics. With practice, complex conscious breathing patterns can be learned and precisely integrated into motor performance. Breathing then becomes automatic and unconscious.

The movement of air into the pulmonary system (inhalation) and the movement out of previously inhaled air (expiration) is regulated in two basic ways, by frequency and by amount or volume. Within seconds after an inhalation, the partial pressure of O_2 in alveoli falls below its optimum, and the partial pressure of CO_2 becomes greater than optimal. Very shortly after beginning to hold a breath, the sensation to breathe again becomes very strong. This sensation is delayed only slightly if we take a large breath and hold it, as opposed to taking a small breath. It is a good thing that we ventilate so frequently, because the O_2 content of our lungs, inflated to maximum, can only sustain us a few minutes even at rest.

Pulmonary Minute Volume

The rate of pulmonary ventilation is usually expressed in terms of volume (liters) per minute. The abbreviation for pulmonary ventilation per minute (pulmonary minute volume) is $(\dot{V})$, where the V and dot refer to volume and per minute, respectively. Pulmonary ventilation can be measured either as volume expired per minute $(\dot{V}_E)$, or volume inspired per minute $(\dot{V}_I)$. Because the respiratory exchange ratio (R) is not necessarily 1.0, and because of the addition of H_2O vapor to inspired air, as well as the warming of inspired air during ventilation, $\dot{V}_I$ and $\dot{V}_E$ are not necessarily equal.

Pulmonary minute volume is equal to the product of frequency of breathing during a minute (f) and the average volume of air moved on each ventilatory excursion $(\bar{V}_T)$ (or tidal volume):

$$\dot{V}_E = (f)\ (\bar{V}_T) \tag{11-1}$$

ENVIRONMENTAL INFLUENCES ON PULMONARY GAS VOLUMES

Environmental conditions have significant effects on pulmonary gas volumes. At this point, the reader is referred to Appendix I for a review of pulmonary symbols and terminology. In addition, it is necessary to mention that there are three sets of conditions, or standards, by which pulmonary gas volumes can be defined. The first of these is the STPD volume, where ST = standard temperature = 0°C, P = standard pressure = 760 mmHg = 1 atmosphere (atm), and D = dry = 0.0 mmHg H_2O vapor pressure. Clearly, the STPD condition is nonphysiological, but it is a norm by which results obtained at different times and places can be compared.

The second pulmonary gas condition is the BTPS volume, where BT = body temperature, P = ambient pressure, S = saturated with H_2O. At approximately 1 atm pressure, the P_{H_2O} depends on temperature; for 37°C, the P_{H_2O} = 47 mmHg pressure. The BTPS volume is the volume that a subject actually exhales.

The third pulmonary gas condition of measurement is the ATPS volume, where AT = the ambient temperature in degrees Celsius, P = the ambient pressure, and S = the ambient P_{H_2O}. The ATPS volume is the volume that a subject actually inhales.

In expressing pulmonary gas volumes, $\dot{V}_{O_2}$ will usually be given in liters $\cdot$ min^{-1} (STPD), $\dot{V}_E$ will usually be given in liters $\cdot$ min^{-1} (BTPS), and $\dot{V}_I$ will usually be given in liters $\cdot$ min^{-1} (ATPS).

ENTRY OF O_2 INTO BLOOD

We ventilate to keep the partial pressure of O_2 in the alveoli at about 105 mmHg. From the alveoli, the diffusion distance for O_2 into erythrocytes (red blood cells) in the blood perfusing the alveolar walls is relatively short (Figure 11-1). This is necessary because the solubility of O_2 in body water at 37°C is low; only 0.3 ml $O_2 \cdot$ dl^{-1} blood is physically dissolved. Fortunately, erythrocytes contain the heme–iron compound hemoglobin, which can bind O_2 according to its partial pressure (Figure 11-2). At an O_2 partial pressure of 100

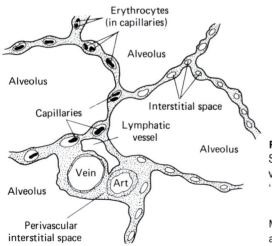

FIGURE 11-1

Schematic of a cross-sectional view through alveolar walls showing the relationship between the "alveolar air" and blood supply.

(SOURCE: Modified from A.C. Guyton, Textbook of Medical Physiology. Copyright W.B. Saunders Co., Philadelphia, 1976, p. 538. With permission.)

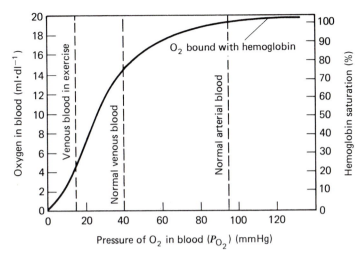

FIGURE 11-2
According to the oxygen–hemoglobin dissociation curve, the O₂ carried in blood in combination with hemoglobin depends mainly on the partial pressure of O₂. At normal arterial partial pressure of O₂ (P_{O_2} approximately 95 mmHg), hemoglobin is 95 to 98% saturated with O₂.

mmHg, as exists in alveolar capillaries at sea level, hemoglobin is nearly 100% saturated with O₂. Because a very small percentage of blood passing through the lungs passes through alveoli that are not ventilated, the saturation of blood returning to the left heart from the lungs is about 96%. This impressive figure is maintained not only in the individual resting at a sea level altitude, but also during maximal exercise (Figure 11-3).

Quantitatively, normal hemoglobin can bind 1.34 ml of $O_2 \cdot g^{-1}$. In the average male, blood hemoglobin is about 15 g $\cdot$ dl^{-1} blood. Arterial O₂ content (C_{aO_2}) is then equal to the sum of the dissolved O₂ plus that combined with hemoglobin:

$$\text{Arterial O}_2 \text{ content} = \text{O}_2 \text{ physically dissolved} + \text{O}_2 \text{ in} \qquad (11\text{-}2)$$
$$\text{combination with hemoglobin}$$

$$C_{aO_2} = 0.3 \text{ ml O}_2 \cdot \text{dl}^{-1} \text{ blood} + (1.34 \text{ ml O}_2 \cdot \text{g}^{-1} \text{ Hb})(15 \text{ g Hb} \cdot \text{dl}^{-1} \text{ blood})$$
$$= 20.4 \text{ ml O}_2 \cdot \text{dl}^{-1} \text{ blood}$$

By convention, this figure can also be referred to as 20.4 vol %, or 20.4 ml O₂/100 ml; note that 1 dl = 100 ml, and vol % = vol/100 ml.

In females where the blood hemoglobin concentration is less than that in males (about 13 g $\cdot$ dl^{-1} in females), the $C_{aO_2} = 17.7$ vol %.

Oxygen Transport

As we have just seen, if one knows the partial pressure of O₂ in arterial blood and the concentration of hemoglobin in the blood, then barring any unusual

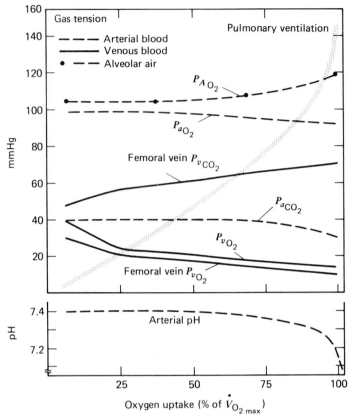

FIGURE 11-3
Oxygen and carbon dioxide partial pressures in alveolar air (P_A) arterial blood (P_a) and mixed venous blood (P_v) during rest and graded exercise. Note that as relative effort increases, the partial pressure of O_2 in arterial blood (P_{aO_2}) remains constant or falls only slightly. Due to the shape of the oxyhemoglobin dissociation curve (Figure 11-2), arterial O_2 content remains close to resting levels of approximately 95 to 98% saturation.

(SOURCE: Modified from P.-O. Åstrand and K. Rodahl, 1977. With permission.)

circumstances, the arterial O_2 content can be calculated. Further, knowing the cardiac output ($\dot{Q}$, the volume of blood ejected from the left ventricle each minute) the maximum O_2 transport capacity (TO$_2$) from heart (or lungs) to the rest of the body can be calculated:

For example, during rest,

$$\dot{T}O_2 = (C_{aO_2})\,(\dot{Q}) \tag{11-3}$$

$$O_2 \text{ transport} = (20 \text{ ml } O_2 \cdot dl^{-1} \text{ blood}) (50 \text{ dl blood} \cdot \text{min}^{-1})$$
$$= 1000 \text{ ml } O_2 \cdot \text{min}^{-1}$$

During maximal exercise in a young fit individual, O_2 transport can be much greater, for example,

$$= (20 \text{ ml } O_2 \cdot dl^{-1} \text{ blood}) (300 \text{ dl of blood } \cdot \text{ min}^{-1})$$
$$= 6000 \text{ ml } O_2 \cdot \text{ min}^{-1}$$

During rest, the actual whole-body O_2 consumption will be much less than the O_2 transport capacity, because mixed venous blood returning to the right heart will contain substantial amounts of O_2. During maximal exercise, as just illustrated for a very fit male, the actual O_2 consumption will approach the limits of the O_2 transport capacity. This is because most of the O_2 present in arterial blood is removed during each circulatory passage.

Effects of CO$_2$ and H$^+$ on O$_2$ Transport

In addition to the partial pressure of O_2 (Figure 11-2), other factors can affect the combination of O_2 with hemoglobin. These factors include temperature, pH, and concentration of 2,3-diphosphoglycerate (2,3-DPG).

The effects of elevated temperature and H^+ (lower pH) on the O_2 dissociation curve are given in Figure 11-4. This shifting of the dissociation curve down and to the right during exercise is termed the Bohr effect, and it facilitates the unloading of O_2 from hemoglobin of blood passing through active muscle beds.

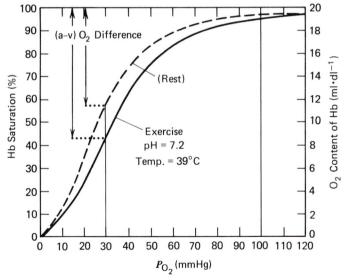

FIGURE 11-4
Decreases in blood pH and temperature such as experienced during exercise cause the oxyhemoglobin curve to shift down and to the right. This Bohr effect facilitates unloading of O_2 from hemoglobin in contracting muscles.

Erythrocytes are unique in that they are specialized for nonoxidative metabolism. In addition to the usual glycolytic enzymes, erythrocytes also possess the enzyme diphosphoglycerate mutase, which catalyzes the formation of 2,3-DPG from 1,3-DPG. Levels of 2,3-DPG are elevated during exercise, particularly during exercise at altitude. The binding of 2,3-DPG to hemoglobin occurs at a site that negatively affects the binding of O_2. Consequently, increased levels of 2,3-DPG cause a rightward shift in the oxyhemoglobin dissociation curve; 2,3-DPG is hydrolyzed to 3-phosphoglcyerate (a common glycolytic intermediate) by diphosphoglycerate phosphatase. Low blood P_{O_2} levels stimulate erythrocyte glycolysis and the formation of 2,3-DPG.

Because of the sigmoidal (S) shape of the O_2 dissociation curve (Figure 11-4), the Bohr effect during exercise will have a minimal impact on the combination of O_2 with hemoglobin in the lung. In active tissues, however, where the P_{O_2} is below 40 mmHg, the binding of O_2 to hemoglobin can be reduced to 10 to 15% as a result of the Bohr effect. This difference represents additional O_2 available in tissues to support metabolism.

THE RED BLOOD CELLS AND HEMOGLOBIN IN CO_2 TRANSPORT

In almost everyone's mind the erythrocyte (red blood cell) and hemoglobin are synonymous with oxygen transport. As described above, O_2 transport is a paramount role for the erythrocyte. In addition, the erythrocyte, which contains hemoglobin and the enzyme carbonic anhydrase, is crucial for CO_2 transport in the blood.

Carbon dixoide is a by-product of cellular respiration. Because the respiratory quotient of metabolism ($R = \dot{V}_{CO_2}/\dot{V}_{O_2}$) approximates unity, quantitatively the problem of CO_2 transport from cells to lungs is as great as transporting O_2 from lungs to body cells. The cellular formation and accumulation of CO_2 results in diffusion out of the cell, mostly in the gaseous form. Once reaching the capillary blood, the reactions diagrammed in Figure 11-5 occur rapidly. Relative to O_2, CO_2 is more soluble in the aqueous phase of the blood; however, only about 5 to 7% of the CO_2 is carried in the dissolved form.

The CO_2 that diffuses from plasma into red blood cells reacts with water in the erythrocyte to form carbonic acid. In the red blood cell this reaction is catalyzed by the enzyme carbonic anhydrase.

$$CO_2 + H_2O \overset{\text{carbonic}}{\underset{\text{anhydrase}}{\rightleftharpoons}} H_2CO_3 \qquad (11\text{-}4)$$

Carbonic anhydrase, the enzyme that wets CO_2, is not present in plasma, so although carbonic acid is also formed there from the physical interaction of CO_2 and H_2O, it is formed at a rate several thousand times slower than in the erythrocytes.

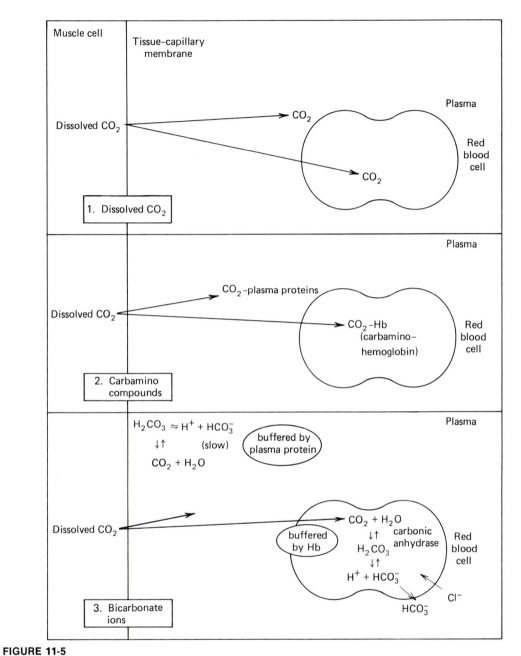

FIGURE 11-5
Carbon dioxide is transported in the blood by three mechanisms: (1) dissolved CO_2, (2) carbamino compounds, and (3) as bicarbonate ion. Hemoglobin and carbonic anhydrase enzyme in red blood cells are essential in the transport of CO_2 from sites of tissue formation to elimination in the lung.

Although the formation of carbonic acid (H_2CO_3) is slow and enzyme limited, the dissociation of carbonic acid to bicarbonate ion (HCO_3^-) and a proton is not enzyme catalyzed and proceeds spontaneously.

$$H_2CO_3 \rightleftharpoons HCO_3^- + H^+ \tag{11-5}$$

The HCO_3^- so formed diffuses from the erythrocyte into the plasma. The resulting chemical–electrical gradient causes a shift of chloride ion (Cl^-) into the erythrocyte (Figure 11-5). About 70% of all CO_2 transported from tissues to lungs is carried in the form of bicarbonate ion (HCO_3^-).

The H^+ so formed as the result of carbonic acid dissociation reacts rapidly with reduced hemoglobin (i.e., hemoglobin that has dissociated its O_2 molecule). The stoichiometry of hemoglobin O_2 and H^+ binding is interesting in that there is almost a one-to-one exchange of H^+ for O_2. In fact, reduced hemoglobin is such a strong buffer that it can take up almost all the H^+ formed as the result of CO_2 transport.

$$Hb^- + H^+ \rightleftharpoons HHb \tag{11-6}$$

The reaction goes to the right as presented here when the P_{O_2} is low (hemoglobin dissociates O_2) and when the P_{CO_2} and $[H^+]$ are high. In fact, elevated CO_2 and H^+ concentrations help to dissociate the oxyhemoglobin complex as part of the Bohr effect. In the lung, where P_{CO_2} is low and P_{O_2} is high, the reactions proceed from right to left. We shall discuss this shortly as part of the Haldane effect.

At present some controversy exists as to whether muscle contains the enzyme carbonic anhydrase. Generally it is believed that muscle and other tissues do *not* contain carbonic anhydrase. Additionally, because CO_2 diffuses through cell membranes much more rapidly than does the negatively charged bicarbonate ion, it would be a disadvantage to form bicarbonate intracellularly. However, on some occasions, such as in heavy exercise, H^+ and HCO_3^- do not appear in or disappear from the blood at identical rates, as would be the case if they were simultaneously formed there. If H^+ and HCO_3^- were formed outside the blood, they might then enter and leave it at different rates. This issue is at present receiving some study, and it is complicated by the fact that during heavy exercise metabolic acids such as lactic acid are formed that affect the kinetics of H^+, HCO_3^-, and CO_2.

Figure 11-5 also illustrates that CO_2 is carried in a third way in the blood. Some of the CO_2 entering the erythrocyte reacts directly with hemoglobin to form carbaminohemoglobin. Carbon dioxide reacts more readily with reduced hemoglobin than with oxyhemoglobin, although both reactions are possible. Some CO_2 also combines with plasma proteins. Together the carbamino compounds account for approximately 25% of CO_2 transported.

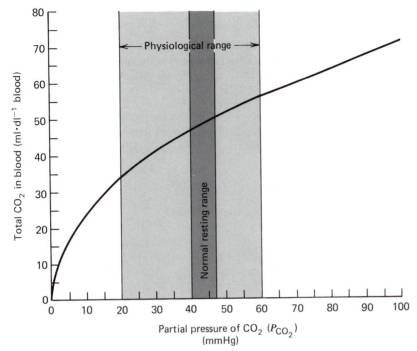

FIGURE 11-6

The CO_2 carried in the blood depends mainly on the partial pressure of CO_2. The carbon dioxide dissociation curve has a positive, almost linear slope within the physiological range.

CO_2 Content of Blood Depends on the P_{CO2}

Just as the combination of O_2 with hemoglobin can be expressed as a function of the P_{O_2} (Figure 11-2), the CO_2 content of blood can be described as a function of the P_{CO_2} (Figure 11-6). The normal resting P_{aCO_2} of 40 mmHg corresponds to an arterial CO_2 content (C_{aCO_2}) of about 48 vol %. During rest the mixed venous P_{CO_2} rises to about 45 mmHg, and the venous concentration of CO_2 (C_{vCO_2}) rises to about 52 vol %.

During rest the C_{aO_2} is about 20 vol %, and mixed venous O_2 content is about 16 vol %. Therefore the $(a - v)O_2$ is about 4 vol %. During maximal exercise the C_{aO_2} remains about constant, but the C_{vO_2} falls to about 4 vol %, so the $(a - v)O_2$ rises to about 16 vol %. During heavy exercise the respiratory exchange ratio ($R = \dot{V}_{CO_2}/\dot{V}_{O_2}$) reaches values of 1.0 or greater. This means that the $(v - a)CO_2$ reaches $\geq$ 16 vol %. As we will see, addition of acid to blood (e.g., a metabolic acid such as lactic acid) causes the P_{aCO_2} to decrease. During hard exercise, the entry of CO_2 into venous blood causes the P_{vCO_2} to increase (Figure 11-3). The physiological range of approximately 40 mmHg corresponds to a $(v - a)CO_2$ of approximately 20 vol % (Figure 11-6).

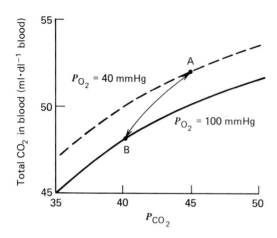

FIGURE 11-7
The effect of oxygenated hemoglobin on CO_2 transport in the blood (Haldane effect). As blood moves from tissues (where CO_2 tension is relatively high and O_2 is low) to the lung (where O_2 tension is high and CO_2 is lower), the CO_2 dissociation curve in effect shifts from A to B. This Haldane effect causes CO_2 to be unloaded at the lungs.

(SOURCE: Redrawn from A.C. Guyton, Textbook of Medical Physiology. Philadelphia: W.B. Saunders, 1976, p. 555, with permission.)

Effects of O_2 on Hemoglobin and CO_2 Transport

As described above for the Bohr effect, the pressure of CO_2 and H^+ can loosen the binding of O_2 with hemoglobin. Oxygen can also act to displace CO_2 and H^+ from hemoglobin. This displacement of CO_2 from the blood by O_2 (a reversal of the Bohr effect) is termed the Haldane effect.

The Haldane effect results from the fact that O_2 causes hemoglobin to dissociate hydrogen ions. Hemoglobin in effect becomes a stronger acid. The increase in acidity in erythrocytes and surrounding plasma brought about by a large increase in P_{O_2} in the alveolar capillary causes reversal of all the reactions diagrammed in Figure 11-5.

Figure 11-7 is an exploded version of Figure 11-6 modified to show the effects of tissue and alveolar O_2 partial pressures on the CO_2 dissociation curve. In actuality there is not one CO_2 dissociation curve, but rather a family of curves depending on the P_{O_2}. As mixed venous blood enters he alveolar capillary, the P_{CO_2} falls and the P_{O_2} rises. Therefore, CO_2 does not dissociate from the blood in a way described solely by the dashed curve in Figure 11-7, but the dissociation is actually described by the drop from point A on the dashed curve, to point B on the solid curve. This shift from one curve to another due to the increased P_{O_2}, as well as the effect of the reduced P_{CO_2}, causes a doubling of CO_2 dissociation from blood. Thus, the Haldane effect has a relatively larger impact on CO_2 transport than does the Bohr effect on O_2 transport.

THE BUFFERING OF METABOLIC ACIDS BY THE BICARBONATE BUFFER SYSTEM

The fact that most CO_2 is transported in forms other than as CO_2 means that the dissolved CO_2 in venous blood is less than might be expected. This is a

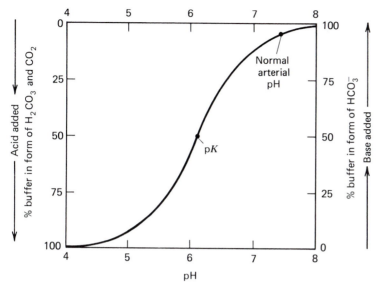

FIGURE 11-8

At its pK, a buffer system has the greatest resistance to additions of base or acid. Despite the fact that the pK of the bicarbonate buffer system (6.1) is removed from the normal arterial pH (7.4), the bicarbonate system works effectively to minimize the effects of acid or base added to the blood. This is because the blood CO_2 content can be varied through ventilation.

highly desirable situation for several reasons. Cellular CO_2 production is greater than the amount that can be transported in physical solution. As hemoglobin acts for O_2, there needs to be a carrier for CO_2. Important also is the fact that dissolution of CO_2 in H_2O causes the formation of an acid (H_2CO_3). However, by converting CO_2 to HCO_3^- and H^+ in the erythrocyte, by buffering the H^+ in hemoglobin, and by shifting the HCO_3^- into plasma, a mechanism is set up whereby the effects of metabolic acids, or bases, on plasma pH can be minimized (buffered).

Figure 11-8 describes the relationship between added acid (or base) to the bicarbonate buffer system that exists in plasma. By definition the system is half dissociated at the pK. Therefore, the pH changes least when acid or base is added to the system when the pH is at the pK. The pK of the bicarbonate system is 6.1. At another pH, such as the normal arterial pH of 7.4, added acid or base will have major effects on pH unless something else is done.

A buffer system consists of a mixture of an acid and its dissociated components. For the bicarbonate (carbonic acid) buffer system we have:

$$H_2CO_3 \overset{K'}{\rightleftarrows} H^+ + HCO_3^- \tag{11-7}$$

where K' is the dissociation constant for the system.

The rate of dissociation of carbonic acid depends on the ratio of products to reactants, and is given by the dissociation constant:

$$K' = \frac{[H^+] [HCO_3^-]}{[H_2CO_3]} \tag{11-8}$$

Therefore, the carbonic acid dissociation constant is given by the ratio of the product of hydrogen ion and bicarbonate concentration divided by the concentration of carbonic acid. Unfortunately, whereas it is practical to measure H^+ and HCO_3^-, it is very difficult to measure the H_2CO_3 concentration. Fortunately, scientists have determined that the H_2CO_3 concentration is directly proportional to the P_{CO_2}. Therefore, because P_{CO_2} is readily measurable, the H_2CO_3 concentration can be accurately estimated, or the P_{CO_2} may be used instead.

$$P_{CO_2} (mmHg) \times 0.03 = [H_2CO_3] \ (mM) \tag{11-9}$$

Because in Equation 11-8 it is the P_{CO_2} that is measured, Equation 11-8 can be rewritten as follows:

$$K = \frac{[H^+] [HCO_3^-]}{[CO_2]} \tag{11-10}$$

Because it is acidity that is usually the variable of interest, Equation 11-10 can be rearranged:

$$H^+ = K \frac{[CO_2]}{[HCO_3^-]} \tag{11-11}$$

Then, taking the logarithm of each side, we get:

$$\log [H^+] = \log K + \log \left(\frac{[CO_2]}{[HCO_3^-]} \right) \tag{11-12}$$

By convention the negative logarithm of the hydrogen ion concentration (i.e., $-\log [H^+]$) is termed the pH. Similarly, the $-\log K$ is termed the pK.

Therefore, by substituting pH and pK into Equation 11-13, and knowing that the pK of the bicarbonate buffer system is 6.1, we get:

$$pH = 6.1 + \log \left(\frac{[HCO_3^-]}{[CO_2]} \right) \tag{11-13}$$

This is called the Henderson–Hasselbalch equation; it describes how plasma pH is regulated by the relative, *not* absolute, quantities of bicarbonate ion and CO_2.

The Control of Blood pH by Ventilation

Inspection of Figure 11-8 reveals that the pK of the bicarbonate–carbonic acid buffer system (6.1) is quite removed from the normal arterial pH (7.4). At pH

7.4, the strength of the buffer system is weak, and small additions of acid or base could markedly affect blood pH. However, despite this deficiency the bicarbonate buffer system is an effective physiological buffer system, because the concentrations of the elements of the system can be regulated.

The main way that blood pH is regulated during metabolic transient periods such as during exercise is by ventilation. As noted at the outset of this chapter, ventilation is important for several reasons, one of them being regulation of blood pH. Ventilation affects pH by changing the CO_2 content of the blood. Arterial CO_2 content is inversely related to the pulmonary minute volume: Too little breathing causes CO_2 to build up; too much breathing causes CO_2 to be eliminated from the blood. For example, if a person volitionally hyperventilates at rest, the alveolar partial pressure of CO_2 (P_{ACO_2}) will fall and the P_{AO_2} will rise. These changes will cause CO_2 to move from blood to alveoli and be expired.

The loss of CO_2 by increased breathing (hyperventilation) will affect pH in two ways. First, CO_2 forms an acid in H_2O (i.e., H_2CO_3). Eliminating CO_2 in effect eliminates an acid from the blood. Additionally, according to the Henderson–Hasselbalch equation (11-13), the pH depends on a constant plus the logarithm of a ratio; CO_2 is in the denominator of that ratio. Therefore, if the denominator decreases, the ratio increases; eliminating CO_2 from the blood causes pH to increase. In actuality it must be realized that this description is a bit simplistic as the concentrations of H^+, HCO_3^- and CO_2 are interrelated. Therefore, if the P_{CO_2} falls as the result of hyperventilation, then HCO_3^- will ultimately also fall.

In the long term, blood pH is determined by those metabolic processes that form or remove metabolic acids, as well as by the kidney, which can regulate blood concentrations of H^+, HCO_3^-, and NH_4^+.

The Buffering of Metabolic Acids

During heavy exercise, production of several metabolic acids is increased, lactic acid in particular. Entry of lactic acid into the blood would cause a large drop in pH if not for the bicarbonate buffer system and the ventilatory regulation of that system. The action of the bicarbonate system in buffering lactic acid is illustrated by the following equations:

$$HLA \longrightarrow H^+ + LA^- \qquad (11\text{-}14)$$

$$\begin{array}{ccc} \text{Lactic} & \text{Hydrogen} & \text{Lactate} \\ \text{acid} & \text{ion} & \text{ion} \end{array}$$

$$H^+ + HCO_3^- \longrightarrow H_2CO_3 \qquad (11\text{-}15)$$

$$H_2CO_3 \longrightarrow H_2O + CO_2 \qquad (11\text{-}16)$$

In summary, lactic acid gives rise to CO_2. This CO_2 is termed "nonmetabolic CO_2," as it did not arise from the immediate combustion of a substrate.

The nonmetabolic CO_2, or its volume equivalent of CO_2, can be eliminated from the blood at the lung. In other words, the efforts of adding a strong acid to blood are lessened by forming a weaker acid (H_2CO_3) and then by eliminating the weaker acid as CO_2. In this way the R ($R = \dot{V}_{CO_2}/\dot{V}_{O_2}$) can exceed 1.0 during hard exercise when lactic acid enters the blood.

BREATHING FOR TALKING

In addition to the exchanges of O_2 and CO_2, and the control of blood pH, another major purpose of ventilation is for vocal communication. The details of control of speech by breathing are not necessarily germane to us here at the present. It is critical to note, however, that the control of speech (a voluntary activity) and breathing (a necessary activity) are tightly integrated. In fact, the control mechanism must be one and the same. This control mechanism will become apparent in Chapter 12.

Summary

Life depends on a constant flow of energy to the body's cells and a controlled conversion of that energy into useful forms. The main cellular mechanism of bioenergetics is respiration. Respiration is the proper term for biological oxidation. Therefore, life also depends on a continuous flow of O_2 to the body's cells. Entry of O_2 into the body begins by breathing O_2 into the lungs. Ventilating the lungs has the effect of raising the partial pressure of O_2 in the alveoli (i.e., the P_{AO_2}). Oxygen then diffuses from alveoli into blood where the O_2 combines with hemoglobin in a way determined mostly by the arterial P_{O_2} (i.e., the P_{aO_2}).

The gaseous by-products of respiration are H_2O and CO_2. It is important to eliminate most CO_2, because it is toxic in concentrations that are not really great. During ventilation of the lungs, the P_{ACO_2} is lowered relative to the tissue and venous P_{CO_2}. Therefore, CO_2 tends to diffuse toward the lungs, where CO_2 can be expired. In the blood, CO_2 is transported in three ways: in simple solution, as bicarbonate ion, and in union with hemoglobin (carbaminohemoglobin).

In addition to CO_2, other by-products of cellular metabolism include strong organic acids such as lactic acid. The effects of metabolic acids on blood pH can be lessened (buffered) by increasing the ventilatory rate to cause diffusion of CO_2 out of the blood. Because CO_2 forms carbonic acid in H_2O, the exit of CO_2 in effect makes room for another acid, such as lactic acid.

Breathing, then, has three critically important metabolic functions for exercise: the consumption of O_2, the elimination of CO_2, and the buffering of metabolic acids.

Selected Readings

American College of Sports Medicine. Symposium on ventilatory control during exercise. Med. Sci. Sports 11:190–226, 1979.

Åstrand, P.-O. and K. Rodahl. Textbook of Work Physiology. New York: McGraw-Hill, 1970. p. 185–254.

Cherniak, R.M. and L. Cherniak. Respiration in Health and Disease, Philadelphia: W.B. Saunders Co., 1961.

Comroe, J. Physiology of Respiration. Chicago: Year Book Medical Publishers, Inc., 1974.

Dempsey, J.A. and C.E. Reed (eds.). Muscular Exercise and the Lung. Madison: University of Wisconsin Press, 1977.

Dejours, P. Control of respiration in muscular exercise. In: Fenn, W.O. and H. Rahn (eds.), Handbook of Physiology, Section 3, Respiration Vol. 1. Washington, D.C.: American Physiological Society, 1964.

Dejours, P. Respiration. New York: Oxford University Press, 1966.

Fenn, W.O. and H. Rahn (eds.). Handbook of Physiology, Section 3, Respiration, Vols. I and II. Washington, D.C.: American Physiological Society, 1964.

Guyton, A.C. Textbook of Medical Physiology. Philadelphia: W.B. Saunders Co., 1976. p. 516–529.

Krogh, A. The Comparative Physiology of Respiratory Mechanisms. Philadelphia: University of Pennsylvania Press, 1941.

Lavoisier, A.L. and R.S. de La Place. Mémoire sur la Chaleur; Mémoires de l'Academie Royal (1780). Reprinted in Ostwald's Klassiker, No. 40. Leipzig: 1892.

McClintic, J.R. Physiology of the Human Body. New York: John Wiley and Sons, 1975. p. 208–215.

Miller, W.S. The Lung. Springfield, Ill.: Charles Thomas, 1947.

Otis, A. The work of breathing. In: Fenn, W.O. and H. Rahn (eds.), Handbook of Physiology, Section 3 Respiration, Vol. 1. Washington, D.C.: American Physiological Society, 1964.

Pappenheimer, J.R. Standardization of definitions and symbols in respiratory physiology. Fed. Proc. 9:602–605, 1950.

Riley, R. Pulmonary function in relation to exercise. In: Johnson, W. (ed.), Science and Medicine of Exercise and Sports. New York: Harper and Brothers, 1960. p. 162–177.

Vander, A.J., J.H. Sherman, and D.S. Luciano. Human Physiology, 3rd ed. New York: McGraw-Hill, 1980. p. 327–365.

West, J.B. Respiratory Physiology—The Essentials. Baltimore: Williams and Wilkins Co., 1974.

12

THE HOW OF VENTILATION

Our lungs represent a protected space wherein O_2 from the atmosphere can gain entry into the blood, and CO_2 in the blood can move to the atmosphere. By continually moving air in and out of the lungs from the atmosphere we can render the partial pressures of O_2 and CO_2 in the alveolar air spaces of the lung more like those in the atmosphere than partial pressures in the systemic venous circulation reaching the lungs from the tissues. Because gases move from areas of higher concentration or partial pressure to areas of lower partial pressure, the exchange of O_2 and CO_2 between "alveolar air" and the blood becomes possible. The process of ventilation, then, becomes a means to set up a diffusion gradient between the respiring cells in various tissues around the body and the atmosphere.

In addition to creating pressure gradients for the movement of respiratory gases (O_2 and CO_2), the ventilatory system and its allied cardiovascular system affect O_2 and CO_2 transport in other ways. One way is to minimize diffusion distances between alveoli and surrounding capillaries, and between capillaries and extra-alveolar cells. The small distance between alveolus and erythrocytes passing through is illustrated in Figure 12-2. Another way to optimize O_2 transport is to have carrier mechanisms whereby larger quantities of both O_2 and CO_2 can be transported in the blood than would be possible on the basis of physical solution.

The rate of ventilation is under control of the respiratory center in the me-

239

John Naber, Montreal Olympic backstroke champion, in action swimming freestyle. In swimming, as well as other muscular activities, ventilatory movements must be integrated with other movements to produce a successful performance. (Gilbert Uzan/Liaison.)

dulla of the brain. This respiratory center receives several different types of input that affect the center and the resulting pulmonary minute ventilation. Some inputs come from synaptic connections within the brain itself and are referred to as *neural* factors. Other inputs are blood borne and are called *humoral* factors.

When exercise starts, pulmonary ventilation accelerates first rapidly and then slowly. When exercise stops, ventilation decelerates first rapidly, then slowly. The rapid control of ventilation is by neural mechanisms; the slower control is by humoral mechanisms. These mechanisms operate to cause rhythmical ventilatory movements. These movements result in the pulmonary exchange of O_2 and CO_2, so that for the sea level resident the partial pressure of O_2 in arterial blood (P_{aO_2}) remains at about 100 mmHg even during maximal exercise, and the resting P_{aCO_2} of 40 mm Hg actually decreases during exercise.

PULMONARY ANATOMY

Gross anatomy of the pulmonary system is diagrammed in Figure 12-1*a*, and the terminal anatomy is diagrammed in Figure 12-1*b*. Although it is only in

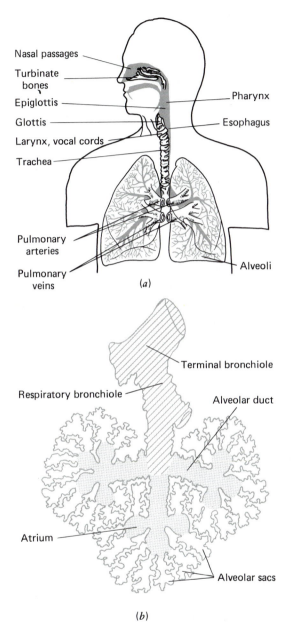

(a)

(b)

FIGURE 12-1
Structure of the ventilatory (respiratory) passages. (a) The gross anatomy.
(b) The respiratory lobule including the alveolar sites of gas exchange.

(SOURCE: (a) From A.C. Guyton, Textbook of Medical Physiology. Copyright W.B. Saunders, Co., Philadelphia, 1976, p. 526. With permission. (b) From W.S. Miller, The Lung, 1947, p. 42. With permission.)

the alveolar air sacs that O_2 and CO_2 are exchanged, the remaining upper respiratory tract (nose, mouth, trachea) performs important functions, including adding H_2O vapor to inspired air, warming (usually, or sometimes cooling) it to body temperature, and trapping particulate material (e.g., dust, yeast, and bacteria) as well as noxious fumes (e.g., smoke, ozone). The upper respiratory tract is extremely efficient in performing these tasks and is adequate to prevent all but the most overwhelming invasions from reaching the delicate alveolar membranes.

As suggested in Figure 12-1, the lower end of the pulmonary tree anatomy provides a large surface area for respiratory gas exchange between air (alveoli) and blood (pulmonary capillaries). The exchange diffusion of O_2 and CO_2 between an alveolus and one of its surrounding capillaries is illustrated in Figure 12-2. Because the cross-sectional area of the capillary is barely adequate to allow for passage of red blood cells (erythrocytes), and because no other structure intervenes between the outer walls of the alveolus and capillary, the diffusion distance is held to a minimum. This short distance facilitates the exchange of O_2 and CO_2. The exchange of respiratory gases, particularly O_2, is further facilitated by the large lipid content of the alveolar and capillary membrane walls, in which O_2 has a greater solubility than in H_2O.

MECHANICS OF VENTILATION

Movement of air into and out of the lungs is caused by changes in thoracic volume, which result in intrapulmonary pressure changes. The structures responsible for this bellowslike action are diagrammed in Figure 12-3. During rest, an inspiration begins with contraction of the diaphragm and external intercostal muscles. These actions "lower the floor" of the thorax and lift the ribs up and out. The volume of the thorax increases and the intrapulmonary pressure momentarily decreases. Atmospheric air moves into the pulmonary system to equilibrate the pressure gradient between lung and atmosphere.

During rest, expiration is a passive action wherein the diaphragm and external intercostals relax. The diaphragm recoils, moving up and "raising the floor" of the thorax, and the lung and ribs recoil to their original positions. The movements decrease volume of the thorax, transiently increasing intrapulmonary pressure and forcing pulmonary air out.

During exercise, inspiratory movements are assisted by *accessory inspiratory muscles,* which include the scalene, sternocleidomastoid, and trapezius muscles. These muscles act to lift the ribs and clavicles vertically and transversely, allowing for large increases in tidal volume (V_T) during exercise.

During exercise, expiration becomes an active (forced) movement. Contractions of the internal intercostals pull the ribs down and in, and contraction of abdominal muscles increases abdominal pressure, forcing the diaphragm up into

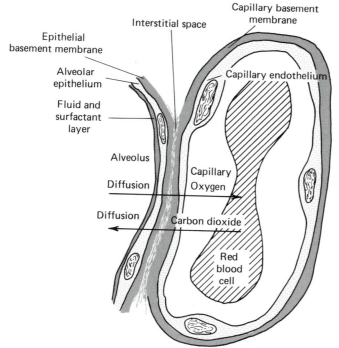

FIGURE 12-2

The ultrastructure of the alveolar capillary provides a minimum distance and mass of tissue between "alveolar air" and red blood cells in the pulmonary circulation.

(SOURCE: Modified from A.C. Guyton, Textbook of Medical Physiology. Copyright W.B. Saunders, Co., Philadelphia, 1976, p. 539. With permission.)

the thorax. The rapid and forceful movements of accessory ventilatory muscles during exercise greatly increase the maximal rate of ventilatory airflow. Consequently pulmonary minute flow ($\dot{V}_E$) can increase tremendously without a dramatic increase in breathing frequency (f). Typical changes in pulmonary minute volume, tidal volume, and breathing frequency during the transition from rest to exercise are given in Table 12-1. Note that by a relatively greater expansion in tidal volume (V_T) than in respiratory frequency (f), sufficient time is allowed for efficient gas exchange in the alveoli, and ventilation of respiratory dead space is minimized.

DEAD SPACE AND ALVEOLAR VENTILATION

Due to the anatomy of the pulmonary system (Figure 12-1), not all the inspired air reaches the alveoli where O_2 and CO_2 are exchanged. Therefore, the alveo-

TABLE 12-1
Pulmonary Ventilation at Rest and During Maximal Exercise in a Large, Healthy, Fit Male. Values in BTPS.

Condition	$\dot{V}_E$ (liters · min^{-1})	$\bar{V}_T$ (liters · breath^{-1})	f (breath · min^{-1})
Rest	6	0.5	12
Maximal exercise	192	4.00	48
Relative increase from rest to exercise	32 × Rest	8 × Rest	4 × Rest

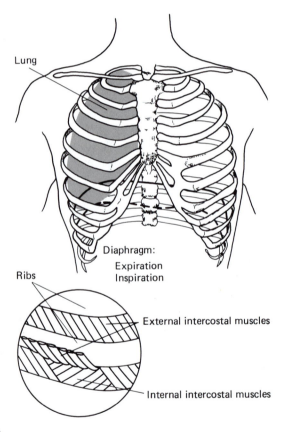

FIGURE 12-3
The major ventilatory muscles for breathing at rest include the diaphragm and external intercostal muscles. During the elevated breathing accompanying exercise (hyperpnea), other thoracic muscles including the sternocleidomastoid, scalene, and trapezius assist in ventilatory movements.

244

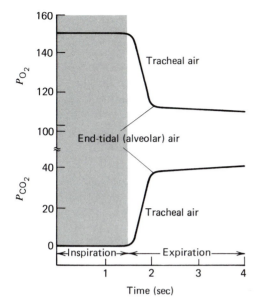

FIGURE 12-4
Partial pressures of O_2 and CO_2 measured in the mouth during a ventilatory cycle (inspiration and expiration) in a resting subject. The last air out during an expiration (i.e., the end-tidal air) represents the alveolar air. The precise values and timing of changes depend on exercise and environmental conditions.

lar minute ventilation ($\dot{V}_A$) will be less than the pulmonary minute ventilation ($\dot{V}_E$). The part of each breath that remains in the upper respiratory tract does not exchange and is called anatomical dead space (DS). Air in the dead space is warmed and humidified, but the relative concentrations of O_2 and CO_2 are like those in atmospheric air. Conversely, as indicated in Figure 12-4, the last air leaving the mouth during an expiration most closely reflects alveolar composition. In general, in a tidal volume the sequence is, last air in, first out—first in, last out. As Figure 12-4 indicates, the fractions of O_2 and CO_2 expired from the mouth during a breath change continuously until the alveolar air streams out. The last air out during a tidal volume (i.e., the *end-tidal air*) has the highest P_{CO_2} and lowest P_{O_2} of gas moved during a breath. This air reflects the alveolar composition.

Recall from Chapter 11 (Equation 11-1) that $\dot{V}_E = (f)\,(\bar{V}_T)$.

However, as just described, not all of the tidal volume will represent air entering the alveoli. The difference between tidal volume and alveolar volume ventilated is the dead space (Figure 12-5).

$$\dot{V}_A = (V_T - DS)\,(f) \qquad (12\text{-}1)$$

The anatomical dead space is not a fixed volume but does increase slightly during exercise as tidal volume increases. Bronchiolar dilatation and greater distances for air to flow within the lung prior to reaching the alveoli increase DS volume during exercise. As opposed to the rather small effect of increasing V_T on DS, the effect of increasing f on ventilation is relatively greater. This is

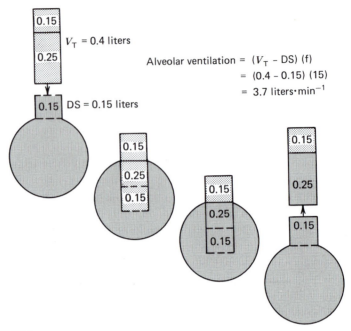

FIGURE 12-5

During each breath (tidal volume, V_T), inspired air enters structures in which no exchange of ventilatory gas is possible. This dead space volume (DS) comprises part of each tidal volume. Consequently, the alveolar minute ventilation equals the pulmonary minute ventilation less the minute dead space ventilation.

illustrated in Table 12-2 for a resting individual. Clearly the effects of rapid, shallow breathing (panting) is to cause $\dot{V}_A$ to be much less than $\dot{V}_E$.

In addition to the anatomical dead space, where respiratory gas exchange is not possible, there in addition exists some "physiological dead space." This space is represented by alveoli that receive blood flow inadequate to effect an equilibration of alveolar and pulmonary capillary blood. In the seated resting

TABLE 12-2

Effect of Panting on Alveolar Ventilation ($\dot{V}_A$) in a Resting Individual

Value	Breathing Normally	Panting
$\dot{V}_E$	6 liters · min^{-1}	6 liters · min^{-1}
V_T	0.4 liters · min^{-1}	0.2 liters · min^{-1}
f	15 breaths · min^{-1}	30 breaths · min^{-1}
$\dot{V}_A$	3.7 liters · min^{-1}	1.5 liters · min^{-1}

Ventilatory volumes in BTPS

person, not all the alveoli are open—particularly those at the top of the lung. Similarly, because blood reaching the lung tends to flow to the bottom of the lung under the influence of gravity, the lower lung tends to be better perfused than the top. Consequently, there can exist alveoli that are ventilated but not circulated. Such areas comprise the physiological dead space.

The pulmonary volumes measured during exercise (i.e., $\dot{V}_E$ and $\bar{V}_T$) are called *dynamic lung volumes*. These are opposed to the "static" lung volumes, which describes the pulmonary dimensions in resting individuals.

STATIC LUNG VOLUMES—PHYSICAL DIMENSIONS OF THE LUNGS

Static lung volumes (Figure 12-6a) are measured with a device called a spirometer (Figure 12-6b). To use the device illustrated in Figure 12-6b, called a displacement spirometer, a person breathes into and out of a chamber that consists of a counterbalanced cylinder suspended over a water seal. On exhalation, the person's breath raises the cylinder. On inhalation the cylinder falls. By means of a pen attached to the cylinder, a written record of ventilatory movements and volumes is obtained. The older displacement-type spirometer illustrated in Figure 12-6 is gradually being replaced with electronic devices called pneumotachometer.

In general, lung volumes are correlated with body size; lung volumes tend to be larger in tall people than in short people and larger in males than in females. Figure 12-6a illustrates that the normal resting tidal volume can be increased or decreased in size by expiring a bit more. This has the effect of increasing or decreasing the inspiratory reserve volume (IRV) or the expiratory reserve volume (ERV). When tidal volume is maximal, it is termed vital capacity (VC). Vital capacity is a commonly measured pulmonary parameter, as is the 1-sec timed vital capacity. In some types of pulmonary problems (e.g., emphysema), distensibility of the lungs is reduced, but the total volume is unaffected. Therefore, VC is normal, but the timed VC, or the percentage of VC achieved in 1 sec, is greatly reduced.

Even after a maximal expiration, some part of the air in the lungs cannot be forced out. This volume is termed the residual volume (RV). Together with the VC, the RV makes up the total lung capacity (TLC).

By means of a much larger spirometer than that illustrated in Figure 12-6, which has been filled with a mixture of 95% O_2 and 5% CO_2, the maximum amount of air a subject can ventilate during a minute can be measured. This volume is termed *maximum voluntary ventilation* (MVV). To perform the MVV test a subject wears a special breathing valve or mask in order to inspire from the spirometer and expire into the atmosphere. The MVV is measured by emptying the spirometer. The high O_2 during the MVV test is to supply O_2 for

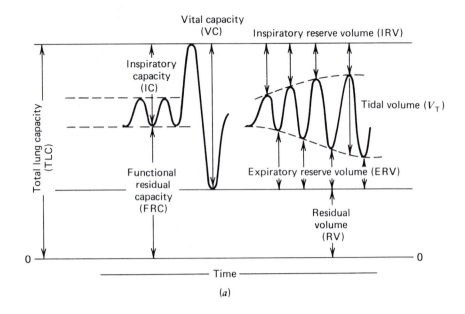

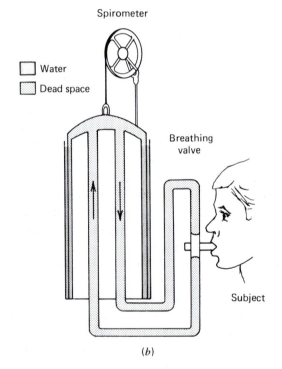

FIGURE 12-6
Lung volumes and capacities (a) can be measured during rest and exercise on devices called displacement spirometers (b). Both tidal volume and ventilatory frequency can range from rest to exercise.

248

respiration, and the high CO_2 is to prevent dizziness from a decrease in the P_{aCO_2}. In most individuals the MVV is much greater than the $\dot{V}_E$ observed during maximal exercise at sea level altitudes. In other words, even during maximal exercise, a ventilatory reserve exists. This ventilatory reserve allows people to exercise intensely and for prolonged periods, to inhabit very high altitudes (14,000 ft), occasionally to sojourn at altitudes exceeding 20,000 ft, and even to abuse their pulmonary systems through habits like smoking without immediate effects.

PHYSICS OF VENTILATORY GASES

Partial Pressures

The purpose of ventilating the pulmonary alveoli is to provide a place where the partial pressure of O_2 can be kept relatively high and the partial pressure of CO_2 can be kept low, so that these respiratory gases exchange with the blood. We have previously defined the partial pressure of a gas as the pressure exerted by that species of gas. The partial pressure of a dry gas depends on the total barometric pressure (P_B) and the fractionated composition (F_{O_2}) of that gas.

$$\text{Partial pressure of gas X in a dry environment} = \qquad (12\text{-}2)$$
$$(P_B \text{ mmHg}) \text{ (fractional composition of gas X)}$$

For oxygen,

$$P_{O_2} = (P_B \text{ mmHg}) (F_{O_2})$$
$$= 760 \text{ mmHg } (0.21)$$
$$= 159.6 \text{ mmHg}$$

This is illustrated in Figure 12-7. In a dry container at 1 atm pressure (760 mmHg) containing 100% O_2, the P_{O_2} will be 760 mmHg (Figure 12-7a). In a container containing 21% O_2 and 79% N_2, the partial pressures of O_2 and N_2 will approximate 160 and 600 mmHg, respectively (Figure 12-7b).

Water Vapor

In studying the movement of gas around the body, it is important to realize that the body is not a dry environment; water is always present and respiratory gases will exist either dissolved or in a gaseous environment saturated with H_2O vapor.

The partial pressure exerted by water is somewhat different from the partial pressure exerted by other ventilatory gases, in that the P_{H_2O} depends on environmental temperature. At body temperature (37°C), the P_{H_2O} is 47 mmHg. Therefore, as illustrated in Figure 12-8, if gas from a tank containing dry 100% O_2 is allowed to flush through and fill a wet container warmed to 37°C, then 47 mmHg of the 760 mmHg total pressure will be occupied by H_2O vapor. The diluting effect of water vapor will be to reduce the partial pressures of other gases present. Therefore, Equation 12-2 must be rewritten to include H_2O vapor.

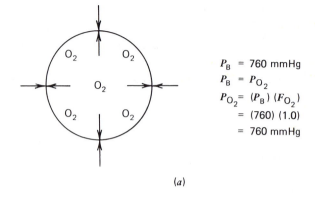

$$P_B = 760 \text{ mmHg}$$
$$P_B = P_{O_2}$$
$$P_{O_2} = (P_B)(F_{O_2})$$
$$= (760)(1.0)$$
$$= 760 \text{ mmHg}$$

(a)

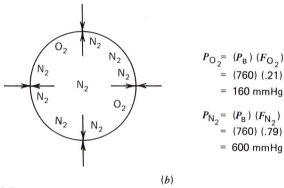

$$P_{O_2} = (P_B)(F_{O_2})$$
$$= (760)(.21)$$
$$= 160 \text{ mmHg}$$

$$P_{N_2} = (P_B)(F_{N_2})$$
$$= (760)(.79)$$
$$= 600 \text{ mmHg}$$

(b)

FIGURE 12-7
In a gas mixture, the total pressure is equal to the total of partial pressures exerted by the individual gases. For dry environments, the partial pressure exerted by a gas equals the total pressure times the fractional concentration of that gas. In environment (a), the partial pressure of O_2 equals the total pressure. In environment (b) the total pressure equals the total of the partial pressures exerted by O_2 and N_2.

$$\text{Partial pressure of gas X in a wet environment} = (P_B - P_{H_2O}) \quad (12\text{-}3)$$
$$\text{(fractional composition of gas X)}$$

The effect of H_2O vapor on the P_{O_2} of inspired tracheal air reaching the alveoli is:

$$P_{O_2} = (P_B - 47 \text{ mmHg})(F_{I_{O_2}})$$
$$P_{O_2} = (760 - 47 \text{ mmHg})(.21)$$
$$= 150 \text{ mmHg} \quad (12\text{-}4)$$

Similarly, for CO_2, which is 0.03% of the inspired air, the tracheal P_{CO_2} will be:

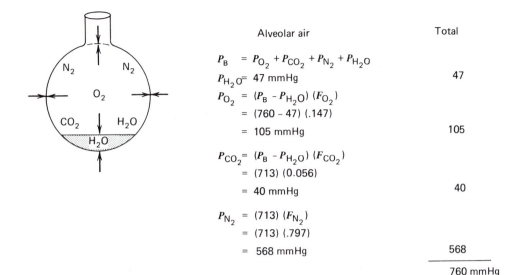

Alveolar air   **Total**

$$P_B = P_{O_2} + P_{CO_2} + P_{N_2} + P_{H_2O}$$

$$P_{H_2O} = 47 \text{ mmHg} \qquad 47$$

$$P_{O_2} = (P_B - P_{H_2O})(F_{O_2})$$
$$= (760 - 47)(.147)$$
$$= 105 \text{ mmHg} \qquad 105$$

$$P_{CO_2} = (P_B - P_{H_2O})(F_{CO_2})$$
$$= (713)(0.056)$$
$$= 40 \text{ mmHg} \qquad 40$$

$$P_{N_2} = (713)(F_{N_2})$$
$$= (713)(.797)$$
$$= 568 \text{ mmHg} \qquad \underline{568}$$

$$760 \text{ mmHg}$$

FIGURE 12-8
In a warm, moist environment (such as the alveolus), H_2O vapor saturates the environment, contributing significantly to the total pressure and diluting the relative fractional concentrations of O_2 and the other ventilatory gases. At a body temperature of 37°C, H_2O exerts a pressure of 47 mmHg. Carbon dioxide delivered to the alveolus from the venous circulation also dilutes the concentration of O_2.

$$P_{CO_2} = (760 - 47 \text{ mmHg})(0.0003) \qquad (12\text{-}5)$$
$$= 0.2 \text{ mmHg}$$

In the alveolus itself, the partial pressure of O_2 will be decreased as a result of the ongoing consumption of O_2 and the diluting effect of CO_2, and the P_{CO_2} will be higher because of the diffusion into the alveolus from the blood. In healthy resting individuals at sea level altitudes, the alveolar partial pressure of O_2 (P_{AO_2}) is about 105 mmHg, and the P_{ACO_2} is about 40 mmHg (Figure 12-8). As we will see, ventilation is controlled to maintain these partial pressures.

In the pulmonary system not only does H_2O move into the gas phase, but the respiratory gases O_2 and CO_2 also dissolve in the aqueous phases of fluids lining the alveoli and into the plasma (Figure 12-2). For movement of respiratory gases, it is important that the partial pressures of O_2 and CO_2 in the gas (alveolar) phase equilibrate with pressures in the aqueous (blood) phase. As illustrated in Figure 12-9, the molecules of a gas continuously move in all directions. Some of the molecules of a gas in contact with water will penetrate the surface and occupy spaces between the molecules as well as (in the case of CO_2) react chemically with the water. Those gas molecules entering the fluid phase dissolve in the fluid. Gas molecules dissolved in a fluid are not trapped

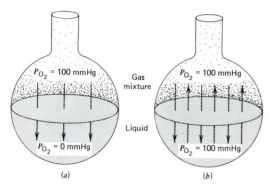

FIGURE 12-9
Given sufficient time, a gas in a closed environment will equilibrate across the gas–liquid interface. The partial pressure of the gas will then be equal in gaseous and liquid phases. O_2 admitted to a closed environment (a) will equilibrate in the liquid phase (b). By analogy, the sphere represents an alveolus, the O_2 is from inspired air, and the water represents body fluids.

(SOURCE: Modified from A.C. Guyton, Textbook of Medical Physiology. Copyright W.B. Saunders Co., Philadelphia, 1976, p. 531. With permission.)

there, but can move around within the fluid or leave it. Therefore the gas exerts a partial pressure both within and around the fluid. Given enough time at a particular temperature and pressure, the gas molecule in liquid and gas phases will come to equilibrium. At that point, the same number of molecules will be entering and leaving each phase for the other. At equilibrium, then, the partial pressures of the gas in liquid and gas phases will be equal. By this means, O_2 in alveolar gas can move through the body fluids to cellular mitochondria, and the reverse pathway for CO_2 is possible (Figure 12-10).

In the pulmonary capillaries, the equilibration of O_2 and CO_2 with alveolar air depends on the pressure gradient and the time the blood is in the capillary. During rest, blood (erythrocytes) are in the pulmonary capillary an average of 0.75 sec. This time period is referred to as the capillary transit time (Figure 12-11). This is adequate time for O_2 and CO_2 to equilibrate between the pulmonary capillary and alveoli. During maximal exercise, capillary transit time has been estimated to decrease to 0.4 to 0.5 sec. At sea level altitudes this is still more than adequate time for equilibration of CO_2, and marginally adequate time for equilibration of O_2. At higher altitudes, where the P_B decreases and with it also the P_{IO_2} and P_{AO_2}, the P_{aO_2} will necessarily decrease in comparison to sea level values of about 100 mmHg.

The amount of O_2 (in ml of O_2 and STPD) diffusing across the pulmonary membranes per minute per mmHg pressure difference between alveolar air and pulmonary capillary blood is defined as lung diffusing capacity (D_L). Diffusing capacity increases from rest to exercise and as the result of endurance training. However, the effect of altitude hypobaria (low pressure) is to decrease D_L.

RESPIRATION, CIRCULATION, AND VENTILATION

Because life depends on energy transduction, and because most of the body's cells depend on O_2 for energy transduction, a constant delivery of O_2 is neces-

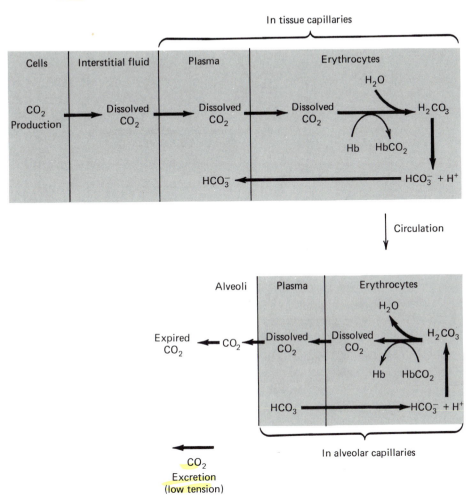

FIGURE 12-10

As CO_2 follows its pressure gradient (high to low), it flows from sites of cellular production to excretion in alveoli. Along the way, CO_2 passes through a number of spatial, fluid, and membrane barriers. Additionally, CO_2 flux through the body involves hemoglobin and the enzyme carbonic anhydrase (see Figure 11-5).

(SOURCE: Modified from A.J. Vander, J.H. Sherman, and D.S. Luciano, Human Physiology, McGraw-Hill Book Co., New York, 1975, p. 305. With permission.)

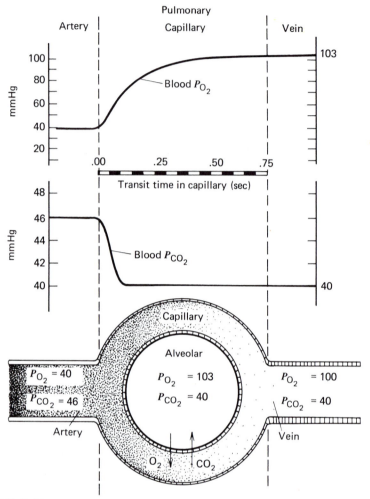

FIGURE 12-11

The exchange of O_2 and CO_2 in alveolar capillaries depends on time as well as partial pressure. Note in the figure that CO_2 equilibrates faster than O_2. During hard exercise at sea level altitudes there is usually sufficient time (0.5 sec) to complete the exchange.

(SOURCE: From P.-O. Åstrand and K. Rodahl, 1970. With permission.)

sary to sustain life. During exercise, energy (ATP) demands are increased and cell respiration must be accelerated. As depicted in Figure 12-12, continuous muscular activity depends on a flux of O_2 from the alveolus to the muscle. Conversely, CO_2 formed in muscle must flow around to the alveolus. Therefore, the circulatory system plays a critical role in sustaining cellular respiration. The pulmonary system provides an equally important role.

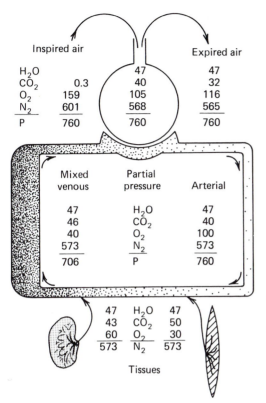

FIGURE 12-12
Typical values of gas tensions in inspired air, alveolar air, the blood, and expired air in a resting individual. Standard conditions are assumed for inspired air.

(**SOURCE**: Redrawn from P.-O. Åstrand and K. Rodahl, 1970. With permission.)

In Figure 12-12 the purpose of ventilating air in and out of the lungs is seen as the means of maintaining the P_{aO_2} slightly above 100 mmHg and the P_{aCO_2} at about 40 mmHg. Maintenance of this partial pressure depends on a continuous ventilation of the pulmonary system.

THE CONTROL OF ALVEOLAR VENTILATION

If a person is forewarned and can anticipate beginning to exercise, the ventilation will start to increase before the exercise starts (Figure 12-13). When the exercise starts, ventilation will increase very rapidly with a half-response time of 20 to 30 sec. After approximately 2 min, ventilation will stop increasing rapidly, but will increase at a slower rate. During submaximal exercise, $\dot{V}_E$ will plateau (stop rising) after 4 to 5 min (Figure 12-13). During maximal exercise, ventilation will continue until either MVV or the point of fatigue is reached.

The pulmonary minute volume reached during exercise will depend on a number of factors, including the work rate, the state of training, and the muscle

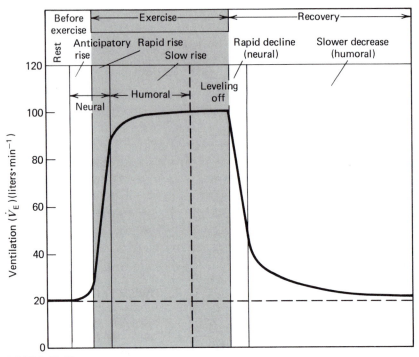

FIGURE 12-13
The neural–humoral control of ventilation during exercise. The elevated pulmonary ventilation during exercise (exercise hyperpnea) is controlled by at least two sets of mechanisms. One set acts very rapidly when exercise starts and stops. This mechanism has a neural basis and even causes ventilation to increase before exercise. The other set of factors that affect ventilation act slowly and result from the effects of blood-borne factors on the ventilatory (respiratory) center.

group used. Typical ventilatory volumes reached during leg exercise for trained and untrained individuals are illustrated in Figure 12-14. The relationship between $\dot{V}_E$ and $\dot{V}_{O_2}$ (or work rate) has two components. These are a linear rise followed by a curvilinear, accelerated increase in response to exercise work rates eliciting more than 50 to 75% of $\dot{V}_{O_2max}$ (Figure 12-14). In general, ventilation is higher in untrained than in trained subjects for given absolute and relative work loads. Ventilation is also higher when small muscles (e.g., arms) perform a given amount of work in comparison to the same work rate performed by larger muscles (e.g., legs).

In the following section, the factors that control ventilation during exercise are described. It would be inaccurate to state that our understanding of ventilatory control is complete. In particular, explaining the elevated ventilation (hyperpnea) during exercise has been one of the major challenges in modern phys-

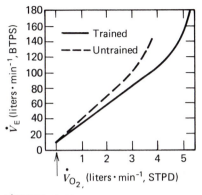 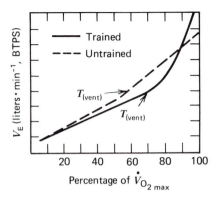

FIGURE 12-14
Relationship between pulmonary minute ventilation ($\dot{V}_E$) and metabolic rate ($\dot{V}_{O_2}$) during rest and exercise. In both untrained (broken line) and trained (solid line) subjects, $\dot{V}_E$ increases linearly with $\dot{V}_{O_2}$ up to about 50 to 75% of $\dot{V}_{O_2max}$. Thereafter, $\dot{V}_E$ increases at a rate disproportionately greater than the change in $\dot{V}_{O_2}$. Note that an effect of endurance training is to delay the ventilatory inflection point (T_{vent}) in trained individuals.

iology. Significant progress has been made, and we look forward to continued research on the regulation of exercise hyperpnea.

CONTROL OF VENTILATION: AN INTEGRATED, REDUNDANT NEURAL—HUMORAL MECHANISM

The neural center that controls ventilation (i.e., the respiratory center) is located in the lower brain (Figure 12-15). The center is designed to alternate inspiration and expiration rhythmically. The rate and amplitude of those ventilatory movements is under direct control of the respiratory center. Impinging on the center are a large number of neural and chemical (humoral) inputs. These inputs operate singularly and in concert to set the frequency and amplitude of output from the respiratory center. Not only are the various imputs into the respiratory center integrated there to produce a given output, but also the center appears to possess redundant mechanisms (Figure 12-16). Following nature's example, when human engineers design sophisticated machines for which there is sometimes little opportunity for repair (such as airliners and the NASA space shuttle), the engineers build in redundant control systems. In this way, if one system fails there will be at least one other—and probably several other— systems to take over. Redundancies in the human respiratory center, as well as its anatomical location, have made it very difficult to study. The way scientists usually study a system is to vary the inputs into the system and then see what happens to the output. Because the various inputs into the respiratory center

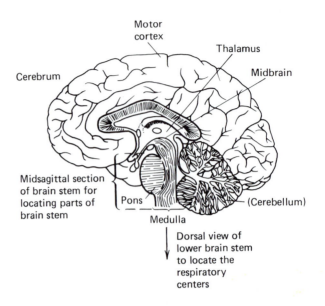

Motor cortex

Thalamus

Cerebrum

Midbrain

Midsagittal section of brain stem for locating parts of brain stem

Pons

(Cerebellum)

Medulla

Dorsal view of lower brain stem to locate the respiratory centers

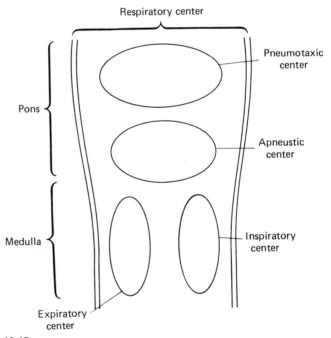

Respiratory center

Pneumotaxic center

Pons

Apneustic center

Medulla

Inspiratory center

Expiratory center

FIGURE 12-15
Schematic of a sagittal (cross) section through the brain showing parts of the respiratory center in relation to the rest of the brain.

(SOURCE: Modified from J.R. McClintic, Physiology of the Human Body, John Wiley & Sons, New York, 1975, p. 209. With permission.)

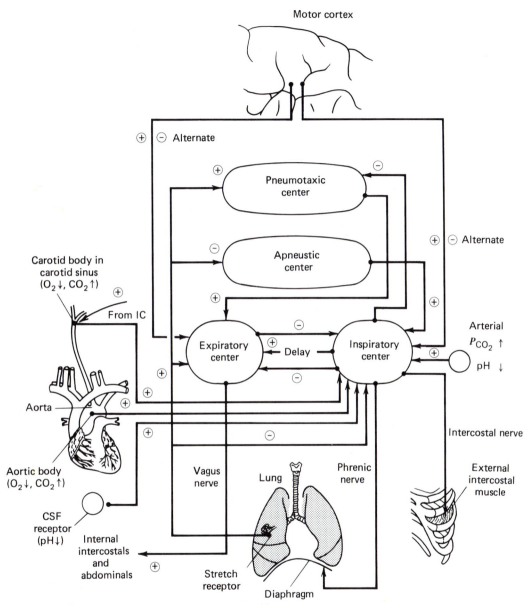

FIGURE 12-16
Schematic representation of the control of ventilation during rest and exercise in various environmental conditions. See text for description.

alter the relationships among the different inputs, and because the center can compensate for loss of a usual input, the output from the center will be different from the sum total actions of the individual inputs. This system of operation is most beneficial for us who must breathe continually under a wide range of circumstances. The complexity of the system, however, makes the respiratory physiologist's job a real challenge.

THE RESPIRATORY (VENTILATORY) CENTER

The respiratory center is divided into four different areas: (1) the medullary expiratory center, (2) the medullary inspiratory center, (3) the apneustic center, and (4) the pneumotaxic center. Both the apneustic and the pneumotaxic centers are located in the pons. Basic rhythmicity of the center is set in the medulla. If the brain of an anesthetized experimental animal is cut (transected) above *and* below the medullary area, breathing will be irregular or spasmodic, but inspirations will alternate with expirations. The apneustic center facilitates (stimulates) the inspiratory center. Transection between the apneustic and pneumotaxic centers (thereby eliminating the pneumotaxic center) results in emphasized (prolonged) inspiration, and short expiration. Conversely, the pneumotaxic center facilitates expiration. Obliteration of the apneustic center allows dominance of the pneumotaxic center. The resulting breathing pattern is characterized by prolonged expiration.

There are two populations of neurons in the medullary rhythmicity center—the inspiratory and expiratory neurons—and they tend to be localized in individual areas. However, the inspiratory and expiratory neurons intermingle as well as have interconnections with the apneustic and pneumotaxic centers. Rhythmicity of the respiratory center is set in the medullary neurons, which spontaneously generate action potentials.

In describing the functioning of any system that acts rhythmically, such as the ventilatory system, it is difficult to select a point at which to begin describing events within that system. Let us begin by describing events leading to an inspiration. Basically, the system works as follows.

Due to facilitatory inputs from other inspiratory neurons, from neurons in the apneustic center, and from central and peripheral sources, increased inspiratory neuronal activity summates to increased phrenic and intercostal nerve activity and contraction of muscles of inspiration (Figure 12-6). Because the inspiratory and expiratory neurons of the medullary center are reciprocally innervated, inspiratory center activity temporarily inhibits expiratory center activity. However, as inspiratory activity begins to wane, reciprocal inhibition of the expiratory neurons decreases also. Because of their own intrinsic rhythmicity, activity of the expiratory neurons increases. This activity is facilitated by synaptic interconnections from other expiratory neurons and from neuronal loops from the inspiratory center. Again, because of the reciprocal inhibitory interconnections between inspiratory and expiratory areas, neuronal discharge in the expiratory area leads to inhibition of inspiratory area activity. During rest, when expiration is passive, inhibition of the inspiratory area is sufficient to result in relaxation of the muscles of inspiration. During exercise when expiration is forced, accessory expiratory muscles are contracted. The ventilatory cycle will begin again as expiratory neuronal activity decreases and inspiratory neuronal activity increases.

CENTRAL INPUTS TO THE INSPIRATORY CENTER

Neural Input—The Motor Cortex

The motor cortex is primarily responsible for stimulating the respiratory center to achieve the elevated ventilatory rates seen in exercise. The motor cortex is responsible for the voluntary control of breathing that allows the individual speaking, singing, or playing a wind instrument to be successful. The motor cortex also allows movement and breathing patterns to be integrated during exercise. For instance, a swimmer must take large inspirations when the mouth and nose are out of the water, and expiration occurs under water. A wide range of breathing patterns are learned by swimmers who ventilate at high rates in a variety of stroke events (freestyle, butterfly, backstroke, breaststroke, etc.). During some sports activities, such as sprint running and weight lifting, ventilation must be inhibited.

The rapid increase in pulmonary ventilation with exercise, which is in fact preceded by an anticipatory rise (Figure 12-13), is the key to the long-standing hypothesis of cortical control over the respiratory center during exercise. The input from the motor cortex into the respiratory center during exercise is an example of a "feed-forward" mechanism in which both ventilation and work rate are proportional to output from the cortical controller. Admittedly there exist little direct data on output of the cortical area to the respiratory center during exercise in humans. The invasive procedures necessary to make such measurements preclude such study. However, there exists considerable experimental evidence obtained on anesthetized animal preparations that *suprapontine* (above the level of the pons) areas have major facilitatory effects on the respiratory center. It may be hoped that in the near future the development of several noninvasive techniques will reach the stage where cortical influences on the respiratory center can be documented on humans during exercise.

In addition to the motor cortex, other suprapontine areas that facilitate the medullary respiratory center include the hypothalamus, the cerebellum, and the reticular formation. The hypothalamus (Chapter 22) is particularly sensitive to increases in body temperature. Increases in body temperature during exercise have been correlated with hypothalamic activity and ventilatory patterns. In some species that cool themselves by panting (e.g., the dog), the hypothalamic effects on ventilation are pronounced. In humans, effects of the hypothalamic mechanisms are apparently not well developed and are probably not observable except during recovery from prolonged hard exercise, which elevates body temperature.

Whereas the cortex exerts feed-forward control over the respiratory center, the cerebellum provides feedback control. Afferents from contracting muscle reach the cerebellum, which then provides facilitatory input to the respiratory center.

The reticular formation is a diffuse subcortical, suprapontine area. The reticular formation is responsible for the general state of CNS arousal. Exercise is associated with heightened activity in the reticular formation. Facilitatory input to the respiratory center from the reticular formation is thought to amplify the ventilatory response to exercise.

Humoral Input—The Medullary Extracellular Fluid

Any substance that circulates in the blood and that has general effects at a site or sites removed from the secretion of the substance in termed a *humor*. The blood-borne substances are thought to be responsible for the slow phase of ventilatory adjustment to exercise (Figure 12-13).

The vast majority of the ventilatory response in resting individuals at sea level altitudes are mediated by the central chemoreceptors. Specialized cells on the ventral surface of the medulla, which are distinct from inspiratory and expiratory neurons, are sensitive to change in hydrogen ion (H^+) concentration. Some of these chemosensitive cells are influenced by the pH of the medullary interstitial fluid, whereas other cells are sensitive to pH of the cerebrospinal fluid (CSF).

Increases in medullary H^+ concentration (decreases in pH) stimulate ventilation. Increases in arterial H^+ concentration decrease medullary interstitial pH; this pH change in turn activates the chemosensitive cells that monitor interstitial pH.

Whereas medullary chemosensitive cells respond to cerebrospinal fluid (CSF) pH, the pH perturbations to CSF come, not from changes in arterial pH, but rather from changes in the P_{aCO_2}. Hydrogen ions diffuse slowly through the "blood–brain barrier" separating CSF and arterial blood. This barrier is rather more permeable to arterial CO_2. Because the buffer capacity of CSF is rather low, CO_2 penetrating the blood–brain barrier results in carbonic acid formation, which lowers the pH.

It should be emphasized that the central chemoreceptors sensitive to H^+ and CO_2 act to control ventilation during rest when the P_{aO_2} is around 100 mmHg. At other times, such as during exercise or acute exposure to high altitude, there is actually an alkalotic response in the medullary extracellular fluid. This increase in pH actually acts to restrain ventilatory response during exercise or acute hypobaric hypoxia.

PERIPHERAL INPUTS TO THE RESPIRATORY SYSTEM

Neural inputs to the medullary respiratory center that originate outside the brain are said to be peripheral to it. Peripheral inputs to the respiratory center include those from chemoreceptors, mechanical receptors, and baroreceptors.

Peripheral Chemoreceptors

The carotid bodies are located at the bifurcation of the common carotid artery into the internal and external carotids (Figure 12-16). Afferent signals from the carotid bodies reach the medullary respiratory center via the carotid sinus and glossopharyngeal nerves. The carotid bodies are stimulated by decreased P_{aO_2}, and the response is heightened by increased P_{aCO_2} and decreased pH. The carotid bodies are in an ideal position to detect a decrease in P_{aO_2} should, for some reason, the partial pressure of inspired O_2 (P_{IO_2}) decrease or pulmonary ventilation be limited. The P_{IO_2} of sea level residents could fall suddenly if they breathed air in which the O_2 content was lowered by the presence of a contaminating gas, or if they went up in altitude. The mechanical requirements of many sports activities such as swimming frequently impair pulmonary ventilation.

Although a fall in P_{aO_2} will rapidly stimulate the carotid bodies, the contribution (if any) of the carotid bodies to stimulating ventilation during exercise has remained a puzzlement. Breathing is so regulated during exercise that mean P_{aO_2} does not decrease (Figure 11-3). During exercise there are a range of normal arterial CO_2 responses, but one frequent response is for P_{aCO_2} to decrease. Therefore, except during transient situations such as at the start of exercise, the carotid bodies would not see an increase in the P_{aCO_2} or a decrease in P_{aO_2}.

During exercise arterial pH does decrease, and the P_{aO_2} does fluctuate subtly with the arterial pulse wave. These factors have been suggested as stimulating the carotid bodies during exercise. It is known that hypoxia exaggerates the CO_2 response of the carotid bodies in resting individuals. Therefore, it is possible that during exercise pulsatile variations in P_{aO_2} increase sensitivity of the carotid bodies to CO_2 and H^+. Additionally, under the influence of the motor cortex, during exercise the medullary respiratory center may be more influenced by input from the carotid bodies than during rest.

In addition to the carotid bodies, other arterial chemoreceptors exist in the aorta and brachiocephalic arteries. These "aortic bodies" are also sensitive to the P_{aO_2}, P_{aCO_2}, and pH; the aortic bodies send afferent signals to the medulla via the vagus nerve.

Other Peripheral Chemoreceptors

In addition to the arterial chemoreceptors, several other peripheral chemoreceptors have been postulated to exist. Because $\dot{V}_E$ and $\dot{V}_{CO_2}$ are so closely related whether at rest or during exercise, increased CO_2 flux to the lungs has long been postulated to be sensed by some receptor in the right heart, pulmonary artery, or lung itself. However, concerted efforts have failed to detect the existence of such a receptor. In the absence of such direct evidence, the high correlation between $\dot{V}_E$ and $\dot{V}_{CO_2}$ while the end-tidal CO_2 is maintained constant during exercise is taken as evidence that ventilation is well coordinated to metabolism. The high correlation between $\dot{V}_E$ and $\dot{V}_{CO_2}$ during exercise may in fact give a misleading impression that a pulmonary CO_2 receptor exists. The recent development of devices to measure change in $\dot{V}_E$, $\dot{V}_{CO_2}$, and $\dot{V}_{O_2}$ breath by breath have allowed investigators such as Hildebrandt and colleagues to study in detail the role of CO_2 flux to the lungs in controlling $\dot{V}_E$ during exercise. In the experiments of Hildebrandt and colleagues, subjects performed bicycle ergometer exercise with venous occlusion of legs with pressure cuffs. Following release of the cuffs, it took 5 to 10 sec for the CO_2 trapped in the legs to reach the lungs. That event was marked by an increase in the partial pressure of CO_2 in end-tidal air (P_{ETCO_2}). The increase in P_{ETCO_2} was followed 10 to 18 sec later by an increase in V_E. This time delay was sufficient for the pH effects of increased CO_2 load to reach and stimulate the arterial chemoreceptors. Had there been a pulmonary CO_2 receptor, V_E would have increased immediately as the P_{ETCO_2} rose.

The existence of muscle chemoreceptors has also been postulated. These muscle receptors are believed to be sensitive to the concentrations of local metabolites. Like the postulated pulmonary chemoreceptors, no anatomical or histological evidence of muscle chemoreceptors exists. Additionally, the great French physiologist Pierre Dejours has shown that ventilatory response following mild exercise with venous occlusion is no greater than ventilation during control rest periods. In that experiment, venous occlusion was achieved by an inflating blood pressure cuff placed around the legs. Those cuffs served to trap metabolites within contracting muscles and prevent their release into the circulation where their presence could be detected by the arterial chemoreceptors.

Peripheral Mechanoreceptors

When the lungs and chest wall expand rapidly, mechanical receptors there are stimulated; these mechanoreceptors transmit to the respiratory center signals that have the effect of inhibiting inspiration. This action is known as the Hering–Bruer reflex. Because mechanoreceptors are known to exist and because of the difficulty in identifying a main peripheral receptor that regulates ventilation-during exercise, scientists have looked for the presence of mechanical receptors that could be activated during exercise. The muscle spindles, Golgi tendon bodies, and skeletal joint receptors are known to send afferent signals to the sensory cortex, which relays information to the respiratory center. Evidence for peripheral mechanoreceptors was provided by Dejours, who used passive limb movement on awake as well as on lightly anesthetized human subjects to demonstrate an increase in $\dot{V}_E$ without an increase in $\dot{V}_{O_2}$. The lack of change in $\dot{V}_{O_2}$ was taken as evidence of no change in muscle metabolism. Furthermore, because the ventilatory response to passive limb movement was not blocked by venous occlusion, a potential role of chemoreceptors was eliminated.

Unfortunately, whereas peripheral mechanoreceptor stimulation has been observed to increase $\dot{V}_E$ in humans, the increase is small in comparison to the large and abrupt changes seen during exercise. Further, although results of some experiments on animal preparations support a role of peripheral mechanoreceptors in controlling ventilation during muscular work, other experiments do not.

CONTROL OF EXERCISE HYPERPNEA

The abrupt and very large change in ventilation that occurs as exercise starts and continues (Figure 12-13) has been and remains a difficult phenomenon for scientists to explain. For the present, the neuro-humoral theory of Dejours remains the best explanation. Basic ventilatory rhythmicity is set by the medullary respiratory center. When exercise starts, output of the respiratory center is increased tremendously by neural inputs from motor cortex, muscles, and joints. Once the broad range of ventilatory response is set by these neural components, the ventilatory response is fine-tuned by humoral factors that affect the peripheral and central chemoreceptors. When exercise stops, the exact opposite happens. Cortical and other neural inputs to the respiratory center cease, and ventilation slows dramatically. Then, as humoral disturbances wane, ventilation slowly returns to resting values.

Summary

Most of the body's cells require O_2 to sustain life. Cellular respiration produces CO_2 as a by-product. Because CO_2 is toxic in high concentrations, it must be removed. The atmosphere is not only the source of O_2; it is also the dump for metabolic CO_2. The sites of exchange of O_2 and CO_2 between the body and atmosphere are the alveoli in the lungs. Through the process of moving air into and out of the lungs, the partial pressure of O_2 is kept relatively high in the alveoli, and the partial pressure of CO_2 is kept low in comparison to the pressures that exist in most tissues. The anatomy of the pulmonary system, its allied circulatory system, and the peripheral tissue anatomy are designed to optimize the opportunity for transport and exchange of respiratory gases.

Ventilation of the lungs is under control of a respiratory center in the medulla of the brain. The respiratory center receives a variety of neural and humoral (chemical) inputs both directly (internally) and indirectly (externally) from receptors in the periphery. These inputs are integrated within the respiratory center to produce an appropriate ventilatory response, but the control circuits within the respiratory center's system of operation contain redundancies. Therefore, the respiratory center can usually produce an appropriate ventilatory response even if one or several inputs are cut off, or if the inputs are contradictory.

Physical exercise results in rapid and large ventilatory adjustments noted by both initial (rapid) and secondary (slow) response characteristics. When exercise stops, ventilation first slows abruptly and then declines slowly to resting levels. This fast–slow ventilatory response to the initiation and cessation of exercise is best explained by the "neural–humoral theory" of ventilatory control. During exercise, neural inputs to the respiratory center come from the motor cortex and from peripheral mechanoreceptors. Of these, the cortical inputs are most important. As exercise continues, humoral (chemical) factors such as fluctuating P_{O_2} and increased P_{CO_2} and $[H^+]$ are sensed in peripheral chemoreceptors. Of these the arterial chemoreceptors have been identified as being the most important for controlling ventilatory response to exercise. Chemoreceptors sensitive mainly to CO_2 have been hypothesized to exist in muscles and in the right heart or lung. Serious efforts to identify these receptors have failed to document their existence.

Selected Readings

Asmussen, E., E.H. Christensen, and M. Nielsen. Humoral or nervous control of respiration during muscular work? Acta Physiol. Scand. 6: 160–167, 1943

Asmussen, E. and M. Nielsen. Experiments on nervous factors controlling respiration and circulation during exercise employing blocking of the blood flow. Acta Physiol. Scand. 60: 103–111, 1964.

Astrand, P.-O. and K. Rodahl. Textbook of Work Physiology. New York: McGraw-Hill, 1970. p. 187–254.

Barcroft, H., V. Basnyake, O. Celander, A.F. Cobbold, D.J.C. Cunningham, M.G.M. Jukes, and I.M. Young. The effects of carbon dioxide in the respiratory response to noradrenaline in man. J. Physiol. London 137: 365–373, 1957.

Band, D.M., I.R. Cameron, and S.J.G. Semple. Oscillations in arterial pH with breathing in the cat. J. Appl. Physiol. 26: 261–267, 1969.

Band, D.M., I.R. Cameron, and S.J.G. Semple. Effect of different methods of CO_2 administration on oscillations of arterial pH in the cat. J. Appl. Physiol. 26: 268–273, 1969.

Barman, J.M., M.F. Moreira, and F. Consolazio. Effective stimulus for increased pulmonary ventilation during muscular exertion. J. Clin. Invest. 22: 53–56, 1943.

Barman, J.M., M.F. Moreira, and F. Consolazio. Metabolic effects of local ischemia during muscular exercise. Am. J. Physiol. 138:20–26, 1942.

Beaver, W.L., K. Wasserman, and B.J. Whipp. On-line computer analysis and breath-by-breath graphical display of exercise function tests. J. Appl. Physiol. 34: 123–132, 1973.

Belmonte, C. and C. Eyzaguirre. Efferent influences on carotid body chemoreceptors. J. Neurophysiol. 37: 1131–1143, 1974.

Biscoe, T.J. Carotid body: structure and function. Physiol. Rev. 51: 427–495, 1971.

Bisgard, G.E., H.V. Forster, B. Byrnes, K. Stark, J. Klein, and M. Manohor. Cerebrospinal fluid acid-base balance during muscular exercise. J. Appl. Physiol. 45: 94–101, 1978.

Byrne-Quinn, E., J.V. Weil, I.E. Sodal, G.F. Filley, and R.F. Grover. Ventilatory control in the athlete. J. Appl. Physiol. 30: 91–98, 1971.

Casaburi, R., B.J. Whipp, K. Wasserman, W.L. Beaver, and S.N. Koyal. Ventilatory and gas exchange dynamics in response to sinusoidal work. J. Appl. Physiol.: Respirat. Environ. Exercise. Physiol. 42: 300–311, 1977.

Chermak, R.M. and L. Cherniak. Respiration in Health and Disease. Philadelphia: W.B. Saunders Co., 1961.

Coles, D.R., F. Duff, W.H.T. Shepherd, and R.F. Whelan. The effect on respiration of infusions of adrenaline and noradrenaline into the carotid and vertebral arteries in man. Brit. J. Pharmacol. 11: 346–350, 1956.

Comroe, J.H. Physiology of respiration. Chicago: Year Book Medical Publishers, 1974, p. 234.

Cropp, G.J.A. and J.H. Comroe, Jr. Role of mixed venous blood PCO_2 in respiratory control. J. Appl. Physiol. 16: 1029–1033, 1961.

Cunningham, D.J.C., E.N. Hey, and B.B. Lloyd. The effect of intravenous infusion of noradrenaline on the respiratory response of carbon dioxide in man. Q.J. Exp. Physiol. 43: 394–399, 1958.

Cunningham, D.J.C., E.N. Hey, J.M. Patrick, and B.B. Lloyd. The effect of noradrenaline infusion on the relation between pulmonary ventilation and alveolar PO_2 and PCO_2 in man. Ann. N.Y. Acad. Sci. 109: 756–770, 1963.

Cunningham, D.J.C., M.G. Howson, and S.B. Pearson. The respiratory effects in man of altering the time profile of alveolar carbon dioxide and oxygen within each respiratory cycle. J. Physiol. London 234: 1–28, 1973.

Dejours, P. Control of respiration in muscular exercise. In: Fenn, W.O. and H. Rahn (eds.), Handbook of Physiology, Section 3, Respiration Vol. 1. Washington, D.C.: American Physiological Society, 1964.

Dejours, P., J.C. Mithoefer, and A. Teillac. Essai de mise en evidence de chemorecepteurs veineux de ventilation, J. Physiol. Paris 47: 160–163, 1955.

Dejours, P., J.C. Mithoefer, and J. Raynaud. Evidence against the existence of specific ventilatory chemoreceptors in the legs. J. App.. Physiol. 10: 367–371, 1957.

Dempsey, J.A., N. Gledhill, W.G. Reddan, H.V. Forster, P.G. Hanson, and A.D. Claremont. Pulmonary adaptation to exercise: effects of exercise type and duration, chronic hypoxia, and physical training. Ann. N.Y. Acad. Sci. 301: 243–261, 1977.

Dempsey, J.A., D.A. Pelligrino, D. Aggarwal, and E.B. Olson. The brain's role in exercise hypernea. Med. Sci. Sports. 11: 213–220, 1979.

Dempsey, J.A., E.H. Vidruk, and S.M. Mastenbrook. Pulmonary control systems in exercise. Fed. Proc. 39: 1498–1505, 1980.

Dutton, R.E. and S. Permutt. Ventilatory responses to transient changes in carbon dioxide. In: Torrance, R.W. (ed.), Arterial Chemoreceptors. Oxford, England: Blackwell, 1968. p. 373–386.

Eccles, J.C. Analysis of electrical potentials evoked in the cerebellar anterior lobe by stimulation of hind and forelimb nerves. Exper. Brain Res. 6: 171–194, 1968.

Eldridge, F.L. Maintenace of respiration by central neural feedback mechanisms. Fed. Proc. 36: 2400–2404, 1974.

Eldridge, F.L. Central neural stimulation of respiration in unanesthetized decerebrate cats. J. Appl. Physiol. 40: 23–28, 1976.

Eldridge, F.L., D.E. Millhron, and T.G. Waldrop. Exercise hypernea and locomotion: Parallel activation from the hypothalamus. Science 211: 844–846, 1981.

Flandrois, R., R. Fravier, and J.M. Pequignot. Role of adrenaline in gas exchange and respiration in the dog at rest and exercise. Resp. Physiol. 30: 291–303, 1977.

Fordyce, W.E., F.M. Bennett, S.K. Edelman, and F.S. Grodins. Evidence in man for a fast neural mechanism during the early phase of exercise hyperpnea. Respir. Physiol. 48: 27–43, 1982.

Freund, P.R., S.F. Hoggs, and L.B. Rowell. Cardiovascular responses to muscle ischemia in man—dependency on muscle mass. J. Appl. Physiol: Respirat. Environ. Exercise Physiol. 45: 762–767, 1978.

Ganong, W.R. Review of Medical Physiology. Los Altos, Cal.: Lange Medical Publications, 1979. p. 517–524.

Gonzalez, F., Jr., W.E. Fordyce, and F.S. Grodins. Mechanism of respiratory responses to intravenous NaHCO3, HCl, and KCN. J. Appl. Physiol.: Respirat. Environ. Exercise Physiol. 43: 1075–1079, 1977.

Guyton, A.C. Textbook of Medical Physiology. Philadelphia: W.B. Saunders, 1981. p. 516–528.

Hagberg, J.M., E.F. Coyle, J.E. Carroll, J.M. Miller, W.H. Martin, and M.H. Brooke. Exercise hyperventilation in patients with McArdle's disease. J. Appl. Physiol.: Respirat. Enivron. Exercise Physiol. 52: 991–994, 1982.

Heistad, D.D., R.C. Wheeler, A.L. Mark, P.G. Schid, and F.M. Abboud. Effects of adrenergic stimultion on ventilation in man. J. Clin. Invest. 51: 1469–1475, 1972.

Henry, J.D. and C.R. Bainton. Human core temperature increase as a stimulus to breathing during moderate exercise. Respir. Physiol. 21: 183–191, 1974.

Henry, J.P. and W.V. Whitehorn. Effect of cooling of orbital cortex on exercise hyperpnea in the dog. J. Appl. Physiol. 14: 241–244, 1959.

Hildebrandt, J.R., R.K. Winn, and J. Hildebrandt. Cardiorespiratory response to sudden releases of circulatory occlusion during exercise. Resp. Physiol. 38: 83–92, 1979.

Hughes, R.L., M. Clode, R.H.T. Edwards, T.J. Goodwin, and N.L. Jones. Effect of inspired O_2 on cardiopulmonary and metabolic responses to exercise in man. J. Appl. Physiol. 24: 336–347, 1968.

Joels, N. and H. White. The contribution of the arterial chemreceptors to the stimulation of respiration by adrenaline and noradrenaline in the cat. J. Physiol. London 197: 1–24, 1968.

Jones, N.L., D.G. Robertson, and J.W. Kane. Difference between end-tidal and arterial PCO_2 in exercise. J. Appl. Physiol.: Environ. Exercise Physiol. 47: 954–960, 1979.

Kao, F.F., C.C. Michel, S.S. Mei, and W.K. Li. Somatic afferent influence on respiration. Ann. N.Y. Acad. Sci. 109: 696–708, 1963.

Krogh, A. and J. Linhard. The regulation of respiration and circulation during the initial stages of muscular work. J. Physiol. London 47: 112–136, 1913.

Krogh, A. and J. Linhard. A comparison between voluntary and electrically induced muscular work in man. J. Physiol. London 51: 182–201, 1917.

Lewis, S.M. Awake baboon's ventilatory response to venous and inhaled CO_2 loading. J. Appl. Physiol. 39: 417–422, 1975.

Linton, R.A.F., R. Miller, and I.R. Cameron. Ventilatory response to CO_2 inhalation and intravenous infusion of hypercapnic blood. Respir. Physiol. 26: 383–394, 1976.

Majcherczyk, S., J.C.G. Coleridge, H.M. Coleridge, M.P. Kaufman, and D.G. Baker. Carotid sinus nerve efferents: properties and physiological significance. Federation. Proc. 39: 2662–2667, 1980.

Majcherczyk, S. and P. Willshaw. Inhibition of peripheral chemoreceptor activity during superfusion with an alkaline c.s.f. of the ventral brainstem surface of the cat. J. Physiol. London 231: 26P–27P, 1973.

Majcherczyk, S. and P. Willshaw. The influence of hyperventilation on efferent control of peripheral chemoreceptors. Brain Res. 124: 561–564, 1977.

Miller, W.S. The Lung. Springfield: C.C. Thomas, 1947.

Mitchell, R.A. and A. Berger. State of the art: Review of neural regulation of respiration. Am. Rev. Respir. Dis. 111: 206, 1975.

Nice, L.B., J.L. Rock, and R.O. Courtight. The influence of adrenaline on respiration. Am.. J. Physiol. 34: 326–331, 1914.

Noguchi, H., Y. Ogushi, I. Yoshiya, N. Itakura, and H. Yambayashi. Breath-by-breath VCO_2 and VO_2 require compensation for transport delay and dynamic response. J. Appl. Physiol. 52: 79–84, 1982.

Pearce, D.H., H.T. Milhorn, G.H. Holloman, and W.J. Reynolds. Computer-based system for analysis of respiratory responses to exercise. J. Appl. Physiol.: Respirat. Environ. Exercise Physiol. 42: 968–975, 1977.

Ponte, J. and M.J. Purves. Frequency response of carotid body chemoreceptors in the cat to changes of $PaCO_2$, PaO_2, and pHa. J. Appl. Physiol. 37: 635–647, 1974.

Purves, M.J. The effect of eliminating fluctuations of gas tensions in arterial blood on carotid chemoreceptor activity and respiration. J. Physiol. London 186: 63P, 1975.

Rowell, L.B., L. Hermansen, and J.R. Blackmon. Human cardiovascular and respiratory responses to graded muscle ischemia. J. Appl. Physiol. 41: 693–701, 1976.

Salmoiraghi, G.C. Functional organization of brain stem respiratory neurons. Ann. N.Y. Acad. Sci. 109: 571–585, 1963.

Sargeant, A.J., M.Y. Rouleau, J.R. Sutton, and N.L. Jones. Ventilation in exercise studied with circulatory occlusion. J. Appl. Physiol. 50: 718–723, 1981.

Segal, S.S. and G.A. Brooks. Effects of glycogen depletion and work load on postexercise O_2 consumption and blood lactate. J. Appl. Physiol.: Respirat. Environ. Exercise Physiol. 47: 514–521, 1979.

Senapati, J. The role of the cerebellum in the hyperpnea produced by muscle afferents. In: Paintal, A.S. and P. Gill-Kumar (eds.), Respiratory Adaptation, Capillary Exchange and Reflex Mechanisms. New Delhi: Navchetan Press (Private) Limited, 1977, p. 115–120.

Strange-Petersen, E., B.J. Whipp, D.B. Drysdale, and D.J.C. Cunningham. Carotid arterial blood gas oscillations and the phase of the respiratory cycle during exercise in man: Testing a model. Adv. Ex. Med. Biol. 99: 335–342, 1978.

Sutton, J.R., R.L. Hughson, R. McDonald, A.C.P. Powles, N.L. Jones, and J.D. Fitzgerald. Oral and intravenous propranolol during exercise. Clin. Pharmacol. Thera. 21: 700–705, 1975.

Sylvester, J.T., B.J. Whipp, and K. Wasserman. Ventilatory control during brief infusions of CO_2-laden blood in the awake dog. J. Appl. Physiol. 35: 178–186, 1973.

Thimm, F. and U. Tibes. Effect of K^+, osmolality, lactic acid, orthophosphate and epinephrine on muscular reseprtors with group I, III, and IV afferents. J. Physiol. London 284: 182P–183P, 1978.

Tibes, U., B. Hemmer, and D. Boning. Heart rate and ventilation in relation to venous K^+, osmolality, pH, PCO_2, PO_2, orthophosphate, and lactate at transition from rest to exercise in athletes and nonatheletes. Europ. J. Appl. Physiol. 36: 127–140, 1977.

Wasserman, D.H. and B.J. Whipp. Coupling of ventilation in pulmonary gas exchange during nonsteady-steady work in man. J. Appl. Physiol.: Respirat. Environ. Exercise Physiol. 54: 587–593, 1983.

Wasserman, K., B.J. Whipp, R. Casaburi, D.J. Huntsman, J. Castagna, and R. Lugliani. Regulation of PCO_2 during intravenous CO_2 loading. J. Appl. Physiol. 36: 651–656, 1975.

Wasserman, K., B.J. Whipp, and J. Castagna. Cardiodynamic hyperpnea: hyperpnea secondary to cardiac output increase. J. Appl. Physiol. 36: 457–464, 1974.

Wasserman, K., B.J. Whipp, S.N. Koyal, and M.G. Cleary. Effect of carotid body resection of ventilatory and acid-base control during exercise. J. Appl. Physiol. 39: 354–358, 1975.

Weissman, M.L., B.J. Whipp, D.J. Huntsman, and K. Wasserman. Role of neural afferents from working limbs in exercise hyperpnea. J. Appl. Physiol.: Environ. Exercise Physiol. 49: 239–248, 1980.

Whelan, R.F. and I.M. Young. The effect of adrenaline infusion on respiration in man. Br. J. Pharmacol. 8: 98–102, 1953.

Whipp, B.J. and J.A. Davis. Peripheral chemoreceptors and exercise hyperpnea. Med. Sci. Sports 11: 204–212, 1979.

Whipp, B.J., D.J. Huntsman, and K. Wasserman. Evidence for a CO_2-mediated pulmonary chemoreflex in dog. (abstract) Physiologist 18: 447, 1975.

Whipp, B.J. and K. Wasserman. Carotid bodies and ventilatory control dynamics in man. Federation Proc. 39: 2668–2673, 1980.

Willshaw, P. Mechanism of inhibition of chemoreceptor activity by sinus nerve efferents. In: Acker, H., S. Fidone, E. Pallot, C., Eyzaguirre, and D.W. Lubbers (eds.), Chemoreception in the Carotid Body. New York: Springer-Verlag, 1977. p. 168–172.

Willshaw, P. and S. Majcherczyk. The effects of changes in arterial pressure on sinus nerve efferent activity. Adv. Exp. Med. Biol. 99: 275–280, 1978.

Yamamoto, W.S. Mathematical analysis of the time course of alveolar CO_2. J. Appl. Physiol. 15: 215–219, 1960.

Yamamoto, W.S. and M.W. Edwards, Jr. Homeostasis of carbon dioxide during intravenous infusion carbon dioxide. J. Appl. Physiol. 15: 807–818, 1960.

Young, M.I. Some observations on the mechanism of adrenalin hyperpnea. H. Physiol. London 137: 374–395, 1957.

Zuntz, N. and J. Geppert. Ueber die Natur der normalen Atomreize un den Ort ihrer Wirkung. Arch. Ges. Physiol. 38: 337–338, 1886.

13

VENTILATION AS A LIMITING FACTOR IN AEROBIC PERFORMANCE

Ventilation is not usually considered to be the factor limiting aerobic performance at altitudes close to sea level. The term *aerobic performance* is sometimes defined as the ability to endure hard and prolonged tasks that are of submaximal intensity. Sometimes aerobic performance is also defined as $\dot{V}_{O_2max}$. Whether aerobic performance is defined in terms of submaximal endurance or short-time capacity to achieve $\dot{V}_{O_2max}$, the process of ventilation is sufficiently robust to continue at high rates for prolonged periods, oxygenating the blood passing through the lungs.

The capacity to increase ventilation during exercise is relatively much greater than the body's capacity to increase cardiac output or oxygen consumption. The alveolar surface area is large in comparison to the pulmonary blood volume; the alveolar partial pressure of O_2 (P_{AO_2}) increases during exercise, and the arterial partial pressure of O_2 (P_{aO_2}) is maintained close to resting levels even during exercise eliciting $\dot{V}_{O_2max}$ at sea level altitudes. A considerable ventilatory reserve exists to oxygenate blood passing through the lungs, and this reserve allows us to perform effectively at altitudes significantly above sea level.

Athletes from several nations attempt to qualify for the Moscow Olympic 5,000 m. final. Aerobic performances, such as this, are seldom limited by ventilatory factors. Excessive heavy breathing during prolonged, hard but sub-maximal efforts likely signal inadequacy of some other system besides ventilation. (Simon/Francolon/Liaison.)

VENTILATORY PERFUSION RATIO ($\dot{V}_E/\dot{Q}$) DURING REST AND EXERCISE

When scientists attempt to understand the limitations in a system such as the O_2 transport system, they frequently attempt to identify the factor or step that limits the system. Finding the rate-limiting step in a series of linked reactions is analogous to finding the weak link in a chain that must support a great weight. In the overall scheme of O_2 transport from the atmosphere to tissues, the process of pulmonary ventilation is not generally considered to be limiting at normally inhabited altitudes near sea level.

As exercise work rate increases from resting levels up through easy to moderate intensities, pulmonary minute ventilation ($\dot{V}_E$) increases linearly (Figure 12-14). However, as exercise intensity becomes more severe, a ventilatory threshold exists beyond which further increments in exercise work rate or $\dot{V}_{O_2}$ result in exaggerated, nonlinear increments in $\dot{V}_E$. Resting values of pulmonary minute ventilatory volume (5 liters $\cdot$ min^{-1}, BTPS) can increase to values of around 190 liters $\cdot$ min^{-1} in a healthy young adult male during exercise. This represents a 35-fold increase. As exercise intensity increases, the volume of blood flowing through (perfusing) the pulmonary vessels also increases. However, this increase in cardiac output is essentially linear from resting levels up through those achieved during maximal exercise (Figure 13-1). In absolute terms, resting cardiac output (5 liters · min^{-1}) can increase five or six times (25 to 30 liters · min^{-1}) during exercise in a young, healthy, and fit adult male.

The ratio of pulmonary minute ventilation to cardiac output (i.e., the ventilation : perfusion ratio, $\dot{V}_E/\dot{Q}$) approximates unity during rest in most individuals. In the robust individual just described, the $\dot{V}_E/\dot{Q}$ may increase five- to sixfold during the transition from rest to maximal exercise. Because the capacity for expanding $\dot{V}_E$ (range 5 to 190 liters · mins^{-1}) is far greater than the capacity for expanding $\dot{Q}$ (range 5 to 30 liters · mins^{-1}), the $\dot{V}_E/\dot{Q}$ increases during exercise. The increase in $\dot{V}_E/\dot{Q}$ seen during exercise is usually presented as one reason pulmonary ventilation is not thought to limit aerobic performances. In individuals less fit than the one just described, the absolute increments in $\dot{V}_E$ and $\dot{Q}$ achieved during exercise will be less, but the $\dot{V}_E/\dot{Q}$ will expand to a similar extent.

THE VENTILATORY EQUIVALENT OF O_2 DURING EXERCISE

An argument similar to that on the $\dot{V}_E/\dot{Q}$ is used with the ventilatory equivalent of O_2 (i.e., $\dot{V}_E/\dot{V}_{O_2}$) to exclude pulmonary ventilation as a factor limiting aero-

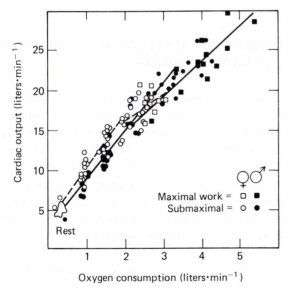

FIGURE 13-1
Relationship between cardiac output and oxygen consumption during rest and exercise. Given physical limits for arterial O_2 content and arterial venous O_2 difference [$(a-v)O_2$], maximal O_2 consumption ($\dot{V}_{O_2max}$) is largely a function of the ability to increase cardiac output.

(**SOURCE:** Redrawn from P.-O. Åstrand and K. Rodahl, 1970. With permission.)

bic performance. During rest, $\dot{V}_E$ approximates 5 liters $\cdot$ min^{-1}, whereas $\dot{V}_{O_2}$ can be as low as 0.25 liters $\cdot$ min^{-1}; the $\dot{V}_E/\dot{V}_{O_2}$ is 20. During maximal exercise, for the individual we are describing, $\dot{V}_{Emax}$ may expand to 190 liters $\cdot$ min^{-1}, and $\dot{V}_{O_2max}$ might be 5 liters $\cdot$ min^{-1}; the $\dot{V}_E/\dot{V}_{O_2}$ increases to 35. Therefore, the ability to expand ventilation is relatively greater than the ability to expand oxidative metabolism.

$\dot{V}_{Emax}$ VERSUS MVV DURING EXERCISE

Pulmonary minute ventilation can increase tremendously from rest to maximal exercise (Figure 12-14). The greatest $\dot{V}_E$ observed in an individual during exercise, however, is usually less than the maximum voluntary ventilatory (MVV, Chapter 12) capacity of that individual. When an athlete turns into the home stretch in finishing a track race, he or she will be breathing very heavily and will probably feel a degree of breathlessness. However, if that athlete were to volitionally attempt to increase $\dot{V}_E$ still further, he or she could probably do that. The fact that a maximal $\dot{V}_E$ observed during exercise is less than the MVV is another reason ventilation is not thought to limit aerobic performance at sea level altitudes.

PARTIAL PRESSURES OF ALVEOLAR (P_{AO_2}) AND ARTERIAL OXYGEN (P_{aO_2}) DURING EXERCISE

The real test of adequacy of pulmonary ventilation rests with the partial pressures of O_2 in alveoli (P_{AO_2}), and arterial blood (P_{aO_2}) during exercise. As described in Chapters 11 and 12, O_2 transport around the body is accomplished only because O_2 (like other gases) moves from areas of high concentration (or partial pressure) to areas of lower partial pressure. Respiratory gas exchange is accomplished through ventilation by maintaining the alveolar partial pressure of O_2 high, and that of CO_2, low. Adequacy of pulmonary ventilation in maintaining the partial pressure of O_2 in alveoli during exercise up to $\dot{V}_{O_2max}$ is illustrated in Figure 11-3. If anything, increased ventilation during exercise raises the P_{AO_2}.

The partial pressure of O_2 (P_{aO_2}) in systemic arterial blood (i.e., blood that has circulated through the lungs and heart, and into the aorta) is also well maintained during exercise (Figure 11-3). This apparently results from several major factors. First, the P_{AO_2} is maintained or increased. Second, the erythrocytes passing through the pulmonary capillaries remain there sufficiently long for equilibration with alveolar O_2 (Figure 12-11). Third, the sigmoid shape of the O_2 dissociation curve (Figure 11-2) is such that when the P_{O_2} is about 100 mmHg, the saturation of hemoglobin with O_2 does not fall off markedly even if the P_{O_2} falls off several millimeters of mercury.

ALVEOLAR SURFACE AREA FOR EXCHANGE

As noted earlier, the alveolar surface area estimated to exist in the average-sized person is 50 m², or 35 times the surface area of the person (Chapter 11). This is the area represented by one half of a singles tennis court. Keeping in mind that the average blood volume is 5 liters, picture the following scene. You are trying to spread the liquid contents of a 5-quart container over one side of the tennis court. How far do you think you could get in spreading the liquid? Could you cover the one side? Now consider that in reality less than 10% of the blood volume is the pulmonary system at any one instant. Therefore, the ratio of alveolar surface area to pulmonary capillary blood volume is enormous. This disproportionality ensures a large capacity for exchange of respiratory gases between blood and alveolar air during ventilation.

FATIGUE OF VENTILATORY MUSCLES AND OTHER LIMITATIONS IN VENTILATION

Any muscle, including the diaphragm and accessory ventilatory muscles, can be made to fatigue. When diaphragm muscle preparations isolated from experimental animals are electrically stimulated to contract at high rates, the muscles demonstrate distinct fatigue characteristics. Additionally, during a MVV test, human subjects will usually demonstrate a degree of fatigue and not be able to ventilate at as high a rate at the end of the test as at the beginning of the test. Maximum voluntary ventilation tests repeated in rapid succession will also usually produce decreasing values. The central question regarding ventilatory muscle fatigue during exercise, however, is whether ventilatory fatigue precedes or coincides with and results in decrements in body exercise performance. The answer to this question is usually no. As noted previously, $\dot{V}_{Emax}$ during exercise is usually less than MVV, even MVV determined after exercise. Further, most athletes can volitionally raise their ventilatory flow at the end of exercise.

The issue of adequacy of the ventilatory system during maximal exercise in very fit individuals has recently been raised by Dempsey and colleagues at the University of Wisconsin. They observed decreases in P_{aO_2} when some subjects were stressed to $\dot{V}_{O_2max}$. Because the P_{aO_2} decreased even though $\dot{V}_E$ approached MVV, the pulmonary system did not adequately oxygenate blood passing through it. The authors suspected that compliance in the ventilatory system prevented $\dot{V}_E$ from rising sufficiently to maintain arterial O_2 concentration. Compliance is a measure of expansibility of the lungs and thorax. These structures do provide a resistance to ventilatory movements and ventilatory flow. Consequently, there is a limitation to ventilatory flow imposed by the internal resistance of the pulmonary tissues. The University of Wisconsin scientists have apparently observed that in some very fit individuals, $\dot{V}_E$ during exercise paproaches a limit imposed by the mechanics of pulmonary compliance. How general this finding is among top aerobic athletes needs to be determined. However, for the present it must be considered unlikely that normal asymptomatic individuals ever achieve a $\dot{V}_E$ during exercise limited by compliance.

Summary

The mechanisms for control of ventilation are redundant in their operation and ensure adequate response under numerous conditions (Chapter 12). Similarly, the robust capacity of the pulmonary system usually ensures that O_2 transport from atmosphere to metabolizing tissues is not limited by ventilation. The ventilatory perfusion ratio ($\dot{V}_E/\dot{Q}$) increases several times during exercise, as does the ventilatory equivalent of O_2 ($\dot{V}_E/\dot{V}_{O_2}$). The alveolar partial pressure of O_2 (P_{AO_2}) is maintained or increased during exercise, and the partial pressure of O_2 in the systemic arterial blood (P_{aO_2}) is maintained at or close to normal resting levels. The rather flat shape of the oxyhemoglobin dissociation curve at partial pressures of 95 to 105 mmHg ensures that even if P_{aO_2} decreases slightly, the O_2 content of arterial blood (C_{aO_2}) will not decrease appreciably.

This robust, "overbuilt" ventilatory capacity allows us not only to sustain high rates of oxidative metabolism at sea level altitudes, but also to withstand additional stresses imposed by unique situations. For instance, the limitations of stroke mechanics may limit ventilatory frequency during swimming; yet swimmers are perhaps the premier aerobic athletes. The ventilatory reserve allows athletes such as soccer and basketball players to converse even during very heavy exercise. Whereas most of the earth's surface is covered by water, and much of the remaining land surface is not much above sea level, significant land masses exist at relatively high altitudes. The ventilatory reserve that we humans possess allows us to inhabit areas as high as 14,000 ft altitude and to sojourn at even higher altitudes. Mt. Everest has recently been climbed without benefit of auxiliary O_2 supplies. In subsequent chapters we will describe other systems and factors that limit O_2 utilization during exercise.

Selected Readings

Anderson, P., and J. Henriksson. Capillary supply to the quadriceps femoris muscle of man: adaptive response to exercise. J. Physiol. London 270: 677–690, 1977.

Åstrand, P.-O, T.E. Cuddy, B. Saltin, and J. Stenberg. Cardiac output during submaximal and maximal work. J. Appl. Physiol. 19: 268–273, 1964.

Åstrand, P.-O. and B. Saltin. Maximal oxygen uptake and heart rate in various types of muscular activity. J. Appl. Physiol. 16: 977–981, 1961.

Åstrand, P.-O. and K. Rodahl. Textbook of work Physiology, New York: McGraw-Hill, 1970. p. 154–178, 187–254.

Bannister, R.G. and C.J.C. Cunningham. The effects on the respiration and performance during exercise of adding oxygen to the inspired air. J. Physiol. London. 125: 118–120, 1954.

Barclay, J.K. and W.N. Stainsby. The role of blood flow in limiting maximal metabolic rate in muscle. Med. Sci. Sports 7: 116–119, 1975.

Bevegård, S., A. Holmgren, and B. Jonsson. Circulatory studies in well trained athletes at rest and during heavy exercise, with special reference to stroke volume and the influence of body position. Acta Physiol. Scand. S7:26, 1963.

Bergh, U., I.-L. Kaustrap, and B. Ekbloom. Maximal oxygen uptake during exercise with various combinations of arm and leg work. J. Appl. Physiol. 41: 191–196, 1976.

Bishop, J.M., K.W. Harold, S.W. Taylor, and P.N. Wormald. Changes in arterial-hepatic venous oxygen content difference during and after supine leg exercise. J. Physiol. London, 137: 309–314, 1957.

Booth, F.W. and K.A. Narahara. Vastus lateralis cytochrome oxidase activity and its relationship to maximal oxygen consumption in man. Pflügers Archiv. 369: 319–324, 1974.

Buick, F.J., N. Gledhill, A.B. Frosese, L. Spriet, and E.C. Meyeis. Effect of induced erythrocythemia on aerobic work capacity. J. Appl. Physiol.: Respirat. Environ. Exercise Physiol. 48: 636–642, 1980.

Byrne-Quinn, E., J.V., Weil, I. Sodal, G.F. Fillez, and R.F. Grover. Ventilatory control in the athlete. J. Appl. Physiol. 30: 91–98, 1971.

Davies, C.T.M. and A.J. Sargent. Physiological response to one- and two-leg exercise breathing air and 45% oxygen. J. Appl. Physiol. 36: 142–148, 1974.

Davies, K.J. A., J.L. Maguire, G.A. Brooks, P.R. Dallman, and L. Packer. Muscle mitochondrial bioenergetics, oxygen supply, and work capacity during dietary iron deficiency and repletion. Am. J. Physiol. 242 (Endocrinol. Metab. 5): E418–E 427, 1982.

Dempsey, J.A., N. Gledhill, W.G. Reddan, H.V. Forester, P.G. Hanson, and A.D. Claremont. Pulmonary adaptation to exercise: effects of exercise type and duration, chronic hypoxia and physical training. Ann. N.Y. Acad. Sci. 301: 242–261, 1977.

Dempsey, J.A., P.E. Hanson, and S.M. Mastenbrook. Arterial hypoxemia during heavy exercise in highly trained runners. Federation Proc. 40: 932, 1981.

Ekbloom, B., A.N. Goldbarg, and B. Gullbring. Response to exercise after blood loss and reinfusion. J. Appl. Physiol. 33:175–189, 1972.

Ekelund, L.G. and A. Holmbren. Circulatory and respiratory adaptation during long-term, non-steady state exercise in the sitting position. Acta Physiol. Scand. 62: 240, 1964.

Ekelund, L.G. and A. Holmbren. Central hemodrynamics during exercise. In: Chapman, C.B. (ed.), Physiology of Muscular Exercise, American Heart Association Monograph No. 15. New York: American Heart Association, 1967. p. I 33–I 43.

Faulkner, J.A., D.E. Roberts, R.L. Elk, and J. Conway. Cardiovascular responses to submaximum and maximum effort cycling and running. J. Appl. Physiol. 30: 457–461, 1971.

Gardner, G.W., V.R. Edgerton, B. Senewirathe, R.J. Barnard, and Y. Ohira. Physical work capacity and metabolic stress in subjects with iron deficiency anaemia. Am. J. Clin. Nutr. 30: 910–917, 1977.

Gleser, M.A., D.H. Horstman, and R.P. Mello. The effects of $\dot{V}O_2$ max of adding arm work to maximal leg work. Med. Sci. Sports 6: 104–107, 1974.

Grimby, G.E. Haggendal, and B. Saltin. Local xenon 133 clearance from the quadriceps muscle during exercise in man. J. Appl. Physiol. 22: 305–310, 1967.

Hanson, P., A. Claremont, J. Dempsey, and W. Reddan. Determinants and consequences of ventilatory responses to competitive endurance running. J. Appl. Physiol.: Respirat. Environ. Exercise Physiol. 52: 615–623, 1982.

Henriksson, J. and J.S. Reitman. Time course of changes in human skeletal muscle succinate dehydrogenase and cytochrome oxidase activities and maximal oxygen uptake with physical activity and inactivity. Acta Physiol. Scand. 99:91–97, 1977.

Holmgren, A. and P.-O. Åstrand. Pulmonary diffusing capacity and the dimensions and functional capacities of the oxygen transport system in humans. J. Appl. Physiol. 21: 1463–1467, 1966.

Holmgren, A. and M.B. McIlroy. Effect of temperature on arterial blood gas tensions and pH during exercise. J. Appl. Physiol. 19: 243, 1964.

Holmgren, A., B. Jonsson, and T. Sjöstrand. Circulatory data in normal subjects at rest and during exercise in the recumbent position with special reference to the stroke volume at different intensities. Acta Physiol. Scand. 49: 343, 1960.

Hughes, R.L., M. Clode, R.H. Edwards, T.J. Goodwin, and N. Jones. Effect of inspired O_2 on cardiopulmonary and metabolic responses to exercise in man. J. Appl. Physiol. 24: 336–347, 1968.

Leith, D.E. and M.E. Bradley. Ventilatory muscle strength and endurance training. J. Appl. Physiol. 41:508–516, 1976.

Martin, B.J., K.E. Sparks, C.W.Z. Willich, and J.V. Weil. Low exercise ventilation in endurance athletes. Med. Sci. Sports 11:181–185, 1979.

Mitchell, J.H., B.J. Sproule, and C.B. Chapman. Physiological meaning of the maximal oxygen uptake test. J. Clin. Invest. 37: 538, 1958.

Pirnay, F., M. Lamy, J. Dujarding. R. Deroanne, and J. M. Petit. Analysis of removal venous blood during maximum muscular exercise. J. Appl. Physiol. 33: 289–292, 1972.

Pirnay, F.R., Marechal, R. Radermecker, and J.M. Petit. Muscle blood flow during submaximum and maximum exercise on a bicycle ergometer. J. Appl. Physiol. 32: 210–212, 1972.

Saltin, B., R.F. Grover, C.G. Blomquist, L.H. Hartley, and R..J. Johnson. Maximal oxygen uptake and cardiac output after 2 weeks at 4,300 m. J. Appl. Physiol. 25: 406–409, 1968.

Sutton, J.R. and W.L. Jones. Control of pulmonary ventilation during exercise and mediators in the blood: CO_2 and hydrogen ion. Med. Sci. Sports 11: 198–203, 1979.

Secher, N.H., N. Ruberg-Larsen, R. Binkhorst, and F. Bonde-Peterson. Maximal oxygen uptake during arm cranking and combined arm plus leg exercise. J. Appl. Physiol. 36: 515–518, 1974.

Taylor, H.E., E.R. Buskirk, and A. Henschel. Maximal oxygen uptake as an objective measure of cardiorespiratory performance. J. Appl. Physiol. 8: 73–84, 1955.

14

THE HEART AND ITS CONTROL

The body, whether resting or exercising, depends on an adequate supply of oxygen and fuels, and must operate within narrow limits of pH and temperature. Additionally, many physiological processes are regulated by hormones that are produced at relatively great distances from their active sites. The cardiovascular system helps to satisfy these biological requirements by serving as the body's internal transportation system. Using blood as the transport medium, the heart and circulation help maintain homeostasis by delivering vital substances to various tissues and eliminating metabolic end products (Table 14-1).

The control of the cardiovascular system is geared toward supplying the tissues' metabolic requirements. At rest, circulation attempts to maintain blood

TABLE 14-1
Functions of the Cardiovascular System

Transportation of O_2 to tissues
Transportation of nutrients
Transportation of CO_2 and metabolites to lungs and kidneys
Distribution of hormones and other substances that regulate cell function
Thermoregulation
Urine formation

Cross-country skier Bill Koch pushes for the finish in a World Cup competition. Aerobic capacities of cross-country skiers compare very favorably with those of other world-class athletes. Aerobic capacity ($\dot{V}O_2$max) depends largely upon cardiovascular capacity. (U.P.I.).

pressure and perfusion to the tissues. During exercise, when the need for oxygen and fuels in working muscles is greater, the system gears itself toward the increased requirement for blood: Cardiac output and muscle blood flow are increased, while blood flow to less active tissues is decreased. However, flow to critical areas such as the brain and heart is either maintained or increased.

The maintenance of systemic arterial blood pressure and the satisfaction of regional tissue demands require coordination of the pumping action of the heart and the optimal distribution of blood flow. These processes are accomplished by a combination of neural, mechanical, and hormonal regulatory mechanisms that serve to maintain circulatory homeostasis at rest and exercise.

This chapter will examine the control of the heart. Chapter 15 will deal with the control of blood flow, and Chapter 16 will discuss the function of the cardiovascular system during exercise. Although basic cardiovascular anatomy and physiology will be reviewed, the reader is referred to basic texts for more in-depth coverage.

THE STRUCTURE OF THE HEART

The heart is a hollow four-chambered muscular organ that serves to pump blood to the lungs and general circulation. It rests in the chest cavity on the diaphragm between the lower parts of the lungs and below the middle and left of the sternum with the apex descending to about the fifth rib (Figure 14-1).

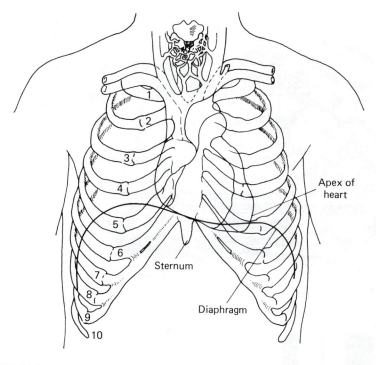

FIGURE 14-1

Location of the heart in the chest cavity.

(SOURCE: Adapted from C.M. Goss, Grey's Anatomy, 1973. With permission.)

The heart is composed of three layers. The outer layer is called the pericardium, which is composed of fibrous tissue interspersed with adipose tissue. The middle layer is the myocardium, which is composed of cardiac muscle. The inner layer is the endocardium, which is composed of squamous endothelial cells and is continuous with the inner lining of the arteries.

The heart is divided into the right and left atria and ventricles. The chambers are distinct, separated by walls called septa. The atria are thin-walled low-pressure chambers that serve as reservoirs for the ventricles. The right and left ventricles pump blood to the pulmonary and systemic circulations, respectively.

The Myocardium

There are two types of cardiac cells: (1) contractile, which makes up the bulk of the myocardium, and (2) electrical, which is made up of specialized excitatory and conductive fibers. The contractile cardiac muscle causes the pumping action of the heart, whereas the electrical tissue is responsible for the rapid conduction of impulses throughout the heart (Figure 14-2).

Cardiac and skeletal muscles are similar in many ways (Chapter 19): Both appear striated, both have myofibrils that contain actin and myosin filaments

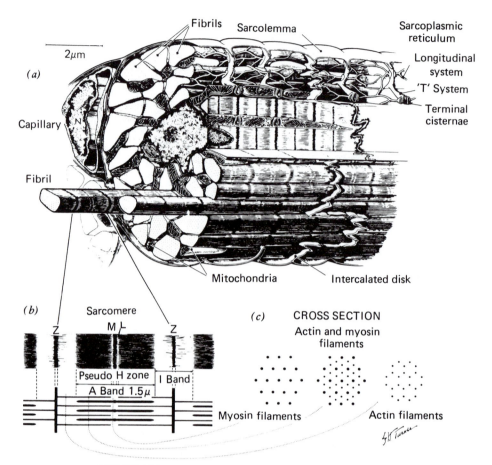

FIGURE 14-2
Schematic diagram of cardiac muscle as seen under the electron microscope. (a) Myocardial cell, showing arrangement of multiple parallel fibrils. (b) Individual sarcomere from a myofibril. A representation of the arrangement of myofilaments that make up the sarcomere is shown below. (c) Cross sections of the sarcomere, showing specific lattice arrangement of myofilaments. N, Nucleus.

(SOURCE: Reproduced, with permission, from E. Braunwald, J. Ross, Jr., and E.H. Sonnenblick, 1967.)

that participate in contraction, and both exhibit an electrical depolarization that precedes shortening.

However, cardiac and skeletal muscle also differ in several significant ways. Cardiac muscle is involuntary. Unlike skeletal fibers, cardiac muscle cells are connected in tight series so that stimulation of one cell results in stimulation of all the cells. Cardiac muscle cells are shorter than skeletal muscle cells and are joined together by intercalated disks. These structures serve as the boundaries between cardiac muscle cells. The electrical resistance between cells is minimal, so that all the cardiac muscle in the atria or ventricles can contract synchronously. So, whereas the "all-or-none" principle applies to a single motor unit in skeletal muscle, it applies to an entire muscle area (called a syncytium) in the heart.

Some cardiac muscle cells have intrinsic rhythmicity and can serve as pacemakers. The most important concentration of these autorhythmic cells is found in the sinoatrial node (SA node), located in the right atrium. Normally, the SA node acts as the pacemaker for the heart. However, there are numerous other pacemakers that can become operative or dominant under certain circumstances.

There is a special electrical conduction system in the heart (Figure 14-3). A nervous-like network connects the SA node with the left atrium. The neuromuscular conduction system of the ventricles consists of the atrioventricular node (AV node), AV bundle, left and right bundle branches, and Purkinje fibers. These systems allow impulses to travel much faster and uniformly than they would if they traveled directly across cardiac muscle.

THE CARDIAC CYCLE

The heart receives blood from the veins and pumps it to the lungs and circulation. This is accomplished by a rhythmic contraction–relaxation process called

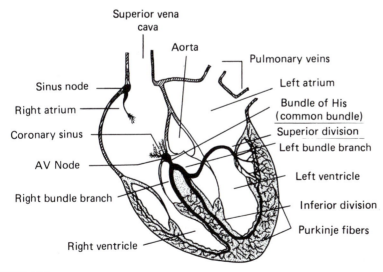

FIGURE 14-3
The electrical conduction system of the heart.

(SOURCE: From E.K. Chung, 1977.)

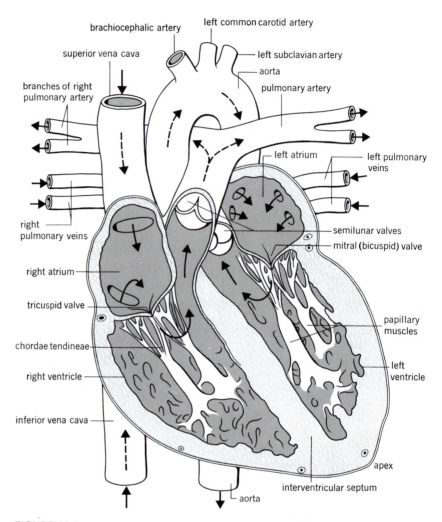

FIGURE 14-4
The chambers of the heart.

(SOURCE: From J.E. Crouch and J.R. McClintic, 1971.)

the cardiac cycle. The cardiac cycle consists of an active phase called *systole* and a relaxation phase called *diastole*. During systole the ventricles contract and push blood from their chambers. During diastole, the ventricles relax and the chambers fill with blood in preparation for the next systole.

Low-oxygenated blood returns to the heart from the circulation via the inferior and superior vena cavae to the right atrium. It then passes through the tricuspid valve into the right ventricle. Blood is then propelled from the right ventricle through the pulmonary valve into the pulmonary arteries, to the lungs where gas exchange occurs. The blood returns to the left atrium via the pulmonary veins. It then passes through the mitral valve into the left ventricle, where it is finally propelled through the aortic valve into the aorta to the systemic circulation (Figure 14-4).

The atria act as reservoirs for the ventricles. At rest, the right and left atria play a small role in helping the heart maintain the necessary cardiac output. However, during exercise the atria are very important in assisting with the large increase in blood returning from the veins. The atria act as primer pumps that greatly facilitate the output of the ventricles.

The ventricles bear the major burden of pumping blood to lungs and circulation. Both ventricles pump essentially the same amount of blood. However, the walls of the left ventricle are thicker than those of the right because of the greater resistance provided by the peripheral circulation. The left ventricle acts as a pressure pump, decreasing mainly in transverse diameter during systole. The right ventricle is more of a volume pump, its large surface-to-volume ratio producing a bellows action.

The amount of blood ejected from the ventricles during systole is called the *stroke volume*. The stroke volume is determined by the difference between ventricular filling (the end-diastolic volume) and ventricular emptying (the end-systolic volume). The percentage of the end-diastolic volume that is pumped from the ventricles is called the *ejection fraction*.

The amount of blood pumped by the heart per unit of time is called the *cardiac output*. Cardiac output ($\dot{Q}$) is a product of the heart rate (f_H) and stroke volume (V_S) (Eq. 14-1). At rest the cardiac output is approximately 5 liters $\cdot$ min^{-1}, but during exercise it can exceed 35 liters $\cdot$ min^{-1} in some world-class endurance athletes.

$$\dot{Q} = (V_S) \times (f_H) \qquad (14\text{-}1)$$

THE ELECTRICAL ACTIVITY OF THE HEART AND THE ELECTROCARDIOGRAM

Resting heart muscle is *polarized*—that is, the cells are negatively charged on the inside and positively charged on the surface. Before heart muscle contracts it must *depolarize;* the inside becomes positively charged and the outside becomes negatively charged. The electrical wave of depolarization causes a progressive contraction of the cardiac muscle. Because it occurs rapidly, the ventricles contract as a unit.

The cardiac cycle involves a series of depolarizations, contractions, and repolarizations. The action potential of depolarization stimulates the release of calcium ions from the T tubules (Figure 14-2) which results in the contraction of cardiac muscle. The T tubules are located at the Z lines of the myocardium. They act with the sarcoplasmic reticulum to initiate contraction. After contraction, the muscle repolarizes back to its resting state.

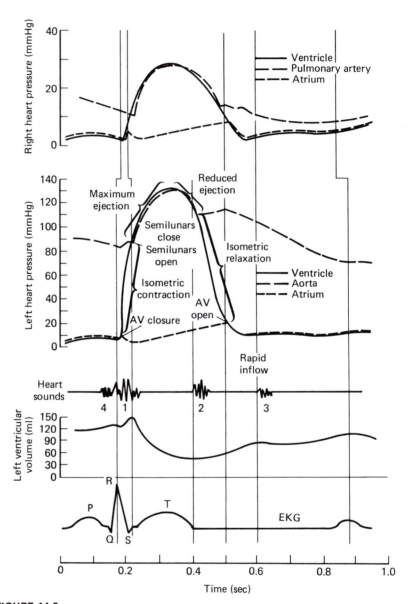

FIGURE 14-5

The elements of the electrocradiogram and the cardiac cycle.

(SOURCE: From J.E. Crouch and J.R. McClintic, 1971.)

Differences in membrane potentials on the heart's surface can be measured by determining differences in potentials between two electrodes on the body's surface. This technique is called electrocardiography (EKG or ECG). The EKG has become an important tool in medicine and physiology. This technique has been used extensively as a noninvasive tool for detecting heart disease and assessing cardiac function.

The elements of the EKG precede the actual mechanical events of the cardiac cycle, as is shown in Figure 14-5. The changing polarity or depolarization stimulates the myocardium to contract. In a normal heart, all the muscle fibers in a given chamber contract at the same time because the depolarization occurs rapidly.

The SA node propagates the initial electrical impulse that initiates the cardiac cycle. This electrical impulse results in a wave of depolarization across both atria called the P wave. The P wave immediately precedes atrial contraction. Conduction tracts from the SA node to the left atrium allow simultaneous contraction of the right and left atria.

A slight delay that occurs when the atrial depolarization wave reaches the AV node allows the blood to pass through the AV valves into the ventricles. After the delay, the AV node is stimulated. This results in the depolarization of the ventricles that produces a QRS complex on the EKG recording. In this process the electrical impulse proceeds across a special neural conduction system consisting of the AV bundle, left and right bundle branches, and Purkinje fibers. This system allows for rapid stimulation of ventricular cardiac muscle. The QRS complex occurs just before the contraction of the ventricles. The repolarization of the atria is usually masked by the QRS complex so cannot be seen on the EKG recording.

The T wave, caused by the repolarization of the ventricles, occurs just after ventricular contraction. The period between the S and T waves is called the ST segment. The ST segment is important in exercise stress testing because it is used to detect deficiencies in coronary blood flow.

There is a *refractory period* after the contraction of cardiac muscle, that is, an interval during which the muscle is incapable of exerting full contraction. An attempt at depolarization during the refractory period will result in a diminished force of contraction (Chapter 17).

CONTROL OF THE HEART

The output of the heart is controlled by the regulation of heart rate and stroke volume. Although the heart has an inherent ability to initiate contraction, the regulation of cardiac output is principally regulated by intrinsic and extrinsic control mechanisms. *Intrinsic* control mechanisms make it possible for the heart automatically to pump all the blood returned to it from the circulation. *Extrinsic* control mechanisms, composed of neural control systems and hormones, work to further increase the rate and strength of cardiac contractions.

The heart is essentially a slave to the circulation. Increased cardiac output is dictated by circulatory demands. When the body's metabolism increases, such as during exercise, the various control mechanisms increase the output of the heart.

Intrinsic Regulation

The normal heart will pump all the blood returned to it from the veins. This is called the Frank–Starling principle of the heart. The ventricles are stretched when they receive more blood, which causes them to contract more forcefully (Figure 14-6). The ability of the heart to contract with increased force as its chambers are stretched is called *heterometric autoregulation.*

The Frank–Starling mechanism is responsible for maintaining the equal output of the right and left ventricles. Any sudden increase in output in the right ventricle is automatically matched by the left ventricle because of this reflex.

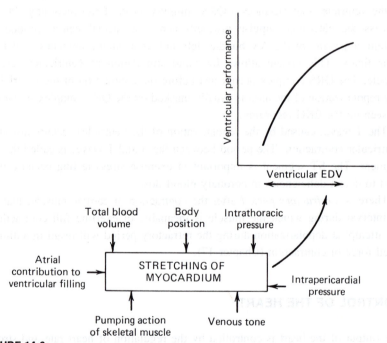

FIGURE 14-6

The Frank–Starling curve and the factors affecting end-diastolic volume (EDV).

(SOURCE: From E. Braunwald, J. Ross, Jr., and E.J. Sonnenblick, 1967.)

Another heterometric autoregulatory mechanism is the atrial stretch reflex. Heart rate is increased by as much as 10 to 15 beats per minute (bpm) when the right atrium is stretched. This, along with the Frank–Starling phenomenon, enables the body to adjust rapidly at the onset of exercise. These mechanical mechanisms operate independently of changes in arterial pressure.

In most people heterometric autoregulation is capable of increasing cardiac output over 100% without assistance from the other mechanisms that control the heart. However, the various control mechanisms operate at the same time. Thus, mechanical and neural controls work together to increase heart rate and stroke volume.

There are limits in the ability of heterometric autoregulatory reflexes to increase cardiac output (Figure 14-6). Any increase in the size of the heart geometrically increases the energy required for contraction. If the linear dimensions of the heart are doubled, then the ventricular muscles must produce a tension four times greater to secure the same systolic pressure. This is expressed in the law of Laplace, which states:

$$\text{Tension} = \frac{\text{Pressure} \times \text{radius}}{2} \qquad (14\text{-}2)$$

During exercise, most of the increase in cardiac output is due to increased heart rate rather than stroke volume. There is a greater mechanical advantage in contracting the smaller ventricular volume.

Additionally, there is a limitation imposed by the contractile characteristics of striated muscle. According to the Frank-Starling principle, an optimal relationship exists between the length of a muscle fiber and the amount of tension it can produce. When a maximum number of myosin cross-bridges have access to actin-receptor sites, the fiber is capable of producing the maximum tension. However, if the fiber is stretched too far, then it will have fewer actin and myosin interactions and will be less capable of exerting force. These limitations can become marked when the heart is affected by disease, acidosis, hypoxia, or pharmacological depressants.

Increased cardiac output by these mechanisms depends on adequate venous return of blood to the right atrium. This is directly affected by cardiac output, circulatory volume, and blood volume. The resulting ventricular filling is called *preload* and will be discussed in Chapter 16.

Extrinsic Regulation

Extrinsic controls consist of nervous reflexes and hormones. These controls involve acute control of chronotropic and inotropic activities of the heart (rate and strength of contraction). As will be seen in the next chapter, many of these mechanisms are also involved in the control of circulation. Activities such as the control of arterial blood pressure, blood volume, and the intake of fluid and electrolytes ultimately have an effect on the heart.

TABLE 14-2

The Autonomic Nervous System and Cardiovascular Function

Sympathetics	Parasympathetics
↑ Heart rate	↓ Heart rate
↑ Strength of contraction	↓ Strength of atrial contraction
Vasodilation of coronary arteries	Vasoconstriction of coronary arteries
Mild vasoconstriction of pulmonary vessels	Dilation of skin blood vessels
Vasoconstriction in abdomen, muscle (adrenergic), skin (adrenergic), and kidneys	
Vasodilation of muscle (cholinergic) and skin (cholinergic)	

Circulatory requirements during exercise easily exceed the ability of the mechanical autoregulatory mechanisms to increase cardiac output. Fortunately, the heart is supplied with both sympathetic and parasympathetic nerves from the autonomic nervous system (Figure 14-7). These nerves affect both the contractile strength of the heart and the heart rate. The sympathetic nerves stimulate the heart, whereas the parasympathetics suppress it.

The autonomic nervous system is an important regulator of circulatory and cardiac physiology. The system provides a number of reflexes that control a wide variety of physiological functions. Sympathetic fibers innervating the heart originate in the spine between the first thoracic and second lumbar vertebrae (Figure 9-7) whereas the parasympathetics innervating the heart originate in the cranial nerves, principally the vagus nerve (Figure 9-6).

The majority of cardiovascular sympathetic and parasympathetic reflexes are controlled by the *vasomotor center* located in the lower third of the pons and the upper two-thirds of the medulla. The center emits constant impulses that maintain the circulation in a sympathetic tone. When the vasomotor center is stimulated, sympathetic fibers, including those to the heart, are stimulated. When the center is inhibited, sympathetic tone is decreased and cardiac function is depressed. The center can be greatly affected by higher brain centers, which explains the effects of emotions on cardiac function.

A summary of the effects of sympathetic and parasympathetic nerves on cardiovascular function is given in Table 14-2. In general, sympathetic stimulation increases cardiovascular function, whereas parasympathetic stimulation inhibits it.

Parasympathetic effects are mediated by the secretion of acetylcholine at the nerve endings, so the parasympathetic nerves are sometimes called cholinergic fibers. Sympathetic fibers secrete both norepinephrine and acetylcholine, al-

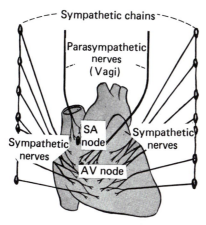

FIGURE 14-7

The cardiac nerves.

(SOURCE: From A.C. Guyton, Textbook of Medical Physiology. Copyright W.B. Saunders Co., Philadelphia, 1976, p. 171. With permission.)

though norepinephrine is predominant. The fibers that secrete norepinephrine are called adrenergic.

Parasympathetic neurons innervate the SA node, the atrial myocardium, and the AV node. Parasympathetic stimulation has a depressant effect on the heart: It results in cardiac slowing, decreased atrial force, and a delay in atrioventricular conduction. At rest, signals through the vagus nerve elicit a parasympathetic tone on the heart that lowers heart rate. This tone seems to be increased with training, which accounts for the low resting heart rates found in endurance athletes.

Sympathetic neurons innervate the SA node, the atrial myocardium, the AV node, and the ventricular myocardium. Sympathetic stimulation is directed at increasing blood pressure by peripheral vasoconstriction of blood vessels and stimulating heart rate and force of cardiac contraction. At the same time, the sympathetics allow more blood to pass through the coronary arteries.

Sympathetic adrenergic effects are classified according to the α and β receptors stimulated (Table 9-1). These receptors can have different effects that sometimes contradict each other. For example, in muscle and skin blood vessels, α-adrenergic receptors cause vasoconstriction, whereas β receptors cause vasodilation. However, one function usually predominates or overrides the other. These receptors are an important consideration in exercise physiology because they are sometimes selectively blocked by drugs in the treatment of heart disease and thus affect exercise tolerance. For example, the drug propranolol, a β-adrenergic blocking agent, is used to reduce myocardial O_2 consumption because it reduces heart rate and myocardial contractility (Appendix III).

Baroreceptors and Chemoreceptors

The baroreceptors, which are stimulated by stretch, detect changes in blood pressure. They are principally located in the carotid arteries and aorta, but exist in the walls of all of the major arteries. When stimulated, they act to inhibit sympathetic centers, which tends to reduce cardiac output. When pressure falls, an opposite reflex occurs: They send fewer impulses to the vasomotor center, which decreases parasympathetic tone to the heart and increases sympathetic tone to the heart and blood vessels. The result is an increase in cardiac output and vasoconstriction, which elevates blood pressure.

Baroreceptors are involved in the control of cardiac output only when arterial blood pressure is affected. They are important for maintaining blood pressure during changes in posture. The baroreceptors act as a buffer to prevent large fluctuations in blood pressure. They are most important in buffering acute changes in pressure rather than long-term regulation. The effective range for the arterial baroreceptor system lies between 60 and 160 mmHg.

Strong stimulation of the carotid baroreceptor, such as occurs when taking the pulse at the neck, can result in a fall in arterial blood pressure and heart rate. In some cases, the heart can even stop. People often use heart rate to assess the intensity of exercise; they should be discouraged from taking their pulse at the carotid artery, because the measurement will probably be inaccurate and they may risk fainting.

Chemoreceptors located in the aortic and carotid bodies can also affect cardiac function. Stimulation of the carotid chemoreceptors produces a parasympathetic effect on the heart, whereas stimulation of the aortic chemoreceptors has a sympathetic effect. Chemoreceptors are discussed in chapters 12 and 15.

Hormonal Controls

Large portions of the sympathetic nervous system are stimulated both during exercise and in anticipation of exercise. The sympathetics are turned on as a unit (except for a few important exceptions) so that the body is ready for the demands of exercise. These sympathetic effects include increases in heart rate, stroke volume, circulatory vasoconstriction in inactive areas, vasodilation in active muscles, metabolic rate, carbohydrate mobilization, muscle strength, and mental acuity.

The mass-action discharge of the system has both specific and general effects on the circulation: Individually innervated areas such as the heart and blood vessels are directly stimulated to increase blood pressure, whereas sympathetic stimulation of the adrenal medulla has general circulatory effects because of the release of the catecholamines epinephrine and norepinephrine. Thus, whereas the heart is directly innervated by sympathetic nerves, it is additionally stimulated by the hormonal action of the adrenal medulla.

The catecholamines are the most important hormones affecting cardiac performance. However, other hormones can also have either a direct or an indirect effect. Both thyroid hormone and glucagon have positive inotropic and positive chronotropic effects on the heart. Hormones that can indirectly affect heart function include the glucorticoids, mineralocorticoids, ACTH, TSH, and growth hormone.

Summary

The heart is a four-chambered organ that pumps blood to the pulmonary and systemic circulations. The control of the heart is geared toward cellular metabolic demands and the maintenance of blood pressure. A combination of neural, mechanical, and hormonal regulation synchronizes the activity of the heart and circulation to ensure the optimal distribution of blood flow.

The heart is composed mainly of involuntary muscle that is unique in its intrinsic rhythmicity and ability to conduct action potentials rapidly. Changes in electrical potential are detected with the electrocardiogram. Exercise EKGs have become extremely useful in predicting some forms of heart disease.

The heart is subjected to both intrinsic and extrinsic control mechanisms. Intrinsic controls include the Frank–Starling mechanism and the atrial stretch reflex, which work largely by adjusting to changes in venous return. Extrinsic controls consist of nervous reflexes and hormones, including sympathetic and parasympathetic nerves, baroreceptors, chemoreceptors, and circulating hormones, such as catecholamines.

Selected Readings

Bern, R.M. and M.N. Levy. Cardiovascular Physiology. St. Louis: C.V. Mosby, 1967.

Bhargava, V. and A.L. Goldberger. Effect of exercise in healthy men on QRS power spectrum. Am. J. Physiol. 243: H964–H969, 1982.

Braunwald, E., J. Ross, Jr., and E.H. Sonnenblick. Mechanisms of contraction of the normal and failing heart. N. Engl. J. Med. 277: 794–799, 853–862, 910–919, 962–971, 1012–1022, 1967

Butler, J., M. O'Brien, K. O'Malley, and J.G. Kelley. Relationship of β-adrenoreceptor density to fitness in athletes. Nature. 298: 60–62, 1982.

Chung, E.K. (ed.). Exercise Electrocardiography: Practical Approach. Baltimore: Williams & Wilkins Co., 1979.

Chung, E.K. Principles of Cardiac Arrhythmias. Baltimore: Williams & Wilkins Co., 1977.

Crouch, J.E. and J.R. McClintic. Human Anatomy and Physiology. New York: John Wiley & Sons, 1971.

Dahlstrom, J.A., B. Lilja, O. Ohlsson, and A. Torp. Left ventricular function at rest and during exercise in coronary insufficiency—a simultaneous evaluation by means of radionuclide angiocardiography and catheterization. Clin. Cardiol. 2: 435–466, 1982.

Devereaux, R.B. and N. Reichek. Echocardiographic determination of left ventricular mass in man. Circulation. 55: 613–618, 1977.

Dubin, D. Rapid Interpretation of EKGs. Tampa, Fla.: Cover Publishing, 1980. p. 1–43.

Ellestad, M.H. Stress Testing. Principles and Practice. Philadelphia: F.A. Davis, 1976.

Feigenbaum, H. Echocardiography. Philadelphia: Lea & Febiger, 1976.

Foster, C., J.D. Anholm, D.S. Dymond, J. Carpenter, M. Pollock, and D.H. Schmidt. Left ventricular function at rest, peak exercise, and postexercise. Cardiology. 69: 224–230, 1982.

Gooch, A.S., M. Kirschbaum, K.B. Mohan, and S.D. Cha. The relationship of carotid sinus stimulation to exercise testing. J. Electrocardiol. 16: 59–72, 1983.

Goss, C.M. Grey's Anatomy. Philadelphia: Lea & Febiger, 1973.

Guyton, A. Textbook of Medical Physiology. Philadelphia: W.B. Saunders, 1976.

Hurst, J.W., R.B. Logue, R.C. Schlant, and N.K. Wenger (eds.). The Heart. New York: McGraw-Hill, 1978.

Ikaheimo, M.J., I.J. Palatsi, and J.T. Takkunen. Noninvasive evaluation of the athletic heart: Sprinters versus endurance runners. Am. J. Cardiol. 44: 24–30, 1979.

Lakatta, E.G., G. Gerstenblith, C.S. Angell, N.W. Shock, and M.L. Weisfeldt. Diminished inotropic response of aged myocardium to catecholamines. Circ. Res. 36: 262–268, 1975.

Longhurst, J.C., A.R. Kelly, W.J. Gonyea, and J.H. Mitchell. Echocardiographic left ventricular masses in distance runners and weight lifters. J. Appl. Physiol. 48: 154–162, 1980.

Morganroth, J. and J. Maron. The athlete's heart syndrome: A new perspective. Ann. N.Y. Acad. Sci. 301: 931–941, 1977.

Mumford, M. and R. Prakash. Electrocardiographic and echocardiographic characteristics of long distance runners. Am. J. Sports Med. 9: 23–28, 1981.

Parker, B.M., B.R. Londeree, G.V. Cupp, and J.P. Dubiel. The noninvasive cardiac evaluation of long-distance runners. Chest. 73: 376–381, 1978.

Port, S., F.R. Cobb, E. Coleman, and R.H. Jones. Effect of age on the response of the left ventricular ejection fraction to exercise. N. Engl. J. Med. 303: 1133–1137, 1980.

Roeske, W.R., R.A. O'Rourke, A. Klein, G. Leopold, and J.S. Karliner. Noninvasive evaluation of ventricular hypertrophy in professional athletes. Circulation. 53: 286–292, 1976.

Ruch, T.C. and H.D. Patton (eds.). Physiology and Biophysics. Philadelphia: W.B. Saunders, 1966. p. 120–167.

Snoeckx, L., H. Abeling, J. Lambregts, J. Schmitz, F. Verstappen, and R.S. Reneman. Echocardiographic dimensions in athletes in relation to their training programs. Med. Sci. Sports Exerc. 14: 428–434, 1982.

Sokolow, M. and M.B. McIlroy. Clinical Cardiology. Los Altos, Calif.: Lange Medical Publications, 1977. p. 1–29.

Sugishita, Y. and S. Koseki. Dynamic exercise echocardiography. Circulation. 60: 743–752, 1979.

Underwood, R.H. and J.L. Schwade. Noninvasive analysis of cardiac function of elite distance runners—echocardiography, vectorcardiography, and cardiac intervals. Ann. N.Y. Acad. Sci. 301: 297–309, 1977.

Wenger, N.K., J.W. Hurst, and M.C. McIntyre. Cardiology for Nurses. New York: McGraw-Hill, 1980.

Wolthuis, R.A., V.F. Froelicher, A. Hopkirk, J.R. Fischer, and N. Keiser, Normal ECG waveform characteristics during treadmill exercise testing. Circulation. 60: 1028–1035, 1979.

Zoneraich, S., J.J. Rhee, O. Zoneraich, D. Jordan, and J. Appel. Assessment of cardiac function in marathon runners by graphic noninvasive techniques. Ann. N.Y. Acad. Sci. 301: 900–914, 1977.

15

THE CIRCULATION AND ITS CONTROL

The control of blood flow is of critical importance during exercise. Blood must be rapidly directed to working muscles to meet their demands for oxygen and fuels. At rest, a relatively large portion of the cardiac output is directed to the spleen, liver, kidneys, brain, and heart. Although muscles comprise over 40% of the body's tissue, they receive only about 20% of the total blood flow. During exercise, however, the muscles can receive more than 85% of the cardiac output.

Although blood volume is only about 5 liters, the circulation has the capacity to hold over 25 liters. So, in order to move blood to where it is needed, circulatory controls must constrict some blood vessels and dilate others. The mechanisms that control blood flow make it possible to maintain or increase blood pressure, to continue supplying blood to vital areas, and to satisfy the metabolic requirements of the working muscles.

THE CIRCULATION

The blood vessels consist of (1) arteries, which serve to transport blood under pressure to the tissues, (2) arterioles, metarterioles, and precapillary sphincters, which act as valves to control blood flow to the various tissues, (3) capillaries, which act as an exchange medium between blood and the interstitial

Cycling activities lasting more than 90 sec. require rates of O_2 consumption and utilization. These, in turn, require a high cardiac output and the ability to circulate blood to active muscles and other metabolically active areas. (Catherine Noren/Photo Researchers.)

spaces, (4) venules, which collect blood from the capillaries, and (5) veins, which return blood to the heart. Another related system, the lymphatics, empties into the veins and is integrally related to circulatory function. The circulation is subdivided into the systemic and pulmonary circulations. Blood is pumped from the right ventricle into the lungs and from the left ventricle into the aorta. The many branches of the aorta (similar to a tree) ultimately deliver blood to most tissues of the body (Figure 15-1).

Systemic arteries are elastic and muscular, which makes them capable of withstanding high pressures. The blood pressure in the arteries is relatively high but diminishes throughout the circulation until it approaches 0 mmHg when it reaches the right atrium. The pressure in the circulation diminishes with increased resistance to flow, unless the resistance is offset by an increased cardiac output (Figure 15-2).

Blood passes from the arteries into the arterioles, metarterioles, and precapillary sphincters. The greatest resistance to flow occurs in these areas, which are critically important for maintaining blood pressure and assuring tissue blood supply, particularly during exercise. The regulation of blood flow occurs in these areas as a result of a combination of sympathetic neural control mechanisms and local chemical vasoconstrictors and vasodilators.

At rest, the arterioles in the inactive muscle remain mostly constricted, which allows the blood to be directed to other areas. However, during exercise, when more blood is needed by the working muscles, the arterioles supplying blood

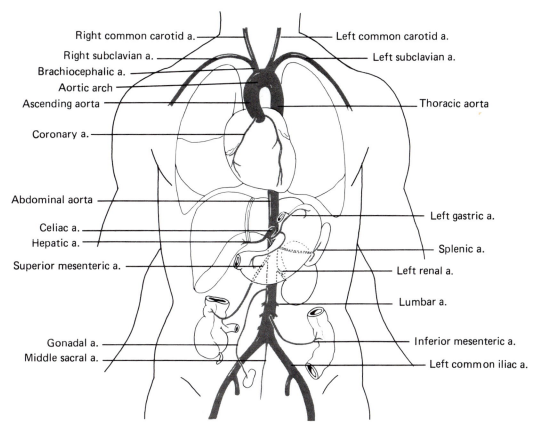

Right common carotid a.

Right subclavian a.

Brachiocephalic a.

Aortic arch

Ascending aorta

Coronary a.

Left common carotid a.

Left subclavian a.

Thoracic aorta

Abdominal aorta

Celiac a.

Hepatic a.

Superior mesenteric a.

Left gastric a.

Splenic a.

Left renal a.

Lumbar a.

Gonadal a.

Middle sacral a.

Inferior mesenteric a.

Left common iliac a.

FIGURE 15-1

Principal branches of the aorta.

(SOURCE: From J.W. Hole, Human Anatomy and Physiology, William Brown & Co., Dubuque, Iowa, 1978.)

to the muscles dilate, thus allowing the transportation of the required O_2 and nutrients.

Blood passes through the arterioles into the capillaries, where gases, substrates, fluids, and metabolites are exchanged. Capillary blood flow is controlled by circulatory driving pressure provided by the heart and by resistance provided by the arterioles, precapillary sphincters, and veins. Because the capillaries contain no smooth muscle, they play no role in regulating their own blood flow.

By far the largest proportion of the blood volume resides in the veins. The veins are sometimes called the "venous capacitance system" because of their storage capacity. The veins can be constricted by sympathetic stimulation and by mechanical compression—both important factors during exercise. Every time muscles are contracted, the veins are compressed, which propels blood toward

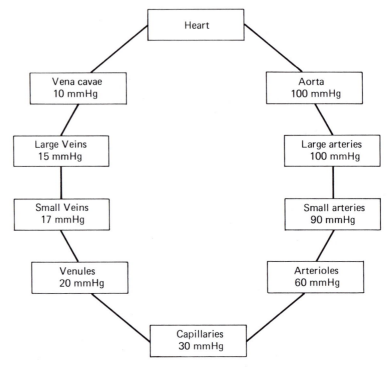

FIGURE 15-2
Mean internal pressure in the circulation.

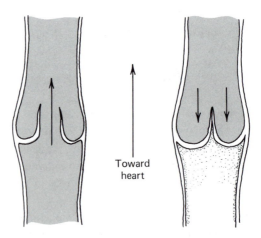

FIGURE 15-3
Venous valves allow blood to flow toward the heart but prevent it from moving away from the heart.

(SOURCE: From J.W. Hole, 1978.)

the heart. Ventilation and negative pressure created by the pumping action of the heart can also stimulate blood flow in the veins. Valves in the veins prevent the blood from flowing away from the heart (Figure 15-3).

DETERMINANTS OF BLOOD FLOW

The rate at which blood flows through the circulation is dictated by the blood pressure. Blood pressure (P) is the product of cardiac output ($\dot{Q}$), the amount of blood pumped from the heart per minute, total peripheral resistance (TPR), and the resistance to blood flow provided by the circulation. At a constant cardiac output, local tissue blood flow is regulated by increasing or decreasing the size or resistance in the local blood vessels.

$$P = Q \times (TPR) \tag{15-1}$$

The control of blood flow by pressure and resistance is expressed in Poiseuille's law, which is shown below. This law states that flow through a tube varies directly with the differences in pressure, varies to the fourth power of the radius of the tube, and varies inversely with the length of the tube and the viscosity of the fluid. Of the four factors in this equation, length is the only one that does not vary in the physiological system.

$$F = \frac{(P_1 - P_2)\pi R^4}{8LN} \tag{15-2}$$

where F is flow rate, $(P_1 - P_2)$ is the drop in pressure, R is the radius of the tube, L is the length of the tube, N is the viscosity of the fluid. and π is pi.

Pressure is perhaps the most important factor that increases blood flow. Blood flow between two points is directly proportional to the driving force between them. During exercise, blood flow to active muscle can be increased as much as 25 times. This is accomplished by increasing cardiac output (the driving force of blood through the circulation) and by directing a greater fraction of the cardiac output to active muscle (shunting).

Blood moves from the left ventricle back to the right atrium very easily, in spite of the effects of gravity in the upright posture. This is accomplished largely because of the principle of the siphon, which states that the flow of fluid between two points depends on the differences in pressures or levels between them. The flow is not affected by the level of the pipes between the points (Figure 15-4a). Thus, although the circulation in the lower extremities is below the heart, blood flow is maintained.

The circulation is not a rigid system, as is assumed by Poiseuille's law. Blood vessels can be distended because of increased pressure on their walls caused by the effects of gravity on body fluids (called hydrostatic pressure; see

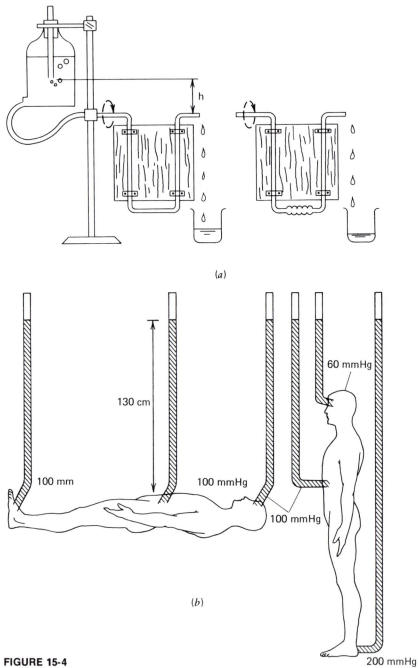

FIGURE 15-4
(a) An apparatus that demonstrates the principle of the siphon in nondistensible (left) and distensible (right) tubes. (b) Hydrostatic pressure in operation on arterial pressure in the supine and erect positions.

(SOURCE: From T.C. Ruch and H.D. Patton, 1966.)

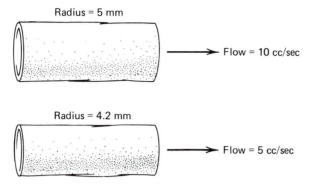

FIGURE 15-5
Effect of decreasing the radius of a tube 16% on flow rate at a constant pressure.

Figure 15-4b). Hydrostatic pressure tends to increase the size of the blood vessels, which increases the volume of the circulation. When there is a volume change in the circulation, there is a momentary interruption in the siphon effect and a delay in the return of blood to the heart. However, this delay is only temporary until the system adjusts to the new volume.

Circulatory control mechanisms help to ensure a constant vascular volume by regulating the size of the blood vessels. Thus, the regulation of blood flow depends on an integration between the control of output from the heart and resistance in the blood vessels.

The distribution of blood is controlled mainly by variations in the size of the arterioles. These controls are extremely sensitive. Because flow varies with the fourth power of the radius, a decrease of 16% in radius will decrease blood flow to $\frac{1}{2}$ of the original value. So, a very small increase in a vessel's radius will result in a significant change in its blood flow (Figure 15-5).

Normally, viscosity is not an important factor affecting blood flow. However, when the *hematocrit* (percentage of red cells in blood) becomes abnormally high, such as in polycythemia vera, the increased viscosity will decrease blood flow. Additionally, extreme cold, such as occurs in frostbite, can increase the viscosity of blood enough to impede blood flow severely.

PERIPHERAL VASCULAR REGULATION AND CONTROL

Vasodilation of blood vessels in active muscle and vasoconstriction in other areas are controlled by intrinsic autoregulation and by extrinsic neurogenic control. Local regulation of blood flow is directed toward meeting tissue requirements, whereas extrinsic neural control is directed toward maintaining blood pressure.

Autoregulation of Blood Flow

Autoregulation or local control of blood flow is perhaps the principal factor in the control of cardiovascular function. At rest, autoregulation of blood flow is directed toward maintaining tissue perfusion at a relatively constant rate despite changes in perfusion pressure. During exercise, it is directed toward increasing blood flow to working muscles. Local autoregulation of blood flow occurs in response to tissue demands for O_2 and fuels, and also responds to CO_2, hydrogen ion, and temperature.

Metabolic activity seems to be the most important factor governing blood flow in skeletal muscle (Figure 15-6). When blood flow is inadequate, vasodilator metabolites accumulate, which stimulates blood flow. Some investigators feel that the process is stimulated by the release of substances such as adenosine and prostaglandins, whereas others feel that hypoxia itself is the stimulus. However, it is possible that muscle contraction itself causes vasodilation. This is the so-called myogenic theory of blood flow. Likewise, the need to eliminate substances such as CO_2, acids, and potassium can also have the same effect. Low O_2 levels in the muscle tissue may have a direct effect on the smooth muscle of the precapillary sphincter. As O_2 levels decline, the sphincters may lose their ability to maintain a contraction, so they relax and blood flow to the area increases.

Muscle blood flow increases exponentially with metabolism. As O_2 requirements of muscles increase, the arterioles dilate to allow more blood flow. At lower levels of metabolism, the precapillary sphincters open and then close. At higher intensities, the vessels tend to stay open. At various times during exercise, the several autoregulatory factors play different roles. For instance, when exercise starts muscle PO_2 falls and this causes vasodilitation. However, as the

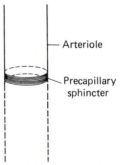

Arteriole

Precapillary sphincter

Condition: Rest on light exercise
Metabolic status O_2, CO_2, and substrate supplies are adequate
Blood flow control: Intermittent vasoconstriction

Condition: Maximal exercise as appears on right
Metabolic status: Hypoxia, hypercapnia, possible inadequate substrates
Blood flow control: Continuous vasodilation

FIGURE 15-6
Effect of local metabolism on blood flow.

local blood flow increases, O_2 delivery increases and PO_2 returns toward normal. Vasodilitation is maintained, however, because of the action of other factors, such as adenosine and K^+.

Changes in arterial pressure do not in themselves have much effect on muscle blood flow because of autoregulation. Rather, local tissue needs are the overriding influence for vasodilation or constriction. This is very important during exercise, because sympathetic stimulation tends to produce a general vasoconstriction. Thus, blood flow can increase in active muscles that need O_2 and nutrients, and decrease in inactive muscles and tissues that have lower metabolic requirements.

Muscle blood flow can be blocked during isometric or heavy isotonic contractions because of the compression of blood vessels. When free flow is restored, blood flow to the area can increase many times above normal, resulting in a phenomenon known as *reactive hyperemia*. A good example of reactive hyperemia is the increase in muscle size following weight-lifting exercise. Many novice weight lifters interpret the increased girth as evidence that the muscle is growing before their eyes. However, it is only the result of reactive hyperemia, and the muscles will shrink back to their normal size in a short time.

Neurogenic Control of Blood Flow

Although autoregulation is a potent means of assuring muscle blood flow during exercise, neural and humoral mechanisms are also vital. Exercise requires a rapid redistribution of blood to the active muscles. Hormones and the sympathetic nervous system can cause rapid vasoconstriction in the viscera and inactive muscles, so that blood pressure can be maintained or increased and blood can be effectively rechanneled to working muscles.

Sympathetic vasoconstrictor fibers are widely distributed in the circulation, but are particularly potent in the spleen, kidneys, skin, and gut. Their distribution in muscle, the brain, and the heart is much less, which is convenient for the circulatory changes that occur during exercise. Sympathetic stimulation that occurs with exercise causes a stimulation of the heart and a progressively increasing vasoconstriction in the spleen, kidneys, gut, and in many instances, the skin. The spleen plays a small role in enhancing venous return of blood to the heart during exercise. It serves as a reservoir of blood that can be released by sympathetically induced vasoconstriction in the organ (Figure 15-7).

During exercise, blood flow to the brain must be maintained at a constant rate, and flow to active muscles and the heart needs to be increased. Because sympathetic stimulation in these areas is less prolific, autoregulatory and other mechanisms facilitate blood flow to these areas.

Both the sympathetics and parasympathetics maintain a constant tone that affects the heart and circulation: The parasympathetic tone acts to suppress heart rate while sympathetic tone maintains a certain amount of circulatory

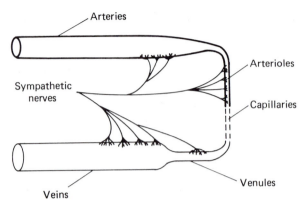

FIGURE 15-7
Sympathetic innervation of circulation.

(SOURCE: From A.C. Guyton, Textbook of Medical Physiology. Copyright W.B. Saunders Co., Philadelphia, 1976. p. 258. With permission.)

vasoconstriction. Sympathetic tone is particularly important during exercise because the autonomic vasodilators are relatively ineffective. Thus, the sympathetic input to the circulation can cause both vasoconstriction and vasodilation—vasoconstriction if they are stimulated and vasodilation if they are inhibited.

During exercise, sympathetic stimulation causes the veins to constrict, resulting in an increased return of blood to the heart. Venous return of blood can also be affected by such factors as increased blood volume, increased tone of large blood vessels, and dilation of small blood vessels.

The sympathetics also have more specific actions that are important during exercise. Blood flow to the skin can be affected independently to cause either vasodilation or vasoconstriction. So, while other inactive areas are vasoconstricted during exercise, the skin can be vasodilated to facilitate body cooling. Muscles also contain sympathetic vasodilators that act to increase blood flow during exercise. Although these vasodilators are relatively weak, they make a definite contribution to increasing muscle blood flow during physical activity.

Hormonal Control Mechanisms

The release of epinephrine and norepinephrine from the adrenal medulla affects the circulation as well as the heart. In the heart both hormones enhance the effect of the specific sympathetic stimulation. However, they have dissimilar effects on circulatory vasoconstriction. Both hormones stimulate the heart by increasing its rate, force and speed of contraction, speed of conduction, and irritability. However, while norepinephrine acts to cause peripheral vasoconstriction, epinephrine causes vasodilation.

The renin–angiotensin mechanism is another hormone control system that exerts a relatively acute effect on blood pressure and blood flow. When blood pressure falls, the enzyme renin is released from the kidneys. The production of renin ultimately leads to the production of angiotensin II, a potent vasoconstrictor of both arteries and veins. The resulting vasoconstriction causes an increase in peripheral resistance and venous return, which raises blood pressure. The system is unidirectional; it acts to raise pressure when it falls too low but has no effect when pressure is too high. There is some evidence that this mechanism plays a role in increasing blood pressure during exercise.

The renin–angiotensin system also causes kidneys to retain fluid and salt and thus increases blood volume. This may be its most potent effect on blood pressure. Angiotensin will also have acute (24-hr) effect on aldosterone secretion, which also increases water and salt retention. Aberrations in this control mechanism are suspected to be related to the development of some types of hypertension.

THE CORONARY ARTERIES

Blood is supplied to the heart by the right and left coronary arteries (Figure 15-8). These arteries originate just past the aortic valve. Thus, although large quantities of blood pass through the chambers of the heart, the myocardium requires a circulatory supply system similar to other tissues in the body.

The left coronary artery supplies mainly the left ventricle, whereas the right coronary supplies the right and part of the left ventricle. These arteries are of obvious importance: If they fail to perform adequately, loss of function or death will ensue.

Increases in stroke volume and heart rate during exercise increase the O_2 consumption of the heart. Thus, increases in cardiac performance must be met by increases in coronary blood flow. Resting coronary blood flow is approximately 260 ml $\cdot$ min^{-1}. During maximal exercise it will rise to about 900 ml $\cdot$ min^{-1}. Endurance training seems to have little or no effect on maximal coronary flow.

Although there is some neurogenic control of blood flow in the coronaries, autoregulation is the principal regulatory mechanism. There is a limited ability to increase O_2 extraction, $(a - v) O_2$, with increased levels of myocardial metabolism. So, blood flow must increase with increasing work loads. If the coronaries are limited in their ability to dilate by disease, then the heart can become hypoxic when its demand for O_2 is high. This can lead to diminished contractile capacity and heart pain called *angina pectoris* (Chapters 24 and 26).

Blood flow through the coronary capillaries of the left ventricle is greatly affected by the cardiac cycle. During systole, these capillaries are compressed

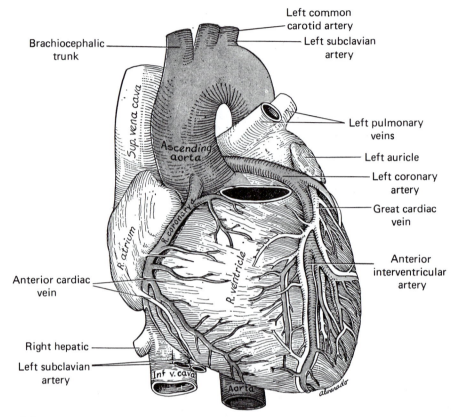

FIGURE 15-8

The coronary blood vessels of the sternocostal aspect of the heart.

(SOURCE: From C.M. Goss, Grey's Anatomy, Lea & Febiger, Philadelphia, 1973. p. 573.)

by the contracting cardiac muscles. Blood flow is restored during diastole when the ventricle is relaxed.

This is an important consideration in exercise physiology and exercise stress testing. If diastolic pressure rises during exercise, blood flow may be impeded at a time when myocardial O_2 demand is greater. Studies have indicated that there is about an 80% chance that an individual has serious stenosis or narrowing of at least one coronary artery if diastolic pressure rises 20 mmHg during exercise. The coronary arteries and disease are further discussed in Chapters 24 and 26.

LONG-TERM REGULATION OF BLOOD FLOW

There are also long-term regulatory mechanisms for controlling blood flow to the tissues. The two most important long-term mechanisms are increased vascularity in response to local hypoxia and control of blood and fluid volumes.

Vascularization

Endurance training and long-term exposure to altitude will lead to increased vascularization (increased blood supply) in skeletal muscle. This adaptation occurs most readily in type I (slow twitch) muscle fibers. Additionally, this vascular response is stronger in the young than in the old. In general, the degree of adaptation reflects the chronic metabolic demands placed on the tissues. The vascularization will also reverse itself if the metabolic demand diminishes.

As a result of coronary narrowing (stenosis) collateral circulation is developed in the heart to improve circulation. Collaterals are accessory vessels that branch from the coronaries to help restore blood flow to an area of the heart. Exercise training seems to enhance the development of these collaterals after a myocardial infarction. Unfortunately, there is presently no evidence to indicate that collateral circulation is developed in a healthy heart in response to exercise training.

Control of Blood Volume

Total blood volume seems to be related to endurance fitness. Athletes have a higher blood volume, plasma volume, total hemoglobin content, and erythrocyte volume. Increased blood volume facilitates venous return of blood to the heart, which enhances the Frank–Starling mechanism of increasing cardiac output.

The control of blood volume is an important consideration in cardiovascular performance at rest and during exercise. As we have seen, cardiac output capacity depends on venous return of blood to the heart. If blood volume is too low, then the capacity of the heart is diminished because it has less blood to pump. If blood volume is high, then cardiac output and arterial blood pressure will increase.

Blood volume control is part of an integrated system involving regulation of extracellular fluid, intracellular fluid, and fluid intake. You cannot affect one of these factors without affecting the others.

Fluid volumes are controlled by (1) the renal pressure diuresis mechanism, (2) ADH stimulation of water reabsorption by the kidneys, (3) control of water reabsorption by thirst, and (4) stimulation of sodium reabsorption and potassium secretion in the renal tubules by aldosterone.

The kidney regulates the rate of fluid excretion by determining the rate of urine formation and electrolyte reabsorption. The hormone ADH (antidiuretic hormone or vasopressin) controls water output in the kidney. Sodium and potassium are controlled by the hormones aldosterone, ADH, and angiotensin.

Fluid intake is extremely important for the maintenance of blood volume. Usually, the thirst mechanism stays abreast with the body's fluid requirements (for important exceptions, see Chapter 22). Fluid intake is an important consideration in exercise physiology. When fluids are ingested there is an increase in both blood volume and interstitial fluid volume.

The capillaries also play an important role in the control of fluid volumes. This function is very important in the regulation of blood volume and the maintenance of cellular homeostasis. Fluid movement in the capillaries is regulated by hydrostatic pressure, interstitial pressure, plasma protein oncotic pressure, interstitial oncotic pressure, and the filtration constant of the capillary membrane, as seen in the following equation.

$$\text{Fluid movement} = K\ [(P_c + O_i) - (P_i + O_p)] \tag{15-3}$$

where P_C is capillary hydrostatic pressure, P_i is interstitial fluid pressure, O_p is plasma protein oncotic pressure, O_i is interstitial fluid oncotic pressure, and K is the filtration constant of the membrane.

Hydrostatic and interstitial pressures refer to the forces exerted against the capillary walls by these fluids. Plasma protein and interstitial oncotic pressures refer to the osmolality of those respective fluid compartments. *Osmolality* refers to the ratio of water to dissolved substances and the tendency of fluids to diffuse into the compartment. The filtration constant of the membrane refers to the ability of the membrane to allow the passage of fluid. Other factors that will affect the movement of fluid in the capillaries are the area of the capillary wall, the distance across the wall, and the viscosity of the fluid.

An increase in capillary pressure tends to cause fluid to move from the circulation to the interstitium, whereas a decrease in pressure results in the osmosis of fluid in the opposite direction. The lymphatics are involved in draining fluid from the interstitium back into the circulation. The lymphatics only return about 100 ml · hr^{-1} of fluid to the circulation, but play an important role in the maintenance of blood flow because the lymphatics supply the only means of returning protein that leaks into the interstitium back to the circulation.

Blood volume undergoes dynamic changes during exercise, largely because of increases in hydrostatic pressure. Plasma volume decreases at least 10% during prolonged exercise, which places an increasing load on the circulation as the physical activity continues.

Fortunately, blood volume increases with exercise training. These adaptations occur rapidly: After a few days of endurance training, plasma volume can be increased by almost 400 ml, even in relatively fit individuals. This adaptation is an important consideration in producing the increase in stroke volume that occurs with endurance training.

There appear to be two mechanisms that produce the increase in blood volume that follows endurance training. The initial increase is due to large increases in plasma renin and ADH levels during exercise. This results in an increased retention of sodium and water by the kidneys. Chronic training leads to an increase in plasma protein, mainly albumin, which increases the osmolality of the blood. This allows the blood to hold more fluid.

Summary

During exercise blood must be rapidly directed to working muscles to meet their demands for oxygen and substrates. The mechanisms that control blood flow make it possible to maintain or increase blood pressure, to continue supplying blood to the tissues, and to satisfy the metabolic requirements of working muscles.

The blood vessels consist of arteries, arterioles, capillaries, venules, and veins. The blood pressure is high in the arteries but diminishes throughout the circulation until it approaches 0 mmHg when it reaches the right atrium. The factors controlling blood flow in individual vessels are described in Poiseuille's law, which states that flow through a tube varies directly with the difference in pressure, varies to the fourth power of the radius of the tube, and varies inversely with the length of tube and the viscosity of the fluid.

Autoregulation is the principal factor that channels blood to muscles and the heart during exercise. Local autoregulation of blood flow occurs in response to tissue demands for O_2 and fuels, and responds to CO_2, hydrogen ion, adenosine, K^+, and temperature. Muscle blood flow increases exponentially with metabolism. At lower levels of metabolism, blood flow through an active muscle capillary is intermittent, but it becomes continuous as metabolic rate approaches maximum.

Neural–hormonal control mechanisms are also vital in facilitating redistribution of blood to active muscles and vasoconstriction in the viscera and inactive muscles. Sympathetically mediated vasoconstriction of the spleen is helpful in maintaining central blood volume during exercise. Circulating catecholamines and the renin–angiotensin system are involved in the acute control of blood pressure and flow, whereas changes in vascularization and fluid volumes are more important as long-term control mechanisms.

Selected Readings

Andersen, P. Capillary density in skeletal muscle of man. Acta Physiol. Scand. 95: 203–205, 1975.

Beiser, G.D., R. Zelis, S.E. Epstein, D.T. Mason, and E. Braunwald. The role of skin and muscle resistance vessels in reflexes mediated by the baroreceptor system. J. Clin. Invest. 49: 225–231, 1970.

Bern, R.M. and M.N. Levy. Cardiovascular Physiology. St. Louis: C.V. Mosby, 1967. p. 41–59, 86–131, and 198–243.

Bezucha, G.R., M.C. Lenser, P.G. Hanson, and F.J. Nagel. Comparison of hemodynamic responses to static and dynamic exercise. J. Appl. Physiol. 53: 1589–1593, 1982.

Brodal, P., F. Ingjer, and L. Hermansen. Capillary supply of skeletal muscle fibres in untrained and endurance trained men. Am. J. Physiol. 232: H705–H712, 1977.

Chevalier, P.A., K.C. Weber, G.W. Lyons, and D.M. Nicoloff. Hemodynamic changes from stimulation of left ventricular baroreceptors. Am. J. Physiol. 227: 719–728, 1974.

Connolly, R.J. Flow patterns in the capillary bed of rat skeletal muscle at rest and after repetitive tetanic contraction. In: Grayson, J. and W. Zingg (eds.). Microcirculation. New York: Plenum Press, 1976.

Convertino, V.A., P.J. Brock, L.C. Keil, E.M. Bernauer, and J.E. Greenleaf. Exercise training-induced hypervolemia: Role of plasma albumin, renin, and vasopressin. J. Appl. Physiol.: Respirat.Environ.Exercise Physiol. 48: 665–669, 1980.

Eriksson, E. and R. Myrhaghe. Microvascular dimensions and blood flow in skeletal muscle. Acta Physiol. Scand. 81: 459–471, 1972.

Falch, D.K. and S.B. Stromme. Pulmonary blood volume and interventricular circulation time in physically trained and untrained subjects. Eur. J. Appl.Physiol. 40: 211–218, 1979.

Fortney, S.M., E.R. Nadel, C.B. Wenger, and J.R. Bove. Effect of acute alterations of blood volume on circulatory performance in humans. J. Appl. Physiol. 50: 292–298, 1981.

Foster, Carl, D.S. Dymond, J. Carpenter, and D.H. Schmidt. Effect of warm-up on left ventricular response to sudden strenuous exercise. Am. J. Physiol. 53(2):380–383, 1982.

Fujita, T., Y. Sato, K. Ando, H. Noda, N. Ueno, and K. Murakami. Dynamic responses of active and inactive renin and plasma norepinephrine during exercise in normal man. Japn. Heart J. 23: 545–551, 1982.

Guyton, A.C. Textbook of Medical Physiology. Philadelphia: W.B. Saunders, 1976, p. 222–339.

Guyton, A.C., A.W. Cowley, D.B. Young, T.G. Coleman, J.E. Hall, and J.W. DeClue. Integration and control of circulatory function. In: Guyton, A. (ed.). Cardiovascular Physiology II. Baltimore: University Park Press, 1976. p. 341–386.

Henry, F.M. Functional tests IV.: Vasomotor weakness and postural fainting. Res. Q. 14: 144–153, 1943.

Hester, R.L., A.C. Guyton, and B.J. Barber. Reactive and exercise hyperemia during high levels of adenosine infusion. Am. J. Physiol. 243:181–186, 1982.

Honig, C.R., C.L. Odoroff, and J.L. Frierson. Active and passive capillary control in red muscle at rest and in exercise. Am. J. Physiol. 243: H196–H206, 1982.

Honig, C.R. and T.E.J. Gayeski. Mechanisms of capillary recruitment: Relation to flow, tissue P_{O_2}, and motor unit control of skeletal muscle. In: Advances in Physiological Sciences. Oxford: Pergamon, 1981.

Ingjer, F. Maximal aerobic power related to the capillary supply of the quadriceps femoris muscle in man. Acta Physiol. Scand. 104: 238–240, 1978.

Johnson, J.M. Regulation of skin circulation during prolonged exercise. Ann. N.Y. Acad. Sci. 301: 195–212, 1977.

Johnson, J.M. and M.K. Park. Effect of heat stress on cutaneous vascular responses to the initiation of exercise. J. Appl. Physiol. 53: 744–749, 1982.

Johnson, J.M. and L.B. Rowell. Forearm skin and muscle vascular responses to prolonged leg exercise in man. J. Appl. Physiol. 39: 920–924, 1975.

Kesselman, R.H. Gravitational effects on blood distribution. Aerospace Med. 39: 162–165, 1968.

Kjellmer, I. Effect of exercise on the vascular bed of skeletal muscle. Acta Physiol. Scand. 62:18–30, 1964.

Konstam, M.A., S. Tu'meh, J. Wynne, J.R. Beck, J. Kozlowski, and B.L. Holman. Effect of exercise on erythrocyte count and blood activity concentration after technetium-99m in vivo red blood cell labeling. Circulation. 66:638–642, 1982.

Liang, I.Y.S. and H.L. Stone. Effect of exercise conditioning on coronary resistance. J. Appl. Physiol.: Respirat. Environ. Exercise Physiol. 53: 631–636, 1982.

Mohrman, D.E. and E.O. Feigl. Competition between sympathetic vasoconstriction and metabolic vasodilation in the canine coronary circulation. Circ. Res. 42: 79–86, 1978.

Ross, G. Adrenergic responses of the coronary vessels. Circ. Res. 39: 461–465, 1975.

Rowell, L.B. Competition between skin and muscle for blood flow during exercise. In: Nadel, E.R. (ed.) Problems with Temperature Regulation During Exercise. New York: Academic Press, 1977. p. 49–76.

Rowell, L.B. Human cardiovascular adjustments to exercise and thermal stress. Physiol. Rev. 54: 75–159, 1974.

Ruch, T. and H. Patton. Physiology and Biophysics. Philadelphia: W.B. Saunders, 1966. p. 523–542.

Sonn, J., A. Mayevsky, B. Acad, E. Guggenheimer, and J. Kedem. Effect of local ischaemia on the myocardial oxygen balance and its response to heart rate elevation. Q. J. Exp. Physiol. 67: 335–348, 1982.

Stone, H.L. Coronary flow, myocardial oxygen consumption, and exercise training in dogs. J. Appl. Physiol. 49: 759–768, 1980.

Vander, A., J. Sherman, and D. Luciano. Human Physiology. New York: McGraw-Hill, 1975. p. 228–282.

Vogt, F.B. An objective approach to the analysis of tilt table data. Aerospace Med. 37: 1195–1204, 1966.

Wenger, C.B., M.F. Roberts, A.J. Stolwijk, and E.R. Nadel. Forearm blood flow during body temperature transients produced by leg exercise. J. Appl. Physiol. 38: 58–63, 1975.

16

CARDIOVASCULAR DYNAMICS DURING EXERCISE

Cardiovascular capacity is critically important for successful performance in sport and physical activity. The ability to sustain an effort in football or basketball, in running long distances, or in skiing down a mountainside, depends on a strong and able oxygen transport system.

The last two chapters showed that cardiovascular control is accomplished by a coordination of mechanical, neural, and hormonal mechanisms. These control systems enable the circulatory system to increase its output greatly in a very short time. This chapter discusses the specific changes that occur in the cardiovascular system with exercise and, in addition, the adaptations that occur in the system with training.

THE MAJOR DETERMINANTS OF CARDIAC PERFORMANCE DURING EXERCISE

The limits within which the heart can increase its output are dictated by its maximum heart rate and maximum stroke volume. These two factors are in turn, governed by the heart's ability to increase its metabolism. A wide variety of factors affect the metabolic load on the heart during exercise (Figure 16-1), including the following:

Sebastian Coe jogs to relax a bit after winning the 1500 m run at the Moscow Olympics. Performances such as these are dependent upon a high cardiac output, and peripheral metabolic adaptations. (Gamma/Liaison.)

- *Preload* The extent of ventricular filling (Frank–Starling factors).
- *Afterload* Impedance or resistance to ventricular emptying.
- *Contractility* The quality of ventricular performance.
- *Heart rate* Frequency of the cardiac cycle.
- *Synergy of ventricular contraction* Coordination of electrical and mechanical components of the cardiac cycle.
- *Distensibility of the ventricles:* the ability of the ventricles to stretch.

Preload

Preload can be defined as the end-diastolic fiber length or the extent of ventricular filling before contraction. Because of the Frank–Starling mechanism (Chapter 14), the heart pumps all the blood it gets (up to a point). However, as we saw in Chapter 14, the heart is subjected to an increasing load as the walls of the ventricles are stretched (due to the law of Laplace).

A number of factors affect preload during exercise, including cardiac output, blood volume, body position, the pumping action of respiration and skeletal muscles, and venous tone. Endurance activities such as distance running significantly contribute to preload stress.

Preload is determined by the venous return of blood to the heart, which, in turn, is directly affected by the cardiac output. Over the course of a few beats, cardiac output will equal venous return. Thus, during endurance exercise ve-

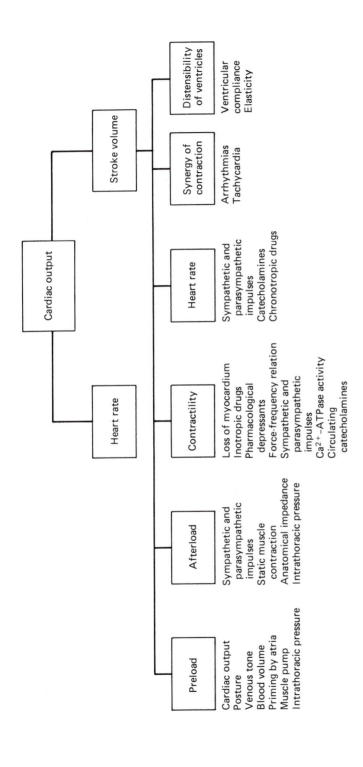

FIGURE 16-1
Factors affecting cardiac output during exercise.

nous return and cardiac output both increase and keep pace with each other. The adaptation to chronic increased preload (called volume load exercise) resulting from endurance training is discussed below.

Blood volume has an important effect on preload. As discussed in the previous chapter, endurance training increases blood volume. Because the circulation is a closed-loop system, increasing blood volume makes more blood available for delivery back to the heart.

A decrease in blood volume can also affect preload. For example, exercise in the heat without adequate fluid replacement can lead to dehydration and thermal injury (Chapter 22). In heat exhaustion, the heart cannot satisfy the demand for blood by the muscles for exercise and by the skin for cooling. The lower blood volume decreases the venous return of blood to the heart and thus impairs cardiac output.

The pumping actions of muscles and ventilation are also important factors in enhancing venous return during exercise. As discussed in Chapter 15, veins have one-way valves that prevent blood from moving away from the heart. Blood enters the capillary beds of muscles when they are relaxed and is pushed toward the heart when they contract. Additionally, the alternating increase and decrease in intraabdominal and thoracic pressures accompanying breathing also act to return more blood to the heart.

Venous tone increases as the intensity of exercise becomes more severe. This decreases the volume of the veins, which facilitates the return of blood to the heart. As discussed in the previous chapter, vasoconstriction occurs as a result of sympathetic stimulation of the circulation.

Posture has an effect on preload, particularly during brief bursts of exertion. At rest, end-diastolic volume is highest in the reclining position and decreases progressively in the seated and standing postures. During exercise in the supine position, end-diastolic volume stays the same, so changes in preload play little role in this type of exercise. In erect exercise, such as cycling and running, end-diastolic volume increases, which increases the preload component of cardiac stress.

The increased preload during exercise in the erect posture is important in increasing stroke volume. This probably occurs because hydrostatic pressure is greater in the lower extremities for erect than for supine exercise. This puts the venous smooth muscle on the stretch, which results in a greater venous tone, thus facilitating the return of blood to the heart.

Prolonged upright exercise at a constant work rate places an increasing load on the heart. Although the metabolic requirement of the exercise does not change, there is a progressive decrease in venous return of blood to the heart. This leads to a decreased stroke volume and a progressive rise in heart rate. This phenomenon is called *cardiovascular drift* and is probably caused by a breakdown in sympathetic blood flow control mechanisms and/or increased distribution of blood to the skin for cooling.

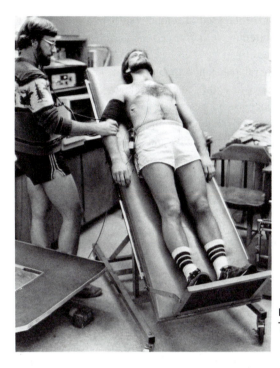

FIGURE 16-2
The tilt table.

(Wayne Glasker.)

The effects of gravity on venous return are particularly acute in static postures such as standing. Soldiers standing at attention for prolonged periods have been known to faint because venous return and cardiac output become inadequate. Because the soldier is standing still, the muscle pump mechanism is not operating to facilitate the return of blood to the heart. Gradually, venous distension increases as compensatory mechanisms by the baroreceptors and sympathetic nervous system become less effective. This phenomenon is accentuated in warmer and more humid weather as a result of the increased shunting of blood to the periphery for purposes of cooling the body.

The tilt table test (Figure 16-2) is often used to measure the capacity of autonomic reflexes to adjust to changes in posture. This device places subjects in a stationary upright posture. Heart rate and blood pressure measurements are made during the test. Gradually, heart rate increases and systolic blood pressure drops until the subject faints.

The tilt table test is frequently used in space and aviation physiology. In space flight, chronic lack of hydrostatic pressure produced by weightlessness results in lower venous tone. This causes a diminished capacity of the cardiovascular system to accommodate to the effects of gravity.

Afterload

Afterload is impedance or resistance to ventricular emptying. Increased afterload has a negative influence on cardiac performance because it creates an increased work load for the heart. Increased afterload results in a reduced ventricular-ejection fraction and shortening velocity, and increased ventricular end-diastolic and end-systolic volumes.

In hypertension, high peripheral resistance to blood flow results in an increased afterload stress that can cause the heart muscle to hypertrophy and sometimes fail. The resulting increase in heart mass increases myocardial O_2 consumption. Because hypertension is often accompanied by coronary atherosclerosis, the combination of cardiac hypertrophy and coronary ischemia places an extraordinary load on the heart.

The destructive effect of chronic afterload stress of hypertension lies partly in the nature of the hypertrophy. The increased heart size is due mostly to an increase in the size and number of connective tissue cells rather than cardiac muscle cells. This results in a marked increase in the load of the heart muscle. The heart has an increased metabolic requirement because of its size but very little increase in its functional capacity.

Exercises that involve high muscle tension, such as weight lifting, can also cause an increased afterload stress. Strength exercises typically involve a *Valsalva maneuver*—expiration against a closed glottis. Valsalva increases the intrathoracic pressure, which is partly responsible for the increased afterload. Also, these "static" exercises (static, as used here, refers to exercise that subjects the heart to an increased pressure load) use smaller muscle groups than dynamic exercise. Muscle contractions in these types of exercises can momentarily impede blood flow, which greatly increases the afterload stress. Brachial arterial pressures of 400 mmHg. have been noted in athletes performing heavy weight lifts.

Static exercise that involves a dynamic motion, such as shoveling snow, causes a much greater stress on the heart than purely isometric exercise, such as squeezing a handgrip dynamometer. "Static–dynamic" exercise builds up more intrathoracic pressure and impedes blood flow with near-maximal contractions in relatively large muscle groups. So, there is a greater afterload stress. Additionally, the dynamic component of the exercise results in more blood returning to the heart than in purely static exercise. Thus, there is a greater preload stress in addition to the greater afterload stress. "Static–dynamic" exercise results in many deaths each year in people with heart disease who are not accustomed to such severe effort.

Repeated static exercise can result in prolonged elevation of heart rate during the course of a workout. This has led some people to believe that weight lifting can produce a significant endurance training effect. However, there is a fundamental difference between endurance and static exercise in their effects on the heart. Endurance exercise places a volume load on the heart, whereas strength exercise exerts a pressure load.

Strength training results in a greater left ventricular mass and wall thickness,

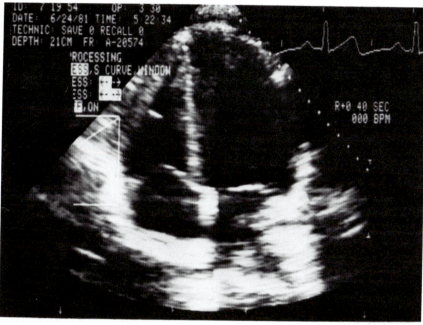

FIGURE 16-3

Two-dimensional echocardiogram of a world-class discus thrower. Weight-trained athletes periodically subject their hearts to an increased pressure load that results in greater ventricular mass and wall thickness.

whereas endurance exercise increases left ventricular dimensions with a lesser amount of ventricular hypertrophy. The pressure load hypertrophy that results from weight lifting is similar to that resulting from hypertension. However, the change due to weight training seems to be functionally benign, even in athletes who have lifted heavy weights for 20 yr or more (Figure 16-3). In fact, most elite weight-trained athletes, such as discus throwers and shot-putters, have better developed cardiovascular systems than the average population.

Contractility

Contractility refers to the quality of ventricular performance at constant conditions of loading and heart rate. This factor is related to the rate of reaction at the contractile sites in the heart. Factors affecting preload are related to the relationship of the changing length of cardiac muscle fibers and their influence on the number of active contractile sites. However, because of the close relationship between the two, it is difficult to discuss contractility without taking preload into consideration.

Myocardial contractility has a tremendous effect on the Frank–Starling curve (Figure 16-4). In heart failure, when the inotropic capacity of the heart is greatly

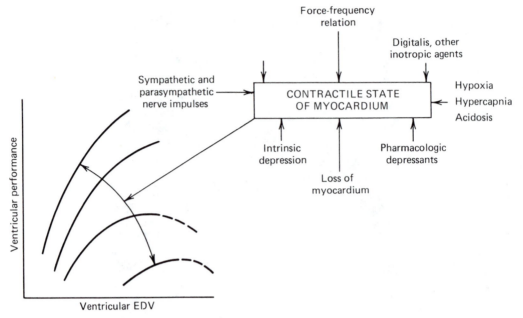

FIGURE 16-4

The effects of changes in myocardial contractility on the Frank–Starling curve and the major factors affecting contractility. EDV, End diastolic volume.

(SOURCE: From E. Braunwald, J. Ross, and E.H. Sonnenblick, 1967.)

reduced, increased preload is much less effective in improving ventricular performance. The Frank–Starling curve shifts to the right (Figure 14-6). Likewise, factors such as sympathetic stimulation, circulating catecholamines, and inotropic agents such as digitalis improve cardiac performance and shift the Frank–Starling curve to the left.

Cardiac muscle depends on increased force by individual fibers to get a stronger contraction. Skeletal muscles, in contrast, rely on recruitment of additional fibers for increased strength. The most important factor dictating the inotropic capacity of the heart is the availability of calcium ion (Ca^{2+}), its rate of release, and its rate of uptake. Calcium ion stimulated myosin-ATPase activity in the heart is integrally related to the excitation–contraction coupling mechanism that determines the strength of contraction. Greater myocardial Ca^{2+} myosin-ATPase activity results in a greater intensity of contraction of the heart.

Endurance training has a marked effect on Ca^{2+} myosin-ATPase activity in the heart. This enhances the contractility of heart muscle and represents one of the major adaptations resulting from endurance training. This enables the heart to increase its stroke volume during exercise. It has also been suggested that training increases extracellular Ca^{2+} influx during depolarization.

Increased stroke volume in the trained person plays a greater role in increasing cardiac output than in the untrained individual. Increased contractility plays an important role in the enhanced stroke volume that occurs with training. This adaptation can increase maximum cardiac output by 15 to 20%.

Coronary blood flow greatly affects contractility. Impaired coronary circulation can lead to decreased contractility and eventual necrosis of cardiac muscle. Loss of cardiac muscle will, in turn, further affect myocardial contractility and exercise performance.

Heart Rate

Heart rate is the major determinant of cardiac output, particularly during moderate to maximal exercise. It is the most important factor affecting myocardial O_2 consumption. Increased heart rate provides a rapid mechanism for increasing O_2 transport during exercise. Even anticipation of exercise results in centrally mediated inhibition of the cardiac parasympathetics and stimulation of the sympathetics acting on the SA node.

Maximal heart rate is fairly consistent under a wide variety of circumstances, but tends to decrease with age. A rough estimate of maximal heart rate can be obtained by subtracting one's age from 220. However, this method is subject to considerable error. Whenever possible, it is better to measure maximal heart rate directly. Maximal heart rate can be accurately determined by measuring the highest heart rate obtained during a treadmill test that pushes the subject to $\dot{V}_{O_2max}$.

There is a high correlation between coronary blood flow and heart rate. In healthy individuals, coronary blood flow seems to be adequate at all levels of exercise. However, in persons with ischemic disease, coronary blood flow may not be adequate during exercise. This can sometimes be detected in the electrocardiogram as a depression of the ST segment (Chapter 24). Coronary ischemia can have a marked effect on cardiac performance by impairing both the chronotropic and inotropic capacity of the heart.

Synergy of Ventricular Contraction

Increases in cardiac output during exercise depend on a coordination of heart rate and ventricular filling. Disturbances in the excitation–contraction coupling mechanisms can upset this coordination. Arrhythmias, such as premature ventricular contractions (PVCs), can have adverse hemodynamic effects, particularly if they occur repeatedly.

A ventricular ectopic focus causing ventricular depolarization results in PVCs. *Ectopic foci* are alternate pacemakers that exist in both the atria and ventricles. When a ventricular ectopic focus fires, depolarization occurs directly across heart muscle. It is usually out of synchronization with the normal cardiac cycle, so there is impairment of end-diastolic volume and stroke volume.

Multiple ventricular premature beats are particularly dangerous because they can result in tachycardia or fibrillation. In tachycardia, the heart may beat so fast that there is no chance for adequate ventricular filling during diastole. In ventricular fibrillation, the normal cardiac cycle ceases to operate.

Distensibility of the Ventricles

Distensibility and compliance refer to the ability of the ventricles to stretch. Normal, healthy ventricles stretch easily and do not resist filling. Lack of distensibility can limit cardiac output in the diseased heart. Stiff fibrous tissue replaces the original fibers following myocardial infarction or when the myocardium is injured by ischemia. This results in a heart that is much less distensible and less capable of dealing with the increased venous return of blood that occurs during exercise. Other pathological conditions that are accompanied by decreased ventricular compliance include chronic overstretching of the myocardium and myocardial hypertrophy.

CARDIOVASCULAR CAPACITY IN EXERCISE

Whole body maximal oxygen consumption ($\dot{V}_{O_2max}$) is the best measure of the capacity of the cardiovascular system, provided there is no pulmonary disease present. The $\dot{V}_{O_2max}$ is the product of maximum cardiac output and maximum arteriovenous (a–v) O_2 difference. It is the point at which O_2 consumption fails to rise despite an increased exercise intensity or work load. After $\dot{V}_{O_2max}$ has been reached, work can be sustained by nonoxidative metabolism in skeletal muscle (Figure 16-5). Whereas $\dot{V}_{O_2max}$ is the best predictor of cardiovascular capacity, biochemical factors, such as oxidative enzyme activity and the size of the mitochondrial reticulum, are better predictors of endurance (endurance is the ability to sustain a particular level of physical effort). $\dot{V}_{O_2}$ is related to cardiac output and arterio-venous O_2 difference by the Fick equation:

$$\dot{V}_{O_2} = \dot{Q} \ (a-v) \ O_2$$

Since the ability to increase $(a-v) \ O_2$ is limited, $\dot{Q}_{max}$ largely determines $\dot{V}_{O_2max}$.

Cardiac output is the most important factor determining $\dot{V}_{O_2max}$. Cardiac output can increase about 20% from endurance training, which accounts for most of the increase in $\dot{V}_{O_2max}$ that occurs. Maximum $(a-v) \ O_2$ difference, a measure of O_2 extraction, changes very little with training. Additionally, there is little difference in $(a-v) \ O_2$ difference between endurance athletes and the sedentary individual during exercise. In contrast, elite endurance athletes have been known to have maximum cardiac outputs in excess of 35 liters $\cdot$ min^{-1}, almost twice that of the average sedentary person. It is easy to see why the limit of $\dot{V}_{O_2max}$ is dictated by the limit of cardiac output.

Work (kg · m/min)	$\dot{V}_{O_2}$ (liters · min^{-1})
100	.5
200	.7
300	.9
400	1.1
500	1.3
600	1.5
700	1.7
800	1.9
900	2.1
1000	2.3
1100	2.5
1200	2.7
1300	2.9
1400	3.1
1500	3.3
1600	3.5
1700	3.7
1800	3.9
1900	4.1
2000	4.3

FIGURE 16-5
Oxygen consumption at various work rates on a bicycle ergometer. $\dot{V}_{O_2}$ (ml · min^{-1}) = work rate (kg · m/min) $\times$ 2 ml O_2/kg · m + sitting $\dot{V}_{O_2}$ (300 ml · min^{-1}).

The measurement of maximal O_2 consumption must satisfy several objective criteria.

1. The exercise must use at least 50% of the total muscle mass continuously and rhythmically for a prolonged period of time.
2. The results must be independent of motivational or skill factors.
3. There must be a leveling of oxygen consumption with increasing intensities of exercise.
4. The measurement must be made under standard experimental conditions. It cannot be made in a stressful environment that exposes the subject to excessive heat, humidity, air pollution, or altitude.

An O_2 consumption measurement made during maximal exercise but not according to the criteria necessary for $\dot{V}_{O_2max}$ is called $\dot{V}_{O_2peak}$. In many instances, $\dot{V}_{O_2peak}$ will equal $\dot{V}_{O_2max}$, but it is important to distinguish between the two because of the inferences made regarding the relationship between $\dot{V}_{O_2max}$ and O_2 transport capacity.

It is not usually possible to measure $\dot{V}_{O_2max}$ during arm exercise. An untrained person will fatigue relatively rapidly during this type of work. Typical $\dot{V}_{O_2peak}$ on an arm ergometer may be 70% of the $\dot{V}_{O_2max}$ measured on a treadmill. However, in a trained rower or canoeist, the difference between the two measurements may be as small as 1 to 2%.

A bicycle ergometer is often used to measure $\dot{V}_{O_2max}$. However, most studies show that these values are 10 to 15% less than those gathered on a treadmill. Additionally, cycling skills and body weight affect the results. So, even though values gathered on bicycle ergometers are commonly called $\dot{V}_{O_2max}$, they are in reality $\dot{V}_{O_2peak}$.

So, an exercise that results in local muscle fatigue before the limits of cardiovascular capacity are reached is not appropriate for measuring $\dot{V}_{O_2max}$ and may not be appropriate for cardiovascular conditioning. An excellent example of this is wheelchair propulsion. Many handicapped individuals use a wheelchair as a means of conditioning their cardiovascular systems (Figure 16-6). However, peak heart rate and $\dot{V}_{O_2peak}$ are much less than what could be attained on other types of exercise. Local muscle fatigue rather than cardiovascular capacity is limiting. Because of this, heavy exercise in a wheelchair may be less useful than other forms of exercise for developing aerobic capacity.

The leveling-off criterion of O_2 consumption is sometimes difficult to achieve in untrained subjects. These people are often unable or unwilling to withstand the pain that accompanies work after maximal O_2 consumption has been reached. However, on a treadmill, $\dot{V}_{O_2peak}$ usually renders an excellent approximation of $\dot{V}_{O_2max}$. A subject typically terminates the test when the limits of cardiovascular capacity have been attained.

For practical purposes, there are several guidelines for judging whether $\dot{V}_{O_2peak}$ is equal to $\dot{V}_{O_2max}$. These include a respiratory exchange ratio (R) greater than 1.0 ($R = \dot{V}_{CO_2}/\dot{V}_{O_2}$), high blood lactate level (>140 mg $\cdot$ dl^{-1}), a peak heart rate similar to the age-predicted maximum, or progressively diminishing differences between successive O_2 consumption measurements.

CHANGES IN CARDIOVASCULAR PARAMETERS DURING EXERCISE AND WITH TRAINING

Changes in cardiovascular function during exercise depend on the type and intensity of exercise. Dynamic exercise requiring a large muscle mass elicits the greatest response from the cardiovascular system. There are large increases in cardiac output, heart rate, and systolic blood pressure, with little change in mean arterial or diastolic blood pressure. Strength exercises, which use less muscle mass, result in marked increases in systolic, diastolic, and mean blood pressure, with more moderate increases in heart rate and cardiac output.

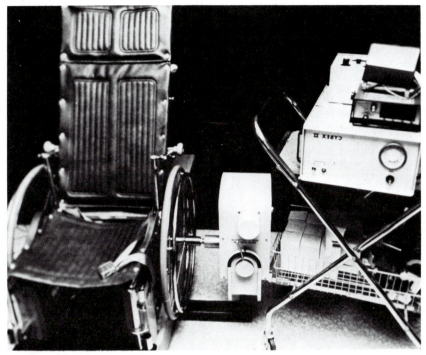

FIGURE 16-6
An isokinetic wheelchair ergometer. Local muscle fatigue may limit the usefulness of this device in developing cardiovascular capacity in some individuals.

Endurance training results in adaptive changes in many aspects of cardio-vascular function. These changes are summarized in Table 16-1. In general, the heart improves its ability to pump blood, mainly by increasing its stroke volume. This occurs because of an increase in end-diastolic volume, left ventricular contractility, and left ventricular muscle mass. At the same time, the metabolic load on the heart decreases at rest and at any submaximal work rate because heart rate decreases.

Endurance or aerobic exercise is best for improving the capacity of the cardiovascular system (provided it meets certain criteria). These exercises require the use of at least 50% of the body's muscle mass in rhythmical exercise, for at least 15 to 20 min, 3 to 5 days a week, and above 60 to 70% of maximum capacity to obtain a significant conditioning effect. Obviously, considerably more training is required to develop the high levels of fitness required of endurance athletes (Appendix VIII).

As discussed in Chapter 21, adaptation to endurance training is somewhat specific. So, swimming, for example, will improve cardiovascular performance

TABLE 16-1

Cardiovascular Adaptations Resulting from Endurance Training

Factor	Rest	Submaximal Exercise	Maximal Exercise
Heart rate	↓	↓	0↓
Stroke volume	↑	↑	↑
$(a-v)$ O_2 difference	0↑	↑	↑
Cardiac output	0↓	0↓	↑
$\dot{V}_{O_2}$	0	0	↑
Work capacity	—	—	↑
Systolic blood pressure	0↓	0↓	0
Diastolic blood pressure	0↓	0↓	0↓
Mean arterial blood pressure	0↓	0↓	0↓
Total peripheral resistance	0	0↓	0↓
Coronary blood flow	↓	↓	↑
Brain blood flow	0	0	0
Visceral blood flow	0	↑	0
Inactive muscle blood flow	0	0	0
Active muscle blood flow	↓0	↓0	↑
Skin blood flow	0	0	0
Blood volume	↑	—	—
Plasma volume	↑	—	—
Red cell mass	0↑	—	—
Heart volume	↑	—	—

Symbols:
↑ increase
↓ decrease
0 no change

in that activity, but will be much less effective in improving endurance in running. Changes in cardiovascular performance with exercise and with training are shown in Figure 16-7*a–j* and Table 16-2.

Interval training, repeated bouts of short to moderate duration exercise, is also used to improve cardiovascular condition. This mode of training manipulates four factors: distance, speed, repetition, and rest. Interval training allows the athlete to train at a higher intensity than is typically employed during competition and thus acts as an overload. Although this type of training optimizes the development of cardiovascular capacity, it is less effective in eliciting the biochemical changes that are critical for optimal endurance performance. Therefore, the endurance athlete must practice both interval and over-distance training (Chapter 21).

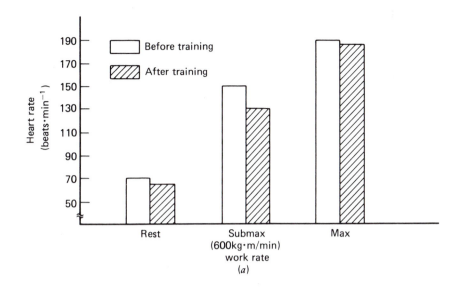

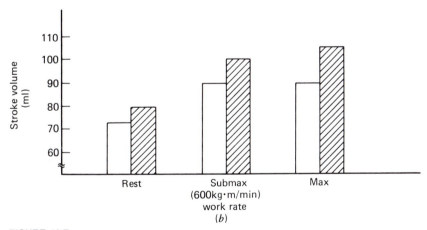

FIGURE 16-7

Cardiovascular responses to exercise before (unshaded columns) and after (shaded columns) endurance training in a 40-yr-old man. (a) Heart rate; (b) stroke volume; (c) cardiac output; (d) a–v O_2 difference; (e) O_2 consumption. In diagrams (f) through (j), open symbols show valuesa before training and closed symbols show values after training. (f) Blood pressure (triangles, systolic; circles, diastolic; (g) total peripheral resistance; (h) coronary (triangles) and brain (columns) blood flow; (i) visceral (triangles) and inactive muscle (columns) blood flow; (j) active muscle (triangles) and skin (columns) blood flow.

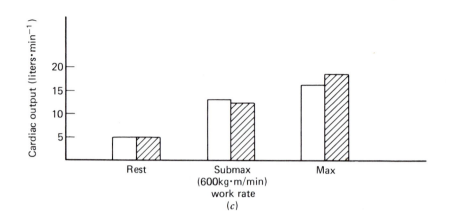

(c)

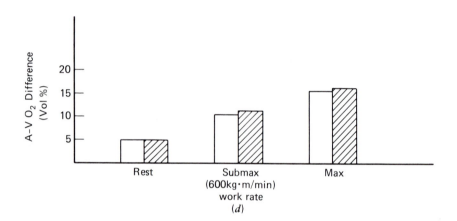

(d)

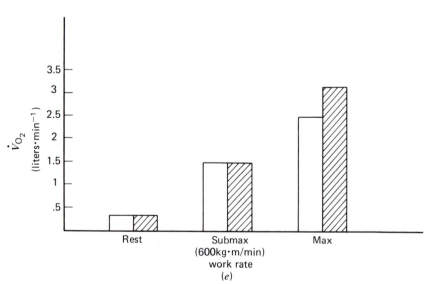

(e)

FIGURE 16-7 *(continued).*

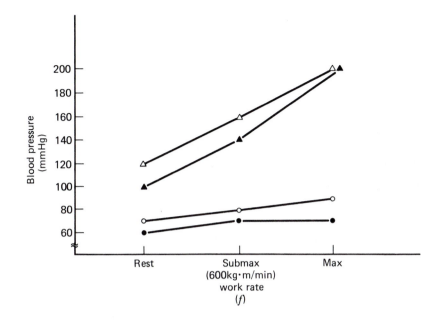

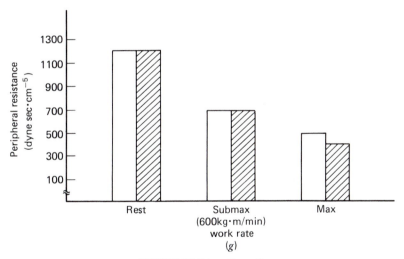

FIGURE 16-7 (continued).

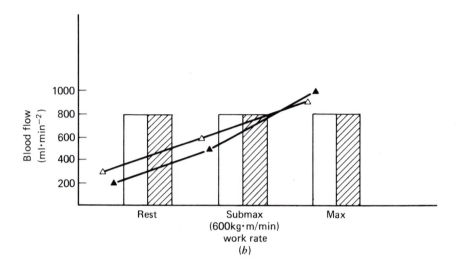

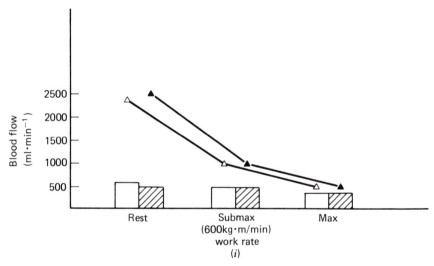

FIGURE 16-7 *(continued)*

Heart Rate

Heart rate response during exercise   Heart rate is the most important factor increasing cardiac output during exercise. In dynamic exercise, heart rate rises with work load and oxygen consumption, leveling off at $\dot{V}_{O_2max}$ (Figure 16-7a). Typically, heart rate will rise from a resting value of approximately 70 beats $\cdot$ min^{-1} to a maximum heart rate of 180 beats or more. The rate and

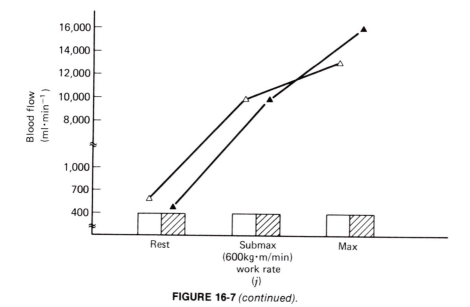

FIGURE 16-7 *(continued).*

extent of the increase depend on the type of exercise and on the fitness, age, and sex of the individual.

During submaximal exercise, heart rate rises and then levels off when the O_2 transport requirements of the activity have been satisfied. At increasing intensities, this leveling off of heart rate takes longer and is more difficult to achieve. In prolonged exercise, as discussed, there is a steady increase in heart rate at the same work rate (Chapter 15).

At rest and during low-intensity submaximal exercise, heart rate is affected by a wide variety of factors. Factors such as anxiety, dehydration, ambient temperature, altitude, and elapsed time after meals can increase heart rate, sometimes markedly. "Resting" heart rates of over 130 beats $\cdot$ min^{-1} are not unusual before anxiety-producing situations such as treadmill tests and sports competitions.

Heart rate response to an identical work load varies with the kind of exercise. Arm-cranking causes a 10% higher heart rate response than either upright or supine cycling (Figure 16-6). Because work cannot be accurately measured on a treadmill, it is difficult to compare submaximal treadmill performance with cycling. However, at the same relative percentage of $\dot{V}_{O_2max}$, cycling produces a higher heart rate than exercise on the treadmill. Differences in heart rate at the same work rate seem to be due to the relative load placed on individual muscle groups. In arm-cranking, for example, small muscle groups are taxed at a higher percentage of their maximum capacity than are the larger leg muscles during treadmill exercise.

TABLE 16-2
Changes in Cardiovascular and Pulmonary Function as a Result of Endurance-Type Physical Training[a]

Measurement	Resting		Upright Submaximal "Steady-state" Exercise		Upright Maximal Exercise	
	Pre	Post	Pre	Post	Pre	Post
Heart rate (beats · min⁻¹)	70	63	150	130	185	182
Stroke volume (ml · beat⁻¹)	72	80	90	102	90	105
Cardiac output (liters · min⁻¹)	5.0	5.0	13.5	13.2	16.6	19.1
$(a-v)$ O₂ Difference (Vol %)	5.6	5.6	11.0	11.3	16.2	16.5
O₂ Uptake (liters · min)	0.280	0.280	1.485	1.485	2.685	3.150
(ml · kg⁻¹ · min⁻¹)	3.7	3.7	19.8	19.8	35.8	42.0
(METS)[b]	1.0	1.0	5.7	5.7	10.2	12.0
Work load (kg · kgm · min⁻¹)	—	—	600	600	1050	1500
Blood pressure (mmHg)						
Systemic arterial systolic BP	120	114	156	140	200	200
Systemic arterial diastolic BP	75	70	80	75	85	75
Systemic arterial mean	90	88	126	118	155	152
Total peripheral resistance (dyne sec · cm⁻⁵)	1250	1250	750	750	450	390
Blood flow (ml · min⁻¹)						
Coronary	260	250	600	560	900	940
Brain	740	740	740	740	740	740
Viscera	2400	2500	900	1000	500	500
Inactive muscle	600	555	500	500	300	300
Active muscle	600	555	10,360	10,000	13,760	16,220
Skin	400	400	400	400	400	400
TOTAL	5000	5000	13,500	13,200	16,600	19,100

Variable	Pre		Post	Pre		Post	Pre		Post
Blood volume (liters)	5.1	+0	5.3						
Plasma volume (liters)	2.8	+0	3.0						
Red cell mass (liters)	2.3	+0−	2.3						
Heart volume (ml)	730	0	785						
Pulmonary ventilation (liters · min⁻¹)	10.2	0	10.3	44.8	0̄	38.2	129	+	145
Respiratory rate (breaths · min⁻¹)	12	0−	12	30	−	24	43	+	52
Tidal volume	850	0+	855	1.5	0+	1.6	3.0	0−	2.8
Lung diffusing capacity (D_L) (ml at STPD)[c]	34.1	0+	35.2	40.6	0+	42.8	48.2	+	50.6
Pulmonary capillary blood volume (ml)	90.1	+	97.2	129.3	+	141.2	124.5	+	220.0
Vital capacity (liters)	5.1	0+	5.2						
Blood lactic acid (mM)	0.7	0	0.7	3.9	−	3.0	11.0	0+−	12.4
Blood pH	7.43	0	7.43	7.41	0+	7.43	7.33	0̄	7.29
Recovery rate		+			+			+	

[a] Estimated for a healthy man, age 45, weighing 75 kg. Pre = Pretraining; Post = posttraining. Minus (−) sign means usually decrease in value with training. Plus (+) sign means usually increase in value with training. Zero (0) sign means usually no change in value with training.

[b] A MET is generally equal to the O_2 cost at rest. One MET is equal to 3.5 ml · kg⁻¹ of body weight per min of O_2 uptake or 1.2 calories · min⁻¹.

[c] STPD, Standard temperature (0° C), pressure 760 mmHg.

SOURCE: Data courtesy of William Haskell

It is not always appropriate to judge physical fitness by the heart rate response to a particular work load. Because maximal heart rate is variable, large fluctuations in submaximal heart rates are possible at identical levels of O_2 consumption. Likewise, postexercise heart rates are a rough measure of fitness. Although fit people do tend to recover faster, the rate at which heart rate decreases after exercise is dependent on such factors as maximum heart rate, the intensity of exercise, and the duration of exercise. There is a very low correlation between $\dot{V}_{O_2max}$ and recovery heart rate. However, changes in physical fitness can be assessed by monitoring heart rate responses to a given load before and after training.

Heart rates are lower during strength exercises such as weight lifting. Heart rate increases in proportion to the muscle mass used and the percentage of maximum voluntary contraction. So, a dynamic lift, such as a clean, and jerk, using heavy weights, will increase the heart rate more than a one-arm biceps curl using minimum resistance.

Heart rate can be valuable in writing individual exercise prescriptions. By measuring the heart rate either during or immediately after exercise, one can easily assess the metabolic load of the activity (70 to 85% of maximal heart rate is equal to about 57 to 78% of $\dot{V}_{O_2max}$). Exercise heart rate can provide a reasonable estimate of cardiac load, providing an index for such factors as coronary blood flow, myocardial O_2 intake, percentage of $\dot{V}_{O_2max}$, and respiratory exchange ratio.

Effect of training on heart rate　Endurance training reduces resting heart rate. It is not unusual to see resting heart rates of less than 40 beats $\cdot$ min^{-1} in champion endurance runners. Although some of these extremely low heart rates are probably genetically determined, aerobic training does induce some of it. It is important to note, however, that a low resting heart rate is not necessarily a sign of physical fitness. On the contrary, bradycardia is often a sign of disease. The reduction in resting heart rate with training is the important sign of fitness, rather than the low heart rate itself.

Training also decreases the heart rate response to a given submaximal level of exercise. Although the metabolic requirement of the work load remains the same (once the movement patterns have been learned), the decline in heart rate does not reduce cardiac output because of a concomitant increase in stroke volume.

Training has only a small effect on maximal heart rate, usually decreasing it by about 3 beats $\cdot$ min^{-1}. Because maximal heart rate is relatively stable, only tending to decrease with age, it can be used as a reference point for judging the relative intensity of exercise. For example, suppose a person achieved a heart rate of 140 beats $\cdot$ min^{-1} running a 10-min mile. The person then returned after 6 months of endurance training and ran another 10-min mile, only this time the heart rate was only 130 $\cdot$ min^{-1}. The work load no longer represented the same percentage of maximum heart rate (which remained the same). Because the heart rate decreased at an identical submaximal work rate, the running speed would have to increase to achieve the same relative intensity of exercise.

The mechanism for decreased heart rate at rest and maximal exercise is not completely clear. There appear to be an increased parasympathetic tone and decreased sympathetic activity. This may be caused by an increased parasympathetic release of acetylcholine in the heart, decreased uptake and turnover of catecholamines, or perhaps a decreased sensitivity to catacholamines. Butler et al. found a high inverse correlation between maximal oxygen consumption and adrenoreceptor density and a decrease in these receptors following endurance training.

The decreased catecholamine uptake has been identified as a protective mechanism against the effects of emotional stress. Stress increases catecholamine secretion, which increases the O_2 consumptioin of the heart, even in the absence of need for an increase in myocardial metabolism. In heart disease, this catecholamine-produced increase in O_2 consumption can result in hypoxia in the myocardium.

Stroke Volume

Stroke volume during exercise   Stroke volume increases during exercise in both the upright and supine positions (Figure 16-7b). Stroke volume increases steadily until about 25% of $\dot{V}_{O_2max}$ and then tends to level off. After peak stroke volume has been achieved, additions in cardiac output occur because of increased heart rate. In a sedentary man, stroke volume can increase from about 72 ml to about 90 ml at maximal exercise.

Although changes in stroke volume appear to be similar in the upright and supine postures, the mechanics of the change are different. Studies using a technique called radionuclide ventriculography indicate that left ventricular end-diastolic volume (EDV) increases during upright exercise but is unchanged during supine exercise. Opposite changes in end-systolic volume result in the same ejection fraction for both types of exercise.

Stroke volume is perhaps the most important factor determining individual differences in $\dot{V}_{O_2max}$. This is readily apparent when the components of cardiac output of a sedentary man with a champion cross-country skier are compared. Both men have maximum heart rates of 185 beats $\cdot$ min^{-1}, yet the maximum cardiac output of the untrained man is 16.6 liters $\cdot$ min^{-1}, whereas that of the skier is 32 liters. The stroke volumes of the skier and the sedentary man were 173 and 90 ml, respectively.

Stroke volume and training   Endurance training increases the stroke volume at rest and during submaximal and maximal exercise. Most studies indicate that training can increase stroke volume by no more than about 20%. The tremendous levels seen in world-class runners and cross-country skiers are probably genetically determined.

The training-induced increase in stroke volume seems to be due to a variety of factors. The decreased heart rate results in increased ventricular filling (increased end-diastolic volume), activating the Frank–Starling mechanism, which enhances ejection. This process is facilitated by an increase of the ventricular volume and a thickening of the ventricular walls, as well as an increase in blood volume (principally the plasma volume). As previously discussed, there is increased Ca^{2+} influx and Ca^{2+} myosin-ATPase activity, which enhances myocardial contractility. Adaptations in heart rate and stroke volume make the energy utilization by the heart more efficient.

Blood Pressure

Blood pressure during exercise   It is very important that blood pressure increase during exercise, because blood flow must be maintained to critical areas such as the heart and brain, as well as satisfying the requirements of the working muscles and skin. Blood pressure is a function of cardiac output and peripheral resistance. There is a large drop in peripheral resistance during exercise, but blood pressure is increased. This occurs because of the large increase in cardiac output and vasoconstriction in nonworking tissues (Figure 16-7c, d, f, g, and h–j).

Systolic blood pressure rises steadily during exercise from about 120 mmHg at rest to 180 mmHg or more during maximal exercise. It follows the same general trend as heart rate. The increase is variable; maximum systolic pressures can vary from as little as 150 to over 250 mmHg in a normal individual. Likewise, mean arterial pressure rises from about 90 mmHg at rest to about 155 mmHg at maximal exercise. A failure to increase systolic and mean arterial blood pressures during exercise is correlated with an increased prevalence of coronary artery disease.

There is debate in the scientific literature at present about the maximum safe systolic pressure. While some experts do not become overly concerned about pressures above even 250 mmHg, others suggest terminating stress tests at levels above 220 mmHg. It is important to look at the characteristics of the subject. A high systolic pressure is very significant in a hypertensive patient, but is probably meaningless in a world-class endurance athlete. During exercise, systolic blood pressure increases with cardiac output. Because endurance athletes have extremely high cardiac output capacities, their systolic blood pressures will naturally be higher.

High systolic pressure during exercise in untrained people is of concern. The combination of a high heart rate and systolic blood pressure indicates that the myocardial O_2 consumption is also high. In fact, the pressure–pulse product (systolic blood pressure $\times$ heart rate) is an excellent predictor of myocardial work. If a person has heart disease, extreme levels of systolic blood pressure could easily result in myocardial hypoxia.

Diastolic pressure changes little during exercise in the normal individual. Typically, there is either no change or a slight decrease of less than 10 mmHg during exercise, and a small decrease during recovery of less than 4 mmHg. A study by Sheps and associates indicated that an increase in diastolic pressure of 15 mmHg or more is associated with a greater prevalence of coronary artery disease, a greater severity of disease, and a greater frequency of left ventricular contraction abnormalities. Exercise and hypertension are discussed in Chapter 26.

Blood pressure and training   Endurance training tends to reduce resting and submaximal exercise systolic, diastolic, and mean arterial blood pressures. Diastolic and mean arterial but not systolic blood pressures are reduced at maximal exercise. The mechanism of reduced blood pressure at rest is not known. However, the fact that it occurs is another reason that exercise training is important in reducing the risk of heart disease.

Blood Flow

Blood flow during exercise   Blood flow control mechanisms were discussed in the previous chapter. During exercise there is a redistribution of blood from inactive to active tissues. Critical areas such as the brain and heart are spared the vasoconstriction that occurs in other areas (Figure 16-7h–j). There is a progressive increase in the sympathetic vasomotor activity with increasing severity of exercise.

Resting blood flow to the spleen and kidneys is about 2.8 liters $\cdot$ min^{-1}. It is reduced to about 500 ml during maximal exercise. After training, blood flow increases to these areas at any absolute value of O_2 consumption. However, when the exercise intensity is expressed as a percentage of $\dot{V}_{O_2max}$, then the blood supply is about the same in the trained and untrained individuals. Training seems to have little effect on the distribution of blood to these areas at high levels of exercise.

Muscle blood flow is lower in the trained person during submaximal exercise but considerably higher during maximal exercise. This points to the importance of the metabolic adaptations (discussed in previous chapters) in improving performance in submaximal exercise, while maximum cardiovascular capacity seems to be paramount in determining $\dot{V}_{O_2max}$.

Coronary blood flow increases with intensity during exercise from about 260 ml $\cdot$ min^{-1} at rest to 900 ml $\cdot$ min^{-1} at maximal exercise. As discussed previously, increases in coronary blood flow occur mainly by autoregulation and mainly during diastole. Therefore, coronary disease, which impedes blood flow, could be catastrophic.

Recent research has pointed to the importance of warm-up in facilitating the increase in coronary blood flow during the early stages of exercise. Ischemia-like electrocardiographic changes (ST segment depression) occurred in the majority of people subjected to sudden strenuous exercise. Although these changes appear to be easily tolerated in healthy individuals, they could be very dangerous in persons with heart disease.

Blood flow and training With training, there is a slight decrease in coronary blood flow at rest and during submaximal exercise. The increased stroke volume and decreased heart rate result in a reduced myocardial O_2 consumption, which decreases the requirements for blood. There is an increase in coronary blood flow at maximal exercise with training that supports the metabolic requirement of increasing cardiac output. Again, blood flow accommodates to the metabolic load in the normal heart. As discussed, training does not seem to increase myocardial vascularity unless there is stenosis or a myocardial infarction.

Arteriovenous Oxygen Difference

Arteriovenous oxygen difference during exercise There is a linear increase in $(a - v)$ O_2 difference with the intensity of exercise (Figure 16-7d). The resting value of about 5.6 volume percent (i.e., ml $O_2 \cdot dl^{-1}$) is increased to about 16 vol % at maximal exercise. There is always some oxygenated blood returning to the heart even at exhaustive levels of exercise. This is because some blood continues to flow through metabolically less active tissues that do not fully extract the oxygen from the blood.

$(a - v)$ O_2 difference and training Arteriovenous O_2 difference increases slightly with training. Possible mechanisms for the change include a rightward shift of the oxyhemoglobin dissociation curve, mitochondrial adaptations, elevated myoglobin concentration, and/or increased muscle capillary density. Increased capillary density results in a shorter diffusion distance between the circulation and muscle. The circulatory and biochemical changes appear to be linked. Stromme and Ingjer have noted that the number of capillaries around each fiber is a good predictor of aerobic power and is related to the mitochondrial content of the fiber. The adaptation in $(a - v)$ O_2 difference does not seem to occur to the same extent in older individuals who train.

Oxygen Consumption

Oxygen consumption during exercise Oxygen consumption increases with increasing exercise intensity until it plateaus near maximal effort (Figure 1-7). Further increases in exercise intensity do not lead to a further increase in O_2 consumption.

There is very little variance among individuals in O_2 consumption performing the same work load. If a world-class cyclist and a sedentary individual ride a bicycle ergometer at the same submaximal work load, then their O_2 consumptions will be very similar. The work load will represent a lower percentage of maximum for the cyclist, but the O_2 cost will be the same. However, the bout of exercise must be conducted under identical conditions. If different combinations of resistance and pedal revolutions are used, then the efficiency of work will be affected and thus the O_2 cost of the exercise.

Oxygen consumption and training The improvement in O_2 consumption with training will depend on current condition, type of training, and age. Most studies find that $\dot{V}_{O_2max}$ can only be improved about 10 to 20%. It appears that an athlete must be endowed with a relatively high-capacity O_2 transport system in order to be successful in endurance events. This is not to say that training is not extremely important, but rather that there is only a limited opportunity to improve aerobic capacity ($\dot{V}_{O_2max}$).

Summary

A number of factors affect cardiac performance during exercise, including preload, afterload, contractility, heart rate, and synergy of ventricular contraction. Preload is the end-diastolic fiber length and is affected by such factors as cardiac output, blood volume, posture, venous tone, and the pumping action of ventilation and skeletal muscles. Afterload is impedance to ventricular emptying and is affected by the degree of circulatory vasoconstriction (peripheral resistance) and by circulatory obstructions. Contractility refers to the quality of ventricular performance and is affected by preload, afterload, training, and disease. Heart rate, the number of cardiac cycles per minute, is the major determinant of cardiac output and myocardial O_2 consumption. Synergy of ventricular contraction refers to the rhythm of the excitation–contraction coupling mechanisms and is an important factor in cardiac performance.

Maximal O_2 consumption is the best measure of the capacity of the cardiovascular system and is the product of maximum cardiac output and maximum arteriovenous O_2 difference. The measurement of $\dot{V}_{O_2max}$ must satisfy objective criteria of standardization and measurement. Most commonly it is measured on a treadmill or bicycle ergometer.

Endurance exercise is best for improving the capacity of the cardiovascular system. Training results in increased maximum cardiac output, stroke volume, and blood volume, and decreased resting and submaximal exercise heart rate.

Selected Readings

Andersen, P. Capillary density in skeletal muscle of man. Acta Physiol. Scand. 95: 203–205, 1975.

Åstrand, P.-O. and K. Rodahl. Textbook of Work Physiology. New York: McGraw-Hill, 1970.

Bhan, A. and J. Scheur. Effects of physical training on cardiac myosin ATPase activity. Am. J. Physiol. 228: 1178–1182, 1975.

Boyer, J. Effects of chronic exercise on cardiovascular function. Phys. Fitness Res. Digest. 2: 1–6, 1972.

Butler, J., M. O'Brien, K. O'Malley, and J.G. Kelly. Relationship of β-adrenoreceptor density to fitness in athletes. *Nature* 298: 60–62, 1982.

Braunwald, E., J. Ross, and E.H. Sonnenblick. Mechanisms of contraction of the normal and failing heart. N. Engl. J. Med. 277: 794–799, 853–862, 910–919, 962–971, 1012–1022, 1967.

Clausen, J.P. Effect of physical training on cardiovascular adjustments to exercise in man. Physiol. Rev. 57: 779–815, 1977.

Consolazio, C.F., R.E. Johnson, and L.J. Pecora. Physiological Measurements of Metabolic Functions in Man. New York: McGraw-Hill, 1963.

Cooper, K., J.G. Purdy, S.R. White, M.J. Pollock, and A.C. Linnerud. Age-fitness adjusted maximal heart rates. BAG Med. Sport. 6: 1–10, 1976.

Cunningham, D.A. and J.S. Hill. Effect of training on cardiovascular response to exercise in women. J. Appl. Physiol. 39: 891–895, 1975.

Ekblom, B. Effect of physical training on oxygen transport system in man. Acta Physiol. Scand. 328(Suppl.): 5–45, 1969.

Fortney, S.M., E.R. Nadel, C.B. Wenger, and J.R. Bove. Effect of acute alterations of blood volume on circulatory performance in humans. J. Appl. Physiol. 50: 292–298, 1981.

Foster, C., J.D. Anholm, C.K. Hellman, J. Carpenter, M.L. Pollock, and D.H. Schmidt. Left ventricular function during sudden strenuous exercise. Circulation. 63: 592–596, 1981.

Ingjer, F. Maximal aerobic power related to the capillary supply of the quadriceps femoris muscle in man. Acta Physiol. Scand. 104: 238–240, 1978.

Keul, J., H.H. Dickhuth, G. Simon, and M. Lehmann. Effect of static and dynamic exercise on heart volume, contractility, and left ventricular dimensions. Circ. Res. (Supp. I). 48: 162–170, 1981.

McDonough, J.R. and R.A. Danielson. Variability in cardiac output during exercise. J. Appl. Physiol. 37: 579–583, 1974.

Mellerowicz, H. and V.N. Smodlaka (eds.). Ergometry: Basics of Medical Exercise Testing. Baltimore: Urban & Schwarzenberg, 1981.

Pollock, M.L., J.H. Wilmore, and S.M. Fox. Health and Fitness Through Physical Activity. New York: John Wiley & Sons, 1978.

Rippe, J.M., L.A. Pape, J.S. Alpert, I.S. Ockene, J.A. Paraskos, P. Kotilainen, J. Anas, and W. Webster. Studies of systolic mechanics and diastolic behavior of the left ventricle in the trained racing greyhound. Basic Res. Cardiol. 77: 619–644, 1982.

Rowell, L.B. Human Cardiovascular adjustments to exercise and thermal stress. Physiol. Rev. 54: 75–159, 1974.

Sanders, M., F.C. White, T.M. Peterson, and C.M. Bloor. Effects of endurance exercise on coronary collateral blood flow in miniature swine. Am. J. Physiol. 234: H614–H619, 1978.

Scheuer, J. and C.M. Tipton. Cardiovascular adaptations to physical training. Ann. Rev. Physiol. 39: 221–251, 1977.

Sheps, D. Exercise-induced increase in diastolic pressure: Indicator of severe coronary artery disease. Am. J. Cardiol. 43: 708–712, 1979.

Slutsky, R. Response of the left ventricle to stress: Effects of exercise, atrial pacing, afterload stress and drugs. Am. J. Cardiol. 47: 357–364, 1981.

Snoeckx, L., H. Abeling, J. Lambregts, J. Schmitz, F. Verstappen, and R.S. Reneman. Echocardiographic dimensions in athletes in relation to their training programs. Med. Sci. Sports Exerc. 14: 428–434, 1982.

Stromme, S.B., F. Ingjer, and H.D. Meen. Assessment of maximal aerobic power in specifically trained athletes. J. Appl. Physiol. 42: 833–837, 1977.

Stromme, S.B. and F. Ingjer. The effect of regular physical training on the cardiovascular system. Scand. J. Soc. Med. 29 (Suppl.): 37–45, 1982.

Suga, H., R. Hisano, S. Hirata, T. Hayashi, and I. Ninomiya. Mechanism of higher oxygen consumption rate: Pressure-loaded vs. volume-loaded heart. Am. J. Physiol. 242: H942–H948, 1982.

Sutton, J.R. Control of heart rate in healthy young men. Lancet. 2: 1398–1400, 1967.

Whipp, B.J. and S. Ward. Cardiopulmonary coupling during exercise. J. Exp. Biol. 100: 175–193, 1982.

Zak, R. Development and proliferative capacity of cardiac muscle cells. Circ. Res. (Suppl. II). 34 and 35: 17–26, 1974.

Zeldis, S.M., J. Morganroth, and S. Rubler. Cardiac hypertrophy in response to dynamic conditioning in female athletes. J. Appl. Physiol. 44: 849–852, 1978.

17

EXCITABLE TISSUES

The ability to generate electrochemical signals (action potentials) is a characteristic that allows nerve and muscle cells to be excitable. This characteristic of excitability depends on two membrane properties: a sodium–potassium (Na^+–K^+) ion exchange pump and a greater membrane permeability of K^+ than for Na^+. These characteristics allow nerve and muscle cells to become polarized in the resting state, with a negative charge on the inside as compared to the outside. Various stimuli can disturb the local membrane charge separation. This membrane depolarization can give rise to an action potential that not only exists locally, but can propagate over the cell membrane surface. Terminal ends of nerve cells release chemical transmitter substances that affect the membrane permeabilities and membrane potentials of adjacent nerve and muscle cells. When nerve cells, or nerve and muscle cells, are arranged such that the terminal ends of a nerve cell (neuron) are in contact with another excitable cell, a synapse is formed. When sufficient neurotransmitter substance is released from the presynaptic neuron to depolarize the post-synaptic cell to a threshold sufficient to initiate an action potential in the postsynaptic cell, then an electrochemical signal is propagated from one cell to the next.

There are many synapses in the central nervous system (CNS—the brain and spinal cord) between neurons. At the end of the peripheral nervous system (PNS), neurons can communicate with skeletal muscle cells. Such a synapse is alternatively called a *neuromuscular junction* or a *motor end plate*. Major mo-

Nadia Comaneci during her balance beam routine at the Montreal Olympics. Performances such as these are dependent upon integrated function of nerves, muscles, spinal cord, brain, and sensory receptors. (Gilbert Uzan/Liaison.)

tor nerve cells (α motor neurons) and their associated muscle cells constitute a *motor unit*. Motor units are the basic functional units of the body's motor control system.

Based on two twitch characteristics (fast and slow) and two metabolic profiles (oxidative and glycolytic), three skeletal muscle fiber types have been identified: (1) fast glycolytic (FG), (2) fast oxidative glycolytic (FOG), and (3) slow oxidative (SO). All the fibers in a particular motor unit will be similar in their biochemical and physiological characteristics. Unique muscle fiber type characteristics have been identified in highly successful athletes at far ends of the athletic spectrum. In sprint runners FG fibers predominate, whereas in distance runners SO fibers predominate.

EXCITABILITY

Muscular movement is always under the direct control of nerves. Not only do nerve cells (neurons) send the signals (action potentials) that cause muscles to contract, but a variety of sensory receptors provide a constant feedback about the movement status. This information is constantly fed to and utilized in the CNS to modify ongoing movement as well as to initiate new movements. The ability of nerve cells to develop signals and to signal or communicate with other nerve or muscle cells depends on their characteristic of excitability. Excitability of nerve and muscle cells is a unique characteristic, dependent on the presence of a resting transmembrane voltage difference in excitable tissues.

DEVELOPMENT OF THE TRANSMEMBRANE VOLTAGE

Development of a transmembrane voltage (electrical potential) across the cell membrane of nerve and muscle cells depends on two unique characteristics: (1) a sodium–potassium ion (Na^+–K^+) exchange pump driven by ATP hydrolysis and (2) a greater permeability to potassium (pK^+) than to sodium (pNa^+). In describing development of the transmembrane voltage, two characteristics of the current flow need to be noted. The first is that the charge developed is of an electrochemical nature. The charge developed depends mainly on separation of Na^+, K^+, and Cl^- ions. Electrical current depends on the movement of these ions. This is different from the usual form of electricity used by machines, which depends on the flow of electrons, not ions. The second characteristic is that current always flows in complete loops. The movement of charge from one area into another will be accompanied by a flow of charge into the area from which the ion flow (or flux) began. This is like current flow due to electron movement; a complete circuit is necessary. The following example will illustrate how these two unique characteristics result in generation of a transmembrane voltage.

Figure 17-1 depicts the starting conditions in a nerve cell for generation of a transmembrane voltage. At the outset, the cell has equal intra- and extracellular ion concentrations and is uncharged. The cellular Na^+–K^+ exchange pumps activate on the first pump cycle, eliminating 100,000 Na^+ ions and bringing in an equal number of K^+ ions. No charge difference will have been developed between the inside and outside of the cell because of this electroneutral ion exchange. However, the concentration of K^+ inside the cell will decrease to the extent that the chemical repulsion of K^+ ions will tend to cause ions to leave the cell. Outside the cell, a similar repulsion among Na^+ ions will tend to cause them to reenter the cell. Because the permeability of K^+ is about 50 times greater than that of Na^+, 50 K^+ will diffuse out of the cell, whereas only 1 Na^+ will reenter. Thus, the relative exit of K^+ (carrying a net 49 $^+$electron charge units) will leave the inside of the cell with a slight negative ($-$) charge. As the result of this developing electronegativity within the cell, chloride ion (Cl^-) will tend to be electrically repelled and exit the cell. According to our model (Figure 17-1), 48 Cl^- will exit. Only 48 Cl^- will exit because the

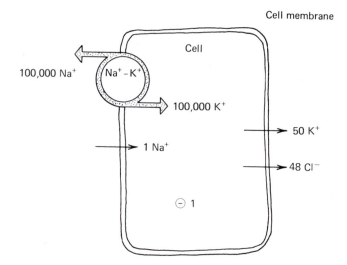

FIGURE 17-1
Two features of the excitable cell membrane allow for the development of a transmembrane electrical potential: (1) a (Na^+–K^+) exchange pump and (2) a greater permeability for K^+ than for Na^+. As shown in the figure and explained in the text, the pumping of relatively large amounts of ions eventually results in charge separation across the cell membrane.

migration will be chemically opposed by Cl^- already present in the extracellular medium. In sum, then, because of the activation of the Na^+–K^+ pump, the relatively greater membrane permeability of K^+ than of Na^+, and the efflux of Cl^-, the inside is 1 electron charge negative relative to the outside.

On the second and succeeding pump cycles, Na^+ and K^+ will continue to be exchanged, and K^+ will accumulate within the cell. Potassium ion will continue to diffuse from the cell because of chemical repulsion, but the developing negative charge will restrain the efflux of K^+. Chloride ion will continue to exit the cell as it becomes more and more electronegative inside. The net exchange of diffusable ions will continue until there is an equilibration of chemical and electrical forces across the cell membrane. This is essentially when the efflux of intracellular K^+ due to chemical repulsion is balanced by the intracellular electrical attraction. The resting membrane potential (V_m) will be the sum of the chemical and electrical potentials across the membrane. V_m will be essentially the diffusion potential of K^+, and V_m will be very close to the resting membrane potential for K^+ (V_K).

Physiologically, a variety of factors can perturb the V_m. Among these are chemical transmitter substances, but mechanical, chemical, and other stimuli can affect nerves, muscles, and specialized receptor organs.

THE ACTION POTENTIAL

The message unit of nerves and muscles is the action potential. An action potential is a wave of depolarization that moves along the surface of a nerve or muscle cell (Figure 17-2). An action potential is due to the sudden increase in permeability of Na^+ and the events that follow. Figure 17-3 displays a glass electrode system used to measure voltage differences across excitable membranes. Figure 17-4 displays the electrical and chemical events that occur at the site of an action potential. As suggested in Figure 17-4, stimulation of an excitable membrane and perturbation of the resting membrane voltage does not necessarily result in an action potential. Subtle changes in the permeability of sodium ion (pNa$^+$) result in some increase in resting V_m (depolarization). However, sufficient perturbation of the V_m results in the achievement of a threshold voltage at which the influx of sodium regenerates itself. During the time that the influx of sodium is regenerative, the pNa$^+$ increases approximately 600 times, and the V_m approaches the equilibrium voltage of Na^+ ($V_{Na}{}^+$).

The influx of sodium ion is short-lived, and the V_m rapidly changes back toward resting levels. The influx of Na^+ ($J_{Na}{}^+$) is followed in time by an efflux of K^+ ($J_K{}^+$). During the time K^+ is leaving the cell (thereby taking away positive charge), the V_m actually becomes more negative than usual. During this time of hyperpolarization, the membrane is in a refractory period. It is either impossible to restimulate the membrane (absolute refractory period) or more difficult than usual to stimulate the cell (relative refractory period) while the efflux of K^+ is taking place.

Although during the course of an action potential there are dramatic changes

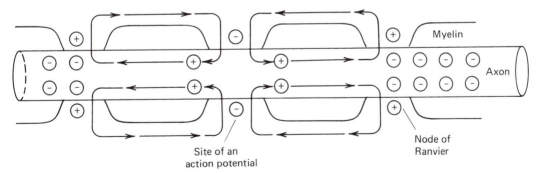

Site of an
action potential

FIGURE 17-2
The movements of charged ions across excitable cell membranes causes the resting membrane voltage to reverse. Because ion current aways flows in a complete circuit, and because current always flows at the point of least resistance, in myelinated nerves such as exist in mammals, the conduction velocity of action potentials is very rapid. Breaks in the myelin are termed nodes of Ranvier.

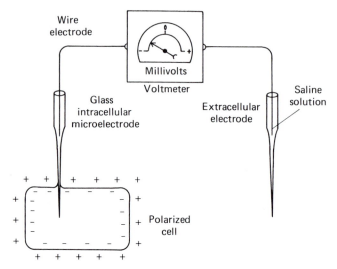

FIGURE 17-3

Microelectrodes are made from glass capillary tubes that are heated and then rapidly stretched to become microscopically thin. These remain patent and can then be filled with a conducting saline solution and inserted into a cell. A voltmeter then records electrical potential differences across the cell membrane.

(SOURCE: Modified from J.A. Vander, J.H. Sherman, and D.S. Luciano, 1980, p. 158. With permission.)

in the permeabilities of Na^+ and K^+, the number of ions that move is relatively small in comparison to the intra- and extracellular populations. Consequently, the resting membrane potential (V_m) recovers rapidly following an action potential. Even if the Na^+–K^+ pump is poisoned, most neurons can continue to function and propagate many action potentials. Neurons are very fatigue resistant.

Whereas most excitable cells need an extrinsic stimulus to develop an action potential, other very important regulatory cells spontaneously and regularly depolarize themselves. This spontaneity is due to the presence of a cell membrane that has a relatively high permeability to Na^+. At regular intervals, the V_m reaches threshold and an action potential results. Specialized cardiac muscle cells comprising the SA and AV nodes in the heart (Chapter 14) and neurons in the medullary respiratory center (Chapter 12) control heart and breathing rates by their spontaneous excitability. As we will see, the frequency of depolarization can be affected by a variety of factors, but the characteristic of rhythmicity is intrinsic to these specialized cells.

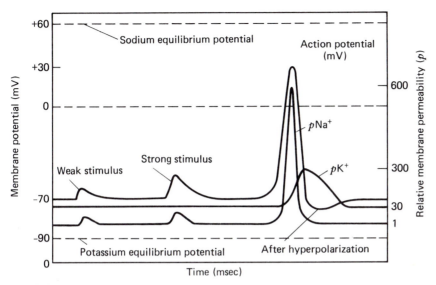

FIGURE 17-4
Weak stimuli can perturb the resting membrane voltage in the locus of the stimulus. Together, several weak stimuli can add up to produce a greater local effect. When the membrane voltage is disturbed to a threshold value (approximately 10 to 15 mV), then an action potential occurs (see text and Figure 17-7).

(SOURCE: Modified from J.A. Vander, J.H. Sherman, and D.S. Luciano, 1980, p. 147.)

THE ANATOMY OF THE NERVE CELL (NEURON)

The anatomy of an α-motor neuron is displayed in Figure 17-5. As is typical of nerve cells (neurons), the α-motor neuron consists of a cell body (or soma), short projections from the soma (dendrites), a long projection (axon), and terminal endings on the axon that contain and can release transmitter substances. In the case of an α-motor neuron, the terminal ending is alternatively called the motor end plate or the neuromuscular junction. The transmitter substance released by the α-motor neuron into the synaptic cleft at the neuromuscular junction is acetylcholine (ACH). The transmitter substance released by neurons varies around the body. In the brain, norepinephrine, serotonin, and γ-aminobutyric acid (GABA) as well as other neurotransmitters are released in addition to ACH.

Characteristic of mammalian axons is the presence of *myelin* (a white substance high in lipid content). The myelin sheath covering axons is intermittently

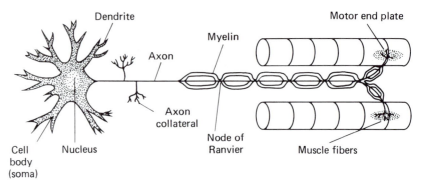

FIGURE 17-5
A motor unit, consisting of an α motor neuron and the muscle fibers (cells) it innervates.

broken up. The break points are termed *nodes of Ranvier*. The electrical properties of myelin are such that it increases the capacitance of the sections of the axon that are sheathed. This property increases the probability that ions will flow first at the unmyelinated nodes of Ranvier during action potentials. Consequently, in myelinated axons action potential conduction velocity is increased as a result of the jumping of action potentials from node of Ranvier to node. This is termed *saltitory conduction* and is illustrated in Figure 17-2. In general, the thicker the myelin sheath sheath around a nerve axon, the faster the conduction velocity.

FACILITATION, INHIBITION, AND ALL OR NONE

When there is an action potential in an α-motor neuron (Figure 17-5), there is always an action potential in each of the postsynaptic muscle cells. These then contract to the best of their ability. The transmission from nerve to muscle is said to be "all or none." However, within the CNS, such as at the soma of an α-motor neuron, presynaptic action potentials do not always result in postsynaptic action potentials. In some instances, just the opposite happens. Figure 17-6 illustrates that presynaptic terminal branches on neuron cell bodies can have either excitatory or inhibitory effects on the postsynaptic cell. Release of one neurotransmitter substance such as ACH increases the postsynaptic permeability of Na^+ so that the V_m rises toward but does not necessarily reach threshold. This localized depolarization at the synaptic site is termed an *excitatory postsynaptic potential* (EPSP). Release of another neurotransmitter into the synaptic cleft from another presynaptic terminal ending can result in a decrease in postsynaptic sodium permeability or an increase in chloride permeability. These events result in a decrease in postsynaptic V_m and reduce the chance of the V_m

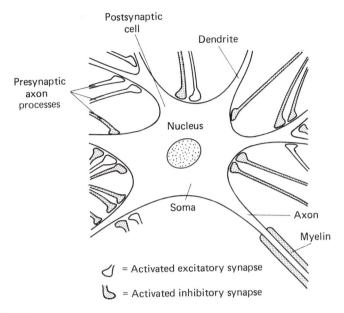

Postsynaptic cell

Dendrite

Presynaptic axon processes

Nucleus

Soma

Axon

Myelin

$\mathcal{J}$ = Activated excitatory synapse

$\mathcal{L}$ = Activated inhibitory synapse

FIGURE 17-6
The α-motor neuron cell body (soma), located in the gray matter of the spinal cord, has many excitatory and inhibitory synapses. Activity from these presynaptic cell endings greatly affects membrane voltage in the soma.

reaching threshold. The localized, nonpropagating decrease in V_m illustrated in Figure 17-7 is termed an *inhibitory postsynaptic potential* (IPSP).

The activity or lack of it in a neuron within the CNS depends on the sum total of EPSPs and IPSPs occurring within a narrow range of time and space. EPSPs can summate to give rise to an action potential in two basic ways. Excitatory presynaptic nerve endings distributed around a postsynaptic soma can release sufficient neurotransmitter to affect a postsynaptic action potential. This is referred to as *spatial summation* (Figure 17-6). Alternatively, one or a few presynaptic endings can be repeatedly active within a short period of time. The summation to threshold is termed *temporal summation* (Figure 17-7).

THE MOTOR UNIT

A motor unit consists of an alpha (α)-motor neuron and its associated muscle cells (fibers). The motor unit (Figure 17-5) is the functional unit of the neural control of muscular activity. All muscle fibers are innervated by at least one motor neuron, and an action potential in the neuron will result in action potentials in all the muscle cells. The number of fibers in a motor unit can vary considerably (from perhaps 10 fibers in an eye muscle to approximately 1000

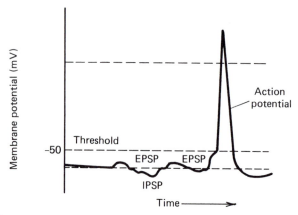

FIGURE 17-7

Recording from inside a α-motor neuron soma reveals the effects of excitatory presynaptic activity and inhibitory presynaptic activity. When the postsynaptic potential reaches threshold, an action potential results.

(SOURCE: Modified from J.A. Vander, J.H. Sherman, and D.S. Luciano, 1980, p. 172. With permission.)

fibers in a gastrocnemius muscle). Normally, all the fibers within a motor unit will have identical biochemical and physiological properties.

Figure 17-8 illustrates that motor units can be classified in two ways on the basis of speed of contraction, and three ways on the basis of metabolic characteristics. Muscle fiber types within motor units are usually identified by histochemical techniques on muscle samples taken during biopsy. In histochemistry, various biochemical characteristics allow reactions with specific stains. The characteristics are then identified in a light microscope.

Histochemically, speed of contraction of a fiber is indicated by the myofibrillar-ATPase (M-ATPase) stain. Myofibrills are the contractile proteins in muscle (Chapter 19). Fast-contracting (fast-twitch) fibers stain dark with alkaline (high pH) myofibrillar ATPase (M-ATPase), and light with acid M-ATPase (Ac-ATPase). Usually there is a good correlation between M-ATPase histochemistry and physiological contraction characteristics of individual muscle fibers. In histochemistry, the metabolic characteristics of a muscle fiber are usually identified with a stain that reacts with an oxidative enzyme or enzyme complex in the fiber. A stain for succinic dehydrogenase (SDHase) is frequently used. If a fiber stains dark for SDHase, it is assumed to be an oxidative fiber. Usually, if a fiber stains weakly with SDHase, it is assumed to have a nonoxidative, glycolytic metabolism. Sometimes the glycolytic characteristics of a fiber are determined by staining specifically for a glycolytic enzyme such as α-glycerol phosphate dehydrogenase (α-GPDase). Because M-ATPase activity, correlates highly with glycolytic capacity, in practice, high M-ATPase activity as indicated by histochemical appearance, is also taken as evidence for a high glycolytic capacity.

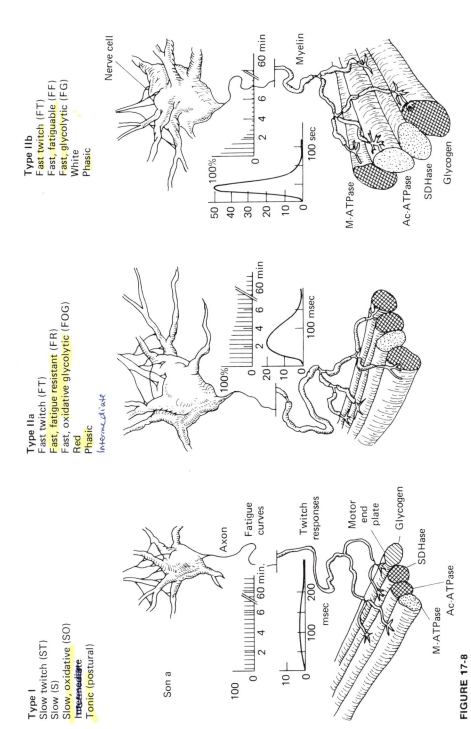

Type I
Slow twitch (ST)
Slow (S)
Slow, oxidative (SO)
~~Intermediate~~
Tonic (postural)

Type IIa
Fast twitch (FT)
Fast, fatigue resistant (FR)
Fast, oxidative glycolytic (FOG)
Red
Phasic
Intermediate

Type IIb
Fast twitch (FT)
Fast, fatiguable (FF)
Fast, glycolytic (FG)
White
Phasic

Son a

Axon

Fatigue curves

Twitch responses

Motor end plate

Glycogen

SDHase

M-ATPase

Ac-ATPase

Nerve cell

Myelin

M-ATPase

Ac-ATPase

SDHase

Glycogen

FIGURE 17-8

In adult mammalian muscles, microscopic examination after transverse (cross) sectioning and histological staining reveals three muscle fiber types (see Figure 5-3). Fibers are classified as either type I (slow) or type II (fast), depending on intensity of staining with alkaline myofibrillar ATPase. Note that type I and II fibers demonstrate an "acid reverse" when stained for myofibrillar ATPase at low pH (Ac-ATPase). With the oxidative marker succinic dehydrogenase (SDH), type I and type IIa fibers stain dark, whereas type IIb fibers stain lightly. Glycogen content is revealed by the periodic acid Schiff (PAS) stain. The three fiber types differ in ease of recruitment, metabolism, twitch characteristics, and rate of fatigue. The alternative terminologies used to classify skeletal muscle fiber types are indicated at the top. Refer to Figure 5-3. (SOURCE: Modified from D.W. Edington and V.R. Edgerton, 1976, p. 53. With permission of the authors.)

On the basis of these twitch and metabolic characteristics, three main skeletal muscle fiber types have been identified: (1) fast glycolytic (FG), the classical phasic or white fiber; (2) fast oxidative glycolytic (FOG), the classical red fiber, and (3) slow oxidative (SO), or classical tonic, intermediate fiber. To the eye, FG fibers are pale or white in color, whereas both FOG and SO fibers are red in color.

Muscle fibers may also be classified by their resistance to fatigue. The FG fiber is alternatively termed fast, fatiguable (FF). The FOG fiber, is also termed fast, fatigue resistant (FR), and the SO fiber is also termed slow (S) fiber. In this classification, the metabolic profile of fibers is implied.

THE SIZE PRINCIPLE IN FIBER TYPE RECRUITMENT

According to the size principle, the frequency of motor unit utilization (recruitment) is directly related to the size and ease of triggering an action potential in the soma. In general, the smaller the soma (neuron cell body), the easier it is to recruit. According to the size principle, those motor units with the smaller cell bodies (S motor units) will be used first and overall most frequently. Those motor units with larger cell bodies (FF and FR) will be used last during a recruitment and, overall, least frequently. Despite the terminology used, fiber recruitment is usually determined not by the speed of a movement but rather by the force necessary to perform a movement. For instance, S motor units may be exclusively recruited while lifting a very light weight fairly rapidly. However, in lifting a very heavy weight, necessarily very slowly, all motor units will be recruited. Dr. P.D. Gollnick of Washington State University and colleagues have shown that fast fibers are used to support low-frequency, high-resistance bicycle ergometer exercises. More rapid cycling at higher speeds but against less resistance was supported by slow muscle fibers.

Evidence that fast motor units are recruited when rapid, unloaded movements are made has been provided by J. Smith and V.R. Edgerton with their colleagues at UCLA. For instance, when a cat steps on a piece of adhesive tape, the cat will attempt to flick the tape off its paw by recruiting fast motor units before slow units.

MUSCLE FIBER TYPES IN ATHLETES

Outstanding performers at far ends of the athletic spectrum have been shown to possess specialized muscle fiber type characteristics. For instance, sprint runners have been found to possess fast-twitch glycolytic fiber types, whereas slow-twitch, high-oxidative fiber types have been found to predominate in distance runners. Figure 17-9 is from work by Gollnick and associates. In that

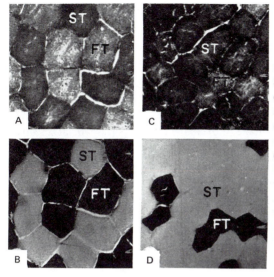

FIGURE 17-9
Serial sections of quadriceps muscle from two different athletes, stained with two different stains; (a) and (b) from an outstanding sprinter, (c) and (d) from an outstanding distance runner. (a) and (c) Stained for succinic dehydrogenase (SDHase); (b) and (d) stained for alkaline myofibrillar-ATPase stain (M-ATPase). Note that fast fibers, which stain dark with M-ATPase, usually are pale and stain weakly with SDHase. FT, Fast-twitch fibers; ST, slow-twitch fibers. Note the two dark FT fibers in (d), which also stain dark for SDHase in (c). These are FOG fibers (Figure 17-8).

(SOURCE: Courtesy of P.D. Gollnick.)

figure, (a) and (b) are from one subject, a sprinter; (c) and (d) are from a distance runner. (a) and (c) are stained for SDHase, whereas (b) and (d) are stained for M-ATPase. Note the pale appearance in (a), the oxidative marker of the sprinter. In the sprinter, however, note the dark appearance of fibers when fixed for the contractibility marker, M-ATPase (b). Results obtained on the distance runner are the inverse: high oxidative capacity is indicated in (c), but low contractile activity is indicated in (d).

It should again be pointed out that results such as those in Figure 17-9 typify histological characteristics in athletes at the far ends of the athletic spectrum. These results are consistent with other observations made in the biosphere where species noted for speed (e.g., cats) possess fast-twitch fiber characteristics, whereas species noted for endurance (e.g. dogs) possess slow-twitch fiber type characteristics. On humans, Costill and associates have observed a significant range of fiber type characteristics in athletes successful in middle-distance competition requiring both speed and endurance. Clearly muscle fiber type is but one factor that affects human motor performance.

Although physical training can significantly affect muscle biochemistry and therefore histological appearance of muscle cells, muscle fiber type is genetically determined. Therefore, it is not possible for the muscle cells of the sprint athlete displayed in Figure 17-9 a and b to assume by training the histological appearance of the endurance athlete displayed in Figure 17-9c and d. Similarly, the endurance athlete cannot by training become like the sprint athlete. Properly used, training can enhance the histochemical and underlying biochemical properties of muscle, but training can not reverse a muscle's intrinsic, genetically determined qualities.

Summary

The ability of nerve and muscle cells to respond to chemical (neurotransmitter) stimuli ultimately governs all human behavior. The ability to think, to breathe, to have a heartbeat, and to move reflexively or volitionally, all depend on excitability in nerve and muscle cells. The characteristic of excitability depends on two cell membrane phenomena: a Na^+–K^+ exchange pump and a greater permeability to K^+ (outward) than to Na^+ (inward). In response to appropriate stimuli, the resting membrane voltage of excitable cells is perturbed sufficiently to generate action potentials. Action potentials represent the basic message unit of communication among neurons and between neurons and muscle fibers. The transmitter substance released from the terminal ends of neurons as the result of action potentials, the frequency of action potentials, their spatial distribution, and related factors determine the recruitment of motor units.

The motor unit, which comprises the basic functional unit of the muscular system, also represents the final means by which the nervous system originates and controls muscular activity. All the muscle fibers innervated by an α-motor neuron in a motor unit will have the same physiological characteristics. On the basis of metabolic (oxidative or glycolytic) and twitch characteristics (fast or slow), three basic muscle fiber types have been identified. Histological examination of human muscle reveals a mosaic appearance. The histological characteristic of muscles vary around the body and from one individual to another. Individuals having exceptional muscular speed and power are characterized by a predominance of fast-twitch muscle fibers. Individuals noted for their endurance capability are observed to possess a preponderance of slow-twitch but highly oxidative muscle fibers. Athletes whose competitive roles require aspects of both speed and endurance have been successful with a wide range of muscle fiber types.

Selected Readings

Barondes, S.H. Cellular Dynamics of the Neuron. New York: Academic Press, 1969.

Brooke, M.H. and K.K. Kaiser. Muscle fiber types: How many and what kind? Arc. Neurol. 23: 369–379, 1970.

Buchthal, F. and H. Schmalbruch. Contraction times and fiber types in intact muscle. Acta Physiol. Scand. 79: 435–452, 1970.

Burke, R.E. and V.R. Edgerton. Motor unit properties and selective involvement in movement. In: Wilmore, J. and J. Keogh (eds.), Exercise and Sports Sciences Reviews. New York: Academic Press, 1975. p. 31–83.

Costill, D.L., J. Daniels, W. Evans, W. Fink, G. Krahenbuhl, and B. Saltin. Skeletal muscle enzymes and fiber composition male and female athletes. J. Appl. Physiol. 40:149–154, 1976.

Eccles, J.C., The Physiology of Synapses. Berlin: Springer Verlag, 1964.

Eccles, J.C. The Understanding of the Brain. New York: McGraw-Hill, 1973.

Edington, D.W. and V.R. Edgerton. The Biology of Physical Activity. Boston: Houghton Mifflin, Co., 1976. p. 51–72.

Evarts, E.V. Brain mechanism in movement. Sci. Amer. 229: 96–103, 1973.

Gollnick, P.D., R.B. Armstrong, C.W. Saubert, K. Piehl, and B. Saltin. Enzyme activity and fiber composition in skeletal muscle of untrained and trained men. J. Appl. Physiology 33: 312–319, 1972.

Gollnick, P.D., R.B. Armstrong, W.L. Sembrowich, R.E. Shepherd, and B. Saltin. Glycogen depletion pattern in human skeletal muscle fibers after heavy exercise. J. Appl. Physiol. 34: 615–618, 1973.

Gollnick, P.D., K. Piehl, C.W. Saubert, R.B. Armstrong, and B. Saltin. Diet, exercise, and glycogen changes in human muscle fibers. J. Appl. Physiol. 33: 421–425, 1972.

Granit, R. Muscular Afferents and Motor Control. Stockholm: Almqvist and Wiksell, 1966.

Granit, R. The Basis of Motor Control. London: Academic Press, 1970.

Guth, L. "Trophic" influences of nerve on muscle. Physiol. Rev. 48: 645–687, 1968.

Guyton, A.C. Textbook of Medical Physiology. Philadelphia: W.B. Saunders, Co., 1981, p. 560–697.

Henneman, E. Skeletal muscle. The servant of the nervous system. In: Mountcastle, V.B. (ed.), Medical Physiology, 14th ed., Vol. 1. St. Louis: Mosby Publishing, 1980. p. 674–702.

Hunt, C.C. Relation of function to diameter in affuent muscle fibers. J. Gen. Physiol. 38: 117–131, 1954.

Hursh, J.B. Conduction velocity and diameter of nerve fibers. Am. J. Physiol. 127: 131–139, 1939.

Kandel, E.R. and J.H. Schwartz (eds.), Principles of Neural Science. New York: Elsevier/North-Holland, 1981.

Katz, B. Nerve Muscle and Synapse. New York: McGraw-Hill, 1966.

Mathews, B.H.C. Nerve endings in mammalian muscle. J. Physiol. London 78: 1–53, 1933.

Nemeth, P. and D. Pette. Succinate dehydrogenase activity in fibers classified by myosin ATPase in hind limb muscles of rat. J. Physiol. London 320: 73–80, 1981.

Pappas, G.D. and D.P. Purpura. Structure and Function of Synapses. New York: Raven Press, 1972.

Phillis, J.W. The Pharmacology of Synapses. Oxford: Pergamon Press, 1970.

Ruch, T.C. and H.D. Patton. Physiology and Biophysics, 19th ed. Philadelphia: W.B. Saunders, Co., 1965.

Sherrington, C. The Integrative Action of the Nervous System, 2nd ed. New Haven: Yale University Press, 1947.

Vander, A.J., J.H. Sherman, and D.S. Luciano. Human Physiology, 3rd ed. New York: McGraw-Hill, 1980. p. 144–190.

Woodbury, J.W. The cell membrane: Ionic and potential gradients and active transport. In: Ruch, T.C. and H.D. Patton (eds.), Physiology and Biophysics, 19th ed. Philadelphia: W.B. Saunders, 1965. p. 1–25.

18

SENSATION, INTEGRATION, AND MOVEMENT PATTERNS

Reflexes represent the simplest movements of which we are capable. They involve the integrated function of receptors, receptor afferent neurons, central nervous system (CNS), α-motor neurons, and skeletal muscle tissue. Reflexes are involuntary and automatically follow the activation of receptors. Reflexes can be very simple in function (such as withdrawal from pain), or a bit more complex, involving movements of a limb on the opposite side of the body.

More complicated and volitional movements involve some part of the brain. Most learned movement patterns, especially those movements that will require sensory input (like hitting a baseball), require participation of one or several areas of the brain. In particular, the motor cortex is in direct control of learned movement patterns. The coordination and modification of various movement patterns depend on participation of various subcortical areas including the reticular formation, cerebellum, and thalamus. The learning of intricate motor tasks involves the programming of the motor cortex. The initiation of motor tasks involves the playing back of stored programs in which motor units in active and supporting muscles are recruited in a precise pattern of time, space, frequency, and amplitude.

A few very essential movement patterns appear to be built into the CNS through the process of evolution. These primitive movements involve locomotion (swimming, walking, and running) and appear to be located in the spinal cord. This segmentation hierarchy of control is advantageous, in that it allows

In ballet exist some of the finest examples of fine control over body movement. Integrated function of nerves, muscles, and several parts of the central nervous system along with years of practice are required. (Cary Wolinsky/Stock Boston.)

more than one function to be performed simultaneously. These primitive spinal cord movement patterns are activated under the control of higher cortical centers.

NEURAL CONTROL OF MOVEMENT

Our lives as we know them are a cascade of movements. In most of human history, movement has played a critical role. Movement determined such basic things as who ate or was eaten, who swam or sank, who froze or overheated, and who migrated successfully and who didn't. In general, movement in large measure determined who coped with the environment. Our great gift as humans in the grand scheme of things is our intellect, but intellect must ultimately be translated into discrete movement patterns. The emphasis in this volume is on explaining performance in sporting activities where the ability for movement is of primary importance. Human movement during sports, like movement during other human endeavors, is under the ultimate control of the brain and central

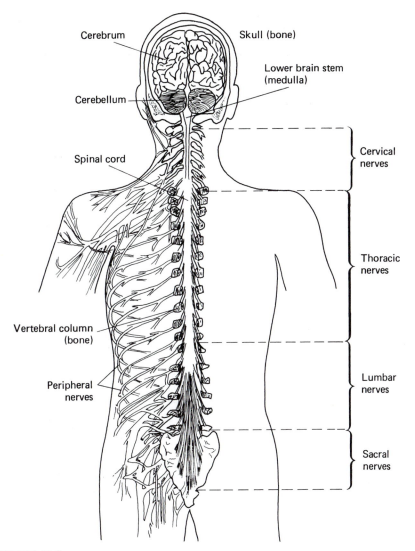

Cerebrum

Skull (bone)

Lower brain stem
(medulla)

Cerebellum

Spinal cord

Cervical
nerves

Thoracic
nerves

Vertebral column
(bone)

Lumbar
nerves

Peripheral
nerves

Sacral
nerves

FIGURE 18-1
Dorsal (back) view of the nervous system. The central nervous system
(CNS) consists of the brain and spinal cord. The peripheral nervous system
(PNS) consists of nerves (bundles of axons) extending from the brain and
spinal cord.

(SOURCE: Adapted from J.A. Vander, J.H. Sherman, and D.S. Luciano, 1980, p. 180
. . . and R.T. Woodburne, Essentials of Human Anatomy, 5th ed., Oxford University
Press, New York, 1973.)

nervous system. Therefore, understanding the neural control of movement is of primary importance. Before describing how muscular activities are controlled by the nervous system, we will first need to consider its gross anatomy.

GROSS ANATOMY OF THE NERVOUS SYSTEM

The central nervous system (CNS) consists of the brain and spinal cord (Figure 18-1). The peripheral nervous system (PNS) consists of all the nerves extending from the brain or spinal cord. Nerves consist of bundles of myelinated neuronal axons that may carry either sensory information to the CNS (afferents) or motor information from the CNS (efferents). In connection with the nervous system, the terms *afferent* (toward) and *efferent* (away from) are used to describe the direction of action potential flow.

Spinal Cord

Seen in cross section, the spinal cord (SC) is divided into white and gray areas (Figure 8-2). The central gray matter consists of neuronal cell bodies with associated dendrites, short interneurons that do not leave the spinal cord, and terminal axon processes from other neurons whose cell bodies are located else-

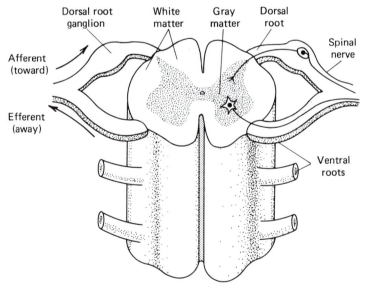

FIGURE 18-2
Cross-sectional view through the spinal cord and joining nerves. Gray matter contains cell bodies for motor nerves, which send axon processes out of the ventral root, as well as cell bodies of interneurons. Cell bodies for sensory nerves exist in the dorsal root ganglion. Sensory neurons also synapse in the gray matter. The SC white matter contains tracts for ascending (sensory) and descending (motor) signals.

where. It is in the gray area that synapses occur and that integration can take place. Some areas in the gray matter can appear especially dark because of the high concentration of cell bodies. Such an area, where cell bodies having related function cluster together, is called a *nucleus*.

The gray matter of the spinal cord is surrounded by white matter consisting of bundles of myelinated neuronal axons running in parallel. These neuronal bundles are called tracts or pathways and are analogous to nerves in the PNS. Nerve tracts run both up (ascending) and down (descending) the spinal cord, and carry sensory and motor signals. No synapses are possible in white matter, but axons may enter the gray matter to receive or transmit information (action potentials).

Groups of afferent fibers carrying sensory information enter the spinal cord on the dorsal (back) side of the body. Cell bodies of these sensory nerves are located immediately outside the spinal cord, in the dorsal root ganglion (Figure 18-2). Groups of efferent fibers carrying motor signals leave the spinal cord on the ventral (belly) side.

Brain

The brain consists of six major areas; the cerebrum, cerebellum, diencephalon, midbrain, pons, and medulla (Figure 18-3).

The *cerebral cortex* (derived from the Greek word meaning "tree bark") covers most of the brain's surface (Figure 18-3*a*). The cortex is responsible for those functions of intellect and motor control that set human beings apart. Four different sections (lobes) of the cortex are associated with specific functions. The *frontal* lobe is responsible for fine intellect and motor control. The motor cortex is at the rear of the frontal lobe. The *parietal* lobe is associated with sensation and interpretation of sensory information. The sensory cortex is at the front of the parietal lobe. The *temporal* lobe is associated with auditory sensation and interpretation, and the *occipital* lobe is associated with visual sensation and interpretation.

The large core of the brain, the *diencephalon,* is made up of two areas: the thalamus and hypothalamus. The *thalamus* is an important integration center through which most sensory inputs pass. From the thalamus, neuronal signals arise for input to the cortex and cerebellum. Additionally, descending signals from the cortex can be integrated with incoming sensory information by synapses in the thalamus.

The *hypothalamus* is an important area where neural and hormonal functions effect a constancy of the internal body environment. It is in the hypothalamus that appetite is governed and body temperature is regulated.

The *cerebellum* is an area especially important for motor control. Motor coordination, balance, and smooth movements are accomplished through action of the cerebellum. Damage to the cerebellum results in impaired motor control.

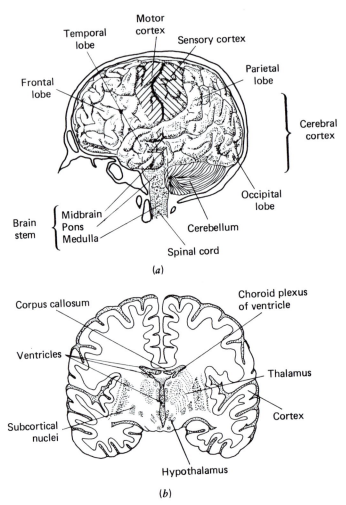

FIGURE 18-3
The human brain. The side view (a) shows major divisions of the brain, and the cross-sectional view (b) reveals the interior structures.

(SOURCE: From J.A. Vander, J.H. Sherman and D.S. Luciano, 1980, p. 181–183. With permission.)

All afferent and efferent signals pass through the brain stem, which consists of the midbrain, pons, and medulla. The *brain stem* is an area that sets rhythmicity of breathing and controls the rate and force of breathing movements and heartbeat. Additionally, within the brain stem is an area called the *reticular formation* that allows us to focus on specific sensory inputs and influences arousal and wakefulness. For instance, in baseball the reticular formation allows a batter to focus on the pitch rather than the taunts of opposing players and spectators.

NEURAL CONTROL OF REFLEXES

Reflexes are the simplest type of movement of which we are capable. The knee jerk reflex (Figure 18-4) involves at least four components: (1) the receptor (in this case a muscle stretch receptor, the *spindle*), (2) the gamma (γ)-afferent neurons, which synapse in the gray matter of the spinal cord (e.g., at A in the

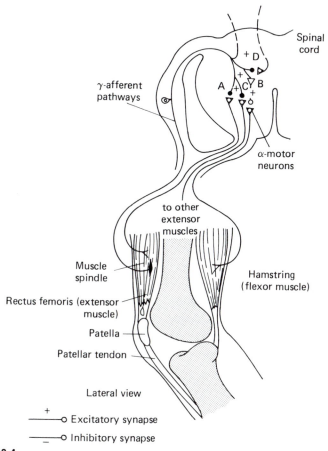

FIGURE 18-4

The knee jerk reflex involves a minimum of four components: the muscle spindle stretch receptor, the γ-afferent neuron, the α-motor neuron for the rectus femoris extensor muscle, and the rectus femoris. Sufficient rapid stretching of the spindle results in reflex contraction of the rectus femoris (A) and other extensors in the quadriceps (B), inhibition of the soma of motor neurons innervating the antagonist hamstrong flexors (C), and ascending signals (D), perhaps to motor units in arms and back.

(SOURCE: From J.A. Vander, J.H. Sherman, and D.S. Luciano, McGraw-Hill, 1980, p. 594. With permission.)

figure), (3) the α-motor neurons with a cell body in the spinal cord, and (4) the muscle fibers within the motor unit. Contrary to immediate appearance, the knee jerk reflex and other reflexes are not all-or-none responses.

Although the knee jerk response is involuntary, control is experienced at several levels. While stretching does depolarize the spindle receptor, an action potential in the γ afferent does not always result. There must be a summation of receptor potentials to generate an afferent action potential or a train (series) of potentials. At the level of synapse with the α-motor neuron, γ-afferent activity results in EPSPs. These must summate to result in an action potential in the α-motor neuron. Because the α-motor neuron resting membrane potential is also subject to IPSPs, a degree of voluntary inhibition is possible.

The simplest reflexes are monosynaptic. As illustrated in Figure 18-4, synapses with other neurons such as interneurons are possible. At B, the interneuron stimulated will result in release of an inhibitory neurotransmitter on the soma of an antagonistic motor neuron—in this case, one that innervates a motor unit in the hamstring muscle. Thus the antagonist hamstring will stay relaxed while the quadriceps agonist contracts to perform the actual knee jerk. In this terminology, the *agonist* is the muscle primarily responsible for movement, and an *antagonist* a muscle that retards movement.

Other synapses possible during a knee jerk reflex are those that activate synergists to the knee jerk (at C in Figure 18-4) or that send ascending signals to the brain or to motor units innervating a contralateral limb (at D).

The Gamma Loop and Muscle Spindle

Muscle spindles (Figure 18-5) respond to the amount and rapidity of stretch. The rapid stretching (lengthening) of spindles in the quadriceps that results when the patellar tendon is struck is a rather extreme example of the spindle's range of responses. In real life, we use the full jerk response only in emergency situations, such as when a foot slips and it is necessary to maintain posture and keep from falling. The spindle, however, works continuously to monitor muscular movement. Muscle spindles exist in parallel with normal extrafusal muscle fibers (Figure 18-6). When a volitional muscular activity is initiated by activation of α-motor neurons, there is also activation (coactivation) of smaller, γ-motor neurons. These efferent γ neurons cause contraction of muscle fibers within the muscle spindle (Figure 18-5). Contraction of the fibers within muscle spindle capsules (intrafusal fibers) in effect takes up slack within the spindle capsule during contraction of the larger extrafusal fibers. In this way, the central receptor portion of the spindle stays at relatively the same length. The receptor can then respond to sudden changes in length throughout almost the full range of limb movement. The system through which muscle spindles participate in monitoring muscular activity is the γ loop (Figure 18-7).

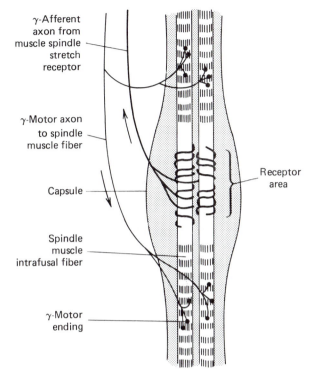

γ-Afferent
axon from
muscle spindle
stretch
receptor

γ-Motor axon
to spindle
muscle fiber

Capsule

Spindle
muscle
intrafusal fiber

γ-Motor
ending

Receptor
area

FIGURE 18-5
Muscle spindles exist in parallel with normal extrafusal skeletal muscle
fibers. Rapid stretching of the muscle also stretches the sensitive receptor
area of the spindle, and action potentials are sent to the spinal cord via
the γ-afferent neuron.

Tension Monitoring and Golgi Organs

Golgi tendon organs (GTO) are receptors that respond to tension rather than to
length as do muscle spindles. Golgi tendon organs are high-threshold receptors
that exert inhibitory effects on agonists and facilitatory affects on antagonists.
When forces of muscle contraction, together with forces resulting from external
factors, sum to the point where injury to the muscle tendon or bone becomes
possible, then the GTO cause IPSPs on the cell body of the agonistic motor
units. Similarly, when shortening during muscle contraction progresses to the
point where continued shortening could damage the joint because of hyperex-
tension or hyperflexion, then GTO act to shut off the agonist (with ISPSs) and
stimulate the antagonist (with EPSPs). In this way, GTO bring about smooth
retardation of muscular contractions.

The GTO mechanism is not, however, a fail-safe mechanism. Because the
GTO influence motor neuron cell bodies with IPSPs, their effects can be counterbalanced by additional EPSPs from higher centers. The process of minimiz-

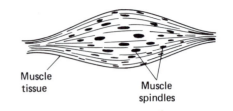

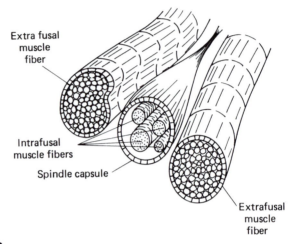

FIGURE 18-6
Muscle spindles exist in parallel with extrafusal muscle fibers. Stretch of the muscle results in stretch of the spindle capsule and its contents.

(**SOURCE**: Modified from D.W. Edington and V.R. Edgerton, 1976, p. 98.)

ing the influence of GTOs is referred to as *disinhibition*. Indeed practicing disinhibition appears to be part of athletic training, the purpose of which is to push performance to the limits of tissue capacity. In the sport of wrist wrestling, there occasionally occur ruptured muscles or tendons and broken bones. In highly motivated and disinhibited individuals, the combination of active muscle contraction plus tension exerted by the opponent can exceed the strength of tissues.

Volitional and Learned Movements

The general scheme by which the brain controls motor activities is illustrated in Figure 18-8. According to this scheme, movement is initiated in the motor cortex, which sends commands (action potentials) directly via a spinal axon to α-motor neurons whose cell bodies are located in the spinal cord. Simultaneously, or perhaps with a slight delay, the cortex signals subcortical nuclei, which receive input from sensory receptors. Meanwhile, the cerebellum con-

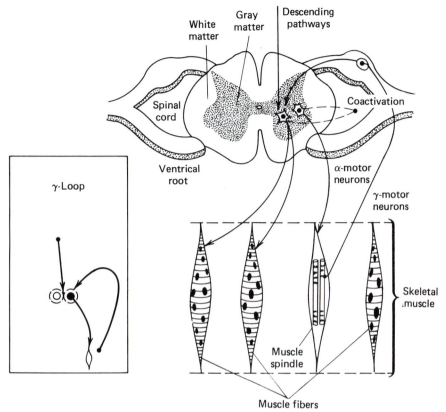

FIGURE 18-7

The γ loop (in heavy lines) allows for fine control of muscle position in complex movement patterns. Descending signals result in activation of both α- and γ-motor neurons (coactivation). Stretch on the spindle receptors provides feedback on muscle position via the γ afferents. Input of the γ loop back to the α-motor neuron cell body affects its firing.

stantly compares the intended with actual movements and integrates each sequential movement component into a coordinated effort.

Corticospinal Tract

The corticospinal tract (Figure 18-9) consists of bundles of neurons whose cell bodies exist in the motor cortex, whose axons extend through the spinal cord, and whose terminal processes communicate with α-motor neurons in the gray matter of the spinal cord. Because portions of the corticospinal tract run from the top of the head to the lower back, the tract can be quite long. The cell bodies from which the corticospinal tract originates are pyramidal in shape, therefore the tract is alternatively referred to as the *pyramidal tract, pathway,* or *system.*

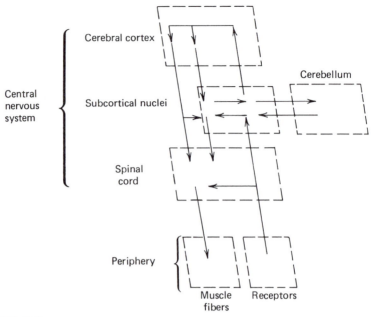

FIGURE 18-8

General scheme of motor control. The cerebral motor cortex initiates movement by sending descending signals to muscle fibers and to subcortical areas, which relay signals to the cerebellum. Various receptors provide sensory feedback, and the cerebellum allows a comparison (integration) of actual and intended movements. Subsequent activity initiated by the motor cortex is varied depending on the initial results.

(SOURCE: From J.A. Vander, J.H. Sherman, and D.S. Luciano, 1980. With permission.)

The corticospinal pathway is the major effector of complicated and rapid volitional movements. Dr. Franklin Henry has compared function of the motor cortex to the running of a computer program. Because at the time that Henry described his hypothesis, computer programs were stored on magnetic drums (rather than on the contemporary tape or disks), the theory is now known as Henry's memory drum theory. According to this theory, intricate movements, which depend on the precise contraction of motor units in different muscles that are ordered specifically in terms of space, time, intensity, and duration, are "programmed" by an incredibly precise order of depolarization of pyramidal cells in the motor cortex. At present, the precise mechanism by which memory is accomplished is not understood. Learning may well involve the synthesis of proteins or neurotransmitters in specific neurons or areas of neurons. The precise imprinting mechanism involved in learning rapid, precise performance of an intricate motor task depends on a computer program-like sequencing of events in the motor cortex.

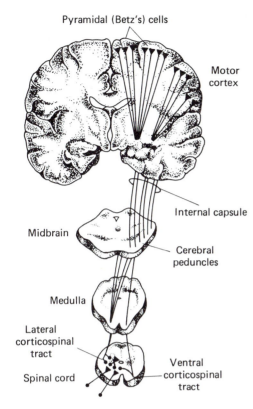

Pyramidal (Betz's) cells

Motor cortex

Internal capsule

Midbrain

Cerebral peduncles

Medulla

Lateral corticospinal tract

Spinal cord

Ventral corticospinal tract

FIGURE 18-9
The pyramidal (corticospinal) tract originates from the pyramid-shaped Betz's cells in the cerebral cortex. Axons from the motor cortex cross over in the medulla and run down in the ventral and lateral white matter of the spinal cord. Bundles of descending axons make up a tract. Synapses occur with motor neurons in the gray area of the spinal cord at appropriate levels.

(SOURCE: Adapted from J.A. Vander, J.H. Sherman, and D.S. Luciano, 1980, p. 598. With permission.)

Multineuronal, Extrapyramidal Pathways

As indicated in Figure 18-8, there exist descending motor pathways in addition to the pyramidal pathway. These extrapyramidal pathways are multineuronal in structure. Extra synaptic connections slow the conduction velocity, but they allow integration of motor programs with sensory inputs including progress on the result of programs originating in pyramidal cells. Additionally, the extrapyramidal system allows the cerebellum to become involved in coordination of motor activities.

SENSORY INPUTS DURING MOTOR ACTIVITIES

During motor activities an individual will use at various times most of the sensory information available. The contributions of various forms of sensory information in leading to successful completion of an activity will vary depending on the activity, but afferent inputs from the eyes, inner ear, muscle spindles, joints, and skin are very important. Various individuals may, in fact, perform identical motor tasks while relying on the different sensory inputs to various degrees. Depending on the activity and sense involved, some individuals can adapt and successfully complete motor tasks even though deprived of a usually important form of sensory input. Sometimes also in the performance of motor tasks, receipt of too much information confuses individuals, distracting them from relevant inputs. Concentration and focusing are tasks governed by the frontal cortex and the reticular formation.

SPINAL MOVEMENTS

The gruesome expression, "running like a chicken with its head cut off," carries for us some useful information concerning spinal control of locomotion. Whereas swimming, walking, and running are usually considered to be learned activities controlled by the pyramidal system, many anesthetized species will sprint when suddenly decapitated. Some of the victims of the guillotine during the French Revolution are reported to have run away from the block, but only after the blade had fallen. In residential areas where swimming pools are common, infants too young to walk are frequently taught to swim.

On the basis of these observations, we must conclude that motor patterns either can be learned by or are built into the spinal cord. These movement patterns are not usually apparent but are probably under the ultimate control of the motor cortex. For these movement patterns, the cortex does not necessarily determine each action potential that will occur in each motor unit involved; rather, the cortex may function to activate a spinal locomotory pattern and to set the frequency and amplitude of the movement pattern. During sudden spinal transection, mechanical stimulation of descending tracts leads to activation of a whole series of responses, one of which occasionally is to activate running motions.

V.R. Edgerton and J. Smith, with their colleagues at UCLA, have been able to induce cats with complete spinal transection to walk. This is done by providing appropriate sensory stimuli to the spinal cord in animals that cannot walk under volitional control. This research holds great promise not only for advancing our understanding of motor control, but also for the rehabilitation of para- and quadriplegics.

Removing the immediate control of locomotion from the cortex to a lower level in the CNS has several theoretical advantages. It may allow for more direct and immediate (faster) control, and it frees higher centers to perform other tasks. When chasing and catching a fly ball, an outfielder in baseball does not have to think about how each muscle is contracted in running; similarly, the relay runner can concentrate on timing his or her start and speed while focusing on receiving the baton.

Summary

The biological control of muscular movement depends on the central nervous system (CNS). Even the simplest form of movement involves at least two neurons and a synapse in the CNS. As motor activities become more complex, the center of neuromuscular control tends to move up the CNS and include the cerebral cortex of the brain. The learning of intricate motor tasks involves programming of the area of the brain's frontal lobe, called the motor cortex. Once the motor cortex is imprinted with a program, the program can be played back very rapidly and accurately.

Perhaps the most difficult motor tasks include those in which locomotion and a fine eye–hand or eye–foot skill are performed simultaneously. For instance, consider running to and hitting a tennis ball. In this situation, programs learned at various levels of the CNS are integrated with sensory input to produce a coordinated response. The CNS represents a fantastic computer that can contain large quantities of information, can simultaneously make decisions and affect actions of differing nature, and can constantly monitor and correct the results obtained.

Selected Readings

Arsharsky, Yu I., M.B. Berkinblit, O.I. Fukson, I.M. Gelfand, and G.N. Orlovsky. Recordings of neurons of the dorsal spinocerebellar tract during evoked locomotion. Brain Res. 272–275, 1972.

Arsharsky, Yu I., M.B. Berkinblit, I.M. Gelfand, G.M. Orlovsky, and O.I. Fukson. Activity of the neurons of the ventral spinocerebellar tract during locomotion. Biophysics 17: 926–935, 1972.

Asanuma, H. Cerebral cortical control of movement. Physiologist. 16: 143–166, 1973.

Betz, V. Anatomischer Nachweis Zweier Gehirncentrabl. Med. Wiss. 12: 578–580; 595–599, 1874.

Burke, R.E., P. Rodomin, and F.E. Zajac, III. The effect of activation history or tension produced by individual muscle units. Brain Res. 109: 515–529, 1976.

Carew, T.J. Descending control of spinal circuits. In: Kandel, E.R. and J.H. Schwartz (eds.), Principles of Neural Science. New York: Elsevier/North-Holland, 1981. p. 312–322.

Carew, T.J. Spinal cord I: Muscles and muscle receptors. In: Kandel, E.R. and J.H. Schwartz (eds.), Principles of Neural Science. New York: Elsevier/North-Holland, 1981. p. 284–292.

Carew, T.J. Spinal cord II: Reflex action. In: Kandel, E.R. and J.H. Schwartz (eds.), Principles of Neural Science. New York: Elsevier/North-Holland, 1981. p. 293–304.

Chambers, W.W., C.N. Liu, and G.P. McCouch. Anatomical and physiological correlates of plasticity in the central nervous system. Brain Behav. Evol. 8: 5–26, 1973.

Crowe, A. and P.B.C. Mathews. The effects of stimulation of static and dynamic fusimotor fibers on the response to stretching of the primary endings of muscle spindles. J. Physiol. London 174: 109–131, 1964.

Evarts. E.V. Relation of pyramidal tract activity to force exerted during voluntary movement. J. Neurophysiol. 31: 14–27, 1968.

Evarts, E.V., E. Bizzi, R.E. Burke, M. DeLong, and W.T. Thach. Central control of movement. Neurosci. Res. Program Bull. 9: 1–170, 1971.

Evarts, E.V. and J. Tanji. Reflex and intended responses in motor cortex pyramidal tract neurons of monkey. J. Neurophysiol. 39: 1069–1080, 1976.

Forssberg, H., S. Grillner, and S. Rossignol. Phase dependent reflex reversal during walking in chronic spinal cats. Brain Res. 85: 103–107, 1975.

Ghez, C. Introduction to the Motor Systems. In: Kandel, E.R. and J.H. Schwartz (eds.), Principles of Neural Science. New York: Elsevier/North-Holland, 1981, p. 272–283.

Ghez, C. Cortical control of voluntary movement. In Kandel, E.R. and J.H. Schwartz (eds.), Principles of Neural Science. New York: Elsevier/North-Holland, 1981. p. 324–333.

Ghez, C. and D. Vicario. The control of rapid limb movement in the cat II. Scaling of isometric force adjustments. Exp. Brain Res. 33: 191–202, 1978.

Grillner, S. and P. Zangger. How detailed is the central pattern generation for locomotion? Brain Res. 88: 367–371, 1975.

Guyton, A.C. Textbook of Medical Physiology. Philadelphia: W.B.Saunders Co., 1981. p. 560–697.

Hams, D.A. and E. Henneman. Feedback signals from muscle and their efferent control. In: Mountcastle, V.B. (ed.), Medical Physiology, 14th ed. Vol. 1. St. Louis: Mosby Publishing, 1980. p. 703–717.

Henneman, E. Organization of the spinal cord and its reflexes. In: Mountcastle, V.B. (ed.), Medical Physiology, 14th ed., Vol. 1. St. Louis: Mosby Publishing, 1980. p. 762–786.

Henry, F.M. Increased response latency of complicated movements and a ''memory drum'' theory of neuromotor reaction. Res. Quart. 31: 448–458, 1960.

Henry, F.M. Influence of motor and sensory sets on reaction latency and speed of discrete movements. Res. Quart. 31: 459–468, 1960.

Henry, F.M. Stability-instability of the motor memory trace; A modification of the 1960 memory drum theory. In: Roberts, G.G. and D.M. Landers (eds.), Psychology of Motor Behavior and Sport. Champaign, Ill.: Human Kinetics Publishing, 1980. p. 60–70.

Henry, F.M. The evolution of the memory drum theory of neuromotor reaction. In: Brooks, G.A. (ed.), Perspectives on the Academic Discipline of Physical Education. Champaign, Ill.: Human Kinetics Publishing, 1981. p. 301–322.

Homma, S. Understanding the stretch reflex. Prog. Brain Res. 44:1–507, 1976.

Houk, J.C. Motor control processes: New data concerning motoservo mechanisms and a tentative model for stimulus-response processing. In: Talbot, R.E. and D.R. Humphrey (eds.), Posture and Movement. New York: Raven Press, 1979, p. 231–241.

Houk, J. and E. Henneman. Responses of Golgi tendon organs to active contractractions of the soleus muscle of the cat. J. Neurophysiol. 30:466–481, 1967.

Hunt, C.C. and S.W. Kaffler. Stretch receptor discharges during muscle contraction. J. Physiol. London 113: 298–315, 1951.

Hunt, C.C. and E.R. Perl. Spinal reflex mechanisms concerned with skeletal muscle. Physiol. Rev. 40: 538–579, 1960.

Johnson, E.J., A. Smith, E. Eldred, and V.R. Edgerton. Exercise-induced changes of biochemical, histochemical, and contractile properties of muscle in cordotomized kittens. Exp. Neurol. 76: 414–427, 1982.

Kaiserman-Abramof, I.R. and A. Peters. Some aspects of the morphology of Betz cells in the cerebral cortex of the cat. Brain Res. 43: 527–546, 1972.

Kandel, E.R. and J.H. Schwartz, (eds.), Principles of Neural Science. New York: Elsevier/ North-Holland, 1981.

Katz, B. Nerve, Muscle and Synapse. New York: McGraw-Hill, 1966.

Kuypers, H.G.J.M. The anatomical organization of the descending pathways and their contributions to motor control especially in primates. In: Desmedt, J.E. (ed.), New Developments in Electromyography and Clinical Neurophysiology, Vol. 3. Basel: Karger Publishing, 1973. p. 38–68.

Lundberg, A. Control of spinal mechanisms from the brain. In: Tower, D.B. (ed.), The Nervous System, the Basic Neurosciences, Vol. 1. New York: Raven Press, 1975. p. 253–265.

Lundberg, A. Integration in the propriospinal motor center controlling the forelimb in the cat. In: Asanuma, H. and V.J. Wilson (eds.), Integration in the Nervous System. Tokyo: Igaku-Shoin, p. 47–64, 1979.

Liddell, E.G.T. and C. Sherrington. Reflex in response to stretch (myotatic reflexes). Proc. Roy. Soc., Series B. Biol. Sci. 96: 212–242, 1924.

Mathews, P.B.C. Muscle spindles and their motor control. Physiol. Rev. 44: 219–288, 1964.

Merton, P.A. How we control the contraction of our muscles. Sci. Amer. 226: 30–37, 1972.

Nashner, L.M. Adaptation reflexes controlling the human posture. Exp. Brain Res. 26: 59–72, 1976.

Nichols, T.R. and J.C. Houk. Reflex compensation for variations in mechanical properties of a muscle. Science 181: 182–184, 1973.

Pearson, K. The control of walking. Sci. Am. 235: 72–86, 1976.

Phillips, C.G. and R. Porter. Cortical Spinal Neurones: Their Role In Movement. London: Academic Press, 1977.

Reback, P.A., A.B. Scheibel, and J.L. Smith. Development and maintenance of dendrite bundles after cordotomy in exercised and non-exercised cats. Exp. Neurol. 76: 428–440, 1982.

Ruch, T.C. and H.D. Patton. Physiology and Biophysics, 19th ed. Philadelphia: W.B. Saunders, Co., 1965.

Sherrington, C. The Integrative Action of the Nervous System, 2nd ed. New Haven: Yale University Press, 1947.

Shik, M.L., F.V. Severin, and G.N. Orlovskii. Control of walking and running by means of electrical stimulation of the mid-brain. Biophysics 11: 756–765, 1966.

Smith, J.L., L.A. Smith, R.F. Zernicke, and M. Hoy. Locomotion in exercised and non-exercise cats cordotomized at two or twelve weeks of age. Exp. Neurol. 76: 393–413, 1982.

Stein, R.B. Peripheral control of movement. Physiol. Rev. 54: 215–242, 1974.

Swett, J.E. and E. Eldred. Distribution and numbers of stretch receptors in medial gastrocnemius and soleus muscles of the cat. Anat. Rec. 137: 453–473, 1960.

Tanji, J. and E.V. Evarts. Anticipatory activity of motor cortex neurons in relation to direction of an intended movement. J. Neurophysiol. 39: 1062–1068, 1976.

Towe, A.L. and E.S. Luschei. Motor coordination. In: Schoff, J. (ed.), Handbook of Behavioral Neurobiology, Vol. 4, Biological Rhythms. New York: Plenum Press, 1970.

Vallbo, Å.B. Muscle spindle response at the onset of isometric voluntary contractions in man. Time difference between fusimotor and skeletomotor effects. J. Phyisol. London 218: 405–431, 1971.

Vallbo, Å.B. Discharge patterns in human muscle spindle afferents during isometric voluntary contractions. Acta Physiol. Scand. 80: 552–566, 1970.

Vander, A.J., J.H. Sherman, and D.S. Luciano. Human Physiology, 3rd ed. New York: McGraw-Hill, 1980. p. 144–190.

19

SKELETAL MUSCLE STRUCTURE AND FUNCTION

Many of the events of daily living, including those involving physical work performed on the job, magnificent physical performances in athletics and the performing arts, as well as migrations, predatory chases, and flights of survival in the animal kingdom, are but a few of the behaviors supported by muscle contraction. These sometimes fantastic feats of strength and endurance are possible because of the conversion of chemical energy to mechanical energy in muscle. In previous chapters (2 through 8) we have described the conversion of the potential chemical energy in foodstuffs to chemical energy forms (glycogen, ATP, CP) used in muscle and other cells. In this chapter we endeavor to explain the process by which this stored cellular energy is used to swim, run, jump, and lift and throw things.

Skeletal, cardiac, and smooth muscles are chemomechanical transducers; they convert chemical energy to mechanical and heat energy. By nature, this process is at best 30% efficient, and so 70% of the energy released appears as heat (Chapter 3). At present, the chemical events of muscle contraction are fairly well understood. Progress has also been made toward understanding how those chemical events lead to muscle shortening.

Contemporary athletes such as Arnold Schwarzeneger have developed muscle size and definition to amazing degrees. (Michael Abramson/Liaison.)

SKELETAL MUSCLE STRUCTURE

Skeletal muscle tissue is composed of individual muscle cells (fibers). These fibers are arranged in bundles (fasciculi) and are subdivided into myofibrils and myofilaments (Figure 19-1). Integrally arranged in the structure of muscle is connective tissue made up largely of the substance collagen. Connective tissue runs from end to end of muscle (i.e., from tendon of origin to tendon of insertion) and exists within muscle tissue surrounding the fibers and giving rise to muscle bundles. In some muscles, the cells run from end to end, whereas in other muscles the fibers attach to connective tissue within the muscle. Consequently, in some muscles (e.g., gastrocnemius) the fibers do not run exactly in the direction of the muscle tissue.

Under a magnifying glass or a light microscope, skeletal muscle fibers appear to have a striped (striated) appearance (Figure 19-2). These striations are due to the presence of actin and myosin, the two main proteins of contraction. The lighter I (isotropic) band allows more polarized light to pass because of

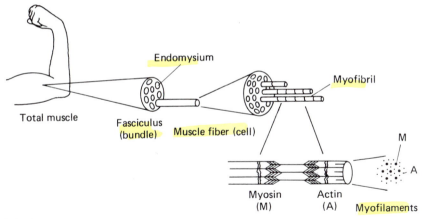

FIGURE 19-1
Muscle tissue is composed of muscle bundles (fasciculi), muscle fibers
(cells), myofibrils, and myofilaments (actin and myosin).

(SOURCE: Reproduced from D.W. Edington and V.R. Edgerton, 1976, p. 16. By permission of the authors.)

the presence of the thin actin filaments. The darker A (anisotropic) bands block
more light because of the presence of both the thick myosin and the thin actin
filaments. Under higher magnification, a thin H band appears in the center of
the A band; and H band is due to the structure of the myosin filament, which
is less dense at the center (Figure 19-3).

How the fine (ultra) structure of skeletal muscle gives rise to the striated
appearance under the light microscope is illustrated in Figure 19-3. This figure
also shows how the thin actin filaments are attached to the Z line. The Z line
is actually a latticelike structure that forms the foundation for the molecular
mechanism of shortening in muscle. The distance between one Z line and the

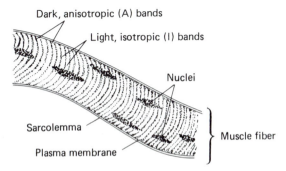

FIGURE 19-2
Skeletal muscle fibers have a striated (striped) appearance due to the
banding of actin and myosin myofilaments.

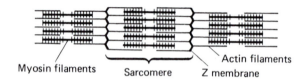

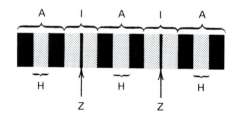

FIGURE 19-3

Dark (anisotropic, A) bands in striated muscle are due to the overlap of thick (myosin) and thin (actin) myofilaments. Light (isotropic, I) bands obtain their appearance due to a lesser density and greater penetration of light.

next makes up the *sarcomere*, which represents the basic functional unit of the muscle.

HUXLEY'S SLIDING FILAMENT MECHANISM OF MUSCLE CONTRACTION

When muscle shortens, the I band shortens, but the A band does not change in length. The mechanism of how this occurs and how the thick and thin filaments operate to cause muscle shortening (contraction) was developed largely by H.E. Huxley. According to this mechanism, the protruding heads of myosin filaments interact with actin filaments, and movement occurs due to a ratchet or oarlike interaction between the two contractile proteins (Figure 19-4). At present, the chemistry of muscle contraction is fairly well understood, but the precise mechanism by which chemical energy is converted to shortening is a subject of some controversy and active investigation.

The structures of myosin (Figure 19-4B) and actin (Figure 19-5) reveal that the filamentous structures are made up of individual subunits. Whereas filamentous actin (F-actin) is a helical structure made up of globular subunits (G-actin), the structure of filamentous myosin is like a double-ended bottle brush (Figure 19-4*b*). Like actin filaments, myosin filaments are also made up of individual myosin subunits, or molecules. In addition to supporting the basic structure of muscle, both actin and myosin subunits possess enzymatic activities. Each has a complementary binding site for the other, and each can bind ADP or ATP.

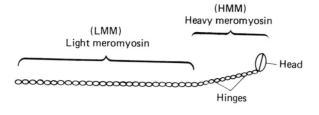

(a)

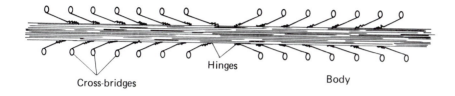

Hinges

Cross-bridges

Body

Myosin filament

(b)

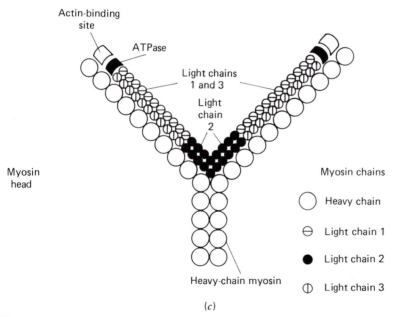

(c)

FIGURE 19-4

Myosin molecules have structural (light) and enzymatic (heavy ends) (a).
The light meromyosin provides a connecting link to similar units in the
myosin filament, which appears as a double-ended bottle brush (b). The
head of heavy meromyosin contains binding sites for actin and ATP.
Myosin in fast-contracting muscle has a higher ATPase activity than does
myosin in slow muscle. These differences in ATPase activity are due to differences in the light- and heavy-chain composition in the myosin head (c).

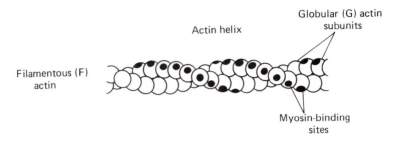

FIGURE 19-5

Filamentous (F) actin is a helix made up of individual globular (G) actin subunits. Each G-actin subunit contains a myosin-binding site. Tropomyosin (not shown) provides structural support to the helical structure and allows for enzymatic control (see Figure 19-8).

Furthermore, F-actin binds two other proteins, troponin and tropomyosin, which are involved in triggering muscle contraction.

Whereas the actin molecule appears to be fairly similar across species, myosin differs not only between species, but also between fast and slow muscles.

EXCITATION—CONTRACTION COUPLING

The process by which nerve signals result in muscular movement is referred to as excitation–contraction coupling. This process (Figures 19-6 and 19-7) involves a series of events by which action potentials in α-motor neurons cause actin and myosin to interact. As described in Chapter 17, the release of acetylcholine at the neuromuscular junction causes an action potential in the postsynaptic muscle cell (Figure 19-6). The muscle action potential spreads over the cell surface and likely also down the transverse (T) tubules (Figure 19-7). As a result, calcium ion (Ca^{2+}) is released from the lateral sacs of the sarcoplasmic reticulum, and Ca^{2+} is released into the sarcoplasm. Calcium ion then binds to troponin and causes tyopomyosin to shift its orientation along the F-actin helix. As indicated in Figure 19-8, the myosin cross-bridge binding sites on actin are exposed, thereby allowing actin and myosin to interact.

The postsynaptic muscle cell membrane (Figure 19-6) possesses two important features in the control of muscle contraction. Acetylcholine (ACH) released into the synaptic cleft from the terminal branches of the motor neuron finds binding sites on the muscle cell membrane. The binding of ACH has the effect of increasing the permeability of Na^+, which depolarizes the postsynaptic muscle cell membrane, giving rise first to an end plate potential (EPP) and then to an action potential (AP), which spreads across the muscle cell surface and down the T tubules. The effect of ACH is short-lived, as it is degraded by the presence of the enzyme acetylcholinesterase. Thus the release of ACH

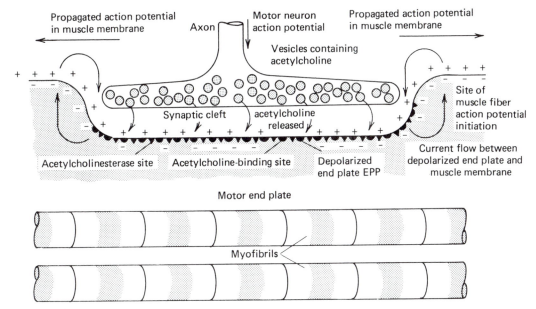

FIGURE 19-6
The motor end plate is the site where α-motor neurons communicate with muscle cells. At the motor end plate (myoneural junction), terminal axon branches contain acetylcholine, which can be released into the synaptic cleft and bind to receptor sites on the muscle cell surface. Enzymatic degradation of ACH by acetylcholinesterase means that continual α-motor neuron activity is required for continued muscle cell contraction.

(SOURCE: From J.A. Vander, J.H. Sherman, and D.S. Luciano, Human Physiology, 1980, p. 224. With permission.)

constitutes the first muscle signal to "switch on" contraction, and the activity of acetylcholinesterase is the signal to "switch off" the contraction.

Whereas muscle contraction depends on the release of Ca^{2+} from the lateral sacs of the sarcoplasmic reticulum (Figure 19-7, upper right), muscle relaxation depends on sequestering Ca^{2+} back into the lateral sacs. The pumping of Ca^{2+} into the lateral sacs is an endergonic process and requires ATP. Fast-contracting muscle fiber types, such as type IIb (Figures 5-3 and 17-8) have high contents of sarcoplasmic reticulum and high activities of sarcoplasmic reticulum ATPase. These characteristics facilitate rapid contraction and relaxation.

THE MOLECULAR MECHANISM OF MUSCLE MOVEMENT

All the essential elements for muscle contraction (ATP, actin, myosin, troponin, tropomyosin, Mg^{2+}, Ca^{2+}) are normally present in muscle. Yet contrac-

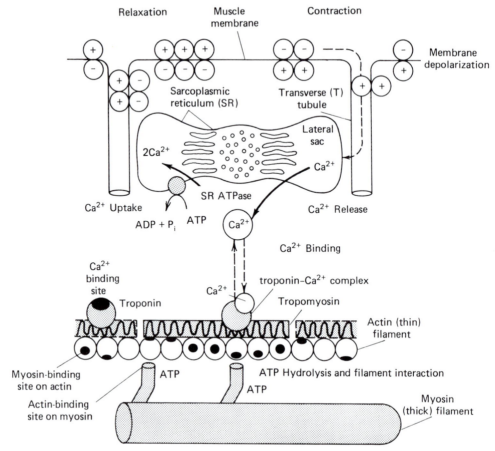

FIGURE 19-7

Excitation–contraction coupling is regulated in several steps. Muscle cell membrane depolarization (Figure 19-6) causes the release of Ca^{2+} from the lateral sacs of the sarcoplasmic reticulum. Calcium ion released into sarcoplasm finds binding sites on troponin, forming a troponin–Ca^{2+} complex. Binding of Ca^{2+} to troponin causes a direct conformational change in troponin and an indirect shift of tropomyosin. The shift in tropomyosin relieves its inhibition between actin and myosin, so that cross-bridges can be formed and contraction can occur. Relaxation depends on pumping Ca^{2+} back into the lateral sacs.

(SOURCE: Modified from J.A. Vander, J.H. Sherman, and D.S. Luciano, 1980, p. 222. Human Physiology. With permission.)

tions result only in particular circumstances. Normally (Figure 19-8), tropomyosin blocks or screens out the possible interaction between actin and myosin. The release of Ca^{2+} from the sarcoplasmic reticulum provides the stimulus for actin–myosin interaction and contraction.

Huxley's ratchet mechanism of muscle filament sliding is diagrammed in Figure 19-9. Although at present there is fair agreement on the fact of actin–

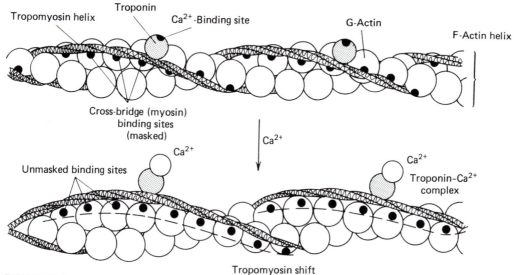

FIGURE 19-8

The binding of Ca²⁺ to troponin causes a conformational (shape) change in that protein and a movement of the tropomyosin helix. The tropomyosin shift causes the myosin-binding sites on actin to become unmasked. Thus Ca²⁺ binding to troponin indirectly allows for actin and myosin to interact.

myosin interaction, as suggested earlier, the precise mechanism by which the energy of ATP is utilized to effect filament sliding is a subject of some uncertainty. As originally proposed by Huxley, a rotation of the myosin head toward the direction of movement could result in filament sliding. Alternatively, movement in the "hinge" area could produce shortening. Additionally, other possibilities exit. Solution of this problem is probably worthy of a Nobel Prize, and some of the best minds in biology, chemistry, and physics are currently working on the problem.

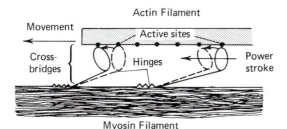

FIGURE 19-9

Muscle contraction (shortening) is thought to occur when cross-bridges extend from myosin to actin and a conformational change occurs in the cross-bridge. The specifics of this schematic model are yet to be worked out.

(SOURCE: Modified from H.E. Huxley, 1958.)

The cyclic events of muscle contraction are summarized in Figure 19-10. Description of this model starts with the resting state (Figure 19-10a). In the presence of magnesium ion (Mg^{2+}), ATP unites with myosin to energize it. Thus, in Figure 10-10a, myosin is like a loaded and cocked revolver ready to discharge. The energized myosin can then combine with actin if Ca^{2+} affects the troponin–tropomyosin complex to reveal (unmask) the actin-binding sites. The binding of actin to myosin triggers the release of potential energy stored in the myosin head. Mechanical work (movement) and heat release are consequences of this energy release. In order for the myosin cross-bridge to be used again (i.e., to recycle), it must be dislodged from the actin-binding site and be reenergized. This is done by ATP in several ways. Ca^{2+} must be pumped back into the lateral sacs of the sarcoplasmic reticulum. This process is endergonic and requires ATP (Figure 19-7). Additionally, ATP must be present to break up (dissociate) the actin–myosin complex and to react with the myosin to reenergize it.

CHARACTERISTICS OF MUSCLE CONTRACTION

Classically, the contractile characteristics of muscle have been studied at several levels. These have been in the intact living, moving organism (*in vivo*), with the muscle in place in an anesthetized animal preparation (*in situ*), and with the muscle in a physiological fluid bath isolated from the individual (*in vitro*). Small muscles from frogs and little mammals have particularly lent themselves to study *in vitro*, but detailed studies on muscles of larger mammals have also provided useful information.

Isometric and Isotonic Contractions

Isolated muscles can be arranged in an apparatus (Figure 19-11), so that the muscle can shorten (*a*) or be prevented from shortening following stimulation (*b*). When the muscle's length is fixed, an isometric (same length) condition exists. A recording of the isometric tension developed can be made following a single electrical stimulus (Figure 19-11b). When the load on the muscle is light enough to be lifted, the rate of shortening is constant (Figure 19-11a). Because the lifting of a constant weight by an isolated muscle produces a linear response, the tension produced is constant. Therefore, the contraction is said to be isotonic (i.e., same tension). In athletic training (Chapter 20) the terms isometric and isotonic are also used, or rather misused. These terms were developed by muscle physiologists and correctly apply to contractile properties of isolated muscles. In the intact individual it is difficult to establish an isometric condition and almost impossible to establish an isotonic condition. In the isotonic situation, increasing the weight lifted decreases both the distance shortened and the velocity of shortening following a single stimulus (Figure 19-

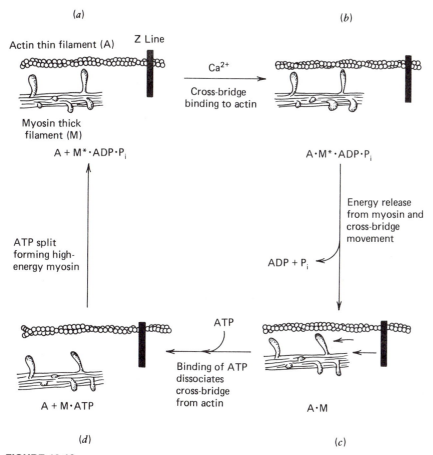

(a)

Actin thin filament (A)   Z Line

Ca^{2+}

Cross-bridge
binding to actin

(b)

Myosin thick
filament (M)

$A + M^* \cdot ADP \cdot P_i$

$A \cdot M^* \cdot ADP \cdot P_i$

Energy release
from myosin and
cross-bridge
movement

ATP split
forming high-
energy myosin

$ADP + P_i$

ATP

$A + M \cdot ATP$

Binding of ATP
dissociates
cross-bridge
from actin

$A \cdot M$

(d)

(c)

FIGURE 19-10
The cyclic process of muscle contraction and relaxation. In the resting
state (a), actin (A) and energized myosin ($M^* \cdot ADP \cdot P_i$) cannot interact because of the effect of tropomyosin. Upon the release of Ca^{2+} from the sarcoplasmic reticulum, actin binds to $M^* \cdot ADP \cdot P_i$ (b). Tension is developed
and movement occurs with the release of ADP and P_i (c). Dissociation of
actin and myosin requires the presence of ATP to bind to myosin and to
pump Ca^{2+} into the SR (d). Myosin is energized upon return to the resting
state (a).

(SOURCE: From J.A. Vander, J.H. Sherman, and D.S. Luciano, 1980, p. 218. With
permission.)

12A). Additionally, the response of the muscle to shorten is delayed as a result
of the extra time required to develop sufficient tension to move the heavier
weight. Also in Figure 19-12, note that although strengthening a muscle does
not affect its maximum rate of unloaded contraction, because of the hyperbolic
nature of the relationship, heavy loads can be lifted at significantly faster rates
(Figure 19-12b).

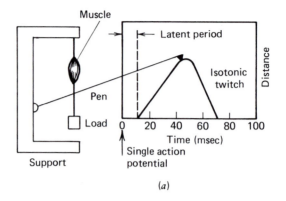

(a)

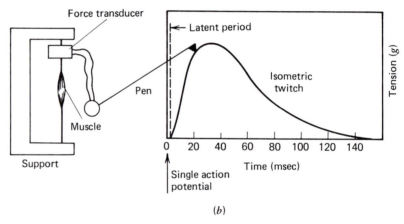

(b)

FIGURE 19-11

Schematics of apparatus to study the contractile properties of isolated muscles. When a loaded muscle shortens against a load, the shortening record produces a linear, isotonic response (a). When a muscle's length is fixed, isometric twitch characteristics can be recorded (b).

(**SOURCE:** Adapted from J.A. Vander, J.H. Sherman, and D.S. Luciano, 1980, p. 226. With permission.)

Two other special characteristics of muscle contraction also bear mention; these are the effects of summation and of initial length. Summation and length effects are usually studied on isometric preparations. When a muscle receives repeated stimuli, at a frequency such that the muscle cannot completely relax from the previous stimulus before the subsequent stimulus arrives, the tension produced adds up (sums) to a value greater than after a single stimulus is re-

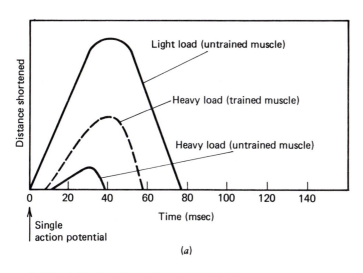

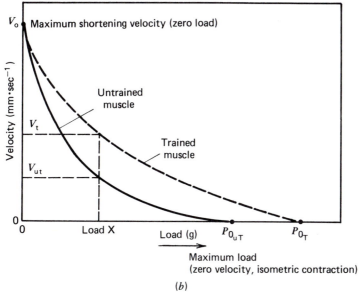

FIGURE 19-12
Characteristics of isotonic contractions. As compared to lifting light loads, isotonic responses to given stimuli when lifting heavy loads results in a greater latent period, slower movement, and less movement (a). The effect of strength training is one that appears to make the load lighter. The force–velocity relationship (b) is hyperbolic in nature. Greater loads produce slower speeds, but greater tension. The effect of strength training is to increase P_0 (maximum isometric tension) but not V_0 (maximum unloaded velocity). However, a stronger trained muscle can move a given isotonic load (x) at a greater velocity ($V_t > V_{ut}$).

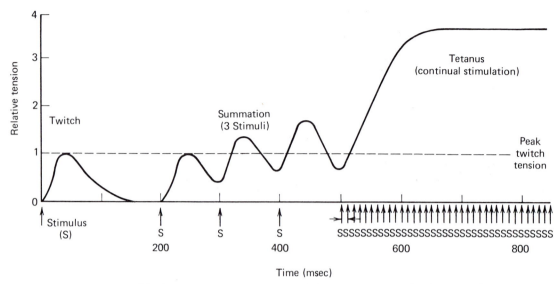

FIGURE 19-13

Repeated stimuli, each of a given strength (S) can produce a tension that sums to greater than the twitch tension. Continual stimulation results in a tetanic contraction three to five times stronger than twitch tension.

(**SOURCE:** Adapted from J.A. Vander, J.H. Sherman, and D.S. Luciano, 1980, p. 228. With permission.)

ceived (Figure 19-13). When repeated shocks are given to a muscle (30 shocks · sec^{-1} for a slow muscle and 120 shocks · sec^{-1} for a fast muscle), the muscle's response fuses into a smooth contraction that yields a tension three to five times greater than that following a single twitch. This response to high-frequency stimulation is termed a *tetanus*. The tetanic response is due to increased time for Ca^{2+} release from the sarcoplasmic reticulum, which increases the probability of myosin cross-bridge interaction with actin.

The initial length at which a muscle begins to contract affects the tension that it develops (Figure 19-14). At the muscle's full resting length (L_0), the overlap between actin and myosin is greatest. Consequently the maximum probability of cross-bridge development exists and the greatest tension is developed. For the biceps, L_0 would exist at full elbow extension. As the muscle's length changes from L_0, the probability of actin–myosin interaction decreases, and the maximal tension developed during an isometric tetanus also decreases.

Properties of Muscle Contraction in the Body

Although studies on isolated muscles provide a great deal of useful information, the properties of isolated muscles do not always predict the properties *in vivo*.

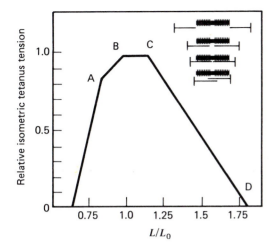

FIGURE 19-14

The length–tension relationship for skeletal muscle. At full resting muscle length (L_0), the probability of actin–myosin interaction (B and C) is maximal, and the greatest isometric tension is recorded. At muscle lengths significantly less than L_0 (A), or greater than L_0 (D), isometric tension declines.

(SOURCE: Adapted from Gordon, Huxley, and Julian, 1966.)

For instance, in the body muscles act across a system of bone levers. Whereas maximal isometric tension of the biceps muscle is developed at L_0, or at full elbow extension, maximal tension during elbow flexion is developed at an elbow angle of approximately 90 degrees. Additionally, the forces developed through muscular contraction will not be the same as those developed in the muscle (Figure 19-15). In real life, external forces placed on a muscle by the circumstances of movement (e.g., through an opponent's efforts or a fall) are additive to the tension developed by the contractile apparatus. These forces can sometimes exceed the tensile strength of a muscle's supporting contractile apparatus and connective tissue. In such cases, injury results.

Muscle Length and Cross Section

The characteristics of muscle contraction *in vivo* will also depend on several other factors. In general, the longer a muscle is, the faster it will contract.

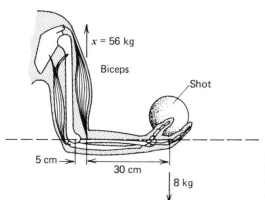

FIGURE 19-15

In the body, muscles act across bone levers so that the force exerted will usually be different from the force of muscular contraction.

Given that in most individuals sarcomere length does not vary much, longer muscles mean more sarcomeres arranged end to end. If these all shorten at the same rate, the individual having the longer muscle will achieve the greatest shortening per unit time.

The maximal tension developed by a muscle is directly related to its thickness, or cross-sectional area. Greater area means more cross-bridges pulling in parallel. The tension developed by human skeletal muscle has been estimated to be 1 to 2 kg/cm².

The maximal tension and speed of muscle contraction will also vary depending on the muscle fiber type (Chapter 17) and the recruitment pattern (Chapter 18). Because few movements in sports require tetanic contractions, maximal force achieved during a movement likely depends on the coordination and skill of the competitor. For instance, the velocity of Jimmy Connors's backhand shot in tennis is likely greater than that of Arnold Schwartzeneger, the body builder. Even in sports such as wrist wrestling, where the individual with the thickest muscle is potentially the strongest, muscular recruitment depends on pyramidal tract activity mediated by the inhibitory influences of Golgi tendon organs (Chapter 18). Here again, muscular performance does not necessarily depend completely on the muscle.

The neural control of muscle recruitment affects speed of movement in another important way. Many movement patterns (e.g., running and swimming) are cyclic in nature. Although muscle strength certainly affects the rate at which a limb can be moved, continued movement depends on both contraction and relaxation of agonists as well as antagonists. Therefore, the individual with the strongest legs is not necessarily the fastest runner.

Summary

Physical activities are the result of chemical to mechanical energy transduction in skeletal muscle. In muscle and other cells there exist two main components to energy transduction. Through the processes of intermediary metabolism, potential energy contained in the by-products of foodstuffs is captured in the form of high-energy phosphate compounds (ATP and CP). The second major component of energy transduction in muscle is the conversion of potential chemical energy in ATP to mechanical work by the two contractile proteins, actin and myosin.

Together, actin and myosin make up a majority of the muscle mass, which in turn can represent a majority of the body's mass. Therefore, in a lean, muscular individual much of the body is composed of these two contractile proteins. Actin and myosin therefore determine not only the body's structure, the enzymatic properties of these proteins also largely determines the body's capacity for movement.

The contractile characteristics of isolated muscle, in terms of speed of movement and tension development, are mediated by a number of factors in the body during muscular exercise. These include the bone lever system, motor unit recruitment pattern, muscle fiber type, and the neural mechanisms of muscular control.

Selected Readings

Abbott, B.C. and D.G. Gordon. A commentary on muscle mechanics. Circ. Res. 36: 1–7, 1975.

Bárány, M. and K. Bárány. A proposal for the mechanism of contraction in intact frog muscle. Cold Spring Harbor Symp. Quant. Biol. 37: 157–167, 1972.

Bornhorst, W.J. and J.E. Minardi. A phenomenological theory of muscular contraction II. Generalized length variations. Biophysical J. 10: 155–171, 1970.

Bornhorst, W.J. and J.E. Minardi. A phenomenological theory of muscular contraction I. Rate equations at a given length based upon irreversible thermodynamics. Biophysical J. 10: 137–154, 1970.

Briskey, E.J., R.G. Cassens, and J.C. Thautman (eds.). The Physiology and Biochemistry of Muscle as a Food, I. Madison: University of Wisconsin Press, 1966.

The Physiology and Biochemistry of Muscle as a Food, II. Briskey, E.J., R.G. Cassens, and B.B. Marsh (Eds.), Madison: University of Wisconsin Press, 1970.

Cohen, C. The protein switch of muscle contraction. Sci. Amer. 233: 36–45, 1975.

Davies, R.E. A molecular theory of muscle contraction: calcium-dependent contractions with hydrogen bond formation plus ATP-dependent extensions of part of the myosin-actin cross bridges. Nature 199: 1068–1074, 1963.

Donovan, C.M. and G.A. Brooks. Muscular efficiency during steady-rate exercise II. Effects of walking speed and work rate. J. Appl. Physiol.: Respirat. Environ. Exercise Physiol. 43: 431–439, 1977.

Ebashi, S. Muscle contraction and pharmacology. Trends Pharmacol. Sci. 1: 29–31, 1979.

Fenn, W.O. The relation between the work performed and the energy liberated in muscle contraction. J. Physiol. London 85: 373–345, 1923.

Fuchs, F. Striated muscle. Ann Rev. Physiol. 36: 461–502, 1974.

Gaesser, G.A. and G.A. Brooks. Muscular efficiency during steady-rate exercise: effects of speed and work rate. J. Appl. Physiol. 38: 1132–1139, 1975.

Gordon, A.M., A.F. Huxley, and F.J. Julian. Tension development in highly stretched vertebrate muscle fibers. J. Physiol. London 184: 143–169, 1966.

Gordon, A.M., A.F. Huxley, and F.J. Julian. The variation in isometric tension with sarcomere length in vertebrate muscle fibers. J. Physiol. London 184: 170–192, 1966.

Greaser, M.L. and J. Gergeley. Purification and properties of the components of troponin. J. Biol. Chem. 248: 2125–2133, 1973.

Guyton, A.C. Textbook of Medical Physiology. Philadelphia: W.B. Saunders, Co., 1981, pp. 122–136.

Harrington, W.F. Muscle proteins and muscle contraction. In: Anfinsen, C.B., R.F. Goldberger, and A.N. Schechter (eds.), Current Topics in Biochemistry. New York: Academic Press, 1972. 135–185.

Hill, A.V. The revolution in muscle physiology. Physiol. Rev. 12: 56–67, 1932.

Hill, A.V. The heat of shortening and the dynamic constants of muscle. Proc. Roy. Soc. London, B. 126: 136–195, 1938.

Hill, A.V. The heat of activation and the heat of shortening in a muscle twitch. Proc. Roy. Soc. London B. 136: 195–211, 1949.

Hill, A.V. Chemical change and the mechanical response in stimulated muscle. Proc. Roy. Soc. London B. 141: 314–321, 1953.

Hill, A.V. Production and absorbtion of work by muscle. Science 131: 897–903, 1960.

Hill, A.V. The effect of tension in prolonging the active state in a twitch. Proc. Roy. Soc. London B. 159: 589–595, 1964.

Hill, T.L., E. Eisenberg, Y.-D. Chen, and R.J. Podolsky. Some self-consistent two-state sliding filament models of muscle contraction. Biophysical J. 15: 335–372, 1975.

Huxley, A.F. Muscle structure and theories of contraction. Progr. Biophys. Biophys. Chem. 7: 255–318, 1957.

Huxley, A.F. The activation of striated muscle and its mechanical response. Proc. Roy. Soc. London B. 178: 1–27, 1971.

Huxley, A.F. A note suggesting that the cross-bridge attachment during muscle contraction may take place in two stages. Proc. Roy. Soc. Lon. B. 183: 83–86, 1973.

Huxley, H.E. The contraction of muscle. Sci. Amer. 1958.

Huxley, H.E. The mechanism of muscular contraction. Sci. Amer. 213: 18–27, 1965.

Huxley, H.E. The mechanism of muscular contraction. Science. 164: 1356–1366, 1969.

Huxley, H.E. Structural aspects of energy conversion in muscle. Ann. N.Y. Acad. Sci. 227: 500–503, 1974.

Lehman, W. and A.G. Szent-Györgi. Regulation of muscular contraction. Distribution of action control and myosin control in the animal kingdom. J. Gen. Physiol. 66: 1–30, 1975.

Morales, M.F. Energy transfer in muscle contraction. Ann. N.Y. Acad. Sci. 227: 183–187, 1974.

Mommaerts, W.F.H.B. Energetics of muscle contraction. Physiol. Rev. 49: 427–508, 1969.

Morel, J.E., I. Pinset-Härström, and M.P. Gingold. Muscular contraction and cytoplasmic streaming: a new general hypothesis. J. Theoret. Biol. 62: 17–51, 1976.

Murray, J.M. and A. Weber. The cooperative action of muscle proteins. Sci. Amer. 230: 58–71, 1972.

Peachey, L. The sarcoplasmic reticulum and transverse tubules in frog's sartorims. J. Cell. Biol. 25: 209–231, 1965.

Podolsky, R.J. and L.E. Teichholz. The relation between the calcium and contraction kinetics in skinned muscle fibers. J. Physiol. London 211: 19–35, 1970.

Rodahl, K. and S. Horvath (eds.). Muscle as a Tissue. New York: McGraw-Hill, 1962.

Szeng-Györgi, A. Chemistry of Muscular Contraction, Academic Press, N.Y., 1951.

The Mechanism of Muscular Contraction. Cold Springs Harbor Symp. Quant. Biol. 37: 1–706, 1972.

Vander, A.J., J.H. Sherman, and D.S. Luciano. New York: McGraw-Hill, Human Physiology. 1980. p. 211–252.

Weber, A. and Murray, J. Molecular control mechanisms in muscle contraction. Physiol. Rev. 53: 612–673, 1973.

Wilkie, D.R. The efficiency of muscular contraction. J. Mechanochem. Cell Motility 2: 257–267, 1974.

Yu, L.C., R.M. Dowlsen, and K. Kornacker. Proc. Nat'l Acad. Sci. 66: 1199–1205, 1970.

20

CONDITIONING FOR MUSCLE STRENGTH

One of the most remarkable aspects of skeletal muscle is its adaptability. If a muscle is stressed (within tolerable limits), it adapts and improves its function. For example, weight lifters exercise their arms and shoulders, so the muscles hypertrophy and improve their strength in order to accommodate the increased load. Likewise, if a muscle receives less stress than it is used to, its function deteriorates. For example, the muscles of a casted leg atrophy in response to disuse.

The purpose of physical training is to stress the body systematically so that it improves its capacity to exercise. Physical training is beneficial only as long as it forces the body to adapt to the stress of physical effort. If the stress is not sufficient to overload the body, then no adaptation occurs. If a stress cannot be tolerated, then injury or overtraining results. The greatest improvements in performance will occur when the appropriate exercise stresses are introduced into the athlete's training program.

Physical fitness is largely a reflection of the level of training. When an athlete is working hard, fitness is high. However, when heavy training ceases, fitness begins to deteriorate. The reader is referred to the discussion of the general adaptation syndrome, appearing in chapter 1, for a better understanding of the relationship between stress and adaptation in physical training.

This chapter explores the adaptation of skeletal muscle to the stresses of strength training, and Chapter 21 discusses the adaptation to endurance train-

Super heavyweight Vasily Alexeiev in action winning a gold medal in the Montreal Olympics. Improvements in training techniques have resulted in fantastic feats of human muscular strength. (Alain Dejean/Sygma.)

ing. Chapters 20 and 21 examine factors involved in muscular adaptation, the physiological effects of training, and examples of athletes' training programs.

STRENGTH TRAINING

Strength training has a long history, beginning with the great Olympic champion Milo of Crotona, who lived in Greece in the sixth century B.C. Milo is said to have hoisted a baby bull on his shoulders to improve his strength. He repeated this every day, and as the bull grew heavier with age, Milo improved his strength. Since then, strength training has become an important part of the training programs of many types of athletes, ranging from american football and track and field to swimmers and figure skaters.

Strength training is a controversial area of exercise physiology as a result of many poorly conducted studies, the difficulty in conducting long-term research, problems associated with extrapolating the results of animal studies to humans and of untrained subjects to athletes, and the rampant commercialism connected with various types of strength training equipment. In many studies, it is difficult to separate the effects of motor learning from strength gains, because the subjects were untrained and the investigations were conducted over a short time period. Additionally, many studies did not allow for adequate recovery from the training program before evaluating the program's effectiveness. Every accomplished athlete knows that a period of rest is necessary for peak performance following a period of heavy training. Yet, strength was often evaluated in the midst of the training program. Consequently, we must look at the results of animal and human studies as well as evaluate the countless empirical observations in order to assess the knowledge in this area.

CLASSIFICATION OF STRENGTH EXERCISES

Strength exercises can be classified into three categories: isometric (static), isotonic, and isokinetic. Isometric exercise involves the application of force without movement, isotonic exercise involves force with movement, and isokinetic exercise involves the exertion of force at a constant speed.

Isometric Exercise

Hettinger and Mueller caused a stir in 1953 when they found that 6 sec of isometric exercise at 75% effort increased strength. However, subsequent research has shown that isometrics have limited applications in the training programs of athletes. Although this type of training received considerable attention

FIGURE 20-1
Isometric squat on the power rack.

(Wayne Glusker.)

FIGURE 20-2
Constant resistance exercise: the power snatch.

(Wayne Glusker.)

in the 1950s and 1960s, it is seldom practiced unless it is included with other techniques in the training program.

Isometric exercise does not increase strength throughout the range of motion of a joint but rather is specific to the joint angle at which the training is being effected. Likewise, isometric training does not improve (and may hamper) the ability to exert force rapidly. Athletes will sometimes use isometrics to help them overcome "sticking points" in the range of motion of an exercise. For example, an athlete who has difficulty lifting a weight from a low position in the squat may perform the exercise isometrically at the point where he or she is experiencing difficulty (Figure 20-1).

Most of the benefits of isometrics seem to occur during the early stages of training. Maximal contraction is essential for the optimal effect, and the duration of contraction should be long enough to recruit as many fibers in a muscle group as possible. The greatest gains in strength occur when isometrics are practiced several times a day. However, as with other strength training techniques, excessive training will eventually lead to a deterioration in performance (overtraining).

Isotonic Exercise

Isotonic exercise is the most familiar strength training technique to most athletes and coaches. Isotonic loading methods include constant, variable, eccentric, plyometric, and speed resistance.

In constant resistance exercise, the load remains constant, but the difficulty in overcoming the resistance varies with the angle of the joint. For example, in the "free weight" bench press, it is easier to move the weight at the end of the range of motion than when the weight is on the chest. Barbells and dumbbells are the best example of constant resistance exercise devices and continue to be the most popular with the majority of athletes who depend on strength and power for maximal performance (Figure 20-2).

Variable resistance exercise is done on specially designed weight machines, which impose an increasing load throughout the range of motion so that a more constant stress is placed on the muscles (Figure 20-3). This is accomplished by changing the relationship of the fulcrum and lever arm in the weight machine as the exercise progresses. Although the concept of placing a muscle group under a relatively uniform stress throughout the range of motion is intellectually appealing, the experimental evidence for the superiority of variable resistance over constant resistance exercise remains equivocal.

FIGURE 20-3
Variable resistance exercise: Nautilus pullover machine.

(Wayne Glusker.)

FIGURE 20-4
Plyometric loading. Subject jumps from bench (a) to floor (b), absorbs the shock, and then jumps
to bench (c and d).

(Wayne Glusker.)

Eccentric loading is tension exerted during the lengthening of a muscle. In a bench press, for example, the muscles work eccentrically by resisting the movement of the bar as it approaches the chest. Several studies have shown that this is an effective means of gaining strength, although it is not superior to other isotonic techniques. One drawback of eccentric training is that it seems to create more muscle soreness than other methods. Eccentrics are not widely practiced by strength athletes except as an adjunct to other training methods.

Plyometric loading is implosion training in which they muscles are loaded suddenly and forced to stretch before they can contract and elicit movement. An example of plyometrics is jumping from a box to the ground, then rebounding to another box (Figure 20-4). Although plyometrics have been shown to increase strength and jumping ability, they carry with them an increased risk of injury. The authors know of an instance when plyometric training by two world-class throwers (former world record holder in the shot put and the American record holder in the discus) resulted in serious knee injuries. Plyometrics have become very popular with track and field athletes, but more research is needed to assess their effectiveness and safety.

Speed loading involves moving the resistance as rapidly as possible. Most studies have found that constant-resistance isotonic exercise is superior to speed loading for gaining strength. Speed loading may not allow sufficient tension to elicit a training effect. This was demonstrated in the 1920s when Hill established that tension diminishes as the speed of contraction increases. Nevertheless, this technique is often practiced by strength athletes at various times in their training schedule, particularly during the competitive period when maximum power is desired.

Proprioceptive neuromuscular facilitation is manual resistive exercise that utilizes a combination of isotonic and isometric loading (Figure 20-5). This technique is widely used by physical therapists and athletic trainers in the treatment and prevention of athletic injuries. Unfortunately, there are few comparative data with traditional loading techniques. However, proprioceptive neuromuscular facilitation appears to be a promising method of resistance training, particularly during the early phase of recovery following an injury.

Isokinetic Exercise

Isokinetic exercise controls the rate of muscle shortening. It is sometimes called accommodating resistance because the exerted force is resisted by an equal force from the isokinetic dynamometer (Figure 20-6). As with variable resistance isotonic exercise, isokinetics require a specially designed machine to produce the isokinetic loading. Isokinetics have become extremely popular with athletic trainers and physical therapists because they allow the training of injured joints with a lower risk of injury. Additionally, isokinetic dynamometers provide a speed-specific indication of the absolute strength of a muscle group.

The most effective strength gains have come from slower training speed (60

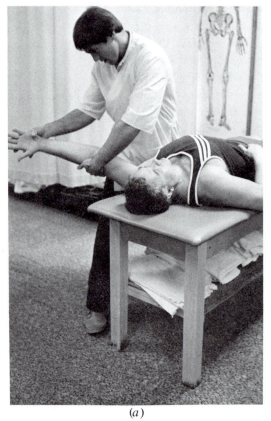
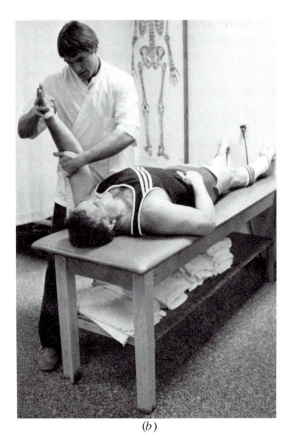

(a)   (b)

FIGURE 20-5
Proprioceptive neuromuscular facilitation.

(Wayne Glusker.)

degrees · sec^{-1} or less). Training at fast speeds of motion has been found to increase the ability to exert force rapidly but no more than traditional isotonic techniques. Although this appears to be a promising method of loading, more research is needed to establish its role in strength training and to determine the ideal isokinetic training protocol.

FACTORS INVOLVED IN MUSCULAR ADAPTATION TO RESISTANCE EXERCISE

Muscles are strengthened by increasing their size and by enhancing the recruitment and firing rates of their motor units. It appears that both of these processes are involved in the adaptive response to resistive exercise. There seem to be age and sex differences in the mechanism involved in effecting strength gains. For example, old men and women of all ages seem to increase strength mainly by neural adaptation (with some hypertrophy), whereas young men rely more

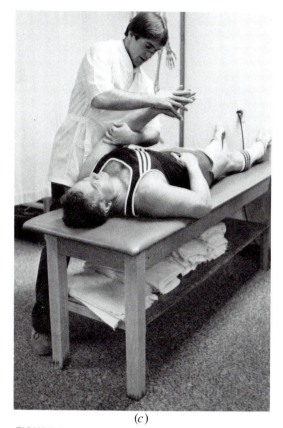 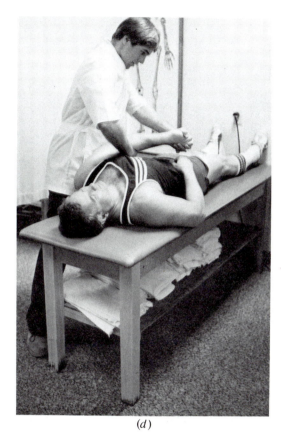

(c) (d)

FIGURE 20-5 (continued).

on increases in muscle size (Figure 31-3). There is a limit to the neural contribution to muscular strength increases. The development of superior strength levels (required by top-level football players, weight lifters, discus throwers, etc.) depends on the ability of increase the contractile properties of muscle.

There are countless exercise devices and training programs available that are hailed as the best way to gain strength. In most instances, as long as a threshold tension is developed, increases in strength will occur. However, the type of strength developed is the important consideration in exercise and sports. Long-distance running up steep hills, for example, will develop a certain amount of muscular strength, but the muscular adaptations that result will differ from those produced from high-resistance, low-repetition squats (knee bends). The distance runner tends to develop sarcoplasmic protein (oxidative enzymes, mitochondrial mass, etc.), whereas the weight lifter tends to develop contractile protein. The nature of the adaptive response must always be considered when designing the training program. There are several factors that determine the rate and type of strength that result from a resistance training program, including overload, specificity, and reversibility.

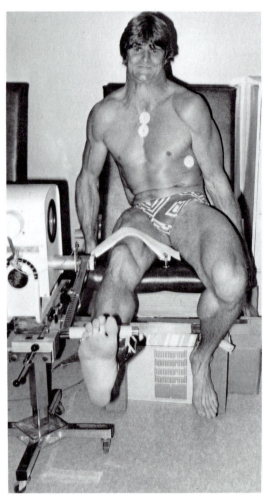

FIGURE 20-6
The Cybex isokinetic dynamometer.

(Wayne Glusker.)

Overload

Muscles increase their strength and size when they are forced to contract at tensions close to their maximum. If muscles are not overloaded, they do not improve their strength, and they do not hypertrophy. A number of experimental and empirical observations have allowed some generalizations concerning the amount of overload necessary for strength gains.

Goldberg demonstrated that the rate of protein synthesis in a muscle is directly correlated with the rate of entry of amino acids into the cells. He also showed that amino acid transport into muscle is influenced directly by the intensity and duration of muscle tension. Uptake of $[^{14}C]\alpha$-amino isobutyric acid by stimulated muscle preparations was more rapid and the intracellular–extracellular concentration gradient much greater than in nonexercised muscles.

Weight training studies and empirical observations of athletes have reinforced the importance of generating muscular tension, at an adequate intensity and duration, for the optimal development of strength. The majority of studies have found that the ideal number of repetitions are between four and eight (repetitions maximum, 4–8 RM), practiced in multiple sets (three or more). Strength gains are less when either fewer or greater numbers of repetitions are used. These findings are consistent with the strength training practices of athletes.

Proper rest intervals are important for maximizing tension, between both exercises and training sessions. Insufficient rest results in inadequate recovery and a diminished capacity of the muscle to exert full force. Unfortunately, the ideal rest interval between exercises has not been determined. Most athletes strength train 3 to 4 days per week, with large muscle exercises such as the bench press and squat seldom practiced more than twice a week. This practice has been empirically derived, as it appears to allow adequate recovery between training sessions.

Athletes involved in speed-strength sports, such as discus throwing and shot-putting, practice low-repetition, high-intensity exercise during or immediately preceding the competitive season. Such training seems to improve explosive strength, while allowing sufficient energy reserves for practicing motor skills. However, the effectiveness of this practice has not been established experimentally.

The overload must be progressively increased for consistent gains in strength to occur. However, because of the high dangers of overtraining in strength building exercises, constantly increasing the resistance is sometimes counterproductive. A relatively new practice among strength-trained athletes is *periodization of training*. This practice varies the volume and intensity of exercises so that the nature of the exercise stress frequently changes. Many athletes believe that this practice produces a faster rate of adaptation. Periodization of training will be discussed further in the section on the strength training programs of athletes.

Specificity

Muscles tend to adapt specifically to the nature of the exercise stress. The strength training program should stress the muscles in a similar manner in which they are to perform. The most obvious example of specificity is that the muscle exercised is the muscle that adapts to training. In other words, when the leg muscles are exercised, it is the leg muscles that hypertrophy and not the muscles of the shoulders.

There is specific recruitment of motor units within a muscle depending on the requirements of the contraction. As discussed in Chapter 19, the different muscle fiber types have characteristic contractile properties. The slow-twitch

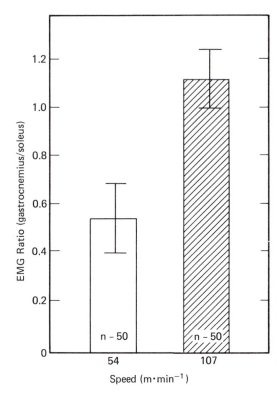

FIGURE 20-7
Electromyogram (EMG) ratio of gastrocnemius to soleus in the cat during treadmill runs of 54 and 107 m · min^{-1}. Fibers from the predominantly fast-twitch gastrocnemius are recruited more frequently as the intensity of exercise increases.

(SOURCE: Adapted from V.R. Edgerton, 1976.)

fibers are relatively fatigue resistant but have a lower tension capacity than the fast-twitch fibers. The fast-twitch fibers have the capacity to contract more rapidly and forcefully, but they also fatigue rapidly.

The use of a motor unit is dependent on the threshold levels of its α motor neuron. The low-threshold, slow-twitch fibers are recruited for low-intensity activities such as jogging (and, for that matter, most tasks of human motion). However, for high-speed or high-intensity activities, such as weight lifting, the fast-twitch motor units are recruited. Figure 20-7 shows the effects of running speed on the selective use of the gastrocnemius (predominantly fast twitch) and the soleus (predominantly slow twitch). As the velocity increases, there is an increased reliance on the predominantly fast-twitch gastrocnemius.

The amount of training that occurs in a muscle fiber is determined by the extent to which it is recruited. As discussed, high-repetition, low-intensity exercise, such as distance running, relies on the recruitment of slow-twitch fibers and results in improvements in the fibers' oxidative capacity. Low-repetition, high-intensity activity, such as weight training, causes hypertrophy of fast-twitch fibers, with some changes to the lower threshold slow-twitch fibers. The training program should be structured to produce the desired training effect.

Muscle fiber type appears to play an important role in determining success in some sports. Successful distance runners have a high proportion of slow-twitch muscles (percentage of slow-twitch fibers is highly related to $\dot{V}_{O_2max}$), whereas sprinters have a predominance of fast-twitch muscles. Several studies have shown that a high content of fast-twitch fibers is a prerequisite for success in strength training. This is understandable, as the fast-twitch fibers experience selective hypertrophy as a result of high-resistance, low-repetition exercise.

However, all sports do not require prerequisite fiber characteristics. Coyle et al., for example, found a surprisingly diverse muscle fiber composition in the gastrocnemius of eight world-class shot-putters. In those athletes, selective adaptation of the fast-twitch fibers, rather than their percentage, accounted for their performance. Although there is tremendous variability in the relative percentage of fast-twitch fibers in explosive strength athletes, they exhibit a higher fast- to slow-twitch fiber area ratio than in sedentary subjects and endurance athletes. It appears that individual differences in training intensity and technique can make up for deficiencies in the relative percentage of fast-twitch fibers in these athletes. It would be interesting to speculate about the performance of a shot-putter with a high percentage of fast-twitch fibers, who developed great strength and had good technique. The high percentage of fast-twitch fibers would probably be a decided advantage.

It appears that simultaneous participation in a training program designed to stimulate both strength and endurance will interfere with the ability to gain strength. Hickson observed that there was more than a 20% difference in subjects who participated in a strength-endurance program compared to subjects who trained only for strength (Figure 20-8). He was unable to demonstrate a

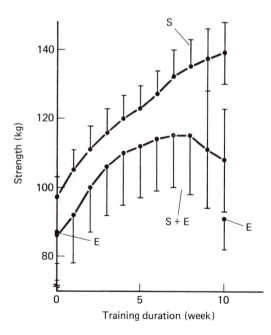

FIGURE 20-8
Simultaneously training for strength (S) and endurance (E) interferes with strength development.

(**SOURCE**: From R.C. Hickson, 1980.)

deleterious effect of strength training on endurance. He suggested that strength athletes may inhibit their ability to gain strength by participating in vigorous endurance activities and hypothesized that muscles may be unable to adapt optimally to both forms of exercise.

Reversibility

Muscles will atrophy as a result of disuse, immobilization, and starvation. Just as muscles adapt to increasing levels of stress by increasing their function, their disuse leads to decreasing strength and muscle mass. Atrophy results in a decrease in both contractile and sarcoplasmic protein.

The muscle fiber types do not atrophy at the same rate. Joint immobilization results in a faster rate of atrophy for slow-twitch than for fast-twitch muscle. This has important implications for rehabilitation. Often, increasing strength is the major consideration following immobilization. Endurance should also be stressed because of the relatively greater loss of slow-twitch muscle capacity.

Immobilization affects muscle length. If a muscle is fixed in a lengthened position, sarcomeres are added; they are lost if the muscle is immobilized in a shortened position. Immobilization also leads to a variety of biochemical changes, including decreased glycogen, adenosine triphosphate (ATP), creatine phosphate (CP), and creatine. All of these factors can affect muscular performance after immobilization has ended.

Deconditioning in strength athletes results in atrophy of fast-twitch and slow-twitch fibers. An interesting phenomenon is that the endurance capacity of muscle improves during cessation of strength training, even in the absence of any endurance training. A study by Staron et al. found an increase in mitochondrial volume and in maximal oxygen consumption (both in liters $\cdot$ min^{-1} and ml $\cdot$ kg^{-1} $\cdot$ min^{-1}) in an elite powerlifter during a 7-month detraining period.

Hypertrophy and Hyperplasia

An ongoing controversy in muscle physiology is whether resistance training induces the muscle to increase the number of muscle cells (hyperplasia) as well as increase their size (hypertrophy). For many years it was believed that muscles increased only in size, and that hyperplasia was only possible until the neonatal period. However, recent research has raised questions about this.

Researchers such as Gonyea, Ho, and Edgerton have reported fiber splitting (longitudinal division of muscle fibers resulting in a new muscle cell), as well as hypertrophy in response to resistive exercise. Fiber splitting seems to require a sufficient tension to result in the hypertrophy of type IIb fibers (fast twitch, glycolytic). Below that intensity, no fiber splitting occurred. Fiber splitting as a means of increasing cell number (hyperplasia) is very controversial, as some researchers have criticized the methodology used in these studies. It is possible that split fibers seen in biopsy samples are damaged and degenerating fibers, not new, budding fibers. So, it will be some time before this question is resolved.

Hypertrophy is the major mechanism involved in enlarging muscle in response to overload stress. Muscle fibers increase in size by increasing the number and size of their myofibrils. As discussed, this probably occurs as a result of increased amino acid transport into the cells (caused by tension) which enhances their incorporation into contractile protein.

MacDougall et al. compared the muscle ultrastructure characteristics of elite powerlifters and bodybuilders with those of control subjects who trained with weights for 6 months. Despite large differences in strength between the two groups, there were no apparent differences in fiber areas or muscle fiber type. The authors hypothesized that the elite group possessed a greater total number of muscle fibers than the controls did. They also found a variety of structural abnormalities in the athletes that may have been due to years of heavy training, anabolic steroids, or a combination of both. Results of this study are illustrated in Figure 20-9.

A number of biochemical alterations accompany hypertrophy during strength training. High-intensity, slow-speed training using isokinetic loading is associated with increases in muscle glycogen, CP, ATP, ADP, creatine, phosphorylase, phosphofructokinase (PFK), and Krebs cycle enzyme activity. Training at faster speeds does not induce these changes. Weight training studies have generally replicated the results of the slow-speed, isokinetic training.

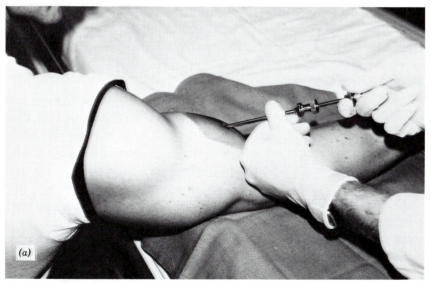

(a)

FIGURE 20-9a
Needle biopsy being taken from biceps brachii of an elite bodybuilder.

(J.D. MacDougall and D.G. Sale.)

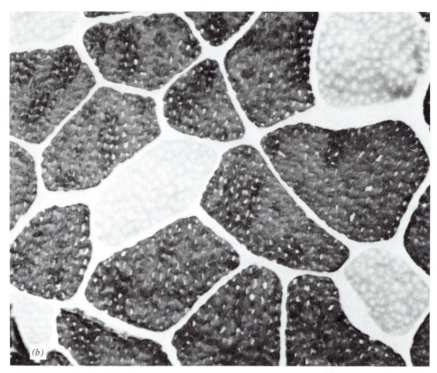

FIGURE 20-9b
Tissue from the biopsy in Figure 20-9a stained for myosin ATPase activity following pre-incubation at pH 10.0. The dark staining fibers are Type II fibers (× 200).

(J.D. MacDougall and D.G. Sale.)

Muscle Soreness

Delayed muscle soreness (muscle soreness that appears 24 to 48 hr after strenuous exercise) is an overuse injury that is a common experience (and a definite factor to contend with) in persons attempting to develop muscular strength. It seems to be associated, in whole or in part, with eccentric muscle contractions. A number of explanations have been proposed for the cause of muscle soreness, including torn tissue, muscle spasm, and connective tissue damage.

Abraham did a series of experiments to determine which of these hypotheses was the most plausible. He used surface electromyograms to test the muscle spasm theory, monitored myoglobin in the urine to test the muscle damage theory, and measured the ratio of hydroxyproline to creatinine to assess disruption of the connective tissue elements. His data support the connective tissue theory of muscle soreness and provide little support for the other hypotheses.

However, other investigators have presented evidence supporting the muscle damage theory of delayed soreness. Armstrong et al. recently compared the

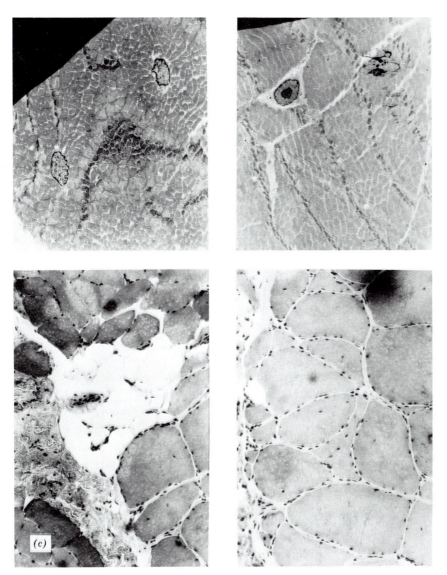

FIGURE 20-9c

The ultrastructure of elite bodybuilders and powerlifters is characterized by an increased absolute amount of contractile protein per fiber but a decrease in the contractile protein density. These athletes exhibited a variety of structural abnormalities including central nuclei (upper left and upper right) and atrophied fibers (lower left and lower right). It is unlcear whether these structural changes are due to prolonged training or anabolic steroids. Upper left: Electron micrograph illustrating two central nuclei in a single muscle fiber from a powerlifter ($\times$ 3800). Upper right: Electron micrograph illustrating a central nucleus and enlarged cytoplasm volume from a bodybuilder ($\times$ 3800). Lower left: Light micrograph of tissue from a bodybuilder, showing atrophied fibers and a proliferation of connective and fatty tissue ($\times$ 200). Lower right: Light micrograph of tissue from a bodybuilder showing atrophied fibers and a proliferation of connective and fatty tissue ($\times$ 200).

(J.D. MacDougall and D.G. Sale.)

degree of muscle damage resulting from high-intensity treadmill exercise in rats who ran at 0°, down a 16° incline, and up a 16° incline. The eccentrically loaded downhill runners demonstrated significant aberrations in serum enzymes that indicate muscle damage (creatine phosphokinase and lactate dehydrogenase) and disruptions in cellular histology 1.5 to 2 days postexercise. Friden et al. have demonstrated a widening and streaming of the Z lines in human soleus muscle following severe eccentric exercise.

It appears that delayed muscle soreness results from muscle injury and the breakdown of connective tissue. These injuries may result in an influx of fluid in the affected area causing pressure and a local inflammatory response causing pain.

STRENGTH TRAINING PROGRAMS OF ATHLETES

Athletes are faced with an incredible array of strength training devices and programs to help them develop strength and power. No doubt all of them will increase ability to exert force to a certain extent. It is beyond the scope of this text to provide in-depth analysis of the many programs available. So, this discussion will deal strictly with techniques that seem to work for weight-trained athletes noted for their great strength.

Although there are undoubtably exceptions, the vast majority of truly strong athletes train with free weights. Although this does not necessarily mean that this is the best way to gain strength, it does infer a de facto recognition of the superiority of this type of training. Athletes are extremely pragmatic and will generally gravitate to the techniques that produce the best results.

The strength training programs of athletes involved in speed-strength sports employ three major types of exercises: presses, pulls, and squats. Examples of presses include the bench press, incline press, jerk, seated press, and behind-the-neck press. These lifts are important for developing the muscles of the shoulders, chest, and arms. Pulling exercises include the clean, snatch, high-pull, and dead lift. These lifts develop the muscles of the legs, hips, back, and arms. Squats include the squat and leg press. These exercises develop the legs and back (leg press develops very little back strength). These three types of exercises represent the most important part of the program and are usually supplemented by auxiliary exercises such as biceps curls, sit-ups, pullovers, and bar dips (exercises designed to develop specific strengths for a sport).

Successful strength athletes usually train 3 or 4 days per week during strength building periods and 1 to 3 days per week during competitive periods. However, the major lifts (press, pulls, and squats), are seldom practiced more than 2 days per week. If these lifts are practiced too frequently and too intensely, overtraining invariably results.

Overtraining results in an increased risk of injury and a decrease in performance, probably because of the inability to train heavily during training sessions. This sounds like a contradiction. However, constant, severe training schedules do not provide adequate recovery, and thus the training stimulus cannot be maximal. As discussed already, the intensity and duration of tension are the most important factors eliciting strength increases.

Periodization of training, made popular by Soviet weight lifters and track and field athletes, seems to prevent overtraining and the problems that go with it. Periodization employs several types of training cycles throughout the year: load cycles, recovery cycles, peak cycles, and conditioning cycles. The load cycle is the building portion of the program and is employed during the off-season and precompetitive period. The recovery cycle is a transition period of active rest that separates the building and competitive periods. The peak cycle is the period for developing maximum strength, while affording time for work on motor skills. The conditioning cycle is also a period of active rest that is usually employed during the few months after the competitive season is over. Examples of load and peak cycles are shown in Tables 20-1 and 20-2.

Load cycles involve relatively high numbers of sets (five to seven) and a moderate number of repetitions (four to seven), using moderate amounts of weight (approximately 80% of one repetition maximum). The intensity of the load is not great because the concentrated volume results in high-intensity training. Within each load cycle are microcycles lasting approximately 2 weeks. Notice in Table 20-1 that the highest levels of resistance for a particular exercise are practiced only once during a microcycle. A typical intensity progression of a microcycle proceeds as follows: heavy, light, moderate, light, and then the cycle is repeated using more weight.

The progression of the load cycle is established to enable the athlete to exert maximum effort during a heavy workout. Even though the microcycles provide a certain amount of rest between heavy workouts, load cycles are exhausting and cannot be practiced for more than 2 to 3 months without the risk of overtraining. A common observation of individuals on this phase of the strength program is that functional indicators of strength begin to decrease toward the end of the load cycle and during the early phase of the recovery period, but they quickly rebound and exceed the initial level. Verkhoshansky has called this phenomenon the delayed-training effect and it is analogous to glycogen supercompensation (see Chapter 27).

The recovery cycle is designed to give the athlete a period of active rest in preparation for the high-intensity low-volume workouts of the peak cycle. This cycle involves low to moderate volume at light to moderate intensity. Recovery cycles typically last approximately 2 to 3 weeks.

The peak cycle is designed to produce maximum strength and employs low volume and high intensity. Peak cycles are employed during or immediately

TABLE 20-1

Four Weeks of an Eight-Week Load Cycle of a World-Class Discus Thrower[a]

Week	Monday			Wednesday			Friday		
1	Power clean			Snatch high pull			Power clean		
	2s	4r	220 lb	3s	8r	198 lb	1s	5r	297 lb
	Squat			Squat				1s	5r	314 lb
	2s	8r	375 lb	1s	3r	435 lb	Hack squat		
	2s	6r	405 lb	Bench Press			3s	10r	220 lb
	Good morning exercise			3s	8r	315 lb	Squats		
	3s	10r	200 lb	Dumbbell bench press			3s	10r	285 lb
	Behind the neck press			4s	8r	100 lb	Behind the neck press		
	3s	5r	195 lb				2s	4r	205 lb
2	Snatch high pull			Squat			Bench press		
	1s	4r	264 lb	2s	10r	220 lb	6s	6r	385 lb
	1s	4r	297 lb	Good morning exercise			Hack squat		
	1s	4r	330 lb	4s	10r	200 lb	3s	10r	220 lb
	Squat						Squat		
	4s	6r	355 lb				3s	10r	325 lb
	Bench						Dumbbell press		
	1s	4r	330 lb				4s	8r	100 lb
	1s	4r	340 lb						
	1s	4r	350 lb						
3	Power clean			Snatch high pull			Snatch high pull		
	1s	4r	264 lb	3s	8r	198 lb	3s	8r	198 lb
	Squat			Squat			Squat		
	1s	4r	335 lb	2s	8r	415 lb	1s	4r	465 lb
	Bench			2s	6r	435 lb	Bench press		
	1s	3r	410 lb	Behind the neck press			1s	6r	365 lb
	Dumbbell flies			3s	3r	225 lb	Dumbbell bench		
	4s	8r	50 lb				4s	10r	100 lb
4	Power clean			Snatch high pull			Bench press		
	1s	5r	308 lb	1s	4r	275 lb	6s	6r	420 lb
	1s	5r	325 lb	1s	4r	308 lb	Squat		
	1s	5r	341 lb	1s	4r	347 lb	1s	6r	315 lb
	Good morning exercise			Squat			Good morning exercise		
	4s	10r	200 lb	3s	10r	340 lb	4s	10r	220 lb
	Bench press			Behind the neck press			Dumbbell bench		
	1s	10r	315 lb	3s	5r	185 lb	4s	10r	110 lb
	1s	4r	335 lb						

[a] s = sets; r = repetitions. Warmup not included.

TABLE 20-2

Four-Week Peak Cycle of World-Class Discus Thrower[a]

Week	Tuesday			Friday			Sunday
1	Power snatch			Power snatch			Competition
	1s	2r	220	1s	1r	198 lb	
	1s	1r	242 lb	Dumbbell flies			
	1s	1r	262 lb	3s	10r	25 lb	
	1s	1r	275 lb	Biceps curls			
	1s	1r	286 lb	3s	10r	125 lb	
	Squat						
	1s	5r	352 lb				
	1s	5r	374 lb				
	1s	5r	396 lb				
	1s	3r	418 lb				
	1s	3r	440 lb				
	Bench						
	1s	5r	365 lb				
	Flies						
	3s	10r	50 lb				

Week	Monday			Wednesday			Friday–Saturday
2	Power snatch			Power snatch			Competition
	1s	3r	198 lb	1s	1r	286 lb	
	Incline dumbbell			1s	1r	303 lb	
	4s	6r	100 lb	1s	1r	314 lb	
	Bench press			1s	1r	319 lb	
	1s	3r	385 lb	Squat			
	1s	3r	415 lb	1s	2r	440 lb	
				1s	2r	484 lb	
				1s	2r	529 lb	
				1s	2r	562 lb	
3	Squat			Power snatch			National championship
	1s	5r	374 lb	1s	1r	198 lb	
	Bench press			Dumbbell flies			
	1s	3r	405 lb	3s	10r	25 lb	
	1s	2r	435 lb	Biceps curls			
	1s	2r	465 lb	3s	10r	125 lb	
	1s	2r	485 lb				
	Dumbbell flies						
	4s	10r	50 lb				

(continued)

TABLE 20-2 (*continued*).
Four-Week Peak Cycle of World-Class Discus Thrower[a]

Week	Monday			Wednesday			Friday–Saturday
4	Squat			Bench Press			
	1s	2r	440 lb	1s	2r	385 lb	
	1s	2r	496 lb	1s	2r	415 lb	
	1s	2r	550 lb	1s	1r	445 lb	
				1s	1r	475 lb	
				Flies			
				3s	10r	30 lb	
5			Olympic trials				

[a] Olympic trials were scheduled during Week 5. The athlete did not lift weights during the week of the trials. s = sets; r = repetitions. Warm-up lifts not included.

preceding the competitive period. As with the load cycle, the peak cycle is also composed of a microcycle lasting 2 to 3 weeks. Again, maximum resistance for each exercise is employed only once during each microcycle.

The load cycle prepares the athlete for maximum performance during the peak cycle. However, the athlete must be aware of two factors that seem to interfere with maximum strength gains during the peak-cycle: The peak cycle must be continued long enough for the athlete to experience maximum strength gains (generally, the duration of the peak cycle should equal that of the load cycle); in addition, the athelete must not introduce high-volume workouts during the peak period, or maximum strength gains will be compromised.

The conditioning cycle is a period of active rest that typically follows the competitive season. The training sessions are characterized by low intensity (60 to 70% of maximum), moderate volume (4 to 6 sets), and high repetition (8 to 10 repetitions). Approximately every 3 weeks, the athlete should employ a slightly higher intensity to maintain strength. This period provides both a physical and mental break from periods of heavy training, while assuring that no significant deconditioning occurs.

These programs are not appropriate for all types of athletes. The strength requirements of each sport must be assessed in order to develop an appropriate, specific program. In general, sports requiring muscular endurance employ strength training schedules involving a higher number of repetitions, whereas those requiring strength use fewer repetitions. Strength training exercises should be chosen to develop the muscles used in the sport.

Computerized Training of Paralyzed Muscles

At Wright State University, Petrofsky and Glaser have utilized computers to control movements of paralyzed leg muscles in paraplegic and quadriplegic individuals. In their system (Figures 1-6, 1-8, and 20-10), a computer regulates electrical stimulation of leg muscles through skin surface electrodes. The resulting movements are then sensed by force and displacement transducers located on the body and exercise apparatus. Movement feedback to the computer is then interpreted and output to the contracting muscles is regulated to produce the intended result by the computer program. To some extent, the subject can vary the frequency and amplitude of muscular contractions. For example, the subject cycling (Figure 1-6) can vary pedaling speed by twisting the handle grip, a control similar to that on a motorcycle.

Glaser and Petrofsky have also utilized a progressive resistance sequence of computer controlled muscle contractions to drastically improve strength. In some instances, computer controlled, progressive resistance stimulation of paralyzed quadriceps muscles (Figure 20-10) has resulted in a doubling or tripling of strength within only a few months. Some individuals paralyzed for 6 to 8 years have been able to develop leg strength comparable to that of normal individuals. By attaching their feet to a unique pedal system, paralyzed subjects are now capable of propelling wheel chairs under their own muscular power. As a result of research at Wright State University, paralyzed individuals can look foward to enhanced levels of muscular and cardiovascular fitness, greater mobility, less obesity, and a lesser feeling of confinement than before.

Recovery of Muscle from Transplantation and Serious Injury

Since 1960 when Studitsky and Bosova first demonstrated the possibility of transplanting muscles from one site in the body to another, the survival of transplants have been reported in rats, cats, dogs, monkeys, and humans. Because a serious injury seldom damages a major proportion of an individual's skeletal musculature, the possibility of transplanting muscle within the body has important implications in the management of a wide variety of injuries and muscular disorders (myopathies). The study of muscle during transplantation injury also provides a basis for management of sports related injuries to the musculature. Muscle transplantation presents fewer problems and the probability of success is potentially greater than heart transplantation.

Terminology of transplantation   When a muscle is transplanted from a member of one species to a member of the same species, the transplantation is said to be homologous. When the transplantation occurs within one individual, the transplant is autologus. When a muscle is removed from, and reinserted into its original site, the transplant is orthotopic. When a muscle is transplanted from one site to another in the body, the transplant is heterotopic. As part of

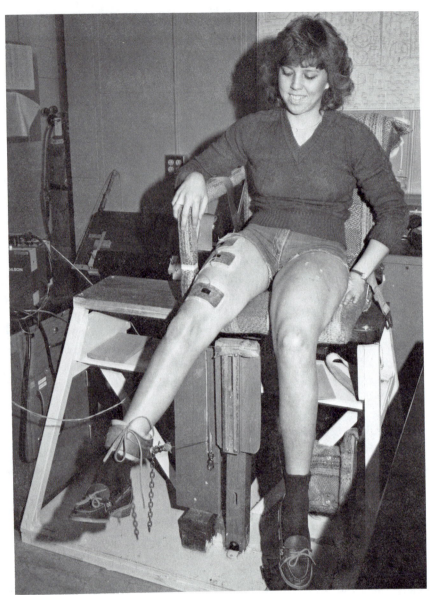

FIGURE 20-10
Computer controlled progressive resistance electrical stimulation of paralyzed muscle.

the transplantation procedure, tendons from the transplanted muscle are sutured (tied) to fixed points in the site of transplantation to maintain some tension. Nerve stumps and sometimes also blood vessels are brought (anastomosed) to the transplanted tissue.

Evolution of a free muscle graft   When a muscle is surgically freed from its bed by severing all connections (nerve, muscle, blood, tendon) to the original site, and placed in some other, or the same site, the muscle will degenerate and then regenerate. Viability of the regenerated fibers depends not only on the procedures of surgery, but also the postoperative exercise procedures used. The sequence of events in a transplanted muscle, such as in the rat extensor digitorum longus (EDL) muscle, has been described in detail by Carlson and associates (Figure 20-11).

Deprived of their blood and nerve supply, fibers in the graft undergo ischemic necrosis (Figure 20-11, B–C). In contrast, satellite cells survive the

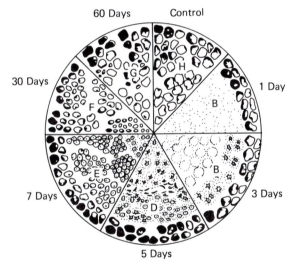

FIGURE 20-11
Schematic representation of cellular reactions within a cross-section of the grafted normal EDL muscle of the rat. The diagram is divided into segments that represent the histological appearance of grafts at various days after transplantation. The letters refer to groups of cells showing similar histological reactions. **A:** Surviving muscle fibers. **B:** Original muscle fibers in a state of ischemic necrosis. **C:** Muscle fibers invaded by macrophages that are phagocytizing the necrotic cytoplasm. **D:** Myoblasts and early myotubes within the basal laminae of the orininal muscle fibers. **E:** Early cross-striated muscle fibers. **F:** Maturing regenerating muscle fibers. **G:** Mature regenerated muscle fibers. **H:** Normal control muscle fibers.

(**SOURCE**: From B.M. Carlson, F.M. Hansen-Smith, and D.K. Magon, 1979. Used with permission.)

transplant and serve as a basis for the regenerating tissue (D–G). In mature muscle fibers, small mononuclear satellite cells can be seen beneath the basement membrane. The function of these satellite cells has long been a question of interest, but they may provide the basis for recuperation after serious injury. When the regenerative response in muscle is activated by serious injury, and after degeneration of the muscle fibers, satellite cells trace a course that is similar to that of embryologic muscle fibers. Mononuclear myoblast cells (from satellite cells) grow and fuse together to form myotubes. These eventually develop into new multinucleated muscle fibers. Thus, in the regenerated muscle, the fibers are different from those in the muscle before transplantation.

Viability of transplanted muscle   Results of several group of investigators, including Faulkner and associates (Figure 20-12) indicate that the functional characteristics of transplanted muscles, while better than no muscles at all, are significantly less than those of the normal muscles they replace. As indicated in Figure 20-12, same characteristics of transplanted muscle are completely normal (*a*). However, other parameters are much less than normal (*b*). Although these parameters are improved by nerve anastomoses to transplanted muscle, complete return to normal is not accomplished for several parameters (*c*). Mainly, the important characteristics of mass, strength, and endurance of transplanted muscle is impaired.

Training of transplanted muscle   Two lines of evidence suggest that exercise training of transplanted muscle can significantly improve its performance. In her work on transplanted rat soleus muscles, Coan also removed the synergestic gastrocnemius muscles, which placed a greater burden of weight bearing on the soleus. In response, both the number and size of fibers in the regenerating soleus was improved.

More recently, White and associates have utilized treadmill running to improve both the biochemical and physiological characteristics of transplanted muscles in rats. Moreover, the work of White et al. suggests that there are critical periods in a muscle's regeneration when exercise therapy should be initiated for maximum benefit.

Therepeutic implications of muscle transplantation studies   At present, autologus muscle transplantation is most frequently used to improve the appearance of individuals after facial injuries or myopathies. Transplantation also offers the possibility of improving locomotion in individuals with lower limb disorders. Results of transplantation studies also have important implications for the management of sports-related muscular injuries.

In sports injury, the muscle usually has several important advantages that

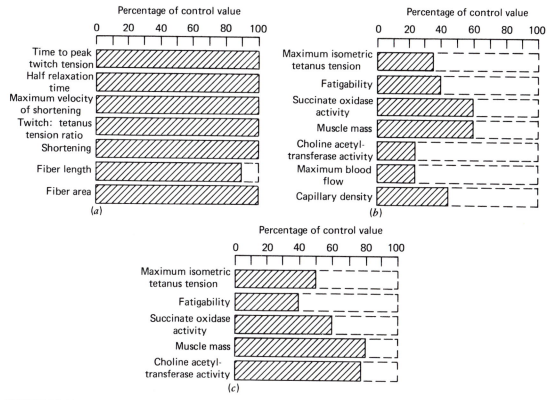

FIGURE 20-12

Pooled data on stabilized standard EDL grafts in cats normalized as a percentage of control value. (a) Variables that are restored to control values. (b) Variables that show incomplete restoration of structure and function. (c) Data 120 days after transplantation on grafts made with nerve-anastomoses. *Note:* For variables displayed in (a), nerve-anastomosed grafts also attained 100 percent of the control values. *Note:* 3a. Fiber area is of *single* fibers.

(SOURCE: From J.A. Faulkner, J.M. Markley, and T.P. White, 1981. Used with permission.)

the transplanted muscle does not. Mainly the important blood and nerve supplies are intact. After the initial inflammatory reaction is minimized by icing the affected part, a variety of procedures can be employed to increase circulation. These procedures include alternate icing and warming of the injured area. Immediately after injury, exercise of the affected part should be minimized. However, when regeneration and recovery is well established, physical therapy should be aggressively pursued. Hopefully, in the near future, research will provide some noninvasive, objective (perhaps EMG) means to assist clinicians in evaluating when physical therapy should begin and training should resume.

Summary

Strength building exercises have become an important component of the training programs of many athletes. Strength exercises include isometrics, isotonic with variable resistance, isotonic with constant resistance, plyometrics, eccentrics, speed loading, and isokinetics. The ideal strength building technique and protocol has yet to be determined.

A variety of factors must be considered when designing a program to build strength, including overload, specificity, reversibility, individual differences, and injury. The programs of proficient strength-trained athletes increasingly emphasize maximum loads but provide enough rest to prevent overtraining.

Selected Readings

Abraham, W.M. Factors in delayed muscle soreness. Med. Sci. Sports. 9: 11–20, 1977.

Andrews, J.G. On the relationship between resultant joint torques and muscular activity. Med. Sci. Sports Exerc. 14: 361–367, 1982.

Armstrong, R.B., R.W. Ogilvie, and J.A. Schwane. Eccentric exercise-induced injury to rat skeletal muscle. J.Appl.Physiol.: Respirat.Environ.Exercise Physiol. 54: 80–93, 1983.

Atha, J. Strengthening muscle. Exerc. Sport Sci. Rev. 9: 1–73, 1981.

Berger, R. Optimum repetitions for the development of strength. Res. Q. 33: 334–338, 1962.

Carlson, B.M., F.M. Hansen-Smith, and D.K. Magon. The life history of a free muscle graft. In: A. Mauro (ed.), Muscle Regeneration. New York: Raven Press, 1979. p. 493–507.

Chaffin, D.B., M. Lee, and A. Freivalds. Muscle strength assessment from EMG analysis. Med. Sci. Sports Exerc. 12:205–211, 1980.

Coan, M.R. and R.J. Tomanek. The growth of regenerating soleus muscle transplants after ablation of the gastrocnemius muscle. Exp. Neurology. 71: 278–294, 1981.

Costill, D.L., E.F. Coyle, W.F. Fink, G.R. Lesmes, and F.A. Witzmann. Adaptations in skeletal muscle following strength training. J. Appl. Physiol. 46: 96–99, 1979.

Coyle, E.F., S. Bell, D.L. Costill, and W.J. Fink. Skeletal muscle fiber characteristics of world class shot-putters. Res. Q. 49: 278–284, 1978.

Currier, D., J. Lehman, and P. Lightfoot. Electrical stimulation in exercise of the quadriceps femoris muscle. Phys. Ther. 59: 1508–1512, 1979.

DeLorme, T.L. Restoration of muscle power by heavy resistance exercises. J. Bone Joint Surg. 27: 645–667, 1945.

Dons, B., K. Bollerup, F. Bonde-Petersen, and S. Hancke. The effect of weight-lifting exercise related to muscle fiber composition and muscle cross-sectional area in humans. Eur. J. Appl. Physiol. 40: 95–106, 1979.

Edgerton, V.R. Mammalian muscle fiber types and their adaptability. Am. Zool. 18: 113–125, 1978.

Edgerton, V.R. Neuromuscular adaptation to power and endurance work. Can. J. Appl. Sport Sci. 1: 49–58, 1976.

Essig, D.A., T.P. White, G.H. Jones, and P.G. Morales. Polysome cell-free protein synthesis of soleus muscle autografts. Physiologist 25(4): 260, 1982.

Fahey, T.D., R. Rolph, P. Moungmee, J. Nagel, and S. Mortara. Serum testosterone, body composition, and strength of young adults. Med. Sci. Sports. 8: 31–34, 1976.

Faulkner, J.A., J.M. Markley, and T.P. White. Skeletal muscle transplantation in cats with and without nerve repair. In G. Freilinger (ed.), Muscle Transplantation. Vienna: Springer-Verlag, 1981. p. 47–54.

Fenn, W.O., H. Brody, and A. Petrelli. The tension developed by human muscles at different velocities. Am. J. Physiol. 97: 1–14, 1931.

Friden, J., M. Sjostrom, and B. Ekblom. A morphological study of delayed muscle soreness. Experientia 37: 506–507, 1982.

Garhammer, J. Energy flow during Olympic weight lifting. Med. Sci. Sports Exerc. 14: 353–360, 1982.

Garhammer, J. Performance evaluation of Olympic weight lifters. Med. Sci. Sports Exerc. 11: 284–287, 1979.

Gettman, L.R., L.A. Culter, and T. Strathman. Physiological changes after 20 weeks of isotonic vs. isokinetic circuit training. J. Sports Med. Phys. Fitness. 20: 265–274, 1980.

Gettman, L.R. and M.L. Pollock. Circuit weight training: A critical review of its physiological benefits. Phys. Sports Med. 9: 44–60, 1980.

Gettman, L.R., P. Ward, and R.D. Hagen. A comparison of combined running and weight training with circuit weight training. Med. Sci. Sports Exerc. 14:229–234, 1982.

Glaser, R.M., J.S. Petrofsky, J.A. Gruner, and B.A. Green. Isometric strength and endurance of electrically stimulated leg muscles of quadriplegics. Physiologist 25: 253, 1980.

Goldberg, A.L. Mechanisms of growth and atrophy of skeletal muscle. In: Cassens, R.D. (ed.), Muscle Biology. New York: Marcel Dekker, 1972.

Goldspink, G. (ed.). Development and specialization of skeletal muscle. Cambridge, England: Cambridge University Press, 1980.

Goldspink, G. Morphological adaptation due to growth and activity. In: Briskey, E.J., R.G. Cassens, and B.B. Marsh (eds.), The Physiology and Biochemistry of Muscle as a Food, 2. Madison, Wisc.: The University of Wisconsin Press, 1972.

Gollnick, P.D., R.B. Armstrong, C.W. Saubertt, K. Piehl, and B. Saltin. Enzyme activity and fiber composition in skeletal muscle of trained and untrained men. J. Appl. Physiol. 33: 312–319, 1972.

Gonyea, W. J. Role of exercise in inducing increases in skeletal muscle fiber number. J. Appl. Physiol. 48: 421–426, 1980.

Gonyea, W.J. and D. Sale. Physiology of weight lifting. Arch. Phys. Med. Rehabil. 63: 235–237, 1982.

Henry, F.M. and J.D. Whitley. Relationships between individual differences in strength, speed, and mass in an arm movement. Res. Q. 31: 24–33, 1961.

Hettinger, T.L. and E.A. Muller. Muskelleistung und muskeltraining. Int. Z. Angew. Physiol. 15: 111, 1953.

Hettinger, T. Physiology of Strength. Springfield, Ill.: Charles C Thomas, 1961.

Hickson, R.C. Interference of strength development by simultaneously training for strength and endurance. Eur. J. Appl. Physiol. 45: 255–263, 1980.

Hill, A.V. The maximum work and mechanical efficiency of human muscles and their most economical speeds. J. Physiol. 56: 19–41, 1922.

Ho, K.W., R.R. Roy, C.D. Tweedle, W.W. Heusner, W.D. Van Huss, and R.E. Carrow. Skeletal muscle fiber splitting with weight-lifting exercise in rats. Am. J. Anat. 157: 433–440, 1980.

Houston, M.E. and P.H. Goemans. Leg muscle performance of athletes with and without knee support braces. Arch. Phys. Med. Rehabil. 63: 431–432, 1982.

Jackson, A., M. Watkins, and R.W. Patton. A factor analysis of twelve selected maximal isotonic strength performances on the Universal Gym. Med. Sci. Sports Exerc. 12: 274–277, 1980.

Johnson, B.L. Eccentric vs. concentric muscle training for strength development. Med. Sci. Sports. 4: 111–115, 1972.

Kamen, G., W. Kroll, and S.T . . Zigon. Exercise effects upon reflex time components in weight lifters and distance runners. Med. Sci. Sports Exerc. 13: 198–204, 1981.

Katch, V.L., F.I. Katch, R. Moffatt, and M. Gittleson. Muscular development and lean body weight in body builders and weight lifters. Med. Sci. Sports Exerc. 12: 340–344, 1980.

Lesmes, G.R., D. Costill, E.F. Coyle, and W.J. Fink. Muscle strength and power changes during maximal isokinetic training. Med. Sci. Sports. 10: 266–269, 1978.

MacDougall, J.D., D.G. Sale, G.C.B. Elder, and J.R. Sutton. Muscle ultrastructure characteristics of elite powerlifters and bodybuilders. Eur. J. Appl. Physiol. 48: 117–126, 1982.

MacDougall, J.D., G.R. Ward, D.G. Sale, and J.R. Sutton. Biochemical adaptation of human skeletal muscle to heavy resistance training and immobilization. J. Appl. Physiol. 43: 700–703, 1977.

Menapace, F.J., et al. Left ventricular size in competitive weight lifters: An echocardiographic study. Med. Sci. Sports Exerc. 14: 72–75, 1982.

Moffroid, M., R. Whipple, J. Hofkosh, E. Lowman, and H. Thistle. A study of isokinetic exercise. J. Am. Phys. Ther. Assoc. 49: 735–746, 1969.

Moritani, T. and H.A. deVries. Potential for gross muscle hypertrophy in older men. J. Gerontol. 35: 672–682, 1980.

O'Shea, P. Effects of selected weight training programs on the development of strength and muscle hypertrophy. Res. Q. 37: 95–102, 1964.

Petrofsky, J.S., R.M. Glaser, C.A. Phillips, and J.A. Gruner. The effect of electrically induced bicycle ergometer exercise on blood pressure and heart rate. Physiologist 25: 253, 1982.

Pipes, T. and J. Wilmore. Isokinetic vs. isotonic strength training in adult men. Med. Sci. Sports. 7: 262–274, 1975.

Romero, J.A., T.L. Sanford, R.V. Schroeder, and T.D. Fahey. The effects of electrical stimulation of normal quadriceps on strength and girth. Med. Sci. Sports Exerc. 14: 194–197, 1982.

Salleo, A., G. Anastasi, G. LaSpada, G. Falzea, and M.G. Denaro. New muscle fiber production during compensatory hypertrophy. Med. Sci. Sports Exerc. 12: 268–273, 1980.

Sharp, R.L., J.P. Troup, and D.L. Costill. Relationship between power and sprint freestyle swimming. Med. Sci. Sports Exerc. 14: 53–56, 1982.

Staron, R.S., F.C. Hagerman, and R.S. Hikida. The effects of detraining on an elite power lifter. J. Neurol. Sci. 51: 247–257, 1981.

Studitsky, A.N. The neural factor in the development of transplanted muscles. In: A.T. Milhort (ed.), Exploratory Concepts in Muscular Dystrophy, II. Exerpta Medica, Amsterdam, 1974, pp. 351–366.

Stull, G.A. and D.H. Clarke. High resistance, low repetition training as a determinant of strength and fatigability. Res. Q. 41: 189–194, 1970.

Surburg, P.R. Neuromuscular facilitation techniques in sportsmedicine. Physician Sports Med. 9: 115–127, 1981.

Thorstensson, A. Muscle strength, fibre types and enzyme activities in man. Acta Physiol. Scand. 443 (Suppl.): 1–45, 1976.

Verkhoshansky, U. How to set up a training program in speed-strength events (part 1). Legkaya Atletika 8: 8–10, 1979. Translated in: Sov. Sports Rev. 16: 53–57, 1981.

Verkhoshansky, U. How to set up a training program in speed-strength events (part 2). Legkaya Atletika 8: 8–10, 1979. Translated in: Sov. Sports Rev. 16: 123–126, 1981.

White, T.P., J.F. Villanacci, and P.G. Morales. Influence of physical conditioning on autografted skeletal muscle. Med. Sci. Sports Exerc. 13(2): 81, 1982.

White, T.P., J.F. Villanacci, P.G. Morales, and O.J. Grossman. Exercise-induced growth of regenerating muscle in rats. Physiologist 25(4): 260, 1982.

21

CONDITIONING FOR RHYTHMICAL ATHLETIC EVENTS

It would seem that studying the physiology of various training methods to identify those methods that produce the greatest results would be a simple matter compared to, for instance, studying the biochemistry of fat metabolism in muscle. However, such is not the case. It has proved extremely difficult to manipulate, in a scientific way, the training regimens of quality athletes. It is very understandable that athletes do not want their training regimens "tampered" with. Therefore, studies of various training regimens frequently utilize only a few trained athletes or more numerous untrained volunteers who train intensively for from 8 to 15 weeks. Such results are not directly applicable to the training of talented individuals engaged in intense training for many years. Moreover, the very large number of training regimens and the possible combinations of these make it very difficult to include in a single study sufficient subjects to analyze benefits of the many different training protocols.

The sports of running and swimming epitomize those rhythmical activities where speed and endurance depend on physiological power. In general, running is the most universal sport, not only because of the worldwide interest in it, but also because skill in running is essential to success in many other sports. It is of interest to describe conditioning for running, because internationally recognized (Olympic) competition includes those events that span the range of pure sprint to pure endurance. In general, many of the same training principles that apply to running also apply to conditioning for swimming. As we will see,

Daly Thompson, Moscow Olympic Decathlon champion and world record holder, performs in a hurdles competition. Improvements in training techniques, intensity and volume have resulted in improved performances in most classes of athletes. (U.P.I.)

however, swimming differs in that Olympic competition does not include pure sprint events. Furthermore, in swimming, stroke technique is exceedingly important, perhaps more important than physiological power.

This chapter is based on three sources of information: (1) the limited studies on training methods, (2) the application of basic physiological principles to athletic training, and (3) the practical experiences of athletes and coaches.

TRAINING FOR ATHLETIC COMPETITION

It should be made clear that the following discussion refers to training for athletic competition and not the training of middle-aged, older, or recreational asymptomatic individuals. The training stimulus necessary to maintain or improve cardiovascular function in the population at large, as described by the American College of Sports Medicine (ACSM), is less in terms of both intensity and volume than is the training of competitive athletes described here. The training stimulus recommended to maintain cardiovascular fitness (Chapter 24) will not develop the exceptional performance levels required for success in competitive athletes, but the ACSM training guidelines are probably no less effective for developing fitness for daily living and for promoting longevity than is training for competitive athletics.

OVERLOAD, STIMULUS, AND RESPONSE

The principle of overload is a rephrasing of the well-known general adaptation syndrome (Chapter 1), wherein physiological adaptations occur in response to appropriate stimuli. The amount of overload to a system can be varied by manipulating two basic factors:

1. Training intensity.
2. Training volume, which is made up of training frequency and training duration.

In general, the greater the overload, the greater the resulting adaptation and increase in functional capacity. Because it takes *time* for physiological responses to occur following application of a training stimulus, the progressive application of a training stimulus must be accomplished within particular constraints. The application of the stimulus (i.e., the increase in training volume and intensity) must be gradual and progressive. Rather than a rapid and continuous increase in training to achieve an increase in athletic performance (Figure 21-1), the training progression should be gradual and discontinuous in nature. Periods of heightened training should be interspersed with recovery periods involving decreased training intensity and volume. Adequate rest each day of training is important. Additionally, hard-training days should be interspersed with easier days. Training should also be scheduled in cycles wherein a period of lesser training intensity and volume follows the peak training of the previous cycle. Therefore, the training schedule of a serious athlete should look more like the broken sawtooth in Figure 21-2 than the smooth blade in Figure 21-1.

FIGURE 21-1
Over the course of a training season, the training volume and intensity should be progressively increased to achieve optional performance. The linear increase in training illustrated here is probably *not* the best protocol of increasing training intensity (see Figure 21-2).

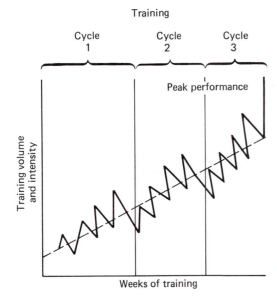

FIGURE 21-2
Over the course of a training season, training volume and intensity should increase in a cyclic, discontinuous manner. Days and weeks of heightened training should be followed by recovery days and weeks of lesser intensity. Similarly, each training cycle begins at an intensity less than that of the previous cycle. In this system the stimulus of progressive training is accompanied by periods of recovery and adaptation.

Recuperation periods at the beginning of a training cycle are to be considered an essential part of the training regimen and are to be followed scrupulously, for its is during recuperation periods that adaptation occurs.

SPECIFICITY, SKILL ACQUISITION, AND DEVELOPING METABOLIC MACHINERY

Increasing performance as the result of training requires that the training be appropriate for the event. Two factors are involved in the specificity of training response: learning the event of interest and developing the metabolic machinery to support the event. In sports such as running, and especially in swimming, where skill level can tremendously affect the performance outcome, long hours of training develop the neuromuscular control mechanisms to optimize results by mimizing the effort necessary to achieve a given result, or by maximizing the work output for a given metabolic power output.

The principle of specificity requires that the training regimen overload the metabolic system that supports the activity. In evaluating the relative contributions of immediate, glycolytic, and oxidative energy sources in supporting an activity, the duration of the activity is of primary importance in determining which energy system is most important (Figures 2-6 and 2-7). Activities lasting a few seconds (e.g., the shot put) depend mainly on immediate energy sources (ATP and CP). Activities lasting from a few seconds to a minute (e.g., 100- to 400-m track running) depend mostly on glycogenolysis and the glycolytic formation of lactic acid. Events lasting longer than a minute become increasingly dependent on oxidative metabolism. Because most athletic events last longer than a minute, training of the cardiovascular mechanism of O_2 transport and the muscular mechanisms of O_2 utilization is of primary importance. For endurance-type activities, training the systems of O_2 delivery and utilization is of such importance that its emphasis can allow athletes to be very successful, even if they overlook and neglect training of the other energy systems. At the conclusion of endurance events, the athletes with the strongest finishing kicks are not necessarily the best sprinters: they are the freshest individuals with the best capabilities of prolonged oxidative metabolism.

Many types of training regimens exist to develop oxidative capacity, but they can be divided into two basic types: over-distance training and high-intensity interval training. Both types of training appear to be important, and clever coaches will work athletes through a range of combinations and variations of these two to optimize adaptation and to keep athletes from becoming bored.

Over-Distance Training

The usually stated objectives of over-distance training (sometimes also called long, slow distance training) are twofold: (1) to increase $\dot{V}_{O_2max}$ and (2) to increase tissue respiratory (mitochondrial) capacity. Of these two, increasing respiratory capacity rather than increasing $\dot{V}_{O_2max}$ appears to be the more important. Although it is true that endurance athletes such as marathon runners display very high values of $\dot{V}_{O_2max}$, $\dot{V}_{O_2max}$ actually correlates poorly with performance in events such as the marathon. Middle-distance runners sometimes record higher values of $\dot{V}_{O_2max}$ than do long-distance runners. Obviously, however, top marathon runners are faster in their event than are top mile runners. Some marathon runners can apparently sustain a pace eliciting over 90% of $\dot{V}_{O_2max}$ for several hours. The basis for this phenomenal endurance ability lies in the muscles and other tissues rather than in the O_2 transport apparatus.

Over-distance training, consisting of running or swimming mile after mile at a speed much less than the competitive pace, causes a proliferation of mitochondrial protein in muscle. In detailed animal studies, muscle mitochondrial density has correlated better with endurance capacity than has $\dot{V}_{O_2max}$. The ability to utilize fats as fuels and the ability to protect mitochondria against damage during prolonged work are qualities developed through over-distance training.

Each training regimen has its advantages and disadvantages. The advantage of over-distance training is that it develops tissue respiratory capacity; the disadvantages come from the fact that the training regimen overlooks the basic principle of specificity. All athletic activities, even the most basic such as running, require practice in technique. Therefore, an endurance athlete must prepare for competition by training at or near the race pace. A mile runner practicing to break 4 min by running at an 8-min-mile pace makes about as much sense as a gymnast attempting to practice a double somersault in slow motion. Athletes must learn to pace and improve skill by performing at high rates in practice. Over-distance training can build tissue respiratory capacity but does not develop sense of pace, skill, or the capacity to achieve a high $\dot{V}_{O_2max}$. However, athletes who attempt to practice at or near race pace cannot possibly sustain much of a training volume. An approach to providing specific skill and pace training with volume training is the interval training regimen.

Interval Training

The interval training regimen is one in which periods of intense training intensity during a workout are interspersed with relief or rest periods. For instance, an interval workout for the runner practicing to dip under 4 min might consist of 10 repeated 60-sec quarter-mile runs with a 2-min relief or rest interval between each run.

The advantage of interval training is that the athlete learns pace; he or she practices the specific competitive skill, and the cardiovascular training stimulus intensity will be greater than in over-distance training. The interval training stimulus will possibly maximize the improvement in $\dot{V}_{O_2max}$ as well as result in significant improvements in mitochondrial density.

In addition to these aerobic training benefits, high-intensity interval training will stress the glycolytic system in muscle and will result in significant lactate accumulation. Because the accumulation of lactate is distressing to a competitor, he or she must practice tolerating its presence by repeated exposure to it. More importantly, radiotracer studies of lactate metabolism in animals indicate that training can improve the pathways of lactate removal (Chapter 10). The sites of lactate removal are heart and red skeletal muscle (which oxidize lactate), and gluconeogenic tissues (liver and kidney), which participate in the Cori cycle. Because the rate of lactate removal is directly dependent on its concentration (i.e., the greater the concentration, the greater the removal), interval training that increases blood lactate levels will stimulate improvement in the capacity to remove lactate.

Training Variation and Peaking

With just two basic aerobic training regimens (over-distance and interval), the permutations and combinations of varying distance, speed, duration of rest interval, and so forth, result in almost an infinite combination of training possibilities. The existence of such a large number of choices can be used to benefit by coaches who must often work to sustain interest and desire in athletes over months if not years of training. Varying training regimen is one way to maintain the interest in athletes.

As the training year progresses and the competitive season approaches, the training regimen should be adjusted to achieve peak performance. Let us illustrate by continuing to describe the training of a miler trying to break 4 min.

Early in the training year (i.e., fall and winter), the athlete should concentrate on over-distance training, with interval training sessions only once or twice a week. The athlete should alternate hard and easy days, with an interval-training day substituted for the hard over-distance day. The hard over-distance day might consist of a 10-mi cross-country run at a pace of 6 min to 6 min, 30 sec $\cdot$ mi^{-1}. An easy over-distance day would consist of a 5-mi run at the same pace. Interval training on the track might focus on developing the 4-min mile rhythm. Therefore, an interval day might consist of either ten quarter-mile runs at 60-sec pace, each with a 5-min rest interval in between, or twenty 30-sec, 220-yd runs, each with 2 min in between, or five 880-yd ($\frac{1}{2}$ mi) runs in 2 min with 5-min rest intervals. During rest intervals, the athlete should walk or jog to stimulate oxidative recovery. By following such a training regimen, the athlete will be accumulating a large volume of training, which will lay a solid base for the competitive season.

As the competitive season approaches, the athlete should gradually convert from a basic over-distance program to an interval program more specifically designed around the competitive event of interest. Instead of one or two interval sessions and five or six over-distance sessions a week, these training protocols should be reversed. Again, hard days should be alternated with easy days. If minor competitions begin to appear on the schedule, such as on every other Saturday, then the heavy volume of training should be centered early in the week (Sunday, Monday, Tuesday), and the training intensity and volume should taper at the end of the week. On hard interval training days the 4-min-mile race pace should be the cornerstone of the interval training. For variety, the rest interval between runs may be shortened, distances lengthened, pace quickened, or number of interval repeats increased. If the distances and numbers of repeats are increased, care should be taken in not letting pace fall much below the intended race pace. In practice, if an athlete can manage 10 quarter-mile runs with only a 2-min interval between, then he or she might expect to break 4 min.

Sprint Training

True sprints are events that last from a few seconds to approximately 30 sec. Sprinting requires an extreme degree of skill, coordination, and metabolic power. This power comes from immediate and nonoxidative energy sources present in muscle before the activity starts. Although proficiency in sprinting certainly can be improved through training, the nature of the activity is such that genetic endowment determines in a major way the success that can be achieved by an individual in sprinting.

Whereas endurance athletes require daily training to develop the cardiovascular and muscle respiratory capacities necessary to be competitive, sprint training requires more intense but less frequent training.

An adequate training frequency for sprinters might be 3 to 5 days per week.

The highly specialized nature of sprinting requires that the training performed must develop the specific skills used in sprinting. Some of these skills might include starting, accelerating, relaxing while sprinting, and finishing. Running and sprinting drills should be all out, but the distances should be kept less than the competitive distances so that repeated bouts at maximal intensity can be practiced.

In addition to high-intensity intervals at maximal speed, analysis of a particular sprinter's performance may reveal a need for exercises to improve other specific aspects of his or her performance. For instance, exercises to develop hip flexion (knee lift) can be particularly important. Weight lifting and isokinetic exercises can be used to develop both quadriceps and hamstring strength. High-speed filming may reveal inefficiencies in form that can be corrected in practice.

Although over-distance training is not strictly necessary for sprinting, under particular circumstances the training volume required by sprinters should be improved. Over-distance training early in the training session may be used to effect a weight loss in over-fat athletes. Long sprints (e.g., 400-m track running) require a significant aerobic component, therefore interval training should be employed. Over-distance training can also be successful if the competitive situation requires repeated bursts of activity. For example, a particular competition may involve trials, quarter- and semifinals, as well as final heats, or a competition may involve participation in several events plus a relay. Even in american football and soccer, repeated sprinting is required. Therefore, the ability to recover rapidly is essential. Recovery is an aerobic process that can be improved through over-distance, interval training. Over-distance conditioning of sprinters may also reduce the incidence of injury.

VOLUME VERSUS INTENSITY OF TRAINING

A perpetual question in athletic training is whether increasing the volume (frequency and duration) of training is more beneficial than increasing the intensity. There appears to be no simple answer to this question, but consideration should be given to both the type of event involved and the phase of training in relation to the competitive season. In general, the more intense (sprint) types of activities require higher intensities of training. This is because of the principle of specificity, which dictates that attention be given to developing the metabolic apparatus and skill levels necessary to compete in the event. Of necessity, more intense training requires reduced training volume.

In general, during the early training season, athletes should focus on increasing the training volume. As the competitive season approaches, training intensity should be emphasized and the volume diminished. In preparing for a major competition, both training volume and intensity should be reduced.

THE TAPER FOR COMPETITION

The period prior to major competition, when athletes rest by decreasing training volume and intensity to very low levels so that peak performances can be achieved, is termed the taper period. The taper can be understood generally in terms of the adaptive response syndrome. It is during the taper that athletes recover from the hard training and adaptive responses peak. The taper period used varies from sport to sport. In track, the most frequent taper period used ranges from 1 to 2 weeks; in swimming, the taper used is frequently twice that used in track. Unfortunately, at present we have an insufficient research basis on which to calculate duration of the optimal taper period. Two to three days after training should be sufficient to result in maximum glycogen supercompensation. Within a week following intense training, minor injuries should have healed, soreness should have disappeared, and nitrogen balance should have returned to zero from positive levels. In other words, the response to training overload should have peaked. On physiological bases, then, a taper period of around a week to 10 days would seem to be ideal. The best information available at present on the half-life of muscle respiratory proteins (mitochondria) indicates the adaptation period to have a half-life of 2 weeks. Therefore, a layoff of longer than 1 to 2 weeks should result in physiological decrements. However, despite the research data available, most athletes and coaches who utilize a taper regimen are of the opinion that a taper period longer than a week is superior. At present we must state, therefore, that the physiology of the taper is not completely understood.

THREE COMPONENTS OF A TRAINING SESSION

Each training session should consist of three components: warm-up, training, and cool-down. There are several objectives to preliminary exercise (warm-up). The athlete attempts to increase temperatures of the tissues to take advantage of the Q_{10} effect (Chapter 2) on raising metabolic rate and the speed of muscle contraction. Preliminary exercise also raises the cardiac output and dilates capillary beds in muscle. In this way circulation of blood and O_2 is raised before hard exercise starts. Additionally, during a warm-up, an athlete attempts to stretch out the active tissues by stretching exercises and warming of the tissues. Through exercise and wearing heavy clothing, prior stretching and warming up is thought to minimize the possibility of injury. Preliminary exercise also provides a "last-minute" practice session. Motor skills are fine-tuned and adjusted for the prevailing conditions. The warm-up procedure also provides the athlete a time when she or he can make psychological preparation for the practice or competition. This neural aspect of preparing for exercise may exceed in importance the other benefits of preliminary exercise.

The nature of the preliminary exercise (warm-up bout) depends on the specific activity to be performed. In general, however, several considerations apply. The exercise performed should utilize the major muscles to be involved in training or competition. The activity performed should be the same or very similar to that to be engaged in, and it should progress from mild to hard intensity. Because about 10 min of exercise at a particular work intensity are required to reach a steady muscle temperature, a warm-up should be at least 10 min long. Because the athlete should avoid fatiguing exercise prior to practice or competition, and because tissues cool down more slowly than they heat up, frequent rest periods will allow warming up without imparting fatigue. Once an athlete has warmed up, a brief recovery period should intervene prior to training or competition. Ideally, this period may be 5 to 10 min, although a well-clothed athlete may remain ''warmed up'' for 20 to 30 min.

Whereas most athletes, by habit or tradition, include time in a training session to warm up, the cool-down period following training is usually overlooked. A cool-down period is very much like the reverse of a warm-up. The exercise intensity is gradually decreased and is followed by a period of passive stretching, wherein heavily used muscles are held in elongated positions for two 1-min intervals. For instance, in runners the hamstrings and gastrocnemius muscles are those usually stretched after training. Particularly in older or less fit individuals, a cool-down period may effectively minimize soreness and stiffness during the days following hard training or competition.

METHODS OF EVALUATING TRAINING INTENSITY

Is a workout schedule too hard, too easy, or just right? At various times, both coaches and athletes desire some objective evaluation of an athlete's training regimen. To answer these questions several approaches can be taken, all of which involve evaluation of both performance and physiological criteria.

Of primary importance is an evaluation of the athlete's performance. If the athlete is meeting or exceeding performance criteria during workouts, time trials, and competitions, then the training regimen is obviously having good results. If the athlete feels good an hour or so after training, and if the athlete feels good the day after training, then the training regimen is probably appropriate.

Application of the above criteria may be able to answer the question whether the training schedule is too easy or too hard, but it cannot identify an optimal-training regimen or provide an objective assessment of training intensity. For this evaluation, determination of exercise heart rate can be useful. During interval training, the exercise intensity should be sufficient to stimulate heart rate to maximum. Maximal heart rate can be determined during an exercise stress test (Chapter 26) or immediately after a time trial by ECG or palpation (counting the pulse). The recovery interval should end and the next training interval should begin when heart rate falls to two thirds of maximum (i.e., about 120 beats · min^{-1}). Such a regimen would constitute a very hard interval training program. Heart rate during submaximal, over-distance training should stabilize at three quarters of maximum and progress to maximum as the training session is completed.

Determination of blood lactic acid level has been purported to be another "objective" means of evaluating intensity of the training stimulus. Lactate level can easily be measured in a tiny drop of blood taken by means of a pinprick to the earlobe or finger tip. The East German swim team has reportedly identified the work intensity that elicits a 4-mM blood lactic acid level as the optimal aerobic (over-distance) training intensity. Unfortunately, no data have been put forward to justify a 4-mM blood lactic acid level as the ideal, and no theoretical basis exists to justify use of the 4-mM value. Certainly, for interval exercise, a 4-mM blood lactate level would lack sufficient intensity. Probably also for over-distance training, a 4-mM blood lactate level might be too easy. For instance, this author (who is now neither young nor trained) has been able to maintain exercise loads for an hour that elicit blood lactic acid levels from 6 to 8mM. Because the ability to remove lactic acid is related to its concentration, an athlete should occasionally experience high circulating lactate levels (10 mM) to develop the mechanisms of lactate removal (Chapters 10 and 32).

The ventilatory threshold (T_{vent}) (Figure 21-3) is that point at which ventilation begins to increase nonlinearly in response to increments in work rate. The T_{vent} is sometimes called the anaerobic threshold and is usually associated with an increase in the blood lactic acid level (Chapter 10). However, other stress factors can elevate ventilation during exercise, and as such, determination of the ventilatory inflection point (T_{vent}) is probably a more appropriate training criterion than is the determination of a blood lactate inflection point (T_{lact}) or a blood lactate level of 4 mM. The problem, however, with using T_{vent} as a training guide is that its measurement requires special apparatus that is not ordinarily available to the athlete and coach. If a laboratory is available

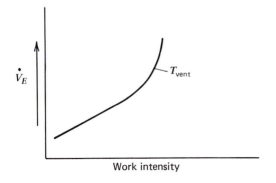

Work intensity

FIGURE 21-3
Pulmonary minute ventilation ($\dot{V}_E$) is a function of work intensity. As work rate increases, V_E increases linearly up until a particular work load, after which V_E increases disproportionally. This break in the ventilation versus work rate curve is termed the "ventilatory threshold" [T(vent)]. Blood lactate level (Figure 10-1b) and other factors affect the T(vent); please refer to Chapter 10.

to a runner, the treadmill pace that elicits the T_{vent} can be determined with precision and then utilized in the field. Repeated laboratory tests will be required, however, as the athlete improves.

COMMON MISTAKES IN TRAINING

In preparing for competition, many athletes expend a great deal of effort but are unsuccessful for some simple reasons. Athletic training is in some cases, at some times, a very important endeavor. Therefore, a system or training strategy should be constructed and written out. A particular training schedule should be prescribed, and a log of training adherence and performance should be kept. Perhaps the most frequent failing of some training regimens is that insufficient overload is applied. Either training volume or intensity is too little to result in training adaptations that will distinguish a particular competitor.

Some training regimens are often too shortsighted in their approach. Many athletes expect rapid improvements and become discouraged even as the basis of real adaptation is being achieved. The progression of an athlete in training and competition should realistically be planned to extend over several seasons. Particularly in American track and field competition, where rapid progress is frequently desired, hard interval training as been used to optimize performance in the short run, but athletes become discouraged when they fail to improve consistently from season to season, because no real basis has been laid to support the improvement. The temptation to develop racers rather than physiologically superior athletes must be suppressed.

Sometimes athletes fail to reach their level of competitive aspiration because they lack an individualized training program. In a group of 10 cross-country runners running interval quarter miles together, perhaps 5 will be running very hard to keep up. Several others will not be training very hard, and at least 1 or 2 will be receiving a minimal training stimulus. Such a predicament happens when there are more athletes to train than can interact with the coach.

Summary

As with strength training (Chapter 20), training for rhythmical events involves application of the basic training principles of overload, specificity, individuality, and reversibility. According to the principle of overload, application of an appropriate stimulus will result in adaptation; the greater the stimulus, the greater the adaptation. According to the principle of specificity, the adaptations will be specific to the type of stimulus provided and will occur only in the tissues and organs stressed. In other words, preparation for particular events involves very specific training regimens. Whereas according to the principles of overload and specificity, two individuals of equal initial ability will respond in the same direction as a result of a particular training regimen, the degree of response will likely be different. This is because of uniqueness of each person and the principle of individuality. The principle of reversibility reminds us that the adaptations due to overload are not permanent and that withdrawl of training will result in regression toward the untrained state.

Selected Readings

American College of Sports Medicine. The recommended quality and quantity of exercise for developing and maintaining fitness in healthy adults. Med. Sci. Sports 10: vii–x, 1978.

Anderson, P. and J. Henrickson. Capillary supply of the quadriceps femoris muscle of man. Adaptive response to exercise. J. Physiol. London 270: 677–690, 1977.

Bevegard, S., A. Holmgren, and B. Jonsson. Circulatory studies in well trained athletes at rest and during heavy exercise, with special reference to stroke volume and the influence of body position. Acta Physiol. Scand. 57: 26–50, 1963.

Bouchard, C., P. Godbout, and C. Leblanc. Specificity of maximal aerobic power. Eur. J. Appl. Physiol. 40: 85–93, 1979.

Brodal, P., F. Inger, and L. Hermansen. Capillary supply of skeletal muscle fibers in untrained and endurance-trained men. Am. J. Physiol. 232: H705–H712, 1977.

Brynteson, P. and W. Sinning. The effects of training frequencies on the retention of cardiovascular fitness. Med. Sci. Sports 5: 29–33, 1973.

Clausen, J.P. Effect of physical training on cardiovascular adjustments to exercise in man. Physiol. Rev. 57: 779–815, 1977.

Costill, D., J. Daniels, W. Evans, W. Fink, G. Krahenbuhl, and B. Saltin. Skeletal muscle enzymes and fiber composition in male and female track athletes. J. Appl. Physiol. 40: 149–154, 1976.

Costill, D., H. Thomason, and E. Roberts. Fractional utilization of the aerobic capacity during distance running. Med. Sci. Sports 5: 248–252, 1973.

Daniels, J.T., R.A. Yarbrough, and C. Foster. Changes in max $\dot{V}O_2$ and running performance with training. Eur. J. Appl. Physiol. 39: 249–254, 1978.

Davies, C.T.M. and A.U. Knibbs. The training stimulus: The effects of intensity, duration and frequency of effort on maximum aerobic power output. Int. Z. Angew. Physiol. 29: 299–305, 1971.

Ekblom, B. Effect of physical training on the oxygen transport system in man. Acta Physiol. Scand. 328 (Suppl.): 11–45, 1969.

Ekblom, B., P. Astrand, B. Saltin, J. Stenberg, and B. Wallstrom. Effect of training on circulatory response to exercise. J. Appl. Physiol. 24: 518–528, 1968.

Ekblom, B. and L. Hermansen. Cardiac output in athletes. J. Appl. Physiol. 24: 619–625, 1968.

Faria, I. Cardiovascular response to exercise as influenced by training of various intensities. Res. Quart. 41: 44–50, 1970.

Fox, E.L. Difference in metabolic alteration with sprint versus endurance interval training programs. In: Howald, H. and J. Poortmans (eds.), Metabolic Adaptation to Prolonged Physical Education. Basel: Birkhauser Verlag, 1975. p. 119–126.

Fox, E.L., R.L. Bartels, J. Klinzing, and K. Ragg. Metabolic responses to interval training programs of high and low power output. Med. Sci. Sports 9: 191–196, 1977.

Fox, E., R. Bartels, C. Billings, D. Mathews, R. Bason, and W. Webb. Intensity and distance of interval training programs and changes in aerobic power. J. Appl. Physiol. 38: 481–484, 1975.

Frick, M., A. Konttinen, and S. Sarajas. Effects of physical training on circulation at rest and during exercise. Am. J. Cardiol. 12: 142–147, 1963.

Fringer, M.N. and G.A. Stall. Changes in cardio respiratory parameters during period of training and detraining in young adult females. Med. Sci. Sports 6: 20–25, 1974.

Gettman, L.R., M.L. Pollock, J.L. Darstine, A. Ward, J. Ayers, and A.C. Linnerud. Physiological responses of men to 1, 3, 5 day per week training programs. Res. Quart. 47: 638–646, 1976.

Gollnick, P., R. Armstrong, B. Saltin, C. Saubert, W. Sembrowich, and R. Shepherd. Effect of training on enzyme activity and fiber composition of human skeletal muscle. J. Appl. Physiol. 34: 107–111, 1973.

Hagberg, J.M., R.C. Hickson, A.A. Ehsani, and J.O. Holloszy. Faster adjustment to and recovery from submaximal exercise in the trained state. J. Appl. Physiol. 48: 218–224, 1980.

Henriksson, J. and J.S. Reitman. Time course of changes in human skeletal muscle succinate dehydrogenase and cytochrome oxidase activities and maximal oxygen uptake with physical activity and inactivity. Acta Physiol. Scand. 99: 91–97, 1977.

Hickson, R.C., H.A. Bomze, and J.O. Holloszy. Faster adjustment of O_2 uptake to the energy requirement of exercise in the trained state. J. Appl. Physiol. 44: 877–881, 1978.

Hickson, R.C., J.M. Hagberg, A.A. Ehsani, and J.O. Holloszy. Time course of the adaptive responses of aerobic power and heart rate to training. Med. Sci. Sports Exercise 13: 17–29, 1981.

Hickson, R.C. and M.A. Rosenkoetter. Reduced trained frequencies and maintenance of increased aerobic power. Med. Sci. Sports Exercise 13: 13–16, 1981.

Hoppeler, H., P. Lüthi, H. Classen, E.R. Weibel, and H. Howard. The ultrastructure of the normal human skeletal muscle. A morphometric analysis of untrained men, women, and well-trained orienteers. Pflügers, Arch. 344: 217–232, 1973.

Inger, F. Capillary supply and mitochondrial content of different skeletal muscle fiber types on untrained and endurance trained men. A histochemical and ultrastructural study. Eur. J. Appl. Physiol. 40: 197–209, 1979.

Jackson, J., B. Sharkey, and L. Johnston. Cardiorespiratory adaptations to training at specified frequencies. Res. Quart. 39: 295–300, 1968.

Jansson, E., B. Sjokin, and P. Tesch. Changes in muscle fiber type distribution in men after physical training. Acta Physiol. Scand. 104: 235–237, 1978.

Karlsson, J., P.V. Komi, and J.H.T. Vitasalo. Muscle strength and muscle characteristics of monozygous and dizgous twins. Acta Physiol. Scand. 106: 319–325, 1979.

Karlsson, J., L.-O. Nordesjö, and B. Saltin. Muscle glycogen utilization during exercise after physical training. Acta. Physiol. Scand. 90: 210–217, 1974.

Kiessling, K., K. Piehland, C. Lundquist. Effect of physical training on ultrastructural features of human skeletal muscle. In: Pernow, B. and B. Saltin (eds.), Muscle Metabolism During Exercise. New York: Plenum Press 1971. p. 97–101.

Klissouras, V. Heritability of adaptive variation. J. Appl. Physiol. 31: 338–344, 1981.

Klissouras, V. Genetic limit of functional adaptability. Int. Z. Angew. Physiol. 30: 85–94, 1972.

Klissouras, V., F. Pirnay, and J. Petit. Adaptation to maximal effort: Genetics and age. J. Appl. Physiol. 35: 288–293,1973.

Knuttgen, H., L. Nordesjö, B. Ollander, and B. Saltin. Physical conditioning through interval training with young male adults. Med. Sci. Sports 5: 206–226, 1973.

Komi, P.V., J.H.T. Viitasalo, M. Havy, A. Thorstensson, B. Sjödin, and T. Karlsson. Skeletal muscle fibers and muscle enzyme activities in monozygous and dizygous twins of both sexes. Acta Physiol. Scand. 100: 385–392, 1977.

Longhurst, J.C., A.R. Kelly, W.J. Gonyea, and J.H. Mitchell. Echocardiographic left ventricular masses in distance runners and weight lifters. J. Appl. Physiol. 210: 154–162, 1980.

Morganroth, J., B. Maron, W. Henry, and S. Epstein. Comparative left ventricular dimensions in trained athletes. Ann. Intern. Med. 82: 521–524, 1975.

Pedersen, P. and K. Jørgensen. Maximal oyxgen uptake in young women with training, inactivity, and retraining. Med. Sci. Sports 10: 233–237, 1978.

Roeske, W.R., R.A. O'Rourke, A. Klein, G. Leopold, and J.S. Karlinger. Noninvasive evaluation of ventricular hypertrophy in professional athletes. Circulation, 53: 286–292, 1975.

Rowell, L.B. Human cardiovascular adjustments to exercise and thermal stress. Physiol. Rev. 51: 75–159, 1974.

Saltin, B., K. Nazar, D.L. Costill, E. Stein, E. Jansson, B. Essen, and P.D. Gollnick. The nature of the training response; peripheral and central adaptations to one-legged exercise. Acta Physiol. Scand. 96: 289–305, 1976.

Shephard, R. Intensity, duration and frequency of exercise as determinants of the response to a training regiment. Int. Z. Agnew. Physiol. 26: 272–278, 1968.

Stromme, S.B., F. Ingjer, and H.D. Meen. Assessment of maximal aerobic power in specifically trained athletes. J. Apply. Physiol. 42: 833–837, 1977.

Wilt, F. Training for competive running. In: Falls, H. (ed.), Exercise Physiology. New York: Academic Press, 1968.

Zeldis, S.M., J. Morganroth, and S. Rubler. Cardiac hypertrophy in response to dynamic coordinating in female athletes. J. Appl. Physiol. 44: 849–852, 1978.

22

EXERCISE IN THE HEAT AND COLD

People have a remarkable ability to exercise in both very hot and very cold environments. It is not unusual to see hundreds of die-hard skiers schussing down the slopes at temperatures well below 0°C. Likewise, marathons are sometimes run in desert climates at temperatures exceeding 37°C (normal internal body temperature).

Humans are often forced to live and work at temperatures considerably hotter and colder than these. The mean annual temperature of much of the Soviet Union and Canada is less than 0°C. In the winter, temperatures are often so cold that exposed skin will freeze within 1 min. Even greater numbers of people are exposed to extreme heat. Summer temperatures regularly surpass 43 to 49°C in places like Australia, the American Southwest, the Middle East, and India (Figure 22-1). In some desert areas, the inhabitants must endure extremely hot temperatures in the daytime and extremely cold temperatures at night.

We can tolerate these hot and cold climates because of a well-developed ability to control body temperature. When the environment is cold, we can maintain our body temperature by increasing the body's heat production and putting on more clothes. When it is hot, we can increase heat dissipation by sweating, increasing blood flow to the skin, and removing clothes.

443

Olympic champion Franz Klammer flies over a bump during practice for a World Cup Downhill competition. Athletes such as Klammer must generate extremely high metabolic rates and display precise neural-muscular control while competing in freezing environments. (Wide World.)

HUMANS AS HOMEOTHERMS

Animal body temperatures either remain constant (homeotherms) or vary with the environment (poikilotherms). Poikilotherms such as lizards and insects are at the mercy of the elements. When the climate is cold, their body temperatures can become so low that their metabolic rate drops to a level that forces inactivity. Likewise, in the heat they must seek shelter or perish.

Advanced animals such as human beings, monkeys, dogs, bears, and birds are homeotherms. They are able to function relatively independently of the environment because of their ability to maintain constant body temperatures.

Processes such as O_2 transport, cellular metabolism, and muscle contraction remain unimpaired in hot and cold environments as long as the internal temperature is maintained. Our mastery of the planet Earth probably would have been impossible if we were not homeotherms.

Various physiological mechanisms, such as neural function, depend on a normal body temperature to function properly. Abnormal increases and decreases in body temperature are catastrophic to the organism. At temperatures above 44°C, the parenchyma of many cells begins to deteriorate and denature. Heat stroke and permanent brain damage can ensue if the body temperature is not quickly brought under control. At temperatures below 34°C cellular metabolism slows greatly, leading to unconsciousness and cardiac arrhythmias (Figure 22-1).

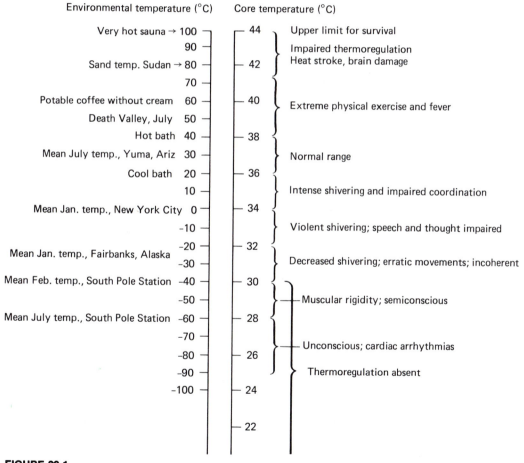

FIGURE 22-1
Environmental temperature extremes and the effects of alterations in core temperature.

Normal Body Temperature

Human beings experience a range of normal resting body temperatures that typically lies between 36.5 and 37.5°C. However, in the early morning the temperature can fall to less than 36°C and during exercise, body temperature can exceed 40°C with no ill effects. There is considerable temperature variation throughout the body. The internal temperature of the core remains relatively constant, while the skin temperature is closer to that of the environment. Body temperature is typically expressed in terms of the core temperature.

Core temperature is usually defined as the temperature of the hypothalamus, the temperature-regulatory center of the body. The most common method of measuring core temperature is orally. However, this has severe limitations, particularly during exercise, as increased ventilation will result in evaporative cooling of the thermometer, which will produce an inaccurate measurement. In research, core temperature is most often measured rectally. Rectal temperature is typically 0.6°C higher than oral temperature. Although rectal temperature is more accurate, it also has its limitations. Vigorous exercise of local muscle groups will produce a higher regional temperature that will render spurious results. Additionally, there are temperature variations in the rectum itself. The thermistor should be inserted to a depth of 5 to 8 cm to ensure an accurate and reproducible measurement of rectal temperature.

Researchers also estimate core temperatures by taking measurements in the auditory canal (tympanic temperature) and in the stomach. Tympanic temperature is advantageous because of its close proximity to the hypothalamus. However, its measurement can cause discomfort to the subject. Stomach temperatures are obtained by telemetry. Subjects swallow a small radio transmitter that signals temperatures to the researcher. Unfortunately, stomach temperatures can vary considerably from the temperature of the hypothalamus. Changes in the ambient temperature or the digestion of food will result in variations of as much as 3°C.

The assessment of mean body temperature (MBT) must take into consideration both skin and core temperatures. This is typically accomplished by measuring the rectal temperature and a series of skin temperatures at various places on the body. Mean body temperature is expressed by the following equation:

$$\text{MBT} = (0.33 \times \text{skin temperature}) + (0.67 \times \text{rectal temperature}) \qquad (22\text{-}1)$$

Mean body temperature is important, because thermal gradients determine whether the transfer of heat will tend to be toward the body or away from it.

TEMPERATURE REGULATION

Heat is not a substance. The temperature of an object is a measure of the kinetic activity of its molecules. If the object is hot, then its molecules move very rapidly, and as it cools, its molecules move more slowly. The temperature of the body is directly proportional to the amount of heat it stores. When heat storage increases, such as in fever or during exercise, body temperature rises. When the heat storage decreases, such as in hypothermia, the body temperature falls.

Body temperature is regulated by controlling the rate of heat production and heat loss. When the rate of heat production is exactly equal to the rate of heat loss, the body is said to be in heat balance. When out of balance, the body either gains or loses heat. Heat balance is controlled centrally by the hypothalamus, with feedback from peripheral heat and cold receptors in the skin. The hypothalamus works like a thermostat by increasing the rate of heat production when body temperature falls and increasing the rate of heat dissipation when it rises. Body temperature conforms to the heat balance as follows.

$$O = M - E \pm C \pm R \pm S \qquad (22\text{-}2)$$

where M = metabolic heat production,
$\quad E$ = evaporative heat loss,
$\quad C$ = conductive heat loss or gain.
$\quad R$ = radiant heat loss or gain, and
$\quad S$ = body heat storage or loss.

The mechanism for temperature regulation can be divided into physical and chemical processes. Physical temperature regulation works principally by changing the resistance to heat flow, while chemical mechanisms work by increasing the body's metabolic rate (Figure 22-2).

Body Temperature, Environment, and Exercise Intensity

Within the broad range of 4°C to 30°C, environmental still air temperature, body core temperature is independent of environmental temperature. Under such

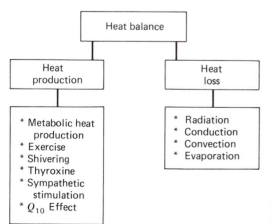

FIGURE 22-2
Factors determining thermal balance.

conditions, core temperature rises in direct proportion to the relative exercise intensity, the greater the relative intensity, the greater core temperature will rise (Figure 22-8b). Under such conditions, temperatures of peripheral tissues (skin and muscle) will reflect the environmental temperature.

Temperatures of tissues in the body's periphery (i.e., outside of the core) will reflect both metabolic rate and environment. For instance, the temperature of contracting muscle usually rises during exercise as muscle is the site of heat production. However, muscle temperature will be higher for a given work load on a warm day than on a cool day.

Skin temperature is a complex resultant of environment, metabolic rate, clothing, and state of hydration. For instance, on a warm day skin temperature might tend to be higher than on a cool day during exercise of a given intensity. However, active sweating in a warm, but dry environment might actually cool the skin in comparison to a cooler, perhaps more humid day. If exercise is prolonged in a hot environment, skin temperature tends to rise as sweat capacity decreases because of dehydration, and because of convection of heated blood to the skin. Skin temperature during heat exposure can also vary considerably around the body due to the presence of clothing and differences in the movement of air across various parts of the body.

Heat Production

Heat is a by-product of all biochemical reactions, because they are not 100% efficient (Chapter 3). Metabolism (M in Eq. 22-2) is the body's source of internal heat production. Even when the body is in deep sleep, there is a certain amount of heat produced. During exercise, the heat production is considerable and increases as a function of exercise intensity.

Heat production can be enhanced by increasing the metabolic rate. The main mechanism for accomplishing this in the face of negative heat balance is shivering. Shivering is an involuntary contraction of muscle. Maximal shivering can increase the body's heat production by up to five times. It is an effective way of increasing body temperature because no work is done by the shivering muscles; thus most of the expended energy appears as heat. However, shivering adds to heat loss by increasing the thermal gradient between the environment and the individual.

Increased thyroxine secretion from the thyroid and catecholamine secretion from the adrenals also increase metabolic rate. Thyroxine increases the metabolic rate of all the cells in the body. The catecholamines, principally norepinephrine, cause the release of fatty acids, which also increase metabolic heat production.

As we saw in Chapter 2, increased body temperature can be self-perpetuating and dangerous. This is because the metabolic rate increases with rising temperature resulting from the Q_{10} effect (see Chapter 2). At high core temperatures, the hypothalamus begins to lose its ability to cool the body. Unfortunately, the rate of temperature increase is faster at these higher temperatures. At core temperatures above 41.5°C, sometimes the only recourse to preventing thermal damage is external cooling, because the hypothalamus may no longer be functional.

Heat Loss

The body loses heat by radiation, conduction, convection, and evaporation. At room temperature, when skin temperature is greater than air temperature, most heat is lost by outward heat flow caused by the negative thermal gradient. In the heat, or during heavy exercise, evaporation becomes the dominant mechanism of heat dissipation.

Radiation Radiation (R in Eq. 22-2) is the loss of heat in the form of infrared rays. At room temperature and at rest it accounts for 60% of the total heat loss. Radiant heat loss will vary with body position and clothing. Any substance not at absolute zero temperature (-273°C) will radiate such waves. However, the body is both radiating and receiving them at the same time. If the body temperature is greater than that of the surrounding environment, then more heat is radiated from the body than to it. If the temperature of the environment is greater, then the net flow of radiation is inward.

The color and texture of an object will affect its ability to absorb radiant heat rays. Light-colored, shiny objects absorb radiant heat less easily than black, rough objects. The human body has been called a perfect blackbody radiator. Human skin, regardless of color, absorbs about 97% of radiant energy that strikes it. So, a person exercising in the hot sun will be better off wearing a light white cotton shirt than going bare-skinned. The environmental radiant heat load is measured with a black globe thermometer.

Conduction Conduction (C in 22-2) is the transfer of heat from the body to an object. About 3% of the total heat loss at room temperature occurs by this mechanism. A good example of conduction is the transfer of heat to a chair while a person is sitting on it. Excretion is related to conduction and represents heat loss in the urine and feces.

Convection The conduction of heat to air or water is called convection. This accounts for about 12% of the heat loss at room temperature. In convection, heat conducted to air or water moves so that other particles can also be heated. As heat is transmitted to the surrounding air, it rises, allowing additional heat transfer to the surrounding air. Heat loss by conduction and convection occurs much more rapidly in water than in air.

Heat loss by convection is greater in the wind, because warmed air is quickly

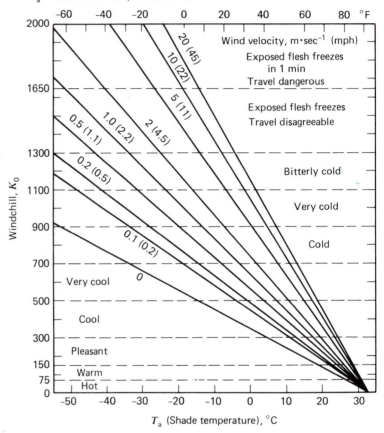

WINDCHILL FACTOR
$$K_0 = (\sqrt{100v + 10.45 - v})(33 - T_a)$$

K_0 = Windchill as kcal·hr^{-1}/m^2 exposed skin surface in shade, ignoring evaporation

v = Wind velocity in m·sec^{-1} = 2.2369 mph)

T_a = Ambient air temperature in °C

FIGURE 22-3

Nomogram for the calculation of windchill.

(SOURCE: Modified from N. Pace, 1972, p. 39.)

replaced by colder air lowering the effective temperature. For example, the heat loss at 10°C in a 2.2-mph wind is the same as that at −10°C in still air (Figure 22-3). The effect of wind on temperature is called the windchill factor (see 22-3) and is expressed in kcal · hr^{-1}/m^2 exposed skin surface.

$$K_o = (\sqrt{100v + 10.45 - v})\,(33 - T_a) \tag{22-3}$$

where K_o = wind chill as kcal/hr/m^2 exposed skin surface in shade, ignoring evaporation.

V = Wind velocity in m/sec

T_a = Ambient air temperature in °C

Wind speeds above 40 mph have no additional effect on heat loss, because of heat transfer to the skin does not occur rapidly enough.

Evaporation At rest in a comfortable environment, about 25% of heat loss is due to evaporation. However, it is the only means of cooling at high environmental temperature and is critically important during exercise. When the environmental temperature is greater than that of the skin, the body gains heat by radiation and conduction. If the body cannot lose heat by evaporation under these circumstances then body temperature rises.

The body loses 0.58 kcal of heat for each gram of water that evaporates (E in 22-2). Sweat is only effective for cooling if it evaporates. If the humidity is high, the rate of evaporation may be greatly reduced or totally prevented, so that the sweat remains in a fluid state. Effective evaporation is hampered by lack of air movement, because the air surrounding the body becomes saturated with water vapor. This explains why fans are desirable on a hot day.

Evaporation occurs as a result of sweating and insensible water loss. Water evaporates insensibly from the skin and lungs at a rate of about 600 ml · day^{-1}. This amounts to a continual heat loss of about 12 to 18 kcal · hr^{-1}. Insensible water loss cannot be controlled, and it occurs regardless of body temperature. However, evaporative sweat loss can be controlled by regulating the rate of sweating.

Except for insensible water loss (i.e., evaporative heat loss through ventilation and sweating under the arms and in the genital region), sweat rates are essentially zero when the skin temperature is low. In hot weather, an unacclimatized individual (person not used to the heat) has a maximum sweat rate of about 1.5 liters · hr^{-1}, whereas an acclimatized person can sweat up to 4 liters · hr^{-1}. During maximum sweating a person can lose 3.6 kg. per hour.

There are two kinds of sweat glands called apocrine and eccrine. The apocrine sweat glands exist primarily under the arms and in the genital region. Their secretions contain a lipid material that produces a characteristic odor when acted on by bacteria. Apocrine sweat glands do not typically become fully operational until adolescence.

Eccrine sweat glands are tubular structures consisting of a deep, coiled portion that secretes sweat and a duct portion that passes outward through the dermis of the skin. They cover most of the body and secrete a clear, essentially odorless sweat that accounts for most of the evaporative heat loss in the body (Figure 22-4). In addition to water, sodium chloride, urea, lactic acid, and potassium ions are also lost in sweat.

Eccrine sweat glands located on the hands and feet can be stimulated by sympathetic adrenergic fibers. The feeling of sweaty palms before competition

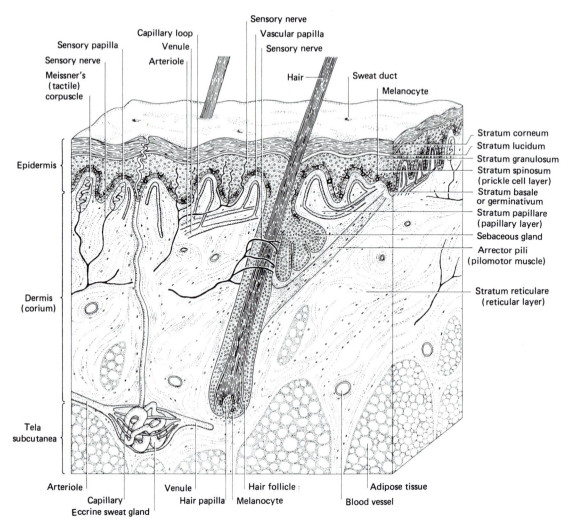

FIGURE 22-4

Structure of human skin and eccrine sweat gland.

(SOURCE: From N. Pace, 1972, p. 28.)

is due to sympathetic mobilization rather than temperature regulation. These sweat glands have the practical effect of providing traction for the feet and improved grip for the hands.

Heat storage If the chemical and physical effects of heat acquisition in the body exceed those of heat loss, then body temperature will increase. In such a case, body heat storage (S in 22-2) will be positive. On the other hand, if the loss of heat exceeds heat acquisition, the body will lose heat, and the S term will be negative.

Body heat balance is one of the best controlled parameters in human physiology. Over the course of a day, a month, an entire lifetime, the heat balance equation (22-2) equals zero. As noted above, even brief imbanaces in mechanisms of heat loss-heat gain can be extremely dangerous.

The Hypothalamus

The temperature-regulatory center is located in the hypothalamus (Figure 22-5). The responses to heat are primarily controlled by heat-sensitive neurons in the preoptic area of the anterior hypothalamus, whereas responses to cold are controlled by the posterior hypothalamus, which receives neural (afferent) input from cold receptors around the body. Although some temperature control is attributed to peripheral receptors, the hypothalamus is by far the most important. Hot and cold skin temperature receptors transmit nerve impulses to the spinal cord and then to the hypothalamus, which then initiates the appropriate response.

Overheating the preoptic area of the hypothalamus results in the stimulation of heat loss mechanisms. The anterior hypothalamus stimulates the sweat glands, resulting in evaporative heat loss from the body. Additionally, the vasomotor center is inhibited. This removes the normal vasoconstrictor tone to the skin vessels, allowing increased loss of heat through the skin. The anterior hypothalamus may also increase the local release of bradykinin, which causes further vasodilation. In animals, such as dogs, there is stimulation of the panting center, which also increases evaporative heat loss.

When cold receptors in the skin and other body areas are stimulated, neural signals are sent to the posterior hypothalamus. The hypothalamus, in turn, initiates various processes that increase heat production and reduce heat loss. The vasomotor center is stimulated, which results in vasoconstriction of blood vessels in the skin. However, there is intermittent vasodilation, particularly in the hands and feet, to maintain the health of the skin. Vasoconstriction can increase the effective insulation of the core by 1 to 2 in. There is stimulation of the shivering center to cause shivering and the pilomotor center to cause piloerection (goose bumps). The posterior hypothalamus also initiates the release of norepinephrine, which results in the mobilization of fatty acids and an increase in metabolic heat production. It indirectly increases thyroxine production by secreting thyrotropin-releasing factor, which in turn stimulates the secretion of thyrotropin by the pituitary gland.

The hypothalamus functions as the body's thermostat by keeping the core temperature within a normal range. When the core temperature goes above or below its ''setpoint'' the hypothalamus initiates processes to increase heat pro-

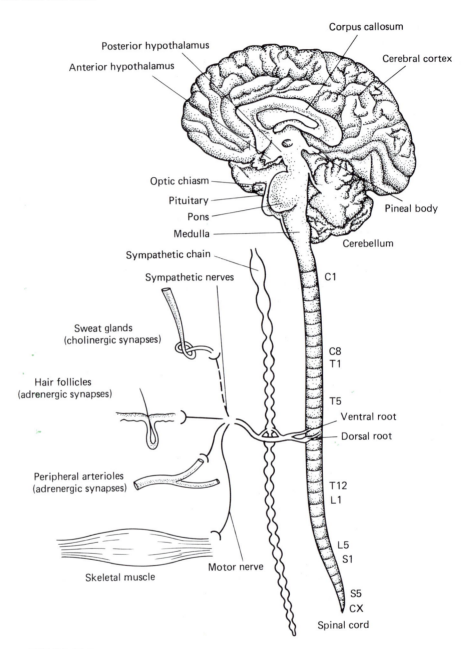

FIGURE 22-5

Temperature regulation schema for humans.

(SOURCE: Modified from N. Pace, 1972, p. 31.)

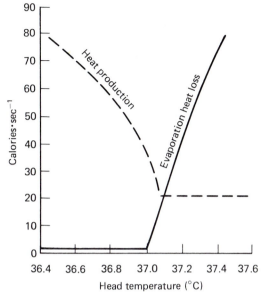

FIGURE 22-6

The effect of hypothalamic temperature on heat production and heat loss. The body's temperature-regulatory system works very much like a thermostat in a house.

(SOURCE: Modified from A.C. Guyton, Textbook of Medical Physiology. Copyright W.B. Saunders Co., Philadelphia, 1976, p. 962. With permission.)

duction or heat loss. Normally, sweating begins at almost precisely 37°C, and thermogenesis mechanisms begin below this point (Figure 22-6).

The setpoint can move either up or down under certain circumstances. Pyrogens (fever-producing agents) and exercise have the effect of raising the setpoint. During exercise, temperatures of over 40°C are often reached without ill effects. This would be very dangerous in the resting state but is easily tolerated when the setpoint is allowed to increase. The setpoint can also be lowered when a person is chronically exposed to cold.

Behavior

Human behavior is an important component in temperature regulation. When the anterior hypothalamus is overheated, a person has the sensation of being hot and will do something about it, such as drinking a glass of cold water, removing some clothing, or turning on the air conditioning. Likewise, if a skier's skin receptors relay the sensation of cold during a chilly day on the slopes, the logical tendency may be to head for the warmth of the nearest bar.

EXERCISE IN THE COLD

It .is rare, except when survival is at stake, that people exercise in a state of hypothermia. The combination of the increased metabolism of exercise and extra clothing make the theoretical problems of hypothermic exercise moot.

However, the cold does present problems for athletes, as is readily apparent during athletic contests played in extremely adverse wintertime weather.

Perhaps the two biggest handicaps of exercising in the cold are the numbing of exposed flesh and the awkwardness and extra weight of protective clothing. Manipulative motor skills requiring finger dexterity such as catching and throwing are tremendously impaired in the cold, because the cold effectively anesthetizes sensory receptors in the hands. Exposed flesh, particularly on the face, is susceptible to frostbite, which can become a serious medical problem.

Clothing is an important consideration during physical activity in the cold. The insulation value of the clothing must be balanced with the increased metabolic heat production of exercise. If too much clothing is worn, the individual risks becoming a "tropical person" in a cold environment. There have been some instances of heat stroke in overclothed persons exercising in extremely cold climates.

Clothes are valuable in the cold because they increase the body's insulation. Clothing entraps warm air next to the skin and decreases heat loss by conduction. The best clothing for exercise in the cold allows for the evaporation of sweat while providing added protection from the cold. Clothing should be worn in layers so that it may be removed as the metabolic heat production increases during exercise. There has been tremendous progress made by clothing manufacturers in recent years in developing lightweight clothing that provides sufficient insulation and freedom of movement during exercise.

Exercise can partially or totally replace the heat production of shivering during exposure to a cold environment. The peripheral blood vessels dilate during physical activity, which effectively decreases the body's insulation to the cold. Heat production from exercise and shivering must be adequate to maintain heat balance, or hypothermia will result.

Maximal O_2 uptake and the O_2 cost of submaximal exercise are unaffected in the cold. However, the O_2 cost of submaximal exercise by subjects in wet clothing exercising in the wind is about 15 to 20% higher than the same exercise practiced in a comfortable environment. The increased O_2 cost is due to shivering.

It is theoretically possible for cold exposure to impair exercise capacity, particularly in short-term, high-intensity exercise. Muscle functions best at a temperature slightly over 40°C. In a start-and-stop sport such as downhill skiing, it's very possible for muscle temperature to drop substantially during the long chairlift rides between ski runs. In addition to impaired function, there may be an increased risk of injury caused by enhanced muscle viscosity or shivering due to cold exposure.

Swimming in cold water can cause marked deterioration in exercise capacity and $\dot{V}_{O_2peak}$ (in this case, $\dot{V}_{O_2peak}$ is the maximal O_2 consumption measured during swimming) (Figure 22-7), because heat conductance is about 25 times

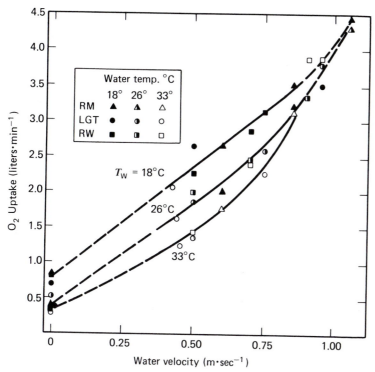

FIGURE 22-7

The effects of swimming speed and water temperature (T_w) on O_2 consumption for three different swimmers.

(SOURCE: Modified from E. Nadel, 1980.)

greater in water than in air. Body fat percentage is an important factor in determining heat loss in cold water. Greater amounts of body fat increase insulation, which slows the transfer of heat to the water. Long-distance channel swimmers typically have high fat percentages, which slows the rate of heat loss during their prolonged swims. It is unclear if this characteristic represents an adaptation to the cold or simply a conscious effort on the part of the swimmer to increase insulation.

Acclimatization and Habituation to Cold

People exposed to environmental stresses, such as heat, cold, and altitude, usually make adjustments to improve their comfort. Adjustments to these environments include acclimatization and habituation. *Acclimatization* is defined as physiological compensation to environmental stress occurring over a period of time. *Habituation* is the lessening of the sensation associated with a particular environmental stressor. Simply stated, in acclimatization there are definite physical alterations that improve physiological function. In habituation, the person learns to live with the stressor.

It is difficult to demonstrate acclimatization even in persons chronically exposed to cold. Except for primitive societies, people do not commonly get the opportunity for chronic exposure to cold. There are three basic tests of acclimatization to cold in humans. The first is the threshold skin temperature that results in shivering. Several studies indicate that shivering occurs later in subjects exposed to several weeks of cold temperatures. Cold-acclimatized individuals maintain heat production with less shivering by increasing nonshivering thermogenesis. They increase the secretion of norepinephrine, which results in uncoupled oxidative phosphorylation—heat is released without the production of ATP. Nonshivering thermogenesis is thought to be most prolific in brown adipose tissue (Chapter 25).

The second test of acclimatization involves measuring the temperatures of the hands and feet. In the unacclimatized person, hand and foot temperatures drop progressively with time during cold exposure. However, the acclimatized person is able to maintain almost normal temperature. Acclimatization results in improved intermittent peripheral vasodilation to make the hands and feet more comfortable. Habituation also seems to play a part. Some individuals seem to lose or learn to tolerate the pain sensations associated with cold feet and hands, even though there is little improvement in circulation or temperature.

The third test is the ability to sleep in the cold. Unacclimatized persons will shiver so much that it is impossible to sleep. Some studies show that it is possible to acclimatize enough to sleep, but these findings have not been consistently replicated. The ability to sleep in the cold seems to depend on the extent of nonshivering thermogenesis induced by increased secretion of norepinephrine. Some primitive peoples, such as the aborigines of Australia, are exceptions to this. They are capable of sleeping in the cold with little or no clothing without an increase in metabolism. These individuals almost resemble poikilotherms in this regard.

Physical conditioning seems to be beneficial in acclimatization to the cold, just as it is in the heat and at altitude. Physical conditioning results in a higher body temperature in sleep tests in the cold. Consequently, the individual can sleep better and is more comfortable.

Cold Injury

Hypothermia As noted in Figure 22-1, the hypothalamus ceases to control body temperature at extremely low core temperatures. Hypothermia depresses the central nervous system, which results in an inability to shiver, sleepiness, and eventually, coma. The lower temperature also results in a lower cellular metabolic rate, which further decreases temperature.

Frostbite   Frostbite is frozen tissue and typically occurs to exposed body parts such as the earlobes, fingers, and toes. It can cause permanent circulatory damage, and sometimes the frostbitten part is lost because of gangrene.

EXERCISE IN THE HEAT

A naked man in still air can maintain a constant body temperature at an ambient temperature of 54 to 60°C. However, exercise in the heat sets the stage for a positive thermal balance. The rate of exercise is the most important factor precipitating this increase. Core temperature increases proportionally with increasing intensities of exercise. While there is considerable variability in core temperature at any absolute work load, there is very little variation when the load is expressed as percentage of maximum capacity (Figure 22-8).

The extent of the effect of environmental temperature on exercise capacity depends on the body's ability to dissipate heat and maintain blood flow to the active muscles. During exercise in the heat, the combined circulatory demands of muscle and skin can effectively impair O_2 transport capacity. An added problem is that the general vasodilation caused by the inhibition of the vasomotor center results in an increase of blood in the venous capacitance vessels. Additionally, there is a decrease in plasma volume during exercise that becomes increasingly acute as the intensity of the effort increases. The decrease in plasma volume can be augmented by the loss of body fluids through sweating. This can become a particularly acute problem during dehydration. During exercise in the heat, there may not be enough blood to go around. There can be a decrease in central blood volume with a resulting decrease in cardiac filling pressure (preload).

Unacclimatized humans have several physiological mechanisms for dealing with heat stress during exercise. Sweating is the primary means of heat dissipation under these circumstances. During submaximal exercise there is an increase in heart rate. However, this mechanism becomes less effective at higher intensities of exercise because of the approach of maximum heart rate. At maximal levels of exercise, there is a peripheral vasoconstriction that overrides the vasodilation originally elicited in response to heat stress. This helps maintain blood pressure and cardiac output. Unfortunately, this response has a negative effect on heat transfer. In this instance, circulatory regulation takes precedence over temperature regulation. These circumstances help to emphasize the danger of heat injury during exercise in hot climates.

Maximal O_2 consumption, as measured during a laboratory treadmill test, is not impaired in the heat unless the subject is experiencing thermal imbalance when the test begins. Athletes rarely have the luxury of going from the comforts of normal room temperature into the heat of an environmental chamber

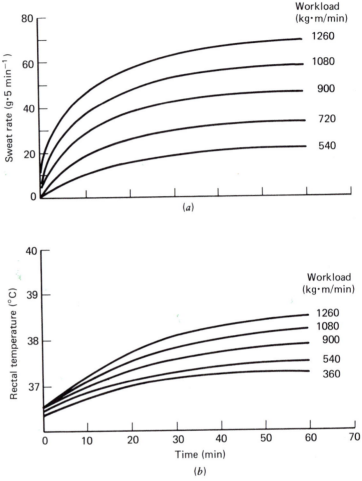

FIGURE 22-8
The effects of work load during prolonged exercise on sweat rate (a) and rectal temperature (b).

(SOURCE: From B. Neilson, Die regulation der korpertemperature ber muskelarbeit. Skand. Arch. Physiol. 79: 193–230, 1938.)

for athletic contests. They usually have to sleep, eat, and train in the adverse environment before they have to perform. Dehydration, lack of sleep, and anxiety can combine to cause a degree of physiological and psychological stress. Most studies show that $\dot{V}_{O_2max}$ is decreased by 6 to 8% in preheated subjects.

Acclimatization to Heat
During the first week of heat exposure, the body makes several adjustments resulting in lower heart rate, core temperature, and skin temperature at rest and

during submaximal exercise. Additionally, there is an increased stability in blood pressure during prolonged exercise. The primary physiological adjustments of acclimatization to heat include increased peripheral conductance, increased plasma volume, increased sweating capacity, a fall in the threshold of skin temperature for the onset of sweating, and a better distribution of sweat over the skin. Acclimatization to heat is not complete unless the exposure is accompanied by exercise training.

Blood flow to the skin decreases with acclimatization. This adjustment helps to restore central blood volume, which is vitally important for maintaining stroke volume and muscle blood flow during exercise. However, core temperature is lower during exercise in acclimatized humans. The decrease in skin blood flow is accompanied by a large increase in sweating and evaporative cooling capacity, resulting in greater peripheral conductance of heat.

Acclimatization to heat induces a 12% increase in plasma volume, provided that heat exposure is accompanied by exercise training. This increase is precipitated mainly by an increase in plasma proteins. For every increase of 1 g of plasma protein, 15g of water are added to the plasma. The increased plasma volume helps to ensure the maintenance of stroke volume, central blood volume, and sweating capacity. Additionally, it enables the body to hold more heat.

Accompanying the increase in plasma volume is an almost threefold increase in sweating capacity from about 1.5 to 4 liters $\cdot$ hr^{-1}. This is accompanied by a more complete and even distribution of sweating, which is an advantage in heat accompanied by high humidity. Sweat losses of sodium chloride decrease because of increased secretion of aldosterone.

The fall in sweating threshold is important for keeping core temperature under control during the early stages of exercise. As discussed, increased temperature tends to cause further increases in temperature because of the Q_{10} effect. The early onset of sweating serves to negate this in part. This phenomenon probably reflects a lower setting of the hypothalamic setpoint in heat acclimitization.

Exercise training is essential for acclimatization to the heat. However, training at normal temperature by itself will not provide a full measure of heat adaptation. An individual from a temperate environment who must compete in the heat can become acclimatized for the effort by exercising (endurance training) in a hot room or while wearing added clothing.

THERMAL DISTRESS

Thermal distress and heat injury are becoming increasingly common with the popularity of distance running and competitive sports for the weekend athlete.

Fortunately, many of the severe effects of heat stress in athletics can be avoided if the necessary precautions are taken. Thermal distress includes dehydration (loss of body fluid), heat cramps (involuntary cramping of skeletal muscle), heat exhaustion (hypotension and weakness caused by an inability of the circulation to compensate for vasodilation of skin blood vessels), and heat stroke (the failure of the temperature-regulatory function of the hypothalamus). The distinction between heat stroke and heat exhaustion is academic as they represent two points along a continuum. The American College of Sports Medicine has issued position statements regarding dehydration in wrestling and the prevention of heat illness during distance running. These appear in Appendix V, VI and XII.

Dehydration

Dehydration is the loss of body fluid that amounts to 1% or more of body weight. Dehydration can decrease sweat rate, plasma volume, cardiac output, maximal O_2 uptake, work capacity, muscle strength, and liver glycogen content. Although this is a common condition during exercise in the heat, it can occur even in thermally neutral environments (Figure 22-9).

A water deficit of 700 ml (approximately 1% of body weight) will cause thirst. At a fluid deficit of 5% of body weight, the person feels discomfort and alternating states of lethargy and nervousness. Irritability, fatigue, and loss of appetite are also characteristic of this level of dehydration. Five percent dehydration is extremely common in athletics such as football, tennis, and distance running. Dehydration levels greater than 7% are extremely dangerous. At these levels, salivating and swallowing food become difficult. At fluid losses above 10%, the ability to walk is impaired and is accompanied by incoordination and spasticity. As 15% dehydration approaches, delirium and shriveled skin are experienced along with decreased urine volume, loss of the ability to swallow food, and difficulty swallowing water. Above 20% dehydration the skin bleeds and cracks. This is the upper limit of tolerance of dehydration, after which death ensues.

Osmoreceptors in the hypothalamus stimulate the drive to drink fluids. Unfortunately, thirst does not keep up with the fluid requirements, so it is very easy to experience fluid deficits of 2 to 4% of body weight. It is very important that dehydrating athletes have regular fluid breaks rather than relying only on their thirst for fluid replacement.

The inadequacy of the thirst mechanism can be compounded by the type of fluid replacement. Fluids with a high carbohydrate content (>2.5 g $\cdot$ dl^{-1}) have been shown to delay the rate of gastric emptying. Some commercially available fluid replacements are actually counterproductive; for quick rehydration their high sugar content delays the replenishment of water while satiating thirst. Cold water is one of the best fluid replacement, although some commercial fluids have

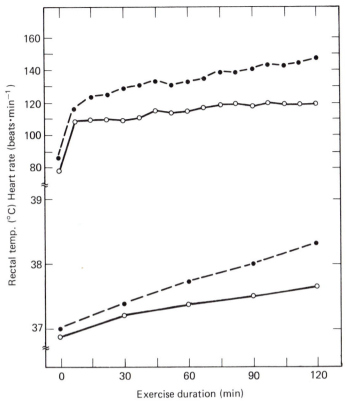

FIGURE 22-9
Effects of dehydration on heart rate and rectal temperature during
prolonged cycling. ○, Hydrated; ●, dehydrated.

(**SOURCE**: Modified from D. Costill, 1977.)

almost equal rates of gastric emptying. Some athletes prefer the commercial
fluids because they are more palatable. As long as a suitable fluid is selected,
the wishes of the athletes should be honored. The important thing is to get
them to drink enough fluid. A common practice among distance runners and
other athletes is to dilute high-glucose commercial preparations, which satisfies
the requirement for good taste and replenishes fluids as well.

Heat Cramps

Heat cramps are characterized by involuntary cramping and spasm in muscle
groups used during exercise. They stem from an alteration in the relationship
of sodium and potassium at the muscle membrane (Chapter 17) and result from
dehydration and salt depletion. They typically occur in people who have exercised and sweated heavily. Often these individuals are conditioned and heat
acclimatized.

The treatment of heat cramps is somewhat controversial. The classical treatment includes oral or intravenous salt replacement and rehydration. Unfortunately, most of the emphasis has been placed on the salt replacement, which can actually compound the problem. Disproportionate salt and water intake will lead to intracellular dehydration. The kidneys and sweat glands have a remarkable ability to conserve salt, particularly in the acclimatized individual. Present evidence suggests that in most instances providing copious quantities of fluid during and after exercise is sufficient for treating and preventing heat cramps. Most people consume more than enough salt in their diets to prevent electrolyte depletion and heat cramps, even in extremely hot, humid climates.

Heat Exhaustion

Heat exhaustion is characterized by a rapid, weak pulse, hypotension, faintness, profuse sweating, and psychological disorientation (Figure 22-10). It results from an acute volume loss and the inability of the circulation to compensate for the concurrent vasodilation in the skin and active skeletal muscles. Although core temperature may be elevated somewhat (usually <39.5°C), it does not reach the extremely high level usually seen in heat stroke (>41°C).

The treatment for heat exhaustion includes having the person lie down in a cool area and administering fluids. Intravenous fluid administration may be appropriate in some instances. The athlete should not participate in any further activity for the rest of the day and should be encouraged to drink plenty of fluids for the next 24 hr.

Heat Stroke

Heat stroke is the failure of the hypothalamic temperature-regulatory center and represents a major medical emergency (Figure 22-10). It is principally caused by a failure of the sudomotor center, (sweating center in the hypothalamus) which results in an explosive rise in body temperature due to the lack of evaporative cooling. It is characterized by a high core temperature (>41°C), hot, dry skin, and extreme confusion or unconsciousness. Complications of heat stroke include coma, central nervous system depression, delirium, renal dysfunction, myoglobinuria, impaired blood coagulation, liver damage, vomiting, and diarrhea. People who are particularly susceptible to heat stroke include those who are obese, unfit, dehydrated, unacclimatized to heat, ill, have a history of heat stroke, and those who are very young or very old. In exercise-related heat stroke, the individual may still be sweating and the core temperature may be below 41°C. The cardiovascular effects are variable, with some experiencing hypotension and others experiencing a full bounding pulse and high blood pressure.

A heat stroke victim should be packed in ice and transported to a hospital as soon as possible. Hospital treatment includes submersion in an ice bath ac-

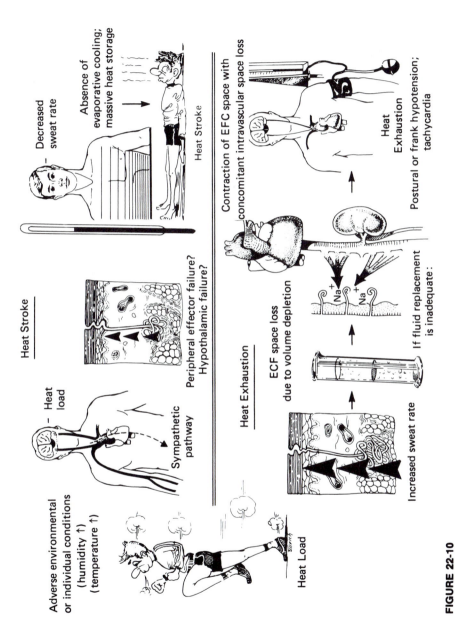

FIGURE 22-10

Pathophysiology of heat stroke and heat exhaustion in the athlete. ECF, Extracellular fluid.

(SOURCE: From T.F. O'Donnell, Management of heat stress injuries in athletes. Orthop. Clin. North Am. 11: 841–855, 1980.)

companied by massage to counteract peripheral cutaneous vasoconstriction. Rectal temperature is monitored to guard against rebound hypothermia. Fluids are administered to counteract hypotension. More drastic measures are sometimes required in the event of continued shock or renal failure.

Preventing Thermal Distress

The problems of thermal distress can be minimized by following a few simple principles:

- Ensure that athletes are in good physical condition. There should be a gradual increase in intensity and duration of training until the athletes are fully acclimatized.
- Schedule practice sessions and games during the cooler times of the day. Sutton has suggested that road race promoters post heat stress warning flags for those who are particularly susceptible to thermal injury.
- Modify or cancel exercise sessions when the wet bulb globe temperature is 25.5°C or greater. Wet bulb temperature is used to determine humidity, and globe temperature is an indication of radiant heat.
- Plan for regular water breaks. Athletes should be encouraged to drink even when they are not thirsty.
- Supply a drink that is cold (8 to 13°C), low in sugar (≤ 2.5 g $\cdot$ dl^{-1}), with little or no electrolytes.
- Athletes should be encouraged to ''tank up'' before practice or games by drinking 400 to 600 ml of water 30 min before activity.
- Splashing water on the skin will facilitate cooling.
- The exercise intensity during competition or practice should reflect the individual's fitness. This point is particularly important for occasional ''fun runners.''
- Athletes should train at the time of day that they plan to compete. This will ensure that the individual is acclimatized.
- Fluid replacement should be particularly encouraged during the early stages of practice and competition. As exercise progresses, splanchic blood flow tends to decrease, which diminishes water absorption from the gut.
- Athletes should be weighed every day before practice. Any athlete showing a decrease of 3% or more in weight should not be allowed to participate until he or she is rehydrated. People who tend to lose a lot of weight in the heat should be identified and closely monitored.
- Salt tablets are prohibited. However, athletes should be encouraged to consume ample amounts of salt at mealtime.

Summary

People can tolerate extremely hot and cold climates because of a well-developed ability to control body temperature. Body temperature is regulated by controlling the rate of heat production and heat loss. Mechanisms of heat production include basal metabolic rate, shivering, exercise, thermogenic hormone secretion, and the Q_{10} effect. The body loses heat by radiation, conduction, convection, evaporation, and excretion. The hypothalamus is the main center for temperature regulation.

Clothing is the most important limiting factor when exercising in the cold. However, swimming in the cold or exercising with wet clothing can result in greatly diminished performance because of an accelerated heat loss. Acclimatization to cold is difficult to demonstrate but is known to occur.

The ability to exercise in high ambient temperatures depends on the ability to dissipate heat and maintain blood flow to active muscles. Sweating is the primary means of cooling in the heat. An increased sweat rate is the most important adaptation that occurs with heat acclimatization. Exercise training is necessary to maximize this process.

Thermal injuries are a serious problem in both competitive and recreational sports. Thermal injuries include heat cramps, heat exhaustion, and heat stroke. These injuries are typically preceded by dehydration. Preventive measures include adequate physical conditioning, exercising during cooler periods of the day, planning regular water breaks, and limiting workout sessions until adequate acclimatization is achieved.

Selected Readings

Adams, W.C. Influence of exercise mode and selected ambient conditions on skin temperature. Ann. N.Y. Acad. Sci. 301: 110–127, 1977.

Baum, E., K. Bruck, and H.P. Schwennicke. Adaptive modifications in the thermoregulatory system of long-distance runners. J. Appl. Physiol. 40: 404–410, 1976.

Branday, J.M., P.R. Fletcher, and R.A. Carpenter. Mortality from thermal injuries at the university hospital of the West Indies (1972–1976). West Indian Med. J. 30: 185–187, 1981.

Brengelmann, G.L. Circulatory adjustments to exercise and heat stress. Ann. Rev. Physiol. 45:191–212, 1983.

Carlson, L.D., H.L. Burns, T.H. Holmes, and P.P. Webb. Adaptive changes during exposure to cold. J. Appl. Physiol. 5: 672–676, 1953.

Carlson, L.D. and A.C.L. Hsieh. Cold. In: Slonim, N.B. (ed.), Environmental Physiology. St. Louis: C. V. Mosby, 1974, p. 67–84.

Coetzee, J.H., J. Hattingh, and D. Mitchell. Effects of heat and cold exposure and exercise on the interstitial fluid proteins of the rat. Comp. Biochem. Physiol. (A) 72: 437–440, 1982.

Cohen, J.S. and C.V. Gisolfi. Effects of interval training on work-heat tolerance of young women. Med. Sci. Sports Exerc. 14: 46–52, 1982.

Convertino, V., J. Greenleaf, and E. Bernauer. Role of thermal and exercise factors in the mechanism of hypervolemia. J. Appl. Physiol. 48: 657–664, 1980.

Costill, D. The New Runner's Diet. Mountain View, Cal.: World Publications, 1977.

Coyle, E., D. Costill, W. Fink, and D. Hoopes. Gastric emptying rates for selected athletic drinks. Res. 49: 119–124, 1978.

Danzl, D. Deep-to-the-core rewarming. Emergency Med. 14: 102–109, 1981.

Erikson, H., J. Krog, K.L. Andersen, and P.F. Scholander. The critical temperature in naked man. Acta Physiol. Scand. 37: 35–39, 1956.

Folk, G. Textbook of Environmental Physiology. Philadelphia: Lea & Febiger, 1974.

Froese, G. and A.C. Burton. Heat losses from the human head. J. Appl. Physiol. 10: 235–241, 1957.

Gisolfi, C. Work-heat tolerance derived from interval training. J. Appl. Physiol. 35: 349–354, 1974.

Gisolfi, C.V. and J. Cohen. Relations between training, heat acclimatization, and heat tolerance in men and women: The controversy revisited. Med. Sci. Sports. 11: 56–59, 1979.

Gisolfi, C. and S. Robinson. Relations between physical training acclimatization and heat tolerance. J. Appl. Physiol. 26: 530–534, 1969.

Golden, F.S.T., I.F.G. Hampton, G.R. Hervey, and A.V. Knibbs. Shivering intensity in humans during immersion in cold water. J. Physiol. (Lond.). 290: 48, 1979.

Greenleaf, J.E., B.L. Castle, and W.K. Ruff. Maximal oxygen uptake, sweating and tolerance to exercise in the heat. Int. J. Biometeorol. 16: 375–387, 1972.

Greenleaf, J. and F. Sargent II. Voluntary dehydration in man. J. Appl. Physiol. 20: 719–724, 1965.

Guyton, A.C. Textbook of Medical Physiology. Philadelphia: W.B. Saunders Co., 1976.

Hanson, P. Heat injuries in runners. Physician Sports Med. 7: 91–96, 1979.

Haymes, E.M., A.L. Dickinson, N. Malville, and R.W. Ross. Effects of wind on the thermal and metabolic responses to exercise in the cold. Med. Sci. Sports Exerc. 14: 41–45, 1982.

Hayward, M.G. and W.R. Keatinge. Progressive symptoms in water: Possible cause of diving accidents. Br. Med. J. 1: 1182, 1979.

Hayward, M.G. and W.R. Keatinge. Roles of subcutaneous fat and thermoregulatory reflexes in determining ability to stabilise body temperature in water. J. Physiol. (Lond.) 320: 229–251, 1981.

Holmer, I. and U. Bergh. Metabolic and thermal response to swimming in water at varying temperatures. J. Appl. Physiol. 37: 702–705, 1974.

Hong, S.K. Patterns of cold adaptation in women divers of Korea (AMA). Fed. Proc. 32: 1614–1622, 1973.

Hong, S. and E.R. Nadel. Thermogenic control during exercise in a cold environment. J. Appl. Physiol. 47: 1084–1089, 1979.

Horvath, S.M. Exercise in a cold environment. Exerc. Sport Sci. Rev. 9: 221–263, 1981.

Keatinge, W.R., M.G. Hayward, and N.K.I. McIver. Hypothermia during saturation diving in the North Sea. Br. Med. J. 1: 291, 1980.

Kobayashi, Y., Y. Ando, N. Oluda, S. Takaba, and K. Ohara. Effects of endurance training on thermoregulation in females. Med. Sci. Sports Exerc. 12: 361–364, 1980.

Kollias, J., L. Bartlett, V. Bergsteinova, J.S. Skinner, E.R. Buskirk, and W.C. Nicholas. Metabolic and thermal responses of women during cooling in water. J. Appl. Physiol. 36: 577–580, 1974.

Morgans, L.F., S.A. Nunneley, and R. Stribley. Influence of ambient and core temperatures on auditory canal temperature. Aviat. Space Environ. Med. 52: 291–293, 1981.

Murphy, R.J. Heat illness and athletics. In: Strauss, R.H. (ed.), Sports Medicine and Physiology. Philadelphia: W.B. Saunders, 1979.

Nadel, E. Circulatory and thermal regulations during exercise. Fed. Proc. 39: 1491–1497, 1980.

Nadel, E., I. Holmer, U. Bergh, P.-O. Åstrand, and J. Stolwijk. Energy exchanges of swimming man. J. Appl. Physiol. 36: 465–471, 1974.

Nilsson, A.L., G.E. Nilsson, and P.A. Oberg. On periodic sweating from the human skin during rest and exercise. Acta Physiol. Scand. 114: 567–571, 1982.

Nunneley, S. Physiological responses of women to thermal stress: A review. Med. Sci. Sports. 10: 250–255, 1978.

O'Donnell, T.F. Management of heat stress injuries in the athlete. Orthop. Clin. North Am. 11: 841–855, 1980.

Oldridge, N.B. and J.D. MacDougall. Cross-country skiing: Precautions for cardiac patients. Physician Sports Med. 9: 64–70, 1981.

Pace, N. Syllabus of Environmental Physiology. Berkeley, Cal.: University of California Press, 1972.

Pirnay, F., R. Deroanne, and J. Petit. Maximal oxygen consumption in a hot environment. J. Appl. Physiol. 28: 642–645, 1970.

Pugh, L. Clothing insulation and accidental hypothermia in youth. Nature. 209: 1281–1286, 1966.

Rennie, D., B. Covino, B. Howell, S. Song, B. Hang, and S. Hang. Physical insulation of Korean diving women. J. Appl. Physiol. 17: 961–966, 1962.

Roberts, M. and C.B. Wenger. Control of skin circulation during exercise and heat stress. Med. Sci. Sports. 11: 36–41, 1979.

Rowell, L. Human cardiovascular adjustments to exercise and thermal stress. Physiol. Rev. 54: 75–159, 1974.

Shibolet, S., M.C. Lancaster, and Y. Danon. Heat stroke: A review. Aviat. Space Environ. Med. 47: 280–301, 1976.

Shvartz, E., V.A. Convertino, L.C. Keil, and R.F. Haines. Orthostatic fluid-electrolyte and endocrine responses in fainters and nonfainters. J. Appl. Physiol. 51: 1404–1410, 1981.

Shvartz, E., Y. Shapiro, H. Brinfeld, and A. Magazanik. Maximal oxygen uptake, heat tolerance, and rectal temperature. Med. Sci. Sports. 10: 256–260, 1978.

Sillau, A.H., L. Aquin, A.J. Lechner, M.V. Bui, and N. Bachero. Increased capillary supply in skeletal muscle of guinea pigs acclimated to cold. Respir. Physiol. 42: 233–245, 1980.

Stromme, S., K.L. Andersen, and R.W. Elsner. Metabolic and thermal responses to muscular exertion in the cold. J. Appl. Physiol. 18: 756–763, 1963.

Sutton, J.R. Heat illness. In: Strauss, R.H. (ed.), Medicine in Sports and Exercise: Nontraumatic Aspects. in press.

Van Someren, R.N.M., S.R.K. Coleshaw, P.J. Mincer, and W.R. Keatinge. Restoration of thermoregulatory response to body cooling by cooling hands and feet. J. Appl. Physiol. 53: 1228–1233, 1982.

Verde, T., R.J. Shephard, P. Corey, and R. Moore. Sweat composition in exercise and in heat. J. Appl. Physiol. 53: 1540–1545, 1982.

Weltman, A. and B. Stamford. Exercising safely in winter. Physician Sports Med. 10: 130, 1981.

Widerman, P.M. and R.D. Hagan. Body weight loss in a wrestler preparing for competition: A case report. Med. Sci. Sports Exerc. 14: 413–418, 1982.

Wyndham, C.H. The physiology of exercise under heat stress. Ann. Rev. Physiol. 35: 193–220, 1973.

23

EXERCISE IN HIGH- AND LOW-PRESSURE ENVIRONMENTS
ALTITUDE AND DIVING

Technological advances and improved transportation have given many people the opportunity to participate in sports in both high- and low-pressure environments. Thousands ski and hike every year at altitudes over 3000 m (9840 ft). With the influx of adventure travel companies, even novice mountaineers can climb some of the peaks in the rarified air of the Andes, Himalayas, Alps, and Rockies.

The invention of scuba (self-contained underwater breathing apparatus) has enabled the common person to explore the ocean's depths. Recreational dives to 30.5 m (100 ft) or more are common; commercial divers working on oil rigs have gone as deep as 457 m (1500 ft) using special diving systems.

High- and low-pressure environments place extraordinary stresses on the human body. In the low pressure of altitude, called *hypobaria,* the body is forced to survive in an atmosphere with less oxygen. Exposure to moderate altitudes above 1524 m (5000 ft) results in decreased maximal O_2 consumption. Prolonged exposure to extremely high altitudes over 6000 m (19,685 ft) leads to progressive deterioration that can eventually cause death unless the person is moved to a lower altitude.

At high pressures (*hyperbaria*), the driving force of gases is increased while their volumes are decreased. Prolonged exposure to depths greater than 10 m (33 ft) leads to an increased retention of nitrogen, which can have disastrous effects if the person is not allowed to decompress. Rapid ascents from the

Mountain climbers must overcome several environmental difficulties including: a thin atmosphere, cold, and blinding sunlight. (R.D. Lamb/ Photo Researchers.)

depths without expiration of air can actually explode the lungs. Additionally, gases that are either essential or benign at sea level can become toxic or narcotic at higher pressures.

This chapter examines the physiology of exercise in hypobaric and hyperbaric environments. Particular emphasis is placed on the effects of these conditions on acute and long-term exercise responses and on acclimatization.

THE GAS LAWS

Gases are composed of individual molecules that can be either compressed or expanded. High pressure compresses the gas so that there are more molecules per unit volume, whereas low pressure allows the gases to expand so there are fewer molecules per unit volume.

As discussed in Chapters 11 to 13, the ability to move O_2 from the ambient environment to the bloodstream depends on ventilation, diffusion, and blood flow. Of these, diffusion is most affected in both the hypobaric and hyperbaric environments. If the pressure is low, as at altitude, the system is stressed because there is less O_2 available. If the pressure is high, as in scuba diving, the driving force of O_2 in the lungs is greater because there is more of it. The rate of diffusion in these environments is profoundly affected by the physics of gases as expressed by the gas laws. For our purposes, the most important of these are the ideal gas law (which includes Boyle's and Charles' laws), Dalton's law of partial pressure, and Henry's law. By necessity, this discussion will be brief, so the reader is referred to basic reference books in physics, chemistry, and biophysics for more detailed descriptions.

The Ideal Gas Law

The ideal gas law states that the volume of a gas is affected by its pressure and temperature. Stated symbolically, $PV = nRT$, where P is the pressure in atmospheres, V is the volume in liters, n is the number of moles of gas, R is a constant, and T is the temperature in degrees Kelvin.

The ideal gas law is based on two more basic laws: Boyle's law and Charles' law. Boyle's law states that at a constant temperature, the volume of a gas is inversely proportional to its pressure (Figure 23-1). Thus the volume of a given mass of gas increases as the altitude increases (decreased barometric pressure) and decreases with increasing depths underwater (increased barometric pressure). Boyle's law is an extremely important consideration in both hypobaric and hyperbaric physiology. Charles' law states that at a constant pressure, the volume of a gas is directly proportional to its temperature.

The Law of Partial Pressures

This law states that in a mixture of gases, the pressure exerted by each gas is the same as it would be if it were alone in the same volume. In other words, each gas exerts an independent pressure, and the sum total of each individual partial pressure equals the total pressure exerted by the combined gases. The percentage of oxygen, nitrogen, and carbon dioxide are the same in the ambient air of high altitude and the compressed air breathed underwater (Figure 23-1). However, the number of molecules in a given volume of each gas in the two environments varies considerably.

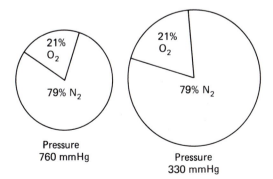

Pressure
760 mmHg

Pressure
330 mmHg

FIGURE 23-1
The volume of a gas varies inversely with pressure, but the relative percentages of oxygen and nitrogen remain the same.

Diffusion from the alveoli to the pulmonary capillary blood largely depends on the movement of gases from a high concentration to a lower one. The partial pressure determines the driving force of the gas moving from one place to another. At altitude, this driving force is diminished; when breathing compressed air underwater, it is increased.

Henry's Law

Henry's law states that the volume of gas dissolved in a liquid varies with the pressure of the gas in which it is at equilibrium. If a gas and a liquid come into contact with each other, the extent that the gas dissolves into the liquid will depend on its pressure. This law is important in both diving and aviation physiology where the pressure can change rapidly. Gases can be rapidly absorbed at higher pressures, then seek to escape at lower pressures. This can lead to either diver's or aviator's bends, depending on the environment.

ALTITUDE

Barometric pressure decreases with increasing altitudes, resulting in less O_2 per volume of air (Figure 23-2). Fortunately, the body has several mechanisms for adjusting to hypoxia. The body's ability to sustain homeostasis is limited, so a deterioration in maximal O_2 consumption begins to manifest itself at about 1524 m (5000 ft). Initially, $\dot{V}_{O_2max}$ decreases about 3% for each increase of 300 m

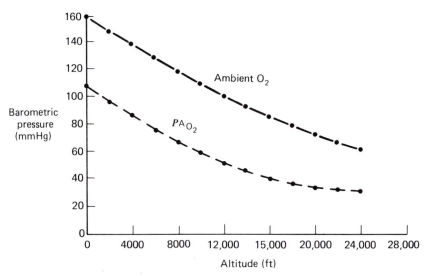

FIGURE 23-2
Ambient and alveolar oxygen partial pressure at altitude.

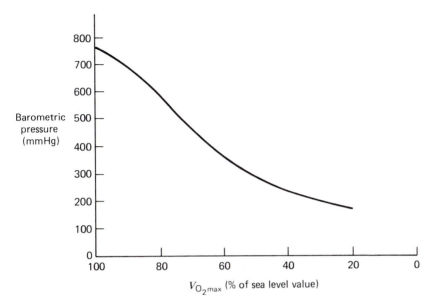

FIGURE 23-3
Maximal O_2 consumption at decreasing barometric pressures.

(1000 ft) of elevation. However, the rate of decrease is more severe at higher altitudes (Figure 23-3). $\dot{V}_{O_2\ peak}$ on Mt. Everest in mountain climbers has been estimated to be as low as 7–10 ml · kg^{-1}.

Limiting Factors at Altitude

Pulmonary diffusion seems to be the major factor limiting performance at altitude. Oxygen and carbon dioxide move across the alveolar membrane by passive diffusion from an area of high partial pressure to one of low partial pressure. At altitude, the partial pressure (and thus the diffusion driving force) is less, which can result in less O_2 transfer from the environment to the blood.

At sea level, blood entering the pulmonary capillaries has a P_{O_2} of about 40 mmHg. At rest, blood takes approximately 0.75 sec to travel the length of a capillary. Oxygen from the alveoli equilibrates with the capillary blood within 0.25 sec, illustrating the tremendous diffusion reserve of the lungs (Figure 23-4). During heavy exercise, blood travels through the capillaries more rapidly (0.25 sec) but still reaches equilibrium with the alveoli. So, diffusion is not a limiting factor of endurance performance in the normal lung at sea level.

In the resting subject at high altitude, relatively large decreases in O_2 partial pressure can be tolerated because of the shape of the oxyhemoglobin dissociation curve. Notice in Figure 23-5 that a P_{O_2} of 50 mmHg (a value found at an altitude of 4300 m), the hemoglobin is about 85% saturated. This explains why

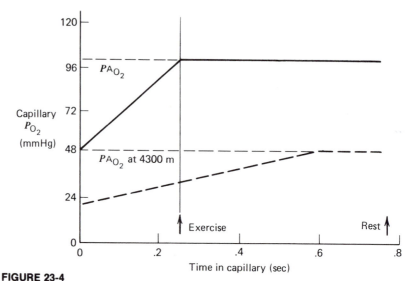

FIGURE 23-4
Resting and exercise pulmonary diffusion at sea level and altitude. Adapted from West, 1974.

some acclimated people are quite comfortable at rest for short periods of time without external O_2 supplies at altitudes as high as 8000 m (26,246 ft.).

Diffusion is severely affected during exercise at high altitudes. While the transit time in the pulmonary capillaries remains at 0.25 sec, the driving force for diffusion is much less than at sea level. As shown in Figure 23-4, the diffusion driving force is 60 mmHg at sea level and only 30 mmHg at 4000 m

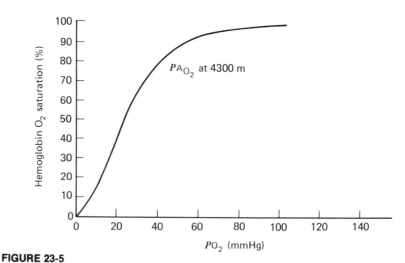

FIGURE 23-5
The oxyhemoglobin dissociation curve. At an altitude of 4300 m the hemoglobin of a resting subject is approximately 85% saturated.

TABLE 23-1
Effects of Acute Exposure to Altitude

Change	Effect
Increased resting and submaximal heart rate	Increased O_2 transport to tissue
Increased resting and submaximal ventilation	Increased alveolar P_{O_2}
	Decreased CO_2 and H^+ in CSF and blood
	Predominance of hypoxic ventilatory drive
	Left shift of oxyhemoglobin dissociation curve
	Acute mountain sickness
Decreased $\dot{V}_{O_2max}$	Decreased exercise capacity
Few acute changes in blood, muscle, or liver	

(13,123 ft). The diffusion driving force is equal to the difference between the alveolar and venous P_{O_2}. This results in a drastic reduction in hemoglobin saturation that produces a decrease in maximal O_2 consumption.

Acute Exposure to Altitude

Slow ascent to about 5486 m (18,000 ft) can be accomplished with few adverse symptoms other than diminished exercise capacity, shortness of breath, and Cheyne-Stokes breathing at night (Cheyne-Stokes is an irregular breathing pattern). If the ascent is rapid, as when going to high altitude in a car or plane, acute mountain sickenss (AMS) will often appear within 2 hr (Table 23-1).

The critical height for the appearance of AMS is about 3353 m (11,000 ft), but it may occur as low as 1828 m (6000 ft) in susceptible individuals or as high as 15,000 ft in the exceptionally tolerant. Symptoms include headache, insomnia, irritability, weakness, vomiting, tachycardia, and disturbance of breathing. It seems to result from hypoxia, which increases water entry into the cerebrospinal fluid (CSF) and brain. This condition usually disappears within a few days but can sometimes evolve into a serious medical emergency. Some people develop high altitude pulmonary edema (HAPE), or cerebral edema, which can be life threatening unless the individual is moved to a lower altitude. High-altitude natives can develop a condition called chronic mountain sickness, characterized by malaise, polycythemia, and pulmonary hypertension.

The individual differences in the onset of and susceptibility to AMS seem to lie in the susceptibility toward fluid retention and the ventilatory response to high altitude. Individuals most prone to AMS and HAPE show decreased sodium retention and increased fluid retention. Those who can ventilate more and maintain a higher P_{O_2} in response to hypoxia seem to have fewer symptoms. Sutton and co-workers were able to demonstrate a relationship between the severity of AMS and the decrement in pulmonary gas exchange. Once experienced, these conditions are more likely to recur (particularly HAPE). These conditions can be prevented to a certain extent by a slow rate of acent and the use of acetazolamide (a diuretic that promotes the loss of bicarbonate and sodium).

Resting respiratory minute volume increases exponentially beginning at an altitude of about 2438 m (8000 ft) (Figure 23-6). Hypoxia, by its effect on the aortic and carotid bodies, is the driving force for this increase in ventilation. Unfortunately, as ventilation increases, P_{CO_2} decreases, which tends to reduce ventilation (Chapter 12). Thus, during acute exposure to altitude there are two regulatory mechanisms working against each other to slow and speed up ventilation at the same time. This process has been implicated in causing AMS. The hypoxic drive to increase ventilation is most predominant in persons with the least susceptibility to mountain sickness.

At sea level, the most important factors in the regulation of ventilation during rest are P_{CO_2} and H^+ because of their effect on the central chemoreceptors located in the medulla (Chapter 12). The central chemoreceptors, which are bathed in CFS, are responsive to H^+. Although the blood–brain barrier, which separates the CSF from blood, is impermeable to H^+, CO_2 moves across it easily. Once in the CSF, the CO_2 releases hydrogen ion as part of the bicarbonate buffer reaction, causing an increase in ventilation. Conversely, if P_{CO_2}

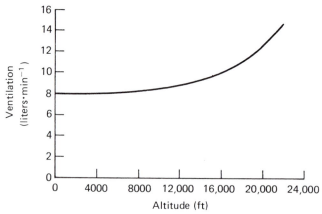

FIGURE 23-6
The effects of acute exposure to altitude on resting ventilation.

decreases, then H^+ levels decrease in the CSF and ventilation slows. Carbon dioxide receptors in the carotid and aortic bodies also affect ventilation, but they are less important than the central chemoreceptors during rest.

Resting and submaximal exercise heart rate increase at altitude. Because heart rate increases as a function of the relative percentage of maximal oxygen consumption, the increase in heart rate makes up for the decreased O_2-carrying capacity of the blood. The O_2 cost of a given work load is identical at sea level and at altitude, but when it is expressed in relation to maximal oxygen consumption, it is higher at altitude. Stated simply, maximal exercise capacity is less at altitude, so a given work load is more difficult.

At higher elevations, maximum heart rate, stroke volume, and cardiac output decrease slightly, particularly in older individuals. This is probably caused by decreased coronary blood flow. As discussed in Chapter 15, autoregulation is the principal mechanism of increasing blood flow in the coronary arteries. The decreased P_{O_2} at high altitude may hamper this process, particularly in the elderly, who sometimes have a decreased capacity for coronary vasodilation due to atherosclerosis.

Acclimatization to Altitude

The body begins to acclimatize to altitude within the first few days of exposure. The initial adaptation includes changes in acid–base balance, which improve the regulation of ventilation and oxygen binding. More long-term adaptations include increases in O_2-carrying capacity, cellular metabolic efficiency, and pulmonary and muscular vascularity (Table 23-2).

During the first few days of exposure to altitude, changes take place in the control of ventilation that enable both the hypoxic and CO_2 drives to control ventilation. The first effect is to decrease the amount of bicarbonate in the CSF and blood. Bicarbonate levels are reduced in the CSF by active transport and reduced in blood through excretion in the kidneys. The second effect is an increased ventilatory sensitivity to CO_2. These changes normalize the pH of the blood and CSF, allowing almost normal respiratory responsiveness to changes in P_{CO_2}. In addition, a higher baseline ventilation is possible in the face of the hypoxia of altitude. This phenomenon is analogous to the resetting of the hypothalamic setpoint during exercise.

The reestablishment of near-normal blood pH also affects the binding of hemoglobin and oxygen. The hyperventilation that accompanies the early stages of altitude exposure has two conflicting effects: a relative increase in blood P_{O_2} and a shift in the oxyhemoglobin dissociation curve to the left (Figure 23-7). The leftward shift of the curve causes hemoglobin to bind more tightly to oxygen, so that a lower P_{O_2} is necessary to release oxygen to the tissues. An added disadvantage is that the lower P_{O_2} decreases the capillary–tissue diffusion gradient, which slows the movement of O_2. The bicarbonate excretion that occurs

TABLE 23-2
Acclimatization to Altitude

Change	Effect
Decreased bicarbonate in CSF and excretion of bicarbonate by kidneys	Increases CO_2–H^+ control of ventilation Shifts oxyhemoglobin dissociation curve to the right
Increased RBC 2,3-DPG	Shifts HbO_2 curve to right
Decreased plasma volume; increased hemoglobin, RBC, and hematocrit	Improved O_2-carrying capacity of blood
Reduction in resting and submaximal heart rate (below increases of early altitude exposure)	Restoration of more normal circulatory homeostasis
Increased pulmonary BP	Improved pulmonary perfusion
Increased pulmonary vascularity	Improved pulmonary perfusion
Increased size and number of mitochondria and in quantity of oxidative enzyme	Improved muscle biochemistry
Increased skeletal muscle vascularity	Improved O_2 transport
Increased tissue myoglobin	Improved cellular O_2 transport

with acclimatization shifts the curve to the right, restoring the normal binding relationship between oxygen and hemoglobin and increasing the diffusion gradient. This improves tissue oxygenation because more O_2 can be delivered at the same or higher O_2 tension.

Also occurring during these first few days is a gradual increase in the concentration of 2,3-diphosphoglycerate (2,3-DPG) in the erythrocytes. This compound decreases the affinity of hemoglobin for oxygen and has the effect of further shifting the oxyhemoglobin dissociation curve to the right without affecting blood pH.

A more long-term mechanism of increasing O_2 supply is to improve the O_2-carrying capacity of blood. There is a gradual increase in hemoglobin that is induced by stimulation of the bone marrow by erythropoietin in response to hypoxia. It is not the number of red blood cells in the blood that controls the

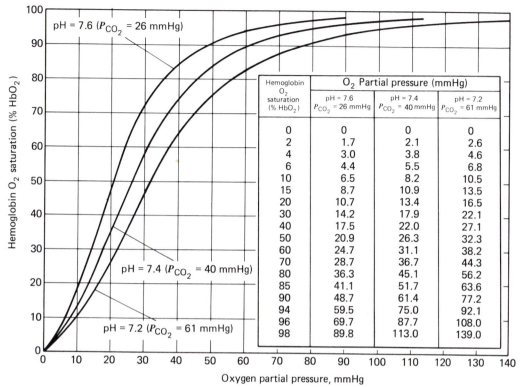

FIGURE 23-7

The effects of pH on the oxyhemoglobin dissociation curve (37°C).

(SOURCE: Modified from N. Pace, 1972)

rate of red blood cell production, but instead the functional ability of the cells to transport O_2 to the tissues. Any condition that causes the quantity of O_2 transported to the tissues to decrease increases the rate of red cell production.

The production of erythropoietin decreases when an individual returns to sea level. Someone who is fully acclimatized to altitude will produce essentially no red blood cells for 24 to 36 hr after returning to sea level. Production will begin again when the demands of O_2 transport require the production of red blood cells.

Hemoglobin can increase from a sea level average of 14 to 16 g · dl^{-1} to as much as 35 g · dl^{-1} at high altitude. Each gram of hemoglobin is capable of combining with approximately 1.33 ml of O_2. Thus, although hemoglobin saturation is impaired at altitude, acclimatization improves the actual O_2-carrying capacity above the initial exposure values. Permanent residents at 4267 to 4572 m (14,000 to 15,000 ft) sometimes have an O_2-carrying capacity of over 30 vol % (i.e., 30 ml O_2 · dl^{-1} of blood).

Hematocrit increases with the increase in red blood cells and the decrease in plasma volume. Although this improves O_2-carrying capacity, it has a mixed

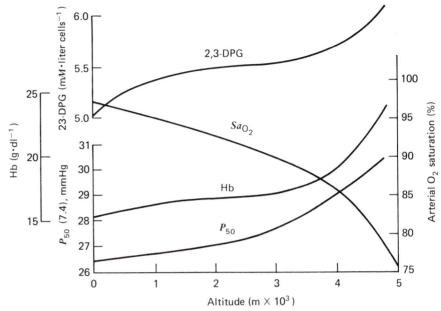

FIGURE 23-8

Long-term adaptation in 2,3-DPG, Sa_{O_2}, Hb, and P_{50}.

(SOURCE: Modified from C. Lenfant, J.D. Torrance, R. Woodson, and C.A. Finch, 1970.)

effect on exercise capacity, particularly prolonged activity. As discussed in Chapter 16, prolonged exercise is accompanied by a gradual increase in cardiovascular load, a phenomenon called cardiovascular drift. This is at least partly caused by a decrease in central blood volume, which impairs venous return of blood to the heart. The combination of an initial decrease in central blood volume and increased viscosity of the blood introduces an added cardiovascular load during prolonged exercise at altitude.

The combined effects of improved O_2 transport and oxygen affinity with altitude acclimatization are shown in Figure 23-8. The higher the altitude at which people live, the higher the values of 2,3-DPG, hemoglobin, and P_{50} (the O_2 tension when the hemoglobin is 50% saturated—the point where the greatest change in hemoglobin saturation occurs). Therefore, acclimatization is altitude specific, which explains why mountain climbers must acclimatize to a particular altitude before moving to a higher elevation.

As O_2-carrying capacity and O_2 affinity improve, heart rate and cardiac output decrease at rest and during submaximal exercise. The initial increase in cardiac output seems to be a stopgap method of improving O_2 transport until the slower acting mechanisms of increased O_2-carrying capacity can engage.

There is a gradual increase in pulmonary diffusing capacity during altitude acclimatization. This is accomplished by increases in pulmonary blood pressure (which makes pulmonary perfusion more uniform), pulmonary vascularity, and pulmonary blood volume, as well as hypertrophy of the right side of the heart. Pulmonary adaptation is a very slow process and is of negligible benefit to the temporary visitor to high altitude. These changes are advantageous to permanent altitude residents and are particularly evident in individuals who have been chronically exposed to altitude since childhood.

Chronic exposure to altitude also induces cellular changes similar to those experienced in endurance training. There are increases in mitochondrial mass, in the amount of oxidative enzymes, and in active tissue vascularity. As would be expected, physically fit individuals acclimatize more readily than the unfit.

Performance at Altitude

Most of the information on physical performance at altitude comes from research associated with the 1968 Olympics held at Mexico City and from various high-altitude mountain-climbing expeditions. These studies have provided valuable information about the helps and hindrances of altitude, and useful techniques for maximizing performance and comfort.

Competitive athletics   Altitude causes marked improvements in events of short duration and high intensity (sprints and throwing events) and deterioration in events of long duration and lower intensity (endurance events). Table 23-3 compares the Olympic performances of the top three athletes in selected sprint events at the Mexico City Olympics with their previous personal bests. High-altitude performances were much better in almost every instance. In the long jump and triple jump (not shown in table), for example, the world records were surpassed by margins of $21\frac{3}{4}$ and $14\frac{1}{4}$ in., respectively. In these types of events, much of the energy cost stems from overcoming air resistance, which is less at altitude.

In throwing events, performances can be helped or hindered by altitude. In events such as the discus or javelin, air mass provides lift to the implements, so performance tends to be hampered. In the shot put and hammer throw, where the implements have minimal aerodynamic characteristics, performances will be improved marginally because of the decreased air resistance, and less weight of a given mass.

At altitude, there is a decline in performance in running events longer than 800. At the Mexico City Games, many athletes who dominated distance running at sea level were soundly defeated by athletes native to high altitude. High-altitude natives placed first and second in the 5000-m run, 3000-m steeple-chase, 10,000-m run, and the marathon.

Athletes who must compete at altitude will benefit from a period of acclimatization of from 1 to 8 weeks. Athletes involved in activities of short dura-

TABLE 23-3

Comparison of Personal Best and Mexico City Olympic Performances of Selected Sprint Athletes

Event, Placement, and Athlete's Name	Mexico City Olympics Time (sec)	Previous Personal Best Time (sec)
100 m—Men		
1st J. Hines (U.S.A.)	9.9	9.9
2nd L. Miller (Jamaica)	10.0	10.0
3rd C. Green (U.S.A.)	10.0	9.9
100 m—Women		
1st W. Tyus (U.S.A.)	11.0	11.1
2nd B. Farrell (U.S.A.)	11.1	11.2
3rd I. Szewinska (Poland)	11.1	11.1
200 m—Men		
1st T. Smith (U.S.A.)	19.8	19.9
2nd P. Norman (Australia)	20.0	20.5
3rd J. Carlos (U.S.A.)	20.0	19.7[a]
200 m—Women		
1st I. Szewinska (Poland)	22.5	22.7
2nd R. Boyle (Australia)	22.7	23.4
3rd J. Lamy (Australia)	22.8	23.1
400 m—Men		
1st L. Evans (U.S.A.)	43.8	44.0
2nd L. James (U.S.A.)	43.9	44.1
3rd R. Freeman (U.S.A.)	44.4	44.6
400 m—Women		
1st G. Besson (France)	52.0	53.8
2nd L. Board (Great Britain)	52.1	52.8
3rd N. Burda (U.S.S.R.)	52.2	53.1
400 m Hurdles—Men		
1st D. Hemery (Great Britain)	48.1	49.6
2nd G. Hennige (Federal Republic of Germany)	49.0	50.0
3rd J. Sherwood (Great Britain)	49.0	50.2
110 m Hurdles—Men		
1st W. Davenport (U.S.A.)	13.3	13.3[a]
2nd E. Hall (U.S.A.)	13.4	13.4[a]
3rd E. Ottoz (Italy)	13.4	13.5

[a] Previous personal best set at altitude

tion, such as sprints, jumps, and throws need only acclimatize long enough to get over the effects of mountain sickness. Although the adjustments in acid–base balance take less than a week, the changes in O_2-carrying capacity can take many months. Athletes in championship form may risk losing their peak condition by too much acclimatization, because they will be unable to train as hard at altitude as at sea level.

Controversy exists over the effects of altitude training on subsequent performance at sea level. Most studies show no improvement in maximal O_2 consumption or maximal work capacity when returning from altitude. In the studies that show an improvement, the subjects may not have been in good condition to start with. At altitude, they improved their exercise capacity, but the improvement was no greater than they would have achieved by training at sea level.

Training intensity and duration are the most important factors in improving exercise performance. Athletes cannot train as hard at high altitude. Even though they can reach the same relative percentage of maximum, their maximum is less. The physiological adaptations to altitude are not necessarily beneficial at sea level. Although the increase in hemoglobin is probably helpful, the decrease in plasma volume and alkaline reserve (bicarbonate) are decided disadvantages. The increased pulmonary ventilation at altitude is worthless at sea level where the O_2 tension is much higher.

Mountain climbing   This activity would be physically exhausting even at the O_2 partial pressure of sea level. The fact that it is sometimes performed at altitudes over 6096 (20,000 ft) is a testimony to the amazing adaptability of human physiology.

Mountain climbers maintain a surprisingly high exercise capacity at altitudes up to 6096 m (20,000 ft). They do this by adjusting the number of climbing hours and the climbing cadence. A man capable of 10 to 14 hr a day at an alpine altitude is capable of only 5 to 7 hr a day at 6096 m. An intermittent climbing cadence allows climbers to exercise at levels for 1 hr that would result in exhaustion in 10 min if done continuously. Mt. Everest climbers typically follow a schedule of a set number of steps followed by rest (e.g., 10 steps, rest, 10 steps, rest). The higher the altitude, the longer the rests and the shorter the work periods. Himalayan porters, who carry loads equal to half their weight, follow this procedure.

Climbing at extreme altitudes (greater than 4500 m) presents severe physiological challenges. Ventilatory equivalent ($\dot{V}_E/\dot{V}_{O_2}$), a measure of ventilatory efficiency, increases steeply (ventilation becomes more costly). Maximum exercise ventilation decreases due to respiratory muscle fatigue. Diffussion capacity is severely limiting due to the extremely low diffusion gradient. Glycogen depletion is often a problem on these climbs due to extremely heavy exercise loads and inadequate caloric intake. $\dot{V}_{O_2max}$ drops to the point where there is little margin for error.

Acclimatization to each range of altitude is important if the climber is to work effectively. However, 5334 m (17,500 ft) seems to be the maximum point where prolonged exposure leads to improved physiological function. Prolonged exposure to higher altitudes can lead to deterioration with symptoms such as loss of weight, increasing lethargy, and weakness. In the first assault on Mt. Makalu in the Himalayas, climbers who were in the field for only 2 months performed better than climbers who had been at the base elevation of 19,000 ft for 6 months.

Lack of oxygen at high altitude can result in blunting emotional response, lack of insight and mental inertia, and the slowing of thought processes. These symptoms are markedly improved by oxygen breathing. Although a climb of Mt. Everest was recently made without oxygen by Austrians Peter Habeler and Reinhold Messner, lack of supplementary oxygen introduces an extraordinary hardship to the climbers of extremely high peaks.

Supplemental oxygen breathing has found its way into other sports as well. It is not unusual to see football players breathing through an oxygen mask during an american football game played at a relatively low altitude such as Denver. This practice is probably of little or no benefit. At that altitude, hemoglobin saturation at rest is essentially the same as at sea level. Unless the players can find some way to breathe pure oxygen while they play, they should think of something else to do with their time on the sidelines.

There are other factors that affect performance at altitude other than low O_2 partial pressure. There is a greater fluid loss from the respiratory tract due to hyperventilation and the low humidity, and from increased sweating due to the higher levels of solar radiation. In addition, ultraviolet (UV) radiation is greater at high altitude, making people more susceptible to sunburn. At sea level, the intensity of the UV rays is less, because a greater percentage are blocked by the denser atmosphere.

HUMANS IN HIGH-PRESSURE ENVIRONMENTS

Our exposure to high-pressure environments is mainly restricted to underwater diving. The exceptions are subaqueous tunnel and caisson workers who are exposed to pressures as high as 4 to 6 atm. Divers' exposures to hyperbaria range from underwater breath-holding to industrial and experimental saturation diving.

The popularity of sports diving has resulted in many accidents because of ignorance of hyperbaric physiology. The high-pressure environments offer some extreme physiological challenges, because gases can become toxic or narcotic,

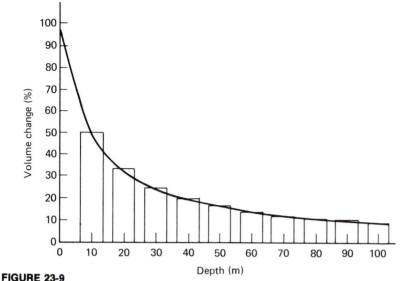

FIGURE 23-9
Percentage change in volume with increasing depth.

and rapid changes in pressure can sometimes be extremely traumatic or fatal. These dangers can be minimized with a consideration of the physics of gases and their physiological effects under increased pressure.

Divers encounter increasing pressure as the depth of the dive increases. Pressure increases by 1 atm (760 mmHg) for every 10 m (33 ft) of depth (Figure 23-9). While the pressure subjects divers to increased air density and partial pressures, they have the added difficulty of exercising in an aqueous environment. There, they must contend with cold water, currents, restrictive equipment, and dangerous marine animals.

Physiological Effects of Exposure to High Pressure

The principal dangers of the hyperbaric environment stem from changes in gas volumes within enclosed spaces and the increased solubility of gases. Medical problems include barotrauma, gas toxicity, and decompression sickness (Table 23-4). Medical contraindications to diving are listed in Table 23-5.

Barotrauma Barotrauma is tissue injury caused by changing pressure. The human body has a limited ability to distend and compress. Trauma results when these limits are exceeded. The volume of a gas decreases or increases as the diver descends or ascends (Boyle's law). On descent, the volume decreases in enclosed spaces such as the middle ear, lungs, sinuses, face mask, or teeth, creating a negative pressure. If the pressure is not allowed to equalize with air from outside the space, tissue injury can result. On ascent, the volume in these areas increase, which can also cause injury if gas is not allowed to escape.

TABLE 23-4
Medical Problems in Scuba Diving

I. Barotrauma
  A. Barotrauma of descent
    Ears
    Sinuses
    Teeth
    Lungs
    Equipment (mask and suit)

  B. Barotrauma of ascent
    Ears
    Sinuses
    Teeth
    Lungs
    Suit

II. Gas toxicity
  A. O_2 toxicity
  B. Inert gas narcosis
    Nitrogen
    Helium
  C. Compressed air impurities
  D. CO_2 intoxication
  E. CO poisoning

III. Decompression sickness
  A. Joints, muscle, bone
  B. Respiratory system
  C. CNS

Ordinarily, descent barotrauma of the ears, teeth, or sinuses is prevented by pain that is relieved by ascending. However, if descent occurs too rapidly, the tympanic membrane and/or round window in the ear can rupture, the sinuses can fill with blood, or a tooth may collapse. Preventive measures, such as slow descent, can minimize the risk of these problems. Patency of the eustachian tube, which is essential for equalizing pressure in the inner ear, is facilitated by increased muscular control or by decongestant medication such as Sudafed. Mask squeeze can be easily remedied by clearing the mask as negative pressure builds up. Individuals with serious nasal obstructions or colds should probably not dive because of the increased difficulty of equalizing pressure.

TABLE 23-5
Contraindications to Diving

Body System	Contraindications
Eyes, ears, nose, and throat	Vision <20/200 Perforated eardrum Sinusitis or severe allergy Residual or prior ear, nose, or throat surgery
Cardiovascular	Recent myocardial infarction Syncope or arrhythmia prone Hypertensive on therapy Cardiac murmurs
Pulmonary	Obstructive airway disease (asthma, COPD, bullae) Prior thoracotomy History of pneumothorax Recurrent pneumonia
Gastroesophageal	Severe gastroesophageal reflux
Musculoskeletal	Chronic disk disease
Central nervous system	Seizure disorder or syncope
Metabolic	Unstable diabetes mellitus
Other	Drug use or heavy smoking Pregnancy Emotional instability or proneness to accidents

SOURCE: From A.P. Spivak. Medical aspects of scuba diving. Compr. Ther. 6: 6–11, 1980.

Pulmonary descent barotrauma (pulmonary squeeze) occurs when the lung volume decreases to the point where negative intrathoracic pressure is created, which increases pulmonary blood volume to the point of pulmonary capillary hemorrhage. This phenomenon can occur in breath-hold diving at extreme depths or in standard diving when the gas supply does not keep pace with the rate of descent. Pulmonary squeeze places a definite limit on the depths that can be achieved in breath-hold diving.

The leading cause of death in scuba diving is pulmonary ascent barotrauma. This occurs when divers hold their breath while ascending, which results in overdistension and rupture of the lungs as the gas volume expands. This can occur at a surprisingly shallow depth: A diver whose lungs were filled to their total capacity at 4 ft would experience lung rupture if he or she failed to expire while ascending to the surface. Problems associated with this serious condition include pulmonary tissue damage, surgical emphysema (gas escaping into interstitial pulmonary tissues), pneumothorax (gas entering the pleural cavity), and air embolism (gas entering the pulmonary veins and systemic circulation).

Gas toxicity   Gases such as oxygen, carbon dioxide, carbon monoxide, nitrogen, and helium can be toxic or narcotic to the diver under certain circumstances.

The sport diver's lifeline is the compressed air carried on his or her back. Contamination of this air supply impairs performance and can be life threatening. Compressed air may contain CO if the air intake of the compressor was contaminated with an impure air source such as automobile exhaust. The danger of CO lies in its great affinity for hemoglobin, which is 200 times greater than that of oxygen. Carbon monoxide poisoning leads to hypoxia, which results in progressive impairment of judgment and psychomotor performance, headache, nausea, weakness, dizziness, impaired vision, syncope, and coma. This problem is compounded if the individual has chronically high levels of CO from cigarette smoking or exposure to air pollution. Other contaminants in the compressed air include dust, oil vapor, and CO_2. Rust may also be a source of contamination by removing O_2 from compressed air.

Oxygen at high pressure is toxic to all life forms, with the degree of toxicity dependent on its concentration and length of exposure. Physical exercise speeds its development. The principal sites of O_2 toxicity are the lungs and CNS. Pulmonary symptoms include substernal distress with soreness in the chest usually accompanied by airway resistance on inspiration, histological changes in the alveoli, pulmonary edema (resembling pneumonia), flushing of the face, and a dry cough that eventually leads to a wet cough. Central nervous system effects include nausea, contraction in the field of vision, convulsions, lack of sphincter control, unconsciousness, and death. Oxygen toxicity can also cause cardiac arrhythmias.

The mechanism of O_2 toxicity is unknown. Hyperbaric oxygen may interfere with CO_2 transport. At high pressure, more O_2 is dissolved in blood, so the hemoglobin does not desaturate and become available for CO_2 transport. This may cause acidosis and increase the P_{O_2} in the brain because of the vasodilation effect of increased CO_2 on cerebral blood vessels. High O_2 concentrations are known to interfere with several metabolic enzymes, which would disrupt cell function. High O_2 concentration also inhibits the action of γ-aminobutyric acid, which is important for neural transmission in the CNS.

Carbon dioxide toxicity is most common in closed-circuit scuba systems and hose-supplied diving helmets, but can occur in conventional scuba. Hypercapnia occurs when there is an inadequate respiratory exchange and has been shown to occur at depths during heavy exercise. Divers often try to suppress their ventilation in an attempt to conserve air. This can lead to a buildup of CO_2 that can have severe consequences. Symptoms of CO_2 toxicity include increased respiratory stimulation, dyspnea (uncomfortable breathing), headache, mental deterioration, violent respiratory distress, unconsciousness, and convulsions. Usually dyspnea will prevent the more serious consequences of CO_2 buildup; however, this may not be the case in an emergency when the diver may not have the opportunity to surface or rest.

Several gases also exert a narcotic or anesthetic effect at high pressure. The anesthetic effect of a gas depends on its partial pressure and its solubility in the body's tissues and fluids. Nitrogen can cause a condition called nitrogen narcosis beginning at about 30 m (100 ft) of depth. Progressive symptoms of nitrogen narcosis include euphoria, impaired performance, weakness, drowsiness, and unconsciousness. It is caused (as is any inert gas anesthesia) by interference of the transfer of electrical potential across neural synapses. This condition limits the use of compressed air to depths of about 65 m (180 ft). Helium, because it is less anesthetic, is usually used as part of the gas mixture on deep dives.

Decompression sickness (the bends)   Decompression sickness is caused by nitrogen bubble formation in the tissues as a result of rapid ascent. It can occur in divers coming from the depths to the surface or in aviators in unpressurized aircraft going to great heights. Symptoms include itching of the skin, fatigue, pain in muscles, joints, and bones, perspiration, nausea, and less frequently respiratory distress, ataxia, vascular obstruction, paralysis, unconsciousness, and death. When the bends affects the lungs, it is sometimes called "the chokes," and when it affects the CNS, it is sometimes called "the staggers." Although the symptoms usually appear about 1 hr after coming to the surface, they can occur either immediately or as long as 12 hr later.

As discussed already, a gas is absorbed into the tissues in proportion to its partial pressure and solubility (Figure 23-10). The uptake is exponential, with some tissues, such as blood, absorbing the N_2 rapidly and other tissues, such as fat, absorbing the N_2 more slowly (Figure 23-11a and b). On decompression (going to the surface or to a shallower depth), the process is reversed—nitrogen leaves the tissue in proportion to the decrease in barometric pressure. It takes longer for N_2 to leave the slower tissues such as fat. If the process of decompression occurs too rapidly, then N_2 returns to the gaseous state and bubbles are formed in the blood and tissues.

Decompression tables have been established to help divers safely reestablish surface N_2 absorption levels (Figure 23-12). In general, decompression time

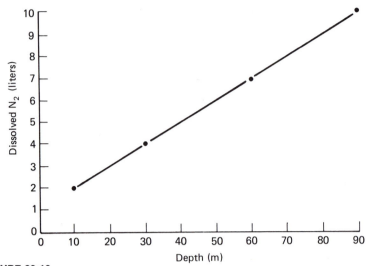

FIGURE 23-10
Nitrogen (STPD) dissolved in tissues at saturation with depth.

increases with the length and depth of the dive. Although these tables provide a margin of safety, factors such as percentage of body fat, age, physical condition, gas mixture, and depth of the dive will affect decompression time. Repetitive diving or flying (airplane, hang glider, etc.) soon after a dive will also affect the accuracy of the table.

Hyperbaric Exercise

Hyperbaric exercise studies are typically conducted in a hyperbaric chamber or underwater. The hyperbaric chamber provides the opportunity to isolate variables such as O_2 partial pressure, temperature, and gas mixtures, whereas the underwater experiments provide a more realistic environment. The hyperbaric chamber sometimes contains wet-pots (small pools about 3 to 4 ft deep) that are used to simulate ocean dives. It is important to understand the differences between these environments when making generalizations about hyperbaric exercise. Although subjects exercising in a hyperbaric chamber may be exposed to high pressure, they are usually not exposed to the added difficulties working in water, where problems of buoyancy, fluid viscosity, temperature, airway obstruction, and awkwardness are more prevalent.

The cycle ergometer is the most common method of administering a reproducible exercise task in the hyperbaric chamber and underwater experimental sessions. This device has the advantage of providing data that are readily comparable with numerous studies done at sea level. However, because cycling is not a common method of underwater transportation, more specific exercise tests have been devised. One of the most useful to date is a drag board designed by Pilmanis and co-workers (Figure 23-13). This device introduces a reproducible

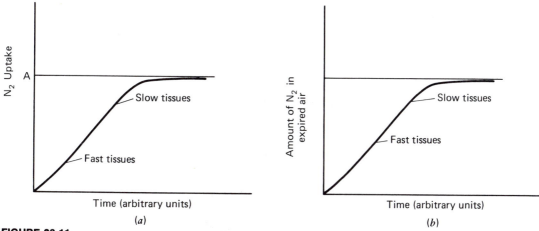

FIGURE 23-11

The uptake and elimination of nitrogen follows a two component curve. The fast tissues, such as blood and extracellular fluid, represent the first component, and the slow tissues, such as fat, represents the second component. N_2 uptake and elimination is expressed in the following equation: $Y = A(1 - e^{-kt})$ where k = rate constant, t = time, A = equilibrium, and $Y = N_2$ uptake or elimination.

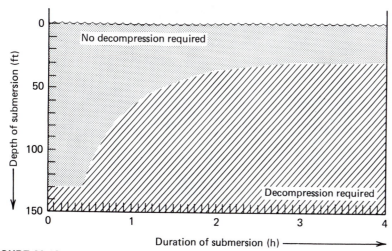

FIGURE 23-12

Decompression requirements in diving.

(**SOURCE**: Modified from G.J. Duffner, Ciba Clinical Symposia 10:99, 1971.)

resistance to the swimming diver. A standard exercise protocol can be administered by regulating the swimming speed and the resistance. More primitive techniques involve measurements while swimming at various speeds.

Biological measurements are exceedingly difficult in the hyperbaric environment. Measurements in hyperbaric chambers require expensive equipment and facilities and are technically exacting. Open-water measurements are complex

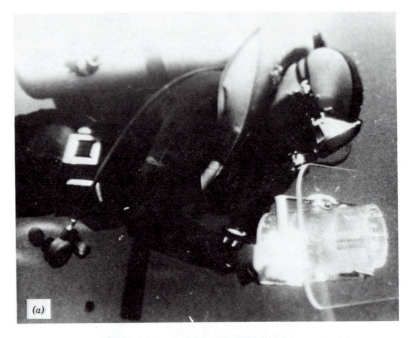

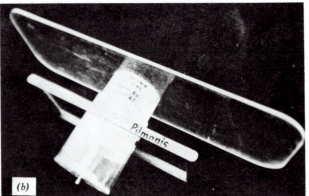

FIGURE 23-13
A diving ergometer.

(J. Dwyer.)

because of the restraints imposed by the aqueous environment. Early open-water studies used pulse rate to study cardiovascular function and trapped air bubbles to study diving energetics. More recently, Dwyer introduced a system allowing accurate measurement of O_2 consumption at different depths and of exercise intensities during the same dive (Figure 23-14). Additionally, he developed a recording system that allows simultaneous measurement of the electrocardiogram, ventilation, breathing frequency, and tidal volume during an open-water dive.

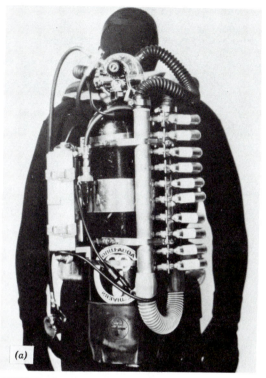
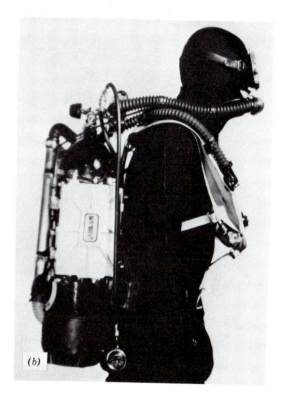

FIGURE 23-14
A device for monitoring $\dot{V}_{O_2}$ in a free-water diver.

(J. Dwyer.)

Divers, are exposed to a number of factors that adversely affects their ability to exercise underwater. These include increased air density, cold, decreased efficiency, CO_2 retention, and inert gas narcosis.

Ventilation may be a limiting factor of physical performance in diving. Maximal voluntary ventilation decreases with depth, which results in a progressively smaller difference between exercise ventilation and maximum voluntary ventilation during heavy exercise (Figure 23-15). The higher density of air increases the flow resistance in the scuba equipment and airways, resulting in hypoventilation. This leads to increased retention of CO_2, increased work of breathing, and dyspnea. The ability to increase expiratory flow rate is limited. After maximum flow rate is reached, further effort only results in partial airway collapse. Maximum expiratory flow rate seems to be independent of effort at depth (Figure 23-16).

Although some studies have shown that the dyspnea accompanying heavy physical activity does not always decrease exercise capacity, it could be extremely dangerous during heavy exercise in an emergency situation. Heavy work done in the presence of CO_2 retention increases the risk of CO_2 intoxi-

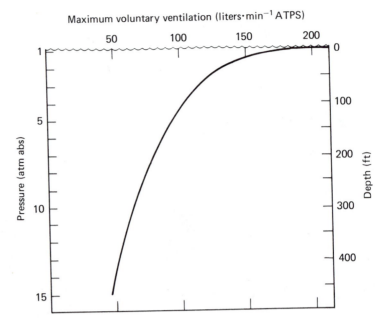

FIGURE 23-15

Maximal voluntary ventilation at depth. ATPS, ambient temperature and pressure saturated.

(SOURCE: Modified from J.N. Miller, O.D. Wangensteen, and E. Lanphier, 1971.)

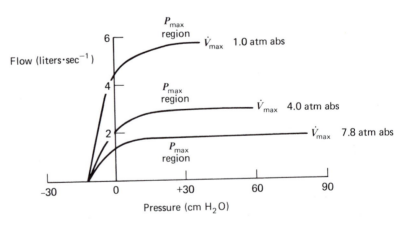

FIGURE 23-16

Respiratory effort and flow rate at depth. P_{max} is the pressure at which flow first becomes effort independent, and $\dot{V}_{max}$ is the maximum effort-independent flow.

(SOURCE: Modified from J.N. Miller, O.D. Wangensteen, and E. Lanphier, 1971.)

cation, which is compounded in the presence of N_2 narcosis. At depths greater than 180 ft, exercise is made possible by using helium–oxygen mixtures, which have a lower density and anaesthetic quality.

Most studies show that O_2 consumption increases in submaximal work with increasing depth (Figure 23-17). Depth affects $\dot{V}_{O_2}$ because of the increased energy cost of breathing, the maintenance of body temperature in colder water, and movement in higher hydrostatic pressures. Water becomes cooler with depth. Because wet suit material is compressed with greater pressure, thus decreasing insulation, O_2 consumption may be increased to maintain body temperature. The increased hydrostatic pressure may hamper mobility by its effects on wet suit compression and by effectively increasing the viscosity of the surrounding water.

Experienced divers can work at as much as 91% of their land-measured maximal O_2 consumption. However, the effective work that a diver can perform is much less because of the greatly reduced efficiency. Probably the most important factors dictating a diver's maximum exercise capacity are the ability to tolerate high levels of CO_2 and the percentage of maximal O_2 consumption that can be attained before reaching the critical P_{CO_2}.

Factors such as the diver's swimming angle and the drag produced by the scuba equipment can make a tremendous difference in the energy cost of swimming underwater. For example, swimming at a moderate speed of $30 \ m \cdot min^{-1}$ in a partial feet-down position increases the O_2 cost by 30%. This can easily occur by improper adjustment of the buoyancy compensator when changing

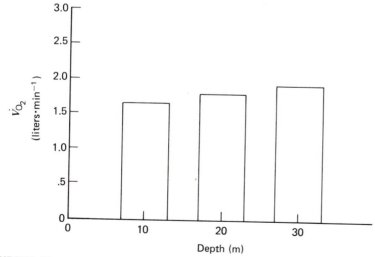

FIGURE 23-17
The effects of depth on oxygen uptake during a swim at $30 \ m \cdot min^{-1}$.
(SOURCE: Data from J. Dwyer, 1975.)

depths. Individual differences in swimming efficiency can also affect energy costs.

The land-measured relationship between O_2 consumption and heart rate cannot be used in diving, and it could be potentially dangerous for the sports diver to do so. Heart rate tends to be reduced as water temperature and pressure decrease, a phenomenon called diving bradycardia. Additionally, individual differences in swimming efficiency will affect this relationship. It appears that heart rate can only be used to estimate energy cost in diving when the heart rate–$\dot{V}_{O_2}$ relationship is known for a particular diver, at a particular depth.

Breath-Hold Diving

Breath-hold diving was the earliest human exposure to the hyperbaric environment. It is of interest to people involved in athletics and exercise physiology because of the popularity of snorkeling, the many associated deaths that occur each year, and the relatively recent attempts to set breath-hold diving performance records.

The danger of breath-hold diving occurs if the dive is preceded by vigorous hyperventilation. This results in a decrease in CO_2, which depresses the urge to breathe. The diver may faint for lack of O_2 before the buildup of CO_2 is again high enough to force him or her to the surface to take a breath. Hyperventilation has little effect on hemoglobin saturation. It serves only to suppress the CO_2 ventilatory drive mechanism. Hyperventilation followed by exercise during the breath-hold dive may compound the dangers because of the more rapid rate of O_2 consumption.

An added danger during breath-hold diving is an increased incidence of cardiac arrhythmias, which increases if the water temperature is low. This is related, in part, to the previously mentioned diver's bradycardia. The effect of cardiac arrhythmias on the death rate in breath-hold diving has not been determined.

The depth limits of the breath-hold diver are determined by the ratio of total lung capacity to residual lung volume. As breath-hold divers descend, their lung volume decreases until residual lung volume is reached. After that point, negative pressure is created in the thoracic cavity, which draws blood into the intrathoracic circulation until the capillaries rupture. Champion breath-hold divers, such as Robert Croft and Jacques Mayol, have extremely large total lung capacities with low residual lung volumes, which partially explains their exceptional abilities.

Summary

Increasing numbers of people are exercising in low- and high-pressure environments, which places increased stress on physiological systems. The decreased O_2 tension of altitude hampers physical working capacity; the increased pressure under the water decreases efficiency.

Maximal O_2 consumption begins to deteriorate at an altitude of approximately 5000 ft, with pulmonary diffusion being the major limiting factor. As the partial pressure of O_2 decreases with altitude, the driving force of O_2 from the environment to the pulmonary circulation is impaired.

Acclimatization to altitude begins within the first few days of exposure. Early exposure to altitude diminishes the role of CO_2 in controlling ventilation. The early effects of acclimatization are to restore the CO_2 ventilatory control mechanisms by reducing bicarbonate levels in the blood and CSF, and by increasing the ventilatory sensitivity to CO_2. Bicarbonate excretion and an increased concentration of 2,3-DPG shift the oxyhemoglobin dissociation curve to the right, which decreases the affinity of hemoglobin for oxygen. Long-term changes include increased hemoglobin, hematocrit, pulmonary diffusion capacity, and skeletal muscle vascularity.

Sport diving constitutes our principle exposure to hyperbaric environments. The sea provides many potential hazards, including gases that can be toxic or narcotic, currents that can tax physical stamina, and rapid changes in pressure that can be traumatic or fatal. Medical problems stemming from exposure to high pressure include barotrauma, gas toxicity, and decompression sickness.

Underwater research has focused on a number of factors that pose difficulty during exercise in this environment, including increased air density, cold, decreased efficiency of movement, CO_2 retention, and inert gas narcosis. The principles of exercise physiology determined on land cannot necessarily be extrapolated to underwater work. For example, pressure and temperature affect the relationship between O_2 consumption and heart rate. Using relationships between these variables calculated on land could be dangerous.

Selected Readings

Anthonisen, N., G. Utz, M. Kryger, and J. Urbanetti. Exercise tolerance at 4 and 6 ATA. Undersea Biomed. Res. 3: 95–102, 1976.

Åstrand, P.-O. and I.O. Åstrand. Heart rate during muscular work in man exposed to prolonged hypoxia. J. Appl. Physiol. 13: 75–80, 1958.

Balke, B. Variations in altitude and its effects on exercise performance. In: Falls, H. (ed.), Exercise Physiology. New York: Academic Press, 1968. p. 240–265.

Balke, B., F.J. Nagel, and J. Daniels. Altitude and maximum performance in work and sports activity. J.A.M.A. 194: 646–649, 1965.

Bjurstedt, H., G. Rosenhamer, C.M. Heser, and B. Lindborg. Responses to continuous negative-pressure breathing in man at rest and during exercise. J. Appl. Physiol. 48: 977–981, 1980.

Bradley, M., N. Anthonisen, J. Vorosmarti, and P. Linaweaver. Respiratory and cardiac responses to exercise in subjects breathing helium-oxygen mixtures at pressures from sea level to 19.2 atmospheres. In: Lambertsen, C. (ed.), Underwater Physiology. New York: Academic Press, 1971. p. 325–337.

Buskirk, E.R., J. Kollias, R.F. Akers, E.K. Prokop, and E.P. Reategui. Maximal performance at altitude and on return from altitude in conditioned runners. J. Appl. Physiol. 23: 259–266, 1967.

Craig, A.B. Underwater swimming and loss of consciousness. J.A.M.A. 1976: 255–258, 1961.

Dempsey, J.A. and H.V. Forster. Mediation of ventilatory adaptations. Physiol. Rev. 62: 262–346, 1982.

Dressendorfer, R., S. Hong, J. Morlock, J. Pegg, B. Respicio, R.M. Smith, and C. Yelverton. Hana Kai II: A 17-day saturation dive at 18.6 ATA. V. Maximal oxygen uptake. Undersea Biomed. Res. 4: 283–296, 1977.

Drinkwater, B., L.J. Folinsbee, J.F. Bedi, S.A. Plowman, A.B. Loucks, and S.M. Horvath. Response of women mountaineers to maximal exercise during hypoxia. Aviat. Space Environ. Med. 50: 657–662, 1979.

Dwyer, J. Energetics of scuba diving and undersea work. Proc. Int. Symp. on Man in the Sea., Honolulu, Hawaii: 32–46, 1975.

Dwyer, J. Measurement of oxygen consumption in scuba divers working in open water. Ergonomics. 4: 377–388, 1977.

Dwyer, J. Physiological studies of divers working at depths to 99 fsw in the open sea. In: Shilling, C. and M. Beckett (eds.), Underwater Physiology VI. Bethesda, Md.: FASEB, 1978. p. 167–178.

Dwyer, J., H. Saltzman, and R. O'Bryan. Maximal physical-work capacity of man at 43.4 ATA. Undersea Biomed. Res. 4: 359–372, 1977.

Elliott, P.R. and H.A. Atterbom. Comparison of exercise responses of males and females during acute exposure to hypobaria. Aviat. Space Environ. Med. 49: 415–418, 1978.

Elsner, R.W., W.F. Garey, and P.F. Scholander. Selective ischemia in diving man. Am. Heart J. 65: 571–572, 1963.

Fagraeus, L., C. Hesser, and D. Linnarsson. Cardiorespiratory responses to graded exercise at increased ambient air pressure. Acta Physiol. Scand. 91: 259–274, 1974.

Goddard, R.F. (ed.). The International Symposium on the Effects of Altitude on Physical Performance. Chicago: The Athletic Institute, 1967.

Green, M.K., A.M. Kerr, I.B. McIntosh, and R.J. Prescott. Acetazolamide in prevention of acute mountain sickness: A double-blind crossover study. Br. Med. J. 283: 811–813, 1981.

Gross, P.M., R.L. Terjung, and T.G. Lohman. Left-ventricular performance in man during breath-holding and simulated diving. Undersea Biomed. Res. 3: 351–360, 1976.

Grover, J. Performance at altitude. In: Strauss, R.H. (ed.), Sports Medicine and Physiology. Philadelphia: Saunders, 1979. p. 327–343.

Hartley, L.H., J.A. Vogel, and J.C. Cruz. Reduction of maximal heart rate at altitude and its reversal with atropine. J. Appl. Physiol. 36: 362–365, 1974.

Hays, P.A. and E.H. Padbury. Respiratory heat transfer and oxygen consumption when breathing warm humidified gas at 250 m (26 bar) following cold-water diving. J.Physiol. 308:110, 1980.

Hong, S.K. and H. Rahn. The diving women of Korea and Japan. Sci. Am. 216: 34–43, 1967.

Hoon, R.S., V. Balasubramanian, O.P. Mathew, S.C. Tiwari, S.C. Sharma, and K.S. Chadha. Effect of high-altitude exposure for 10 days on stroke volume and cardiac output. J. Appl. Physiol. Respirat. Environ. Exercise Physiol. 42: 722–727, 1977.

Hultgren, H. Circulatory adaptation to high altitude. Ann. Rev. Med. 19: 119–130, 1968.

Kaijser, L. Limiting factors for aerobic muscle performance: The influence of varying oxygen pressure and temperature. Acta Physiol. Scand. 346 (Suppl.): 1–96, 1970.

Kizer, K.W. Corticosteroids in treatment of serious decompression sickness. Ann. Emergency Med. 10: 485–488, 1981.

Kizer, K.W. The role of computer tomography in the management of dysbaric diving accidents. Radiology. 140: 705–707, 1981.

Kizer, K.W. Women and diving. Physician Sports Med. 9: 84–92, 1981.

Klausen, K. Cardiac output in man in rest and work during and after acclimatization to 3,800 m. J. Appl. Physiol. 21: 609–616, 1966.

Kollias, J. and E.R. Buskirk. Exercise and altitude. In: Johnson, W.R. and E.R. Buskirk (eds.), Structural and Physiological Aspects of Exercise and Sport. Princeton, N.J.: Princeton Book Co., 1980. p. 211–227.

Lahari, S. Physiological responses and adaptations to high altitude. In: Robertshaw, D. (ed.), Environmental Physiology II. Baltimore: University Park Press, 1977. p. 217–251.

Lally, D.A., F.W. Zechman, and R.A. Tracy. Ventilatory responses to exercise in divers and non-divers. Respir. Physiol. 20: 117–129, 1974.

Lanphier, E. Man in high pressures. In: Handbook of Physiology. Section 4: Adaptation to the Environment. Dill, D.B., E.F. Adolph and C.G. Wilber (eds.), Washington, D.C.: American Physiological Society, 1964.

Lenfant, C. and K. Sullivan. Adaptation to high altitude. N. Engl. J. Med. 284: 1298–1309, 1971.

Lenfant, C., J.D. Torrance, R. Woodson, and C.A. Finch. Adaptation to hypoxia. In: Brewer, G.J. (ed.), Red Cell Metabolism and Function. New York: Plenum Press, 1970. p. 203–212.

Levin, L., G.J. Stewart, P.R. Lynch, and A.A. Bove. Blood vessel wall changes induced by decompression sickness in dogs. J. Appl. Physiol. 50: 944–949, 1981.

Lin, Y.-C. Breath-hold diving in terrestrial mammals. Exerc. Sport Sci. Rev. 10: 270–307, 1982.

Luce, J. Respiratory adaptation and maladaptation to altitude. Physician Sports Med. 7: 55–69, 1979.

Malhotra, M.S. and W.S. Murthy. Changes in orthostatic tolerance in man at an altitude of 3,500 meters. Aviat. Space Environ. Med. 48: 125–128, 1977.

McFadden, D.M., C.S. Houston, J.R. Sutton, A.C.P Poowles, G.W. Gray, and R.S. Roberts. High altitude retinopathy. J.A.M.A. 245: 581–586, 1981.

Miller, J.C. and S.M. Horvath. Cardiac output during sleep at altitude. Aviat. Space Environ. Med. 48: 621–624, 1977.

Miller, J.N., E.H. Lanphier, and O.D. Wangensteen. Respiratory limitations to work in diving and their significance to the design of underwater breathing apparatus. Med. Dello Sport 24: 231–237, 1971.

Mori, S., M. Sakakibara, A. Takabayashi, S. Takagi, and G. Mitari. Cardiac output responses in rest and work during acute exposure to simulated altitudes of 3,000, 4,500, and 6,000 m, and during overnight sleep at 4,500 m. Jpn. J. Physiol. 32:337–349, 1982.

Neuman, T.S., R.F. Goad, D. Hall, R.M. Smith, J.R. Claybaugh, and S.K. Hong. Urinary excretion of water and electrolytes during open-sea saturation diving to 850 fws. Undersea Biomed. Res. 6: 291–302, 1979.

Openshaw, P.J.M. and G.M.F. Woodroof. Effect of lung volume on the diving response in man. J. Appl. Physiol. 45: 783–785, 1978.

Patajan, J.H. The effects of high altitude on the nervous system and athletic performance. Semin. Neurol. 1: 253–261, 1981.

Pilmanis, A.A., J. Henriksen, and H.J. Dwyer. An underwater ergometer for diver work performance studies in the ocean. Ergonomics. 20: 51–55, 1977.

Schneider, E.C. Respiration at high altitude. Yale J. Biol. Med. 4: 537–550, 1932.

Shilling, C.W., M.F. Werts, and N.R. Schandelmeir. The Underwater Handbook, A Guide to Physiology and Performance for the Engineer. New York: Plenum Press, 1976.

Slichter, S.J., P. Stegall, K. Smith, T.W. Huang, and L.A. Harker. Dysbaric osteonecrosis: Consequences of intravascular bubble formation, endothelial damage, and platelet thrombosis. J. Lab. Clin. Med. 98: 568–590, 1981.

Spivack, A. Medical aspects of scuba diving. Comp. Ther. 6: 6–11, 1980.

Squires, R.W. and E.R. Buskirk. Aerobic capacity during acute exposure to simulated altitude, 914 to 2286 meters. Med. Sci. Sports Exerc. 14: 36–40, 1982.

Strauss, R. Diving Medicine. New York: Grune & Stratton,1976.

Sutton, J., A.C. Bryan, G.W. Gray, E.S. Horton, A.S. Rebuck, W. Woodley, I.D. Rennie, and C.S. Houston. Pulmonary gas exchange in acute mountain sickness. Aviat. Space Environ. Med. 47: 1032–1037, 1976.

Sutton, J.R. and N.L. Jones. Exercise at altitude. Ann. Rev. Physiol. 45: 427–437, 1983.

Thalmann, E., D. Sponholtz, and C. Lundgren. Effects of immersion and static lung loading on submerged exercise at depth. Undersea Biomed. Res. 6: 259–289, 1979.

Vogel, J.A. and C.W. Harris. Cardiopulmonary responses of resting man during early exposure to high altitude. J. Appl. Physiol. 22: 1124–1128, 1967.

Vogel, J.A., L.H. Hartley, J.C. Cruz, and R.P. Hogan. Cardiac output during exercise in sea-level residents at sea level and high altitude. J. Appl. Physiol. 36: 169–172, 1974.

Ward, S.A., B.J. Whipp, and C.-S. Poon. Density-dependent airflow and ventilatory control during exercise. Respir. Physiol. 49: 267–277, 1982.

West, J.B. Respiratory Physiology—the Essentials. Baltimore: Williams & Wilkins Co., 1974.

Williams, C.I. and E.M. Bernauer. Diving reflex during exercise in man. Med. Sci. Sports. 8: 63–64, 1976.

Wohns, R.N.W. High-altitude cerebral edema: A pathophysiological review. Crit. Care Med. 9: 880–882, 1981.

24

CORONARY HEART DISEASE

Coronary heart disease (CHD) is the leading cause of death in Western countries, resulting in staggering economic and human costs. Every year billions of dollars are lost in wages, productivity, and medical expenditures. Even worse is the toll in human suffering as scores of relatively young men and women die prematurely, resulting in an irretrievable loss to their families and a waste of human potential. What is really unfortunate is that most of these deaths were probably preventable.

The disease process involves a steady buildup of atherosclerotic plaques in the coronary arteries leading to diminished myocardial blood flow. The effect on the heart depends on the degree of ischemia. If blood flow is not totally blocked but is insufficient to satisfy the myocardial oxygen demand, then angina (heart pain) or arrhythmias (aberrant ECG) can develop. If, however, blood flow is totally cut off or reaches a critical level of ischemia, the damage is irreparable, because anoxia does not just hamper cardiac function, it destroys it. The death of a portion of the myocardium is commonly called a heart attack. Survivability, or the degree of impairment, depends on the amount of tissue destroyed.

Atherosclerosis refers to the development of fibrotic, lipid-filled plaques in the walls of the larger arteries. Although its deadly effects are often centered in the heart, it can impair blood flow to all vital organs, including the brain, kidneys, and liver. Arteriosclerosis is a more general term to describe the lipid-

503

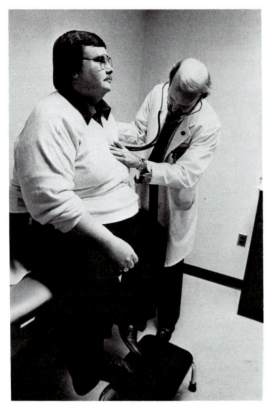

Obesity is a problem of overeating relative to activity level. The obese condition not only affects life style, but it also predisposes individuals to degenerative diseases, such as cardiovascular disease and diabetes. Mike McCanham has lost 100 pounds and is checked by his physician, Dr. James Lawson, in a Southfield, Michigan clinic. (Wide World.)

fibrous process in any blood vessel (e.g., small arteries, arterioles, and veins). The process of arteriosclerosis can result in extensive circulatory damage that is ultimately impervious to medical intervention. The most logical approach is to prevent the disease before it becomes life threatening.

Current research indicates that prevention is possible and that, to some extent, the process may be reversible. Although there is still much to be learned, the incidence of heart disease can be decreased by a systematic reduction in factors that increase the risk of heart disease. This chapter discusses the role of various risk factors in the development of coronary artery disease with specific emphasis on the importance of exercise. Additionally, the role of the exercise stress test in detecting coronary heart disease and measuring physical fitness will be explored.

THE DEVELOPMENT OF ATHEROSCLEROSIS

Atherosclerosis is thought to begin as a small injury to the arterial endothelium. The injury, which might be caused by a variety of factors, including ''wear

and tear'', and toxic chemicals, allows substances in the plasma, such as cholesterol, to penetrate into the intima of the artery. Ordinarily, the endothelial cells
act as a barrier to cholesterol and other lipids. However, an injury to the endothelial cells results in their inundation with blood fats. This process causes a
proliferation of the arterial smooth muscle that eventually results in its en-
croachment into the intima and arteriovascular space. Circulating platelets and
monocytes are thought to release substances at the injury site that stimulate the
migration of smooth muscle cells from the media to the intima. The proliferating smooth muscle cells also fill with fats. In addition, blood cells called
platelets cluster around the exposed smooth muscle and release substances that
are thought to result in further proliferation of the arterial smooth muscle (Figure 24-1).

The smooth muscle cells have difficulty removing cholesterol, so this sub-

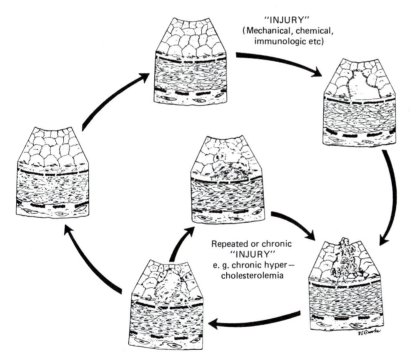

FIGURE 24-1
The response-to-injury hypothesis of the production of atheroma. The figure illustrates the response to a single (outer cycle) and repeated (inner
cycle) injury. Endothelial injury results in smooth muscle proliferation and
connective tissue formation. In the single event, the lesion heals and re-
gresses. During chronic insult, the process accelerates and is accompanied by lipid deposition in the lesion.

(SOURCE: From R. Ross and J.A. Glomset. The pathogenesis of atherosclerosis. N.
Engl. J. Med. 295: 369–377, 420–425, 1976.)

stance tends to accumulate, gradually increasing the size of the lesion. This process is accelerated by a variety of conditions such as hyperlipidemia and hypertension. Eventually, the plaque calcifies and connective tissue forms, producing a narrowed, rigid blood vessel. Once the vessel has been narrowed, blood flow may be impaired or the vessel can be blocked by a clot, resulting in a coronary thrombosis (one kind of heart attack). In summary, atherosclerosis develops because of smooth muscle proliferation, lipid accumulation, and connective tissue formation in the arterial intima.

Factors other than atherosclerosis, such as coronary vasospasm, may play a role in interfering with coronary artery blood flow and result in angina or myocardial infarction. Maseri et al. have suggested that coronary vasospasm with accompanying platelet aggregation may be responsible for the onset of perhaps a majority of cases of acute myocardial infarction and coronary thrombotic obstructions. They cite that coronary spasm has been shown to cause ventricular fibrillation and cardiac arrest and may be a leading cause of angina (angina caused by coronary spasm is often called Prinzmetal angina, after the man who first identified the phenomenon). The triggering stimuli of coronary vasospasm is not known, nor has the relationship between coronary spasm and atherosclerosis been established. However, while coronary spasm has been shown to occur in persons with largely patent coronary arteries, spasm is most common in persons with a significant degree of atherosclerosis.

THE RISK FACTORS IN THE DEVELOPMENT OF CORONARY HEART DISEASE

Heart disease is the number one killer in countries like the United States and Great Britain, while it is practically unknown in over half the world. In most Western countries, heart disease begins to manifest itself during childhood. By age 20, it is estimated that 75% of males have heart disease to a significant degree. Although females are less likely to suffer from coronary heart disease, they are experiencing an increase in incidence.

Although it is difficult to explain fully the reasons for the differences in heart disease between countries with high incidence (such as the United States) and those with low incidence (such as Japan), epidemiological (population disease studies) and experimental studies have identified several factors that are associated with (and may promote) coronary heart disease (Table 24-1). The three most important risk factors appear to be hyperlipidemia, hypertension, and cigarette smoking. A person who smokes two packs of cigarettes a day, or who has a cholesterol level above 260 mg $\cdot$ dl^{-1} (normal, 180), or has a blood pressure greater than 150/90 mmHg (normal, $\leq$ 120/80), is more than three times more likely to develop the disease.

TABLE 24-1

Risk of Developing CHD

Risk Factor	Relative Level of Risk				
	Very Low	Low	Moderate	High	Very High
Blood pressure (mmHg)					
Systolic	<110	120	130–140	150–160	170>
Diastolic	<70	76	82–88	94–100	106>
Cigarettes (per day)	Never None in 1 yr	5	10–20	30–40	50>
Cholesterol (mm · 100 cc^{-1})	<180	200	220–240	260–280	300>
Triglycerides (mm · 100 cc^{-1})	<80	100	150	200	300>
Glucose (mg · 100 cc^{-1})	<80	90	100–110	120–130	140>
Body fat (%)					
Men	<12	16	22	25	30>
Women	<15	20	25	33	40>
Stress and tension	Almost never	Occasional	Frequent	Nearly constant	
Physical activity minutes above 6 cal · min^{-1} (5 METS)[a] per week	240	180–120	100	80–60	30<
ECG stress test abnormality (ST depression-mm)[b]	0	0	0.5	1	2>
Family history of premature heart attack[c] (blood relative)	0	1	2	3	4>
Age	<30	35	40	50	60>

[a] A MET is equal to the oxygen cost at rest. One MET is generally equal to 3.5 ml · kg^{-1} of body weight per minute of oxygen uptake or 1.2 calories · min^{-1}.

[b] Other ECG abnormalities are also potentially dangerous and are not listed here.

[c] Premature heart attack refers to <60 yr of age.

SOURCE: From Pollock, M.L., J.H. Wilmore, and S.M. Fox. Health and Fitness Through Physical Activity. New York: John Wiley & Sons, 1978.

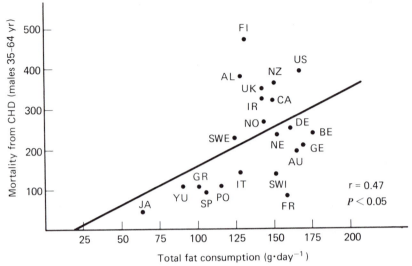

FIGURE 24-2

The correlation between total fat consumption and mortality in selected countries. AL, Australia; AU, Austria; BE, Belgium; CA, Canada; DE, Denmark; FI, Finland; FR, France; GE, Germany (West); GR, Greece; IR, Ireland; IT, Italy; JA, Japan; NE, the Netherlands; NO, Norway; NZ, New Zealand; PO, Portugal; SP, Spain; SWE, Sweden; SWI, Switzerland; UK, United Kingdom; US, United States; YU, Yugoslavia.

(SOURCE: Adapted from O. Turpeinen, 1979 by permission of the American Heart Association.)

Other risk factors, such as gender, family history, and inactivity, also increase the risk of heart disease, particularly when they are combined with other factors. Factors such as physical fitness and obesity are considered particularly important, because they often have a direct effect on other risk factors as well as exerting an independent effect.

There has been a reduction in heart disease in the United States in recent years (Figure 24-5). Although the mechanisms of this decrease are unclear, the increased awareness of CHD risk factors and the blossoming interest in preventive medicine may have played a role.

Hyperlipidemia

Hyperlipidemia is a highly controversial risk factor. For many years, researchers have been divided in their opinions about the role of blood fats, such as triglycerides and cholesterol, in the development of CHD. Epidemiological studies indicate a general trend toward a greater incidence of heart disease in countries with a high consumption of fat (Figure 24-2), particularly dairy fats (Figure 24-3). Countries that derive most of their fats from vegetable sources show a lower rate of the disease (Figure 24-4).

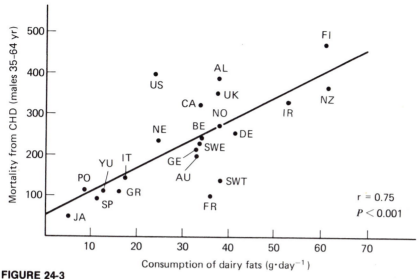

FIGURE 24-3

Correlation between consumption of dairy fats and mortality from CHD in selected countries. For country abbreviations, see Figure 24-2.

(SOURCE: Adapted from O. Turpeinen, 1979 by permission of the American Heart Association.)

The evidence is far from being clear-cut. Whereas the rate of heart disease has fallen in the United States, where the consumption of animal and dairy fats has decreased, it has also fallen in Switzerland, where the consumption of these fats has risen. In spite of a vigorous heart disease risk reduction campaign, the heart disease rate in Sweden has been rising by about 2% a year. Much of the controversy apparently stems from the failure to distinguish between lipids that promote atherosclerosis and those that do not.

Because lipids are not water soluble, they are transported as components of lipoproteins, which contain combinations of fat molecules and protein. Cholesterol, which is a precursor to steroids, is considered a lipid because it is synthesized from fatty acids and has many of the properties of other lipids. There are five types of lipoproteins that are classified according to density, composition, and size. In order of increasing density, the classes of lipoproteins are very-low-density lipoproteins (VLDL), low-density lipoproteins (LDL), intermediate-density lipoproteins (IDL), high-density lipoproteins (HDL, which are further subdivided into HDL1, HDL2, and HDL3), and chylomicrons.

The liver produces VLDL, which contain mostly triglycerides and some phospholipids and cholesterol, and carry these substances away from the liver. The catabolism of VLDL, results in the formation of IDL. The principal cholesterol carriers, LDL, are formed in the catabolism of IDL and are mostly made up of cholesterol. Chylomicrons are formed in the intestinal wall from

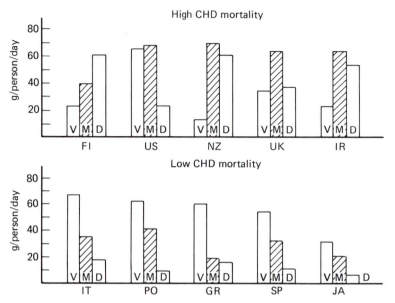

FIGURE 24-4

Consumption of major types of food fats in selected countries with high and low CHD mortalities. V, Vegetable fats; M, meat fats; D, dairy fats. For country abbreviations, see Figure 24-2.

(SOURCE: Adapted from O. Turpeinen, 1979 by permission of the American Heart Association.)

the diet and reach the circulation from the thoracic duct. They carry dietary triglycerides and cholesterol. HDL, consisting of about 50% protein and a mixture of lipids, are formed as a result of the enzyme lipoprotein lipase acting on the surface components of chylomicrons and VLDL. Approximately 17 different apolipoproteins, the protein constituents of lipoproteins, have been identified. Apolipoproteins seem to play an important role in lipoprotein metabolism by acting as recognition sites for cell membrane receptors and as cofactors in enzymatic reactions. Total cholesterol is the sum total of cholesterol carried by VLDL, LDL, HDL, and chylomicrons.

Most people in the United States eat approximately 100 to 120 g of fat per day. About two-thirds of this is transported to the liver where it forms VLDL and LDL. The liver forms cholesterol even if the dietary intake of cholesterol is low. The rate of formation is greater if the fats are saturated. The level of LDL is far more important than the total cholesterol in the development of CHD, because it is the primary substance involved in the development of the atherosclerotic plaque.

HDL are also formed in the liver, but they tend to retard the development of atherosclerosis. HDL contain an enzyme called lecithin cholesterol acid transferase (LCAT), which gathers free cholesterol and returns it to the liver where it is released as cholesterol in the bile or is converted to bile salts. HDL act as a reverse cholesterol transfer system that resists the development of atherosclerosis.

The ratio of cholesterol to HDL seems to be an extremely important factor in the development of heart disease. The average ratio of heart attack victims in the Framingham, Massachusetts heart study (a long-term epidemiological study) was 5.4. It is recommended that the ratio be less than 4.0. People with total cholesterols of 200 mg · dl^{-1}, generally regarded as favorable, may still be at risk if the level of LDL is high and the level of HDL is low. Likewise, populations such as the Eskimos that have high levels of LDL may have a low risk of heart disease, because they also have high levels of HDL.

HDL levels are considered by many researchers to be an extremely important risk factor. It may help to explain why some populations are at higher risk than others and may provide added evidence for the benefits of exercise. HDL2 levels are higher in women until menopause (approximately three times higher), at which time HDL2 levels decrease and become similar to those of men. The heart disease rates follow a similar course (Figure 24-5). Low levels of HDL

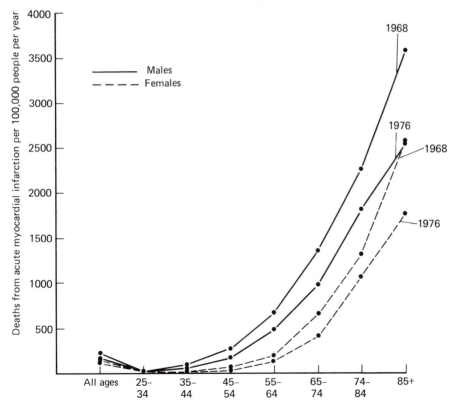

FIGURE 24-5

Chances in 100,000 of dying from acute myocardial infarction in 1968 and 1976 (all races).

are also seen in hypertriglyceridemia, obesity, diabetes, and cigarette smokers.

The constituents of HDL, such as apolipoproteins A-I and A-II, phosphatidylcholine, and sphingomyelin, may also be related to the risk of CHD. Currently, a great deal of research has been directed at these factors, but the findings have been equivocal.

Testosterone, the male sex hormone, is known to depress HDL levels. In athletes, administration of high doses of exogenous testosterone derivatives in the form of anabolic steroids has been shown to depress HDL levels to as low as 5 mg $\cdot$ dl^{-1} (20- to 40-year-old males typically average 40 mg $\cdot$ dl^{-1}). Other factors affecting HDL levels include diet, alcohol consumption, and exercise.

Endurance training has been shown to increase the levels of HDL, LCAT, and lipoprotein lipase, while decreasing LDL and triglycerides. This results in an increased elimination of cholesterol in the bile and may be the principal mechanism through which exercise reduces the risk of CHD. Low-fat, low-cholesterol diets, popularized by Nathan Pritikin, appear to be promising in reducing serum cholesterol, hypertension, and glucose intolerance (all of which are related to an increased risk of CHD). The lipid management program should be aimed at reducing LDL and increasing HDL through proper diet, exercise, cessation of smoking, and perhaps moderate alcohol consumption. (Note: high levels of alcohol consumption can be toxic to the heart and liver.)

Hypertension

Hypertension is discussed extensively in Chapter 26. It is sometimes called the silent killer because it usually has no apparent symptoms and accounts for over 1.5 million deaths a year in the United States alone. Blood pressure above 150/90 mmHg increases the risk of heart disease by three times in men and by six times in women who are between the ages of 40 and 60 yrs.

Hypertension affects the development of atherosclerosis, increases myocardial O_2 consumption at rest and exercise, and causes cellular changes in the heart and blood vessels. Hypertension has two principal effects on the development of atherosclerosis: (1) It is thought to be directly involved in the process of damage to the vascular endothelial cells by increasing shear forces, torsion, and lateral wall pressure; (2) the higher pressure is thought to increase the filtration of lipids into the intimal cells.

Hypertension increases the myocardial O_2 consumption by increasing the afterload of the heart, causing the heart to pump harder to deliver the same quantity of blood. This process usually results in cardiac hypertrophy, which in itself increases myocardial O_2 consumption. Myocardial hypertrophy, together with the enhancement of atherosclerosis, increases the likelihood of the development of coronary ischemia.

The chronic afterload stress imposed by hypertension can affect the ultrastructure of the myocardial cells. A recent study by Singh et al. demonstrated that the chronically pressure-overloaded heart exhibited myofibril disorganization, disintegration and broadening of Z bands, swelling and aggregation of mitochondria, intracellular edema, and decreased regional myocardial blood flow to the subendocardium during exercise.

An exercise-induced increase in diastolic pressure has been shown to be a good indicator of severe coronary artery disease. In a study by Sheps et al., 83% of subjects demonstrating an increase in diastolic pressure of 15 mmHg or more during exercise had coronary artery disease.

Smoking

There is a definite relationship between smoking and heart attack in the United States. In the Framingham study, 60% of the men and 40% of the women who had heart attacks were smokers. However, no such relationship has been demonstrated in Finland, the Netherlands, Italy, Greece, Yugoslavia, Japan, or Puerto Rico.

The role of smoking in the development of atherosclerosis has not been clearly established. Smoking may be involved by increasing blood pressure, decreasing high-density lipoproteins, and mobilizing free fatty acids. Additionally, it may be involved in the development of fatal cardiac arrhythmias by lowering the ventricular fibrillation threshold and causing coronary vasoconstriction. Although the mechanism of accelerated atherosclerosis in smokers has not been clearly established, the statistical reality of a high number of heart attacks among smokers makes it a serious risk factor.

Family History, Sex, and Age

These are complex risk factors with both genetic and sociological implications. Although the factors themselves cannot usually be altered, the relative risk of any individual can probably be reduced by a vigorous risk factor management program.

A person who had a close male relative (father or grandfather) die from a heart attack before age 60 has a three to six times greater risk of contracting CHD. The chances are even greater in people with familial hypercholesterolemia, as 50% of these die before age 60.

Although some people have a genetic tendency toward hypertension, diabetes, and obesity, many familial risk factor patterns are learned. It is likely that some of the dietary and psychological characteristics of families could be amenable to change under the right circumstances. Such individuals would undoubtably be worthy candidates for high risk factor management programs.

Premenopausal females have a lower risk of heart attack than men. The reason may lie in their higher levels of HDL (60 mg $\cdot$ dl^{-1} in women versus 40 mg $\cdot$ dl^{-1} in men), and perhaps the higher level of stress in men. As discussed, the higher levels of testosterone in males may be a factor in their lower levels of HDL.

Age is somewhat of a paradox, because in general, the older people become, the closer they are to death. Heart disease is a long-term degenerative disease, so with fewer infectious diseases to die from, it is not surprising that more people die from a disease that tends to get progressively worse with age. Even though it is extremely common in Western countries, atherosclerosis is not inevitable, as demonstrated by the low rate of heart disease among many peoples of the world. A risk factor reduction program will not allow people to live forever but may prevent them from dying prematurely from heart disease.

Obesity

Life insurance statistics and results of the Framingham study have drawn a link between excessive weight (weighing more than the average weight for persons of the same sex, age, and height) and heart disease. Unfortunately, these statistics are extremely misleading. They fail to differentiate between those of higher weight but normal body fat and those with excess body fat. Many individuals have normal weight but are really obese because they have low lean body mass and excess body fat.

Obesity does not appear to be a risk factor of coronary artery disease unless it is associated with hypertension or hyperlipidemia (which is often the case). When obese and normal-weight subjects who have the same blood pressure and serum lipids are compared, there is no difference in the risk of heart disease. However, because of cosmetic considerations and the relationship of obesity with other diseases, it remains a serious problem.

Emotional Stress

For many years, clinicians have identified the "heart attack personality" as aggressive, restless, impatient, and hard driving. Rosenman and Friedman coined the term "Type A" personality to describe this behavior pattern and defined it as an action/emotional behavior complex exhibited by an individual in a life-long struggle to obtain too much from the environment in too short a time, against the opposing efforts of objects or people in the environment. The less cardiac-prone behavioral type is the Type B personality, who is characterized as less competitive and more relaxed. Extensive subsequent research has drawn a definite relationship between the Type A personality and an increased risk of heart disease.

Although there is a statistical relationship between Type A personality and heart disease, there is no evidence directly relating it to the process of coronary artery disease. It is possible that the adrenergic drive produced by emotional stress may induce mechanical trauma to the intima of the coronary arteries, but this has not been demonstrated. There is some evidence to suggest that the Type A person releases more norepinephrine in the face of stress. This may enhance platelet aggregation and lipid mobilization, which could accelerate the development of the disease. However, there is no relationship between the degree of Type A behavior and the degree of coronary occlusion.

Jenkins has identified a number of psychosocial factors that have been associated with an increased risk of CHD, in addition to the Type A personality. These factors include lower social class, indices of neuroticism such as anxiety and depression, sleep disturbances, and excessive work load. These factors appear to exert their pathogenic influence through atherosclerosis, rather than by precipitating sudden coronary events.

Physical Activity

Until recently, lack of physical activity was classified as a minor risk factor in the development of CHD. However, recent evidence suggests that regular physical exercise is extremely important in reducing the risk of the disease. Exercise training may result in regression of atheroma and reduce the magnitude of other risk factors such as hypertension, obesity, and hyperlipidemia.

Since the early 1950s, epidemiological studies have demonstrated the relationship between physical activity and a reduced risk of heart disease. A num-

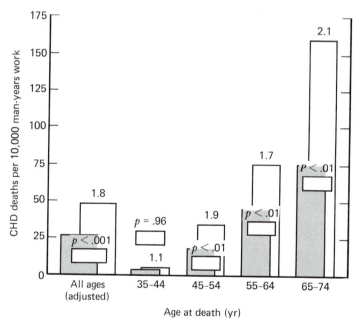

FIGURE 24-6

Death from CHD in longshoremen according to physical activity of work (range in $kcal \cdot min^{-1}$) and age at death. Shaded bars, heavy activity (5.2–7.5); unshaded bars, moderate and light activity (1.5–5.0). Relative risk (moderate and light/heavy) given above bars.

(SOURCE: Adapted from R.S. Paffenbarger and W.E. Hale. Work activity and coronary heart disease. N. Engl. J. Med. 292: 545–550, 1975.)

ber of studies compared active and sedentary populations: London bus drivers (sedentary) and bus conductors (active), postal clerks (sedentary) and postal deliverers (active), and active versus sedentary longshoremen (Figure 24-6). These studies have been criticized for being ex post facto (after the fact) investigations. It is possible that less coronary-prone individuals self-selected themselves into more active occupations. In addition, not all epidemiological studies demonstrated the reduced risk of CHD in more active occupations.

A fundamental problem in all of these population profile studies is the failure to define adequately the level of physical activity in the active and sedentary groups and to distinguish between the fit and the unfit. A study by Cooper demonstrated an inverse relationship between the level of physical fitness, as measured on a treadmill test, and the risk of heart disease. Extremely fit people tended to be less fat, with lower blood pressure, cholesterol, triglycerides, uric acid, and glucose. Morris and co-workers showed that vigorous exercise during leisure time resulted in a substantial reduction in the risk of heart disease.

Recent studies using animals provide direct experimental evidence that moderate endurance exercise can result in a reduction of coronary atherosclerosis. Kramsch et al. demonstrated that exercising monkeys who were fed an atherogenic diet had substantially reduced overall atherosclerotic involvement, lesion size, and collagen accumulation, than did sedentary controls. Further, the trained

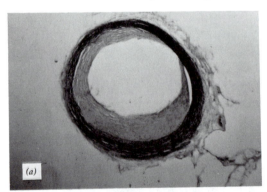

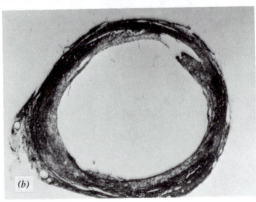

FIGURE 24-7
Micrographs of comparable sections of the left main coronary artery in sedentary (a) and active (b) monkeys on an atherogenic diet.

(SOURCE: D.M. Kramsch, A.J. Aspen, B.M. Abramowitz, T. Kreimendahl, and W.B. Hood. Reduction of coronary atherosclerosis by moderate conditioning exercise in monkeys on an atherogenic diet. N. Engl. J. Med. 305: 1483–1489, 1981.)

monkeys had larger hearts and wider coronary arteries, which further reduced the degree of luminal narrowing (Figure 24-7). In contrast, the control animals, who were fed the same diet, suffered significant narrowing of the coronary arteries and, in one case, sudden death.

Autopsies on marathon runners, who died from causes other than CHD, consistently find enlarged and widely patent coronary arteries. This does not actually prove that endurance training provides a measure of protection against the disease. It does, however, add to existing evidence that points in that direction. Reports of heart attacks among serious runners have usually been unsubstantiated by autopsy and have perhaps been confused with other disorders. Whereas atherosclerotic, cardiovascular deaths are not uncommon in people involved in exercise, most victims are usually relatively untrained and possess predisposing risk factors such as cigarette smoking, hypertension, or hyperlipidemia.

A summary of the mechanisms by which physical activity may reduce the occurrence or severity of CHD appears in Table 24-2. Endurance training aids

TABLE 24-2

Mechanics by Which Physical Activity May Reduce Occurrence or Severity of CHD

Increase	Decrease
Coronary collateral vascularization (?)	Serum lipid levels:
Vessel size	Triglycerides
HDL Cholesterol	LDL Cholesterol
Myocardial efficiency	Glucose intolerance
Efficiency of peripheral blood distribution and return	Obesity–adiposity
	Platelet stickiness
Electron-transport capacity	Arterial blood pressure
Fibrinolytic capability	Heart rate
Red blood cell mass and blood volume	Vulnerability to dysrhythmias
	Neurohormonal overreaction
Thyroid function	"Strain" associated with psychic "stress"
Tolerance to stress	
Prudent living habits	
"Joie de vivre"	
Width of coronary artery lumina	Reduction of size of atheroma
Metabolic turnover of collagen in the heart	Collagen accumulation in coronary arteries

SOURCE: Adapted from Fox, S.M., J.P. Naughton, and W.L. Haskell. Physical activity and the prevention of coronary disease. Ann. Clin. Res. 3: 404–432, 1971.

in a reduced risk of CHD by improving the metabolic capacity and affecting factors involved in the development of the disease. Metabolic function is improved by such adaptations as increased electron-transport capacity, thyroid function, coronary collateral vascularization, myocardial efficiency, efficiency of peripheral blood distribution and return, and decreased resting heart rate, submaximal exercise heart rate, arterial blood pressure, and obesity. These factors lessen the load on the heart, thus decreasing the chance of developing ischemia.

Other changes may affect the disease process itself. Adaptations such as increased fibrinolytic capability, growth hormone production, tolerance to stress, increased HDL and LCAT, and decreased LDL, triglycerides, glucose intolerance, and platelet stickiness may retard or reverse the process of atherosclerosis.

HOW MUCH EXERCISE DO PEOPLE NEED?

A high maximal O_2 consumption is associated with reduced risk of CHD. Studies by Owen et al., and Bjurstrom and Alexiou show that improving fitness reduces the risk of heart disease. The American College of Sports Medicine has issued a position statement regarding the quantity and quality of exercise necessary for achieving the desired levels of fitness (Table 24-3).

The mode, intensity, frequency, and duration of the exercise should be considered. Endurance exercises, such as running, walking, cycling, swimming, and cross-country skiing are best. These continuously use at least 50% of the muscle mass in a rhythmical manner. Other exercises, such as tennis, racquetball, handball, basketball, and squash are acceptable but do not usually develop endurance capacity to the same extent as the more continuous activities. Additionally, these ''stop-and-start'' sports usually require a prerequisite level of skill in order to achieve a training effect.

The frequency of training should be 3 to 5 days per week. Participation of 2 days or less does not produce sufficient improvement in maximal O_2 consumption. Sporadic participation may result in an increased risk of soft tissue injury.

The exercise intensity should be between 60 and 90% of maximum heart rate reserve (maximum heart rate − resting heart rate) plus resting heart rate or 50 to 85% of maximal O_2 consumption. This is an extremely important consideration, often neglected by the novice. Activities such as swimming and walking, can be practiced at intensities very close to resting heart rate. Intensity can be easily gauged by taking the heart rate during the first 10 sec of recovery and multiplying the value by six. This heart rate closely approximates the rate during exercise.

The duration of exercise should be between 15 and 60 min of continuous

TABLE 24-3
The Recommended Quantity and Quality of Exercise for Developing and Maintaining Fitness in Healthy Adults

Mode of Activity
Aerobic or endurance exercises, such as running-jogging, walking-hiking, swimming, skating, bicycling, rowing, cross-country skiing, rope skipping, and various game activities

Frequence of Training
3 to 5 days per week

Intensity of Training
60% of capability range plus resting pulse rate, or 50–85% of maximum O_2 uptake

Duration of Training
15 to 60 min of continuous aerobic activity. Duration is dependent on the intensity of the activity

SOURCE: Adapted from a position statement issued by the American College of Sports Medicine.

activity depending on the intensity of exercise. A low-intensity activity such as walking requires more time than a high-intensity activity such as interval training on a running track.

THE EXERCISE STRESS TEST

The stress test measures the physiological response to a standard bout of exercise. Although the treadmill is the most popular device for stress testing, others include the bicycle ergometer, supine cycle ergometer, arm ergometer, wheelchair ergometer, and bench step. The use of the classic "stimulus—response" method of measurement (an exercise stimulus is introduced and the physiological response is measured) allows subjects to be compared accurately with their own past performances and with other peoples'.

The factors measured during a stress test may include exercise or work capacity, ECG, heart rate, blood pressure, O_2 consumption, ventilation, blood lactate, and perceived exertion. The principal uses of the exercise stress test are the measurement of endurance fitness, the diagnosis of CHD, and the classification of the degree of impairment in cardiac patients. The ability to perform endurance exercise depends on the functional capacity of the cardiovascular system. Endurance fitness is also related to the relative risk of heart disease. The exercise stress test is valuable because it provides a method of objectively measuring fitness in a safe environment.

The stress test has proved valuable in the detection of coronary artery disease. The stress test is used to provoke a coronary perfusion deficit in those with CHD. Myocardial O_2 demand increases with the intensity of exercise, enhancing the requirement for coronary blood flow. Ischemia may develop if the coronaries are burdened with atherosclerosis. In many instances, this "relative" coronary ischemia will not manifest itself at rest.

The number of diagnostic tools used during exercise stress testing are growing and include multiple-lead electrocardiography, radioisotope myocardial perfusion, and echocardiography. Although the ECG is the most popular and widely used method of CHD detection, the use of the other methods is becoming more prevalent because they afford a more direct view of cardiac function.

Echocardiography uses ultrasound to get a picture of the heart. Two-dimensional echocardiography allows the study of wall motion, valve function, and stroke mechanics. The infusion of isotopes, such as thallium-201, allows an accurate estimation of myocardial perfusion during exercise.

Coronary artery disease can be predicted during the treadmill test from indirect measures such as effort-induced angina pectoris, changes in cardiac electrical repolarization, cardiac arrhythmias, heart rate impairment, and ventricular dysfunction (Table 24-1). These can be determined by monitoring the subjective sensations of the subject, the ECG, and blood pressure. Other possible contributing factors include symptom-limited duration of the exercise test and maximal O_2 consumption relative to sex and age.

Angina pectoris is characterized by a pressure or constricting sensation anywhere in the upper body. It is typically experienced as a burning sensation under the sternum, which may radiate to the jaw, neck, left shoulder, or left arm. Angina may occur at predictable levels of exertion, at rest, or during sleep. It is caused by coronary ischemia due to atherosclerosis or coronary spasm and is the most common clinical symptom of coronary artery disease.

ST-segment depression of the electrocardiogram, an aberration of ventricular repolarization, is considered as one of the best measures for detecting coronary ischemia (Figure 24-8). Although it may appear in persons with coronary artery disease, it may also occur whenever there is excess ventricular strain, such as in hypertension. The appearance of ST depression depends on the number of coronary vessels significantly affected by disease (greater than 70% occluded). It will appear in about 65% of those with one-vessel disease, 85% with two-vessel disease, and 90% with three-vessel disease.

The exercise electrocardiogram is called a false negative when CHD is present without the appearance of ST-segment depression, and a false positive when there is no disease but there is ST-segment depression. A false-positive result may occur because of hyperventilation, the effects of some drugs, or poor response time in the instrument. Multiple-lead ECG systems have been shown to improve considerably the ability of the electrocardiogram to predict CHD.

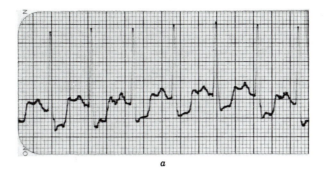

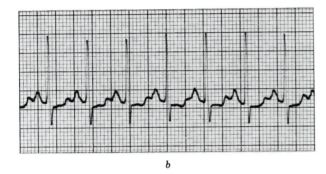

FIGURE 24-8
ST Segment on ECG tracing before (a) and after (b) coronary bypass surgery (V₅ lead).

Exercise tends to increase the irritability of the heart and may produce arrhythmias. These arrhythmias include premature atrial or ventricular contractions (Figure 24-9), intraventricular conduction disturbances (bundle branch blocks), atrial or ventricular tachycardia, and atrial or ventricular fibrillation. The incidence of arrhythmias tends to increase with the intensity of exercise in those with CHD. Frequent ventricular arrhythmias during exercise suggest the presence of coronary artery disease.

Impairment of the chronotropic capacity of the heart may be associated with

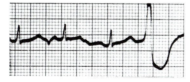

FIGURE 24-9
Example of a premature ventricular contraction (PVC) on an
ECG tracing. PVC's arise from ventricular ectopic foci.

(SOURCE: P. Dubin, 1974, p. 91.)

CHD. The chances of sustaining a fatal myocardial infarction increase dramatically as the maximum heart rate falls below that predicted for the age and sex of the individual. Depressed maximum heart rate is usually accompanied by ST-segment depression and diminished work capacity, so it usually is of little additional value in predicting the existence of CHD. However, by itself, it is of considerable prognostic significance.

Ventricular dysfunction is characterized by a diminished inotropic or contractile capacity of the myocardium. The inotropic capacity of the heart can be estimated by examining wall motion with echocardiography, especially during isometric exercise, and by monitoring the pressure–rate product (heart rate × systolic pressure) during a treadmill test. Deviation of the pressure–rate product from that of people of the same age, sex, and habitual level of physical activity, or a failure to increase systolic blood pressure with increasing exercise intensity suggests the existence of coronary artery disease.

Low exercise capacity and low maximal O_2 consumption also suggest CHD. People unable to complete stage 2 on a Bruce test (2.5 mph, 12% grade), or who have a maximal O_2 consumption of less than 20 ml $O_2 \cdot kg^{-1}$ are more likely to have the disease. Bruce estimated the incidence of CHD mortality in males with extremely low endurance at about 225 in 1000 men at risk per year.

Bruce has found that the three best predictors of CHD are an exercise test duration of less than 6 min (stage 2 on Bruce treadmill test), chest pain, and ST-segment depression of −0.15 mV. Mortality from CHD is 10 in 1000 if none of these variables are present, 96 in 1000 if one is present, 252 in 1000 if two are present, and 884 in 1000 if all three are present.

Summary

Coronary heart disease (CHD) is the leading cause of death in Western countries. CHD involves a steady buildup of atherosclerotic plaques in the coronary arteries. The atheroma probably begins as an injury to the arterial endothelium, which causes a migration and proligeration of smooth muscle cells in the intima. Eventually, the lesion fills with fat, connective tissue forms, and the area becomes calcified.

A number of risk factors have been associated with CHD, including hypertension, cigarette smoking, hyperlipidemia, male sex, family history, diabetes, psychosocial stress, obesity, and physical inactivity. A great deal of effort has been directed at increasing the public's awareness of these risk factors, and there has been an increased emphasis on preventive medicine.

Exercise has received increasing attention as a factor that may decrease the risk of CHD. Regular physical activity is important because it modifies the severity of other risk factors as well as exerting an independent effect. A program that may provide a measure of protection from CHD includes endurance-type exercise such as walking, jogging, swimming, and cycling, practiced three to five times a week for 15 to 60 min at between 60 and 85% of maximum heart rate reserve.

The exercise stress test is a valuable procedure for measuring exercise capacity as well as noninvasively screening for CHD. A number of procedures can be utilized during various exercise tests, including electrocardiography, O_2 consumption, blood pressure, radioisotope myocardial perfusion, and echocardiography.

Selected Readings

Assman, G., H. Funk, and H. Schriewer. The relationship of HDL-apolipoprotein A-I and HDL-cholesterol to risk factors of coronary heart Disease. J. Clin. Chem. Clin. Biochem. 20: 287–289, 1982.

Ball, K.P. and R.W.D. Turner (eds.). Dietary prevention of coronary heart disease. Postgrad. Med. J. 56: 537–615, 1980.

Basser, T. J. Marathon running and immunity to atherosclerosis. Ann. NY. Acad. Sci. 301: 579–592, 1977.

Bjurstrom, L.A. and N.G. Alexiou. A program of heart disease intervention for public employees. J. Occup. Med. 20: 521–531, 1978.

Blankenhorn, D.H. Will atheroma regress with diet and exercise? Am. J. Surg. 149: 644–645, 1981.

Bowyer, A., J.F. Schwalbach, A.C. Jain, H. Asato, and B.C. Pakrashi. The logic of exercise stress testing to predict the presence of significant coronary stenosis. Proc. San Diego Biomed. Symp. 16: 127–134, 1978.

Braunwald, E., J. Ross, and E.H. Sonnenblick. Mechanisms of contraction of the normal and failing heart. N. Engl. J. Med. 277: 794–799, 853–862, 910–919, 962–971, 1012–1022, 1967.

Bruce, R.A. and T.A. DeRouen. Exercise testing as a predictor of heart disease and sudden death. Hosp. Pract. 14: 69–75, 1978.

Cooper, K.H., M.L. Pollock, R.P. Martin, S.R. White, A.C. Linnerud, and A. Jackson. Physical fitness vs. selected coronary risk factors. J.A.M.A. 236: 166–169, 1976.

Cooper, T. Coronary-prone behavior and coronary heart disease: A critical review. Circulation. 63: 1199–1215, 1981.

Detry, J.M., S. Abouantoun, and W. Wyns. Incidence and prognostic implications of severe ventricular arrhythmias during maximal exercise testing. Cardiology. 68: 35–43, 1981.

Dubin, P. Rapid Interpretation of EKG's. Tampa, Fla.: Cover Publishing Co., 1974. p. 91.

Dufaux, B., G. Assmann, and W. Hollmann. Plasma lipoproteins and physical activity: A review. Int. J. Sports Med. 3: 123–136, 1982.

Fox, S.M. coronary Heart Disease, Prevention, Detection, Rehabilitation with Emphasis on Exercise Testing. Denver, Colo.: Department of Professional Education, International Medical Corporation, 1974.

Friedman, M., R.H. Rosenman, R. Straus, M. Wurm, and R. Kositchek. The relationship of behavior pattern A to the state of the coronary vasculature. Am. J. Med. 44: 525–537, 1968.

Froelicher, V.F. Exercise testing and training. New York: Le Jacq Publishing Co., 1983.

Galen, R.S. HDL cholesterol: How good a risk factor? Diagnostic Med. Dec. 1979. p. 39–58.

Gobel, M.F., R.R. Nelson, L.A. Nordstrom, and F.L. Gobel. Predictors of coronary artery disease from exercise stress testing. Minn. Med. 64: 143–148, 1981.

Goldstein, J.L. and M.S. Brown. Atherosclerosis: The low-density lipoprotein receptor hypothesis. Metabolism. 26: 1257–1275, 1977.

Kramsch, D.M., A.J. Aspen, B.M. Abramowitz, T. Kreimendahl, and W.B. Hood. Reduction of coronary atherosclerosis by moderate conditioning exercise in monkeys on an atherogenic diet. N. Engl. J. Med. 305: 1483–1489, 1981.

Maseri, A., S. Chierchia, and A. L'Abbate. Pathogenetic mechanisms underlying the clinical events associated with atherosclerotic heart disease. Circulation 62(suppl V): 3–13, 1980.

McNeer, J.F., J.R. Margolis, K.L. Lee, J.A. Kisslo, R.H. Peter, Y Kong, V.S. Behar, A.G. Wallace, C.B. McCants, and R.A. Rosati. The role of the exercise test in the evaluation of patients for ischemic heart disease. Circulation. 57: 65–70, 1978.

Morris, J.N., R. Pollard, M.G. Everitt, and S.P.W. Chave. Vigorous exercise in leasure-time: Protection against coronary heart disease. Lancet i:1207–1210, 1980.

Morris, J.N., J.A. Heady, P.A.B. Raffle, C.G. Roberts, and J.W. Parks. Coronary heart disease and physical activity of work. Lancet. 2: 1053–1057, 1111–1120, 1953.

Paffenbarger, R.S. and W.E. Hale. Work activity and coronary heart disease. N. Engl. J. Med. 292: 545–550, 1975.

Pollock, M.L. A comparative analysis of four protocols for maximal treadmill stress testing. Am. Heart J. &§: 39–46, 1976.

Pollock, M.L., J.H. Wilmore, and S.M. Fox. Health and fitness through physical activity. New York: John Wiley & Sons, 1978.

Rosenman, R.H., M. Friedman, R. Straus, M. Wurm, R. Kositchek, W. Hahn, and N.T. Werthessen. A predictive study of coronary heart disease. J.A.M.A. 189: 103–110, 1964.

Ross, R. and J.A. Glomset. The pathogenesis of atherosclerosis. N. Engl. J. Med. 295: 369–377, 420–425, 1976.

Sheps, D.S., J.C. Ernst, F.W. Briese, and R.J. Myerburg. Exercise-induced increase in diastolic pressure: Indicator of severe coronary artery disease. Am. J. Cardiol. 43: 708–712, 1979.

Singh, S., F.C. White, and C.M. Bloor. Effect of acute exercise stress in cardiac hypertrophy: I.Correlation of regional blood flow and qualitative ultrastructural changes. Yale J. Biol. Med. 53: 459–470, 1980.

Simko, V. Physical exercise and the prevention of atherosclerosis and cholesterol gallstones. Postgrad. Med. J. 54: 270–277, 1978.

Sotobata, I., T. Kondo, and N. Kawai. Present status of exercise testing in the evaluation of coronary artery disease. Jpn. Circ. J. 45: 381–393, 1981.

Stromme, S.B., H. Frey, O.K. Harlem, O. Stokke, O.D. Vellar, L.E. Aaro, and J.E. Johnsen. Physical activity and health. Social Medicine 29 (Suppl.): 9–36, 1982.

Tharp, G.D. and C.T. Wagner. Chronic exercise and cardiac vascularization. Eur. J. Appl. Physiol. 48: 97–104, 1982.

Turpeinen, O. Effect of cholesterol-lowering diet on mortality from coronary heart diseases and other causes. Circulation 59: 1, 1979.

Wainwright, R.J., M.N. Maisey, A.C. Edwards, and E. Sowton. Functional significance of coronary collateral circulation during dynamic exercise evaluated by thallium-201 myocardial scintigraphy. Br. Med. J. 43: 47–55, 1980.

Wood, P. and W.L. Haskell. The effect of exercise on plasma high density lipoproteins. Lipids. 14: 417–427, 1979.

25

OBESITY AND BODY COMPOSITION

Obesity refers to excess body fat and is the direct result of consuming more energy than is expended. It affects nearly 30% of the adults in Western countries and constitutes a serious health problem. The etiology of this disorder is complex and the problem is often impervious to cure. Overfatness runs in families as a result of both genetic and sociological factors, so the disorder is often passed from one generation to the next.

Obesity is associated with many diseases that together account for a very large fraction of the total morbidity and mortality. High levels of body fat are associated with an increased risk of coronary heart disease, stroke, hypertension, hyperlipidemia, diabetes, osteoarthritis, gallstones, gallbladder disease, renal disease, hepatic cirrhosis, accident proneness, surgical complications, and back pain.

Obesity seriously affects physical performance. Because increased energy is required to move a larger mass, the overweight person is forced to work harder at most tasks. Unfortunately, the overweight are usually less fit, which compounds their handicap. The obese are usually chronically inactive, which deprives them of a potent means of weight loss.

The obese are also socially stigmatized. Fat people are often ridiculed and subtly discriminated against. This social rejection often causes them to develop emotional problems. Consequently, many people are involved in a constant battle to lose weight in an attempt to escape their obesity.

Rachel McLish winning the first U.S. Women's Body Building championship. Anyone who is not convinced that proper exercise and nutrition positively affects body composition ignores the obvious. (U.P.I.)

Huge industries have been established to cater to the obsession with weight control. Millions of dollars are spent on various diet books, low-calorie foods, and appetite suppressants in the quest for leanness. These remedies usually provide only a short-term solution to the problem and result in large fluctuations in weight that can be extremely unhealthy.

Chronic dieters usually overemphasize excessive weight rather than excessive fat. They often see the loss of water or lean mass as desirable and concentrate on the rate of weight loss rather than its permanence. They fail to realize that food provides nutrients as well as energy and that an adequate food intake is necessary for good health. The long-term solution lies in achieving a balance between energy intake and expenditure.

This chapter explores the causes and remedies of obesity. Emphasis will be placed on the role of exercise in reducing body fat and on the relationship between body composition and physical performance. Additionally, techniques used in measuring body composition will be discussed.

OBESITY AND HEALTH

The association between obesity and increased mortality is controversial. Although the higher incidence of many diseases among the obese is well established, obesity as a cause of premature death has been questioned. In the absence of serious risk factors such as hypertension and diabetes, obesity seems to pose a much less serious risk to overall mortality. However, there exists the possibility that obesity may be a cause of diabetes and hypertension, and thus be vitally linked with coronary heart disease (CHD), the number one killer.

Hypertension (both systolic and diastolic pressure) is twice as common in overweight persons as it is in those with normal weight, and three times more common than in those who are underweight (Figure 25-1). This relationship is most striking in white women and blacks of both sexes, and it exists in all age groups, even in children as young as 5 yr of age. In Western countries, the incidence of obesity and hypertension increases with age. However, hypertension does not increase with age in populations that do not experience the age-related increase in body fat, such as the Masai and New Guinea tribal natives. It is important to point out that the relationship between hypertension and obesity is not a result of measurement artifact (i.e., the use of small blood pressure cuffs on people with large arms often presents a spuriously high blood pressure).

The increased incidence of hypertension in the obese seems to be due to

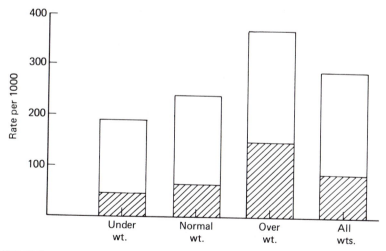

FIGURE 25-1

The frequency of hypertension in various weight categories (men and women). Darker bars represent 20- 39-year age group, and lighter bars represent 40 to 64 years.

(SOURCE: Data from P. Berchtold, V. Jorgens, C. Finke, and M. Berger, 1981.)

their higher salt consumption. The higher dietary salt content causes an increased blood volume due to expansion of erythrocyte mass and greater plasma volume. Weight loss in the obese almost always results in a reduction in blood pressure, even when no restriction is placed on salt intake. This is probably due to the high sodium content of most diets, which decreases proportionately with the caloric content of the meals eaten during a period of weight loss. However, there is good evidence that there are individual differences in the sensitivity to sodium in increasing blood pressure.

Obese individuals have a greater risk of developing diabetes, with the risk increasing with the severity and duration of the problem. Obesity often fosters resistance to insulin by decreasing the number of active insulin receptors. This seems to be caused by the chronically elevated oral glucose loads associated with a positive energy balance (higher energy intake than expenditure). The pancreas is sometimes unable to handle the increased insulin requirement, resutling in glucose intolerance. Obesity-related pancreatic islet failure may take as long as 15 to 20 yr to manifest itself. This type of diabetes seems to be related more to dietary overload than to the increased adiposity. Dietary restriction often increases the number of insulin receptors even before the ideal weight is achieved.

Hyperlipidemia is common in obesity, with over 31% of obese persons exhibiting this disorder. In obese persons with hyperlipidemia, 75% have high triglyceride levels, 25% have high cholesterol, and 50% have both. Additionally, there is a strong negative correlation between plasma high-density lipoproteins (HDL) and relative weight (proximity to ideal weight). HDL and triglyceride levels increase during weight reduction in the obese. Exercise reduces triglycerides and increases HDL, which reinforces the importance of physical activity in weight reduction.

Obesity greatly increases the risk of a variety of musculoskeletal disorders, particularly osteoarthritis and backache. The increased mass places chronic stress on joints, eventually leading to arthritic changes. A sagging abdomen, especially prominent in obese men, results in an increased lumbar lordosis that has been implicated as a cause in many incidents of back pain. This problem is compounded by generally weak and inflexible abdominal, spinal, and leg muscles.

THE THERMODYNAMICS OF OBESITY

The body's energy balance determines whether the amount of fat increases, decreases, or remains the same. The metabolism is in energy balance when energy intake equals energy expenditure (Figure 25-2). Obesity occurs when this balance becomes positive (more energy is consumed than expended.). Once

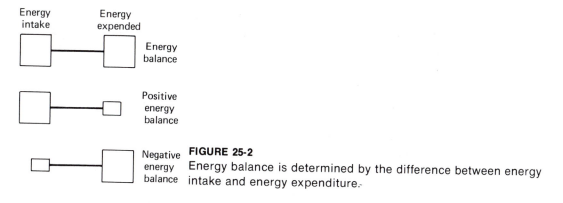

FIGURE 25-2
Energy balance is determined by the difference between energy intake and energy expenditure.

gained, the body fat will remain unless there is also a period of negative energy balance. Reducing obesity appears to be simply a matter of putting a rein on gluttony and burning up more calories through physical activity. Unfortunately, there are individual differences in the desire for energy consumption and the rate of energy expenditure.

Control of Food Intake

The control center for food intake is found in the hypothalamus; the hunger center lies in the lateral regions of the ventral hypothalamus, whereas the satiety center is located primarily in the ventromedial area of the hypothalamus. Normally, the satiety center dominates, except when the nutrient status declines. Then, the activity of the satiety center decreases, while that of the hunger center increases. Aberrations in the hunger–satiety center are involved in some types of obesity. However, demonstrable hypothalamic lesions are rare.

A number of factors have been suggested as triggering the hypothalamic hunger center, including glucose, stored triglycerides, plasma amino acid levels, and hypothalmic temperature. The *glucostatic theory of hunger regulation* proposes that glucose controls hunger and satiety with glucose sensors located in the hypothalamus. These sensors are thought to be sensitive to the effects of insulin and elicit satiety signals as the utilization of glucose increases within the sensor cells.

The *lipostatic theory* proposes that the long-term control of appetite is controlled by the amount of stored triglycerides. A faulty lipostatic mechanism may be involved in obesity. This theory is only conjecture at present and requires considerable research to substantiate it. Likewise, other researchers have proposed an *aminostat*—that is, control of hunger by plasma amino acids—as a regulator. Likewise, more experimental support is required for this theory.

Hypothalamic temperature may be involved in the control of appetite. This theory, called the *thermostatic theory,* proposes that the specific heat of food, the heat given off during digestion, causes a stimulation of the satiety center.

This is consistent with the often-observed phenomenon of increased appetite in the cold and decreased appetite in the heat. Likewise, it may help to explain the mechanism by which exercise exerts a short-term depressant effect on appetite.

A number of other factors have been implicated in the hunger–satiety mechanism. These include gastrointestinal (GI) stretch receptors, catacholamines, nutrients, and osmotic changes, GI hormones such as cholecystokinin, hepatic nutrient receptors, hormones sensitive to nutrient status such as insulin, glucagon, and growth hormone, and plasma levels of nutrients.

Exercise is thought to have a depressing effect on appetite. In rats, appetite is suppressed following high-intensity exercise but not low-intensity exercise. This is thought to be due to the greater release of catecholamines in response to the more intense level of physical activity, or, as discussed, because of a larger increase in core temperature. In humans, Mayer et al. has found a lower appetite in moderately active Indian mill workers compared to workers who were sedentary. However, as the energy requirements of the workers increased above moderate, the energy intake also increased. This may indicate that a minimum amount of physical activity may be necessary for appetite control and the maintenance of energy balance. Several exercise training studies using human subjects have found that physical activity had no effect on appetitie. Wilmore has interpreted this finding as evidence for a net decrease in appetite because caloric intake did not increase in spite of an increased caloric expenditure.

Psychological factors can affect food intake. This is extremely important as a cause of obesity and perhaps the most difficult to alter. Psychosocial factors such as food-centered social gatherings, structured mealtime, and anxiety-stimulated eating binges are often independent of physiological hunger-stimulatory mechanisms. A summary of hunger–satiety mechanisms appears in Figure 25–3.

Energy Expenditure

The components of energy expenditure include basal metabolic rate (BMR), thermal effect of food, facultative thermogenesis, and physical activity. There can be individual differences in each of these factors, which accounts for the considerable variation in body fat between individuals on diets with similar caloric content.

Basal metabolic rate is the energy requirement of an awake person during absolute rest. It is measured under stringent laboratory conditions and requires that the subject (1) had not eaten in 12 hr, (2) had a restful night's sleep, (3) performed no strenuous exercise after sleeping, and (4) was reclined in a comfortable, nonstressful environment [68 to 80°F (20 to 27°C)] for 30 min prior to the measurement.

Although BMR tends to decrease with age and is higher in men than women,

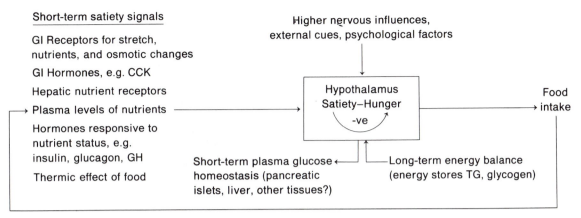

FIGURE 25-3
A simplified diagram of the mechanisms of hunger and satiety. The hypothalamus receives information on the amount and composition of the nutrients absorbed. This information is coordinated in relation to short-term
energy requirements and the status of the body's energy stores, and appropriate adjustments to food intake are initiated. Higher nervous influences reflecting external cues and psychological factors can exert important influences on this process in man. G-I. = gastrointestinal, CCK =
cholecystokinin, GH = growth hormone, TG = triglyceride, -ve = inhibitory
effect.

(SOURCE: From C.J. Bailey, 1978.)

it varies less than 10%. Most studies fail to demonstrate a difference in BMR
between fat and lean individuals. However, the increased body fat in older
individuals may be partially caused by gradual decrease in BMR with age. A
small deficit in caloric expenditure may become quite substantial when accumulated over a period of many years.

There is an increase in metabolic rate during the digestion and absorption of
food. The energy cost of digestion is highest for protein, lower for carbohydrates, and lowest for fats. Some fad diets advocate the consumption of high
amounts of protein because of its higher energy loss during digestion. However, close examination of these diets demonstrate that their ability to help
people lose weight stems from a low caloric content rather than the paltry
number of calories dissipated during digestion.

Facultative thermogenesis is heat production not accounted for by the BMR,
the thermic effect of food, or physical activity. When food is ingested, the
portion of energy not required for metabolism is either stored as fat or dissipated as heat. It appears that humans vary in their efficiency of utilizing food
energy. Some are very efficient and store most of the unneeded energy in adipose tissue, whereas others lose a portion as heat. This phenomenon may ac-

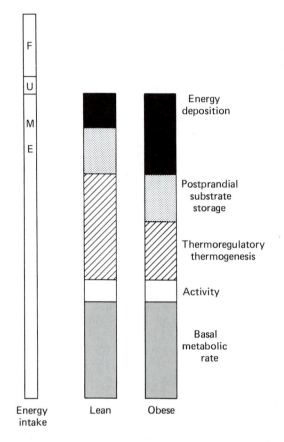

FIGURE 25-4

Components of energy intake and expenditures in lean and obese subjects. Energy intake provides the metabolizable energy (ME) with appreciable fecal (F) and urine (U) losses.

(SOURCE: From W.P.T. James and P. Trayburn, 1981.)

count for individual differences in fat deposition among people with seemingly identical caloric intake (Figure 25–4).

Brown adipose tissue (BAT) seems to be an important center for facultative thermogenesis in humans. There are small amounts of this tissue in the interscapular, axillary, and perirenal regions and around the great vessels in the thorax. The thermogenic property of BAT stems from the ability of its mitochondria to perform controlled uncoupling of oxidative phosphorylation (Figure 25-5). When the uncoupling pathway is activated, the utilization of substrate results in the production of heat rather than adenosine triphosphate (ATP).

Although BAT has been recognized as a heat-producing tissue in several animal species and in human infants, its relatively small amount excluded it from serious consideration as an important contributor to thermogenesis in the

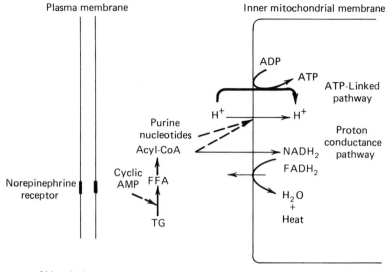

Plasma membrane

Inner mitochondrial membrane

Abbreviations:
FFA: free fatty acids
TG: triglycerides

FIGURE 25-5
Thermogenesis in brown adipose tissue.

(SOURCE: Modified from W.P.T. James and P. Trayburn, 1981.)

adult human. However, recent studies using thermography, a technique used to determine thermal gradients in the body, and direct thermistor readings following norepinephrine infusion, suggest that BAT may play a far more important role in thermogenesis than previously suspected. The obese show a much lower thermogenic response to infusion of norepinephrine, which may indicate an abnormality in the thermogenic pathway in BAT.

The amount of physical activity is probably the most significant and variable factor of energy expenditure. Basal metabolic rate is less than 1000 kcal $\cdot$ day^{-1} (depending on body size), but the additional caloric cost of physical activity can vary tremendously. Whereas a sedentary person typically expends approximately 1800 kcal $\cdot$ day^{-1}, the endurance athlete may use as much as 6000 kcal.

Many health experts place little emphasis on the importance of exercise in caloric expenditure. However, observing the threefold difference in caloric expenditure between the sedentary and extremely active should dispel this notion. Even small differences in caloric expenditure can become quite substantial when accumulated over many months.

During exercise, approximately 5 kcal are utilized for the consumption of every 1 liter of O_2. However, there is an increase in metabolic rate after exercise, due to the Q_{10} effect (heat increases metabolic rate) and to the anabolic processes elicited by the exercise stimulus, which can almost equal the caloric cost of the exercise itself. These postexercise effects add to the potent effect of physical activity in caloric expenditure.

Many studies have demonstrated that the amount of physical activity is the most important difference between the lean and obese. Obese individuals typically lead a sedentary life-style and when involved in sports, pursue them less vigorously than lean people. These patterns were evident in studies of children as well as adults.

HYPERCELLULARITY OF ADIPOSE TISSUE

Obesity can occur as a result of an increase in adipose cell size, number, or both (Figure 25-6). Severely obese individuals have been shown to have three times as many fat cells that can be 40% larger than in lean people. Weight loss results in a decrease in the size of the fat cells but not the number, so it is important to prevent the initial development of fat cells during their growth period.

Fat cells increase in both size and number during childhood, and to a certain extent during adolescence. In the adult the number of cells becomes fixed, and they increase only in size (except in extreme obesity).

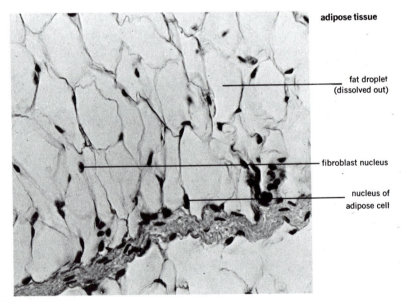

adipose tissue

fat droplet
(dissolved out)

fibroblast nucleus

nucleus of
adipose cell

FIGURE 25-6
Micrograph of adipose tissue.

(SOURCE: From J.E. Crouch and J.R. McClintic, 1971, p. 112.)

Obesity that begins in childhood is much more difficult to manage than if it develops in later life. In obese adult, severe triglyceride depletion of the adipocytes during weight reduction seems to upset their homeostasis, which may result in stimulation of the appetite center in the hypothalamus to reinstitute the cellular status quo. This is the basis of the lipostatic theory of hunger and satiety discussed earlier. This process makes it extremely difficult for people with increased fat cell number to maintain a lower body weight after a period of weight reduction.

Animal studies show that the development of fat cells can be reduced by endurance exercise and diet during growth. Additionally, suppression of fat cell growth during childhood makes the development of obesity during adulthood much less of a problem, even if physical activity is discontinued. It is unfortuntate that so many of our medical resources are devoted to fighting obesity in adults, rather than preventing the problem in children, where the effort will do much more good.

THE TREATMENT OF OBESITY

The positive energy blance of obesity may be caused by a number of factors, including genetic tendency, hypothalamic hunger–satiety defects, endocrine abnormalities, lipid or carbohydrate metabolism aberrrations, and inappropriate social attitudes toward eating and physical activity. It is not surprising that a variety of treatments and remedies have been proposed to deal with the problem (Table 25-1).

TABLE 25-1
Treatment of Obesity

Diet
Exercise
Anorectic drugs
Behavior modification
Jejunoileal bypass
Lipectomy
Hypnosis
Acupuncture
Miscellaneous agents: thyroid hormones, drugs that uncouple oxidative
  phosphorylation (e.g., dinitrophenol), human chorionic gonadotrophin,
  growth hormone, glucagon, progesterone, and biguanides

Diet

Caloric restriction is the most prominent treatment of obesity and is an essential part of any weight control program. Most quick-loss fad diets stress weight loss rather than fat loss and fail to provide a regimen that can be followed for life. The goal of a dietary program should be to lose body fat and then maintain the loss. Unfortunately, the composition of these fad diets is often so unpalatable and unhealthy that a rebound weight gain is inevitable.

Many of these diets promote low carbohydrate intake, which results in dehydration as muscle and liver glycogen stores are depleted. Although the weight loss appears impressive, most of it is in the form of water rather than fat. Additionally, glycogen depletion greatly diminishes exercise capacity, which almost eliminates physical activity as a source of caloric expenditure (Figure 25-7). Low-carbohydrate diets are usually high in fats, which could increase the risk of developing CHD.

The success or failure of any diet depends on its effect on energy balance. The loss of 1 pound of fat requires a caloric deficit of 3500 calories. A sedentary person of average size does not even expend that many calories in a day, so it is impossible to lose more than a few pounds of fat in a week. Diets that promise large decreases in fat in a short time are misleading the public.

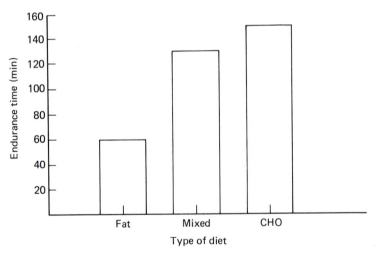

FIGURE 25-7
Endurance time on bicycle ergometer exercising at 75% of $\dot{V}_{O_2max}$ in subjects consuming high-fat, mixed, and high-carbohydrate (CHO) diets.

[SOURCE: From data presented in E. Hultman and L.H. Nilsson. Liver glycogen in man. Effects of different diets and muscular exercise. In: B. Pernow and B. Saltin (eds.), Muscle Metabolism During Exercise, Plenum Press, New York, 1971.]

Dehydration

Diuretics and impermeable clothing have been advocated from time to time for losing weight. Again, these result in weight loss and not fat loss. The resulting dehydration can be extremely dangerous. There have been numerous cases of severe cardiac arrhythmias due to dehydration-weight loss programs. Although diuretics are often used in the treatment of hypertension in obese individuals, they have no place in the treatment of obesity itself.

Sauna belts and rubber suits are often used by the uninitiated in their quest to lose weight and inches. The resulting dehydration of the cells under the clothes does cause a temporary loss of circumference. However, the change is short-lived as normal fluid balance is quickly restored.

Medical Procedures

Medical intervention in the hospital is sometimes employed in extreme cases of obesity. Hospital procedures include prolonged fasting, jejunoileal bypass, and lipectomy. These techniques have rendered mixed results and are often accompanied by undesirable side effects.

Prolonged starvation has resulted in weight losses of over 100 pounds. However, this procedure results in the loss of a considerable amount of lean body mass and can result in serious side effects, such as gout, anemia, hypotension, and various metabolic disturbances. In addition, starvation does little to modify eating habits that will help the person avoid regaining weight that has been lost.

The jejunoileal bypass is a resection of a portion of the small intestine to decrease the absorption of nutrients. This is used as a last resort in the grossly obese and is usually accompanied by side effects such as chronic diarrhea and kidney stones. Lipectomy, the surgical removal of adipose tissue, is seldom successful because of compensatory hypertrophy of the remaining cells. Additionally, there is risk from the surgical procedure itself.

A number of agents have been used to alter metabolic rate, curb appetite, uncouple oxidative phosphorylation, or alter lipid metabolism. Thyroid hormones were in common use a number of years ago, but they caused an undesirable reduction in lean body mass and an increased incidence of heart arrhythmias. Anorectic drugs, such as amphetamine, although somewhat effective, have serious side effects such as habituation and cardiac arrhythmias. Uncoupling agents, such as dinitrophenol, are toxic at the dosage required to render the desired effect. Other agents with questionable effectiveness and desirability include human chorionic gonadotrophin, growth hormone, glucagon, progesterone, and biguanides.

Acupuncture, a quasi-medical technique involving the puncture of the body with small needles, has been suggested as a technique in the management of obesity. Little objective experimental evidence is available, and its effectiveness remains unestablished.

Behavior Modification

Behavior modification programs focus on the elimination of behavior associated with poor eating habits. The success rate of this technique has been mixed. Whereas some studies have demonstrated weight losses of 9 to 18 kg in over 80% of their subjects, others showed considerably less success than this.

The long-term success rate after weight reduction through behavior modification is somewhat dismal. Follow-ups indicate that at least 70% of subjects regained their lost weight after a year. On a more positive note, Currey et al. noted that almost one-third of those involved continued to be successful after a year and even continued to lose more weight. So, even though this procedure of weight reduction through behavior modification will not help everyone, it can be helpful for some.

At present there is little known relationship between such factors as personality type, onset of obesity, age, or weight at entry into the program and predictability of success in behavior modification programs. Unfortunately, the biggest failures, as a group, have been medically referred older obese persons with numerous health problems. These are the people who need to reduce the most.

Hypnosis can be thought of as a kind of behavior modification and is widely used. Unfortunately there is no clinical research that clearly substantiates its effectiveness. Most available evidence is restricted to clinical observations and one-subject anecdotal reports. There is no standardization of techniques or of hypnotic suggestions, and little regard for scientific method. Although hypnosis may be a valuable tool in the treatment of obesity, it usefulness has not been established.

Exercise

Exercise is a vital part of a lifelong weight control program, because it allows the consumption of enough calories to supply the body with adequate nutrients as well as energy. Caloric restriction alone often leads to malnourishment, because the diet may not contain enough food to provide the necessary vitamins and minerals. Chronic caloric restriction may eventually have serious health consequences.

Although a single session of exercise results in little fat loss, regular training can make a substantial difference in the weight control program. The expenditure of 300 calories during exercise, three or four times a week, can result in ther loss of 13 to 23 pounds of fat in a year, provided the caloric intake remains the same. Although that may not seem like much to a crash dieter, the weight loss consists of fat and not a combination of water, lean tissue, and fat, which is commonly lost on most fad diets. When weight is lost soley through caloric restriction, lean mass may represent 35 to 45% of the loss. Exercise training spares body protein as it mobilizes lipids. Provided that the total weight loss is not too great and the rate of weight loss is not too rapid, a weight-loss program that stresses exercise and diet will result in the loss of body fat and the sparing (retention) of lean body mass.

Whereas dieting is a drudgery, exercise can represent an enjoyable form of caloric expenditure. For this reason it is important to link a person to a compatible form of physical activity. Whereas some people enjoy exercising in a group, others like to train alone. For others, competition is an important component of physical activity. These individual differences in motivation should be considered or discovered if physical activity is to become a permanent and important part of the weight management program.

As fitness improves, exercise has a more potent effect on caloric utilization. A change in maximal O_2 consumption from 3 to 4 liters $\cdot$ min^{-1} has increased the ability to burn calories by one third. Exercise for weight control should center on long-term endurance activity for a minimum of 20 min.

The amount of fat loss due to exercise is directly proportional to the duration and intensity of the activity. At the same intensity of exercise, in activities that are greatly affected by gravity, the obese lose more fat than the lean because they do more work.

The most successful weight loss programs seem to be those that use a combination of diet and exercise. Exercise complements diet because of its negative effect on appetite, making a program of dietary restriction more bearable. In addition, the calories expended during exercise often allow the dieter to consume more calories and still lose weight, which presents a more acceptable long-term weight control program.

Unfortunately, exercise does not result in the spot reduction of fat. Although the improved muscle tone that results from training will usually make a particular area of the body look better, the subcutaneous adipose layer that lies over the muscles is unaffected (except as it is affected by any negative caloric balance).

BODY COMPOSITION

Body composition can be divided into two components: (1) lean body mass or fat-free weight, and (2) body fat. The lean body mass encompasses all of the body's nonfat tissues including the skeleton, water, muscle, connective tissue, organ tissues, and teeth. The body fat component includes both the essential and the nonessential lipid stores. Essential fat includes lipid incorporated into organs and tissues such as nerves, brain, heart, lungs, liver, and mammary glands. The storage or nonessential fat exists primarily within adipose tissue.

The morning weighing ritual on the bathroom scale is really an attempt at estimating body composition. Using this method, it is impossible to determine accurately if a fluctuation in weight is due to a change in muscle, body water, or fat. The method cannot distinguish between overweight and over-fat. A 260-pound muscular football player may be overweight according to population height–weight standards yet actually have much less body fat than average. Likewise, a 40-year-old woman may weight exactly the same as when she was in high school, yet have a considerably different body composition.

Several more precise methods have been developed that can provide a more precise estimation of body composition. Laboratory methods include densitometry and potassium-40 counting. Field test methods include ultrasound, anthropometry, and skinfolds. The field tests are usually validated against standard laboratory techniques.

Hydrostatic Weighing

Hydrostatic or underwater weighing is considered the most accurate indirect means of measuring body composition and serves as a standard for others indirect techniques such as skinfolds (Figure 25-8). This procedure was popular-

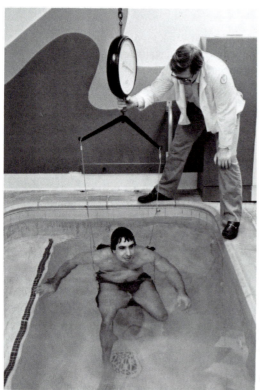

FIGURE 25-8
The underwater weighing procedure. A) Subject maximally expires from lungs. B) Subject submerges and is weighed. C) Data is corrected for residual lung volume (O_2 washout method).

(a)   (Wayne Glusker.)

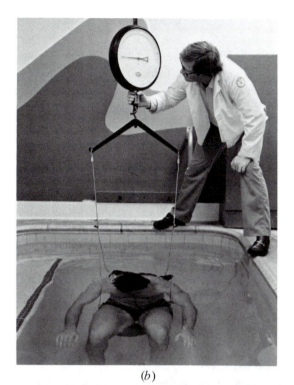

(b)

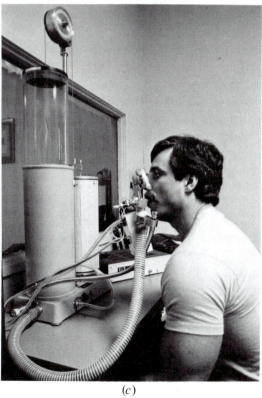

(c)

FIGURE 25-8 *(continued)*.

ized by researchers such as Behnke and Pace in the early 1940s and has become an important tool in exercise physiology and medicine. The equations used in this method stem from the direct chemical analysis of human cadavers.

Chemical analyses of human cadavers have demonstrated several universal characteristics of body composition that have made possible the indirect measurement of body fat. These studies have shown that the density of fat (density is a measure of compactness) and muscle, and the ratios of skeletal weight and body water to lean body weight are extremely constant. Furthermore, numerous animal studies showed an inverse relationship between the density of the animal and its percentage of fat.

Density is equal to mass divided by volume. Unfortunately, the irregular shape of the human body makes a simple geometric estimation of its volume impossible. The volume of the body can be measured by using Archimedes' principle of water displacement. This principle states that "a body immersed in water is buoyed up with a force equal to the weight of the water displaced." The volume of the body can be measured by determining the weight loss by complete immersion in water. The density of the body, and thus its percentage of fat, can be calculated by dividing the body weight (scale weight) by the body volume (calculated by underwater weighing).

In this procedure, the subject is submerged and weighed underwater. Muscle has a higher density and fat a lower density than water (1.1 g/cm³ for muscle, 0.91 g/cm³ for fat, and 1 g/cm³ for water). Therefore, fat people tend to float and weigh less underwater, whereas lean people tend to sink and weigh more underwater. At a given body weight, a fat person has a larger volume than a thin one, and thus a smaller density.

A number of errors are possible even in this relatively precise laboratory procedure. Failure to consider factors such as residual lung volume, intestinal gas, and water density will decrease the underwater weight and result in an overestimation of volume. During the measurement, the subject is weighed completely submerged in water while attempting to expel as much air as possible from the lungs. A small but variable amount of air called the residual lung volume remains, which must be taken into account in the equation. While residual lung volume can be estimated, accurate assessment of body composition requires that it be measured directly. Furthermore, intestinal gas will also increase buoyancy and confound the results. Finally, water has a density of 1 g/cm³ only at a temperature of 39.2°F (4°C). The calculation of volume must be corrected for a difference in water density if the water temperature is other than 39.2°F.

The two most widely used equations for the calculation of lean body mass and percentage of fat were derived by Brozek et al., and Siri. Their slight variance stems from different estimations of the density of fat and muscle. These equations, together with sample calculations of density, percentage fat, and lean mass appear in Table 25-2.

TABLE 25-2

Equations for the Calculation of Body Density, Percentage Fat, Body Fat, and Lean Body Mass

---

1. Body density $= \dfrac{BW}{\dfrac{BW - UWW}{D_{H_2O}} - RLV}$

   where BW = body weight (kg), UWW = underwater weight, D_{H_2O} = density of water (at submersion temperature), and RLV = residual lung volume.

2. Percentage fat $= \left(\dfrac{4.950}{D_b} - 4.50 \right) \times 100$
   (Siri equation)
   where D_b = body density.

3. Percentage fat $= \left(\dfrac{4.570}{D_b} - 4.142 \right) \times 100$
   (Brozek equation)

4. Total body fat (kg) = Body weight (kg) $\times \% \dfrac{Fat}{100}$

5. Lean body weight = Body weight − fat weight

Although the hydrostatic weighting technique is considered the "gold standard," it is subject to error due to violations of the underlying assumptions of the method. Wilmore summarized some of the possible errors in a recent review, including individual variability in the density of the lean mass component, variations in bone mineral and total body water, and differences in density and proportions between the target population (or individuals within that population) and the "reference man." The "reference man" is a composite of the human body composition based on direct chemical analysis of relatively few cadavers.

Biochemical Techniques

A number of biochemical techniques have been employed in the measurement of body composition and are based on biological constants observed during the direct chemical analysis of the body. These methods include potassium-40, total body water, and inert gas absorption.

The lean body mass contains a relatively constant proportion of potassium. Part of this potassium is in the form of potassium-40, a naturally occurring isotope. The γ rays emitted by the potassium-40 can be measured with a whole-body scintillation counter, which allows the prediction of total body potassium and lean body mass. The results of this method are very similar to those of hydrostatic weighing. Because of the expense of the scintillation counter, this method is almost completely restricted to research.

Various diffusion techniques rely on the property of various substances to diffuse into specific tissues or compartments. Tracers such as deuterium oxide, tritium oxide, antipyrine, and ethanol have been used to estimate total body water. Lean body mass can thus be estimated, because body water represents an almost constant 73.1% of body weight and the water is contained almost entirely within the lean body mass. Fat-soluble, inert gases such as krypton and cyclopropane have been used to estimate percentage fat by measuring their rate of absorption into the body. As discussed in Chapter 23, body fat is an important factor in the uptake and release of inert gases in the body.

Anthropometric and Skinfold Techniques

Anthropometric assessment of body composition utilizes various superficial measurements such as height, weight, and anatomical circumferences. Of these, height–weight is by far the most popular. Height–weight tables, periodically produced by insurance companies, are inadequate, because they are subject to individual interpretation by requiring people to decide if they are of small, medium, or large frame and because they do not take into consideration individual differences in lean body mass and relative fat.

Sheldon devised a rating system for assessment of body composition based on three components, rated on a seven-point scale. The three components were (1) *endomorphy*, relative predominance of corpulence and roundness, (2) *mesomorphy*, relative predominance of muscularity, and (3) *ectomorphy*, relative predominance of linearity and fragility of body build. This method is very subjective and requires a test administrator who is well trained in the photographic protocol and rating system.

Behnke presented an anthropometric technique for the estimation of body composition that compares various circumferences of a subject with those of a reference man or woman. Like other field test methods, however, it requires experience to obtain the precise measurements necessary for accurate results.

Skinfold measurements are probably the most popular ''scientific'' means of assessing body composition (Figure 25-9). This method is inexpensive, rapid, and takes little time to learn. Skinfold equations are derived using a statistical technique called multiple regression, which predicts the results of the hydrostatic weighing procedure from the measurement of skinfolds taken at various sites. It is absolutely essential that a subject be measured using an equation derived from a similar population. For example, it would be inappropriate to use a skinfold equation to estimate the body fat of a 40-year-old man that was derived from 18-yr-old college females.

Several models of skinfold calipers are available. The ideal caliper should have parallel jaw surfaces and a constant spring tension, regardless of the de-

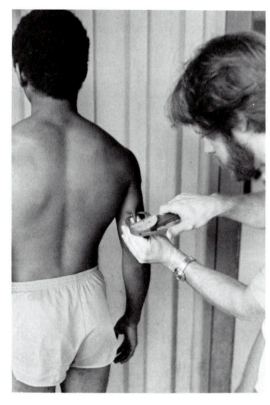

FIGURE 25-9
The skinfold procedure.

(Wayne Glusker.)

gree of opening. Many inexpensive plastic calipers have become available recently, but these should be avoided because of the potential for error.

The skinfold method is subject to severe limitations and should only be used
in field studies or in circumstances where only gross estimations of body composition are required. This technique has been shown to render considerable
measurement error, even among experienced observers. Dehydration will decrease a skinfold thickness by as much as 15%, so would likely cause variability between measurements taken in the morning and evening. In addition, skinfold measurements have been shown to be of no value in predicting changes in
body composition following weight loss. The accuracy of this method can be
improved through the use of multiple measurements by a single experienced
observer.

THE IDEAL BODY COMPOSITION

Perhaps the three most important considerations determining the ideal body
composition are health, aesthetics, and performance. The average person tends
to be most concerned with the first two factors, whereas the athlete is concerned about all three.

Although the absolute range of the ideally "healthy fat percentage" has not been clearly established, it is generally agreed to be between 16 and 25% for women and less than 20% for men. Figure 25-10a and b presents the 90th, 50th, and 10th percentiles of percentage fat for men and women in different age groups. Notice that the average man and woman fall above the recommended fat percentage in almost all age categories.

The ideal aesthetic body composition is even more difficult to establish. These days, the lean, athletic look is prized, whereas the more corpulent look of the turn of the century is disdained. Unfortunately, the quest for the "lean look" often leads to unhealthy dietary habits. This is especially disturbing among athletes, who usually have very high caloric requirements because of heavy training. Young teenage female athletes seem to be especially prone to overzealous caloric restriction that can be dangerous.

Body composition is an extremely important consideration in many sports. In sports with weight categories such as wrestling, boxing, and weight lifting, serious abuses have taken place in an attempt to lose weight, which have sometimes compromised the health of the athletes. These athletes are subjected to severe dehydration in attempts to "make weight." These practices caused the American College of Sports Medicine to issue a position statement on weight loss in wrestlers (appendix VI). Two of their recommendations were particularly relevant to the importance of body composition in sport: (1) "Assess the body composition of each wrestler several weeks in advance of the competitive season. Individuals with a fat content less than 5 percent of their certified body weight should receive medical clearance before being allowed to compete." (2) "Discourage the practice of fluid deprivation and dehydration."

Successful athletes in various sports usually possess a characteristic body composition (Table 25-3). The variability in body fat seems to depend on the metabolic requirement of the activity and the relative disadvantage of carrying an extra load. For example, successful male distance runners almost always have less than 9% fat. For these athletes, excess fat is a decided disadvantage, and the tremendous caloric cost of running long distances makes the deposition of large amounts of fat extremely difficult.

Football linemen, in contrast, almost always have more than 15% fat. This may be advantageous because of the added mass and padding provided by the subcutaneous fat and the increase in lean mass that accompanies excess weight (muscle mass accompanies gains in fat to support the extra weight). Unfortunately, many younger athletes gain too much fat in an attempt to attain the high body weights of the professional football player.

The use of relative fat (the expression of body fat as a percentage) can be misleading in large athletes because the high body weight "dilutes" the fat pool. For example, suppose two 6-ft 4-in. athletes each had 10% fat. However, the first athlete weighed 200 pounds and carried 20 pounds of fat, whereas the

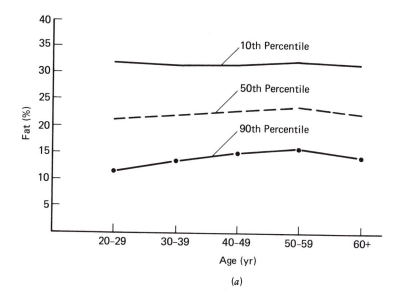

(a)

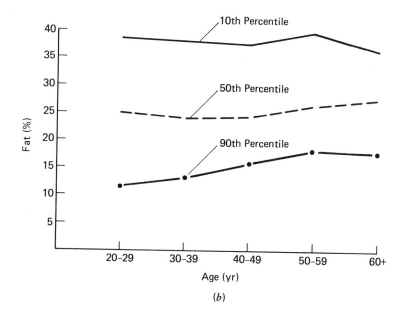

(b)

FIGURE 25-10
Fat percentage in (a) males and (b) females as a function of age.

(SOURCE: From data presented in M. Pollock, J.H. Wilmore, and S.M. Fox, Health and Fitness Through Physical Activity, John Wiley & Sons, New York, 1978.)

TABLE 25-3
Body Composition Values in Athletes

Athletic Group or Sport	Sex	Age (yr)	Height (cm)	Weight (kg)	Relative Fat (%)
Baseball	Male	20.8	182.7	83.3	14.2
	Male	—	—	—	11.8
	Male	27.4	183.1	88.0	12.6
Basketball	Male	26.8	193.6	91.2	9.7
	Female	19.1	169.1	62.6	20.8
	Female	19.4	167.0	63.9	26.9
Football	Male	20.3	184.9	96.4	13.8
	Male	—	—	—	13.9
Defensive backs	Male	17–23	178.3	77.3	11.5
Offensive backs	Male	17–23	179.7	79.8	12.4
Linebackers	Male	17–23	180.1	87.2	13.4
Offensive linemen	Male	17–23	186.0	99.2	19.1
Defensive linemen	Male	17–23	186.6	97.8	18.5
Defensive backs	Male	24.5	182.5	84.8	9.6
Offensive backs	Male	24.7	183.8	90.7	9.4
Linebackers	Male	24.2	188.6	102.2	14.0
Offensive line	Male	24.7	193.0	112.6	15.6
Defensive line	Male	25.7	192.4	117.1	18.2
Quarterbacks, kickers	Male	24.1	185.0	90.1	14.4
Gymnastics	Male	20.3	178.5	69.2	4.6
	Female	19.4	163.0	57.9	23.8
	Female	20.0	158.5	51.1	15.5
	Female	14.0	—	—	17.0
	Female	23.0	—	—	11.0
	Female	23.0	—	—	9.6
Ice hockey	Male	26.3	180.3	86.7	15.1
Jockeys	Male	30.9	158.2	50.3	14.1
Skiing	Male	25.9	176.6	74.8	7.4
Swimming	Male	21.8	182.3	79.1	8.5
	Male	20.6	182.9	78.9	5.0
	Female	19.4	168.0	63.8	26.3

Athletic Group or Sport	Sex	Age (yr)	Height (cm)	Weight (kg)	Relative Fat (%)
Track and field	Male	21.3	180.6	71.6	3.7
	Male	—	—	—	8.8
Runners	Male	22.5	177.4	64.5	6.3
Distance	Male	26.1	175.7	64.2	7.5
	Male	40–49	180.7	71.6	11.2
	Male	50–59	174.7	67.2	10.9
	Male	60–69	175.7	67.1	11.3
	Male	70–75	175.6	66.8	13.6
	Male	47.2	176.5	70.7	13.2
	Female	19.9	161.3	52.9	19.2
	Female	32.4	169.4	57.2	15.2
Sprint	Female	20.1	164.9	56.7	19.3
Discus	Male	28.3	186.1	104.7	16.4
	Male	26.4	190.8	110.5	16.3
	Female	21.1	168.1	71.0	25.0
Jumpers and hurdlers	Female	20.3	165.9	59.0	20.7
Shot put	Male	27.0	188.2	112.5	16.5
	Male	22.0	191.6	126.2	19.6
	Female	21.5	167.6	78.1	28.0
Tennis	Male	—	—	—	15.2
Volleyball	Female	19.4	166.0	59.8	25.3
Weight lifting	Male	24.9	166.4	77.2	9.8
Power	Male	26.3	176.1	92.0	15.6
Olympic	Male	25.3	177.1	88.2	12.2
Body builders	Male	29.0	172.4	83.1	8.4
Wrestling	Male	26.0	177.8	81.8	9.8
	Male	27.0	176.0	75.7	10.7
	Male	22.0	—	—	5.0
	Male	19.6	174.6	74.8	8.8
	Male	15–18	172.3	66.3	6.9

SOURCE: From Wilmore, J.H. and J.A. Bergfeld. A comparison of sports: physiological and medical aspects. In: Strauss, R.J. (ed.), Sports Medicine and Physiology. Philadelphia: W.B. Saunders, 1979.

second athlete weighed 300 pounds and carried 30 pounds of fat. The second athlete carried 50% more fat even though he was of equal stature and fat percentage.

Summary

Obesity results from a positive energy balance and is associated with an increased risk of a variety of diseases. The causes of obesity are complex and possibly include aberrations in appetite control, socially mediated abnormal eating patterns, physical inactivity, and impaired facultative thermogenesis.

Gross obesity has been associated with adipocyte hypercellularity. Fat cell development can probably be affected by manipulation of diet and physical activity patterns during growth. The number of fat cells appears to be fixed in adults, which could explain the extreme difficulty in dealing with obesity.

A variety of treatments and remedies have been proposed to deal with obesity, including extreme diets, dehydration, fasting, jejunoileal bypass, exercise, acupuncture, behavior modification, and hypnosis. Although some of these techniques have been somewhat successful, the long-term prognosis has been dismal. Exercise seems to be an essential component of any successful long-term weight control program.

The measurement of body composition has become important in sports medicine. These measurements can help the individual develop the ideal body composition for health, aesthetics, and physical performance. Methods include densitometry, potassium-40, anthropometry, and skinfolds.

Selected Readings

Alliston, J.C. The use of a danscanner ultrasonic machine to predict the body composition of Hereford bulls. Anim. Prod. 35: 361–365, 1982.

American College of Sports Medicine. Position statement on weight loss in wrestlers. Med. Sci. Sports 7: vii, 1975.

Angel, A. Pathophysiologic changes in obesity. Can. Med. Assoc. J. 119: 1401, 1406, 1978.

Bailey, C.J. On the physiology and biochemistry of obesity. Sci. Prog. 65: 365–393, 1978.

Behnke, A.R. Anthropometric fractionation of body weight. J. Appl. Physiol. 16: 949–954, 1961.

Behnke, A.R. Quantitative assessment of body build. J. Appl. Physiol. 16: 960–968, 1961.

Behnke, A.R. and J. Royce. Body size, shape, and composition of several types of athletes. J. Sports Med. Phys. Fitness. 6: 75–88, 1966.

Behnke, A.R. and J.H. Wilmore. Evaluation and Regulation of Body Build and Composition. Englewood Cliffs, N.J.: Prentice-Hall, 1974.

Berchtold, P., V. Jorgens, C. Finke, and M. Berger. Epidemiology of obesity and hypertension. Int. J. Obesity. 5: 1–7, 1981.

Berg, K. and C.W. Bell. Physiological and anthropometric determinants of mile run time. J. Sports Med. Phys. Fitness. 20: 390–396, 1980.

Bray, G.A., J.A. Glennon, L.B. Salans, E.S. Horton, E. Danforth, and E.A. Sims. Spontaneous and experimental human obesity: Effects of diet and adipose cell size in lipolysis and lipogenesis. Metabolism. 26: 739–747, 1977.

Brown, C.H. and J.H. Wilmore. The effects of maximal resistance training on the strength and body composition of women athletes. Med. Sci. Sports. 6: 174–177, 1974.

Brozek, J., F. Grande, J.T. Anderson, and A. Keys. Densitometric analysis of body composition: Review of some quantitative assumptions. Ann. N.Y. Acad. Sci. 110: 113–140, 1963.

Brozek, J. and A. Henschel (eds.). Techniques for Measuring Body Composition. Washington, D.C.: National Academy of Sciences, National Research Council, 1961.

Buskirk, E.R. Obesity: A brief overview with emphasis on exercise. Fed. Proc. 33: 1948–1951, 1974.

Carter, J.E.L. (ed.). Physical structure of Olympic athletes: The Montreal Olympic Games Anthropological Project. Basel: S. Karger, 1982.

Carter, J.E.L. The somatotypes of athletes—a review. Hum. Biol. 42: 535–569, 1970.

Clarke, H.H. Exercise and fat reduction. Physical Fitness Research Digest. 5: 1–27, 1975.

Consolazio, C.F., R.E. Johnson, and L.J. Pecora. Physiological Measurements of Metabolic Functions in Man. New York: McGraw-Hill, 1963.

Costill, D.L., R. Bowers, and W.F. Kammer. Skinfold estimates of body fat among marathon runners. Med. Sci. Sports. 2: 93–95, 1970.

Crouch, J.E. and J.R. McClintic, Human Anatomy and Physiology. New York: John Wiley & Sons, 1971. p. 112.

Currey, H., R. Malcolm, E. Riddle, and M. Schachte. Behavioral treatment of obesity. JAMA 237: 2829–2831, 1977.

Dubois, D. and E.F. Dubois. A formula to estimate the approximate surface area of the body if height and weight be known. Arch. Intern. Med. 17: 863–871, 1916.

Epstein, L.H. and R.R. Wing. Aerobic exercise and weight. Addict. Behav. 5: 371–388, 1980.

Fahey, T.D., L. Akka, and R. Rolph. Body composition and $\dot{V}_{O_2max}$ of exceptional weight-trained athletes. J. Appl. Physiol. 39: 559–561, 1975.

Franklin, B.A. and M. Rubenfire. Losing weight through exercise. J.A.M.A. 244: 377–379. 1980.

Himms-Hagen, J. Brown adipose tissue as an energy buffer: A role in energy balance and obesity. J. Can. Dietetic Assoc. 44: 36–48, 1983.

Himms-Hagen, J. Obesity may be due to a malfunctioning of brown fat. Can. Med. Assoc. J. 121: 1361–1364, 1979.

James, W.P.T. and P. Trayburn. Thermogenesis and obesity. Br. Med. Bull. 37: 43–48, 1981.

Katch, F.I. Apparent body density and variability during underwater weighing. Res. Q. 39: 993–999, 1968.

Katch, F.I. and W.D. McArdle. Nutrition, Weight Control, and Exercise. Boston: Houghton-Mifflin Co., 1977. p. 101–134.

Kleiber, M. The Fire of Life. New York: John Wiley & Sons, 1961. p. 41–59.

Lewis, S., W.L. Haskell, H. Klein, J. Halpern, and P.D. Wood. Prediction of body composition in habitually active middle-aged men. J. Appl. Physiol. 39: 221–225, 1975.

Malcolm, R., P.M. O'Neil, A.A. Hirsch, H.S. Currey, and G. Moskowitz. Taste hedonics and thresholds in obesity. Int. J. Obesity. 4: 203–212, 1980.

Malina, R.M., B.W. Meleski, and R.F. Shoup. Anthropometric, body composition, and maturity characteristics of selected school-age athletes. Ped. Clin. North Am. 29: 1305–1323, 1982.

Malina, R.M., W.H. Mueller, C. Bouchard, R.F. Shoup, and G. Lariviere. Fatness and fat patterning among athletes at the Montreal Olympic Games, 1976. Med. Sci. Sports Exerc. 14: 445–452, 1982.

Mayer, J., P. Roy, and K.P. Mitra. Relation between caloric intake, body weight, and physical work: Studies in an industrial male population in West Bengal. Am. J. Clin. Nutr. 4: 169–175, 1956.

Miles, C.A. and G.A. Fursey. A note on the velocity of ultrasound in living tissue. Anim. Prod. 18: 93–96, 1974.

Mott, T. and J. Roberts. Obesity and hypnosis: A review of the literature. Am. J. Clin. Hypnosis. 22: 3–7, 1979.

Oscai, L.B., S.P. Babirak, F.B. Dabach, J.A. McGarr, and C.N. Spirakis. Exercise or food restriction: Effect on adipose tissue cellularity. Am. J. Physiol. 277: 901–904, 1974.

Oscai, L.B., S.P. Babirak, J.A. McGarr, and C.N. Spirakis. Effect of exercise on adipose tissue cellularity. Fed. Proc. 33: 1956–1958, 1974.

Oscai, L.B. and J.O. Holloszy. Effects of weight changes produced by exercise, food restriction, or overeating on body composition. J. Clin. Invest. 48: 2124–2128, 1969.

Pace, N. and E. Rathbun. Studies on body composition. J. Biol. Chem. 158: 685–691, 1945.

Pascale, L.R., M.I. Grossman, H.S. Sloan, and T. Frankel. Correlations between thickness of skinfolds and body density in 88 soldiers. Hum. Biol. 28: 165–176, 1956.

Pollock, M.L., L.R. Gettman, A. Jackson, J. Ayres, A. Ward, and A.C. Linnerud. Body composition of elite class distance runners. Ann. N.Y. Acad. Sci. 301: 297–309, 1977.

Pollock, M.L., T. Hickman, Z. Kendrick, A. Jackson, A.C. Linnerud, and D. Dawson. Prediction of body density in young and middle-aged men. J. Appl. Physiol. 40: 300–304, 1976.

Pollock, M.L., E.E. Laughridge, B. Coleman, A.C. Linnerud, and A. Jackson. Prediction of body density in young and middle-aged women. J. Appl. Physiol. 38: 745–749, 1975.

Powers, P.S. Obesity. Psychosomatics. 23: 1028–1039, 1982.

Rathbun, E.N. and N. Pace. Studies on body composition. J. Biol. Chem. 158: 667–676, 1945.

Severson, D.L. Regulation of lipid metabolism in adipose tissue and heart. Can. J. Physiol. Pharmacol. 57: 923–937, 1980.

Sharp, J.T., M. Barrocas, and S. Chokroverty. The cardiorespiratory effects of obesity. Clin. Chest Med. 1: 103–117, 1980.

Sheldon, W.H., C.W. Dupertuis, and E. McDermott. Atlas of Men. New York: Harper Brothers, 1954.

Siri, W.E. Gross composition of the body. In: Lawrence, J.H. and C.A. Tobias (eds.), Advances in Biological and Medical Physics, IV. New York: Academic Press, 1956.

Sloan, A.W. Estimation of body fat in young men. J. Appl. Physiol. 23: 311–315, 1967.

Sloan, A.W., J.J. Burt, and C.S. Blyth. Estimation of body fat in young women, J. Appl. Physiol. 16: 967–970, 1962.

Stern, J.S. and M.R.C. Greenwood. A review of development of adipose cellularity in man and animals. Fed. Proc. 33: 1952–1955, 1974.

Tanaka, K. and Y. Matsuura. A multivariate analysis of the role of certain anthropometric and physiological attributes in distance running. Ann. Hum. Biol. 9: 473–482, 1982.

Tanner, J.M. The Physique of the Olympic Athlete. London: George Allen & Unwin, 1964.

Weltman, A., S. Matter, and B.A. Stamford. Caloric restriction and/or mild exercise: Effects on serum lipids and body composition. Am. J. Clin. Nutr. 33: 1002–1009, 1980.

Wilcox, A.R. The effects of caffeine and exercise on body weight, fat-pad weight, and fat-cell size. Med. Sci. Sports Exerc. 14: 317–321, 1982.

Wilmore, J.H. A simplified method for determination of residual lung volumes. J. Appl. Physiol. 27: 96–100, 1969.

Wilmore, J.H. Body composition in sport and exercise: Directions for future research. Med. Sci. Sports Exerc. 15: 21–31, 1983.

Wilmore, J.H., C.H. Brown, and J.A. Davis. Body physique and composition of the female distance runner. Ann. N.Y. Acad. Sci. 301: 764–776, 1977.

Wilmore, J.H. and J.A. Bergfeld. A comparison of sports: Physiological and medical aspects. In: Strauss, R.J. (ed.), Sports Medicine and Physiology. Philadelphia: W.B. Saunders, 1979.

Wilmore, J.H., R.N. Girandola, and D.L. Moody. Validity of skinfold and girth assessment for prediction of alterations in body composition. J. Appl. Physiol. 29: 313–317, 1970.

Wilmore, J.H. and W.L. Haskell. Body composition and endurance capacity of professional football players. J. Appl. Physiol. 33: 564–567, 1972.

26

EXERCISE AND DISEASE

The human body has a high tolerance for exercise. Heart rate can increase by 150%, stroke volume by 50%, and pulmonary venous blood remains almost fully saturated at maximal exercise, even in the most unfit individual. Metabolism can increase as little as 3 to 4 times above rest in the debilitated person to as much as 25 times above rest in the endurance athlete. However, in disease the body is often under intolerable stress. An additional load could lead to further deterioration or, in some cases, death.

Exercise is currently being used as an adjunct treatment for many diseases. Training provides the body with the ability to meet higher metabolic demands that otherwise would lead to further deterioration from disease. The stress of exercise, however, must be applied cautiously and conservatively in disease states, if at all. This chapter presents an overview of the indications and contraindications of exercise in the face of disease.

CONTRAINDICATIONS OF EXERCISE

Relative contraindications to exercise are described in Table 26-1. In some instances, exercise is allowable in the presence of some of these conditions under medical supervision.

Studies of sudden death during exercise almost always indicate the existence

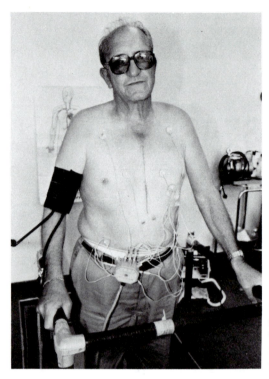

Exercise stress testing is valuable means to evaluate cardio-pulmonary status in many individuals, including those recovering from open heart surgery. (Courtesy M. Harvey, Stanford University Heart Disease Prevention Center.)

of acute cardiovascular disease. Because thorough cardiovascular screening is not commonplace (and perhaps impractical), many of these deaths are probably not preventable. However, in cases where contraindications to exercise are known, prudent behavior dictates cessation or modification of the activity program.

Conditions that pose dangers to the heart itself are obvious contraindications for exercise training. Vigorous exercise within 6 to 8 weeks of a myocardial infarction (MI) is usually considered dangerous because of the greater risk of reinfarction and arrhythmias. There is a trend toward earlier ambulation, but this is done only after careful screening of the individual case. Patients are now started on programs in the hospital as soon as they are stable (2 to 5 days post-MI) and are given home programs to continue to increase their exercise tolerance during recovery.

Unstable or crescendo angina also poses dangers in an exercise program, as it may indicate that an infarction is imminent. *Angina pectoris* is cardiac pain that is usually felt beneath the upper sternum, often radiating to the left arm and shoulder, and to the neck and side of the face. Angina is caused by ischemic heart muscle due to an imbalance between oxygen supply and oxygen demand, or coronary spasm. In stable angina, the pain begins at a predictable level of exertion or environmental temperature. However, in crescendo angina

TABLE 26-1
Contraindications to Exercise and Exercise Stress Tests

Congestive heart failure
Unstable angina
Severe aortic stenosis
Pulmonary hypertension
Myocarditis, pericarditis, bacterial endocarditis, acute rheumatic fever or
  cardiomyopathy within the past year
Uncontrolled hypertension
Serious arrhythmias
  Second- and third-degree AV blocks
  Uncontrolled atrial fibrillation
  Excessive or complex PVCs
  Ventricular tachycardia
Marked bradycardia (other than in endurance-trained individuals)
Fixed-rate artificial pacemaker
Significant cardiac enlargement
Valvular disease, moderate to severe
Recent pulmonary embolism
Severe anemia
Uncontrolled metabolic disease (diabetes mellitus, thyrotoxicosis,
  myxedema)
Transient illness accompanied by fever
Certain orthopedic disabilities
Inappropriate blood pressure response to exercise testing
Overdose of cardiac drugs such as digitalis, quinidine, lidocaine, procain-
  amide, propranolol, and verapamil
Mental instability

---

the pain may develop at rest or during sleep, at variable levels of exertion, and tends not to be relieved by nitroglycerine (a medication used to provide relief of symptoms). The exercise prescription in stable angina calls for training below the pain threshold.

Heart failure, as described in Chapter 14, occurs when the heart cannot adequately accommodate to the load placed on it by the circulation. Characteristically, there is an increased ventricular volume and end-diastolic pressure with a decreased cardiac output. Exercise is contraindicated because it puts additional stress on the heart. Additionally, conditions that predispose the heart to failure, such as pulmonary hypertension, aortic stenosis, severe valvular disease, and cardiac enlargement, may also make exercise training dangerous.

Exercise is not recommended in the presence of electrical disturbances in the heart such as ventricular tachycardia, multiple premature ventricular contractions, atrioventricular nodal blocks, and atrial tachycardia. Such disturbances lead to diminished cardiac output that is impaired further by the demands of exercise. Most important, exercise can also make the existing arrhythmias worse. Exercise may also be dangerous in the presence of marked sinus bradycardia (resting heart rate below 45 beats · min^{-}1 in the untrained individual). The individual with sinus node disease may not be able to meet the demands of exercise, because the heart rate will not increase.

Myocarditis is inflammation of the heart muscle. It can impair the metabolism of the myocardium, interfere with ventricular filling, induce myocardial ischemia, and cause dilation and hypertrophy of the heart. Myocarditis can be associated with a variety of diseases and disorders such as infections, rheumatic fever, metabolic diseases such as myxedema, thyrotoxicosis, and acromegaly, damage from drugs such as corticosteroids and digitalis, and connective tissue diseases such as rheumatoid arthritis and systemic lupus erythematosus. Exercise is contraindicated in the presence of myocarditis or in conditions where the possibility of myocarditis exists.

Even the flu, accompanied by fever, may present the possibility of myocarditis. A person should rest an additional day for each day of a viral flu or infection to guard against the possibility of subclinical symptoms of the illness. (You feel good, but you still have the flu.)

Many disease states are self-limiting: If you feel terrible, a 2-mi run may be the last thing on your mind. However, because many people today seem to be addicted to exercise, they may attempt to train in spite of the possible risks. It is best to have individual cases evaluated by a physician knowledgeable about the disease and exercise physiology. Although exercise is contraindicated in some instances, in others it may prove to be essential.

DECONDITIONING AND BED REST

Just as training results in improvement in physical working capacity, inactivity leads to deterioration. Until quite recently, treatment for cardiac and postoperative patients involved prolonged bedrest. However, the forced inactivity resulted in additional difficulties that were sometimes worse than the original problem. The negative effects of bedrest were aptly summed up by R.A.J. Asher:

> Bedrest is anatomically and physiologically unsound. Look at a patient lying long in bed. What a pathetic picture he makes! The blood clotting in his veins, the lime draining from his bones, the scybala stacking up in his colon, the flesh rotting from his seat, the urine leaking from his distended bladder, and the spirit evaporating from his soul. Teach us to live that we may dread unnecessary time in bed. Get people up and we may save our patients from an early grave.

TABLE 26-2

Effects of Prolonged Bedrest on Physiological Function

Decreases	Increases	No Effect
Maximum stroke volume	Maximum heart rate	Mean corpuscular hemoglobin concentration
Orthostatic tolerance	Diastolic blood pressure	Forced vital capacity
Arterial vasomotor tone	Resting heart rate	Resting or exercise arterio-venous O_2 difference
Systolic time interval	Extra vascular and intravascu-lar IgG[a]	Vital capacity
Coronary blood flow	Submaximal exercise heart rate	Maximum voluntary ventilation
Maximal O_2 consumption	Submaximal exercise cardiac output	Total lung capacity
Pulmonary capillary blood vol-ume	Sleep disturbances	Proprioceptive reflexes
Plasma volume	Diuresis	
Skin blood flow	Incidence of urinary infection	
Total diffusing capacity	Incidence of deep vein throm-bosis	
Cerebrovascular tone	Urinary excretion of calcium and phosphorus	
Sweating threshold tempera-ture	Nitrogen excretion	
Vasomotor heat loss capacity	Serum corticosteroids	
Red blood cell production	Cultured staphylococci in nasal mucosa	
Red cell mass	Extracellular fluid	
Hemoglobin	Tendency to faint	
Serum proteins	Incidence of constipation	
Serum albumin	Cholesterol	
Intracellular fluid volume	Low-density lipoproteins	
Serum electrolytes	Growth hormone	
Coagulating capacity of blood	Electrocardiographic ST-seg-ment depression	
Bone calcium	Renal diurnal rhythms	
Bone density		
Insulin sensitivity		
Acceleration tolerance		
Blood flow to extremities		
Catecholamines		
Serum androgens in males		
Muscular strength and mass		
Muscle tone		
Leukocyte phagocytic func-tion		
Visual acuity		
Resistance to infection		
Systolic blood pressure		
Balance		

[a] Immunoglobulin G

As shown in Table 26-2, bed rest has far-reaching effects on most aspects of physiological function. Studies overwhelmingly show that decreased function or inactivity of an organ leads to serious changes in its functional state as well as its histochemical state. In many instances, disease can be made worse because of bed rest, and sometimes irreversible problems can develop. It is little wonder that current treatment for the postoperative patient and for most diseases calls for early ambulation and physical activity.

There is a large body of research on deconditioning and bed rest stemming largely from the space programs of the United States and the Soviet Union. Well-controlled and well-funded studies have examined the effects of extreme deconditioning on most aspects of physiological function. Some of these studies were conducted for several months and involved almost complete restriction of movement.

The negative effects of bed rest seem to be due primarily to the decrease in hydrostatic pressure within the cardiovascular system, the lower energy expenditure due to inactivity, decreased pressure on the skeleton, and psychological stress.

The most profound changes from bed rest occur in the cardiovascular system. Impairments include diminished capacity of the heart, reduced plasma and blood volumes, and impaired automaticity of the blood vessels (Figure 26-1).

Maximal O_2 uptake and work capacity decrease from as little as 1% to as much as 26%, depending on the type and duration of confinement. Work capacity is affected more in the upright than the supine posture because of the added effects of orthostatic intolerance developed during bed rest.

The changes in O_2 transport capacity are the result of reduced function in many parts of the system. Stroke volume and cardiac output decrease, both in upright and supine exercise. This is a result of impaired venous return of blood to the heart and decreased myocardial contractility. There is a decrease in tissue-oxidative enzymes that affects submaximal exercise capacity. Diffusing capacity in the lungs decreases because of reduced pulmonary blood volume.

Bed rest leads to reduced size and contractility of the heart. In addition, there have been some changes noted in the electrocardiogram (ECG): Heart rate increases at rest, conductivity through the atrioventricular node and bundle of His is slowed, ST-segment depression develops, and there is an increased incidence of sinus arrhythmia. These changes seem to be caused by changes in water and salt metabolism that impair the K^+ and Na^+ cellular gradient and thus the ECG.

Bed rest also produces *orthostatic intolerance*, the inability of the circulation to adjust to the upright posture. The reduction in hydrostatic pressure seems to be the primary stimulus for this phenomenon. When a bedridden patient assumes an upright posture, there is a sudden decrease in venous return of blood to the heart (Figure 26-2). This is caused by a reduction of vasomotor tone,

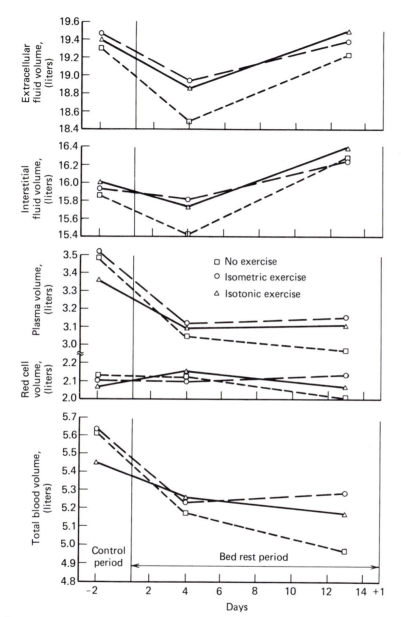

FIGURE 26-1
Fluid and blood changes during bed rest in subjects who practiced isometric exercise (circles), isotonic exercise (triangles), no exercise
(squares).

(SOURCE: Modified from J.E. Greenleaf and S. Kozlowski, 1982.)

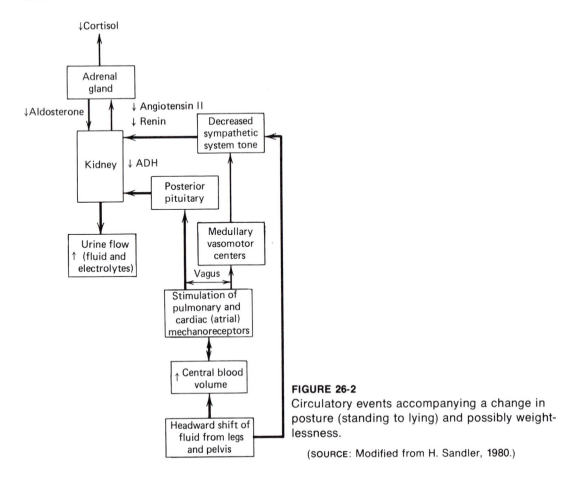

FIGURE 26-2
Circulatory events accompanying a change in posture (standing to lying) and possibly weightlessness.

(SOURCE: Modified from H. Sandler, 1980.)

blood volume, and muscle tone. The heart rate increases rapidly in an attempt to increase cardiac output. However, blood pressure falls and cerebral blood flow is impaired, leading to dizziness, ataxia, and fainting.

Orthostatic intolerance is a good example of how bedrest can worsen a disease state. In a heart patient, the decreased stroke volume, coronary blood flow, and myocardial contractility can sometimes lead to heart failure or reinfarction.

Orthostatic tolerance is improved in direct proportion to the time spent in the upright posture. Exercise in the horizontal position during bed rest has no effect on orthostatic intolerance, but it does help prevent the deterioration in $\dot{V}_{O_2max}$ and work capacity. Likewise, sitting for 8 hr a day helps maintain orthostatic tolerance, but has no effect on fitness. Interestingly, static exercise has been found to be effective in preventing some of the decrease in $\dot{V}_{O_2max}$ during bed rest.

Plasma and blood volumes decrease about 15%. The loss of red cell mass is due to a decrease in erythropoiesis. The decrease in plasma volume is not totally understood. It may be due to the increase in the extravascular protein distribution, which changes the extracellular osmolarity, thus decreasing fluid in the vascular compartment. The increased extravascular distribution of proteins is caused by impairment of the capacity of lymph to transport proteins due to reduced muscular function of the inactive patients.

The decrease in vasomotor tone is due primarily to the low hydrostatic pressure in the recumbent position. Bed rest appears to have a direct effect on the vasomotor center. There also seem to be effects on local smooth muscle reflexes. Ordinarily, the smooth muscle contracts in response to the stretch elicited by the hydrostatic pressure characteristic of the upright posture. This pressure is much less in bed rest, so the smooth muscle of the precapillary sphincters loses some of its tone. Additionally, depletion of endogenous epinephrine decreases venous vascular responsiveness, which contributes to the decrease in venous return of blood to the heart.

Bed rest has a far-reaching effect on metabolism, including decreased insulin sensitivity. Lowered insulin effectiveness seems to be due to a decrease in the number or sensitivity of cell membrane insulin receptors. The extent of effect of bed rest on carbohydrate metabolism seems to depend on the length and degree of inactivity. There is also a decrease in free fatty acids and an increase in triglycerides in blood caused by a depletion of norepinephrine. Cholesterol and low-density lipoproteins increase, which has possible implications in heart disease.

There is a marked decrease in muscle tone that is due to the loss of intracellular fluid and contractile protein. Disuse atrophy is accompanied by a loss of muscular strength and endurance. Almost all studies show that the bedridden patient goes into negative nitrogen balance very early during convalescence. There is, however, little change in body weight because the loss of lean body mass is usually accompanied by an increase in fat.

Bone demineralization occurs at a rapid rate during bed rest. This is a result of reduced longitudinal stress on the bones rather than inactivity. The dissolution occurs at different rates in the various parts of the skeleton. The weight-bearing bones are particularly vulnerable; because they are usually under the most stress, they have the most to lose during deconditioning.

The effects of prolonged bed rest are usually reversible with adequate ambulation. The physiological effects, if not too advanced, can be reversible with an appropriate training program. One problem is that the negative effects of bed rest lead to a vicious circle, particularly in the elderly: Disuse leads to debilitation, which leads to a further desire to stay in bed or remain inactive.

Psychologists have described bed rest as a form of sensory deprivation. Bed rest produces a greater incidence of intellectual inefficiency, bizarre thoughts, exaggerated emotional reactions, time distortions, changes in body image, unusual body sensations, and an array of physical discomforts. As one researcher put it, it may be better to burn out than to rust out.

The negative adaptability resulting from inactivity certainly has implications for the athlete or individual interested in a high level of fitness. Although the atrophy in muscles of a casted leg or the debilitation experienced by the post-operative patient are extreme examples of physiological deterioration, the essence is similar in a highly conditoned individual who suddenly assumes a sedentary life-style; physiological processes adapt to the lower stresses with reduced functional capacity.

CARDIAC REHABILITATION

Exercise therapy represents a radical change in philosophy and treatment of the cardiac and cardiac-prone patient. It has gained widespread acceptance as an important tool in returning these individuals to physiological and psychological competence. Myocardial infarction patients involved in a rehabilitation program have been found to experience an improved sense of well-being, reduced anxiety and depression, improved exercise capacity, reduced ST-segment depression on the ECG, reduction in blood pressure, reduced heart rate at rest, reduced serum cholesterol and triglycerides and elevated high-density lipoprotein cholesterol, and perhaps a reduction in mortality rate.

Candidates for the program include people with documented myocardial infarction or stable angina pectoris, postoperative cardiovascular surgical patients following procedures such as myocardial revascularization, peripheral arterial obstructive surgery, valve replacement, and repair of congenital heart defects, and those who are at high risk to develop heart disease because of significant predisposing factors. Patients with unstable conditions, such as those described in Table 26-1, are generally excluded from the program.

The heart patient suffers both physical and emotional distress. After a myocardial infarction or heart surgery, there is a loss of cardiac function due to injury, and a loss of working capacity (fitness) because of deconditioning. In addition, the patient is faced with psychological difficulties stemming from fear of death or disability, and the perception of a changing life-style that could affect family, friends, economic well-being, and employment.

A primary goal of the cardiac rehabilitation program is to help patients return to their former quality of life and develop optimal physiological function. Initially, this consists of simple tasks such as dressing and showering. Ultimately, this may include resumption of physically demanding occupations and recreational activities. Ideally, rehabilitation will return the individual to an acceptable level physiologically, mentally, socially, vocationally, and economically.

Exercise training has been shown to play an important role in the rehabilitation of most coronary patients. Physical activity should be integrated with a total program that includes cessation of cigarette smoking, weight control, proper diet, return to work, and normal social life. Interestingly, some heart patients actually improve their fitness compared to before their heart attack. The cardiac event provides potent motivation to become more active and increase awareness of coronary risk factors.

Education is an important part of the rehabilitation program. The program should prepare patients and their families for healthy alternatives in life-style that might reduce the risk of recurrence of CHD. Most studies have shown that the prognosis is poor in patients who continue to remain inactive, to smoke, or maintain elevated blood lipids. In addition the educational program helps teach the family about emergency procedures and alert them of warning signs of further problems.

Cardiac rehabilitation reduces the cost of health care through shortened treatment time and prevention of disability. In addition, it may help to reduce occupational losses caused by cardiovascular disease.

The ideal cardiac rehabilitation program consists of three phases: the inpatient program, the outpatient therapeutic program, and the exercise maintenance program. During cardiac rehabilitation the patient will deal with a team that may include physicians, coronary care nurses, exercise physiologists, physical therapists, occupational therapists, dietitians, social workers, clinical psychologists, exercise specialists, and exercise technologists.

The Inpatient Program (Phase I)

The inpatient program is conducted in the hospital during the early period following the myocardial infarction or cardiovascular surgery. Early activity has been shown to reduce anxiety and depression, and prevent some of the adverse effects of bed rest such as muscle atrophy, thromboembolic complications, negative nitrogen balance, orthostatic intolerance, decreased aerobic capacity, tachycardia, and pulmonary atelectasis. In addition, the program helps to provide for both the physiological and psychological climate for the resumption of normal activities.

The inpatient program includes appropriate medical treatment, patient and family education, and graded physical activity. By the end of this phase the patient should be able to meet low-level demands of daily activity such as dressing, showering, toilet, and walking up a flight of stairs.

The Outpatient Therapeutic Program (Phase II)

The second phase is also usually conducted in the hospital and includes supervised exercise, vocational rehabilitation, and behavioral counseling. The exercise program is typically held at least 3 days per week and is closely monitored by a physician. Occupational task simulation is conducted to ensure that the individual is physiologically prepared to resume work.

In the past, endurance exercise training has been delayed for at least 6 weeks after myocardial infarction or cardiac surgery because of the fear that working above the 50% of maximum capacity required for the training effect might be dangerous. However, clinical trials indicate that active training can begin as early as 3 weeks provided that patients are adequately screened to rule out serious ventricular arrhythmias and evidence of heart failure.

Symptom-limited treadmill testing is important in exercise prescription in early exercise programs and is typically administered prior to discharge from the hospital. In addition, patients should be retested periodically to detect late-occurring ischemia and arrhythmias, and to reassess the exercise prescription.

A principal goal of the high-level outpatient program is to develop a functional capacity of 8 METS or 28 to 30 ml $O_2 \cdot kg^{-1}$. This will produce a fitness level that will serve the needs of a sedentary life-style and act as a minimum criterion for discharge from the outpatient program. A person typically remains in most outpatient programs for about 3 months after a myocardial infarction.

The outpatient program allows the physician to ensure satisfactory medical status by reviewing resting and exercise ECG, blood pressure response to various stimuli, and symptoms. Additionally, patients can obtain the necessary education about the nature of heart disease and the signs and symptoms associated with their problem.

The Exercise Maintenance Program (Phase III)

This program is ideally conducted in an environment especially organized for cardiac and risk-prone individuals. Many fine programs have been established through hospitals, the YMCA, university physical education departments, medical schools, and private clinics. The purpose of this phase of the cardiac rehabilitation program is to prevent recurrence and to improve physical working capacity. This program can be used to evaluate the status and effectiveness of treatment regimens in the patient, maintain patient compliance with life-style changes, and provide a safer environment for the exercise program. Exercises may include endurance activities such as walking, jogging, cycling, or swimming, resistive exercise, and arm exercise. The design of the exercise program should reflect the physical requirements of the patient's job.

Because of the dangers of reinfarction, it is particularly important that attention be directed at resuscitation and safety procedures. Personnel, patients, and patients' families should be trained in cardiopulmonary resuscitation (CPR), and key personnel should be certified in advanced cardiac life support. Proper equipment and drugs should be available for evaluating and correcting life-threatening cardiac dysrhythmias.

Exercise Prescription for the Cardiac Patient

The exercise training program should be based on an exercise test that includes measurement of heart rate, blood pressure, and ideally O_2 consumption during graded exercise. Patients should be evaluated for ischemia, arrhythmias, and adequate blood flow. Medications such as β-blocking agents, digitalis, diuretics, vasodilators, and antiarrhythmics should be considered in the exercise prescription. Exercise tests designed to aid in the exercise prescription should be conducted while patients are taking their normal medications.

Exercise testing can begin as early as 1 to 2 days prior to discharge from the hospital. The exercise prescription should be evaluated regularly. A training diary, ideally kept by the exercise leader, should be used to record the nature, intensity, and duration of each exercise session. Clinical studies have shown that consideration of intensity of the exercise is vital to improved capacity. Training above 50 to 60% of $\dot{V}_{O_2max}$ will result in improved work capacity and lowered resting and submaximal exercise heart rates, whereas recreational activities generally will not.

Myocardial oxygen consumption ($M\dot{V}_{O_2}$) is an important consideration when prescribing calisthenic and resistive exercises to cardiac patients. An acceptable indirect index of myocardial O_2 consumption is heart rate $\times$ systolic blood pressure (i.e., called the pulse-pressure product or double product). Exercises of the same metabolic cost can vary considerably in $M\dot{V}_{O_2}$ (Figure 26-3). Exercises that require use of the upper extremities, or those that induce Valsalva's

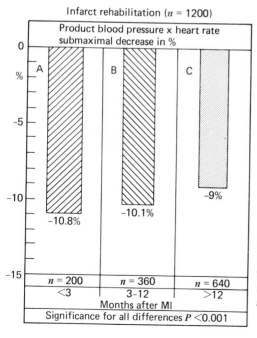

FIGURE 26-3
Reduction in pressure product (heart rate $\times$ systolic blood pressure) during identical work loads after 4 to 6 weeks of exercise training. A) $<$ 3 mo. after MI. B) 3–12 mo. after MI. C) $>$ 12 mo. after MI.

(SOURCE: Modified from K. Konig, 1977.)

maneuver should be administered with care. Likewise, exercise should be prescribed with discretion in the presence of angina. An exercise that might be easily tolerated by one patient may produce an ischemic response in a patient with angina.

The amount of improvement in cardiovascular capacity in the cardiac rehabilitation program depends on adherence to the program over months and years. Adherence to the program depends on program design, attitude of the spouse, nature of the activities, and the accessibility of facilities and personnel. Reinforcers such as percent fat, body weight, reduced heart rate and blood pressure at rest and during exercise, decreased angina, and improved working capacity are potent motivators. Realistic short-term goals can also be very effective in maximizing adherence to the program.

There has been considerable interest in marathon training for the postcardiac patient. It has even been suggested that such training may offer immunity from coronary disease. However, marathon training has definite limitations as the vast majority of postcardiacs simply do not have the aerobic capacity to run a marathon safely. In addition, most occupations require training in the upper extremities rather than the legs. There are minimal crossover benefits between lower and upper body exercise. Muscle specificity must be considered in the exercise prescription of the cardiac patient.

Cardiac patients involved in heavy exercise should take precautions and avoid unnecessary risks: Over 60% of coronary patients who die, do so before they get to the hospital. Care should be taken to avoid competition and exercising in adverse environments. Patients should run with a group of people trained in CPR and ideally in a supervised, monitored environment.

Cardiac Rehabilitation and Reinfarction

In studies of dropouts versus adherents to the training program, the majority of recurrences seem to happen to the dropouts. Their relative risk toward CHD is higher, and their mortality rate has been reported to be 50% greater than patients who were faithful to their rehabilitation programs. The reason for this is not clear. It may be due to the actual effects of the program or to personality or physiological characteristics of the dropouts. Factors that increase the risk of reinfarction are discribed in Table 26-3.

Cardiac Drugs and Exercise

A variety of drugs are used in cardiovascular disease that alter such things as the rate, contractility, conduction, preload, and afterload of the heart. These drugs can alter the electrocardiogram and the response to exercise. They can also have far-reaching effects on other physiological functions. Exercise response can help gauge the effectiveness of a drug, so it is important for people involved in exercise prescription and training to understand the interactions

TABLE 26-3

Factors That Increase the Risk of Reinfarction

Nonadherence to exercise program
Prior history of angina or hypertension
Poor left ventricular function
Large arteriovenous O_2 difference
Ventricular arrhythmias—multifocal
Depression
Left atrial enlargement
Cardiac enlargement
Smoking
ST-Segment depression >2 mV
Use of digitalis or diuretic agents
Cholesterol >270 mg · dl⁻¹
Age, after age 55
Low systolic pressure during exercise
Increased ST-segment depression with exercise
Diastolic pressure rise of 15 mmHg or more during
   exercise

between physical activity and drugs. A summary of drugs used in heart disease, their actions, and effects on exercise and the ECG appears in Table 26-4.

Digitalis drugs are used to treat or prevent heart failure, and/or to suppress supraventricular arrhythmias. They can improve exercise capacity by increasing cardiac output. This occurs because the contractility and blood flow of the heart improves, and fluid and electrolyte imbalances normalize. Digitalis frequently induces ST-segment depression in the ECG, which can lead to false-positive treadmill tests or mask ischemia if not taken into consideration (a false-positive test incorrectly indicates ischemia). Digitalis can be toxic, which can lead to heart block and almost any arrhythmia, as well as weakness, depression, and anorexia.

Diuretics are used primarily in the treatment of hypertension, fluid retention states, pulmonary edema, and heart failure. In addition, they are sometimes misued by people seeking to lose weight. Diuretics that cause hypokalemia (low potassium) can render false positives on the ECG and an increased incidence of arrhythmias. There is also the danger of hypotension, particularly when more potent diuretics are used or when diuretics are combined with vasodilators. Work capacity may be impaired because of reduced plasma volume. Particular care should be taken when exercising in the heat—severe fluid diuresis will impair the capacity to sweat. Carbohydrate metabolism is further

TABLE 26-4
Cardiac Drugs and Their Effects[a]

Digitalis (cardiac) Glycosides

Examples:

Oral: digitalis, digitoxin, digoxin, lanatoside C, acetyldigitoxin, gitalin.

Parenteral: ouabain, deslanoside, digitoxin, digoxin

Use: In heart failure—slows heart rate, increases cardiac contractility (inotropic effect), suppresses supraventricular antiarrhythmias, and reduces ventricular rate during atrial flutter or atrial fibrillation by increasing the automaticity of the secondary pacemakers in the atrioventricular node.

Mechanism of action: (1) CNS autonomic activity (2) Inhibition of (Na^+, K^+)-ATPase exchange pump

Effects on exercise or ECG: May increase work capacity and produce S T-segment depression on ECG.

Diuretics

Examples:

High-potency diuretics (site of action is the ascending limb of the loop of Henli): organomercurials, ethacrynic acid, furosemide, bumetamide

Moderately potent (site of action is the proximal portion of the distal convoluted tubule): chlorothiazide, thiazides, phthalimidines, quinazolinones, benzene disulfonamides, and chlorobenzamides

Potassium-sparing diuretics (site of action is the distal portion of the distal convoluted tubule): spironolactone, triamterene, and amiloride

Use: Hypertension, heart failure, fluid retention, pulmonary edema

Mechanism: Generally work by inducing sodium diuresis (increased sodium excretion)

Effects on exercise and the ECG: May induce hypotension, S T-segment depression on ECG (if there is potassium depletion), and hyperglycemia. Will increase the frequency of arrhythmias in digitalis toxicity.

Catecholamines (sympathomimetic amines)[b]

Examples: epinephrine, norepinephrine, isoproterenol, dopamine, and dobutamine—administered intravenously or intramuscularly, or in aerosol (epinephrine or isoproterenol)

Use: To regulate heart rate and rhythm, arterial blood pressure; to increase cardiac contractility and peripheral organ perfusion (particularly in the lungs). Clinically, used in treatment of chronic obstructive pulmonary disease, allergic reactions (anaphylaxis), low cardiac output, and cardiac arrest.

Mechanism of action: Acts directly on α and β receptors of various tissues (smooth muscle, gland, or heart), which can produce mixed results.

Stimulated α receptors cause vasoconstriction, whereas β receptors result in vasodilation, stimulation of heart rate and myocardial contractility, and relaxation of nonvascular smooth muscle.

Effects on exercise or ECG: May improve work capacity by increasing cardiac output and myocardial contractility, decrease the work of breathing in asthmatics, cause arrhythmias and electromechanical dissociation.

Vasodilators

Examples: Nitroglycerin, isosorbide dinitrate, hydralazine, isordil, prazocin, phentolamine, phenoxybenzamine, and trimethaphan—administered intravenously, sublingually, orally, or topically

Use: Angina, chronic congestive heart failure, hypertension, valvular heart disease, congenital heart diseases, chronic pulmonary hypertension

Mechanism: Some exert a direct and general effect on smooth muscle (nitroglycerine), whereas others act as an α-adrenergic blockade (phenoxybenzamine). Work principally by reducing the afterload of the heart, thus lowering myocardial O_2 consumption.

Effects on exercise and the ECG: Nitrites increase work capacity by raising the angina threshold. May cause hypotension and fainting. May increase orthostatic intolerance in bed rest.

β-Adrenergic Blocking Drugs[b]

Examples: propranolol, alprenolol, lopressor, tenformin, metoprolol, atenolol

Use: Hypertension, angina, hypertrophic cardiomyopathy, congestive cardiomyopathy, cardiac arrhythmias, and thyrotoxicosis. Reduces heart rate and blood pressure.

Mechanism: β-adrenergic blocker

Effects on exercise and ECG: Decreases heart rate and blood pressure at rest and during exercise, decreases cardiac contractility, decreases work capacity, and may cause false negative on ECG.

Antiarrhythmic Drugs

Examples: Quinidine, procainamide, lidocaine, phenytoin, mexiletine, tocainide, and aprindine—administered orally or intravenously

Use: Suppression of arrhythmias; type of arrhythmia will dictate type of drug.

Mechanism: Individual drugs in this category have different mechanisms, including decreased automaticity, decreased membrane conductance to sodium, increased membrane conductance to potassium, decreased membrane conductance to calcium, and antisympathetic activity.

Effect on exercise or ECG: May cause false negative on ECG and may increase submaximal exercise heart rates (dose dependent)

(continued)

TABLE 26-4 *(continued).*
Cardiac Drugs and Their Effects[a]

Calcium Channel Blockers

Examples: Nifedipine, verapamil, diltiazem.

Use: Variant angina, classic angina, unstable angina, coronary spasm, reentrant supraventricular tachycardia, idiopathic hypertrophic subaortic stenosis.

Mechanism of action: Inhibits calcium ion influx from sarcoplasmic reticulum and T-tubules into myocardial cells, preventing contraction of calcium-dependent muscle.

Effects on exercise or ECG: Increases exercise tolerance by raising angina threshold and decreasing peripheral vascular resistance. ECG effects include: increased atrial refractory period, reduced A-V node conduction velocity, decreased automaticity and excitability of Purkinje fibers, and decreased ventricular conduction rate resulting in a broadened QRS complex.

---

[a]Note: This chart is a generalization. Particular drugs within a group may have different effects, particularly in abnormal physiological states, or when other drugs are taken concurrently.
[b]Note: see Appendix III

impaired in diabetics on diuretics, so the exercise leader should be extremely careful with these individuals.

Vasodilators, such as nitroglycerine, have come to play an important part in the exercise program of cardiac patients. By reducing the afterload of the heart, they lower myocardial O_2 consumption and allow the patient to do more work before the onset of angina. Vasodilators have no effect on heart rate but tend to reduce blood pressure, systemic vascular resistance, and pulmonary wedge pressure. They also increase cardiac output in patients with acute myocardial infarction complicated by left ventricular failure.

Nitroglycerine is used both to relieve an angina attack and to prevent one from occurring. This substance can induce hypotension, which increases the risk of fainting during exercise. Nitroglycerine has a variable effect on the ECG. In many cases it will reduce ST-segment depression and the incidence of arrhythmias. However, these changes are poorly correlated with changes in pain threshold.

Although catecholamines are used primarily in emergencies such as cardiac arrest and shock, they are also administered as an aerosol in chronic obstructive pulmonary disease (epinephrine or isoproterenol). Because they can cause arrhythmias and increase blood pressure, adequate supervision is important for the heart patient using them. Ephedrine, another sympathomimetic drug that is used in asthma, can also increase heart rate and blood pressure.

β-Adrenergic blocking drugs such as propranolol are used to decrease myocardial O_2 consumption. This is accomplished by depressing heart rate and blood pressure. Because of the chronotropic effect of the β blockers, the exercise prescription must be based on an exercise stress test in which the drug was present in the typical therapeutic dosage.

Antiarrhythmic drugs such as procainamide and quinidine are used to suppress ventricular arrhythmias that can be particularly dangerous during exercise. These drugs can decrease myocardial contractility and cause slight muscle weakness. In some people they can cause nausea and vomiting, which could affect the desire to exercise. Because of their effects on suppressing arrhythmias, they can cause false negatives on the ECG during the exercise stress test.

HYPERTENSION

Hypertension is high arterial blood pressure, usually defined as above 140 to 160 mmHg systolic and/or above 90 to 100 mmHG diastolic. (Note: The acceptable upper normal level of diastolic blood pressure is considered by many to be 85 to 88 mmHg.) It is considered to be a major health hazard, with approximately 12% of the population dying as a direct result from stroke, congestive heart failure, kidney failure, and heart attack. Although there is presently no cure, it can usually be controlled by medication, diet, weight control, and exercise. It should be emphasized that even ''marginal'' hypertension should be closely monitored. The classifications of hypertension are described in Table 26-5.

As discussed in Chapter 15, blood pressure is the force exerted by the blood against any unit area of the vessel wall. When the heart contracts, it expels blood from its chambers, which elevates the arterial blood pressure rapidly. As the heart muscle relaxes, there is a reduction in this pressure. The peak pressure of this cycle is called systolic blood pressure, whereas the minimum pressure is called diastolic blood pressure.

As with other physiological functions, the body has many mechanisms of maintaining adequate blood flow to the tissues. Blood pressure can be affected by such factors as increased smooth muscle contraction, baroreceptor reflexes, autonomic discharge from the medulla and hypothalamus, arteriolar autoregulation, changes in blood volume, and neurogenic and humoral influences from hormones and the kidney (such as the renin–angiotensin–aldosterone system). Although presently unknown, the cause of essential hypertension undoubtedly lies among these control mechanisms.

TABLE 26-5
Classification of Hypertension

I. Primary or essential hypertension—undetermined cause

II. Secondary hypertension or hypertension associated with other
   disorders
   A. Renovascular disease
   B. Endocrine disorders such as Cushing's syndrome, acromegaly, and
      primary aldosteronism
   C. Coarctation (narrowing) of aorta
   D. Certain enzymatic defects
   E. Neurological disorders—increased intracranial pressure from brain
      tumors or cardiovascular accident
   F. Drug-induced hypertension from such agents as corticosteroids,
      desoxycorticosterone, amphetamines, thyroxine, or oral contra-
      ceptives
   G. Hypercalcemia
   H. Neurogenic and psychogenic disorders
   I. Deficiency of vasodilating tissue enzymes such as prostaglandins
      and bradykinin

SOURCE: Adapted from Sokolow, M. and M.B. McIlroy. Clinical Cardiology. Los
Altos, Calif.: Lange Medical Publications, 1977.

There is no absolute value of blood pressure below which the mortality rate
is unaffected and above which it is increased. Symptoms in people suffering
from hypertension may include headache, heart failure, renal disorders, neurological problems, claudication, and chest pain. However, during the early phases
of hypertension, there are usually no symptoms (thus it is sometimes called the
silent killer).

Labile Hypertension

Blood pressure increases naturally in a variety of circumstances where circulatory homeostasis is perceived to be threatened. Examples include exercise,
emotional responses such as anger, anxiety, or fear, and hostile environments
such as extreme cold. Blood pressure will typically increase before treadmill
tests, sometimes dramatically, in inexperienced subjects or in athletes before
competition.

Increased catecholamine secretion due to psychologically induced excessive
sympathetic stimulation may produce a labile hypertension (blood pressure
$>140/90$ mmHg) that subsides with relaxation. Labile hypertension results from
an increase in cardiac output rather than an increase in peripheral vascular re-

sistance. In general this is often seen in young athletes and young adults. Its significance is unknown.

Blood pressure response in labile hypertension is usually normal during exercise, with blood pressure often dropping to normal values during recovery. In true labile hypertension, no modification of the exercise program is required.

Long-term exposure to sports that induce significant pressure loads on the heart, such as weight lifting, may increase the risk of developing established hypertension (blood pressure that does not return to normal during relaxation). Chronic periodic distension of smooth muscle by autoregulation (see Chapter 15) that occurs during these activities may ultimately lead to structural changes that could result in a permanent increase in peripheral vascular resistance.

Systolic Blood Pressure

Although much emphasis has been placed on the importance of diastolic pressure and health, epidemiological studies have shown that high systolic blood pressure is a significant risk factor for many diseases. Systolic pressures above 140 to 160 mmHg are considered hypertensive.

Blood meets considerable resistance to its flow in the small capillaries and arterioles; therefore it must be pumped at high pressure to meet the tissue demands. Blood pressure increases even more during exercise, because the tissue demands for blood are higher. Normally, the body can easily tolerate increasing blood pressure during exercise. However, blood pressure can reach dangerous levels in the hypertensive individual.

In laboratory animals, systolic blood pressures of 300 mmHg have been shown to tear the intimal layer of the arterial wall. Presumably, in a vessel affected by arteriosclerosis, which would be less distensible than normal, less pressure would be needed to cause damage. Normally, maximal systolic pressure during exercise ranges between 150 and 220 mmHg. It is not uncommon for hypertensive individuals to exceed these levels at rest and to go considerably above them during exercise.

Diastolic Blood Pressure

Diastolic blood pressure is particularly important because of its effect on coronary blood flow. Diastolic blood pressures above 90 mmHg are considered hypertensive. During systole when the heart muscle is contracting, there is a constriction of the intramuscular vessels that leads to a reduction in blood flow. So, the most significant amount of coronary blood flow occurs during diastole. During exercise, myocardial O_2 demand may increase five or six times above resting demand. If, in the hypertensive individual, diastolic blood pressure increases during exercise, then coronary blood flow could be compromised at a time when more blood is needed. This would create a relative ischemia—more blood is needed than is being delivered.

Increases in diastolic blood pressure during exercise of 15 mmHg above resting pressure have been found to be highly predictive of coronary artery disease and to indicate greater severity of disease and more frequent left-ventricular contraction abnormalities.

The Treatment of Hypertension

Drugs are currently the front line of defense against hypertension. The introduction of this mode of treatment in the 1950s has led to a 40% decrease in mortality from the disease. In general, the aim of hypertension management is to bring the standing blood pressure below 140/90 mmHg, provided the side effects can be tolerated. In addition to medication, treatment for hypertension includes limiting sodium, cholesterol, and fat intake in the diet, weight reduction, regular exercise, cessation of smoking, and relaxation.

Hypertensives are often put on a low-sodium diet. When body sodium levels exceed the kidney's ability to excrete it, blood pressure increases. Hypernatremia tends to increase blood volume and adversely affect kidney function.

Three types of drugs are commonly employed in the treatment of hypertension, either singly or in combination (Table 26-4): diuretics, adrenergic inhibitors, and vasodilators. Diuretics work by decreasing body sodium, potassium, and fluid volume, and eventually peripheral vascular resistance. Examples of diuretic drugs include hydrochlorothiazide, chlorothiazide, bendroflumethiazide, and metolazone. Vasodilators decrease peripheral vascular resistance. Examples include prazosin and hydralazine. Adrenergic inhibitors interfere with sympathetic transmission to the heart. They include such drugs as propranolol, methyldopa, reserpine, metoprolol, guanethidine, and clonidine.

Control of hypertension is often a trial-and-error procedure that must balance the effects of the drugs with their side effects. The exercise leader can help in this process by observing differing responses to physical activity. Because of the effect of exercise in lowering blood pressure, some people with hypertension may unilaterally cease their medication. The exercise leader should be aware that this may cause a rebound phenomenon, which results in intense vasoconstriction with more serious hypertension that may be difficult to reverse.

Exercise and Hypertension

In general, endurance exercise training reduces resting blood pressure, probably by reducing body fat, which is a contributor to the disease. However, in most cases exercise alone will not reduce high blood pressure to normal levels. In addition, care must be taken when prescribing exercise and administering exercise stress tests to hypertensives. Even though exercise is sometimes used as an adjunct treatment of the disease, the utmost caution should be used in its application: Although exercise training tends to lower resting blood pressure, exercise itself raises systolic blood pressure. During or after exercise these individuals may be more susceptible to heart failure, coronary ischemia, angina, claudication, and possibly stroke.

Heart failure can occur because of an increased load on the heart resulting from hypertension and from the effects of coronary disease. During exercise stress tests, it is particularly important to take frequent measurements so that falling systolic blood pressure can be detected immediately.

Unaccustomed exercise such as shoveling snow, water skiing, or carrying heavy suitcases that induce Valsalva's maneuver (expiring against a closed glottis) are particularly dangerous for hypertensive individuals. The combination of isotonic and isometric exercise, especially in the cold, may produce an imbalance between the O_2 supply and O_2 demand of the myocardium.

Valsalva's maneuver initially causes a fall in arterial blood flow because of the increase in intrathoracic pressure, and as the venous return of blood is cut off, the heart rate increases rapidly along with an increase in peripheral resistance. When the strain of the activity subsides, there is an overshoot in blood pressure because of the sudden increase in stroke volume and the continued high peripheral resistance. In addition, local release of K^+ resulting from the muscle contraction elicits a pressor response that also raises blood pressure. Activities involving upper body work above shoulder level will also raise the blood pressure because of their effects on restricting blood flow through these relatively small muscle groups. Inverted positions, such as ''standing on your head'' are also not recommended.

During exercise stress testing, care should be taken to prevent patients from tightly grasping the supports of the treadmill or bicycle ergometer. This may increase systolic and diastolic blood pressure, because an isometric contraction is introduced.

Persons with hypertension should avoid high-intensity, low-repetition weight lifting. However, moderate resistive exercise using relatively high repetitions could be beneficial. If strength can be improved, then the blood pressure response to possibly dangerous situations may be reduced. During strength-building exercises hypertensives should be particularly careful to breathe normally, rather than holding their breath.

Warm-up is very important for this group. Sudden, high-intensity exercise can result in explosive increases in blood pressure that compromise coronary artery blood flow. Hypertensives should increase the intensity of exercise very gradually. They should stress low-intensity, long-duration endurance activities rather than short-term, high-intensity exercise. Walking, jogging, or swimming is more appropriate for persons with hypertension than interval sprint training or start-and-stop sports such as racquetball.

Drug side effects should be considered in the exercise prescription and the patient warned of possible symptoms. The diuretics can cause hypovolemia and muscle cramps, and could conceivably precipitate or augment heat injury during heavy exercise in hot environments. Certainly if the diuretics cause significant dehydration, they could hamper work capacity and increase the incidence of arrhythmias.

Adrenergic inhibitors, depending on the drug, produce side effects such as orthostatic intolerance, bradycardia, asthma, fatigue, and drowsiness. Propranolol (a commonly prescribed β blocker), for example, reduces heart rate at rest and during exercise, and its effects should be taken into consideration in the exercise prescription.

Vasodilators, again depending on the drug, can produce hypotension, tachycardia, and postural weakness. With all hypertension medication, particular attention should be directed toward warm-down, or continued low level of exercise after the workout, because of the dangers of hypotension, which can produce dizziness or fainting.

DIABETES

Diabetes mellitus is a disease characterized by a relative lack of the hormone insulin, which is produced by the β cells of the pancreas (Chapter 9). There are two types of diabetes: Type I and Type II. Type I is insulin dependent, which requires injections of insulin. Type II is insulin independent, which can usually be controlled by diet alone, or diet and oral hypoglycemic drugs.

Insulin plays a pivotal role in fuel homeostasis. Many tissues require insulin for the uptake of glucose. In the normal individual when more glucose is available in the blood than is needed, insulin is secreted and acts to store glucose in muscle tissue and the liver as glycogen. Glycogen can later be broken down in the liver to provide glucose when blood levels supplies are low. This process of glycogen breakdown becomes very important during exercise when a lot of glucose is needed for optimum performance.

High blood sugar (hyperglycemia) is a prominent characteristic of diabetes. High blood sugar results from a lack of insulin in Type I diabetes, or from tissue insensitivity to insulin in Type II diabetes. A major problem in Type I diabetes is an increase in the breakdown of fats and protein in an effort to meet the body's energy need.

Diseases of large and small blood vessels are common in diabetics. In fact, diabetes is considered a major risk factor of cardiovascular disease. Diabetes is typically accompanied by arteriosclerosis, which greatly increases the risk of heart attack and stroke. Diabetes also produces abnormalities in the capillaries. The disease seems to make the blood platelets more adhesive or sticky, allowing for a greater possibility of intravascular thrombosis. Diabetes also often causes nerve damage (neuropathy) and deterioration of myelin, an important covering of the axons of myelinated nerve fibers. The specific effects of the disease include heart attack, gangrene, kidney failure, blindness, cataracts, muscle weakness, and ulceration of the skin.

Control of the disease involves maintaining normal or near normal blood glucose level (glycemia) through the appropriate therapy. Depending on the type of diabetes being treated, therapy may include insulin, oral hypoglycemic agents, diet, and exercise. It is also important that the metabolism of fats and proteins is regulated. Diet is very important for the diabetic. Food intake must be enough to satisfy the needs of growth (in children) and metabolism, but not so much as to produce obesity. The American Diabetes Association recommends that diabetics eat a balanced diet composed of about 50% complex carbohydrates, 35% fat, and 15% proteins. It is recommended that dietary cholesterol and fats be curtailed to reduce the risk of heart disease.

A fundamental problem in the treatment of diabetes is that an injection of insulin is not the same physiologically as the natural secretion of insulin by the pancreas. Normally, insulin is released both continuously (basal secretion) and in response to blood-borne substrates. For example, after a meal, blood sugar rises and insulin is secreted to utilize the fuel effectively. In the diabetic, insulin is administered all at once. Many of the side effects of the disease manifest themselves when there is a chronic imbalance between insulin levels and glucose; in other words, diabetes becomes very dangerous when it is not controlled.

Exercise seems to be very effective in reducing the side effects of the disease, providing three major benefits for the diabetic: (1) Endurance exercise reduces the diabetic's need for insulin. The lower the insulin dosage, the closer to normal physiology. Therefore, there is less of a metabolic roller coaster, allowing for easier control of the disease. (2) Endurance exercise reduces platelet adhesiveness. The blood platelets, which are more sticky in diabetics, achieve normal adhesiveness for about 24 hr after exercise such as running, swimming, or cycling. (3) Regular endurance exercise reduces the severity of the risk factors of coronary artery disease such as hypertension, obesity, blood lipids, and serum uric acid.

Exercise is not recommended unless the diabetes is under control; that is, the blood sugar levels can be predictably regulated. Participation in physical activity presents control problems that have to be worked out on a trial-and-error basis. The most immediate problem is hypoglycemia—low blood glucose. Glucose utilization is much greater during exercise than at rest. The diabetic who is taking insulin is involved in a daily and often hourly juggling act trying to balance energy intake (food), energy output (exercise and resting metabolism), and insulin (energy regulator). A variation in any one of these three factors requires adjustment in the others.

Many active diabetics choose to counterbalance the increased glucose requirements of exercise by consuming extra food just before exercise, usually a high-sugar type such as candy, orange juice, or graham crackers. This works particularly well if the time or amount of exercise is irregular. If exercise habits are more predictable, then a change in the amount of insulin injected each day might help prevent hypoglycemia. The body works a lot like a machine: Exercise at a given intensity for a given amount of time has a predictable fuel requirement. Regular, predictable exercise habits make the control of diabetes a lot easier.

It is very important to protect against hypoglycemia—particularly for the athletically active diabetic. Hypoglycemia can result in a severe impairment of judgment and loss of coordination, which could lead to injury. A person in a state of hypoglycemia can easily overestimate his or her capacity, with potentially tragic results.

A diabetic who is in poor control of the disease and has hyperglycemia may also be in a state of ketosis. *Ketosis* is a condition characterized by the accumulation of large amounts of acetoacetic acid, β-hydroxybutyric acid, and acetone in the blood (Chapter 7). Ketosis occurs when the intracellular availability of carbohydrates is severely diminished and fats must be used as the predominant substrate. The blood concentration of ketones can rise to as high as 30 times above normal, leading to extreme acidosis. If left to progress, ketosis can lead to diabetic coma. Exercise is not recommended when a person has ketosis or hyperglycemia. Under certain conditions exercise can make the problem worse by increasing blood sugar levels even higher. So, diabetes should be controlled before undertaking an exercise program.

Another problem that occurs when a poorly controlled diabetic exercises is an excessive release of growth hormone. Growth hormone is very important in the regulation of blood lipids, in addition to its role in the growth processes during childhood and adolescence. Excessive secretion of growth hormone in uncontrolled diabetics may contribute to blood vessel disease. Again, the solution is to get diabetes well under control and then begin exercising.

Diabetics who take insulin have a choice of where to inject it. Injections are usually given beneath the skin covering the thigh, upper arms, abdomen, or buttocks. The site of injection is not terribly important unless exercise is to follow. It then becomes critically important. If insulin is injected subcutaneously in an area adjacent to a muscle involved in exercise, then the insulin goes into the bloodstream much more rapidly and may lead to hypoglycemia. For example, if a runner injected insulin into a leg and then went out to run, the insulin would begin to act more quickly and more powerfully than normal. The problem can be avoided by injecting the insulin in the abdomen or arm.

Because of their increased risk of heart disease, diabetics of any age should get a treadmill test before starting an exercise program. Heart disease, which most frequently manifests itself in the nondiabetic population beginning at about

35 to 45 yr of age, appears sooner among diabetics. Therefore, diabetics should be thoroughly evaluated so that they can safely participate in their exercise program.

CHRONIC OBSTRUCTIVE PULMONARY DISEASE

Chronic obstructive pulmonary disease (COPD) is progressive and is characterized by airway destruction of alveoli, retention of mucous secretions, narrowed airways, and respiratory muscle weakness. Categories of this disease include emphysema, asthma, and bronchitis, although they are seldom distinct.

Emphysema is characterized by a loss of alveoli and their related vasculature. In addition to the loss of functional lung tissue, the hypoxia created by the reduction of blood supply results in pulmonary vasoconstriction that tends to further reduce the surface area of gas exchange. The disease increases pulmonary artery pressure, first during exercise and then at rest, which leads to structural hypertrophy and hyperplasia of the smooth muscle pulmonary vascular bed, followed by fibrosis and atherosclerosis. Ultimately, the pulmonary hypertension leads to right heart failure.

Emphysema patients often exhibit chest deformities resulting from their use of accessory muscles for respiration. Their diaphragm is fixed in an inspiratory position, their chest is overexpanded, and their respiratory muscles, as a whole, are weakened. There is a decrease in adenosine triphosphate and creatine phosphate in both intercostals, which seems to be related to increasing airway obstruction. They show a higher than normal residual lung volume and have difficulty expiring rapidly.

Chronic bronchitis is described as a persistent productive cough, with episodes of infected sputum and shortness of breath (dyspnea). Bronchitis patients typically show blood gas abnormalities caused by aberrations in ventilation and perfusion of the lungs. At the later stages of the disease, there is often extensive peripheral edema resulting from pulmonary hypertension and right heart failure.

Ventilation and diffusion of respiratory gases are limiting in obstructive lung disease because of weak respiratory muscles, narrowed obstructed airways, and destroyed or compromised alveoli. Physical work capacity is usually low, and patients often complain of dyspnea at rest.

Because of the overriding effects of the work of breathing during exercise to exhaustion, these subjects will exhibit low maximal heart rates, maximal respiratory exchange ratios of less than 1.0, and lower blood lactate levels than for healthy individuals (Table 26-6). These people are often unable to push themselves during exercise because of anxiety produced by shortness of breath. Maximal exercise ventilation will often reach or exceed that of maximal vol-

TABLE 26-6

Exercise Factors in Chronic Obstructive Pulmonary Disease

Factor	Effect[a]
Maximum heart rate	↓
$\dot{V}_{O_2 max}$	↓
Oxygen pulse	No effect
PA_{O_2}	↓
Pa_{O_2}	↓
$\dot{V}_E/\dot{V}_{CO_2}$	↑
V_D/V_T	↑
Expiratory airflow pattern	Abnormal
MVV	↓
$\dot{V}_{E max}/MVV$	↑
Maximal cardiac output	↓
Δ Cardiac output/$\Delta\dot{V}_{O_2}$	No effect
Mean pulmonary artery pressure	↑
Pulmonary artery wedge pressures	No effect
Pulmonary vascular resistance	↑
Left ventricular fraction	No effect
Right ventricular fraction	↓

[a] Ascending and descending arrows indicate increase and decrease, respectively. N, Normal.

SOURCE: Modified from Brown, H.V. and K. Wasserman. Exercise performance in chronic obstructive pulmonary diseases. Med. Clin. North Am. 65(3): 525–547, 1981.

untary ventilation (MVV), which causes dyspnea and fatigue at low exercise intensities. People with normal lung function normally ventilate about 60% of MVV during maximal exercise.

Unfortunately, lung function does not seem to improve with training in these individuals. However, there is evidence that endurance training may delay the deterioration of pulmonary function. Maximal O_2 consumption and work capacity improve, provided the patients are motivated and the exercise prescription is not overly strenuous.

Exercise prescription for these individuals, as well as for those with other pulmonary diseases such as cystic fibrosis, should include progressive endurance exercise, such as walking, and breathing exercises. Ideally, the exercise pace should be determined by monitoring arterial P_{O_2}. Noninvasive instruments, such as the transcutaneous ear oximeter, make this practical for most pulmonary laboratories. Breathing exercises are aimed at increasing airflow to obstructed, atelectatic, and restricted areas, improving respiratory muscle endurance, decreasing the use of accessory breathing muscles, facilitating secretion removal, and maintaining chest mobility.

ASTHMA

Asthma is technically an obstructive lung disease. However, because it is frequently experienced without emphysema or bronchitis, particularly in the young, it will be treated independently. Asthma is a lung disorder characterized by edema in the walls of the small bronchioles, secretion of thick mucus into their lumens, and spasm of their smooth muscle walls. Symptoms include choking, the sensation of shortness of breath, wheezing, tightness in the chest, increased mucus production, and fatigue. When an asthma attack occurs, breathing becomes labored, particularly during expiration, as a result of a reduction in bronchiolar diameter. An attack can be caused by such things as emotional upset, dust, pollen, cold and damp weather, smoke, and exercise.

Because exercise is one of the most important factors precipitating an asthma attack, it is understandable that many asthmatics shun physical activity. However, medications developed for asthma not only make exercise possible, but actually decrease the incidence and severity of asthmatic attack. Athmatics involved in exercise programs have improved lung function, muscle coordination, and emotional adjustment. Usually there is a decrease in the frequency and severity of asthmatic attacks.

Although exercise can be beneficial, it is not without risk to the asthmatic, as overexertion is a prominent precipitator of asthma. It is particularly important that these individuals increase the intensity of their programs gradually and avoid environmental conditions (e.g., air pollution and extreme cold) that will complicate their asthma. Exercise should be modified or curtailed if the asthmatic is too tired or under emotional stress, or if it is too cold, humid, or smoggy. Under adverse environmental conditions, exercise bouts should be kept short, less than about 3 min at a time (or as tolerated). These people should avoid swimming in excessively cold water, because this can induce an asthma attack.

Asthma is often associated with allergies. Some degree of allergic control is desirable before beginning a vigorous endurance program. One possible suggestion is to undergo a series of allergy shots to desensitize against aggravating substances. At the very least, the asthmatic can prevent allergic reactions by avoiding dusty areas, pollen, and smoke.

An asthmatic should take a treadmill test before beginning an exercise program. Ideally, this test should include arterial blood gas analysis. In people with normal lung function, the lungs are capable of almost completely oxygenating the blood during even the heaviest levels of exercise. However, asthmatics may not be able to supply enough O_2 to the blood at the heavier levels of exertion; diffusion becomes a limiting factor. The treadmill and blood gas measurements make it possible to select a safe upper limit of exercise.

If exercise seems to bring on an attack, asthmatics should make sure to take their medication before the workout. Modern medications have enabled asthmatics to participate in even the most vigorous sports, including endurance events in the Olympic Games. In the 1972 Games, an asthmatic won a gold medal in swimming, only to be disqualified for taking his asthma medication. An excellent prescription drug for exercise-induced asthma is albuterol. This drug is available in aerosol form and is acceptable for use in international competition. These drugs can prevent exercise-induced asthma or reduce the effects once they have occurred. The control of chronic recurrent wheezing is usually accomplished using sympathomimetic drugs such as ephedrine or epinephrine in the form of an aerosol inhalant.

Asthmatics should be encouraged to drink a lot of water when they participate in endurance exercise. Ingesting water reduces the thickness of lung secretions and thus facilitates breathing.

These individuals should participate in self-paced activities and avoid overly competitive situations when they are beginning their exercise programs. Endurance interval training (30-sec to 3-min bouts of exercise followed by a short rest) is well tolerated by most asthmatics and is recommended as a conditioning method during the early stages of physical training. Beginners should ideally exercise in a supervised environment. An inhaler and 100% oxygen should be available, along with somebody to assist if a difficulty develops. The asthmatic should be instructed in relaxation positions and breathing exercises in the event of shortness of breath or distress.

OBESITY

Obesity, although technically not a disease state, requires modification of the exercise prescription because of the greater possibility of musculoskeleltal injuries and overexertion. Although obesity in itself is not considered a major risk factor of coronary heart disease, it is seldom isolated from more serious conditions. Obesity is often accompanied by hypertension, hyperlipidemia, diabetes, and cigarette smoking. When several risk factors are combined, the overall risk is many times greater. This in itself requires caution in prescribing exercise to these individuals.

Relative to their weight, obese people have lower muscle masses than those of normal weight. In addition, the cardiovascular demands of an activity are greater in the obese. Excess body fat forces the obese to do work when they have to climb a flight of stairs, run, or lift their body weight. Usually, these individuals also have low aerobic capacities, which can make even ordinary physical activities very demanding.

Preexercise screening is very important for this group. Screening should include a medical evaluation to assess other health problems such as diabetes, hypertension, and coronary heart disease, which will complicate the exercise prescription. Screening should also include a work capacity test and a determination of body fat.

The best exercises are those that minimize weight bearing and excessive shock to the joints, such as swimming, cycling, and walking. Jogging is usually not a good starting activity because it requires an excessive percentage of aerobic capacity and puts too much stress on the ankles, knees, hips, and back.

It is more important to emphasize duration over intensity. Long-duration endurance exercise is more effective for weight control than high-intensity short-term work. As fitness improves, the intensity can be increased.

This group is extremely prone to weight-reducing fads such as rubber suits, neoprene abdominal bands, and steam baths. The resulting dehydration has no effect on body fat and may lead to heat stroke, heat exhaustion, or electrolyte depletion.

Because the obese have more difficulty dissipating heat because the subcutaneous fat acts as a layer of insulation, they should avoid exercising in the heat. Lower physical capacity also contributes to a lower ability to sweat and an even greater tendency to exceed safe exercise intensities.

ARTHRITIS

Arthritis is an inflammatory disease of a joint. The Arthritis Foundation estimates that over 31 million people in the United States have this disorder to some extent and that the economic impact exceeds $13 billion per year. The most common forms of the disease include osteoarthritis, rheumatoid arthritis, juvenile rheumatoid arthritis, ankylosing spondylitis, systemic lupus erythematosus, and gout. Arthritis is chronic but may go periodically into remission.

Osteoarthritis

Osteoarthritis is the most common form of arthritis. It is caused by wear and tear on the joints and is a common effect of the aging process. Individual differences in the presence of arthritis can be accounted for by such factors as age, previous injury or trauma, heredity, previous joint disease, and metabolic diseases. This type of arthritis is specific to individual joints, rather than spreading to other joints. Pain can vary from none at all to debilitating.

Osteoarthritis damages the articular cartilage of joints. This is often accompanied by bone spurs and adhesions in the membranes and ligaments of the joint. The associated pain usually results in decreased range of motion and disuse atrophy caused by the person's avoidance of any motion that is uncomfortable. The limitation of activity results in the formation of further adhesions, which further limits motion.

Exercise prescription for people with osteoarthritis includes range-of-motion and strength exercises, and activities that minimize weight bearing, such as swimming. Often a reduction of body fat is recommended to help take pressure from joints. Analgesics will often help relieve pain so that the necessary exercises can be accomplished.

Rheumatoid Arthritis

Arthritis is a common manifestation of rheumatoid diseases such as rheumatoid arthritis, lupus erythematosus, and polyarteritis nodosa. These are autoimmune diseases in which aberrant immunoglobulins combine to form anti-immunoglobulin antibodies called rheumatoid factor. The rheumatoid factor binds with antigens and immune complexes in tissue and produces an inflammatory response. Scar tissue is eventually formed with repeated inflammatory episodes, so the inflammation itself is a potent precipitator of tissue damage.

Exercise is contraindicated during an inflammatory period, because it makes the inflammation and the tissue damage more severe. Because excessive bed rest and deconditioning can also lead to deterioration, there has to be the right balance of rest and exercise. Nordemar and co-workers recently studied 23 patients with rheumatoid arthritis who were involved in physical training for 4 to 8 yr. They found less pronounced degenerative changes, improved exercise tolerance, and fewer sick days from work in the active patients. They concluded that although there is a risk of overuse during physical activity in rheumatoid arthritis patients, it is better to be overactive than the reverse.

Range-of-motion exercises and minimal weight-bearing endurance exercises such as swimming or walking may be practiced when appropriate. Drugs such as aspirin, fenoprofen, indomethacin, tolmetin, corticosteroids, and phenylbutazone are usually used to relieve pain and facilitate mobilization exercises.

Another potentially serious problem associated with rheumatoid diseases is cardiopulmonary involvement. Sixty percent of patients with systemic lupus erythematosus will have pleuritis and/or pericarditis. Inflammation of small or medium-sized blood vessels is relatively common in all rheumatoid diseases. Caution should be used, as excessive exercise could be destructive to the heart.

Summary

Disease can pose severe stresses to the body; the additional stress of exercise may be intolerable. Exercise is contraindicated in some diseases and should be modified in others. Conditions where exercise is contraindicated include congestive heart failure, unstable angina, serious arrhythmias, cardiac inflammation, uncontrolled hypertension, and uncontrolled metabolic disease.

Inactivity and bed rest lead to physiological deterioration. Impairments include diminished capacity of the heart, reduced plasma and blood volumes, impaired blood vessel automaticity, decreased maximal O_2 consumption, muscle atrophy, orthostatic intolerance, and bone demineralization. Decreases in maximal O_2 consumption and work capacity range from 1 to 26%, depending on the type and duration of confinement.

Exercise training has become an important part of rehabilitation following myocardial infarction and cardiac surgery. Cardiac rehabilitation programs have been found to improve the patient's sense of well-being, exercise capacity, and HDL cholesterol, while reducing anxiety, ST-segment depression, blood pressure, cholesterol, triglycerides, and resting and submaximal exercise heart rates. There are three phases of cardiac rehabilitation: the inpatient program (early activity following infarction or surgery), the outpatient therapeutic program (supervised exercise in the hospital), and the exercise maintenance program (exercise training conducted in an environment especially organized for cardiac and cardiac-prone individuals).

A variety of drugs are used in the treatment of cardiac and cardiac-prone patients, which can alter the response to exercise. These drugs can alter such factors as the rate, contractility, conduction, preload, and afterload of the heart. Drug effects must be taken into consideration when evaluating the electrocardiogram and prescribing exercise.

Hypertension is a disease with no known cure, but it can usually be controlled by medication, diet, weight control, and physical activity. Exercise causes increases in systolic and sometimes diastolic blood pressure, so it should be used cautiously in individuals with hypertension. High "pressure-load" exercises such as shoveling snow and water skiing should be avoided by these individuals.

Exercise is beneficial in diabetes, because it reduces the need for insulin, platelet adhesiveness, and the severity of risk factors of coronary artery disease. Exercise is not recommended unless blood sugar levels can be predictably regulated with insulin and diet. Control of diabetes requires a balance between energy intake, energy expenditure, and insulin.

Exercise training does not improve but may delay deterioration of pulmonary function in patients suffering from chronic obstructive pulmonary disease. Maximal O_2 consumption improves, resulting in a higher absolute level of exercise-induced dyspnea.

Asthmatics involved in exercise programs have improved lung function, muscle coordination, and emotional adjustment, and a decreased frequency and severity of asthmatic attacks. Because exercise can precipitate an asthmatic attack, the exercise program should be carefully prescribed following determination of a safe upper limit of exercise.

Obesity, although technically not a disease, requires modification of the exercise prescription because of the greater possibility of musculoskeletal injuries and overexertion. Preexercise training is important to rule out any accompanying health problems. The best exercise are the endurance type that minimize weight bearing, such as swimming, cycling, and walking.

Training may prevent joint deterioration in osteoarthritis and rheumatoid arthritis. Exercises that develop range of motion and minimized weight bearing are best. Exercise is contraindicated during an inflammatory period in rheumatoid arthritis, because it makes the inflammation and tissue damage more severe.

Selected Readings

Asher, R.A.J. The dangers of going to bed. Br. Med. J. 4: 967–968, 1947.

Barnard, B.J., H.W. Duncan, K.M. Baldwin, G. Grimditch, and G.D. Buckberg. Effects of intensive exercise training on myocardial performance and coronary blood flow. J. Appl. Physiol.: Respirat. Environ. Exercise Physiol. 49: 444–449, 1980.

Baroldi, G., G. Falzi, and F. Mariani. Sudden coronary death. A postmortem study in 208 selected cases compared to 97 "control" subjects. Am. Heart J. 98: 20–31, 1979.

Bjorntorp, P. Hypertension and exercise. Hypertension. 4(Suppl.): 56–59, 1982.

Boyer, J.L. and F.W. Kasch. Exercise therapy in hypertensive men. J.A.M.A. 211: 1668–1671, 1970.

Braun, S.R., R. Fregosi, and W.G. Reddan. Exercise training in patients with COPD. Postgrad. Med. 71: 163–173, 1982.

Brooke, J.D., A. Chapman, L. Fisher, and P. Rosenrot. Repetitive skill deterioration with fast and exercise-lowered blood glucose. Physiol. Behav. 29: 245–251, 1982.

Brown, H.V. and K. Wasserman. Exercise performance in chronic obstructive pulmonary diseases. Med. Clin. North Am. 65(3): 525–547, 1981.

Bruce, R.A., F. Kusumi, and R. Frederick. Differences in cardiac function with prolonged physical training for cardiac rehabilitation. Am. J. Cardiol. 40: 597–603, 1977.

Bundgaard, A., F. Rasmussen, and L. Madsen. Pretreatment of exercise-induced asthma in adults with aerosols and pulverized tablets. Allergy. 35: 639–645, 1980.

Campbell, J.A., R.L. Hughs, V. Sahgal, J. Frederiksen, and T.W. Shields. Alterations in intercostal muscle morphology and biochemistry in patients with obstructive lung disease. Am. Rev. Respir. Dis. 122: 679–686, 1980.

Cerny, F.J., T.P. Pullano, and G.J.A. Cropp. Cardiorespiratory adaptations to exercise in cystic fibrosis. Am. Rev. Respir. Dis. 126: 217–220, 1982.

Chapman, E.A., H.A. DeVries, and R. Swezey, Joint stiffness: Effects of exercise on young and old men. J. Gerontol. 27: 218–221, 1972.

Chase, G.A., C. Grave, and L.B. Rowell. Independence of changes in functional and performance capacities attending prolonged bed rest. Aerospace Med. 37: 1232–1238, 1966.

Cheitlin, M.D., C.M. DeCastro, and H.A. McAllister. Sudden death as a complication of anomalous left coronary origin from the anterior sinus Valsalva. Circulation. 50: 780–787, 1974.

Convertino, V.A., R.W. Stremel, E.M. Bernauer, and J.E. Greenleaf. Cardiorespiratory responses to exercise after bed rest in men and women. Acta Astronautica. 4: 895–905, 1977.

Convertino, V.A., R. Bisson, R. Bates, D. Goldwater, and H. Sandler. Effects of antiorthostatic bedrest on the cardiorespiratory responses to exercise. Aviat. Space Environ. Med. 52: 251–255, 1981.

Costill, D.L., P. Cleary, W.J. Fink, J.L. Ivy, and F. Witzmann. Training adaptations in skeletal muscle of juvenile diabetics. Diabetes. 28: 818–821, 1979.

Cropp, G.J. Exercise-induced asthma. Pediatr. Clin. North Am. 22: 63–76, 1975.

Cropp, G.J., T.P. Pullano, F.J. Cerny, and I.T. Nathanson. Exercise tolerance and cardiorespiratory adjustments at peak work capacity in cystic fibrosis. Am. Rev. Respir. Dis. 126: 211–216, 1982.

DeBusk, R., N. Houston, W. Haskell, G. Fry, and M. Parker. Exercise training soon after myocardial infarction. Am. J. Cardiol. 44: 1223–1229, 1979.

Duvernoy, W.F.C. Sudden Death in Wolff-Parkinson-White syndrome. Am. J. Cardiol. 39: 472, 1977.

Edmunds, A.T., M. Tooley, and S. Godfrey. The refractory period after exercise-induced asthma. Am. Rev. Respir. Dis. 117: 247–254, 1978.

Fardy, P.S., J.L. Bennett, N.L. Reitz, and M.A. Williams. Cardiac Rehabilitation. St. Louis: C.V. Mosby, 1980.

Froelicher, V., D. Jensen, E. Atwood, M.D. McKirnan, K. Gerber, R. Slutsky, A. Battler, W. Ashburn, and J. Ross. Cardiac rehabilitation: Evidence for improvement in myocardial perfusion and function. Arch. Phys. Med. Rehabil. 61: 517–522, 1980.

Fye, K.H., H.M. Moutsopoulos, and N. Talal. Rheumatic diseases. In: Fudenberg, H. (ed.), Basic and Clinical Immunology. Los Altos, Calif.: Lange Medical Publications, 1978, p. 422–451.

Gibbons, L.W., K.H. Cooper, B.M. Meyer, and R.C. Ellison. The acute cardiac risk of strenuous exercise. J.A.M.A. 244: 1799–1801, 1980.

Godfrey, S., M. Silverman, and S.C. Anderson. Problems of exercise-induced asthma. J. Allergy Clin. Immunol. 52: 199–204, 1973.

Godfrey, S., M. Silverman, and S.D. Anderson. The use of the treadmill for assessing exercise-induced asthma and the effect of varying the severity and duration of exercise. Pediatrics. 56: 893–898, 1975.

Greenleaf, J.E. and S. Kozlowski. Physiological consequences of reduced physical activity during bedrest. Exerc. Sport Sci. Rev. 10: 84–119, 1982.

Greenleaf, J.E. and J. Reese. Exercise thermoregulation after 14 days of bed rest. J. Appl. Physiol. 48: 72–78, 1980.

Greenleaf, J., E. Bernauer, L. Juhos, H. Young, J. Morse, and J.W. Staley. Effects of exercise on fluid exchange and body composition in man during 14-day bed rest. J. Appl. Physiol. 43: 126–132, 1977.

Gyntelberg, F., L. Lauridsen, and K. Schubell. Physical fitness and risk of myocardial infarction in Copenhagen males aged 40–59. Scand. J. Work Environ. Health. 6: 170–178, 1980.

Hanson, P., M.D. Giese, and R.J. Corliss. Clinical guidelines for exercise training. Postgrad. Med. 67: 120–138, 1980.

Haskell, W.L. Cardiac complications during exercise training of cardiac patients. Circulation. 57: 920–925, 1978.

Hellerstein, H. Limitations of marathon running in the rehabilitation of coronary patients: Anatomical and physiological determinants. Ann. N.Y. Acad. Sci. 301: 484–494, 1977.

Hilsted, J., H. Galbo, B. Tronier, N.J. Christensen, and T.W. Schwartz. Hormonal and metabolic responses to exercise in insulin-dependent diabetics with and without autonomic neuropathy and in normal subjects. Int. J. Sports Med. 2: 216–219, 1981.

Holbreich, M. Exercise-induced bronchospasm in children. Fam. Physician. 23: 185–188, 1981.

Hughes, G.R.V. Connective Tissue Diseases. Oxford: Blackwell Scientific Publications, 1977.

Kavanagh, T., R.J. Shephard, A.W. Chisholm, S. Qureshi, and J. Kennedy. Prognostic indexes for patients with ischemic heart disease enrolled in an exercise-centered rehabilitation program. Am. J. Cardiol. 44: 1230–1240, 1979.

Koffler, D. The immunology of rheumatoid diseases. Clin. Symp. 31: 2–36, 1980.

Koivisto, V.A. Diabetes in the elderly: What role for exercise? Geriatrics. 36: 74–83, 1981.

Koivisto, V. and R.S. Sherwin. Exercise in diabetes. Postgrad. Med. 66: 87–96, 1979.

Konig, K. Changes in physical capacity, heart size and function in patients after myocardial infarction, who underwent a 4- to 6-week physical training program. Cardiology. 62: 232–246, 1977.

Korhonen, U.R. Dynamic and static exercise haemodynamics after acute myocardial infarction. Ann. Clin. Res. 12: 152–159, 1980.

LeBlanc, J., A. Nadeau, M. Boulay, and S. Rousseau-Migneron. Effects of physical training and adiposity on glucose metabolism and ^{125}I-Insulin binding. J. Appl. Physiol. 46: 235–239, 1979.

Ludvigsson, J. Physical exercise in relation to degree of metabolic control in juvenile diabetics. Acta Paediatr. Scand. 283(Suppl.): 45–49, 1980.

Mansfield, L., J. McDonnell, W. Morgan, and J.F. Souhrada. Airway response in asthmatic children during and after exercise. Respiration. 38: 135–143, 1979.

McFadden, E.R. and R. Ingram. Exercise-induced asthma. N. Engl. J. Med. 301: 763–769, 1979.

McHenry, P.L. The risks of graded exercise testing. Am. J. Cardiol. 39: 935–937, 1977.

Miller, P.B., R.L. Johnson, and L.E. Lamb. Effects of four weeks of absolute bed rest on circulatory functions in man. Aerospace Med. 35: 1194–1200, 1964.

Morales, A.R., R. Romanelli, and R.J. Boucek. The mural left anterior descending coronary artery, strenuous exercise and sudden death. Circulation. 62: 230–237, 1980.

Noakes, T.D., L.H. Opie, A.G. Rose, and P.H.T. Klynhans. Autopsy-proved coronary atherosclerosis in marathon runners. N. Engl. J. Med. 301: 86–89, 1979.

Nordemar, R., B. Ekblom, L. Zachrisson, and K. Lundqvist. Physical training in rheumatoid arthritis: A controlled long-term study (I and II). Scand. J. Rheumatol. 10: 17–30, 1981.

Patruno, D. Asthma and physical exercise. Curr. Ther. Res. 32: 257–264, 1982.

Petersen, G.E., M. Lorenzi, and T.H. Forsham. The effects of exercise upon platelet adhesion in diabetes mellitus. Diabetes. 28: 360, 1979.

Podolsky, S. (ed.). Clinical Diabetes: Modern Management. New York: Appleton-Century-Crofts, 1980.

Rudderman, N., O. Ganda, and K. Johansen. The effect of physical training on glucose tolerance and plasma lipids in maturity onset diabetes. Diabetes. 28(Suppl.): 89–92, 1979.

Sandler, H. Effects of bedrest and weightlessness on the heart. Hearts and Heart-like Organs. 2: 435–524, 1980.

Sergysels, R., A. van Meerhaeghe, G. Scano, M. Denaut, R. Willeput, R. Messin, and A. DeCoster. Respiratory drive during exercise in chronic obstructive lung disease. Bull. Eur. Physiopathol. Respir. 17: 755–766, 1981.

Shepard, R.J. Exercise-induced bronchospasm. A review. Med. Sci. Sports. 9: 1–10, 1977.

Sokolow, M. and M.B. McIlroy. Clinical Cardiology. Los Altos, Calif.: Lange Medical Publications, 1977.

Stremel, R.W., V.A. Convertino, E.M. Bernauer, and J.E. Greenleaf. Cardiorespiratory deconditioning with static and dynamic leg exercise during bed rest. J. Appl. Physiol. 41: 905–909, 1976.

Sugiura, T., Y.L. Doi, B.G. Haffty, T. Fitzgerald, R.L. Bishop, and D.H. Spodick. Effect of oral propranolol on left ventricular performance in patients with ischemic heart disease. Chest. 82: 576–580, 1982.

Sutton, J.R. Exercise in the post coronary patient: Unanswered question. Med. Sci. Sports. 4: 362–363, 1979.

Taylor, H.L., A. Henschel, J. Brozek, and A. Keys. Effects of bed rest on cardiovascular function and work performance. J. Appl. Physiol. 2: 223–239, 1949.

Thompson, P.D. Cardiovascular hazards of physical activity. Exerc. Sport Sci. Rev. 10: 208–235, 1982.

Vranic, M. and M. Berger. Exercise and diabetes mellitus. Diabetes. 28: 147–163, 1979.

Vuori, I., M. Makarainen, and A. Jaaskelainen. Sudden death and physical activity. Cardiology. 63: 287–304, 1978.

Wahren, J., P. Felig, and L. Hagenfeldt. Physical exercise and fuel homeostasis in diabetes mellitus. Diabetologia. 14: 213–222, 1978.

Walberg-Henriksson, H., R. Gunnarsson, J. Henriksson, R. Defronzo, P. Felig, J. Ostman, and J. Wahren. Increased peripheral insulin sensitivity and muscle mitochondrial enzymes but unchanged blood glucose control in type I diabetics after physical training. Diabetes. 31: 1044–1050, 1982.

Wilson, C.H. Exercise for arthritis. In: Basmajian, J.V. (ed.), Therapeutic Exercise. Baltimore: Williams & Wilkins Co., 1978. p. 514–530.

Wilson, P.K., P.S. Fardy, and V.F. Froelicher. Cardiac Rehabilitation, Adult Fitness, and Exercise Testing. Philadelphia: Lea & Febiger, 1981.

Wilson, P.K., E.R. Winga, J.W. Edgett, and T.T. Gushiken. Policies and Procedures of a Cardiac Rehabilitation Program. Philadelphia: Lea & Febiger, 1978.

Zeppilli, P., M.M. Pirrami, M. Sassara, and R. Fenici. T wave abnormalities in top-ranking athletes: Effects of isoproterenol, atropine, and physical exercise. Am. Heart J. 100: 213–222, 1980.

Zinman, B., M. Vranic, A. Albisser, B. Leibel, and E. Marliss. The role of insulin in the metabolic response to exercise in diabetic men. Diabetes. 28(Suppl.): 76–81, 1979.

27

NUTRITION AND ATHLETIC PERFORMANCE

Because diet affects metabolism, especially resting metabolism, the question arises whether diet can be manipulated to increase an individual's ability to perform at high metabolic rates for extended periods in athletic training and competition. It is apparent that proper nutrition will enable an individual to perform at his or her optimum in both practice and competition. However, the body's metabolism during exercise is directly controlled by biochemical processes, and the previous diet has only indirect influences. Therefore, extreme dietary manipulation immediately before a competition is more likely to affect a performance negatively than to improve it. The normal, balanced diet has not been surpassed for training, competition, or daily living.

NUTRITIONAL PRACTICE IN ATHLETICS

Perhaps no greater mythology exists in sports than in the area of athletic nutrition. To gain an immediate competitive edge, athletes have engaged in all sorts of odd dietary habits that are perpetuated if the athlete is successful because of genetic endowment. Sound nutritional practices of what to eat, how much to eat, and when to eat it will allow an athlete to perform up to his or her potential in practice and competition. Sound nutrition combined with habitual overload training can enable athletes to achieve incredible feats. At present there is no

Grete Waitz, world's premier women's long-distance runner on the way to another marathon victory. Proper training and sound nutrition allow individuals to achieve their maximal performance capabilities. (Ira Berger/Woodfin Camp.)

known foodstuff that will allow a mediocre, moderately conditioned athlete to become an Olympic champion. It is more likely that the miracle diet consumed before a competition will result in problems rather than in outstanding performance. In order to perform to its maximum, the body requires specific nutritional elements. However, those elements are available in a wide variety of foods. The body even has some flexibility in terms of interconversions of materials. With regard to nutrition and sports, it is *not* true that you are what you eat, and it is *not* true that you metabolize what you consume. A balanced diet involves at least three square meals per day, wherein the daily protein content is 15 to 20% of total calories, carbohydrate content is 45 to 55%, and fat consumption is 35% or less. These calories should be consumed in the form of dark green or yellow fresh vegetables, fruits, breads, and whole-grain products or potatoes, as well as milk and lean meats (including fish, poultry, eggs, and cheese). The consumption of fats and refined sugar products should be held to a minimum. If necessary, the reader should review Chapters 5 to 8 before proceeding to the following sections on sports nutrition.

MUSCLE GLYCOGEN

Figure 27-1 illustrates the relationship between the respiratory gas exchange ratio ($R = \dot{V}_{CO_2}/\dot{V}_{O_2}$) and relative work load. From the R, a crude estimate can be made of the relative contribution to energy yield of carbohydrates (glycogen and hexoses) and fats. Once exercise starts, carbohydrates contribute a majority of the energy released to support the exercise. The relative contribution provided by carbohydrates increases as relative work load increases. Endurance training has the effect of shifting the curve slightly down and to the right.

With introduction of the muscle biopsy technique in the late 1960s by Scandinavian scientists, significant contributions to our understanding of exercise metabolism were made. In the biopsy procedure, a hollow needle is inserted into a muscle through a small incision in skin and fascia. A stylet is then inserted into the bipsy needle and a small piece of muscle is excised within the needle. The needle is then removed and the muscle specimen is quick-frozen for biochemical assays or for microscopy. Information contained in the now famous one-legged experiments (Figure 27-2) summarizes much of what was learned by the biopsy technique. In these experiments, two subjects pedaled

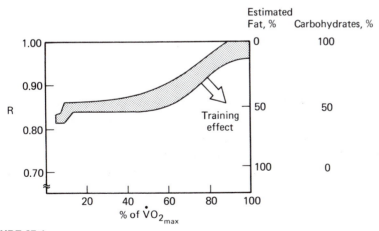

FIGURE 27-1
Relationship between the ventilatory respiratory exchange ratio ($R = \dot{V}CO_2/\dot{V}O_2$) and relative exercise intensity in well-nourished individuals. Utilization of fat and carbohydrate can be estimated from the R. As relative exercise intensity increases the proportional use of carbohydrate increases. Training shifts the curve to the right.

(Based on tables of Zuntz and Schumberg and other sources.)

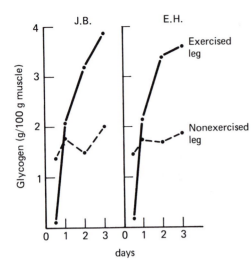

FIGURE 27-2

Effects exhausting exercise and diet on quadriceps muscle glycogen content. Exhausting exercise depletes glycogen content in active muscle. Rest and a high carbohydrate diet results in a glycogen overshoot (supercompensation) in the exercised muscle only.

(Based on data of Bergstrom and Hultman.)

the same bicycle ergometer; each subject used one leg to pedal, and exercise continued until exhaustion. During the days following the exercise, subjects rested and ate a high-carbohydrate diet. These one-legged and subsequent studies indicated the following:

1. Prolonged exercise of submaximal intensity can result in glycogen depletion of the active muscle.
2. Glycogen level is little affected in inactive muscle.
3. Following exhausting exercise and a high-carbohydrate diet, glycogen level in the exercised muscle is restored to a higher level than before exercise.
4. Glycogen level in the inactive muscle is little affected by the exercise and diet procedure.

The relationship between muscle glycogen content and endurance during hard but submaximal exercise (at about 75% $\dot{V}_{O_2max}$) is illustrated in Figure 27-3. In these experiments muscle glycogen content was manipulated by diet and previous exercise. Subjects then attempted to maintain a given work load on a bicycle ergometer for as long as possible. From these studies, it is apparent that for exercises such as cycling, the amount of glycogen present when exercise begins can determine the endurance time.

CARBOHYDRATE LOADING AND GLYCOGEN SUPERCOMPENSATION

The process by which glycogen concentration is raised to levels two or three times greater than normal is called glycogen supercompensation. Glycogen su-

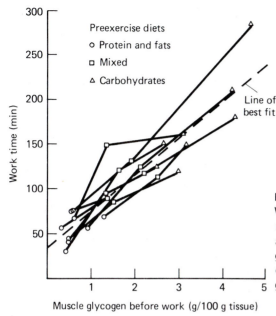

FIGURE 27-3

Work time on a bicycle ergometer at a set work rate for subjects on three different diets. Endurance (work time) depends on the muscle glycogen content before exercise. The diet several days prior to exercise affects the muscle of glycogen content, and therefore exercise endurance.

(Based on data from Bergström et al.)

percompensation results from a program of exercise (submaximal exercise to exhaustion) and a high-carbohydrate diet. This procedure is sometimes called carbohydrate (carbo) loading and is performed as follows. Three or four days prior to competition, the athlete exercises continuously to exhaustion. For some sports it may also be advisable to include some very hard intervals of exercise during and after a brief recovery from the prolonged exercise. These sprints deplete FG and FOG fibers of glycogen. The athlete then rests or trains mildly during the intervening days until competition, and eats a normal, balanced diet that contains substantial carbohydrate. The athlete will then be well prepared to compete in activities requiring hard continuous exercise (e.g., cycling, marathon running) or activities where activity is not continuous but contains intervals of very hard exercise (e.g., soccer, american football).

Problems with Carbohydrate Loading

In the more than 10 years of experience with carbohydrate loading, it appears that the practice is not universally successful. Some individuals apparently do not respond to the procedure. Further, the precise biochemical mechanism by which the practice works is not understood. Obviously glycogen synthase activity in the affected muscles is increased, but the practice in effect fools the biochemical regulation of this enzyme to produce an abnormal effect. Apparently after repeated attempts to glycogen-supercompensate, the biochemistry "smartens up," and the supercompensation effect is diminished.

Another problem encountered with carbohydrate loading is that, with inexperienced individuals, malnutrition can result. A carbohydrate loading diet should be one rich in whole-grain cereals and potatoes. Care should also be taken to include fresh vegetables (for vitamins and minerals) as well as for meats or meat substitutes (for protein). Refined sugar products (cakes, pies, candies) and alcohol should be avoided. In manipulating an athlete's diet immediately prior to competition, care should be taken to include necessary nutrients as well as calories. Too little protein or calorie input will result in lean tissue wasting, and too much carbohydrate will result in fat gain. Moreover, a preliminary trial is necessary to ensure that the diet is agreeable and palatable to the athlete.

Another potential problem or asset in glycogen supercompensation, depending on the activity, is that glycogen storage requires H_2O retention. Because each gram of glycogen requires almost 3 g of H_2O for storage, each gram in effect adds 4 g of body weight. In some sports (e.g., gymnastics or wrestling), added weight would be a liability. For endurance exercise, the added H_2O would be useful for preventing dehydration.

It has sometimes been advocated that during the interim between exercise to exhaustion and carbohydrate loading, a 2-day period of carbohydrate starvation be imposed. Supposedly, this period when proteins and fats are allowed, but no sugars or starches, results in a heightened glycogen supercompensation effect, a super-supercompensation. However, it has never been convincingly demonstrated that the insertion of a carbohydrate starvation period is beneficial. To the contrary, carbohydrate starvation after exhausting exercise has often been reported to produce serious side effects. Inducing a period of depression and lethargy immediately prior to a competition is perhaps not the best way to prepare an athlete psychologically. Furthermore, there is a natural tendency to restore glycogen reserves after exhausting exercise. If adequate carbohydrate is not consumed to accomplish this, lean tissues will be catabolized in the attempt to supply gluconeogenic precursors. The dietary input of proteins may be inadequate to prevent this.

Liver Glycogen and Blood Glucose

The release of glucose from liver during exercise as the result of glycogenolysis and gluconeogenesis is very important during prolonged exercise. Relatively, however, muscle glycogen is more important as a carbohydrate fuel than blood-borne glucose. On the basis of radiotracer studies at the University of California, Donovan and Brooks have estimated that muscle glycogen contributes three to five times as much fuel as does blood glucose during prolonged submaximal exercise. The maintenance of adequate blood glucose levels by hepatic function is essential to allow normal functioning of nonworking areas of the body during exercise. Of critical importance is the maintenance of fuel supply to the brain and nerve tissues, which rely almost exclusively on glucose delivered by the blood.

During sprinting and very intense exercises lasting less than 2 min, glycogenolysis in muscle supplies a very significant proportion of the ATP utilized. For these short-duration types of activities, there is little opportunity for liver glycogen to be mobilized and to contribute significantly to muscle metabolism.

AMINO ACID PARTICIPATION IN EXERCISE

The initial attempts to evaluate the roles of amino acids and proteins in supporting exercise considered only their potential roles as fuels. These experimental attempts were based on urinary nitrogen excretion determinations that did not take into account the substantial nitrogen loss in sweat or the complete balance of nitrogen input versus loss resulting from exercise.

On the basis of radiotracer studies performed on experimental animals by White and Brooks and by Henderson et al., it now appears that oxidation of particular amino acids, including essential amino acids such as leucine, is increased in proportion to the increment in $\dot{V}O_2$ during prolonged submaximal exercise. The contribution that amino acids make in supplying fuels is not great (approximately 5% to 10%) and may seem insignificant. However, the performances by winners of athletic competitions are seldom superior to those of "also rans" by a margin of 5%.

Consideration of the role of amino acids only as substrates (i.e., their deamination and oxidation) would be a mistake. Amino acids participate in a number of transamination reactions, not only between themselves but also with glycolytic and TCA-cycle intermediates. This is illustrated in Figure 27-4, which is a two-dimensional radiochromatogram of rat muscle following injection of [^{14}C]lactate and exhausting exercise. Obvious in the chromatogram is an exchange of carbon between a variety of metabolites. The amino acid alanine (a participant in the glucose–alanine cycle) plays an important role in maintaining blood glucose homeostasis. The calories returning to muscle during exercise as the result of glucose–alanine cycle activity are not great, but the effect on maintaining blood glucose level is more important.

As the result of muscular contraction, skeletal muscle produces glutamine (Figure 27-4) in significant amounts. The carbon source for this amino acid is α-ketoglutarate (α-KG), a TCA-cycle product, and the ammonia source is from the purine nucleotide cycle (Chapter 8). In order for the TCA cycle to continue functioning [citrate cannot be formed unless oxaloacetic acid (OAA) is present], material must be added to the TCA cycle to compensate for the loss of α-KG. When material is lost from the TCA cycle, it is said to be cataplerotic; when material is added to ("fills up") the TCA cycle to sustain its function, the material is said to be anaplerotic. The anaplerotic addition of glycolytic and amino acid derivatives is probably one reason that fat cannot be utilized as the single substrate in humans and most other mammals.

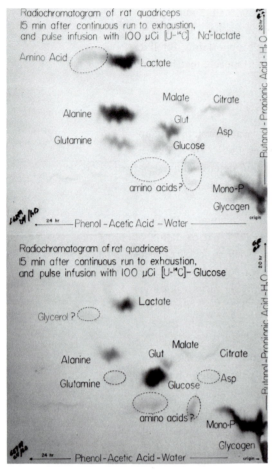

FIGURE 27-4

Two-dimensional radiochromatogram of rat skeletal muscle 15 minutes after exercise to exhaustion and injection with [U-^{14}C] lactate (A), and -glucose (B). Note the incorporation of label from injected tracer into a variety of metabolic intermediates including amino acids (alanine, glutamate, and glutamine).

(From Brooks and Gaesser with permission of the authors and journal.)

The supporting roles played by amino acids in metabolic responses to exercise are probably more important than their use as fuels.

BULKING UP

For most individuals, including those who are moderately active, a dietary protein content of from 0.8 to 1.0 g of protein $\cdot$ kg^{-1} body weight $\cdot$ day^{-1} is recommended to ensure nitrogen balance. On a normal, balanced North American or European diet, this figure is easily met, especially in active individuals who tend to eat more.

Unfortunately, there are few studies that have evaluated nitrogen balance in athletes during training. Determination of nitrogen loss involves collection and analysis of all nitrogen-containing materials from the body (including feces, urine, whiskers, semen, menses, and phlegm). This is a difficult task, to say the least. Studies by Gontzea et al. indicate that individuals on a normal recommended diet for protein content ($1 \text{ g} \cdot \text{kg}^{-1} \cdot \text{day}^{-1}$) go into negative nitrogen balance when initiating a program of exercise training (Figure 27-5). However, within several days to a week, nitrogen balance is restored on the normal diet. It has also been found that the period of negative nitrogen balance can be eliminated if the protein intake is raised to 1.5 to $2 \text{ g} \cdot \text{kg}^{-1} \cdot \text{day}^{-1}$ when training starts. Therefore, when initiating training or when increasing training intensity, it is advisable to increase protein intake.

On the basis of experimentally determined protein requirements for normal-sized individuals engaged in endurance activity, it is usually recommended that $1 \text{g} \cdot \text{kg}^{-1} \cdot \text{day}^{-1}$ is also adequate for heavy-weight athletes and those attempting to maintain or increase lean body mass ("bulk up"). The unfortunate fact is that there is inadequate information on the dietary protein content required to bulk up. Increasing lean body mass is often a requirement for success in some track and field events, american football, wrestling, and weight lifting. In one

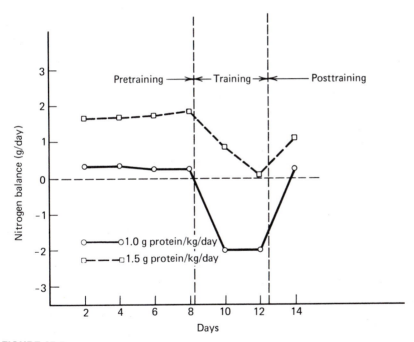

FIGURE 27-5
The initiation of endurance training causes individuals on a diet containing "normal" amounts of dietary protein (1.0 g Ptn/kg body weight/day) to go into negative nitrogen balance. The negative nitrogen balance condition, which occurs when training intensity increases, may be prevented by increasing dietary protein content.

(Based on data from Gontzea et al.)

study on Polish Olympic weight lifters preparing for competition, some athletes were unable to maintain nitrogen balance despite extremely high meat and vegetable protein consumptions. As will be discussed in Chapter 28, dietary protein supplementation by means of consuming concentrated protein materials may result in superior strength and weight gains in those undergoing very heavy weight training and taking anabolic steroids. The testimonials by athletes who believe that their success depends on consumption of large amounts of protein and calories, and the examples of such athletes as Polish weight lifters and Japanese sumo wrestlers suggest further laboratory investigation. The possibility exists that consumption of high-protein, high-calorie diets coupled with extremely high resistance training can positively affect lean body mass. In Chapter 28, the negative side effects of such practices are described.

FAT UTILIZATION DURING EXERCISE

Although the ability to store glycogen in the body is limited, for practical purposes during athletics the stores of lipid in adipose tissue are essentially inexhaustible. In addition to adipose tissue, significant amounts of lipid are also deposited in muscle and liver. On the basis of dry weight, more than twice as many calories are stored in fatty acids ($9 \text{ kcal} \cdot \text{g}^{-1}$) than in glycogen ($4 \text{ kcal} \cdot \text{g}^{-1}$). Because the storage of lipid entails far less additional water weight than glycogen does, the kilocalories stored per gram of lipid plus associated water may exceed that in glycogen by six to eight times. Consequently, storing calories in adipose tissue rather than in glycogen is far more efficient in terms of weight and space. Adipose tissue should be considered as a great fuel reserve. As described in Chapter 7, the mobilization of calories from adipose tissue, their delivery to muscle, as well as their uptake and utilization in muscle, is a complicated and lengthy process. Furthermore, untrained subjects seldom have the circulatory capacity or the muscle enzyme activity to utilize fats as the predominant fuels during endurance exercise. Consequently, glycogen becomes the preferred fuel (Figure 27-1). Through training, however, both circulatory capacity (Chapter 16) and muscle oxidative capacity (Chapter 6) improve, so that the curve in Figure 27-1 shifts down and to the right.

Given the fact that even the leanest endurance athlete has stored away more lipids than he or she will ever use in a competition, athletes do not need to eat extra fat during the days prior to a competition. However, because the uptake of fatty acids by muscle depends to a great extent on circulating levels, anything that raises those levels could potentially benefit endurance. As we will see, some lipid content in the precompetition meal might be desirable. Additionally, caffeine present in coffee, tea, and other beverages, or an oil meal and heparin-injection may enhance lipid mobilization (Chapter 28).

THE PRECOMPETITION MEAL

The timing, size, and composition of the pregame meal are all important considerations in performing optimally. Ideally, the athlete will enter competition with neither feelings of hunger or weakness (from having eaten too little or not at all) nor fullness from having eaten too much or too recently. Depending on the event, an athlete may want to be glycogen loaded. Ideally also, blood glucose level will be in the high-normal range (100 mg $\cdot$ dl^{-1} or 5.5 mM) and will not be falling. Blood insulin levels should be constant or falling slightly, but not rising.

It is important for athletes to eat a moderate-sized meal 2½ to 3 hr prior to competition. The nervous athlete who cannot eat or the athlete with an unfavorable competition schedule could be at a severe disadvantage in terms of maintaining blood glucose homeostasis. Even in well-nourished, glycogen-loaded athletes during the morning after 8 hr of sleep, the liver will essentially be empty of glycogen, and the blood glucose level will be falling. The hepatic glucose output will be the result of protein catabolism and gluconeogenesis. Therefore, a nutritious breakfast is necessary to stop the fall in blood glucose and begin replenishing liver glycogen content.

The practice of carbohydrate loading has unfortunately encouraged the habit, among some athletes, of consuming large amounts of simple starches, syrups, and sugars immediately prior to competition. This writer was appalled when he saw Frank Shorter on television prior to the 1976 Olympic marathon in Montreal eating a heap of pancakes topped with a chocolate bar, all immersed in syrup. If Frank really ate all of that and washed it down with soda pop as he pretended, his failure to repeat his 1972 Olympic victory may be explained. The reason is that a glycogen-supercompensated athlete has no real place, except in adipose tissue, to store much of the glucose in a high-sugar, high-carbohydrate diet. Worse than that, simple sugars and refined carbohydrates are rather rapidly digested. The rush of glucose into the bloodstream results in a tremendous release of insulin. The insulin clears the glucose from the circulation, but the hormone lingers, causing a continuous fall in blood glucose and increased ultilization of muscle glycogen. The results of such a meal are illustrated in Figure 27-6. As illustrated in the figure, soon after a glucose jolt to the circulation, glucose falls lower than if no glucose were consumed. This is no way to be entering competition.

Prior to the advent of carbohydrate loading, athletes were frequently advised to consume a moderate-sized meal of steak or eggs and toast or potatoes. This precompetition meal remains an excellent choice and is perhaps the best way to top off a carbohydrate-loading regimen. A lean steak and eggs present minimal problems with digestibility. Furthermore, their digestion results in the release of some fat and amino acids into the circulation. Amino acids do serve

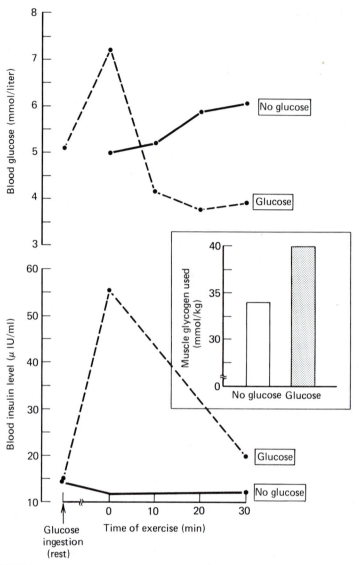

FIGURE 27-6

The effect of consuming a solution containing large amounts of simple sugar (glucose) 45 minutes prior to exercise on blood glucose and insulin during exercise. Because of the exaggerated insulin response, blood glucose eventually falls and places a greater dependence on muscle glycogen as a fuel. Low blood glucose and elevated glycogen utilization can lead to earlier fatigue in endurance exercises.

(Based on data of Costill et al.)

as substrates for exercise, and more importantly, they in effect represent "glucose timed-release capsules." A significant portion of the amino acids will be deaminated in the liver and converted into glucose. A protein meal is in effect a way to supply glucose and avoid an exaggerated insulin response. It has long been recommended that people who suffer from hypoglycemia (low blood sugar) consume protein-rich meals and avoid sugars and simple, highly refined carbohydrates. Hypoglycemics are those with exaggerated insulin responses to elevations in blood sugar (glucose) and resemble the athlete packing away sugar at the last minute.

The complex carbohydrates in whole-grain bread, cereals, and potatoes are digested more slowly than are sugars or simple carbohydrates. Consequently,

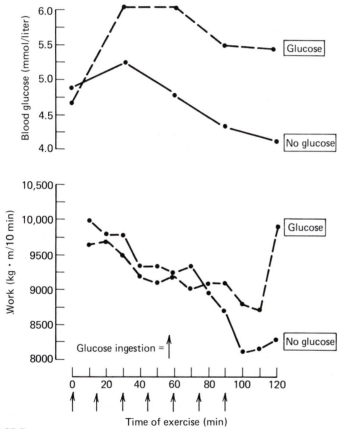

FIGURE 27-7

Injection of aqueous glucose solutions during prolonged exercise can result in greater work output and endurance. This procedure avoids an insulin response and the result is much different from that obtained when glucose is taken shortly before exercise (see Figure 26-6).

(Based on data from Ivy et al.)

glucose is released more slowly from the GI tract. Blood glucose rises more slowly and does not peak as high, but it remains elevated longer as a result of the slower release and lower insulin response. Circulating free fatty acid levels after a meal high in complex carbohydrates are most likely to be higher than after a meal containing sugars and simple carbohydrates of the same calorie value. Free fatty acid uptake by muscle is concentration dependent.

The old standard pregame meal of steak, eggs, toast, and potatoes, consumed 2½ to 3 hr before competition, raises blood glucose and fatty acids and keeps them elevated longer than either no meal or one rich in sugars and simple carbohydrates.

Glucose Ingestion During Exercise

Rather than consuming glucose or simple carbohydrates shortly before competition, a practice that may actually hamper performance (Figure 27-6), the consumption of dilute glucose solutions during exercise may enhance performance (Figure 27-7). When dilute, aqueous solutions are taken during exercise, the fuel can enter the circulation without eliciting a large insulin response. Thus the glucose fuel becomes available to working muscle, and "normal" blood glucose levels are maintained to benefit other glucose-requiring tissues such as the brain. As discussed in Chapter 28, glucose should be taken in dilute aqueous solution to maximize the rate of entry into the blood.

THE NORMAL BALANCED DIET

In 1957 the U.S. Department of Agriculture (USDA) published a report that suggested that an individual's daily diet should be made up of meals containing foods from four basic groups: (1) milk, (2) meat, (3) bread and cereals, and (4) vegetables and fruit. It must be realized that it is very difficult to make specific recommendations for the general population, because individual dietary needs can be quite unique. Periods of growth and physical activity will require extra calorie and protein intake, whereas in older individuals the dietary essentials should be present among fewer calories. During unique physiological situations such as pregnancy, proper nutrition becomes especially important. Another reason it is so difficult to make dietary recommendations is that our knowledge of essential dietary constituents as well as their proportions is inadequate. Therefore, in making dietary recommendations today, it is believed that individuals should eat a wide variety of foods to increase the probability that all the essentials will be included.

In recognition of the need to include more fruits, vegetables, and whole-grain cereals in the diet, the basic four of 25 yr ago have been broadened to today's basic eight: (1) milk, (2) meat (including fish, poultry, cheese, eggs),

(3) dark green or deep yellow vegetables, (4) citrus fruits, (5) other fruits and vegetables, (6) bread (consisting of whole-grain or enriched flour), (7) cereal and potatoes, and (8) fats (butter, margarine, etc.). The effect of these recommendations is to "lighten up" our diet by diluting the input from meats and meat products containing cholesterol and saturated fats, and to include more fruits, vegetables, and cereals. Fats are included as a separate group to ensure assimilation of fat-soluble vitamins. In this balanced diet, the contributions toward the caloric intake (not the contribution by weight) should be approximately as follows:

Protein: 15 to 20%

Carbohydrate: 45 to 55%

Fat: 35% maximum

THE ATHLETE'S DIET AND TRACE ELEMENTS

In eating, the athlete should be regulated by appetite once the competitive weight is achieved. Eat when hungry, but eat from the basic eight. At present we simply do not know if active people require more of the trace dietary elements. Furthermore, we do not know with certainty what the trace elements are. Vitamin and mineral pills contain only what the manufacturers put into them, not what the body's unique needs are. The best way to meet calorie, protein, and other dietary requirements is to eat a broad diet to ensure purposefully that all the essentials are accidentally consumed.

The number of meals consumed by an athlete will be influenced not only by the appetite and caloric requirements, but also by the training schedule. For instance, highly competitive collegiate swimmers often train twice a day for 2 hr at a time (e.g., from 6 to 8 A.M. and 3 to 5 A.M.). Consequently, a light breakfast before the first workout should be followed by a more substantial breakfast after the workout. Snacks in midafternoon and late evening may even be appropriate. With a training schedule of this type, malnutrition could easily result from missed meals.

In Chapter 25, we saw that in active individuals, the mechanisms of appetite result in a stable body weight. Young, active athletes often discover that they can eat "all they want," "whenever they want" and not get fat. Unfortunately, the consumption of "junk foods" can become a habit. Junk foods are usually highly processed foods—composed primarily of sugar, salt, starch, or fat—that supply calories without providing other essential nutrients. Two sets of problems can result from consumption of too much junk food. Immediately, meeting caloric requirements with junk foods lessens the probability of providing essential trace elements, and junk food consumption upsets the balance of nutrients in the diet. This may or may not be a problem to an athlete in heavy training who is consuming many other nutritional foods. Over the long term, however, the persistence of poor nutritional habits in the retired athlete could work to his or her disadvantage.

Summary

The normal balanced diet has not been surpassed for supporting daily life or athletic training and competition. Eating a wide variety of foods ensures the input of necessary calories, proteins, carbohydrates, fats, vitamins, minerals, and trace elements. Further, consumption of such a diet develops sound nutritional habits necessary to sustain the individual throughout his or her life.

Extreme manipulation of the diet immediately prior to competition can more likely upset or handicap the athlete than trigger outstanding performance. Two exceptions to this rule bear mention, however. Carbohydrate loading can potentially benefit athletes engaged in sports requiring endurance and activities causing problems of dehydration. In heavyweight athletes concerned with increasing body mass and lean body mass, the dietary protein content may have to be emphasized to improve deposition of lean tissue. The precompetition meal following carbohydrate loading should be either a balanced meal or one high in protein.

Selected Readings

Ahlborg, G. and P. Felig. Influence of glucose ingestion on fuel-hormone response during prolonged exercise. J. Appl. Physiol. 41: 683–688, 1976.

American Association for Health, Physical Education, and Recreation. Nutrition for the Athlete. Washington, D.C.: AAHPER, 1971.

Bergstom, J. and E. Hultman. Muscle glycogen synthesis after exercise: An enhancing factor localized to the muscle cells in man. Nature 210: 309–310, 1966.

Bergstrom, J. and E. Hultman. Synthesis of muscle glycogen in man after glucose and fructose infusion. Acta Med. Scand. 182: 93–107, 1967.

Bergstrom, J., L. Hermansen, E. Hultman, and B. Saltin. Diet, muscle glycogen and physical performance. Acta Physiol. Scand. 71: 140–150, 1967.

Bergstrom, J. and E. Hultman. A study of the glycogen metabolism during exercise in man. Scand. J. Clin. Lab. Invest. 19: 218–228, 1967.

Bergstrom, J., E. Hultman, L. Jorfeldt, B. Pernow, and J. Wahren. Effect of nicotinic acid on physical working capacity and on metabolism of muscle glycogen in man. J. Appl. Physiol. 26: 170–177, 1969.

Bergstrom, J. and E. Hultman. Nutrition for maximal sports performance. J. Am. Med. Assoc. 221: 999–1006, 1972.

Buskirk, E.R. Diet and athletic performance. Postgrad. Med. 61:229–236, 1977.

Buskirk, E.R. Some nutritional considerations in the conditioning of athletes. Ann. Rev. Nutr. 1:319–350, 1981.

Butterfield, G. and D. Calloway. Physical activity improves protein utilization in young men. Unpublished manuscript.

Cathcart, E.P. and W.A. Burnett. Influence of muscle work on metabolism in varying conditions of diet. Proc. Roy. Soc. (Biol.) 99: 405, 1926.

Celejowa, I. and M. Homa. Food intake, nitrogen and energy balance in Polish weight lifters, during a training camp. Nutrition and Metabolism 12: 259–274, 1970.

Christensen, E.H. and O. Hansen. Arbeitsfähigkeit und Ehrnàhrung. Sand. Arch. Physiol. 81: 160–163, 1939.

Consolazio, C.F., H.L. Johnson, R.A. Dramise, and J.A. Skata. Protein metablism during intensive physical training in the young adult. Am. J. Clin. Nutr. 28: 29–35, 1975.

Costill, D.L., A. Bennett, G. Brahnam, and D. Eddy. Glucose ingestion at rest and during prolonged severe exercise. J. Appl. Physiol. 34: 764–769, 1973.

Costill, D.L. and B. Saltin. Factors limiting gastric emptying during rest and exercise. J. Appl. Physiol. 37: 679–683, 1974.

Costill, D.L., E. Coyle, G. Dalsky, W. Evans, W. Fink, and D. Hopes. Effects of elevated plasma FFA and insulin on muscle glycogen usage during exercise. J. Appl. Physiol.: Respirat., Environ. Exercise Physiol. 43: 695–699, 1977.

Costill, D.L. and J.M. Miller. Nutrition for endurance sport: Carbohydrate and fluid balance. Int. J. Sports Med. 1: 2–14, 1980.

Darling, R.C., R.E. Johnson, G. Pitts, R.C. Consolazio, and R.F. Robinson. Effects of variations in dietary protein on the physical well-being of men doing manual work. J. Nutr. 28: 273–281, 1955.

DuBois, E.F. A graphic representation of the respiratory quotient and the percentage of calories from protein, fat, and carbohydrate. J. Biol. Chem. 59: 43–49, 1924.

Durnin, J.V.G.A. Protein requirements and physical activity. In: Parizkova, J. and V.A. Rogozkin (eds.), Nutrition, Physical Fitness and Health. Baltimore: University Press, 1978. p. 53–60.

Essen, B. Intramuscular substrate utilization. N.Y. Acad. Sci. 301: 30–44, 1977.

Fox, E.L. Sports Physiology. Philadelphia: W.B. Saunders Co., 1979. p. 242–281.

Foster, C., D.L. Costill, and W.J. Fink. Effects of preexercise feedings on endurance performance. Med. Sci. Sports. 11: 1–5, 1979.

Gollnick, P.D., K. Piehl, I.V. Saubert, C.W. Armstrong, and B. Saltin. Diet, exercise, and glycogen changes in human muscle fibers. J. Appl. Physiol. 33: 421–425, 1972.

Gontzea, I., R. Sutzescu, and S. Dumitrache. The influence of adaptation to physical effort on nitrogen balance in man. Nutr. Rep. Int. 11: 231–236, 1975.

Henderson, S.A., A.L. Black, and G.A. Brooks. Leucine turnover and oxidation in trained and untrained rats during rest and exercise. Med. Sci. Sports Exercise. 15: 98, 1983.

Hickson, R.C., M.J. Rennie, W.W. Winder, and J.O. Holloszy. Effects of increased plasma fatty acids on glycogen utilization and endurance. J. Appl. Physiol. 43: 829–833, 1977.

Hultman, E. Muscle glycogen in man determined in needle biopsy specimens. Method and normal values. Scand. J. Clin. Lab. Invest. 19: 209–217, 1967.

Ivy, J.L., D.L. Costill, W.J. Fink, and R.W. Lower. Influence of caffeine and carbohydrate feedings on endurance performance. Med. Sci. Sports. 11: 6–11, 1979.

Jansson, E. Diet and muscle metabolism in man. Acta Physiol. Scand. (Suppl.) 487: 1–24, 1980.

Jansson, E. and L. Kaijser. Effect of diet on the utilization of blood-borne and intramuscular substrates during exercise in man. Acta Physiol. Scand. 115: 19–30, 1982.

Jette, M., O. Pelletier, L. Parker, and J. Thoden. The nutritional and metabolic effects of a carbohydrate-rich diet in a glycogen supracompensation training regimen. Am. J. Clin. Nutr. 31: 2140–2148, 1978.

Karlsson, J. and B. Saltin. Diet, muscle glycogen and endurance performance. J. Appl. Physiol. 31: 203–206, 1971.

Koivisto, V.A., S.-L. Kavonen, and E.A. Nikkila. Carbohydrate ingestion before exercise: Comparison of glucose, fructose, and sweet placebo. J. Appl. Physiol: Respirat. Environ. Exercise Physiol. 51: 783–787, 1981.

Krause, M.V. and L.K. Mahan. Food, Nutrition, and Diet Therapy. Philadelphia: W.B. Saunders, Co. 1979.

Lemon, P.W.R. and J.P. Mullin. Effect of initial muscle glycogen levels on protein catabolism during exercise. J. Appl. Physiol.: Respirat. Environ. Exercise Physiol. 48: 624–629, 1980.

Lusk, G. Analysis of the oxidation of mixtures of carbohydrate and fat. J. Biol. Chem. 59: 41–42, 1924.

Page, L. and E. Phippard. Essentials of an adequate diet. Home Economics Research Report No. 3, U.S. Department of Agriculture, Washington, D.C., 1957.

Pirnay, F., M. Lacroix, F. Mosora, A. Luyckx, and P. Lefebvre. Glucose oxidation during prolonged exercise evaluated with naturally labeled (13 C) glucose. J. Appl. Physiol. 43: 258–261, 1977.

Mayer, J. and B. Bullin. Nutrition, weight control and exercise. In: Johnson, W. and E.R. Buskirk (eds.), Science and Medicine of Exercise and Sport. New York: Harper & Row, 1974.

Merkin, G. Carbohydrate loading: A dangerous practice. J. Am. Med. Assoc. 223: 1511–1512, 1973.

Morgan, W. (ed). Ergogenic Aids and Muscular Performance. New York: Academic Press, 1972.

National Dairy Council. Nutrition and human performance. Dairy Counc. Dig. 51: 13–17, 1980.

Olsson, K. and B. Saltin. Diet and fluids in training and competition. Scand. J. Rehabil. Med. 3: 31–38, 1971.

Pernow, B. and B. Saltin (eds.). Muscle Metabolism During Exercise. New York: Plenum Press, 1971.

Sherman, W.M., M.J. Plyley, R.L. Sharp, P.J. Van Handle, R.M. McAllister, W.J. Fink, and D.L. Costill. Muscle glycogen storage and its relationship to water. Int. J. Sports Med. 3: 22–24, 1982.

Wahren, J. Glucose turnover during exercise in man. Ann. N.Y. Acad. Sci. 301: 45–55, 1977.

White, T.P., and G.A. Brooks. [U-^{14}C]glucose, -alanine, and -leucine oxidation in rats at rest and two intensities of running. Am. J. Physiol. (Endocrinol. Metab. 3) E155–165, 1981.

Zuntz, N. Betrachtungen über dre Bezichungen zwischen Währstoffen und Leistungen des Körpers. Oppenheimers Handuch der Biochemie 4: 826, 1911.

Zuntz, N. and Schumburg. Studren zu einer physiologie des Marsches, Berlin, 1901. p. 361.

28

ERGOGENIC AIDS

The extreme competitiveness within the sports world has forced athletes to grasp at any substance or technique that might provide an edge over the competition. Considering the intensity of athletic training, it is little wonder that athletes consume substances to improve performance, such as amphetamines and anabolic steroids (which are potentially dangerous) or employ far-fetched devices such as appliances to separate teeth.

Substances or techniques (other than training) that improve athletic performance are called ergogenic aids. Their use began in antiquity, when Aztec warriors would eat the hearts of particularly brave fallen adversaries in the hope of increasing their prowess in battle. The ancient Greeks were known to espouse the consumption of red meat for the development of muscle tissue. To this day there seems to be an unending array of ergogenic aids promising improved performance.

The vast majority of ergogenic aids provide no benefit other than the placebo effect. A *placebo* causes an improvement through the power of suggestion (if a person wants it to work, it does). The importance of placebos has long been recognized in both medicine and athletics. There are anecdotes of coaches administering "super pills" to athletes to improve performance. Even though the pills contained only sugar, the athletes believed that their performance would be improved, so it was. The placebo effect was demonstrated experimentally when Ariel administered placebos that resembled Dianabol, an anabolic steroid,

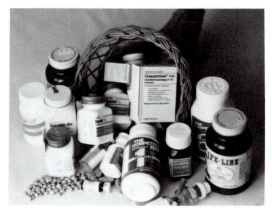

The ergogenic aids in one athlete's medicine chest more than fills a cornucopia. Athletic performances attributed to many substances and inadequate policing by sports authorities leads athletes to jeopardize their immediate and long term health through the consumption of banned and illegal substances.

and was able to produce dramatic increases in strength.

The popularity of specific ergogenic aids usually stems from their use by top athletes. The successful Japanese swim teams of the 1930s used supplemental oxygen during competition, so the practice spread. Many champion weight-trained athletes use anabolic steroids, encouraging their use by younger athletes. Testimonials for vitamin and protein supplements by Mr. America or Mr. Universe in a body-building magazine perpetuate the use of sometimes dubious products, even in the face of conflicting evidence presented in prestigious research journals.

Ergogenic aids research is sometimes fraught with contradiction and uncertainty stemming from poor experimental controls and the use of few subjects. Often the only information available on a specific agent is from studies using animal models or untrained subjects (who show wide variability in response to training), making it difficult to extrapolate the results to highly competitive athletes.

The most popular ergogenic aids used to improve athletic performance are listed in Table 28-1. Even though most of these have been shown to be worthless, the mystique of ergogenic aids continues to captivate almost everyone associated with sport. This chapter explores the most popular ergogenic aids as well as discussing possible side effects.

BANNED SUBSTANCES

The International Olympic Committee (IOC) has passed legislation banning substances ingested or consumed for the purpose of unfairly and artificially improving performance in competition (Table 28-2). The banned list appears to be based on the attempt to discourage any pretext of unfair competition,

TABLE 28-1
Some Substances or Techniques Used as Ergogenic
Aids

Alcohol
Alkalies
Amino acids
Amphetamines
Anabolic steroids
Aspartates
Bee pollen
Caffeine
Camphor
Cocaine
Cold
Digitalis
Electrical stimulation
Epinephrine
Gelatin
Growth hormone
Heat
Human chorionic gonadotrophin
Hypnosis
Marijuana
Massage
Mineral supplements
Negatively ionized air
Nicotine
Nitroglycerine
Norepinephrine
Organ extracts
Oxygen
Periactin
Protein supplements
Sulfa drugs
Strychnine
Vitamin supplements
Wheat germ oil
Yeast

TABLE 28-2
Substances Banned by the International Olympic Committee

Psychomotor Stimulant Drugs

Amphetamine

Benzphetamine

Cocaine

Ethylamphetamine

Fencamfamin

Fenproporex and related compounds

Methylamphetamine

Methylphenidate

Norpseudoephedrine

Pemoline

Phenmetrazine

Phentermine

Pipradol

Prolintane

Sympathomimetic Amines

Ephedrine

Methylephedrine

Methoxyphenamine and related compounds

Miscellaneous CNS Stimulants

Amiphenazole

Bemegride

Leptazole

Nikethamide

Strychnine and related compounds

Narcotic Analgesics

Dextromoramide

Dipipanone

Heroin

Methadone

Morphine

Pethidine and related compounds

Anabolic Steroids

Methandrostenolone (Dianabol)

Nandrolone phenpropionate (Durabolin)

Nandrolone decanoate (Deca-Durabolin)

Oxandrolone (Anavar)

Oxymesterone (Oranabol)

Oxymetholone (Anadrol)

Stanozolol (Winstrol) and related compounds

Testosterone Esters

because any beneficial effect that many of these substances may have has not been demonstrated.

Attempts at detecting the use of banned substances in international sport have escalated from relatively modest efforts at the 1960 Olympics in Rome, to the multimillion-dollar project at the 1984 Olympics in Los Angeles. The increased level of sophistication among athletes and in the field of sports medicine has resulted in a steadily increasing list of banned substances.

Unfortunately, many banned substances are found in a variety of over-the-counter and prescribed medications. For example, banned substances are contained in some nonprescription medications such as decongestants, throat lozenges, topical nasal decongestants, and eyedrops. This has presented and continues to present a tremendous problem for athletic coaches and administrators. In one instance, unwitting consumption of substances found on the list has resulted in the disqualification of an Olympic gold medal winner. Consequently, many of the larger countries involved in international sports competition have engaged pharmacology consultants to help them sift through the complications involved in the consumption and detection of drugs.

ANABOLIC STEROIDS

Anabolic steroids are drugs that resemble androgenic hormones such as testosterone (Figure 28-1) and androstenedione. Athletes consume them in the hope of gaining weight, strength, power, speed, endurance, and aggressiveness. They are widely used by athletes involved in such sports as track and field (mostly the throwing events), weight lifting, and american football. However, in spite of their tremendous popularity, their effectiveness is controversial.

The American College of Sports Medicine has issued a position statement regarding the use and abuse of anabolic–androgenic steroids in sports. They state that studies of the effects of anabolic steroids on athletic performance are contradictory, and that for many individuals, any benefits are likely to be small and not worth the health risks involved. Yet, almost all athletes who consume these substances extol their beneficial effects and feel that they would not have been as successful without them (see Appendix VII).

There are several possible reasons for the large differences between experi-

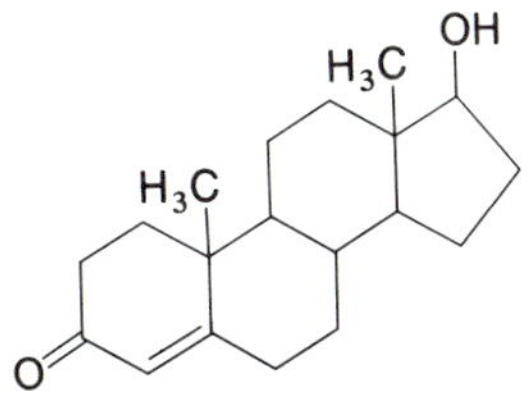

FIGURE 28-1
Chemical structure of testosterone.

mental findings and empirical observations. First, an incredible mystique has arisen around these substances, providing fertile ground for the placebo effect. Second, the use of anabolic steroids in the "real world" is considerably different from that in rigidly controlled, double-blind experiments (in a double-blind study, neither the subject nor experimenter knows who is taking the drug). Most studies have not used the same drug dosage used by athletes, as institutional safeguards prevent administration of high dosages of possibly dangerous substances to human subjects. Additionally, subjects in research experiments seldom resemble accomplished weight-trained athletes. Under these conditions, we must assess the results of sound research studies, as well as clinical and empirical observations, in order to obtain a realistic profile of the use, effects on performance, and side effects of these substances.

How Anabolic Steroids Work

Male hormones, principally testosterone, are partially responsible for the tremendous developmental changes that occur during puberty and adolescence. Male hormones exert both androgenic and anabolic effects. Androgenic effects are characterized by changes in primary and secondary sexual characteristics such as enlargement of the penis and testes, changes in the voice, hair growth on the face, the axilla, and the genital areas, and increased aggressiveness. The anabolic effects of androgens are characterized by accelerated growth of muscle, bone, and red blood cells, and enhanced neural conduction.

Anabolic steroids have been manufactured to enhance the anabolic properties of the androgens and minimize the androgenic properties. However, no steroid has completely eliminated the androgenic effects because the so-called androgenic effects are really anabolic effects in sex-linked tissues. That is, the effects of male hormones on accessory sex glands, genital hair growth, and oiliness of the skin are a reflection of protein anabolism in those specific tissues. The androgens with the most potent anabolic effect are also those with the greatest androgenic effect.

Although the process is not completely understood, male hormones work by stimulation of receptor molecules in muscle cells, which activate specific genes to produce proteins (Figure 28-2). They also affect the activation rate of enzyme systems involved in protein metabolism, thus enhancing protein synthesis and inhibiting protein degradation (called an anticatabolic effect). Additionally, they may affect the permeability of cell membranes to amino acids, which would make more material available for protein synthesis. Souccar et al. have demonstrated a trophic effect of testosterone on neuromuscular transmission, possibly by a facilitated release of acetylcholine from the nerve terminal.

Heavy resistance training seems to be necessary for anabolic steroids to exert any beneficial effect. Some researchers have speculated that the real effect of anabolic steroids is the creation of a "psychosomatic state" characterized

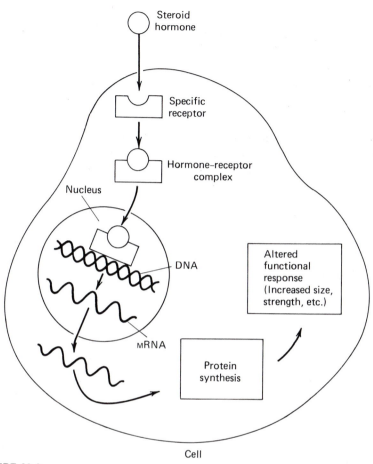

FIGURE 28-2
Theoretical model of how steroid hormones affect cell growth. Steroid hormone enters the cell and binds to a receptor molecule. The bound hormone can then enter the nucleus and activate specific genes to produce proteins. These proteins in turn bring about the cellular changes triggered by the hormone.

(SOURCE: From J.E. Wright, 1978.)

by sensations of well-being, euphoria, increased aggressiveness, and tolerance to stress, allowing the athlete to train harder.

Heavy training has been shown to increase testosterone binding capacity. Coupled with increased availability of the hormone (the anabolic steroid) and the enhanced transport of amino acids into the muscle (due to the training stimulus), it is not surprising that protein synthesis in the muscle might be enhanced. A diet high in protein may also be important to maximize the effectiveness of anabolic steroids.

The effects of anabolic steroids on factors important to physical performance are unclear, as the well-controlled, double-blind studies have rendered conflicting results. In studies showing beneficial effects, body weight increased by about 4 pounds, lean body weight by about 6 pounds (fat loss accounts for the discrepancy between gains in lean mass and body weight); bench press increased by about 15 pounds, and squats by about 30 pounds. Some of the reported weight gain may be due to increased water retention. Anabolic steroids are structurally similar to aldosterone, which causes Na^+ retention, and, ultimately, water retention. Almost all studies have failed to demonstrate a beneficial effect on maximal O_2 consumption or endurance capacity. These studies have typically lasted 6 to 8 weeks and have usually used relatively untrained subjects.

The gains made by athletes in uncontrolled observations have been much more impressive. Weight gains of 30 or 40 pounds, coupled with 30% increases in strength, are not unusual. Such case studies lack credibility because of the absence of scientific controls. However, it would be foolish to completely disregard such observations.

Side Effects of Anabolic Steroids

Anabolic steroids have several side effects, including alterations in liver function, lipid chemistry, and cardiovascular function, suppression of the gonadotrophin–testicular axis, and a variety of miscellaneous effects. The severity of these side effects seems to be affected by dosage and duration of drug therapy. The major side effects are summarized in Table 28-3.

Athletes using anabolic steroids typically exhibit elevated blood levels of liver enzymes such as serum glutamic oxaloacetic transaminase (SGOT), serum glutamic pyruvic transaminase (SGPT), and alkaline phosphatase, all of which indicate liver toxicity. Elevated levels of blood glucose, creatine phosphokinase (CPK), and bilirubin have also been noted. These changes are usually reversible upon withdrawal from the drug. Prolonged administration in some groups of patients has been linked to severe liver disorders such as: peliosis hepatis, hepatocellular carcinoma, and cholestasis.

Oral steroids, such as methandrostenolone (Dianabol), are particularly hepatotoxic because their structure has been altered to make them digestible by adding an alkyl or alkynyl group in the α configuration at position 17 of the sterol. This causes the steroid to become concentrated in the liver much earlier and in greater quantity than the injectable varieties.

A particularly disturbing observation occurs in several factors that are linked to increased risk of coronary heart disease. These are high levels of cholesterol and triglycerides, elevated blood pressure, and decreased levels of high-density lipoproteins (HDL). Cholesterol and triglyceride levels above 300 mg · dl^{-1} and HDL levels below 10 mg · dl^{-1} have been noted in athletes taking

TABLE 28-3
Major Side Effects of Anabolic Steroids

Liver toxicity
    Elevated levels of SGOT, SGPT, alkaline phosphatase, and bilirubin
    Increased bromsulphalein (BSP) retention
    Hepatocellular carcinoma
    Peliosis hepatitis
    Cholestasis
Elevated CPK and LDH
Elevated blood pressure
Edema
Alterations in clotting factors
Elevated cholesterol and triglycerides
Decreased HDL
Elevated blood glucose
Increased nervous tension
Altered electrolyte balance
Depressed spermatogenesis
Lowered testosterone production
Reduced gonadotrophin production (LH and FSH)
Increased urine volume
Premature closure of epiphyses in children
Masculinization
Increased or decreased libido
Sore nipples
Acne
Lowered voice in women and children
Clitoral enlargement in women
Increased aggressiveness
Nosebleeds
Muscle cramps and spasms
GI distress
Dizziness
Disturbed thyroid function
Wilm's tumor
Prostatic hypertrophy
Prostatic cancer
Increased activity of apocrine sweat glands

large doses of anabolic steroids (respective ideal levels for cholesterol, triglycerides, and HDL are 200, 100, and 60 mg · dl^{-1} in males). Although it appears that high cholesterol and triglyceride levels may be partially related to diet (many weight-trained athletes consume high-protein, high-fat, high-cholesterol diets), the low HDL level seems to be directly related to the anabolic steroids. Many weight-trained athletes compete for as long as 10 to more than 20 yr, subjecting themselves to the real possibility of premature death from atherosclerosis. Hypertension is also a common observation, probably due to the fluid-retaining properties of these drugs.

Anabolic steroids have been shown to affect the regulation of the hypothalamic–gonadotrophin–testicular (hypothalamic–pituitary–gonadal) axis, which controls normal reproductive processes (Figure 28-3). The anabolic steroids suppress luteinizing hormone (LH) and follicle-stimulating hormone (FSH), resulting in a decreased production of testosterone by the testes. This process can result in testicular atrophy and decreased sperm production. These changes also reverse themselves after withdrawal from the medication; however, the possibility exists that prolonged use of these substances may permanently disturb this delicate regulatory system.

A relatively new ergogenic aid is human chorionic gonadotrophin (HCG), which acts like LH to stimulate testosterone production in the testes. Studies by Fukutani et al. of normal men (control subjects) and by Jean-Faucher et al. on rats show dramatic increases in serum testosterone and dihydrotestosterone following administration of HCG. However, the effects on performance have not, as yet, been studied. This drug is often used by athletes to counteract the gonadal-suppressant effects of anabolic steroids.

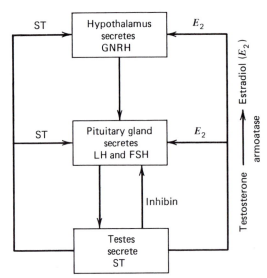

FIGURE 28-3
The feedback control system of the hypothalamic–pituitary–gonadal axis. Testosterone (ST) exerts independent control over the hypothalamus, which secretes gonadotrophin-releasing hormone (GNRH), and pituitary, which secretes luteinizing hormone (LH) and follicle-stimulating hormone (FSH). The aromatization of testosterone to estradiol provides additional regulation of the hypothalamus and pituitary. Inhibin, which is produced in the testes, also controls the production of FSH.

Virilization is a possible side effect of anabolic steroid use, particularly in women and sexually immature children, because the androgenic effects have not been completely eliminated from any of these drugs. Anabolic steroids can cause increased facial and body hair, deepening of the voice, oily skin, prostatic hypertrophy, increased activity of the apocrine sweat glands, acne, baldness, and increased or decreased sex drive. Women may also experience clitoral enlargement and menstrual irregularity. In women, some of these masculine changes are irreversible. Children will initially experience precocious development followed by premature closure of the epiphyseal growth centers in the long bones, and stunted growth.

Long-term use of anabolic steroids may induce changes in the ultrastructure of skeletal muscle. MacDougall and co-workers recently studied the muscle cell characteristics of elite powerlifters and bodybuilders. They found a variety of abnormalities, including central nuclei and atrophied fibers. It is unclear whether these findings are a result of anabolic steroids or the effects of training.

A variety of miscellaneous side effects have been reported, including muscle cramps, gastrointestinal distress, headache, dizziness, sore nipples, and abnormal thyroid function. Some of these side effects have become manifest in individuals who took low doses for short periods of time. To summarize, anabolic steroids present possible grave health risks that could be life threatening.

Use of Anabolic Steroids by Athletes

In some sports, anabolic steroids are used by the majority of accomplished athletes. Unfortunately, the administration of these dangerous substances is often accomplished without medical supervision. Self-administration has resulted in sometimes ridiculously high doses. A report in a body-building magazine mentioned an individual who was taking 100 times the recommended dosage and was spending $150 a week on anabolic steroids.

Table 28-4 presents the most popular oral and injectable anabolic steroids, listed in order of androgenicity. In general, athletes tend to take the less androgenic steroids during building periods (off-season conditioning) and administer the more androgenic substances during competitive periods. Additionally, they will often take combinations of steroids (a practice called stacking) to obtain the most beneficial effects. Table 28-5 presents an example of the anabolic steroids taken by a world-class athlete during a transition period between conditioning and competition.

Screening for Anabolic Steroid Use

Testing for anabolic steroids began with the 1976 Olympic Games, held in Montreal, using radioimmunoassay and gas chromatography techniques. It is estimated that oral steroids, such as Dianabol, can be detected up to 8 weeks following administration, and injectables such as Deca-Durabolin, up to 6 to 8

TABLE 28-4
Anabolic–Androgenic Steroids Used by Athletes Listed in Order of Androgenicity[a]

Orals	Injectables
Maxibolin (ethyloestrenol)	Deca-Durabolin (nandrolone deca-noate)
Anavar (oxandrolone)	
Winstrol (stanozolol)	Durabolin 50 (nandrolone phenpro-pionate)
Dianabol (methandrostenolone)	
Anadrol (oxymetholone)	Delatestryl (testosterone enanthate)
	Testosterone propionate
	Depo-Testosterone (testosterone cy-pionate)
	Aqueous testosterone

[a] Trade names are given with generic names in parentheses, except that testosterone propionate and aqueous testosterone are generic names.

months. Individual differences in detection time depend on such factors as amount and duration of steroid therapy, percentage body fat, and blood volume.

In the past, athletes would switch to testosterone esters such as testosterone cypionate or testosterone enanthate as an international competition neared, in order to escape detection. However, in 1982 testosterone was added to the list of banned substances, because a technique was developed to determine if exogenous testosterone had been injected. The test is based on the natural ratio of testosterone to epitestosterone. Therefore, part of future Olympic competitions will involve athletes balancing the administration of anabolic steroids, testosterone, and related hormones. Some athletes have resorted to the use of diuretics in an attempt to speed the clearance of anabolic steroids.

GROWTH HORMONE

Growth hormone (GH), a polypeptide hormone produced by the adenohypophysis (anterior pituitary gland), has been used by some powerlifters, body builders, and throwers to increase muscle mass and strength. To date there have been no research studies of its effect on athletic performance, but testimonials from athletes have claimed weight gains of 30 to 40 pounds in 10 weeks accompanied by equally amazing but questionable gains in strength. Nitrogen balance becomes highly positive when GH is administered to adult humans, which lends some credence to these claims.

TABLE 28-5
Anabolic Steroid Therapy Program of a World-Class, Weight-Trained Athlete[a]

Week	Injectable Steroid	Daily Oral Steroid	Training Program
1–2	100 mg Deca-Durabolin every 5 days	8 mg Winstrol	Heavy weight training—high volume, medium intensity
3–4	100 mg Deca-Durabolin every 5 days	10 mg Winstrol 10 mg Dianabol	Heavy weight training—high volume, medium intensity
5	100 mg Deca-Durabolin every 5 days	10 mg Winstrol 20 mg Dianabol	Heavy weight training—high volume, medium intensity
6	100 mg Deca-Durabolin every 5 days	Decreasing dosage	Low volume, low intensity—rest week
7–8	No injectables	No orals	Low volume, high intensity
9	No injectables	10 mg Dianabol	Low volume, high intensity Lift heavy once during week
10	200 mg Delatestryl every 5 days	15 mg Dianabol	Low volume, high intensity Lift heavy once during week
11	200 mg Delatestryl every 5 days	20 mg Dianabol	Low-volume, low-intensity weight training
12	200 mg Delatestryl every 5 days 100 mg testosterone propionate day before meet	25 mg Dianabol	Competition, no lifting

[a] This table should not be considered an endorsement for use of anabolic steroids. It is merely intended to demonstrate use patterns in strength-trained athletes.

Growth hormone is a potent anabolic agent that facilitates the transport of amino acids into cells. As discussed, an increased rate of amino acid transport into muscle cells is associated with muscle hypertrophy. Growth hormone is also involved in the formation of connective tissue and exerts a stimulatory effect on somatomedin, which is an important hormone in chondrocyte metabolism. As discussed in Chapter 9, GH is also involved in carbohydrate and fat metabolism.

The use of this substance may have severe consequences. Growth hormone has well-documented diabetogenic effects that could become permanent if large doses are administered. Because of its general anabolic effect, GH could cause cardiomegaly, which could result an increased myocardial O_2 demand. Combining an enlarged heart induced by administration of GH with the possibility of accelerated atherosclerosis resulting from administration of anabolic steroids could drastically increase the risk of myocardial infarction and heart failure. It could also cause symptoms of acromegaly, characterized by enlarged bones in the head, face, and hands, osteoporosis, arthritis, and heart disease. The skeletal changes are irreversible.

In the past, GH has been very expensive and in short supply, because it was only obtained from cadavers. However, recent advances in genetic engineering have made this substance more widely available. Even with such severe side effects, the possibility that GH may allow dramatic increases in performance in a short time may render its use irresistible to some athletes.

CYPROHEPTADINE HYDROCHLORIDE (PERIACTIN)

Periactin is a serotonin and histamine antagonist with anticholinergic and sedative effects. It is not an anabolic steroid. In the 1960s, some athletes began taking this substance because it increased appetite and weight. Clinical observations on over 1000 athletes indicated that weight gains averaged about 1 pound a week. Unfortunately, to date there have been no studies on the effect of Periactin on body composition. There are fewer side effects with Periactin than with anabolic steroids, the most frequent being drowsiness, listlessness, and irritability.

The mechanism of action of Periactin is unclear. It has been hypothesized that it works by changing the hypothalamic appetite setpoint that is sensitive to glucose (see Chapter 25). An alternative explanation is that it stimulates the release of LH-releasing factor from the hypothalamus, ultimately stimulating the production of testosterone.

Richardson studied the effects of Periactin, Dianabol, and a placebo on the strength of monkeys involved in a strength-training program. The monkeys in the two drug groups gained more weight than the control group. All groups gained in strength as a result of training, but the Periactin group gained more than the others. In addition, the Periactin group maintained their strength when the drug was withdrawn, while strength tended to decrease in the steroid group.

AMPHETAMINES

Amphetamines are perhaps the most abused drug in sports, being particularly popular in football, basketball, track and field, and cycling. Athletes use them to prevent fatigue and to increase confidence, cardiovascular endurance, muscle endurance, speed, power, and reaction time. Generic examples of amphetamines include benzedrine, dexedrine, dexamyl, and methedrine.

These drugs act as both central nervous system and sympathomimetic stimulants (sympathomimetic effects mimic the action of the sympathetic nervous system). Amphetamines stimulate the CNS by directly affecting the reticular activating system and postganglionic nerves. Central effects include increased arousal, wakefulness, confidence, and the feeling of an enhanced capability to make decisions. Sympathomimetic effects include increases in blood pressure, heart rate, O_2 consumption in the brain, and glycolysis in muscle and liver, vasoconstriction in the arterioles of the skin and spleen, and vasodilation in muscle arterioles.

The effectiveness of amphetamines in improving athletic performance is controversial. Many studies were poorly controlled, used low dosages, and did not allow enough time for absorption of the drug. Administration of approximately 15 to 50 mg of D-amphetamine is a common dose in athletics, but some studies used a dosage as low as 5 mg. Although these drugs are readily absorbed orally, they take 1½ to 2 hr to reach peak levels in the body (peak levels are reached in 30 min if the amphetamine is administered im or ip). Yet some studies began performance testing within a half hour of administration. Additionally, amphetamines create a euphoric sensation that is easily identifiable, which makes a true double-blind study almost impossible.

Amphetamines became popular during World War II with soldiers who used them to ward off fatigue. Studies have generally supported the effectiveness of amphetamine as a psychotropic drug that masks fatigue, but have been equivocal on their ability to improve endurance performance. Many studies have demonstrated enhanced feelings of well-being and improved exercise capacity in fatigued subjects. Fatigued animal and human subjects have been shown to improve endurance (time to fatigue) in marching, cycling, swimming, and treadmill exercise, and also to improve simple reaction time. Amphetamines have no effect on reaction time in rested subjects.

Most studies have failed to demonstrate an effect on cardiovascular function, even though exercise time to exhaustion was often improved. Maximal values for O_2 consumption, heart rate, minute volume, respiratory exchange ratio, respiratory rate, oxygen pulse, CO_2 production, and ventilatory equivalent were unaffected by amphetamines. Chandler and Blair, in a well-controlled study, demonstrated a 4.5% increase in time to exhaustion on a treadmill (treadmill protocol unclear) and a 10% increase in peak lactate, with no change in maximal O_2 uptake. Their data reinforce the notion of amphetamine as a potent psychotropic agent.

The effects of amphetamines on strength and power seem to depend on the number of motor units recruited. Chandler and Blair demonstrated a 22.5% increase in knee extension strength but no significant change in elbow flexion strength. In sprinting, acceleration, but not top speed, is enhanced by the drug. This points to an increased excitability of the muscles but not to an increase in their maximal capacity. Most studies show increases in static strength but mixed results in muscle endurance.

The effects of amphetamines on sports performance are also unclear. They appear to aid power-oriented movement skills in activities that employ constant motor patterns such as shot putting and hammer throwing. They are probably less effective in sports requiring the execution of motor skills in an unpredictable order, such as football, basketball, and tennis. In these sports, amphetamines may be deleterious because they may interfere with the body's fatigue alarm system, cause confusion, impair judgment, and, in high concentrations, cause neuromuscular blockade and loss of effective motor control.

Amphetamines can cause a variety of severe side effects. They increase the risk of hyperthermia because of their vasoconstricting effect on the arterioles of the skin. Numerous deaths in endurance sports have been reported in athletes competing in the heat while under the influence of these drugs. These drugs can also cause tremulousness, psychic distress, insomnia, dry mouth, addiction, loss of appetite, and cardiac arrhythmias.

COCAINE

Cocaine is an alkaloid derived from the leaves of the coca plant and acts as a CNS stimulant. It has become very popular with certain segments of the population. Cocaine has also become popular with some athletes, with reports often surfacing in the news media of its rampant use by professional football and basketball players. The use of cocaine as a stimulant has a long history beginning with the Incas in Peru. Its use was advocated by Sigmund Freud, and it was an important ingredient in Coca-Cola during the early days of the product.

The drug produces a feeling of exhilaration and an enhanced sense of wellbeing, and it depresses fatigue. It works by inhibiting the reuptake of norepinephrine in sympathetic neurons and has a direct sympathomimetic effect. Cocaine has been shown to increase work capacity in several poorly controlled studies. However, cocaine had no effect in rats, on swim time to exhaustion. Cocaine is an extremely dangerous drug, particularly when it is administered intravenously or converted to the free base and smoked.

CAFFEINE

Caffeine is a xanthine that acts as a cerebrocortical stimulator and may stimulate the adrenal medulla to release epinephrine. It also stimulates the heart (increasing both rate and contractility of the heart at rest), causes peripheral vasodilation, and acts as a diuretic by blocking renal tubular reabsorption of sodium. In athletics, caffeine is used as a stimulant and as a fatty acid mobilizer. It is found in a variety of food products such as coffee, tea, and chocolate.

Caffeine is a much weaker stimulant than amphetamine, yet it is widely used by weight lifters and throwers (discus, shot, javelin, and hammer) to enhance strength and power. These athletes take the caffeine in the form of strong coffee or through over-the-counter medications such as Vivarin or No-Doze. Although some older, poorly controlled studies found improvements in strength and power from this substance, these findings have not been replicated by well-controlled studies.

Caffeine appears to enhance performance in prolonged endurance exercise by mobilizing free fatty acids and sparing muscle glycogen. A study by Ivy et al. found that a total of 500 mg of caffeine, administered before and during a 2-hr ride on an isokinetic bicycle ergometer, resulted in a 7.4% greater work production. It appears that caffeine is an effective ergogenic aid in events such as marathon running (Chapter 27).

The use of caffeine as an ergogenic aid is not without danger. The diuretic and cardiac stimulatory properties of this substance can combine to increase the risk of ECG arrhythmias, such as ventricular ectopic beats and paroxysmal atrial tachycardia. This is particularly alarming for older, less conditioned individuals. Caffeine can also cause delayed or lightened sleep and is addicting.

NUTRITIONAL SUPPLEMENTS

Athletes spend an absolute fortune on an endless variety of dietary supplements such as protein, vitamins, and weight gain products. An overwhelming body of literature has demonstrated that as long as an athlete is receiving a balanced diet, dietary supplements have no effect on performance. However, if the diet is deficient in any essential nutrient, then supplementation may very well be beneficial. Diet and performance are discussed in Chapter 27, so this section will focus on those nutritional supplements that are specifically used as ergogenic aids.

Carbohydrate (CHO) feeding, often in the form of glucose, dextrose, or honey, has long been used as an ergogenic aid to increase strength, speed, and endurance. CHO Feeding has no effect on strength, power, or high-intensity short-term exercise, and has been shown to decrease performance in endurance activity (see Figure 28-4).

Wheat germ oil has been highly touted as an ergogenic aid that increases endurance. The beneficial constituents of this product are purported to be vita-

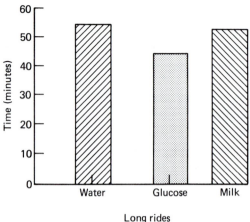

FIGURE 28-4
Endurance capacity at 80% of $\dot{V}_{O_2max}$ following ingestion of water, glucose, and milk.

(SOURCE: From C. Foster, D.L. Costill, and W.J. Fink, 1979.)

min E (α-tocopherol) and octacosanol. The proposed mechanism of its beneficial effect is that it reduces the O_2 requirement of the tissues and improves coronary collateral circulation. Numerous studies by Cureton have shown a beneficial effect of wheat germ oil on endurance performance; however, these findings have generally not been replicated by others. Neither wheat germ nor vitamin E has been shown to increase maximal O_2 consumption, although one study found that vitamin E increased $\dot{V}_{O_2max}$ by 9% at 1524 m (5000 ft) and 14% at 4572 m (15,000 ft) altitude.

A variety of vitamins have been used as ergogenic aids. As discussed in Chapter 27, these substances, although essential, are required in extremely small quantities. Buskirk and Haymes have hypothesized that because the size and number of mitochondria increase with endurance training, more vitamins may be needed to support the increased metabolic activity. However, vitamin deficiency in athletes has not been consistently demonstrated. In fact, most athletes take many times the minimum daily requirement for these substances.

Vitamin C has been used to improve both cardiovascular and muscle endurance. Although vitamin C has been shown to speed up the process of acclimatization to heat, most studies have not demonstrated any effect on factors important to endurance performance. B Vitamins such as thiamine, riboflavin, and niacin have become extremely popular with athletes for improving endurance, strength, and recovery from fatigue. Most studies have failed to find any beneficial effects.

Pangamic acid, sometimes called vitamin B_{15}, has been hailed as a miracle substance that improves endurance and fights off fatigue. This product has come under increasing criticism from the U.S. Food and Drug Administration (FDA), because it is not an identifiable substance (not a vitamin or provitamin), and because it has no established medical or nutritional usefulness. The active ingredient in this product is attributed to be *N,N*-dimethylglycine (DMG), which is said to increase oxygen utilization. Although some studies support the claims of improved endurance, the most stringently controlled investigations generally have not. The use of this substance may be dangerous because of its potential for mutagenesis.

The administration of buffering substances such as sodium bicarbonate has been suggested as an ergogenic aid in preventing fatigue during endurance exercise. This stemmed from the belief that the inducement of alkalosis would prevent the accumulation of blood lactate and thus fatigue. Research studies have been equivocal on the effects of alkaline salts on performance. Kindermann, Keul, and Huber found that infusion of bicarbonate and Tris buffer had no effect on performance in a 400-m run. In addition, they found that in spite of increased buffering capacity, blood lactate levels were unaffected. In opposition to these results are those of Jones and co-workers who have reported that alkalyzing agents improve bicycle ergometer exercise performance. Two mechanisms may be involved. First, alkalyzing agents buffer the effects of metabolic acids (mainly lactic acid) in the blood. Second, and perhaps more important, lactic acid efflux from muscle may be promoted by keeping blood pH higher than muscle pH. In this way bicarbonates and other bases in the circulation may help to minimize the effect of exercise on reducing muscle pH. As discussed in Chapter 32, low muscle pH could cause fatigue. Unfortunately, no definitive study has reported their effects on muscle pH or lactate.

Aspartate, which is potassium and magnesium salts of aspartic acid, has been used to reduce fatigue. It has been hypothesized that they work by accelerating the resynthesis of ATP and CP in muscle, and by sparing glycogen. As with many so-called ergogenic aids, beneficial effects have only been demonstrated in poorly controlled studies that used few subjects.

A variety of nutritional substances have been proposed as ergogenic aids, including bee pollen, gelatin, lecithin, phosphates, and organ extracts. Although little research is available concerning them, there is little reason to support their effectiveness. The case for organ extracts seems to almost be a throwback to the times that warriors ate the hearts or livers of brave adversaries to obtain some of their prowess in battle.

BLOOD DOPING

Blood doping, or induced erythrocythemia, is an increase in blood volume by transfusion for the purpose of increasing O_2 transport capacity. Although the transfusion can utilize blood from a matched donor, it is typically accomplished by the removal, storage, and subsequent reinfusion of the subject's own blood. This procedure received considerable publicity with the publication of a paper by Ekblom, Goldbarg, and Gullbring, who reported increases in work capacity and $\dot{V}_{O2max}$ of 23% and 9%, respectively. It also caught the interest of the media when several long-distance runners were suspected of using this technique in the 1972 and 1976 Olympics.

The majority of studies show improvements in maximal O_2 consumption (ranging from a 1 to 26% increase) and endurance exercise capacity (ranging from a 2.5 to 37% increase) following induced erythrocythemia. The increase in blood volume induces a decreased heart rate and cardiac output, with no change in stroke volume during submaximal exercise, and an increase in stroke volume and cardiac output, with no increase in maximal heart rate, during maximal exercise.

Training may be an important prerequisite for the beneficial effects of blood doping to manifest themselves. In untrained rats, maximal O_2 consumption increases with hematocrit until the hematocrit reaches 40%. After that, further increases in hemoconcentration do not result in increased $\dot{V}_{O2max}$.

Blood doping appears to be a relatively safe procedure in normal, healthy individuals. Reinfusion of autologous blood causes no abnormal changes in the exercise electrocardiogram or in blood pressure. However, homologous transfusions carry with them the risk of hepatitis, bacterial communication, and blood-type incompatibility.

Presently there is no test to detect blood doping. The changes in hemoconcentration are very similar to those induced by altitude acclimatization. Although this practice clearly falls under the spirit of the antidoping regulations of the International Olympic Committee, there is no effective means for enforcing a ban against this procedure.

OXYGEN

Supplemental oxygen has been used before and during exercise in an attempt to enhance performance, and after exercise to hasten recovery. The first use of this technique was by the Japanese in the 1932 Olympic Games held in Los Angeles. Because they were successful, the practice became institutionalized in several sports. The sight of a gasping football player on the sidelines with an oxygen mask over his face has become common.

Oxygen breathing immediately prior to exercise will increase the O_2 content of the blood by 80 to 100 ml. In exercise lasting less than 2 min, supplemental O_2 will improve maximal work capacity by approximately 1% and decrease submaximal exercise heart rate. Most studies (but not all) have shown that O_2 could be beneficial in short-term maximal efforts in swimming and running. This is not a very practical ergogenic aid because of the time lag between the administration of O_2 and the start of performance. Runners and swimmers must take the time to prepare themselves in the starting blocks, and football players must align themselves in formation. During this time, the increase in the O_2 content of blood has dissipated.

Although supplemental O_2 could be of theoretical benefit in recovery from repeated maximal efforts, several studies have been unable to demonstrate an accelerated decrease in ventilation, heart rate, or blood pressure following O_2 breathing. However, these investigations used submaximal exercise, where an effect would be less likely to occur.

Oxygen breathing during exercise improves work capacity, increases maximal O_2 consumption, and decreases submaximal heart rate. Supplemental O_2 effectively increases O_2 transport capacity by increasing the O_2 dissolved in the blood, and perhaps by increasing maximal myocardial O_2 consumption. With the exception of mountain climbing, supplemental oxygen breathing during exercise is of little practical significance, as few sports allow an athlete to carry an oxygen tank on the field.

INTEROCCLUSAL SPLINTS

Interocclusal splints, or mandibular orthopedic repositioning appliances (MORAs), are corrective mouth orthotics designed to correct malocclusion and temporomandibular joint (TMJ) imbalance. The theory behind this device is that TMJ dysfunction causes muscle tension, which impairs muscle strength. It has been hypothesized that the splint changes position of the TMJ, which relieves pressure on the auriculotemporal nerve, thus relieving tension in the neck and shoulder muscles. Such devices have proven to be effective in the treatment of particular dental-related problems. The effectiveness of MORAs in the practice of dentistry has led to their promotion as ergogenic aids.

Although there are numerous personal testimonials and clinical observations testifying to the beneficial effects of MORAs for increasing strength and endurance, there is scant experimental evidence to support such contentions. In a well-controlled investigation, Burkett and Bernstein found that MORAs had no effect on strength and muscular endurance.

Whereas proponents of this device claim that 90% of the population has some degree of TMJ imbalance, critics point out that every joint has a normal range of deviation and that few people experience TMJ pain when clenching their teeth. Some individuals are apparently promoting these devices for use in athletics because of their income-generating potential, so their enthusiasm has to be viewed with a jaundiced eye. Further, it is difficult to run a double-blind investigation with the splint, which makes the placebo effect a prominent factor in any result. It appears that the interocclusal splint is another possible ergogenic aid that requires considerable investigation before it becomes an accepted part of the athletic cornucopia.

ELECTRICAL STIMULATION

Electrical stimulation has been used for many years as a treatment modality in the rehabilitation of injured muscle, and to prevent denervation atrophy, decrease spasticity and muscle spasm, reduce contractures, and prevent deep vein thrombosis. Recently, healthy athletes have begun using this technique to increase strength and power.

There is little evidence that electrical muscle stimulation is a beneficial supplement to traditional strength-training techniques. Romero et al., using untrained females as subjects, demonstrated substantial increases in isometric knee extension strength (31% in the nondominant leg), but were unable to demonstrate such changes isokinetically. Electrical stimulation appears to do little to develop dynamic strength, which is most important in athletics.

This finding is consistent with the time-honored concept of specificity of training. The muscle is stimulated isometrically during electrical stimulation, so gains in strength will tend to be isometric. Although studies of the effects of electrical stimulation on athletes have generally been negative, they have usually been conducted during a 6-week experimental period, which may not be enough time to render statistically significant results.

Summary

Ergogenic aids are substances or techniques used in the hope of improving athletic performance. Although the use of ergogenic aids is rampant in sports, few have proved to be of any significant value. The International Olympics Committee has issued an extensive list of banned substances that appears to be based on an attempt to discourage any pretext of unfair competition.

Perhaps the two most popular substances used as ergogenic aids are anabolic steroids and amphetamines. Anabolic steroids are drugs that resemble testosterone and androstenedione. They are used by athletes in the hope of gaining weight, strength, power, speed, endurance, and aggressiveness. They act mainly by stimulating receptor molecules in muscle cells, which activate specific genes to produce proteins. Although athletes almost universally extol their benefits, research has not confirmed their observations. Besides the equivocally beneficial effects of anabolic steroids, almost all athletes can expect to experience some side effects, including liver toxicity, depressed HDL cholesterol, gonadal suppression, masculinization in women, and premature closure of epiphyses in children.

Amphetamines act as central nervous system and sympathomimetic stimulants. Their effects include increased arousal, wakefulness, confidence, blood pressure, heart rate, muscle and liver glycolysis, vasodilation in the skin arterioles, and vasoconstriction in muscle arterioles. Like anabolic steroids, their effectiveness is questionable. They appear to improve power in large muscle groups and acceleration in sprinting. Additionally, they appear to aid the performance of power-oriented movement skills in activities employing constant motor patterns but may be deleterious in sports requiring the execution of motor skills in an unpredictable order. Amphetamines are known to cause a number of severe side effects, including hyperthermia, addiction, insomnia, and cardiac arrhythmias.

Numerous other ergogenic aids have been employed, including growth hormone, cocaine, oxygen, caffeine, wheat germ oil, vitamins, induced erythrocythemia, interocclusal splints, and electrical stimulation of muscles.

Selected Readings

Aigner, A. Anabolic agents in sport. Oesterr. J. Sportmed. 9: 25–29, 1979.

American College of Sports Medicine. Position statement on the use and abuse of anabolic-androgenic steroids in sports. Reprinted in: Physician Sports Med. 6: 157–160, 1978.

Ariel, G. The effects of anabolic steroids on reflex components. Med. Sci. Sports. 4: 124–126, 1972.

Bender, K.J. and D.H. Lockwood. Use of medications by U.S. olympic swimmers. Physician Sports Med. 5: 63–65, 1977.

Bentivegna, A., E.J. Kelly, and A. Kalenak. Diet, fitness, and athletic performance. Physician Sports Med. 7: 99–105, 1982.

Brooks, R.V., G. Jeremiah, W.A. Webb, and M. Wheeler. Detection of anabolic steroid administration to athletes. J. Steroid Biochem. 11: 913–917, 1979.

Burkett, L.N. and A.K. Bernstein. Strength testing after jaw repositioning with a mandibular orthopedic appliance. Physician Sports Med. 10: 101–107, 1982.

Buskirk, E.R. Some nutritional considerations in the conditioning of athletes. Ann. Rev. Nutr. 1: 319–350, 1981.

Celejowa, I. and M. Homa. Food intake, nitrogen and energy balance in Polish weight lifters, during a training camp. Nutr. Metab. 12: 259–274, 1970.

Chandler, J.V. and S.N. Blair. The effect of amphetamines on selected physiological components related to athletic success. Med. Sci. Sports Exerc. 12: 65–69, 1980.

Consolazio, C.F., R.A. Nelson, L.O. Matoush, and G.J. Isaac. Effects of aspartic acid salts (Mg and K) on physical performance of men. J. Appl. Physiol. 19: 257–261, 1967.

Crist, D.M., P.J. Stackpole, and G.T. Peake. Effects of androgenic-anabolic steroids on neuromuscular power and body composition. J. Appl. Physiol: Respirat. Environ. Exercise Physiol. 54: 366–370, 1983.

Cureton, T.K. The Physiological Effects of Wheat Germ Oil on Humans in Exercise. Springfield, Ill.: Charles C Thomas, 1972.

Currier, D., J. Lehman, and P. Lightfoot. Electrical stimulation in exercise of the quadriceps femoris muscle. Phys. Ther. 59: 1508–1512, 1979.

Dohm, G.L., S. Debnath, and W.R. Frisell. Effects of commercial preparations of pangamic acid (B_{15}) on exercised rats. Biochem. Med. 28: 77–82, 1982.

Drug abuse in sports: Denial fuels the problem. Physician Sports Med. 10: 114–123, 1982.

Ekblom, B., A.N. Goldbarg, and B. Gullbring. Response to exercise after blood loss and reinfusion. J. Appl. Physiol. 33: 175–180, 1972.

Ekblom, B., G. Wilson, and P.O. Åstrand. Central circulation during exercise after venesection and reinfusion of red blood cells. J. Appl. Physiol. 40: 379–383, 1976.

Elinwood, E.H. and R.L. Balster. Rating the behavioral effects of amphetamine. Eur. J. Pharmacol. 28: 35–41, 1974.

Fahey, T.D. and C.H. Brown. The effects of an anabolic steroid on the strength, body composition, and endurance of college males when accompanied by a weight training program. Med. Sci. Sports. 5: 272–276, 1973.

Foster, C., D.L. Costill, and W.J. Fink. Effects of preexercise feedings on endurance performance. Med. Sci. Sports. 11: 1–5, 1979.

Frischkorn, C.G.B. and H.E. Frischkorn. Investigations of anabolic drug abuse in athletics and cattle feed. J. Chromatogr. 151: 331–338, 1978.

Fukutani, K., H. Ishida, M. Shinohara, S. Minowada, T. Niijima, and K. Isurugi. Responses of serum testosterone levels to human chorionic gonadotropin stimulation in patients with Kinefelter's syndrome after long-term androgen replacement therapy. Int. J. Andrology 6: 5-11, 1983.

Gilbert-Dreyfus, M. The use of hormones in sports. Bull. Nat. Acad. Med. 163: 73–78, 1979.

Girandola, R.N., R.A. Wiseman, and R. Bulbulian. Effects of pangamic acid (B_{15}) ingestion on metabolic response to exercise (abstract). Med. Sci. Sports Exerc. 12: 98, 1980.

Gledhill, N. Blood doping and related issues: A brief review. Med. Sci. Sports Exerc. 14: 183–189, 1982.

Golding, L.A. and J.R. Barnard. The effects of D-amphetamine sulfate on physical performance. J. Sports Med. Phys. Fitness. 3: 221–224, 1963.

Gray, M.E. and L.W. Titlow. B_{15}: Myth or miracle? Physician Sports Med. 10: 107–112, 1982.

Gray, M.E. and L.W. Titlow. The effect of pangamic acid on maximal treadmill performance. Med. Sci. Sports Exerc. 14: 424–427, 1982.

Ivy, J.L., D.L. Costill, W.J. Fink, and R.W. Lower. Influence of caffeine and carbohydrate feedings on endurance performance. Med. Sci. Sports. 11: 6–11, 1979.

Jean-Faucher, C., M. Berger, M. deTurckheim, G. Veyssiere, and C. Jean. Plasma and testicular testosterone and dihydrotestosterone in mice: Effects of age and HCG stimulation. IRCS Med. Sci. 11:26–27, 1983.

Karpovich, P.V. Effects of amphetamine sulfate on athletic performance. J.A.M.A. 170: 558–561, 1969.

Kindermann, W., J. Keul, and G. Huber. Physical exercise after induced alkalosis (bicarbonate or Tris-buffer). Eur. J. Appl. Physiol. 37:197–204, 1977.

MacDougall, J.D., D.G. Sale, G.C.B. Elder, and J.R. Sutton. Muscle ultrastructural characteristics of elite powerlifters and bodybuilders. Eur. J. Appl. Physiol. 48:117–126, 1982.

Meyers, F.H., E. Jawetz, and A. Goldfein. Review of Medical Pharmacology. Los Altos, Calif.: Lange Medical Publications, 1980.

Moore, M. Corrective mouth guards: Performance aids or expensive placebos? Physician Sports Med. 9: 127–132, 1981.

Morgan, W. (ed.). Ergogenic Aids and Muscular Performance. New York: Academic Press, 1972.

O'Shea, J.P. Anabolic steroids in sport: A biophysiological evaluation. In: Schriber, K. and E.J. Burke (eds.), Relevant Topics in Athletic Training. Ithaca, N.Y.: Mouvement Publications, 1978.

Percy, E.C. Ergogenic aids in athletics. Med. Sci. Sports. 10: 298–303, 1978.

Richardson, J.H. A comparison of two drugs on strength increase in monkeys. J. Sports Med. Phys. Fitness. 17: 251–254, 1977.

Rogozkin, V. and B. Feldkoren. The effect of retabolil and training on activity of RNA polymerase in skeletal muscles. Med. Sci. Sports. 11: 345–347, 1979.

Romero, J.A., T.L. Sanford, R.V. Schroeder, and T.D. Fahey. The effects of electrical stimulation of normal quadriceps on strength and girth. Med. Sci. Sports Exerc. 14: 194–197, 1982.

Sawyer, D.A., H.L. Julia, and A.C. Turin. "Caffeine and human behavior: Arousal, anxiety, and performance effects. J. Behavioral Med. 5: 415–439, 1982.

Smith, G.M. and H.K. Beecher. Amphetamine, secobarbital and athletic performance. J.A.M.A. 172: 1502–1514, 1623–1629, 1960.

Smith, G.M. and H.K. Beecher. Amphetamine sulfate and athletic performance. J.A.M.A. 170: 542–557, 1959.

Souccar, C., A.J. Lapa, and R.B. do Valle. The influence of testosterone on neuromuscular transmission in hormone sensitive mammalian skeletal muscles. Muscle Nerve. 5: 232–237, 1982.

Stang-Voss, C. and H.J. Appell. Structural alterations of liver parenchyma induced by anabolic steroids. Int. J. Sports Med. 2: 101–105, 1981.

Steadward, R.D. and M. Singh. The effects of smoking marihuana on physical performance. Med. Sci. Sports. 7: 309–311, 1975.

Sumner, J. Nutrition and athletic performance. Food Nutr. Notes Rev. 37: 125–130, 1980.

Videman, T. and T. Rytomaa. Effect of blood removal and auto-transfusion on heart rate response to submaximal workload. J. Sports Med. Phys. Fitness. 17: 387–390, 1977.

Williams, M.H. Drugs and Athletic Performance. Springfield, Ill.: Charles C Thomas, 1974.

Williams, M.H. Nutritional Aspects of Human Physical and Athletic Performance. Springfield, Ill.: Charles C Thomas, 1976.

Williams, M.H., M. Lindhejm, and R. Schuster. The effect of blood infusion upon endurance capacity and ratings of perceived exertion. Med. Sci. Sports. 10: 113–118, 1978.

Williams, M.H. and J. Thompson. Effect of variant dosages of amphetamine upon endurance. Res. Q. 44: 417–421, 1973.

Williams, M.H., S. Wesseldine, T. Somma, and R. Schuster. The effect of induced erythrocythemia upon 5-mile treadmill run time. Med. Sci. Sports Exerc. 13: 169–175, 1981.

Wright, J. Anabolic steroids and athletics. Exerc. Sport Sci. Rev. 8: 149–202, 1980.

Wright, J. Anabolic Steroids and Sports. Natick, Mass.: Sports Science Consultants, 1978.

Wright, J. Anabolic Steroids and Sports, Vol II. Natick, Mass.: Sports Science Consultants, 1982.

29

SEX DIFFERENCES IN PHYSICAL PERFORMANCE

Until recently, women's sports programs in the United States have been almost nonexistent. Although some of this can be attributed to prejudice against women, much is a result of ignorance of female exercise physiology. In the past, women have been thought incapable of competing in vigorous exercise. Women's sports were modified to make them less strenuous, and competition was limited to "playdays." The exceptions were the few sports, such as tennis or swimming, that were considered appropriately feminine.

Today, even though we know that women can develop an extremely high level of fitness, the legacy of past beliefs persists. For example, the longest women's race in Olympic track and field is the 1500-m run, even though thousands of women compete very successfully in marathons every year with no more ill effects than male runners. (The marathon was included in the 1984 Olympics.)

There are sex differences in physical performance. Males tend to be larger with more muscle mass and larger hearts, which gives them more strength, power, and endurance. Sex differences in performance are much less than once thought. Today's top female marathoners equal the performance of their male counterparts of only 3 decades ago, whereas female swimmers are equaling the times of the top male athletes who competed in the 1960s (Figure 29-1). However, regardless of differences in performance and work capacity, sports and physical activity affect and benefit both sexes.

Shirly Babashoff, a contemporary female Olympic athlete. (Wayne Glusker.)

In 1972, Title IX of the Educational Amendments Act dictated that women receive equal opportunities in education. ''No person in the United States shall, on the basis of sex, be excluded from participation in, be denied the benefits of, or be subjected to discrimination under any educational program or activity receiving federal financial assistance.''

The increased sports participation resulting from such legislation raised many fears among educational authorities. Although they want to provide opportunities for participation, they also want the experience to be safe and beneficial. A rational sports and exercise program for women should take into consideration sex differences in anatomy and physical performance capacity. Differences do exist, and it is naive and unfair to consider the physical capacities of men and women to be equal. However, whereas it might be inappropriate to expect a 100-pound girl to play football against a 225-pound boy, it may be perfectly acceptable to provide coeducational competition (or, at least coeducational participation) in tennis, track and field, or skiing. This chapter explores physiological sex differences with special reference to pregnancy and menstruation.

PHYSIOLOGICAL SEX DIFFERENCES

Sex Chromosomes

Although genetic or hormonal abnormalities may alter the development of the sex organs, the sex chromosomes are the ultimate test of whether a person is

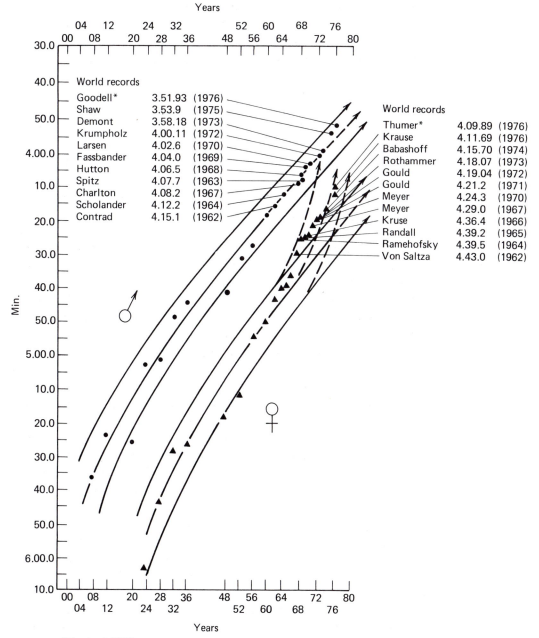

FIGURE 29-1

Progression in the world record in 400-m freestyle swimming for men and women.

(SOURCE: From P. Jokl and E. Jokl, 1977.)

male or female. Human cells have 46 chromosomes distributed in 23 pairs. In both sexes, 22 of the pairs are alike. The last pair, the sex chromosomes, are different in men and women. Females have an identical pair of X chromosomes, whereas males have one X chromosome and a smaller Y chromosome. The sex chromosomes are of major importance in sexual differentiation and act as the ultimate imprintation of a person's sex.

Sex chromosomes are responsible for the anatomical differences that give males an advantage in physical performance. The genetic material contained in the chromosomes act as a blueprint for respective sexual characteristics in muscle mass, heart size, body fat, and probably even psychological traits such as aggression.

The determination of sex chromosomes serves as the basis for the identification of the gender of athletes in major athletic competitions. This test involves obtaining a buccal smear and histologically identifying the sex chromosomes. In the female, one of the X chromosomes contains Barr bodies that stain darkly during histological examination, which provides a definitive means of identifying gender.

Sex tests have proved necessary over the years because of scandals involving men posing as women athletes. In 1968, for example, a group of athletes (including two former gold medal winners) withdrew from the Olympics rather than take the sex tests. In 1938, a man posing as a woman set the world's record in the high jump. There are several instances of Olympic medals won by women who subsequently had surgery to change their sex.

Unfortunately, the legal system has not definitively supported the chromosomal method of sex identification in athletics. In a recent celebrated case, a male tennis player had his sex changed and sought to play women's professional tennis. The courts supported her in her quest even though she possessed male chromosomes. Although this athlete's outward appearance was female, she still had male physical traits provided by the Y chromosomes.

Growth and Maturation

Growth rates in both sexes proceed in four major stages: (1) an accelerated growth during the prenatal and infant period, (2) a plateau in the rate of growth during childhood, (3) an acceleration in growth during puberty, and (4) a plateauing in the rate during late adolescence. Girls tend to mature faster than boys. It is not unusual for some girls to be bigger and more physically skilled than some boys during childhood. The pubertal growth spurt begins between 10 and 13 yr in girls, and between 12 and 15 yr in boys. Young adolescent girls are usually taller than boys of the same age. However, boys catch up and eventually surpass girls (see Chapter 30).

The hormonal changes of puberty affect the body composition of both sexes, with girls showing a greater growth of fat and boys a greater growth of lean tissue. The body composition changes during puberty seem to be caused by increased secretion of gonadotrophic hormones from the pituitary (controlled by their releasing factors in the hypothalamus). These hormones act to increase estrogen levels in girls and androgen levels in boys. Estrogens tend to increase adipose tissue and have a slight retardant effect on lean body mass, whereas androgens tend to increase lean tissue and inhibit the development of body fat. These hormone-controlled changes of puberty ultimately cause the evident sex differences in work capacity and physical performance.

Boys tend to have slightly more muscle mass than girls throughout growth. At puberty, boys experience pronounced muscle growth, whereas girls experience much less. The difference in muscle mass between the sexes is due largely to the male hormones. For example, levels of testosterone (the principal male hormone) are similar in prepubertal boys and girls (about 20 to 60 ng $\cdot$ dl^{-1} of blood). During adolescence, male testosterone levels progress to adult levels of approximately 600 ng $\cdot$ dl^{-1}, whereas females remain at prepubertal levels. In females, the adrenal gland secretes low levels of androgens (principally andros-tenedione).

Anthropometric characteristics, with a few exceptions, are similar in boys and girls before puberty. Until about age 10, the sexes are similar in leg length, upper arm circumference, body weight, sitting height, and ponderal index (i.e., height divided by the cubed root of weight). Boys are greater throughout growth in biepicondylar diameter of the femur and humerus, arm length, and chest and shoulder circumference. Relative to height, girls have larger hips throughout growth.

Anthropometric characteristics change considerably during puberty. Boys develop larger shoulders whereas girls develop larger hips. The smaller shoulder girdle makes it much more difficult for women to develop upper body strength. The wider hips makes the angle of the femur much more pronounced than in males; thus it is more difficult for women to develop the same running speed as males. However, the larger hips give women a slightly lower center of gravity, so that they have an advantage in activities requiring balance.

Body Composition

Adult males tend to have less fat and more muscle mass than adult females. These sex differences remain in athletes who participate in high-energy sports that tend to reduce body fat, such as distance running. Sex differences in fat distribution are caused by a combination of natural and social factors. Active women tend to have lower fat percentages than average (Table 29-1), generally falling below 22% in recreational runners. Women in higher social classes also tend to be leaner. Ideal fat percentages are dictated as much by individual interpretation of fashion as by natural fat deposition patterns.

Differences in body composition affect the work capacity and performance

TABLE 29-1

Sex Differences in Fat Percentage
(50th percentile)

Age (yr)	Males	Females
20–29	21.6	25.0
30–39	22.4	24.8
40–49	23.4	26.1
50–59	24.1	29.3
Over 60	23.1	28.3

SOURCE: Data from Pollock, J.H. Wilmore and
S.M. Fox, Health and Fitness Through Physical
Activity. New York: John Wiley & Sons, 1978.

capability of women. Women are at a distinct disadvantage in sports where they must lift or move their mass against gravity. In activities such as running, climbing, and jumping, they must propel more body fat with less muscle mass than men. The greater body fat also gives women a disadvantage in releasing body heat during exercise.

Women's higher body fat is an advantage in swimming. The higher percentage of body fat allows women to swim higher in the water with less body drag. In swimming, sex differences in performance are smaller than in most other sports. For example, in the 400-m freestyle, the winning time for women in the 1976 Olympics was faster than for men in the 1964 Games (Figure 29-1).

Body fat can be reduced to extremely low levels in endurance-trained women. Body fats of less than 12% have been reported for top marathon runners, whereas levels of less than 16% are typical for women involved in metabolically taxing activities. As discussed on p. 646, strenuous endurance programs resulting in substantial reduction of fat affect the estrous cycle.

Oxygen Transport System

Men exceed women in maximal O_2 consumption by about 20%. However, training can either equalize or reverse these differences, so that some women will exceed the endurance capacity of some men. Differences in $\dot{V}_{O_2max}$ can be attributed to greater cardiac output, blood volume, and oxygen-carrying capacity. Sex differences in maximal O_2 consumption with age are shown in Table 29-2.

Males have a larger heart size and heart volume. The larger heart size enables greater myocardial contractility, while the larger heart volume allows a greater end-diastolic volume. These advantages give males a larger maximal stroke volume and a larger maximal cardiac output.

TABLE 29-2

Sex Differences in Maximal Oxygen Consumption (ml $O_2 \cdot kg^{-1} \cdot min^{-1}$) (50th percentile)

Age (yr)	Males	Females
20–29	40.0	31.1
30–39	37.5	30.3
40–49	36.0	28.0
50–59	33.6	25.7
Over 60	30.0	22.9

SOURCE: Data from Pollock, J.H. Wilmore and S.M. Fox, Health and Fitness Through Physical Activity. New York: John Wiley & Sons, 1978.

Resting, submaximal exercise, and maximal exercise heart rates tend to be higher in women (Table 29-3). Oxygen requirements for a given work load when expressed as $ml \cdot kg^{-1}$, are similar in men and women, so the female heart must beat faster to make up for its lower pumping capacity. In similarly trained individuals, there is little difference in maximal heart rate between the sexes. Because of their larger maximal cardiac output, males tend to have a higher systolic blood pressure during maximal exercise. Heart rate recovery is slower in women because their hearts must beat faster to produce a particular cardiac output.

TABLE 29-3

Sex Differences in Resting and Maximal Heart Rate (50th percentile)

Age (yr)	Heart Rate (beats $\cdot$ min^{-1})			
	Resting		Maximal	
	Males	Females	Males	Females
20–29	64	67	192	188
30–39	63	68	188	183
40–49	64	68	181	175
50–59	63	68	171	169
Over 60	63	65	159	151

SOURCE: Data from Pollock, J.H. Wilmore, and S.M. Fox, Health and Fitness Through Physical Activity. New York: John Wiley & Sons, 1978.

Pulmonary ventilation is higher in men, mostly because of differences in body size. Until puberty maximal ventilation is similar, but it becomes disproportionate during adolescence. There are no sex differences in pulmonary diffusion capacity and hemoglobin saturation, either at rest or during exercise.

Both the amount and concentration of hemoglobin are higher in males, thus giving their blood a greater oxygen-carrying capacity. Women average about 13.7 g Hb $\cdot$ dl^{-1} blood, whereas men average 15.8 g Hb $\cdot$ dl^{-1} blood. The difference is attributed to the potent effect of the androgens on hemoglobin production and the effects of menstrual blood loss.

Some women are prone to iron deficiency and iron deficiency anemia because of the combined effects of low dietary iron intake, limited rates of iron absorption, and iron loss during menstruation. Such women can benefit from dietary iron supplementation. However, universal administration of iron supplements does not appear to be warranted. Studies by Pate et al. and Cooter and Mowbray indicate that the administration of iron supplements to nonanemic women athletes had no effect on hemoglobin or iron status.

Physical Performance

Males have about a 20% greater capacity for both endurance and short-duration, high-intensity exercise. This is largely because of factors already mentioned: body size and oxygen transport capacity. Sex differences in exercise capacity tend to be somewhat greater in activities that benefit from strength and body mass, such as weight lifting and sprinting. Sex differences in performance are reduced to 7 to 13% in swimming because of the greater buoyancy of women.

Boys tend to be slightly superior to girls in sprint running, long jump, and high jump, and considerably superior in throwing skills. At puberty, males accelerate their development of motor skills, whereas females change very little. Sex differences in sports participation probably account for much of this phenomenon.

Males and females seem to experience similar effects from most types of physical training. However, the literature often provides conflicting results. When evaluating these studies, it is important to consider the relative fitness of the male and female subjects. Although some studies show that females improve more than males, this is usually because of a relatively higher initial fitness in males (i.e., males in these studies are usually closer to their maximum physical potential). Eddy et al. demonstrated that the sexes responded similarly to different intensities of interval and continuous endurance training. This was a well-controlled study that equated such factors as initial fitness, frequency, duration and intensity of training, and total caloric expenditure.

The subject of sex differences in temperature regulation is a somewhat controversial one. Numerous early studies found that males were better able to tolerate exercise in the heat. However, these early investigations usually employed fit male subjects and sedentary female subjects. In studies using fit women, such as that by Weinman et al., sweat rates during exercise in the heat are generally less in women, but the ability to acclimatize and control body temperature was similar. Woman may rely on circulatory mechanisms, such as altering vascular tone, to achieve the same degree of thermoregulatory control as men. Relative fitness rather than sex differences seem to be more important in determining heat tolerance and the ability to acclimatize to heat.

Muscle Metabolism

In equally trained male and female subjects, muscle glycogen, blood lactate levels (at the same relative percentage of $\dot{V}_{O_2max}$), fat metabolism capacity, and muscle fiber composition are similar. There is some evidence that certain aspects of female muscle biochemistry (such as the ability to synthesize muscle glycogen and the ability to oxidize fats) do not respond to training as readily as they do in males. Costill et al., comparing similarly trained male and female runners, demonstrated that muscle succinate dehydrogenase and carnitine palmitoyl transferase activities were higher in males. Muscle glycogen production is stimulated by testosterone, which is found in much greater abundance in the male.

It has been hypothesized that women are better endurance athletes because they have more fat and can metabolize it more effectively than men. This notion was derived from empirical observations of female runners who claimed not to experience the extreme fatigue that is common in males during the late stages of the marathon. This extreme fatigue, known as "hitting the wall," is characterized by muscle and liver glycogen depletion. There appears to be no basis for this hypothesis. Even the leanest endurance athletes have ample fat stores, and in fact, most studies indicate that males have a slightly greater ability to metabolize fats than women. A likely explanation for this observation is that women, or for that matter, men who do not experience an extreme degree of glycogen depletion during a marathon, are not running at a fast enough pace or at a high enough percentage of their $\dot{V}_{O_2max}$ to burn up excessive quantities of their carbohydrate stores. These athletes are certainly not among the winners of the race.

Strength

Males are stronger than females throughout childhood, with the gap widening during adolescence. Body weight accounts for much but not all of the sex differences in strength. Even during childhood, when body weight and muscle mass are similar in boys and girls, male muscle can develop more tension per unit volume. In the adult, males are 50% stronger than females in most muscle groups.

Muscle fibers in female athletes and nonathletes are similar to those in males, both histochemically and in their distribution. However, all fiber types in females have a smaller cross-sectional area than males.

Higher androgen levels in males account for the large strength differences between the sexes. Androgens are potent anabolic hormones that are responsible for much of the muscle hypertrophy seen in males during the adolescent growth spurt as well as from strength training. Because women have only low levels of androgens, they experience little muscle hypertrophy from strength training. They can greatly improve their strength, but they just do not develop big muscles. Fahey et al. have found no relationship between serum testosterone and muscle strength in college women. The subjects in this study had normal levels of testosterone, which does not rule out the possibility that women with high levels of testosterone also possess greater strength. Women apparently gain strength by improving their ability to recruit motor units rather than significantly altering the contractile structures of the muscles. Old men, who have low levels of testosterone, also rely principally on enhanced motor unit recruitment to increase strength (see Chapter 31).

EFFECTS OF EXERCISE ON THE FEMALE ESTROUS CYCLE

Sexual maturation in the female is marked by the onset of menarche, the first menstruation. Menarche typically occurs after the peak of the adolescent growth spurt and the appearance of pubic hair, breast development, and mature patterns of fat deposition. The onset of menarche may be brought on by such factors as the attainment of a critical body weight or the concentration of a prerequisite amount of gonadotrophic hormones.

Athletes, particularly in high-energy sports such as swimming and running, tend to reach menarche later than nonathletes. Whereas nonathletes typically experience their first menstruation at 12.9 yr, this does not occur in athletes until 13.1 to 14.0 yr. In addition, there seem to be cultural differences in the attainment of menarche in both athletes and nonathletes. Eastern European girls tend to attain menarche later than their Western counterparts, whereas Latin American females attain menarche earlier than Caucasian Americans. These differences may be attributed to such factors as genetics and diet.

World-class gymnasts often seem to delay the onset of puberty by several years. The low fat and lighter weight of the preadolescent girl is an advantage in a sport that demands a lean body composition for increasingly difficult techniques. The administration of puberty-suppressing drugs to certain successful gymnasts has been suspected. However, the most likely explanation of this phenomenon lies in the protein composition of the diet. Puberty could easily be delayed by combining a diet chronically low in protein with the rigorous training of gymnastics. The age of menarche was much later in Europe after both world wars, periods of protein shortages in the diet. It might be possible to manipulate the diet to delay puberty and yet have a minimal effect on performance in gymnastics. The long-term effects of such a practice are unknown.

Amenorrhea (absence of menstruation) and oligomenorrhea (reduced or irregular menstruation) with concomitant anovulation (failure to ovulate) is common in endurance athletes. This phenomenon may be related to low levels of body fat, weight loss, or chronically high energy expenditure. In a study of track and cross-country runners, for example, 45% of women who ran more than 80 mi a week experienced irregular menstrual cycles. A study by Dale et al. indicates that the mileage run per week was directly proportional to the degree and incidence of menstrual dysfunction. This condition is also common in women experiencing body fat losses from other causes such as dieting and anorexia nervosa. Menstrual irregularities associated with rigorous physical training appear to be benign and reversible.

The mechanism of exercise-induced amenorrhea or oligomenorrhea is unclear. Heavy exercise training stimulates adrenal androgen production. It is possible that increased androgen production in females involved in heavy training could affect hypothalamic and pituitary hormones that control the estrous cycle. However, female endurance athletes do not commonly experience progressive virilization, so it is unlikely that they experience a hyperandrogenic state.

More likely endurance training probably acts to reduce estrogen levels, which directly affect menstruation. Dale et al. showed a decrease of both pituitary gonadotrophins and ovarian steroids in amenorrheic distance runners compared to nonrunners with normal menstrual cycles (Figure 29-2). In addition to their secretion from the ovaries, estrogens are produced by peripheral aromatization of androgens catalyzed by the enzyme aromatase found in fat cells. A reduction in fat may decrease the peripheral production of estrogens, which are thought to be important in stimulating the hypothalamic–pituitary axis during the early follicular phase of the normal menstrual cycle.

Dysmenorrhea (painful menstruation) is of more practical significance than oligomenorrhea or amenorrhea to women involved in sports. Various studies have estimated that, to some extent, as many as 90% of female athletes suffer from dysmenorrhea. Symptoms range from irritability, backache, abdominal cramps, and headache to nausea and vomiting. Because dysmenorrhea detracts from feelings of well-being, it very probably has a negative effect on physical performance that may be psychological in nature.

In a 1975 review, Ryan stated:

"Dysmenorrhea is apparently neither aggravated nor cured by sports participation, although it appears to be less common among women athletes and physical education professional students than for those who are physically inactive. Its appearance and severity appear to depend more on psychological factors, and it seldom prevents a high level woman athlete from competing."

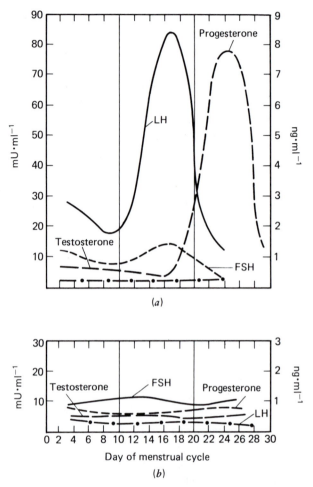

FIGURE 29-2

Hormone profile tf healthy nonrunner (a) and anovulatory runner (b).

(SOURCE: From E. Dale, D.H. Gerlach, and A.L. Wilkite, 1979. Reprinted by permission of The Physician and Sportsmedicine, a McGraw Hill Publication.)

Although researchers such as Billig and Clow have suggested exercises to relieve this condition, dysmenorrhea is often treated successfully by the administration of hormonal antifertility agents and, most recently, antiprostaglandins. Prostaglandin F has been implicated in dysmenorrhea because of its role in initiating gastrointestinal and uterine contractions. Drugs such as Motrin (ibuprofen), which are widely used in athletic medicine as antiinflammatory agents, have been used with considerable success in treating dysmenorrhea.

In most sports (other than endurance activities), participation seems to have no effect on menstruation. Conversely, menstruation has only a minimal effect on performance. Women have achieved record-breaking and Olympic medal-winning performances during every period of the menstrual cycle, but the best performances seem to occur between the immediate postmenstrual period until the fifteenth day of the cycle. The bulk of evidence suggests that participation in sports should be allowed during menstruation by women who wish to do so.

PREGNANCY AND EXERCISE

Pregnancy has traditionally been treated almost as an illness. Pregnant women have been advised to rest and avoid unnecessary physical activity. However, evidence strongly suggests that moderate exercise is beneficial to the mother and does not place the fetus at increased risk (as long as the intensity and duration are not excessive).

Regular exercise during pregnancy combats the effects of deconditioning and thus combats one of the causes of fatigue. Muscle strength is better maintained, which may speed delivery and help the pregnant woman to withstand the effects of a distended abdomen.

Energy Cost of Pregnancy

Pregnancy tends to increase the energy cost of common daily activities. Likewise, pregnant women tend to become deconditioned as they approach term. The result is often chronic fatigue and backache. The higher energy costs of pregnancy are due to a combination of fulfilling the energy needs of the conseptus, increased basal metabolic rate and the increased cost of moving a larger body mass.

There is considerable controversy surrounding the added caloric requirement of pregnancy. Whereas some researchers estimate an added requirement of 80,000 calories during the entire pregnancy (to build new tissues and to maintain the higher metabolism of increased energy cost of daily activities), others believe the added demand is only about 27,000 calories. They point out that pregnant women typically decrease their activity levels to compensate for the increased difficulty of movement.

The management of caloric balance (calories consumed versus calories expended) depends on the extent of daily activities. The caloric cost of these activities is higher during pregnancy, provided the same techniques are used in the pregnant and nonpregnant conditions. A sample of the caloric cost of common household activities during pregnancy is shown in Table 29-4.

Most obstetricians recommend an average weight gain of about 24–35 pounds (Table 29-5). However, there are large individual differences and considerable controversy concerning the ideal weight gain. Some authorities point to an increased risk of toxemia of pregnancy, the difficulty in reducing excess body fat

TABLE 29-4

Energy Consumed in Normal Activities During Pregnancy

Activity	Energy Consumed ($C \cdot min^{-1}$)	
	Pregnant	Not Pregnant
Lying quietly	1.11	.95
Sitting	1.32	1.02
Sitting, combing hair	1.36	1.22
Sitting, knitting	1.55	1.47
Standing	1.41	1.12
Standing, washing dishes	1.63	1.33
Standing, cooking	1.66	1.41
Sweeping with broom	2.90	2.50
Bed making	2.98	2.66

SOURCE: Data from Blackburn, M. and D. Calloway. Energy expenditure of pregnant adolescents. J. Am. Dietetics. 65:24–30, 1974.

after delivery and the possible medical complications introduced by extreme obesity. Others, such as Shanklin and Brewer, point to the dangers of fetal and maternal malnutrition that could result from rigidly adhering to predetermined weight standards. These researchers cite malnutrition fostered by attempts at weight control during pregnancy to be a prominent culprit in low birthweight, toxemia of pregnancy (a controversial viewpoint) (toxemia is also called eclampsia; characterized by hypertension and proteinuria), and fetal malnutrition (i.e., intrauterine growth retardation, where the fetus stops growing). They favor emphasis on sound nutritional practices rather than caloric restriction during pregnancy.

TABLE 29-5

Components of Typical Weight Gain During Pregnancy

Component	Weight Gain (lb)
Baby	7
Amniotic fluid, placenta, and fetal membranes	4
Breasts	3–4
Uterus	2
Mother's tissue (protein 3 lb, fluid 3 lb, fat 2 lb)	8–18
TOTAL	24–35

Effects of Physical Fitness on Childbirth

Clinical observations differ as to the relationship of physical fitness and the ease of delivery. This undoubtedly stems from a lack of objective definition of fitness. Some studies consider a woman physically fit and active if she considers herself to be. This is unsatisfactory, because most people are poor judges of their own physical condition. However, studies of extremely fit women indicate that superior physical fitness does speed delivery and make the experience less tiring.

In a study by Zaharieva of Olympians who became pregnant, the first stage of labor was the same as nonathletes. This is understandable, as this phase of childbirth involves involuntary uterine contractions. However, the second or expulsive stage of labor was about 50% shorter among the sportswomen, probably because they had stronger abdominal muscles. Erdelyi found that athletes exhibited decreased incidences of toxemia, threatened abortions, necessity for forceps delivery, and frequency of cesarean section.

Overall fitness seems to be more important in the ease of delivery than innate physical superiority. Active women seem to get the same benefits as Olympians in their deliveries. Clinical observations of physically fit mothers by researchers such as Erkkola indicate that they do not experience fatigue during labor to the same extent and tend to recover faster after childbirth than mothers with low physical working capacities.

Fit mothers also experience a lower incidence of stretch marks (striae gravidarum), which often occur in the skin over the womb. In one study, stretch marks were found in 26.3% of the athletes compared to 90% of the sedentary nonathletes. Strong muscles distend less in response to the growing abdomen, and thus less stress is placed on the skin. Fit mothers also tend to have less subcutaneous fat, which plays a role in the development of stretch marks.

Effects of Pregnancy on Fitness

Although Olympic medals have been won by women in the early stages of pregnancy, in general, pregnant athletes are unable to maintain their vigorous level of training. The level of Olympic performance is so high today that preparation for peak performance is incompatible with the state of pregnancy.

Pregnancy itself probably has no effect on subsequent athletic performance, although the increased time commitment of child care might. One study noted that 27% of the Olympians who became pregnant participated in at least one more Olympic Games, usually with improved performance. The most likely explanation for the improved performance is that the most highly motivated athletes continued to participate in sports after childbirth, and their improvement merely represented the normal progression in their event.

It has been hypothesized that increased flexibility following pregnancy accounts for some of the increased success of these women. During pregnancy,

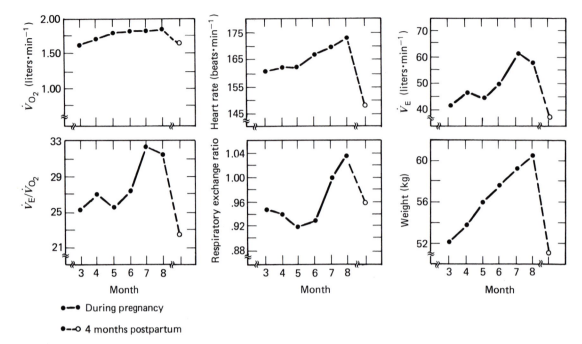

●—● During pregnancy

●--○ 4 months postpartum

FIGURE 29-3

Metabolic and circulatory responses (measured during the tenth minute of treadmill running at 6 mph and 0% grade) and weight changes from the third month through the eighth month of pregnancy and 4 months post-partum.

(SOURCE: From P.L. Hutchinson, K. Cureton, and P. Sparling, 1981. Reprinted by permission of The Physician and Sportsmedicine, a McGraw Hill Publication.)

a hormone called relaxin is released that facilitates the opening of the birth canal during delivery. Relaxin increases the flexibility of all of the body's soft tissue, an effect that seems to persist after pregnancy. However, presently it is extremely difficult to quantify accurately the effect of increased flexibility on performance, so one can only speculate on the contribution of relaxin to subsequent success in athletics. In addition, relaxin is withdrawn after delivery, so it is doubtful that it would have any long-term effects on flexibility.

Pregnant women lose very little capacity for nonweight-bearing exercise such as cycling or swimming during the first two trimesters. However, the capacity for exercise affected by gravity decreases. At a given intensity of exercise there is a progressive increase in O_2 consumption, heart rate, ventilation, ventilatory equivalent ($\dot{V}_E/\dot{V}_{O_2}$), and respiratory exchange ratio (Figure 29-3). The drop in aerobic capacity seems to be proportional to the increase in body weight, the degree of deconditioning of the pregnant woman and proportional to the increase in uterine size, which affects ventilation during exercise.

Erkkola has found that pregnancy does not alter the fact that physical training can actually increase the endurance capacity of previously sedentary pregnant women and minimize the effects of deconditioning in those who were previously trained. In spite of the many positive findings, many women, even the highly motivated, find it difficult to continue their training during the first few months of pregnancy because of morning sickness.

Effects of Exercise on the Fetus

The female anatomy provides reasonable protection to the fetus during pregnancy. The pelvis, uterine wall, and amniotic fluid can absorb considerable shock and thus enable safe participation in most activities. During the embryonic phase (early pregnancy), the amniotic fluid provides a buoyant medium for the delicate tissues. During the later stages of pregnancy it allows the fetus to move freely and permits it to change body position.

Blows to the abdomen resulting from falls or contact sports could be hazardous and should be avoided. Further, sudden rushes of water through the vagina that could occur during water skiing, or changes of pressure that occur during scuba diving, could also place the fetus and mother at risk.

Pregnancy places an added load on the cardiovascular system of the mother. There are increases in resting heart rate, stroke volume, and cardiac output, which result in left ventricular hypertrophy. Blood volume increases by up to 40%, thus adding significant preload stress. During exercise, the needs of the fetus must be added to those of the maternal physiology.

Analogous to exercise in the heat, there is an increased competition for blood in the pregnant woman during physical activity because of the circulatory needs of the placenta and of the working muscles. In general, the response of the fetus to maternal exercise tends to follow those of the mother. During maternal exercise, fetal heart rate increases, while arterial pH and P_{O_2} decrease. Nelson et al. have demonstrated that in guinea pigs, uterine blood flow decreases as a function of the duration and intensity of exercise. The literature is unclear regarding the possible effects of this on fetal well-being but it appears that it makes no difference.

In humans, Erkkola has found that women with higher physical capacities tended to give birth to heavier babies. However, in that study (Int. J. Gynaecol. Obstet 14: 153–159, 1976), the amount of physical exercise practiced by the subjects during pregnancy was not specified. Animal studies using the African pygmy goat, mouse, sheep, and guinea pig reported that chronic exercise resulted in a reduction in fetal weight at term. In the guinea pig, birthweight, placental diffusion capacity, and placental weight decreased with increasing intensity and duration of exercise (exercise duration greater than 30 min caused the most serious effects) (Figure 29-4). If the results apply to humans, all of these factors may have negative effects on the quality of fetal and subsequent child development.

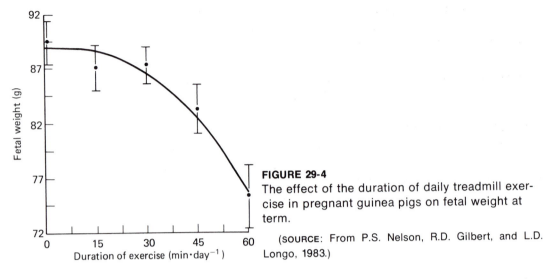

FIGURE 29-4
The effect of the duration of daily treadmill exercise in pregnant guinea pigs on fetal weight at term.

(SOURCE: From P.S. Nelson, R.D. Gilbert, and L.D. Longo, 1983.)

Exercise stress tests have been suggested as a means of early detection of inadequate gas exchange in the uterus and placenta. In uteroplacental insufficiency, fetal heart rate response is irregular during maternal exercise, followed by bradycardia during recovery. Strembera and Hodr have concluded that the higher the degree of potential fetal distress from hypoxia, the more likely is fetal bradycardia to occur after maternal exercise. In the normal fetus, heart rate goes up during exercise, then gradually subsides during recovery. Certainly, under circumstances of fetal distress, exercise is contraindicated.

There are several factors that appear to protect the fetus from hypoxia during exercise. Clapp, using pregnant ewes, showed that although placental blood flow decreased during exercise, uterine and umbilical O_2 uptakes were unchanged as a result of increased O_2 extraction. Further, Curet et al. demonstrated a redistribution of blood flow to the placenta after exercise, resembling reactive hyperemia, that might represent a compensatory mechanism for the fetus. The fetal circulation is well prepared to meet the stresses of hypoxia created by the exercising mother. Fetal hemoglobin can carry 20 to 30% more O_2 than maternal hemoglobin, and fetal blood is about 50% more concentrated. The fetus has a relatively small demand for O_2, because its muscular activity is minimal and its body temperature is kept fairly constant by its fluid environment.

Exercise-induced fetal hyperthermia is a risk, particularly during the first trimester of pregnancy. Rectal temperatures of 40° C are not unusual in recreational marathoners and road racers. A woman could easily mistake pregnancy for amenorrhea that is so common among distance runners. Unfortunately, little is known of the effects of exercise-elevated maternal core temperature on embryonic or fetal well-being.

Numerous clinical studies have appeared in the literature on exercise and pregnancy. In general, these studies have been positive (easier delivery, no apparent fetal defects, normal birthweight, etc.). However, many factors are involved in producing a successful outcome of a pregnancy, possibly including diet, genetics, stress, exercise, and smoking. It is possible that the negative effects of excessive exercise during pregnancy, demonstrated in animal studies, may be overshadowed by positive nutritional and health practices, and thus result in a successful outcome. Certainly in light of the negative evidence, prudence should be used in prescribing endurance exercise. It appears that a little exercise is probably beneficial (probably at the level suggested by the ACSM guidelines of exercise prescription for healthy adults—see Chapter 24), but excessive exercise could compromise fetal well-being.

THE SAFETY OF EXERCISE FOR WOMEN

Safety is a primary concern of educators and physicians when considering sports for women. Women sustain different injuries than men only because they participate in different sports. The higher injury rate in women can usually be attributed to lack of physical conditioning. The evidence overwhelmingly points to sports and exercise being safe and healthy for girls and women. As in boys, sports injuries can be prevented by stressing physical conditioning and by restricting play in most sports to the athletes' size and ability level.

The breasts should be supported during sports. During running, the movement of unsupported breasts against the chest creates between 50 and 100 pounds of force. Prolonged periods of training without adequate breast support could result in inflamed nipples and eventually lead to sagging breast tissue. Sports bras are currently available that adequately support the breasts during physical activity.

Summary

Widespread participation by women in sports is a relatively new phenomenon. Although the nature of exercise training and adaptation are the same in both sexes, there are differences in performance and physiology that present problems unique to women.

Sex chromosomes are responsible for differences between males and females. They act as blueprints for respective sexual characteristics in factors such as muscle mass, heart size, body fat, and hemoglobin concentration. The determination of sex chromosomes serves as a basis for the identification of gender in major athletic competitions.

Hormonal changes during puberty are important in producing sex differences in physical performance. Increased estrogen secretion enhances the deposition of fat and slightly retards lean body mass. Large increases in androgen secretion in boys, in contrast, result in a large acceleration in body size and physical performance.

Adult males tend to have less fat and more muscle mass than adult females. Active women tend to have lower fat percentages than the average, generally falling below 22% in recreational runners. Fat percentages of between 10 and 15% are not uncommon in competitive female runners.

Men exceed women in maximal O_2 consumption by about 20% because of their greater cardiac output, blood volume, and oxygen-carrying capacity. Heart rate, at rest and during exercise, tends to be higher in women, but the O_2 requirements for a given work load are similar between the sexes. Men and women experience similar effects from most types of training. Muscle metabolic processes are also similar in men and women of equal training status.

Exercise training appears to affect the female estrous cycle. Menarche tends to occur later in athletes. Amenorrhea and oligomenorrhea are common in endurance athletes; the incidence in runners seems to be related to the number of miles run per week. Menstrual irregularities associated with rigorous physical training appear to be benign and reversible. Menstrual irregularity appears to be related to the suppression of pituitary gonadotrophins and ovarian steroids. Dysmenorrhea (painful menstruation), although common in athletes, appears to be neither aggravated nor cured by sports participation.

Exercise training is becoming increasingly popular among women during pregnancy. Training appears to combat the effects of deconditioning, facilitate delivery, and have no apparent negative effects on the fetus. However, several studies have shown that placental blood flow decreases with increasing intensity and duration of exercise; so excessive exercise should be avoided. Moderate exercise seems to be important for most women during pregnancy.

Selected Readings

Allsen, P.E., P. Parsons, and G.R. Bryce. Effect of the menstrual cycle on maximum oxygen uptake. Physician Sports Med. 5: 53–55, 1977.

Artal, R., L.D. Platt, M. Sperling, R.K. Kammula, J. Jilek, and R. Nakamura. Exercise in pregnancy. I. Maternal cardiovascular and metabolic responses in normal pregnancy. Am. J. Obstet. Gynecol. 140: 123–127, 1981.

Åstrand, P-O, Human physical fitness with special reference to sex and age. Physiol. Rev. 36: 307–335, 1956.

Baker, E.R. Menstrual dysfunction and hormonal status in athletic women: A review. Fertil. Steril. 36: 691–696, 1981.

Billig, H.E. Dysmennorrhea: The result of a postural defect. *Arch. Surg.* 46:611–613, 1943.

Brewer, T.H. Metabolic Toxemia of Late Pregnancy: A Disease of Malnutrition. Springfield, Ill.: Charles C Thomas, 1966.

Brown, C.H., J.R. Harrower, and M.F. Deeter. The effects of cross-country running on pre-adolescent girls. Med. Sci. Sports. 4: 1–5, 1972.

Buchsbaum, H.J. Trauma in Pregnancy. Philadelphia: W.B. Saunders, 1979.

Bullard, J.A. Exercise and pregnancy. Can. Fam. Physician. 27: 977–982, 1981.

Clapp, J.F. Acute exercise stress in the pregnant ewe. Am. J. Obstet. Gynecol. 136: 489–494, 1980.

Clarke, H.H. (ed.). Physical activity during menstruation and pregnancy. Phys. Fitness Res. Digest. Series 8, No. 3: 1–25, 1978.

Clow, A.E.S. Treatment of dysmenorrhea by exercise. *Brit. Med. J.* 1:4–5, 1962.

Cooter, G.R. and K.W. Mowbray. Effects of iron supplementation and activity on serum iron depletion and hemoglobin levels in female athletes. Res. Q. 49: 114–117, 1978.

Costill, D.L., W.J. Fink, L.H. Getchell, J.L. Ivy, and F.A. Witzmann. Lipid metabolism in skeletal muscle of endurance-trained males and females. J. Appl. Physiol. 47: 787–791, 1979.

Curet, L.B., J.A. Orr, H.G. Rankin, and T. Ungerer. Effect of exercise on cardiac output and distribution of uterine blood flow in pregnant ewes. J. Appl. Physiol. 40: 725–728, 1976.

Dale, E., D.H. Gerlach, and A.L. Wilkite. Menstrual dysfunction in distance runners. Obstet. Gynecol. 54: 47–53, 1979.

Dhindsa, D.S., J. Metcalfe, and D.H. Hummels. Responses to exercise in the pregnant pygmy goat. Respir. Physiol. 32: 299–311, 1978.

Doolittle, T.L. and J. Engebretsen. Performance variations during the menstrual cycle. J. Sports Med. Phys. Fitness. 12: 54–58, 1972.

Dressendorfer, R.H. Physical training during pregnancy and lactation. Physician Sports Med. 6: 74–80, 1978.

Drinkwater, B.L. Physiological responses of women to exercise. Exerc. Sport Sci. Rev. 1: 125–153, 1973.

Drinkwater, B.L., S.M. Horvath, and C.L. Wells. Aerobic power of females, ages 10 to 68. J. Gerontol. 30: 385–394, 1975.

Eddy, D.O., K.L. Sparks, and D.A. Adelizi. The effects of continuous and interval training in women and men. Eur. J. Appl. Physiol. 37: 83–92, 1977.

Emmanouilides, G.C., C.J. Hobel, K. Yashiro, and G. Klyman. Fetal responses to maternal exercise in the sheep. Am. J. Obstet. Gynecol. 112: 130–137, 1972.

Erdelyi, G.J. Gynecological survey of female athletes. *J. Sports Med. Phys. Fitness* 2: 174–179, 1962.

Erkkola, R. The influence of physical training during pregnancy on physical working capacity and circulatory parameters. Scand. J. Clin. Invest. 36: 747–754, 1976.

Erkkola, R. The physical work capacity of the expectant mother and its effect on pregnancy, labor, and the newborn. Int. J. Gynaecol. Obstet. 14: 153–159, 1976.

Fahey, T.D., R. Rolph, P. Moungmee, J. Nagel, and S. Mortara. Serum testosterone, body composition, and strength of young adults. Med. Sci. Sports 8: 31–34, 1976.

Frisch, R.E. and J.W. McArthur. Menstrual cycles: Fatness as a determinant of minimum weight for height necessary for their maintenance or onset. Science. 185: 949–951, 1974.

Frisch, R.E., G. Wyshak, and L. Vincent. Delayed menarche and amenorrhea in ballet dancers. N. Engl. J. Med. 303: 17–19, 1980.

Gilbert, R.D., L.A. Cummings, M.R. Juchau, and L.D. Longo. Placental diffusing capacity and fetal development in exercising or hypoxic guinea pigs. J. Appl. Physiol. 46: 828–834, 1979.

Haycock, C.E. and J.V. Gillette. Susceptibility of women athletes to injury: Myths vs. reality. J.A.M.A. 236: 163–165, 1976.

Hermansen, L. and K.L. Andersen. Aerobic work capacity in young Norwegian men and women. J. Appl. Physiol. 20: 425–431, 1965.

Hoffman, P.G. Menstrual irregularities in women athletes. J.A.M.A. 242: 1539, 1979.

Hutchinson, P.L., K. Cureton, and P. Sparling. Metabolic and circulatory responses to running during pregnancy. Physician Sports Med. 9: 55–61, 1981.

Jenkins, R.R. and C. Ciconne. Exercise effect during pregnancy on brain nucleic acids of offspring in rats. Arch. Phys. Med. Rehabil. 61: 124–127, 1980.

Jokl, P. and E. Jokl. Running and swimming world records. J. Sports Med. Phys. Fitness. 17: 213–228, 1977.

Knuttgen, H.G. and K. Emerson. Physiological responses to pregnancy at rest and during exercise. J. Appl. Physiol. 36: 549–553, 1974.

Kolata, G. Exercise during pregnancy reassessed. *Science* 219: 832–833, 1983.

Lehtovirta, P., J. Kuikka, and T. Pyorala. Hemodynamic effects of oral contraceptives during exercise. Int. J. Gynaecol. Obstet. 15: 35–37, 1977.

Malina, R., A. Harper, H. Avent, and D. Campbell. Age at menarche in athletes and non-athletes. Med. Sci. Sports. 5: 11–13, 1973.

Miles, D.S., J.B. Critz, and R.G. Knowlton. Cardiovascular, metabolic, and ventilatory responses of women to equivalent cycle ergometer and treadmill exercise. Med. Sci. Sports Exerc. 12: 14–19, 1980.

Mostardi, R.A., N.R. Woebkenberg, and M.T. Jarrett. Oral contraceptives and exercise. Ohio Acad. Sci. 80: 3–7, 1980.

Nelson, P.S., R.D. Gilbert, and L.D. Longo. Fetal growth and placental diffusing capacity in guinea pigs following long-term maternal exercise. J. Dev. Physiol. 5:1–10, 1983.

Nesheim, B-I. and P. Bergsjo. Physical activity and reproductive function in women. Scand. J. Soc. Med. 29(Suppl.): 77–81, 1982.

Parizkova, J. Impact of daily work load during pregnancy and/or postnatal life on the heart microstructure of rat male offspring. Basic Res. Cardiol. 73: 433–441, 1978.

Pate, R.R., M. Maguire, and J. Van Wyk. Dietary iron supplementation in women athletes. Physician Sports Med. 7: 81–101, 1979.

Pomerance, J.J., L. Gluck, and V.A. Lynch. Physical fitness in pregnancy: Its effect on pregnancy outcome. Am. J. Obstet. Gynecol. 119: 867–876, 1976.

Powers, S.K., W. Riley, and E. Howley. Comparison of fat metabolism between trained men and women during prolonged aerobic work. Res. Q. Exerc. Sport. 51: 427–431, 1980.

Raven, P.B., B.L. Drinkwater, and S.M. Horvath. Cardiovascular responses of young female track athletes during exercise. Med. Sci. Sports. 4: 205–209, 1972.

Ryan, A.J. Gynecological considerations. *J. Phys. Ed. Rec.* 46: 39–41, 1975.

Shanklin, D.R. and J. Hodin. Maternal Nutrition and Child Health. Springfield, Ill.: Charles C Thomas, 1979.

Sinning, W.E. and G.D. Lindberg. Physical characteristics of college age women gymnasts. Res. Q. 43: 226–234, 1972.

Sloan, A.W. Effects of training on physical fitness of women students. J. Appl. Physiol. 16: 167–169, 1961.

Smith, A.D., R.D. Gilbert, R.J. Lammers, and L.D. Longo. Placental exchange area in guinea pigs following long-term maternal exercise: A sterological analysis. *J. Dev. Biol.* 5: 11–21, 1983.

Terada, J. Effect of physical activity before pregnancy on fetuses of mice exercised forcibly during pregnancy. Teratology. 10: 141–144, 1974.

Thomas, C.L. Factors important to women participants in vigorous athletics. In: Strauss, R.H. (ed.), Sports Medicine and Physiology. Philadelphia: W.B. Saunders, 1979, p. 204–319.

Ueland, K., M.J. Novy, E.N. Peterson, and J. Metcalfe. Maternal cardiovascular dynamics. Am. J. Obstet. Gynecol. 104: 856–864, 1969.

Vellar, O.D. and L. Hermansen. Physical performance and hematological parameters. Acta Med. Scand. 522: 1–40, 1971.

Wardle, M.G. Women's physiological reactions to physically demanding work. Psychol. Women Q. 1: 151–159, 1976.

Weinman, K.P., Z. Slabochova, E.M. Bernauer, T. Morimoto, and F. Sargent. Reactions of men and women to repeated exposure to humid heat. J. Appl. Physiol. 22: 533–538, 1967.

Wells, C.L. Sexual differences in heat stress response. Physician Sports Med. 5: 79–90, 1977.

Wells, C.L., L.H. Hecht, and G.S. Krahenbuhl. Physical characteristics and oxygen utilization of male and female marathon runners. Res. Q. Exerc. Sport. 52: 281–285, 1981.

Williams, C.A., J.S. Petrofsky, and A.R. Lind. Physiological responses of women during lifting exercise. Eur. J. Appl. Physiol. 50: 133–144, 1982.

Wilmore, J.H. and C.H. Brown. Physiological profiles of women distance runners. Med. Sci. Sports 6: 178–181, 1974.

Woodward, S.L. How does strenuous maternal exercise affect the fetus? A review. Birth Fam. J. 8: 17–24, 1981.

Zaharieva, E. Survey of sportswomen at the Tokyo Olympics. J. Sports Med. Phys. Fitness. 5: 215–219, 1965.

Zaharieva, E. Olympic participation by women: Effects on pregnancy and childbirth. J.A.M.A. 221: 992–995, 1972.

30

GROWTH AND DEVELOPMENT

Physical performance in children and adolescents must always be assessed in light of the growth process. Growth involves a series of developmental stages that are remarkably similar in all people. Individual differences in diet, exercise, and health may affect these stages to a certain extent, but the basic pattern remains the same.

Each of these stages has a profound influence on individual capability for physical performance. For example, the ability to throw a ball in the adult fashion of weight transfer and trunk rotation requires a prerequisite neurological development. Likewise, prepubertal boys have a limited ability to gain strength because they do not produce male hormones in sufficient quantities to induce significant muscle hypertrophy. Attempts at trying to rush the developmental process are futile and may lead to physical and emotional harm.

It is extremely important to recognize the capabilities and limitations imposed by growth and maturation. For instance, a contact sport such as tackle football could pose a severe risk to an immature prepubertal child but might be perfectly appropriate for a mature and muscular 16-yr-old high school student. Likewise, the apparent excess body weight in an 8-yr-old boy may not reflect extreme deconditioning, but simply a common characteristic of a prepubertal youngster.

The purpose of this chapter is to examine the relationship between growth, exercise capacity, and training. Emphasis is placed on the importance of exer-

Tommy, Danny, Timmy and Mikey typify the energy and potential for human development. (Courtesy Suzanne Brooks and Kilty Fahey).

cise during growth and the relative importance of environmental and genetic factors on development.

NATURE OF THE GROWTH PROCESS

Growth involves the transformation of nutrients into living tissue. It implies the development of the organism in an orderly fashion and represents a predominance of anabolic over catabolic processes. Growth is characterized by the progressive transformation of the organism into the adult form.

Growth proceeds along an **S**-shaped curve. During the first 2 yr of life there is a rapid increase in height and weight. This is followed by a progressively declining growth rate during childhood. At puberty there is an abrupt increase in the growth rate called the adolescent growth spurt (Figure 30-1).

The different tissues and organs of the body have their own rate of growth. People grow from the head down. The head grows more rapidly than the chest and arms, which grows more rapidly than the legs. The growth of skeletal muscle, heart, liver, and kidney is slower than the rate of growth of the skeleton. As we shall see, differences in these growth rates will affect physical performance at different stages of development.

Although all people adhere to the basic growth curve, individual differences

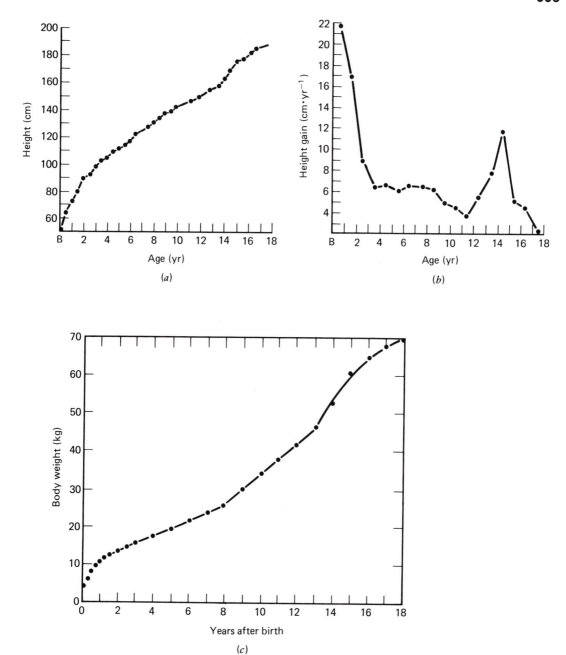

FIGURE 30-1

(a) and (b) Data from an eighteenth-century longitudinal growth study. (a) Height attained at each year. (b) Growth rate. (c) Increase in body weight with age in males.

(SOURCE: From R.E. Scammon, 1927 and A.K. Laird, 1967.)

in environment can affect the rate of change in height, weight, and physiological development. There has been a progressive increase in the rate of growth in height and weight of children living in the Western industrial nations over the past 100 yr. This increased growth rate has been accompanied by earlier maturation.

Although the explanation for this phenomenon is not clear, it seems to be due to improvements in public health, better nutrition, and fewer social stresses. It is interesting to note that these trends are not evident in underdeveloped countries. It is very likely that adverse environmental conditions will also affect the growth of an individual living in an country with a high standard of living.

These secular changes in growth have been particularly evident in Olympic athletes. Performances in sports such as track and field and swimming have improved tremendously since the early days of the competition. However, the relative contribution of greater height and weight is unclear. Most of the increases are probably associated with improved training methods and conditions.

GROWTH IN INFANCY AND CHILDHOOD

The first 2 yr of life are called infancy. During this period there are tremendous changes in body proportions. In the neonate, the head represents about one fourth of the total height. The trunk is slightly longer than the legs. After about 6 months, the rate of growth in the skull slows, while that of the legs and trunk proceeds more rapidly. The fastest rate of growth is in the legs, so that by the time the child is 2, the legs and trunk are about equal in length (Figure 30-2).

Males tend to be taller and heavier than females. However, females have relatively longer legs, which indicates a greater level of maturity. Although the growth rate is faster in males, the relative changes are similar in both sexes.

During childhood, the period between about 2 yr of age and the onset of puberty, there is a steady but gradual increase in height and weight with height increasing faster than weight. The legs continue to grow faster than the trunk, and there is a proportional increase in the growth of the pelvis and shoulders.

The rate of growth in boys and girls is similar, but boys tend to be slightly larger. The skeletal age of girls continues to be more advanced throughout childhood. Between the ages of 6 and 10 yr, girls gain in pelvic width faster than boys, while boys tend to have larger thoraxes and forearms. Although there are some sex differences in growth during childhood, the anthropometric characteristics are almost the same until puberty.

The gradual growth rate during childhood is conducive to learning motor skills. The relatively constant ratio between height and lean body mass provides a stable environment for developing coordination and neuromuscular skill. This is an important period for the introduction and development of gross motor

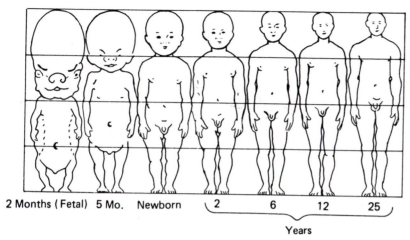

FIGURE 30-2
The changes in the human body's proportions during growth. During growth the head becomes smaller in proportion to the rest of the body, while the legs become relatively longer.

(SOURCE: From R.E. Scammon, 1927.)

activities such as running, jumping, hopping, and throwing. However, the limited muscle mass and the fragile skeletal epiphyseal growth centers make vigorous strength training less appropriate.

GROWTH AT PUBERTY AND ADOLESCENCE

Adolescence is the final period in the growth process leading to maturity. It is a time of rapid increases in height and weight that is accompanied by puberty, the time when the sex organs become fully developed (Figure 30-3). The adolescent growth spurt generally begins at about 10½ to 13 yr in girls and 12½ to 15 yr in boys. Girls will often be taller than boys of the same age, but when the boys' growth spurt occurs they generally catch up and then surpass the girls.

Because puberty and the growth spurt occur at the same time, the degree of sexual maturation is often used to assess the position on the growth curve (Table 30-1). In girls, the growth spurt generally occurs when the breasts and pubic hair first appear. This is followed by menarche, the first menstruation. Menarche provides a definite landmark for the assessment of maturation that is not available in boys. The closest indicators of maturity in boys, in addition to pubic hair, are the appearance of facial hair and the change in voice.

Adolescent growth characteristics are different in late and early maturers. The rate of growth is more intense the earlier the growth spurt occurs (Figure

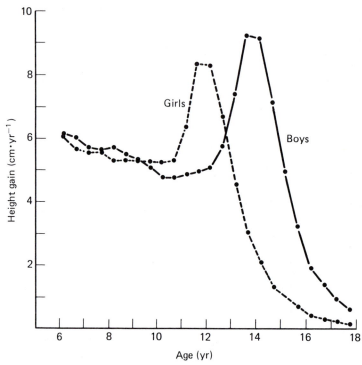

FIGURE 30-3

The adolescent growth spurt in height for girls and boys.

(SOURCE: From J.M. Tanner, 1962.)

30-4). Boys who mature earlier tend to be more muscular, with shorter legs and broader hips. Girls who mature earlier have shorter legs and narrower shoulders than those who mature later. Additionally, late maturers tend to become slightly taller adults. Even though the growth spurt is less sudden in late maturers, they are growing over a longer period of time.

The characteristic sexual differences in anthropometric measures arise during puberty. Males develop greater height and weight with larger musculature and broader shoulders, while females develop larger hips. The growth spurt is slightly over 3 in. $\cdot$ yr^{-1} in girls and about 4 in. $\cdot$ yr^{-1} in boys.

The adolescent growth spurt has a profound effect on physical performance. Males make marked improvements in endurance, strength, speed, power, and various motor skills. Females, in contrast, often experience a leveling off in skills.

SKELETAL CHANGES DURING GROWTH

Increases in height occur because of skeletal growth, principally in the long bones. Bone growth occurs at the epiphyseal growth plates. These are located

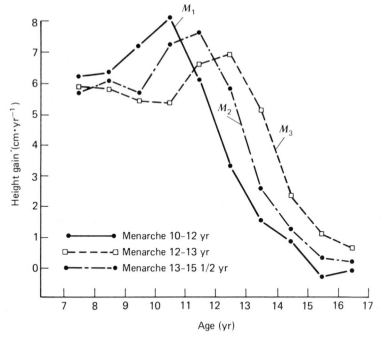

FIGURE 30-4
Peak velocities in height for early-, average-, and late-maturing girls. M_1,
M_2, M_3, Average time of menarche for each group.

(SOURCE: From J.M. Tanner, 1962.)

at both ends of long bones between the articular epiphysis and the central diaphysis (Figure 30-5). The process of bone formation is called ossification.

Bone is a matrix composed of collagen fibers and ground substance. Imbed-
ded in the matrix are osteocytes, osteoblasts, and osteoclasts. Bone grows as a
result of the secretion of collagen by the osteoblasts. During this process, some
of the osteoblasts are trapped in the tissue (now called osteoid). These trapped
osteoblasts, called osteocytes, begin to collect calcium and phosphorus. This
process continues until the growth plates themselves are finally ossified.

The nature of the bone matrix gives it both great compressional and great
tensile strength. The collagen fibers are arranged along vectors of the bone
receiving stress. The bone salts form crystals that are arranged to provide compressional strength greater than that of concrete.

As in most processes in physiology, bone formation is extremely dynamic.
Bone is deposited by the osteoblasts and absorbed by the osteoclasts. This
process allows precise control of calcium and phosphate metabolism and the
maintenance of healthy bone. Additionally, it enables the bones to adapt to
stress. Bone can be deposited when subjected to increased loads or be absorbed
when stresses are decreased.

TABLE 30-1
Pubertal Stage Ratings for Boys and Girls[a]

Pubertal Stage	Pubic Hair	Genital Development	Breast Development
1	None	Testes, scrotum about same size and proportions as in early childhood	Elevation of papilla only
2	Sparse growth of long, slightly pigmented downy hair, straight or only slightly curled, appearing chiefly at base of penis or along labia	Enlargement of scrotum and testes; skin of scrotum reddens and changes in texture; little or no enlargement of penis at this stage	Breast bud stage; elevation of breast and papilla as small mound; enlargement of areolar diameter
3	Considerably darker, coarser, and more curled; hair spreads sparsely over junction of pubes	Enlargement of penis, which occurs at first mainly in length; further growth of testes and scrotum	Further enlargement and elevation of breast and areola, with no separation of their contours
4	Hair now resembles adult in type, but area covered is still considerably smaller than in adult; no spread to medial surface of thighs	Increased size of penis with growth in breadth and development of glans; further enlargement of testes and scrotum; increased darkening of scrotal skin	Projection of areola and papilla to form a secondary mound above the level of the breast

During growth, bone formation exceeds absorption. Exercise, which stresses bones and stimulates bone growth, does not seem to enhance linear bone growth but does increase bone density and width. So, in general, exercise during growth creates a skeleton composed of denser, stronger bone that is better able to withstand stress.

Linear growth can continue as long as the ossification centers are open. Epiphyseal–diaphyseal union begins at puberty in some bones but is not completed until 18 yr of age or later. Disease or trauma can cause injury to the growth plates that could affect growth. For this reason it is extremely important to avoid situations that could adversely damage these areas. For example, studies have shown that excessive baseball pitching by youngsters can cause epi-

TABLE 30-1 (*continued*)

Pubertal Stage	Pubic Hair	Genital Development	Breast Development
5	Adult in quantity and type with distribution of horizontal (or classically "feminine" pattern); spread to medial surface of thighs but not up the linea alba or elsewhere above the base of the inverse triangle. In about 80% of Caucasian men and 10% of women, pubic hair spreads further, but this takes some time to occur after stage 5 is reached. This may not be completed until the mid-twenties or later.	Genitalia adult in size and shape; no further enlargement after stage 5 is reached	The mature stage; projection of papilla only, due to recession of areola to general contour of breast

[a] Pubertal stage is rated on a scale of 1 to 5. Stage 1 represents prepubertal development, whereas stage 5 represents the characteristics of the adult. Stages are determined by the degree of development of pubic hair in both sexes, genital development in boys, and breast development in girls.

SOURCE: Adapted from Larson, L. Fitness, Health, and Work Capacity. New York: MacMillan Co., 1974. p. 516–517.

physeal damage. It is possible that heavy weight lifting could have the same effect (although this had not been consistently demonstrated).

CHANGES IN BODY COMPOSITION

The ratio of fat to fat-free weight is particularly important during growth, because overweight children tend to become overweight adults. Physical activity is vitally important in the maintenance of ideal body composition. Research shows that patterns of inactivity are evident in obese youngsters and that these patterns tend to persist throughout life. However, programs of vigorous physical activity have been shown to reverse the process.

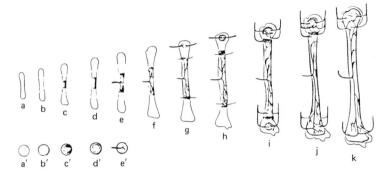

FIGURE 30-5

Diagram of the development of a typical long bone and its blood supply. a, cartilage model; b, development of the bone collar; c, development of calcified cartilage in primary center; d, extension of the bone collar; e, f, the invasion of the cartilage by vascular sprouts and mesenchyme with formation of two areas of bone formation toward the bone ends; g, h, i, secondary centers develop in the bone ends as the central area expands; j, k, epiphyseal plates disappear and the blood vessels of the diaphysis and epiphyses intercommunicate.

(SOURCE: From W. Bloom and D.W. Fawcett, 1975, p. 266.)

Body weight is regulated by the balance between caloric intake and energy output. Increases in body fat occur because of increases in fat cell size, cell number, or both. Fat cells increase in number until early adolescence. After that, increased body fat occurs mainly by increasing cell size.

Once developed, fat cells can only decrease in size, not in number. This may be a reason for the dismal prognosis for long-term weight control in obese adults. It may be possible to affect the development of fat cells during growth by means of diet and physical activity. Failure to do this could result in a lifelong weight problem (Figure 30-6).

During childhood, females have slightly more fat than males. Typical body fat percentages are about 16% for an 8-yr-old boy and 18% for an 8-yr-old girl. At puberty there are marked changes in body composition. Boys make rapid increases in lean weight and decreases in percentage of fat. Percentage of fat typically drops 3 to 5 percentage points between ages 12 and 17 yr. Females also experience increases in lean mass, although less than boys, and increases in body fat. An untrained 17-yr-old female has typically around 25% fat. However, an athlete will often have less than 16 to 18% (Figure 30-7).

Physical activity will not change the essence of these growth-related stages of body composition. In females, for example, vigorous training will not prevent the increase in fat that occurs during adolescence. However, the gains are less than in sedentary girls. Studies comparing active versus inactive youngsters show that training consistently results in higher lean body mass and lower body

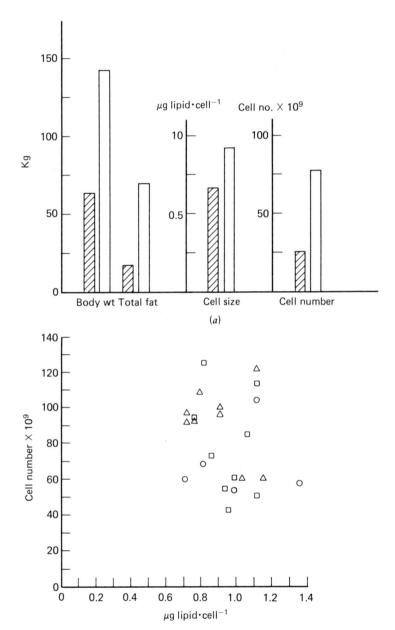

FIGURE 30-6
(a) Adipose cellularity of 20 obese (unshaded) and 5 nonobese (shaded) subjects (mean ± S.D.). (b) Adipose cellularity of 20 obese subjects according to age of onset. Triangles, before 10 yr; □, 10 to 20 yr; circles, >20 yr.

(SOURCE: Modified from J. Stern and M. Greenwood, 1974.)

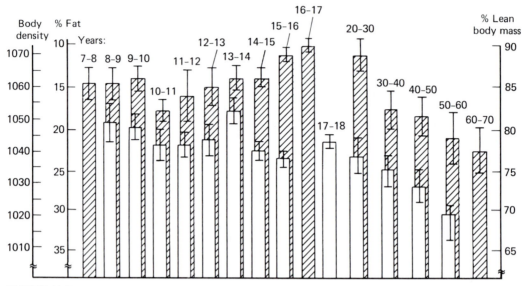

FIGURE 30-7
Body composition during growth and aging in males (shaded) and females (unshaded).

(SOURCE: From J. Parizkova, 1968.)

fat. However, these advantages do not persist into adulthood unless the training is continued. Inactivity will result in decreased muscle mass and increased fat.

Body composition is an important consideration in the development of motor skills. The relatively slow growth period is an important period for learning gross motor skills. These efforts may be compromised if the child is overweight. The relatively weak muscles could have considerable difficulty overcoming the burden of excess body mass.

MUSCLE

Muscle growth accounts for a considerable portion of weight gains during growth. There is a steady growth of muscle tissue during the first 7 yr of life followed by a slowing trend in the years immediately preceding puberty. In the normal child, muscle and skeletal growth keep pace with one another. During the pubertal growth spurt, muscle grows at a rapid rate, particularly in boys. The increase in muscle tissue typically occurs slightly after the greatest increases in height. This explains the awkward gangliness in children that age.

In boys, the increases in muscle size are related to improvements in strength. On the average, rapid increases occur at about 14 yr and continue throughout adolescence. However, there are considerable individual differences that can be attributed to such factors as maturational level, body build, and amount of

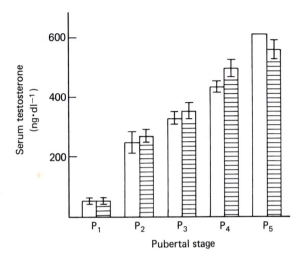

FIGURE 30-8
Serum testosterone before (unshaded) and after (shaded) exercise in pubertal stages 1 to 5.

(SOURCE: Modified from T.D. Fahey, A. DelValle-Zuris, G. Oehlsen, M. Trieb, and J. Seymour, 1979.)

physical activity. Muscle strength and size are of considerable concern to children because they are the most important predictors of athletic success.

The extent of the development and performance of muscle is dependent on the relative maturation of the nervous system. High levels of strength, power, and skill are impossible if the child has not reached neural maturity. Myelination of nerves is not complete until sexual maturity has been achieved, so an immature child cannot be expected to respond to training or reach the same level of skill as adults.

Significant training-induced muscle hypertrophy does not occur until adolescence in boys. This is because of the low levels of male hormones (principally testosterone) found in sexually immature children. Male hormones, called androgens, are important regulators of protein synthesis. Adult males have about 10 times the amount of androgens as prepubertal children and adult women (Figure 30-8). Maximal strength-gaining potential is not possible until adult levels of androgens are achieved.

CARDIOVASCULAR AND METABOLIC FUNCTION

There is a progressive decline in resting and maximum heart rate during childhood and adolescence. Heart rates tend to be the same in girls and boys during childhood but about 3 to 4 beats higher in females during and after adolescence. At puberty in boys, there is a tendency for submaximal exercise and recovery heart rates to be lower, even in the absence of vigorous training. The decreased heart rate is accompanied and perhaps caused by a greater resting and exercise stroke volume. This occurs because of increased heart size and blood volume.

Resting and exercise systolic blood pressure rise progressively, particularly during puberty. Blood pressure assumes adult values shortly after the adolescent growth spurt. The pubertal acceleration in blood pressure occurs in females before males, but the boys quickly catch up and surpass the girls.

Exercise capacity and maximal O_2 consumption (liters $\cdot$ min^{-1}) increase gradually throughout childhood. During puberty, there is a dramatic increase in boys and a leveling off in girls. The improvements in endurance capacity occur because of enhanced oxygen transport and metabolic capacities, and increased muscle mass.

Prepubertal youngsters do not seem to respond to endurance training as well as adolescents and adults. Some studies show no improvement in $\dot{V}_{O_2max}$ from training beyond that expected in normal growth. There may even be a difference between children and adults in the way they adapt to training. For example, studies using interval training as the exercise stimulus have sometimes failed to improve maximal O_2 consumption significantly in children. However, studies using prolonged endurance exercise, such as cross-country running, have resulted in gains.

It appears that the training threshold may be higher in immature children. Prepubertal youngsters have a limited ability for cardiac hypertrophy and metabolic enzyme synthesis, both important factors in improving endurance performance. This may be a result of their low androgen levels. Androgens are potent stimulators of muscle hypertrophy, protein synthesis, and red blood cell production. These factors have far-reaching effects on O_2 transport and the aerobic production of ATP.

The pubertal increase in $\dot{V}_{O_2max}$ corresponds to the time of the greatest increase in height (Figure 30-9). At the same time, androgen secretion increases dramatically, which results in hypertrophy of cardiac muscle, stimulation of red cell and hemoglobin production, and the proliferation of metabolic enzymes. These maturationally induced changes make possible large induced increases in endurance, which are even greater if there is involvement in a vigorous endurance-training program.

The adolescent growth period has been called a critical time for the development of maximal aerobic power. The reasons for this are probably as much sociological as physiological. Adolescence provides the first opportunity for large increases in physical capacity. Those who take advantage of this may develop a superior capacity for exercise that may persist into adulthood. The notion of adolescence as a critical period for the development of physical ca-

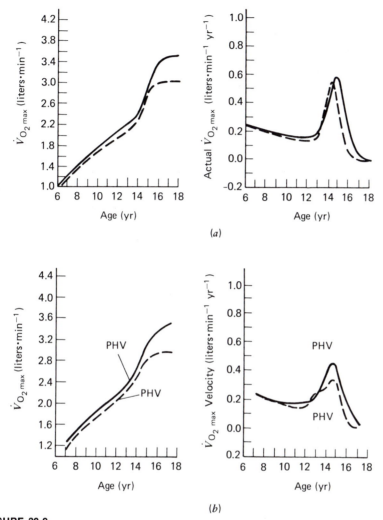

FIGURE 30-9
(a) $\dot{V}_{O_2max}$ with age in active (solid lines) and inactive (broken lines) children. (b) Same curves as in (a), aligned on peak height velocity (PHV).

(SOURCE: Modified from R. Mirwald, D.A. Bailey, N. Cameron, and R.L. Rasmussen, 1981.)

pacities is controversial. Whereas some researchers have presented promising evidence to support this hypothesis, others have claimed just the opposite: A temporary stagnation occurs in the effectiveness of training during the maximum growth spurt.

Successful participation in sport may result in a habit of exercise training that may be continued throughout life. However, the benefits of adolescent exercise are clearly not maintained unless training is continued. A sedentary

40-yr-old man will likely have a low capacity for exercise regardless of his activity level as an adolescent.

There has been a great deal of controversy about young children participating in prolonged exercise such as marathon running. It is not unusual to see 7- or 8-yr-old boys and girls running along with their parents in the "weekend warrior's" test of endurance and athletic prowess. There are three basic considerations: (1) What are the effects on the cardiorespiratory system and metabolism? (2) What are the effects on the bones and joints? and (3) What are the effects on the child's emotional well-being?

The available research indicates that children can tolerate prolonged exercise remarkably well. Heart rate responses to a given level of exercise are about the same as adults. In fact, they do not experience the same degree of cardiovascular drift (increase in heart rate with time at a constant level of exercise).

Prepubertal children have immature temperature-regulatory systems, which limits their ability to sweat. However, there is no evidence that their ability to dissipate heat is of extraordinary concern. In a study comparing rectal temperatures of children and adults entered in a 10,000-m race, the children had consistently lower temperatures. It is possible that the children did not exercise at the same intensity as the adults. Muscle and liver glycogen levels are lower than those in adolescents and adults. This may prevent them from running for extended periods at high percentages of their maximum capacities, which would increase their core temperatures.

Children have a large body surface area compared to their muscle mass. This gives them a very large area from which to dissipate heat. It is doubtful, however, that the same results would occur if the study had been repeated in extreme heat and humidity. There, conduction and convection are useless for heat dissipation. So, great care should be given to preventing heat injury under those circumstances.

Less is known about the possible skeletal damage that may result from extensive early endurance training. Prolonged regular exercise training in rats has been shown to induce epiphyseal growth plate injury and impaired growth. However, no such study has been done on humans. Because of the uncertainties and the possible tragic consequences, it is best to be extremely cautious.

The psychology of the child is beyond the scope of this book. Suffice it to say that marathon running goes beyond the realm of spontaneous play of elementary schoolchildren.

GENETIC CONSIDERATIONS

The question of the relative contribution of environment and genetics has been highly controversial for over 40 yr. Are champions made or born? If a child is

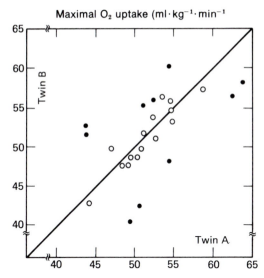

FIGURE 30-10
Intrapair values of V_{O_2max} for identical (white circles) and nonidentical (black circles) twin boys between 7 and 13 yr.

(SOURCE: Modified from V. Klissouras, 1976.)

subjected to the proper training at a young age, can he or she reach high levels of performance? Researchers are beginning to find the answers to these questions.

The determination of the genetic component of performance usually involves the comparison of identical and nonidentical twins. Because identical twins have the same genetic material, variability in performance will be due solely to environmental considerations. Variability in the nonidentical twins is due to a combination of genetic and environmental factors.

Most of these studies indicate that the genetic component is the most important factor determining individual differences in performance (Figure 30-10). In relatively sedentary populations, there are small if any differences in maximal O_2 consumption between identical twins, whereas the differences in nonidentical twins are often considerable. Klissouras (1976) cited two interesting case studies:

Case 1
In one case of non-identical twins, aged 21 years who lived apart since 16, one twin had trained strenuously for competitive middle distance running, whereas his brother had never participated in sports of any kind. It was therefore surprising to find that the untrained twin had a maximal oxygen uptake of 56.0 ml $\cdot$ kg^{-1} $\cdot$ min^{-1} as compared with a value of 52.8 for his trained counterpart. One cannot escape the inference that if it were not for the physical training the intrapair difference between this twin pair would have been greater.

Case 2
Two identical brothers, 40 years of age, had been separated at age 12 and had different lifestyles. More important, one twin had engaged in vigorous training for competitive basketball, whereas his brother was only moderately active during

the same period. For the last ten years neither of them had been involved in regular physical exertion. When tested, their maximal oxygen uptake was closely similar— the absolute values being 37.8 and 41.7 ml · kg^{-1} · min^{-1} for the trained and untrained twin, respectively.

Genetic studies are important for emphasizing the trained state as a temporary level of adaptation. Training studies with identical twins do demonstrate that the active twin improves while the sedentary twin does not. However, when the levels of activity are similar, heredity asserts itself and physiological characteristics in the genetically identical individuals move closer together.

The adaptability of the human body is limited. Maximal O_2 consumption, for example, can only be improved by about 20%. So, an athlete must start out with a high oxygen transport capacity if he is ever to reach Olympic levels of performance. Studies of the characteristics of athletes seem to reinforce this.

Many performance characteristics are determined by the relative percentages of fast- and slow-twitch muscle fibers. There is a high positive correlation between the percentage of slow-twitch fibers and maximal O_2 consumption. Studies of champion endurance athletes show a disproportionately greater amount of slow-twitch fibers in the muscle groups they use in their sport (Figure 30-11).

These same results apply to fast-twitch muscle fibers for sports requiring speed, such as sprinting. A high percentage of fast-twitch fibers is a prerequisite for performing fast muscle contractions. However, many world-class discus and shot putters have been shown to have a normal balance between fast- and slow-twitch fibers (approximately 50-50). So, in many sports, particularly those demanding exacting technique, training may overcome the lack of some genetic prerequisites.

Can champion athletes be selected at a young age? The top few athletes during childhood do tend to become top athletes in adolescence and adulthood. However, except for these, prediction of superior athletic prowess becomes tenuous at best. The best approach seems to be to provide a wide range of activities and provide extra help for those who show interest and talent. This approach has worked very well in many small countries, such as East Germany and Cuba. Although authoritarian techniques of athletic development are repugnant to many people in Western democracies, they are not necessarily harmful if the welfare of the individual child is always considered. Certainly we could do a better job of providing more athletic opportunities for the bulk of our children.

Summary

The growth process is similar in all people and must be considered when assessing exercise capacity and designing training programs. The growth curve follows an **S** shape, with rapid increases in height and weight occurring during

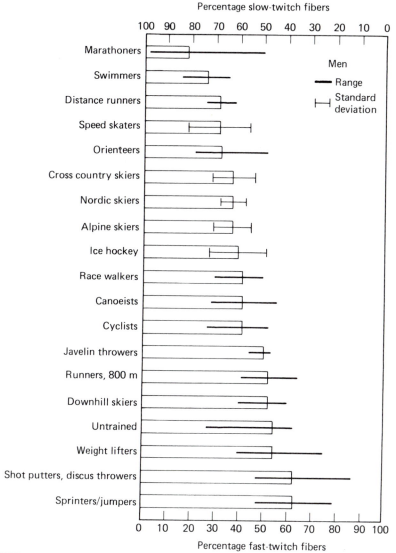

Percentage slow-twitch fibers

FIGURE 30-11

Percentage of fast- and slow-twitch muscle fibers in selected athletes.

(SOURCE: From E. Fox, 1979, p. 106.)

the first 2 yr, followed by a declining growth rate during childhood, and an accelerated rate during puberty.

During childhood the relative changes are similar in both sexes, although girls tend to be more mature than boys. The gradual growth rate during childhood is conducive to learning gross motor skills.

Adolescence is a period of rapid growth during which the child attains sexual maturity. Males, who attain puberty later than females, develop greater height and weight with larger musculature and broader shoulders, while females develop larger hips. Males experience large improvements in all areas of physical performance, while females tend to level off in most areas. The reasons for these sex differences in performance seem to be based as much in sociology as physiology.

During growth, bone formation exceeds absorption. Exercise appears to be important for optimal skeletal development. Trauma can cause injury to the bone growth centers, which could affect growth. Caution should be exercised when prescribing activities that are potentially dangerous to these bone growth centers.

The prevention of obesity is particularly important during periods of rapid growth, because fat cells are proliferating during these times. It may be that regular exercise and proper diet during growth may prevent the development of some fat cells and thus be a factor in lifelong weight control.

Adolescence has been called a critical period for the development of many types of physical abilities. Although this hypothesis is controversial, it is generally agreed that this is an important period of life for developing positive attitudes toward physical activity.

Selected Readings

Albinson, J.G. and G.M. Andrew. Child in sport and physical activity. Baltimore: University Park Press, 1976.

Åstrand, P-O. Human physical fitness with special reference to sex and age. Physiol. Rev. 36: 307–335, 1956.

Bar-or, O., R. Shepard, and C.L. Allen. Cardiac output of 10- to 13-year-old boys and girls during submaximal exercise. J. Appl. Physiol. 30: 219–223, 1971.

Bloom, W. and D.W. Fawcett. A Textbook of Histology. Philadelphia: W.B. Saunders Co., 1975.

Clarke, H.H. (ed.). Individual differences, their nature, extent and significance. Phys. Fitness Res. Digest. Series 4, No. 4, p. 1–23, 1973.

Dallman, P.R. and M.A. Siimes. Percentile curves for hemoglobin and red cell volume in infancy and childhood. J. Pediatr. 94: 26–31, 1979.

Daniels, J., N. Oldridge, F. Nagle, and B. White. Differences and changes in $\dot{V}_{O_{2}max}$ among young runners 10 to 18 years of age. Med. Sci. Sports. 10: 200–203, 1978.

Ekblom, B. Effect of physical training in adolescent boys. J. Appl. Physiol. 27: 350–355, 1969.

Eriksson, B. Muscle metabolism in children—a review. Acta Paediatr. Scand. 283 (Suppl.): 20–28, 1980.

Eriksson, B., I. Engstrom, P. Karlberg, A. Lundin, B. Saltin, and C. Thoren. Long-term effect of previous swim training in girls. A 10-year follow-up of the ''girl swimmers.'' Acta Paediatr. Scand. 67: 285–292, 1978.

Eriksson, O. and B. Saltin. Muscle metabolism during exercise in boys aged 11 to 16 years compared to adults. Acta Paediatr. Belg. 28(Suppl.): 257–265, 1974.

Fahey, T.D., A. DelValle-Zuris, G. Oehlsen, M. Trieb, and J. Seymour. Pubertal stage differences in hormonal and hematological responses to maximal exercise in males. J. Appl. Physiol. 46: 823–827, 1979.

Fox, E. Sports Physiology. Philadelphia: W.B. Saunders Co., 1979. p. 106.

Klissouras, V. Prediction of athletic performance: Genetic considerations. Can. J. Appl. Sport Sci. 1: 195–200, 1976.

Klissouras, V., F. Pirnay, and J. Petit. Adaptation to maximal effort: Genetics and age. J. Appl. Physiol. 35: 288–293, 1973.

Komi, P.V. and J. Karlsson. Physical performance, skeletal muscle enzyme activities, and fibre types in monozygous and dizygous twins of both sexes. Acta Physiol. Scand. (Suppl.) 462: 1–28, 1979.

Laird, A.K. Evolution of the human growth curve. Growth 31: 345–355, 1967.

Larson, L. (ed.). Fitness, Health, and Work Capacity. New York: Macmillan Co., 1974. p. 435–450, 516–524.

Macek, M., J. Vavra, and J. Novosadova. Prolonged exercise in prepubertal boys I. Cardiovascular and metabolic adjustment. Eur. J. Appl. Physiol. 35: 291–298, 1976.

Malina, R.M. Growth and development: The first twenty years in man. Minneapolis, Minn.: Burgess, 1975.

Malina, R.M. Growth, maturation, and human performance. In: Brooks, G.A. (ed.), Perspectives on the Academic Discipline of Physical Education. Champaign, Ill.: Human Kinetics, 1981. p. 190–210.

Meen, H.D. and S. Oseid. Physical activity in children and adolescents in relation to growth and development. Scand. J. Soc. Med. (Suppl.) 29: 121–134, 1982.

Mirwald, R., D.A. Bailey, N. Cameron, and R.L. Rasmussen. Longitudinal comparison of aerobic power in active and inactive boys aged 7 to 17 years. Ann. Hum. Biol. 8: 405–414, 1981.

Murase, Y., K. Kobayashi, S. Kamei, and H. Matsui. Longitudinal study of aerobic power in superior junior athletes. Med. Sci. Sports Exerc. 13: 180–184, 1981.

Parizkova, J. Longitudinal study of the development of body composition and body build in boys of various physical activity. Hum. Biol. 40: 212–223, 1968.

Rarick, G.L. The emergence of the study of human motor development. In: Brooks, G.A. (ed.), Perspectives on the Academic Discipline of Physical Education. Champaign, Ill.: Human Kinetics, 1981. p. 163–184.

Rarick, G.L. (ed.). Physical Activity: Human Growth and Development. New York: Academic Press, 1973.

Rutenfranz, J., K. Andersen, V. Seliger, F. Klimmer, I. Berndt, and M. Ruppel. Maximum aerobic power and body composition during the puberty growth period: Similarities and differences between children of two European countries. Eur. J. Pediatr. 136: 123–133, 1981.

Rutenfranz, J., K.L. Andersen, V. Seliger, J. Ilmarinen, F. Klimmer, H. Kylian, M. Rutenfranz, and M. Ruppel. Maximal aerobic power affected by maturation and body growth during childhood. Eur. J. Pediatr. 139: 106–112, 1982.

Sady, S.P., W.H. Thompson, M. Savage, and M. Petratis. The body composition and physical dimensions of 9- to 12-year-old experienced wrestlers. Med. Sci. Sports Exerc. 14: 244–248, 1982.

Scammon, R.E. The first seriatim study of human growth. Am. J. Phys. Anthropol. 10: 329–336, 1927.

Stern, J. and M. Greenwood. A review of development of adipose cellularity in man and animals. Fed. Proc. 33: 1952–1955, 1974.

Tanner, J.M. Growth at Adolescence. Oxford, England: Blackwell Publications, Ltd., 1962.

Timaris, P.S. Developmental Physiology and Aging. New York: Macmillan Co., 1972.

Vaccaro, P., C.W. Zauner, and W.F. Updyke. Resting and exercise respiratory function in well trained child swimmers. J. Sports Med. Phys. Fitness. 17: 297–306, 1977.

Weber, G., W. Kartodihardjo, and V. Klissouras. Growth and physical training with reference to heredity. J. Appl. Physiol. 40: 211–215, 1976.

31

AGING

It is extremely difficult to quantify fully the effects of aging on physiological function and physical performance. Individual outlooks and expectations can greatly affect life-style, which can appear either to accelerate or to delay the aging process. Some people consider themselves old by the age of 30, whereas others refuse to give up when they are well past their fifties and sixties (and, indeed, sometimes accomplish remarkable athletic performances).

Disease can further complicate the picture. Various systemic disorders, such as osteoarthritis and atherosclerosis, are so common in the aged that they are sometimes considered a normal part of the aging process. So, although it is readily apparent that age has a deteriorating effect on physiological capacity, it is extremely difficult to separate its effect from those of deconditioning and disease.

There has been considerable research in recent years on aging and physical activity. This topic is of particular concern to students of exercise physiology, because the aging adult population is steadily increasing, making contact with these individuals very likely. Because exercise is more hazardous in this population, it is important to know the limitations imposed by advancing age.

Proper exercise maintained throughout life maintains a more youthful physiology and allows for an active and vigorous life style. (Nancy Kaye/Leo de Wys.)

THE NATURE OF THE AGING PROCESS

In a recent review, Shepard characterized aging as the diminished capacity to regulate the internal environment (homeostasis), resulting in a reduced probability of survival. Simply stated, the physiological control mechanisms do not work as well in old people. Reaction time is slowed, resistance to disease is impaired, work capacity is diminished, recovery from effort is prolonged, and body structures are less resilient.

As a student of physiology, you will likely have noticed the large capacity and margin of safety in most of the body's biological systems. For example, in the heart the maintenance of cardiac output is possible even in the face of extreme vascular disease and drastically impaired cardiac function. Likewise, the pulmonary system has a large reserve capacity for ventilating the alveoli. Its function is limited only by extreme environmental conditions or disease. Unfortunately, the margin of safety in these systems diminishes with age (Table 31-1).

It appears that no matter how well people take care of themselves, their physiological processes will eventually fall prey to the ravages of old age. This fact is consistent across all animal species. Cultured bacteria, for example, can only reproduce a limited number of times. Animals such as cats, dogs, horses, and aardvarks have life spans that are characteristic of their species.

TABLE 31-1
Physiological Effects of Aging

Effect	Functional Significance
Cardiovascular	
Capillary/fiber ratio ↓	Decreased muscle blood flow
Cardiac muscle and heart volume ↓	Decreased maximal stroke volume and cardiac output
Elasticity of blood vessels ↓	Increased peripheral resistance, blood pressure, and cardiac afterload
Elasticity of blood vessels ↓	Increased peripheral resistance, blood pressure, and cardiac afterload
Myocardial myosin-ATPase ↓	Decreased myocardial contractility
Sympathetic stimulation of SA node ↓	Decreased maximum heart rate
Respiration	
Condition of elastic lung support structures ↓	Increased work of breathing
Elasticity of support structures ↓	Decreased lung elastic recoil
Size of alveoli ↑	Decreased diffusion capacity and increased dead space
Number of pulmonary capillaries ↓	Decreased ventilation/perfusion equality
Muscle and joints	
Muscle mass ↓	
Number of type II a and b fibers ↓	
Size of motor units ↓	
Action potential threshold ↓	Loss of strength and power
(Ca^{2+}, Myosin)-ATPase ↓	
Total protein and N_2 concentration ↓	
Size and number of mitochondria ↓	Decreased muscle respiratory capacity
Oxidative enzymes: SDH, cytochrome oxidase, and MDH ↓	Decreased muscle respiratory capacity
Lactate dehydrogenase ↓	Slows glycolysis
Stiffness of connective tissue in joints ↑	Decreased joint stability and mobility
Accumulated mechanical stress in joints ↑	Stiffness, loss of flexibility, and osteoarthritis

TABLE 31-1 (*continued*)

Effect	Functional Significance
Muscle and joints	
Water content in intervertebral cartilage ↓	Atrophy and increased chance of compression fractures in spine
Bone	
Bone minerals ↓	Osteoporosis—increased risk of fracture
Body composition and stature	
Body fat ↑	Impaired mobility and increased risk of disease
Kyphosis ↑	Loss of height

Obviously, all animals of a species do not live to be precisely the same age—there are individual differences. Likewise, the quality of life can be considerably different. Some people can be fit and alert at 100 yr, whereas others are invalids in their sixties.

Genetic considerations seem to be the most important factor dictating the length of life, whereas a combination of environmental and genetic factors govern the quality of life. The maximum life span in humans is slightly over 110 yr, a figure that has remained unchanged for 300 yr in spite of tremendous advances in public health. Genetic research reinforces this. Studies of identical and nonidentical twins show that the life spans of the former are remarkably similar (identical twins have identical genes). The identical twins usually died within 2 to 4 yr of each other, whereas the interval in the nonidentical twins averaged 7 to 9 yr.

Although little is known of the process, aging seems to be caused by aberrations in the genetic functions of cells. With time there is a progressive buildup of abnormal genetic material that gradually impairs the ability of the cells to reproduce and function normally. This process undoubtedly affects the cellular communications systems, which control metabolic processes such as protein synthesis. This can result in the formation of tissues that do not function as well, such as stiff and brittle cartilage and inactive enzymes. This process can also affect the immune systems.

The aging process also seems to represent an accumulation of insults and wear and tear that results in the gradual loss of the ability to adapt to stress. As we have seen, most physiological control mechanisms are highly adaptable.

Joints, for example, adapt to mobilization by maximizing their range of motion to the extent of the stress. If the joint is injured repeatedly over a lifetime, the joint capsule thickens, resulting in a gradual loss of range of motion. Likewise, microtrauma in the circulation leads to the proliferation of smooth muscle and the formation of plaque and calcification that eventually impairs the ability of the vessels to dilate and thus adequately supply the tissues with blood. The nature of these insults can be far-reaching and may include such things as environmental radiation, chemical additives in food, and viruses.

Although the maximum life span is finite, the quality of life is extremely variable. Exercise seems to be an important factor in maximizing physiological function throughout life. Research has shown that physical activity has significant effects on many biological activities. In fact, for some processes, it almost appears to "make time stand still."

THE AGING PROCESS AND THE EFFECTS OF EXERCISE

The effects of aging touch both cellular function and systemic regulation (Table 31-1). Peak physiological function, for the most part, occurs at about 30 yr of age. After that, most factors decline at a rate of about 0.75 to 1% a year.

The decline in physical capacity is characterized by a decrease in maximal O_2 consumption, maximal cardiac output, muscle strength and power, neural function, flexibility, and increased body fat. All of these factors can be positively affected by training. In fact, remarkable levels of performance are possible, particularly if physical training has been maintained throughout life.

Exercise training does not retard the aging process; it just allows the individual to perform at a higher level. Comparisons of trained versus sedentary individuals indicate a similar decrease in work capacity with age. Of course, the trained subjects achieved a higher level of performance at all ages. In some cases the decrease in performance is greater in the trained than untrained individual, indicating the difficulty of maintaining a high physical capacity with advancing age (Figure 31-1).

Cardiovascular Capacity

Maximal O_2 consumption decreases approximately 30% between the ages of 20 and 65. This occurs because of decreases in maximal heart rate, stroke volume, and arteriovenous O_2 difference. Thus, aging produces decreases in both O_2 transport and O_2 extraction capacities.

During submaximal exercise, heart rate is higher at any given work rate (Figure 31-2). Additionally, cardiovascular drift is greater in the aged. As discussed in Chapter 15, cardiovascular drift is the tendency of physiological factors such as heart rate, core temperature, and ventilation to rise at a constant

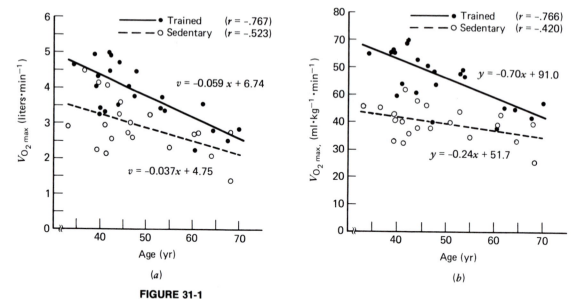

FIGURE 31-1

Effects of age on $\dot{V}_{O_2max}$ in trained and sedentary men. (a) $v = \dot{V}_{O_2max}$ l·min⁻¹. (b) $y = \dot{V}_{O_2max}$ ml·kg·min⁻¹; x = age.

(SOURCE: Adapted from H. Suominen, E. Heikkinen, T. Parkatti, S. Forsberg, and A. Kiiskinen, 1980.)

work rate. There is also a longer recovery rate for both submaximal and maximal exercise. The effects of aging on functional capacity are summarized in Table 31-2.

Aging impairs the heart's capacity to pump blood. There is a gradual loss of contractile strength caused, in part, by a decrease in $(Ca^{2+}$, myosin)-ATPase activity. The heart wall stiffens, which delays ventricular filling. This delay decreases maximum heart rate and perhaps stroke volume. The impairments in maximal heart rate and stroke volume result in substantial decreases in cardiac output.

Numerous cellular changes occur in the cardiovascular system that help explain its diminished capacity to transport gases and substrates (Table 31-1). There is a decreased elasticity in the major blood vessels and in the heart. This is usually accompanied by narrowing of the blood vessels in the muscles, heart, and other organs. Heart mass usually decreases, and there are sometimes fibrotic changes in the heart valves. The fiber/capillary ratio is reduced, which impairs the blood flow to the muscles. Also, there is a deterioration of the venous valves. These serve to keep blood flowing toward the heart.

These changes have several consequences for cardiovascular performance. Vascular stiffness increases the peripheral resistance to blood flow, which in-

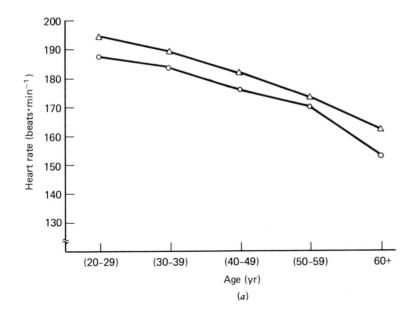

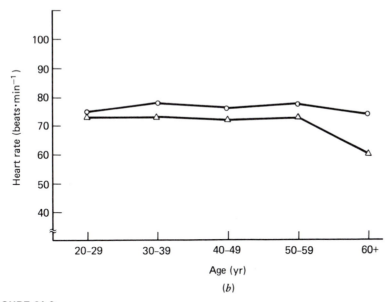

FIGURE 31-2
50th Percentile of maximal (a) and resting (b) heart rate with age. Triangles, males; circles, females.

(SOURCE: Data from M. Pollock, J.H. Wilmore, and S.M. Fox, 1978.)

TABLE 31-2

Changes in Functional Capacity and Body Composition with Age in Males

Effect	Age (yr) 20	Age (yr) 60
$\dot{V}_{O_2 max}$ (ml · kg^{-1} · min^{-1})	39	29
Maximal heart rate (beats · min^{-1})	194	162
Resting heart rate (beats · min^{-1})	63	62
Maximal stroke volume (ml)	115	100
Maximal $(a-v)$ O_2 difference (ml · liter^{-1})	150	140
Maximal cardiac output (liters · min^{-1})	22	16
Resting systolic blood pressure (mmHg)	121	131
Resting diastolic blood pressure (mmHg)	80	81
Total lung capacity (liters)	6.7	6.5
Vital capacity (liters)	5.1	4.4
Residual lung volume (liters)	1.5	2.0
Percentage fat	20.1	22.3

creases the afterload of the heart. This forces the heart to work harder to push blood into the circulation and increases myocardial O_2 consumption at a given intensity of effort. Cardiac hypoxia can ensue because of atherosclerotic changes in the coronary arteries. The higher peripheral resistance also raises systolic blood pressure at rest and during maximal exercise. However, there is little or no increase in diastolic blood pressure.

Arteriovenous O_2 difference decreases, which contributes to the diminished aerobic capacity. This occurs because of a reduction in the fiber/capillary ratio, total hemoglobin, and the respiratory capacity of muscle. Muscle mitochondrial mass decreases along with decreases in several oxidative enzymes.

There also seems to be a diminished capacity of autonomic reflexes that control blood flow. At rest, circulation to the skin is often poor, which can make peripheral body parts uncomfortably cold. During physical activity, a disproportionate amount of blood is directed to the skin, which can further hamper O_2 extraction. Additionally, the elderly are more subject to orthostatic intolerance; they sometimes have difficulty maintaining blood pressure when going from a horizontal to vertical posture.

Endurance training induces improvements in aerobic capacity in the aged that are similar to those in young people (in relative terms). Most studies indicate that gains of about 20% in maximal O_2 consumption can be expected in a 6-month endurance exercise program. As in the young, individual differences can be accounted for by initial fitness, motivation, intensity and duration of the program, and genetic characteristics.

Endurance exercise results in a decrease in submaximal heart rate at a given work load, decreased resting and exercise systolic blood pressure, and faster recovery heart rate. Abnormal electrocardiographic findings, such as ST-segment depression, have also been shown to improve. Although the reason for this is unclear, it could be from reduced O_2 consumption in the heart, or perhaps the development of coronary collateral circulation.

Pulmonary Function

The normal lung enjoys a large reserve capacity that can meet ventilatory requirements even during maximal exercise. The reserve capacity begins to deteriorate gradually between 30 and 60 yr of age, with an acceleration after that. This process may be faster if the individual is a smoker or is chronically subjected to significant amounts of airborne contaminants.

The three most important changes that occur in this system with aging are a gradual increase in the size of the alveoli, the disintegration of the elastic support structure of the lungs, and a weakening of the respiratory muscle. These changes can interfere with the ventilation and perfusion of the lung, both of which can impair the O_2 transport capacity.

Enlargement of alveoli also occurs in chronic obstructive pulmonary disease. Thus, it is difficult to separate the effects of disease from those of aging. Alveolar enlargement is accompanied by a decrease in pulmonary vascularization. Both of these changes decrease the effective area available for diffusion.

The loss of pulmonary elasticity and the weakening of respiratory muscles can have a marked effect during exercise. Both of these changes make expiration more difficult and increase the work (O_2 cost) of breathing. The loss of elastic recoil results in premature closing of airways, which impairs ventilation in some of the alveoli. Because of the restrictions of flow, there is an increased dependence on breathing frequency, rather than tidal volume, for ventilation during increasing intensities of exercise.

The deterioration of pulmonary function is similar in magnitude to that in the cardiovascular system. So, unless the decline in pulmonary function is accentuated by disease, ventilation remains adequate during exercise in the aged. Ventilation does not seem to limit endurance performance in the young or old. Rather, cardiac output is probably the most important limiting factor.

Training will increase maximum ventilation, but the improvements parallel those of cardiac output. Although the breathing muscles can be strengthened through exercise, most of the changes are irreversible. However, because of the large reserve capacity of the lungs, the changes can be tolerated quite well.

Skeletal System

Bone loss is a serious problem in older people, particularly in women. Women begin to lose bone mineral at about 30 and men at about 50 yr of age. Bone loss, called osteoporosis, results in bone with less density and tensile strength.

Osteoporosis increases the risk of fracture, which drastically increases the short-term mortality rate. One study found a 50% death rate within 1 yr after hip fractures in elderly women.

Although the cause of bone loss in the aged is not completely understood, it seems to be related to a combination of factors including inactivity, diet, skeletal blood flow, and endocrine function. These factors may induce a negative calcium balance that steadily saps the bones of this important mineral.

Exercise has been shown to be important in the prevention and treatment of osteoporosis. Bones, like other tissues, adapt to stresses placed on them. They become stronger when stressed, and weaker when not stressed. Even extremely fit astronauts show some bone loss when their bones are not subjected to the stresses of gravity. Studies have indicated that bone mineral content can be improved by exercise, even in people over 80 yr of age.

Joints

Joints become less stable and mobile with age. Aging is often associated with degradation of collagen fibers, fibrous synovial membranes, joint surface deterioration, and decreased viscosity of synovial fluid. Joint stiffness and loss of flexibility are common in the elderly. In fact, some researchers feel that osteoarthritis is a natural result of the aging process.

It is difficult to separate aging from accumulated wear and tear. Trauma to joint cartilage results in the formation of scar tissue, characterized by the buildup of fibrous material that makes the connective tissue stiffer and less responsive to stress. This can result in a thickened joint capsule that often contains debris, which impairs range of motion. Osteoarthritis in the elderly is principally located in areas receiving the most mechanical stress. So, it is unclear if the restricted range of motion found in the joints is principally the result of aging or repeated trauma.

Range-of-motion exercises have been shown to increase flexibility dramatically in a relatively short time. One study by Munn showed increases in flexibility ranging from 8% in the shoulder to 48% in the ankle in only 12 weeks of stretching and dance exercises. However, it is improbable that exercise can significantly undo damage to joints that have undergone extensive degeneration.

Skeletal Muscle

Marked deterioration in muscle mass occurs with aging. It is characterized by decreases in the size and number of muscle fibers, decreases in the muscles' respiratory capacity, and increases in connective tissue and fat. These changes can have severe consequences in the elderly: Mobility may be hampered, incidents of soft tissue pain are more frequent, and work capacity is impaired. Aging results in decreases in isometric and dynamic strength, and speed of movement.

There is a loss of fibers from individual motor units. This results in less available contractile force when a motor unit is recruited. Additionally, the nature of the motor units changes: There is a selective loss of type II fibers (fast-twitch muscle), which diminishes available strength and power. However, in elderly endurance athletes there is a higher percentage of type I (slow-twitch) muscle, which reflects the metabolic requirement of endurance exercise.

The loss of the muscle's biochemical capacity is characterized by decreases in succinate dehydrogenase (SDH), pyruvate and malate dehydrogenase, and cytochrome oxidase. In very old age, there is evidence of the formation of incomplete or inactive enzymes. Some researchers have found decreases in the mitochondrial mass. All of these changes will affect ATP production and thus impair physical working capacity.

The mechanisms involved in muscle contraction are also impaired, which contributes to the loss of strength and power. Aging muscle is less excitable and has a greater refractory period. Thus a greater stimulus is needed for contraction, and a longer period of time is required before the muscle can respond to another stimulus. Myosin-ATPase activity, ATP, and CP are also reduced, particularly in fast-twitch muscle, which further impairs muscle function.

Relative strength changes from training are similar in the young and old, at least in short-term programs. Because of the difficulty conducting such research, available data are restricted to observations of a few months. These studies indicate a fundamental difference in way the elderly increase strength. Young men increase strength primarily by hypertrophy, whereas old people (and perhaps young women) increase strength by increased neural stimulation. Thus, the young improve the contractile capacity of the fibers, whereas the elderly rely on improved motor unit recruitment (Figure 31-3). The diminished capacity for hypertrophy in older males may be related to decreases in testosterone.

It is extremely doubtful that the parity in strength gains between young and old would continue for more than a few months. Because the elderly have a limited ability for muscle hypertrophy, they would probably quickly reach their maximum strength potential because of the finite number of motor units available for recruitment.

Body Composition and Stature

There are marked changes in body composition and stature with age. Body weight increases steadily beginning in the twenties and continues to increase until about 55 or 60 yr, when it begins to decline. Height decreases gradually. Weight gain is accompanied by increased body fat and decreased lean body mass. Males advance from an average of 15% fat at 17 yr to about 28% at 60. Women change from about 25% at 17 yr to about 39% at 60 yr. These values are highly variable and tend to be much less in the trained individual.

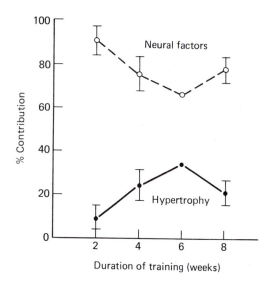

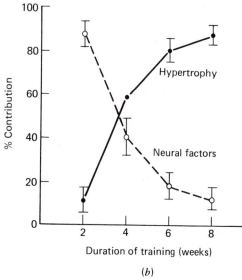

(b)

FIGURE 31-3
Relative contributions of neural factors and hypertrophy toward strength gains in old (a) and young (b) male subjects during an 8-week strength-training program.

(SOURCE: Adapted from T. Moritani and H.A. De Vries, 1980.)

Table 31-3 includes longitudinal body composition measurements made on Dr. Albert Behnke, who is widely regarded as the father of body composition. While Dr. Behnke has remained physically active all of his life, his lean body weight has tended to decrease and his percent fat has tended to increase with age. Of particular interest are the measurements made during periods of weight loss (3-12-40 to 10-9-40, 12-20-55 to 10-16-56, and 1967 to 1971). Note that Dr. Behnke seemed to be able to lose fat and maintain lean body mass much easier when he was younger. While it is difficult to draw any conclusions from

TABLE 31-3

Serial Body Composition Measurements on Dr. A. R. Behnke

Year	Age (yr)	Weight (kg)	Density (g/cm³)	Lean Body Weight (kg)	Percent Fat
1940(3/12)	36	92.0	1.056(SG)	72.7	21.0
1940(7/1)	36	88.4	1.060(SG)	71.6	19.0
1940(8/13)	37	85.0	1.066(SG)	71.3	16.1
1940(10/9)	37	83.2	1.071(SG)	71.9	13.6
1944(5/2)	40	90.7	1.056(SG)	71.7	21.0
1949(—)	46	91.6	1.054(SG)	69.5	24.1
1953(—)	50	95.5	1.044(SG)	69.7	27.0
1953(12/9)	50	95.5	1.045	69.0	27.8
1955(12/20)	52	101.5	1.035	69.9	31.1
1956(3/30)	52	86.3	1.047	66.6	22.8
1956(6/29)	52	84.4	1.053	66.4	21.3
1956(10/16)	53	86.5	1.058	67.5	22.0
1956(11/12) [a]	53	83.0	1.058	64.6	22.2
1967(—)	64	95.6	1.032	67.2	29.7
1970(—)	67	87.9	1.043	66.3	24.6
1971(—)	68	84.3	1.049	65.8	21.9
1971(—)	68	81.8	1.047	65.9	19.4
1973(4/6)	69	87.9	1.040	65.0	26.1
1973(10/31)	70	89.2	1.041	66.5	25.4
1977(3/8)	73	83.2	1.042	62.4	25.0
1977(—)	74	86.2	1.038	63.7	26.1

[a] Measured after week on Pemmican, 1000 kcal/day diet (45% fat, protein, 10% carbohydrate).

(SG) = Specific gravity.

SOURCE: Data courtesy of Dr. A. R. Behnke.

these data, they certainly conform to cross-sectional studies of body composition in older adults.

The distribution of fat tends to change in aging. There is a tendency for a greater proportion of the total body fat to be located internally rather than subcutaneously. Therefore, if body composition is measured by a skinfold technique, an equation developed using elderly subjects must be employed.

Increased body fat with age is of concern because of its possible relationship with disease and premature mortality. Exercise is extremely important in managing body composition in the elderly. Metabolic rate slows with age, which necessitates a low-calorie diet in order to maintain a normal body composition. This diet is often low in the necessary vitamins and minerals. Regular exercise enables the elderly to consume more food and calories, which allows them to satisfy their nutritional requirements.

Stature decreases with age because of an increased kyphosis, or rounding of the back, compression of intravertebral disks, and deterioration of vertebrae. A cross-sectional study by Durin and Womersley showed a 6-cm decrease in height between 17- and 60-yr-olds. This, of course, may represent modern trends of increased height. However, loss of height with age is well documented.

Neural Function

Many neurophysiological changes occur with aging. Again, it is often difficult to separate the changes resulting from aging from those caused by disease. The principal changes include decreased visual acuity, hearing loss, deterioration of short-term memory, inability to handle several pieces of information simultaneously, and decreased reaction time.

Physical training seems to have little effect on the deterioration of neural function. A study by Suominen and co-workers could find no difference in neurobiological factors between extremely fit elderly endurance athletes and sedentary men. They concluded that the effects of endurance training in the elderly are largely limited to functions that are apparently relevant to physical performance. In other words, exercise training will improve performance in the elderly, but it will not affect the aging process itself.

EXERCISE PRESCRIPTION FOR THE ELDERLY

The principles of exercise prescription (see Chapter 24) apply to people of all ages. However, because of the increased risk of exercise for the elderly, caution is required.

Electrocardiographic abnormalities, particularly during exercise, are regular findings in the elderly. Aberrations such as ST-segment depression, heart blocks, and dysrhythmias are common enough to warrant routine exercise stress testing for any elderly person wishing to undertake an exercise program.

Great care must be taken when determining the type and intensity of exercise. In a sedentary person the exercise should be one that minimizes soft tissue injuries. Good choices are walking and swimming. More vigorous exercise, such as running and racquetball, should only be attempted when the person has achieved sufficient fitness.

Ideally, the intensity of exercise should coincide with the target heart rate. As discussed in Chapter 24, this is calculated from the maximal heart rate, which is often calculated from the formula 220 − age. However, this is extremely misleading in the elderly and can lead to problems with their exercise prescription. Although it is fairly accurate for estimating maximal heart rate at the 50th percentile of the population, it does not take into consideration the tremendous variability observed in the older age group. Maximal heart rates in people over 60 yr can range from highs of over 200 to lows of 105 beats $\cdot$ min^{-1}. Thus, target heart rates predicted on the basis of age can often either underestimate or overestimate the ideal exercise intensity. Whenever possible, use an accurate measurement of maximal heart rate rather than a predicted one.

Very often older people have misperceptions about physical conditioning. They sometimes overestimate their stamina, and their exercise program consists of extremely low intensity activities such as gardening. In some elderly persons, it is important to increase the frequency of exercise until sufficient fitness is developed to increase the intensity as well.

De Vries has suggested several considerations for exercise prescription in the elderly.

- The exercise program should follow a careful progression in intensity and duration.
- Slow and careful warm-up should be particularly emphasized in activities that may suddenly overload local muscle groups (such as handball and racquetball).
- Cool-down must be done slowly. Before showering, elderly persons should continue very light exercise, such as walking or unloaded pedaling, until heart rate is below 100 beats $\cdot$ min^{-1}.
- Static stretching is helpful in preventing soft tissue pain that is often associated with exercise training in older persons.

Summary

The effects of aging are difficult to separate from those of degenerative diseases and deconditioning. However, it is clear that aging takes its toll on almost every facet of physiological function. Although the life span of any species is identifiable within relatively narrow limits, the quality of life can be extremely variable. It is clear that regular exercise training can improve the quality of life by increasing, or at least slowing deterioration in, physical capacity.

After 30 yr of age, most physiological functions decline at a rate of approximately 0.75 to 1% a year. The decline in physical capacity is characterized by a decrease in maximal O_2 consumption, maximal cardiac output, muscle strength and power, neural function, flexibility, and increased body fat.

Exercise training does not retard the aging process; it merely allows the individual to perform at a higher level. Although older individuals cannot expect to reach the same absolute capacity as the young, relatively they can improve about the same amount.

Care should be taken when prescribing exercise to older people because of possible complications from degenerative diseases such as atherosclerosis, arthritis, and osteoporosis. Physical capacity should be carefully measured and considered. The duration and intensity of training should be mild at the beginning of the program and then progress slowly.

Selected Readings

Adams, G. and H. DeVries. Physiological effects of an exercise training regimen upon women aged 52–79. J. Gerontol. 28: 50–55, 1973.

Aloia, J.F. Exercise and skeletal health. J. Am. Geriatr. Soc. 19: 104–107, 1981.

Barnard, J., G. Grimditch, and J. Wilmore. Physiological characteristics of sprint and endurance runners. Med. Sci. Sports. 11: 167–171, 1979.

Bortz, W.M. Effect of exercise on aging—effect of aging on exercise. A. Am. Geriatr. Soc. 28: 49–51, 1980.

Clarke, H.H. (ed.). Exercise and aging. Phys. Fitness Res. Digest. Series 7, No. 2: 1–25, April 1977.

Coquelin, A. and C. Desjardins. Luteinizing hormone and testosterone secretion in young and old male mice. Am. J. Physiol. 243(E): 257–263, 1982.

Crilly, R., A. Horsmann, D. Marshall, and B. Nordin. Prevalence, pathogenesis and treatment of post-menopausal osteoporosis. Aust. N.Z. J. Med. 9: 24–30, 1979.

Dehn, M.M. and R.A. Bruce. Longitudinal variations in maximal oxygen uptake with age and activity. J. Appl. Physiol. 33: 805–807, 1972.

De Vries, H. Physiological effects of an exercise training program regimen upon men aged 52 to 88. J. Gerontol. 25: 325–336, 1970.

De Vries, H. Tips on prescribing exercise regimens for your older patient. Geriatrics 35: 75–81, 1979.

Durin, J. and J. Womersley. Body fat assessed from total body density and its estimation from skinfold thicknesses: Measurement on 481 men and women aged from 16 to 72 years. Br. J. Nutr. 32: 77–97, 1974.

Frolkis, V.V., O.A. Martynenko, and V.P. Zamostyan. Aging of the neuromuscular apparatus. Gerontology. 22: 244–279, 1976.

Gordon, T., W.B. Kannel, M.C. Hjortland, and P.M. McNamara. Menopause and coronary heart disease, The Framingham study. Ann. Intern. Med. 89: 157–161, 1978.

Hanzlikova, V. and E. Gutmann. Ultrastructure changes in senile muscle. Adv. Exp. Med. Biol. 53: 421–429, 1975.

Kavanagh, T. and R.J. Shephard. The effects of continued training on the aging process. Ann N.Y. Acad. Sci. 301: 656–670, 1977.

Moritani, T. and H.W. De Vries. Potential for gross muscle hypertrophy in older men. J. Gerontol. 35: 672–682, 1980.

Munn, K. Effects of exercise on the range of motion in elderly subjects. In: Smith, E. and R. Serfass (eds.), Exercise and Aging: The Scientific Basis. Hillside, N.J.: Enslow Publishers, 1981. p. 167–186.

Niinimaa, V. and R.J. Shephard. Training and oxygen conductance in the elderly. J. Gerontol. 33: 354–367, 1978.

Nilsson, S. Medical examination of and advice to middle-aged persons starting physical training. Scand. J. Soc. Med. 29(Suppl.): 161–169, 1982.

Orlander, J. and A. Aniansson. Effects of physical training on skeletal muscle metabolism and ultrastructure. Acta Physiol. Scand. 109: 149–154, 1980.

Pollock, M.W., J.H. Wilmore, and S.M. Fox, Health and Fitness Through Physical Activity. New York: John Wiley & Sons, 1978.

Saltin, B. and G. Grimby. Physiological analysis of middle-aged and former athletes. Circulation. 38: 1104–1115, 1968.

Saltin, B., L. Hartley, A. Kilbom, and I. Astrand. Physical training in sedentary middle-aged and older men. Scand. J. Clin. Lab. Invest. 24: 323–334, 1969.

Shepard, R. Physical Activity and Aging. Chicago: Year Book Publishers, 1978.

Sidney, K.H. and R.J. Shephard. Activity patterns of elderly men and women. J. Gerontol. 32: 25–32, 1977.

Sidney, K. and R.J. Shepard. Frequency and intensity of exercise training for elderly subjects. Med. Sci. Sports. 10: 125–131, 1978.

Smith, E. and R. Serfass (eds.). Exercise and Aging. Hillside, N.J.: Enslow Publishers, 1981.

Stamford, B.A. Physiological effects of training upon institutionalized geriatric men. J. Gerontol. 27: 451–455, 1972.

Stromme, S.M., H. Frey, O.K. Harlem, O. Stokke, O.D. Vellar, L.E. Aaro, and J.E. Johnsen. Physical activity and health. Scand. J. Soc. Med. 29(Suppl.): 9–36, 1982.

Suominen, H., E. Heikkinen, T. Parkatti, S. Forsberg, and A. Kiiskinen. Effects of lifelong physical training on functional aging in men. Scand. J. Soc. Med. 14(Suppl.): 225–240, 1980.

Timaris, P.S. Developmental Physiology and Aging. New York: MacMillan Co., 1972. p. 408–615.

Waaler, H.T. and P.F. Hjort. Physical activity, health, and health economics. Scand. J. Soc. Med. 29: 265–269, 1982.

Wilson, P.D. and L.M. Franks. The effects of age on mitochondria ultrastructure and enzymes. Adv. Exp. Med. Biol. 53: 171–183, 1975.

32

FATIGUE DURING MUSCULAR EXERCISE

Muscular fatigue is usually defined as the inability to maintain a given exercise intensity. As we will see, there is no one cause of fatigue. Fatigue is task specific, and its causes are multifocal. Fatigue during muscular exercise can often be due to impairment within the active muscles themselves, in which case muscular fatigue is peripheral to the CNS and is due to muscle fatigue. Muscular fatigue can also be due to more diffuse, or central factors. For example, for psychological reasons an athlete may be unable to bring his or her full muscle power to bear in performing an activity. Alternatively, environmental factors such as hot, humid conditions may precipitate a whole series of physiological responses that detract from performance. In such cases, the cause of muscular fatigue resides outside the muscles.

The cause of fatigue varies with the nature of the activity, the training and physiological status of the individual, and the environmental conditions. Fatigue can be due specifically to depletion of key metabolites in muscle or to the accumulation of other metabolites, which can affect the intracellular environment and also spill out into the circulation and affect the general homeostasis.

The study of fatigue in exercise has occupied the attention of many of the best biological scientists. Identification of a fatigue cause is not a simple matter, as it is often difficult to separate causality from concurrent appearance.

Walker holds off Van Dam for the Montreal Olympic gold medal. Fatigue and elation expressed simultaneously in the competitors. (Alain Dejean/Sygma.)

IDENTIFICATION OF FATIGUE

Before the cause and site of fatigue can be identified, the definition of fatigue must be decided on. Usually, in athletic competition, fatigue means the inability to maintain a given exercise intensity. For instance, in distance running or swimming, when the competitor is unable to keep the pace, he or she is considered to have fatigued for the particular event. However, the athlete is infrequently completely fatigued and can usually maintain a lesser power output for some time. In a few instances, such as during wilderness hiking, individuals will push themselves until they are completely unable to move. In such cases, death can result from exposure.

Sometimes the cause of a decrement in work performance (fatigue) can be identified as to its specific cause and site. For instance, the depletion of a particular metabolite in a particular fiber type within a specific muscle may be identified. At other times, as in dehydration, the causes of fatigue are diffuse and involve several factors that are contributory to a disturbance in homeostasis.

At particular times it is much easier to identify a factor whose presence is correlated with the onset of fatigue than it is to determine that the presence (or absence) of a factor and fatigue are causally related. For instance, heavy muscular exercise has long been observed to be associated with lactic acid accumulation. Naturally, lactic acid was assumed to be the cause of fatigue. However, whenever lactate is high, other metabolites (glycogen, ATP, CP) can be low. Is fatigue the result of lactate accumulation and CP depletion, or is there some other factor that causes lactate and CP levels to change? As we will see, many of the data relating lactic acid accumulation to the onset of fatigue are often circumstantial.

Compartmentalization in physiological organization, involving the division of the body into various systems, organs, tissues, and cells, and the subdivisions of cells into various subfractions and organelles, can make it difficult to identify the fatigue site. For instance, ATP may be depleted at a particular site within a cell (e.g., on the head of myosin cross-bridges) but may elsewhere be adequate. In such a case, it would be extremely difficult to identify ATP depletion as the cause of fatigue, even if a muscle biopsy were performed or NMR technique used. Compartmentalization can mask the site of fatigue.

The effect of exercise at given absolute or relative values of $\dot{V}_{O_2max}$ can be more severe on untrained than on trained individuals. For instance, fatigue will occur sooner in an untrained person exercising at 75% of $\dot{V}_{O_2max}$ than in an endurance-trained individual exercising at the same work rate, or at a higher work rate that elicits 75% of the trained individual's $\dot{V}_{O_2max}$.

It is well known that environment can affect exercise endurance. For example, endurance is reduced during exercise in the heat. This reduced endurance is due to the redistribution of cardiac output from contracting muscles and hepatic areas, to include greater cutaneous circulation as well. High muscle temperatures can also loosen the coupling between oxidation and phosphorylation in mitochondria. In this case, $\dot{V}_{O_2}$ is unchanged or increased, but ATP production is decreased. To the extent that the exercise is submaximal and there exists a reserve for cardiac output expansion, exercise under conditions of heat can be continued. The stress level will, however, be greater. If the need to circulate exercising muscles, cutaneous areas, and other essential areas exceeds the cardiac output, then endurance will be reduced.

During exercise in hot environments, the rate of sweat loss is increased and body heat is gained over time. Severe sweating results in dehydration and shifts both fluid and electrolytes among body compartments. These shifts, as well as the increased body temperature, represent direct irritants to the CNS that can additionally affect an individual's subjective perception of the exercise.

Physiological status of the individual can easily affect exercise tolerance. For instance, if an individual exercises to fatigue in the heat on Monday, his or her ability to repeat that performance on Monday, Tuesday, or perhaps even Wednesday may be impaired. Similarly, individuals may have less endurance when they are glycogen depleted than when glycogen levels are normal or supercompensated.

METABOLITE DEPLETION

ATP and CP

Adenosine triphosphate (ATP) is the high-energy intermediate that is utilized to support muscular contraction as well as most other cellular endergonic processes. The immediate source of ATP rephosphorylation is creatine phosphate (CP) (Chapter 4). Because the catalyzing enzyme, creatine kinase, functions so rapidly, the muscular concentration of ATP is little affected until the CP level is significantly depleted.

Endurance and its converse, fatigue, can be likened to a competition between two opposing forces: those that utilize ATP and those that restore it. Because the quantities of ATP and CP on hand in a resting muscle are fairly small, any significant utilization must be immediately matched with an equivalent restoration. If the rates of ATP and CP restoration are even a little less than utilization, exercise cannot continue very long.

Biopsies of human quadriceps muscle during cycling exercise have demonstrated that the CP level in muscle declines in two phases (Figure 32-1a). When exercise starts, the CP level first drops rapidly, and then it drops slowly. Both the severity of the initial drop (Figure 32-1a), and the extent of the final drop (Figure 32-2) appear to be related to the relative work intensity for the subject. In general, the greater the relative work load, the greater will be the CP depletion. In subjects willing to push themselves to their limit, the point of fatigue in isometric exercise and in supermaximal cycling exercise coincides with CP depletion in muscle. Detailed studies on isolated muscles stimulated to contract indicate that tension development is related to CP level. Thus it is clear that CP depletion leads to muscle fatigue.

Although ATP level declines somewhat when exercise starts (Figure 32-1b), ATP level is apparently well maintained (largely at the expense of CP) until the level of CP is greatly reduced. Therefore, at the fatigue point, both CP and ATP become depleted.

Glycogen

Glycogen depletion in skeletal muscle is associated with fatigue during prolonged submaximal exercise (approximately 75% of $\dot{V}_{O_2max}$) to exhaustion

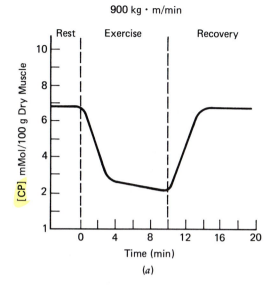

900 kg · m/min

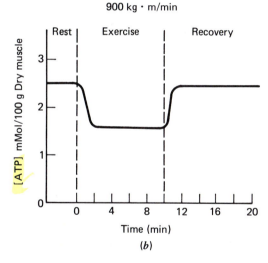

900 kg · m/min

FIGURE 32-1
Creatine phosphate [CP] and adenosine triphosphate levels [ATP] during steady-rate exercise. When exercise starts, both [CP] (a) and [ATP] (b) decline. Although [CP] may continue to decline if the exercise intensity is difficult for the subject, ATP levels will be well maintained until CP is exhausted. In recovery, both CP and ATP levels recover rapidly.

(**SOURCE:** Modified from Bergström, 1967.)

(Chapter 27). When the pedaling rate is moderate (i.e., 60 to 70 revolutions · min^{-1}), glycogen appears to be depleted uniformly from the different fiber types. However, according to the work of Gollnick and associates, when subjects perform at a given work rate, rapid cycling at 100 revolutions · min^{-1} (low resistance) results in selective recruitment and depletion of glycogen from slow-twitch fibers (Chapter 17). Maintaining the same work rate at a slow high-force pedaling frequency (i.e., 50 revolutions · min^{-1}) will result in recruitment of fast-twitch fibers. Thus it is possible for an athlete to exercise to exhaustion and fatigue because of glycogen depletion from specific muscle fibers, while glycogen remains in adjacent fibers within the tissue.

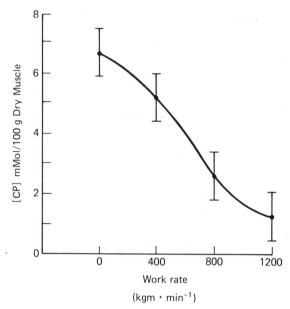

FIGURE 32-2

Creatine phosphate levels ([CP]) in quadriceps muscles after five minutes of cycling. A strong negative relationship is indicated between work rate and CP levels. Extremely hard exercise can lead to CP and ATP depletion, a situation that causes fatigue.

(SOURCE: Modified from Bergström, 1967.)

Blood Glucose

During short, intense exercise bouts, blood glucose rises above preexercise levels because of an autonomic nervous system stimulation of hepatic glycogenolysis. The ability of the liver to maintain a high rate of glucose release over time is limited by the amount of glycogen stored and by the activities of hepatic glycogenolytic and gluconeogenic enzymes. During prolonged exercise, glucose production may be limited to gluconeogenesis because of hepatic glycogen depletion; thus glucose production may fall below that required by working muscle and other essential tissues such as the brain. In this case of falling blood glucose, the exercise becomes subjectively more difficult because of CNS starvation and difficulty in oxidizing fats in muscle due to the absence of anaplerotic substrates (Chapter 27).

METABOLITE ACCUMULATION

Not long ago the story of muscle fatigue was easy for scientists, textbook authors, and students to explain. The explanation went like this: When exercise was too difficult, an athlete went into "O_2 deficit." The athlete then built up lactic acid, which caused fatigue. During recovery the "O_2 debt" was repaid, and lactate was reconverted to glycogen. Unfortunately, the lactic acid explanation is *not* now universally acceptable as an explanation of either the "O_2 debt" or fatigue (Chapter 10). Inspection of most of the data implicating lactic acid and fatigue reveals that the relationship is circumstantial at best. Certainly during prolonged exercise, glucose, glycogen, and lactate levels are low. In this situation, an injected dose of lactate might actually enhance performance.

Lactic Acid

During short-term, high-intensity exercise, lactate accumulates as the result of greater lactic acid production than removal. At a physiological pH, lactic acid, a strong organic acid, dissociates a proton (H^+). It is the H^+ rather than the lactate ion that causes pH to decrease. The H^+ formation as the result of lactic acid production can have several negative effects. Within the muscle, the lower pH may inhibit phosphofructokinase (PFK) and slow glycolysis. Additionally, H^+ may act to displace Ca^{2+} from troponin, thereby interfering with muscle contraction. Further, the low pH may stimulate pain receptors.

Hydrogen ion liberated into the blood and reacting in the brain will cause severe side effects, including pain, nausea, and disorientation. Within the blood itself, H^+ will inhibit the combination of O_2 with hemoglobin in the lung. Some species will actually run themselves to the point where O_2 delivery is reduced by lactic acid formation and the blocking of oxyhemoglobin formation. High circulating H^+ levels will also inhibit hormone-sensitive lipase activity in adipose tissue, thereby limiting the release of free fatty acids (FFA) into the circulation. Fat oxidation in muscle is directly dependent on circulating FFA levels.

As debilitating as high levels of H^+ from lactic acid dissociation may be in the muscles and blood, it is uncertain whether the pH decrement actually stops exercise. Because of a muscle's gross and microanatomy, muscle biopsies actually yield little information on the pH at critical sites of metabolism. Many active sites on enzymes are hydrophobic, and the environmental pH has minimal effect. In theory, a lowered cytoplasmic pH should benefit mitochondrial function (Chapter 6). Recent studies utilizing nuclear magnetic resonance (NMR) technique to look within muscles noninvasively during exercise and recovery suggest that fatigue is due to CP depletion rather than to lactic acid accumulation.

Calcium Ion

The accumulation of Ca^{2+} within muscle mitochondria during prolonged exercise may be more debilitating than the decrease in cytoplasmic pH from lactic acid formation. Some of the Ca^{2+} liberated in muscle from the sarcoplasmic reticulum during excitation–contraction coupling may be sequestered by mitochondria. This process results in O_2 consumption and saps the mitochondrial energy potential for phosphorylating ADP to ATP. When muscle mitochondria are allowed to respire in a test tube in the presence of Ca^{2+}, they will take up so much Ca^{2+} that oxidative phosphorylation will eventually be uncoupled. The extent to which this occurs in the performing athlete remains to be determined. However, such a Ca^{2+} effect could account for the situation in which athletes can repeatedly reach the same $\dot{V}_{O_2max}$ during interval exercise but not match the initial work output.

O_2 DEPLETION AND MUSCLE MITOCHONDRIAL DENSITY

The depletion of muscular O_2 stores, or rather the inadequacy of circulating O_2 delivery to muscle, can result in fatigue. Those with impaired circulatory or ventilatory function, those engaged in exercise at high altitudes, or those engaged in strenuous exercise at sea level can fall short in the balance between muscle respiratory requirement and the actual O_2 supply. Because most of the ATP required to perform any activity lasting 90 sec or more will be from respiration, adequate O_2 supply is essential to support maximal aerobic work. Maximal aerobic work is defined as the situation in which the ability to work is directly dependent on $\dot{V}_{O_2max}$, or the required work rate elicits close to 100% of $\dot{V}_{O_2max}$.

The effects of inadequate O_2 supply or utilization can be represented by increased lactate production or decreased CP levels, or both. Thus inadequate oxygenation of contracting muscle can result in at least two fatigue-causing effects.

Skeletal muscles, even untrained muscles, contain a greater mitochondrial respiratory capacity than can be supplied by the circulation. Therefore, the maximal rate of muscle oxidation is ultimately limited by the cardiac output. For activities that require very high metabolic rates for several minutes, the capacities for a high $\dot{V}_{O_2max}$ and cardiac output are important.

The doubling in muscle mitochondrial activity that takes place in response to endurance training benefits endurance capacity by a means other than increasing the $\dot{V}_{O_2max}$. Twice as much mitochondrial content doubles the capacity to oxidize fatty acids as fuel. This results in a glycogen-sparing effect. In addition, it now appears that more than the minimal number of mitochondria needed to achieve a circulatory-limited $\dot{V}_{O_2max}$ are necessary to minimize the effects of mitochondrial damage during exercise. The utilization of O_2 in mitochondria is associated with liberation of free radicals, which present a real threat to mitochondria. Indeed, Davies et al. at the University of California have supplied evidence of mitochondrial damage due to free-radical accumulation at the point of fatigue. In this sense, having more mitochondria or a larger mitochondrial reticulum to sustain prolonged submaximal exercise bouts is analogous to having more pawns with which to play the game of chess. More pawns (mitochondria) can lessen the effect of an opponent's attack (free radicals).

DISTURBANCES TO HOMEOSTASIS

Previously in Chapter 22, we have illustrated the fatiguing effects of thermal dehydration. It should be emphasized that the continuance of exercise depends on an integrated functioning of many systems, each containing many elements. Any factor or set of factors that upsets this integrated functioning can cause fatigue. Levels of K^+, Na^+, and Ca^{2+}, compartmentalization of these ions within a cell, levels of blood glucose and FFA, plasma volume, pH and osmolality, body temperature, and hormone levels are a few of the factors involved in maintenance of homeostasis.

CENTRAL AND NEUROMUSCULAR FATIGUE

In the linkage between afferent input to the performance of an appropriate motor task, there are several sites that require adequate function. A performance decrement at any site will result in a decrease in motor performance. Thus it is possible to have muscular fatigue when the muscle itself is not impaired. Specifically, the proper functioning of receptors, CNS integrating centers, sensory cortex, spinal cord, α-motor neurons, γ loop, motor end plates, muscle cells, and cerebellum are frequently required to perform motor tasks. As discussed in Chapter 18, it has not been possible to obtain many direct data on CNS function during exercise. Therefore, researchers have tended to focus on the musculature itself as a point of fatigue. It may well be that in many if not most cases, the cause of fatigue lies in the musculature, but the interrelationships between central and peripheral functions should not be overlooked in particular situations. At the very least, painful afferent inputs from muscle and joints might negatively affect an athlete's willingness to continue competition. In such a situation, the physiology makes the psychology.

In a now classic series of experiments, Merton determined that the site of fatigue in the adductor pollicis muscle was peripheral to the neuromuscular junction (i.e., it was within the muscle itself and not at the NMJ, or some site more centrally located). When the voluntary tension developed by the adductor pollicis decreased to indicate significant fatigue, Merton applied electrical shocks to the ulnar nerve to increase stimulation of the muscle (Figure 32-3). As in-

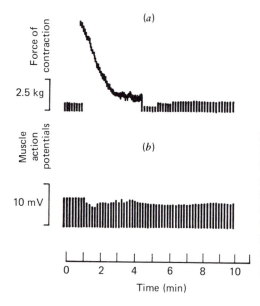

FIGURE 32-3

Maximum voluntary isometric contraction force of thumb adductor muscles (a) and corresponding muscle action potentials (b). Superimposed are electrical shocks to the controlling motor nerve. Results indicate that voluntary effort produces the maximal contraction possible in the muscle, and that the motor nerve and motor endplate are not fatigued when muscle performance declines. Fatigue is then peripheral to the nervous system, in the muscle.

(SOURCE: Modified from Merton, 1954.)

dicated in the figure, muscle action potentials (bottom) indicated clear conduction of the signal from nerve to muscle, but there was no tension response in the fatigued muscle (top).

When the electrical potentials of muscle are recorded with electromyography (EMG), there frequently occurs a distinct change in the point at which fatigue approaches during both dynamic and static muscular exercise. The so-called fatigue EMG can have one or perhaps two characteristics. The first is a large increase in the integrated EMG signal wherein all action potentials are weighted to represent a single voltage (Figure 32-4). Failure of the muscle to respond to increased stimulation indicates that fatigue is peripheral. The second characteristic is a shift to the left in the EMG power frequency spectrum (PFS) (Figure 32-5). The PFS shows the relative electrical activity contributed by slow (on the left) and fast (on the right) motor units. The leftward shift indicates increased stimulation of the smaller, slow, fatigue-resistant motor units.

Perhaps the best evidence for a central basis of fatigue follows from the work of the Russian scientist, Setchenov. He observed that the exhausted muscles of one limb recovered faster if the opposite limb was exercised moderately during a recovery period. The so-called Setchenov phenomenon, wherein an exhausted muscle recovers more rapidly if a diverting activity (either physical or mental) is performed, has been repeated several times, most notably by Asmussen and Mazin. In Figure 32-6, the work performed following "active pauses" (dark bars) is clearly greater than after "passive pauses" (light bars). This greater recovery following a diverting activity is apparently not associated with a greater recovery muscle blood flow. Rather, the Setchenov phenomenon

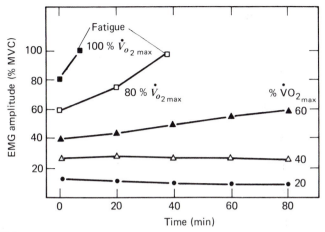

FIGURE 32-4
Integrated EMG signal in a rectus femoris muscle during cycle exercise of graded increasing intensity. The integrated EMG represents the sum total (frequency and amplitude) of all muscle action potentials. Values are given as a percentage of that recorded during a maximal voluntary contraction (MVC). At higher work rates, the EMG signal increases disproportionally, reaching a maximum as fatigue approaches. The so-called "fatigue EMG" represents the attempt to recruit additional motor units as the work output of fatiguing motor units declines.

(SOURCE: Modified from J.S. Petrofsky, 1979.)

is attributed to afferent input having a facilitatory effect on the brain's reticular formation and motor centers.

PSYCHOLOGICAL FATIGUE

The Setchenov phenomenon mentioned above is one example of the overlapping of the sciences of physiology and psychology. Eventually our understanding of the brain will advance to the point where its underlying physiology and biochemistry are understood and the effects of afferent input can be predicted. For the present, however, we can really only begin to address the question of how afferent input during competition (pain, breathlessness, nausea, audience response) can influence physiology of the CNS. For the present, a behavioral (psychological) approach to understanding these questions may be beneficial. Through training or intrinsic mechanisms, some athletes can learn to minimize the influences of distressing afferent input and can therefore approach performance limits set in the musculature. At times, athletes will slacken their pace to reduce discomforting inputs to a tolerable level. Consequently, work output will decrease but not because of a muscular limitation. On the opposite end of

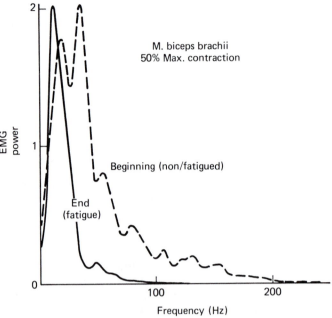

FIGURE 32-5

An EMG power frequency spectrum represents a plot of muscle action potential amplitude versus frequency. Slow, type I, motor units fire at lower frequency than do fast, type II motor units. In the experiment illustrated subjects were asked to hold 50% of their maximum voluntary contraction force. The nonfatigued EMG is bimodal in its distribution. The EMG power frequency spectrum shifts to the left and has only one peak after fatigue. The shift probably indicates recruitment of slow motor units in preference to fast units.

(SOURCE: Modified from Naeije and Zorn, 1982.)

the spectrum, we frequently see examples of inexperienced or foolhardy athletes who set a blistering pace for most of a race, and then experience real muscular fatigue before the end. In track events, these are the competitors who, in the home stretch, look like someone reading the newspaper. In order to perform optimally in an activity, the experiences of prior training and competition are often necessary for an athlete to evaluate afferent inputs properly and to utilize them in determining the maximal rate at which physiological power can be meted out during competition.

THE HEART AS A SITE OF FATIGUE

In healthy individuals there is no direct evidence that exercise, even prolonged endurance exercise, is limited by fatigue of the heart muscle. Because the Pa_{O_2}

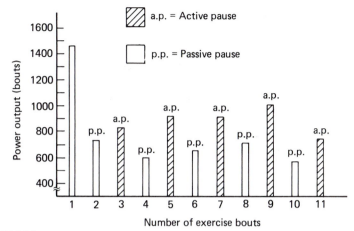

FIGURE 32-6

An illustration of the Setchenov phenomenon. Work output on a Musso, single-arm ergometer is recorded after two minutes of maximal work (bout 1). Thereafter, subjects rested for two minutes prior to each subsequent work bout. Between exercise bouts, subjects rested by means of either an active pause, a.p. (consisting of mental activity or light physical activity), or a passive pause, p.p. (consisting of minimal activity). Work output appears to be significantly greater after an active pause than after a passive pause.

(SOURCE: Modified from Asmussen and Mazin, 1978.)

is well maintained during exercise, and because the heart in effect gets first choice at the cardiac output, the heart is well oxygenated and nourished even at maximal heart rate. Additionally, because the heart is "omnivorous" in its appetite for fuels, it can be sustained by either lactic acid (which rises in short-term work) or fatty acids (which rise in long-term work). In individuals free of heart disease, the ECG does not reveal signs of ischemia (inadequate blood flow) during exercise. If ischemic symptoms are observed (Chapter 24), there is, in fact, evidence of heart disease.

During prolonged work that leads to severe dehydration and major fluid and electrolyte shifts, or other situations in which exercise is performed after thermal dehydration or diarrhea, changes in plasma, Na^+, K^+, or Ca^{2+} could affect excitation–contraction coupling of the heart. In these cases cardiac arrhythmias are possible, and exercise is not advised.

$\dot{V}_{O_2max}$ AND ENDURANCE

Given the obvious importance of maximal O_2 consumption ($\dot{V}_{O_2max}$) for setting the upper limit for aerobic metabolism, many scientists have been interested in studying the factors that limit $\dot{V}_{O_2max}$, and how $\dot{V}_{O_2max}$ may affect exercise performance. As far as $\dot{V}_{O_2max}$ is concerned, researchers have considered two possibilities. One is that $\dot{V}_{O_2max}$ is limited by O_2 transport from lungs to contracting muscles (i.e., by cardiac output and arterial O_2 content). The other possibility is that $\dot{V}_{O_2max}$ is limited by respiratory capacity of contracting muscles. As far as endurance is concerned, some researchers have believed that endurance is directly related to $\dot{V}_{O_2max}$ (i.e., the greater the $\dot{V}_{O_2max}$ the greater the endurance), while other researchers have supported the possibility that endurance is determined by peripheral factors, such as the capacity of muscle mitochondria to sustain high rates of respiration. For the present we must conclude that $\dot{V}_{O_2max}$ is a parameter set by maximal cardiac output, but endurance is determined also by muscle respiratory (mitochondrial) capacity. Several important lines of evidence lead to this conclusion.

Muscle Mass

While it is true that the mass of contracting muscle can influence $\dot{V}_{O_2}$ during exercise, once a critical mass of muscle is utilized, $\dot{V}_{O_2max}$ is independent of muscle mass. The influence of muscle mass is revealed in the observations that $\dot{V}_{O_2max}$ during arm cranking, or bicycling with one leg, is less than cycling with two legs. Furthermore, $\dot{V}_{O_2max}$ during running is greater than that bicycling or swimming, probably because of the greater muscle mass involved in running. However, $\dot{V}_{O_2max}$ while simultaneously arm cranking and leg cycling, or cross-country skiing, is not greater than during running. Thus, it appears that in healthy individuals $\dot{V}_{O_2max}$ increases as active muscle mass increases up to a point beyond which O_2 delivered through the circulation is inadequate to supply the active muscle mass.

Muscle Mitochondria

Several groups of investigators including Booth and Narahara and Ivy et al., have observed the correlation between $\dot{V}_{O_2max}$ and leg muscle mitochondrial activity to approximate 0.8 (a perfect correlation is 1.0, while no correlation is 0.0). Although these results suggest a direct relationship between $\dot{V}_{O_2max}$ and muscle mitochondrial content, the results do not directly address the issue of whether muscle mitochondrial respiratory capacity limits $\dot{V}_{O_2max}$. To address this question, Henriksson and Reitman followed the course of changes in $\dot{V}_{O_2max}$ and leg muscle mitochondrial activity during training and detraining. The results of these investigators indicate that differential responses during training and detraining. Compared to $\dot{V}_{O_2max}$, muscle mitochondrial capacity increases relatively more during training (130% versus 19%), but improvements in $\dot{V}_{O_2max}$ persist longer than does muscle respiratory capacity during detraining. These results demonstrate an independence of $\dot{V}_{O_2max}$ and muscle respiratory capacity.

In their work at the University of California, Davies et al. approached issues

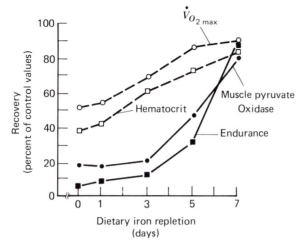

FIGURE 32-7
Recovery of $\dot{V}_{O_2max}$, hematocrit, leg muscle pyruvate oxidase activity, and running endurance in rats restored to a diet containing normal levels of iron following a zero iron-containing diet. Results indicate that hematocrit and $\dot{V}_{O_2max}$ respond in parallel, whereas pyruvate oxidase (a mitochondrial marker) and endurance respond in parallel. Results suggest that $\dot{V}_{O_2max}$ is a function of circulatory, O_2 transport capacity, whereas running endurance is related to muscle mitochondrial content.

(SOURCE: From Davies et al., 1982.)

related to $\dot{V}_{O_2max}$, O_2 transport, and muscle respiratory capacity in several ways. In endurance trained and untrained control animals, $\dot{V}_{O_2max}$ correlated only moderately well (0.74) with endurance capacity, but muscle respiratory capacity correlated very well (0.92) with running endurance (Table 6-3). Endurance training elicited a 100% increase in muscle mitochondrial content and a 400% increase in running endurance. $\dot{V}_{O_2max}$ increased only 15% in response to endurance training. On the other hand, sprint training, which also increased $\dot{V}_{O_2max}$ 15%, did not improve muscle mitochondrial content or running endurance.

Other approaches by Davies et al. to assess whether $\dot{V}_{O_2max}$ or muscle mitochondrial content better predicted exercise endurance involved dietary iron deficiency. In one study, animals were raised on an iron deficient diet so that blood hemoglobin and hematocrit and muscle mitochondrial content were depressed. When iron was restored in the diet (Figure 32-7), hematocrit and $\dot{V}_{O_2max}$ responded rapidly and in parallel. In contrast, muscle mitochondrial capacity (pyruvate oxidase) and running endurance improved in parallel, but more slowly. In still another series of experiments, iron-deficient animals received cross-transfusions of packed red blood cells in exchange for equal volumes of anaemic blood. This procedure raised blood hemoglobin and hematocrit immediately and restored $\dot{V}_{O_2max}$ to within 90% of normal. However, despite the fact that hematocrit, hemoglobin, and $\dot{V}_{O_2max}$ were improved in iron deficient animals by cross-transfusion of red blood cells, running endurance was unimproved.

Taken together, these results strongly suggest that $\dot{V}_{O_2max}$ is a function of O_2 transport (a cardiovascular parameter) whereas endurance is more dependent on muscle mitochondrial capacity. *The maximal ability of mitochondria in muscle to consume O_2 is apparently several times greater than the ability of the circulation to supply O_2. Hence, $\dot{V}_{O_2max}$ is probably limited by arterial O_2 transport.*

Arterial O_2 Transport

Researchers have also utilized several other techniques to study the relationship of arterial O_2 transport and $\dot{V}_{O_2max}$. Arterial O_2 transport ($\dot{T}_{O_2}$) is equal the product of cardiac output ($\dot{Q}$) and arterial O_2 content (CaO_2):

$$\dot{T}_{O_2} = \dot{Q}\,(Ca_{O_2}) \tag{32-1}$$

In general, attempts to raise arterial O_2 content, by having subjects breathe gases high in O_2 content, or by inducing an increase in red blood cell mass (blood doping), raise $\dot{V}_{O_2max}$. Conversely, $\dot{V}_{O_2max}$ is depressed when subjects breathe low O_2 content gas mixtures, have depressed cardiac function, are anaemic, or are exposed to carbon monoxide (CO), a gas that binds to hemoglobin and interferes with arterial O_2 binding and transport.

Muscle Oxygenation at $\dot{V}_{O_2max}$

Despite the large volume of evidence that $\dot{V}_{O_2max}$ is limited by O_2 transport, in fairness it must be said that under some circumstances, scientists have reported that $\dot{V}_{O_2max}$ is not limited by O_2 transport. In dog gastrocnemius muscle preparations stimulated to contract to $\dot{V}_{O_2max}$, Jöbsis and Stainsby have made spectrophotometric measurements on mitochondrial cytochromes and found them to be oxidized. Thus, mitochondrial O_2 content was adequate. Furthermore, Barclay and Stainsby raised $\dot{V}_{O_2}$ and contractile tension of dog muscle preparations by assisting arterial blood flow with a pump. In humans during bicycling, Pirnay et al. noted significant O_2 content of femoral venous blood draining active muscles. The critical O_2 tension for mitochondrial O_2 content (i.e., the O_2 tension below which mitochondrial O_2 consumption declines) is very low. Based on studies of isolated mitochondria, it is probable that mitochondria are capable of achieving maximal rates of respiration when the PO_2 is as low as 2 to 5 mmHg. Although the possibility of venous contamination with shunted blood was not evaluated, the interpretation of studies by Pirnay et al. on exercising humans is similar to the interpretations of results from experiments on dog muscles contracted to contract *in situ*. When relatively small muscle masses are made to work maximally, $\dot{V}_{O_2max}$ may be limited by factors within the muscle, and not by O_2 transport.

THE FUTURE OF FATIGUE

Through the wonders of contemporary science and technology, a whole series of devices and techniques are becoming available to researchers. Many of these devices will be of assistance in understanding muscle fatigue, and two deserve special mention: nuclear magnetic resonance (NMR), and positron emission tomography (PET).

NMR

The technique of nuclear magnetic resonance (NMR) spectroscopy is in theory simple, but is actually very difficult to perform. The technique takes advantage of the fact that nuclei with odd numbers of nucleons (neutrons + protons) are slightly magnetic. Thus atoms containing naturally occurring, nonradioactive elemental isotopes such as ^{31}P, ^{13}C, and ^{1}H can be aligned in a uniform and powerful magnetic field. If a brief pulse of oscillating, radiofrequency magnetic field is applied at a right angle to the fixed field, the influenceable nuclei will resonate between the directions of the two fields. This resonance will persist briefly after the radio frequency field is switch off, and the nuclei will emit a radio frequency signal. Since the radio frequency signal emitted by a particular atomic species is influenced by the nature of the compound in which that species exists, the resulting signal will vary and produce an NMR spectrum (Figure 32-8). Thus, it is possible not only to identify the various compounds in which the susceptible nuclei exist, but also the area under each peak (i.e., the integral) is proportional to the concentration of each compound.

Through the technique of NMR spectroscopy it is now possible to determine concentrations of ATP, CP, Pi, H^+ (pH), and concentrations of water, fat, and particular metabolites without breaking the skin, continuously, and painlessly. In some instances, a cross-sectional image through a body part (i.e., arm, leg, or brain) can be obtained through NMR spectroscopy. The technique of NMR spectroscopy takes advantage of naturally occurring isotopes in the body, or in some cases nonradioactive isotopes such as ^{13}C can be infused or injected to enhance the NMR signal.

Figure 32-8a displays the NMR spectrum from a flexor forearm muscle in a resting individual. In this procedure, a radio frequency (RF) sensor coil is applied to the skin directly over the muscle, and the procedure is called topical nuclear magnetic resonance (TNMR). Initial attempts at utilizing NMR to study muscle metabolism have focused on ^{31}P containing compounds. Table 32-1 presents a comparison of phosphate compound results in resting human muscle obtained by means of muscle biopsy and NMR spectroscopy. Although ATP levels are similar, creatine phosphate (CP) levels are significantly higher in NMR. Thus the CP/ATP in NMR is approximately 5 to 6, while that from the biopsy is only 3 to 4. The difference is likely due to CP degradation during the biopsy procedure.

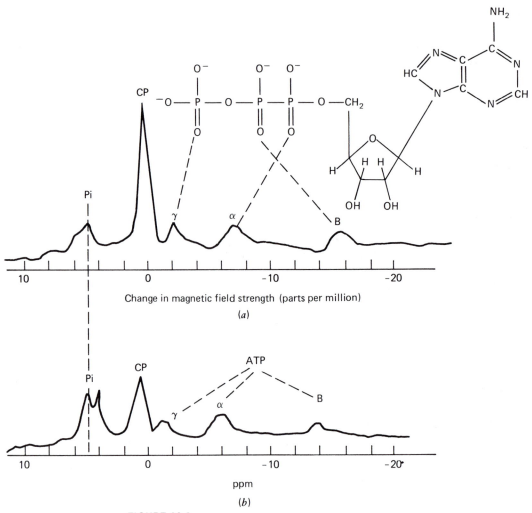

FIGURE 32-8

Tropical NMR (TNMR) phosphate spectrum from a human forearm before (a), and after (b) fatiguing isometric exercise. The area under each peak is proportional to the concentration of each metabolite. After fatigue the concentrations of free phosphate increase and the levels of CP and ATP decrease. Additionally, the phosphate peak splits and shifts to the right.

(SOURCE: From Nunnally, Brooks, and Budinger, 1983.)

In part (b) of figure 32-8, recorded after intermittent isometric contraction to fatigue, the TNMR spectrum has changed dramatically. Both ATP and CP peaks have decreased, Pi has increased, and the Pi peak has split and shifted right toward the CP peak indicating a decrease in pH. Experiments such as these have begun and they will probably make the muscle biopsy procedure obsolete for measuring muscle phosphates and pH during rest, exercise, and in

TABLE 32-1

Comparison of Phosphate Compounds in Healthy, Resting Human
Muscle Obtained by Muscle Biopsy and NMR Spectroscopy

P Metabolite	Needle Biopsy Mean ± SE mmol · kg⁻¹ Wet	^{31}P NMR Mean ± SE mmol · kg⁻¹ Wet
ATP	5.5 ± 0.07	5.11 ± 0.12
PC	17.4 ± 0.19	28.52 ± 0.43
Pi	10	4.27 ± 0.17
CP + Cr	28.6 ± 0.28	—
CP + Pi	31.6 ± 3.27	32.79 ± 0.41
Phosphodiester		0.83 ± 0.17
NAD		0.88 ± 0
pH		
Homogenates	7.08 ± 0.008 (12)	—
Intracellular	7.00 ± 0.02 (13)	7.13 ± 0.02 (20)

SOURCE: Edwards, R.H.T, D.R. Wilkie, M.J. Dawson, R.E. Gordon, and D. Shaw. Lancet, March 1982: 3.

the clinical laboratory. This is because the NMR signal represents a larger, more representative sample, and because the result probably more accurately reflects the situation *in vivo*. Of course, the biopsy procedure will be retained for important reasons, such as when there is a need to do histochemistry. NMR imaging may allow the degree of atherosclerosis and the presence of tumors to be evaluated noninvasively.

PET

The technique of positron emission tomography (PET) offers great potential for studying regional blood flow and metabolism. Like NMR, in principle, PET is conceptually not difficult to understand, but its successful execution requires enormous theoretical and technical competence. Nuclei of particular short half-life isotopes such as ^{11}C and ^{82}Rb emit positions when they decay. These positrons (B$^+$) have the same mass as electrons, but an opposite or positive charge. When an emitted B$^+$ (a particle of antimatter) strikes an electron (matter), the two particles annihilate each other. The result is two photons, roughly equivalent to gamma-rays, which fly off in opposite directions. Thus, if the atom decays within a ring of crystal and photomultiplier sensors, two crystals (on opposite sides of the ring) may be struck simultaneously as the result a nuclear decay. By means of some extremely complex electronics and computer programming, the point within the ring where the nucleus has decayed can be pinpointed.

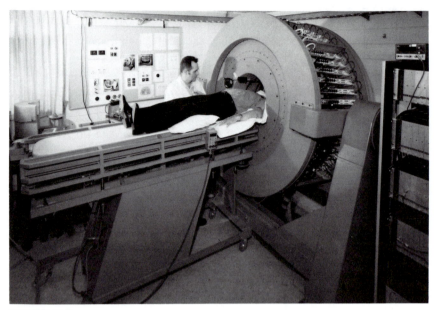

FIGURE 32-9
The University of California, Donner Research Medicine (LBL) positron emission tomograph. When sensors on opposite sides of the ring are simultaneously struck by photons, the locus of the disintegrating nucleus can be pinpointed. See text for a description of how PET can be used to image heart, brain, and other body parts, and to study regional blood flow and metabolism.

(**SOURCE**: Courtesy of T.F. Budinger, Donner Research Medicine Group, Lawrence Berkeley Laboratory.)

Figure 32-9 illustrates the University of California, Donner Research Medicine PET. The device is large enough to insert the head of thorax of an adult. When a positron-emitting isotope is injected into the circulation supplying a tissue, such as the heart, then the isotope flows where the blood flows. The PET device can then develop an image of where the blood flows through the myocardium, and pinpoint areas that receive little or no flow.

Compared to NMR technique, the PET technique has the disadvantage of requiring a radioactive isotope. These B^+ emitting isotopes, however, have short half-lives (on the order of minutes) so that the radioactivity decays away rapidly. Thus repeated measurements can be made in a relatively short period of time. Furthermore, PET is useful not only in imaging tissues of interest, but also for obtaining quantitative data on tissue uptake, metabolism, and washout. The short half-life of B^+ emitters makes it possible to follow the injection of one isotope with another. For instance, in the heart regional blood flow can be determined with injection of ^{82}Rb. This can then be followed with injection of

[11]C labeled metabolite (e.g., lactate, pyruvate, fatty acid) and regional myocardial metabolism can be studied. In this way detailed studies of myocardial flow and metabolism can be performed. Positron emission tomography can be used to study blood flow and metabolism in any tissue that can fit in a tomograph. To date most studies have focused on the heart and brain, but it will also be possible to study exercising limb skeletal muscles.

Summary

The reasons people are unable to continue muscular work and experience fatigue depend on several general factors, including the nature of the activity, the training and physiological status of the individual, and the environmental conditions. The continuance of exercise sometimes depends on availability of some key metabolite in a particular tissue. Sometimes also, the continuance of exercise depends on the maintenance of functional integrity of several individual systems, as well as their proper interaction. Because physiological systems are often matched in their capacities, during exercise stress it is sometimes difficult to identify a single factor that failed in an attempt to maintain exercise intensity.

Fatigue may result from the depletion of key metabolites such as muscle creatine phosphate and ATP, muscle glycogen, liver glycogen, blood glucose, and arterial as well as muscular O_2. Without adequate levels of each of these, high rates of muscular work cannot continue.

The accumulation of particular metabolites can also hinder the continuance of exercise. Historically, the accumulation of lactic acid (which dissociates a proton and decreases pH) has been thought to be the main fatigue-causing agent. It is unlikely, however, that lactic acid accumulation always causes the muscle fatigue, pain, or soreness usually attributed to it. Other potential fatigue factors include the accumulation of Ca^{2+} in mitochondria (which affects oxidative phosphorylation) and heat. Overheating can lead to a number of general as well as local effects, including blood flow redistribution away from muscle and liver, dehydration, fluid and electrolyte redistribution, CNS dysfunction, and mitochondrial uncoupling of oxidative phosphorylation. In the case of overheating, we have an example of how a single specific factor (overheating) can result in a generalized disturbance to homeostasis.

During contraction in athletic competitions as we know them, muscular fatigue usually appears to be a peripheral phenomenon. The central nervous system and the associated motor neurons and neuromuscular junctions appear to be far superior to skeletal muscle in maintaining function. In some instances, however, CNS factors may operate to limit performance of the muscles. In healthy individuals, the heart is not thought to fatigue and cause reduced exercise power output.

Selected Readings

Alpert, N. Lactate production and removal and the regulation of metabolism. Ann. N. Y. Acad. Sci. 119:995–1001, 1965.

American Heart Association, monograph no. 15. Physiology of Muscular Exercise. New York: American Heart Association, 1967.

Anderson, P. and J. Henriksson. Capillary supply to the quadriceps femoris muscle of man: Adaptive response to exercise. J. Physiol. (London) 270: 677–690, 1977.

Asmussen, E. Muscle Fatigue. Med. Sci. Sports 11:313–321, 1979.

Asmussen, E. and B. Mazin. Recuperation after muscular fatigue by "diverting activities." Europ. J. Appl. Physiol. 38: 1–8, 1978.

Asmussen, E. and B. Mazin. A central nervous component in local muscular fatigue. Europ. J. Appl. Physiol. 38:9–15, 1978.

Åstrand, P.-O. and K. Rodahl. Textbook of Work Physiology. New York: McGraw-Hill, 1970. pp. 154–178 and 187–254.

Åstrand, P.-O. and B. Saltin. Maximal oxygen uptake and heart rate in various types of muscular activity. J. Appl. Physiol. 16: 977–981, 1961.

Bang, O. The lactate content of blood during and after exercise in man. Skand. Arch. Physiol. 74(Supp.): 10: 51–82, 1936.

Bannister, R.G. and C.J.C. Cunningham. The effects on the respiration and performance during exercise of adding oxygen to the inspired air. J. Physiol. (London) 125: 118–121, 1954.

Barclay, J.K. and W.N. Stainsby. The role of blood flow in limiting maximal metabolic rate in muscle. Med. Sci. Sports 7: 116–119, 1975.

Barcroft, H. and J.L.E. Millen. The blood flow through muscle during sustained contraction. J. Physiol. 97: 17–31, 1939.

Basmajian, J.V. Muscles Alive. Baltimore: Williams and Wilkins, 1962.

Bergh, U., I.-L. Kaustrap, and B. Ekblom. Maximal oxygen uptake during exercise with various combinations of arm and leg work. J. Appl. Physiol. 41: 191–196, 1976.

Bergström, J. Local changes of ATP and phosphorylcreatine in human muscle tissue in connection with exercise. In: Physiology of Muscular Exercise. New York: American Heart Association monograph no. 15, 1967. pp. 191–196.

Bergström, J. and E. Hultman. The effects of exercise on muscle glycogen and electrolytes in normals. Scand. J. Clin. Lab. Invest. 18: 16–20, 1966.

Bigland-Ritchie, B., D.A. Jones, and J.J. Woods. Excitation frequency and muscular fatigue: Electrical responses during human voluntary and stimulated contractions. Exp. Neurol. 64: 414–427, 1979.

Booth, F.W. and K.A. Narahara. Vastus lateralis cytochrome oxidase activity and its relationship to maximal oxygen consumption in man. Pflugers Archiv. 349: 319–324, 1974.

Bowie, W. and G.R. Cumming. Sustained handgrip-reproducibility; Effects of hypoxia. Med. Sci. Sports 3: 48–52, 1971.

Brooks, G.A., K.E. Brauner, and R.G. Cassens. Glycogen synthesis and metabolism of lactic acid after exercise. Am. J. Physiol. 220: 1053–1059, 1971.

Brooks, G.A., K.J. Hittelman, J.A. Faulkner, and R.E. Beyer. Tissue temperatures and whole-animal oxygen consumption after exercise. Am. J. Physiol. 221: 427–431, 1971.

Brooks, G.A., K.J. Hittelman, J.A. Faulkner, and R.E. Beyer. Temperature, liver mitochondrial respiratory functions, and oxygen debt. Med. Sci. Sports. 2: 71–74, 1971.

Brooks, G.A., R. Nunnally, and T.F. Budinger. Human forearm muscular fatigue investigated by phosphorus nuclear magnetic resonance. Unpublished manuscript, 1983.

Buick, F.J., N. Gledhill, A.B. Frosese, L. Spriet, and E.C. Meyers. Effect of induced erythrocythemia on aerobic work capacity. J. Appl. Physiol: Respirat. Environ. Exercise Physiol. 48: 636–642, 1980.

Chance, B., S. Eleff, J.S. Leigh, D. Sokolow, and A. Sapega. Mitochondrial regulation of phosphocreatine/phosphate ratios in exercised human limbs: Gated ^{31}P NMR study. Proc. Natl. Acad. Sci. USA. 78: 6714–6718, 1981.

Christensen, E.H. Fatigue of the working individual. In: Bourne, G.H. (ed.), Structure and function of Muscle II. Biochemistry and physiology. New York: Academic Press, 1960. p. 455–465.

Ciba Foundation. Symposium 82nd on Human Muscle Fatigue: Physiological Mechanisms. London: Pitman Medical, 1981.

Costill, D.L., R. Bowers, G. Branam, and K. Sparks. Muscle glycogen utilization during prolonged exercise on successive days. J. Appl. Physiol. 31: 834–838, 1971.

Davies, K.J.A., C.M. Donovan, C.J. Refino, G.A. Brooks, and P.R. Dallman. Iron deficiency and work capacity: Distinguishing the effects of anaemia and muscle iron deficiency on exercise bioenergetics in the rat. Unpublished manuscript.

Davies, K.J.A., J.L. Maguire, G.A. Brooks, P.R. Dallman, and L. Packer. Muscle mitochondrial bioenergetics, oxygen supply, and work capacity during dietary iron deficiency and repletion. Am. J. Physiol. 242 (Endocrinol. Metab.): E418–E427, 1982.

Davies, K.J.A., L. Packer, and G.A. Brooks. Biochemical adaptation of mitochondria, muscle, and whole-animal respiration to endurance training. Arch. Biochem. Biophs. 209: 539–554, 1981.

Davies, K.J.A., L. Packer, and G.A. Brooks. Exercise bioenergetics following sprint training. Arch. Biochem. Biophys. 215: 260–265, 1982.

Davies, C.T.M. and A.J. Sargent. Physiological response to one- and two-leg exercise breathing air and 45% oxygen. J. Appl. Physiol. 36:142–148, 1974.

Dawson, M.J., D.G. Gadian, and D.R. Wilkie. Contraction and recovery of living muscles studied by 31 P nuclear magnetic resonance. J. Physiol. (London) 267: 703–735, 1977.

Dawson, M.J., D.G. Gadian, and D.R. Wilkie. Muscular fatigue investigated by phosphorus nuclear magnetic resonance. Nature 274: 861–866, 1978.

Dill, D.B. Fatigue and physical fitness. In: Johnson, W.R. and E.R. Buskirk (eds.), Structural and Physiological Aspects of Exercise and Sport. Princeton, N.J.: Princeton Book Co., 1980.

Edwards, R.H.T., D.K. Hill, and P.A. Merton. Fatigue of long duration in human skeletal muscle after exercise. J. Physiol. (London). 272: 769–778, 1977.

Edwards, R.H.T., D.R. Wilkie, M.J. Dawson, R.E. Gordon, and D. Shaw. Clinical use of nuclear magnetic resonance in the investigation of myopathy. Lancet, March 1982: 725–731.

Ekblom, B., A.N. Goldbarg, and B. Gullbring. Response to exercise after blood loss and reinfusion. J. Appl. Physiol. 33: 175–189, 1972.

Ekblom, B., R. Hout, E. Stein, and A. Thorstensson. Effect of changes in arterial oxygen content on circulation and physical performance. J. Appl. Physiol. 39: 71–75, 1975.

Ekblom, B., G. Wilson, and P.-O. Åstrand. Central circulation during exercise after venesection and reinfusion of red blood cells. J. Appl. Physiol. 40: 379–383, 1976.

Ekelund, L.G. and A. Holmgren. Circulatory and respiratory adaptation during long-term, non-steady state exercise in the sitting position. Acta Physiol. Scand. 62: 240, 1964.

Ekelund, L.G. and A. Holmgren. Central hemodynamics during exercise. In: C.B. Chapman (ed.), Physiology of Muscular Exercise. New York: American Heart Association monograph no. 15, American Heart Association, 1967. pp. 133–143.

Faulkner, J.A., D.E. Roberts, R.L. Elk, and J. Conway. Cardiovascular responses to submaximum and maximum effort cycling and running. J. Appl. Physiol. 30: 457–461, 1971.

Gadian, D.G. and G.K. Radda. NMR studies of tissue metabolism. Ann. Rev. Biochem. 50: 69–83, 1981.

Gardner, G.W., V.R. Edgerton, R. Senewirathe, R.J. Barnard, and Y. Ohira. Physical work capacity and metabolic stress in subjects with iron deficiency anaemia. Am. J. Clin. Nutr. 30: 910–917, 1977.

Gleser, M.A. Effect of hypoxia and physical training on hemodynamic adjustments to one-legged exercise. J. Appl. Physiol. 34: 655–659, 1973.

Gleser, M.A., D.H. Horstman, and R.P. Mello. The effects on $\dot{V}_{O_2max}$ of adding arm work to maximal leg work. Med. Sci. Sports 6: 104–107, 1974.

Gordon, R.E., P.E. Hanley, and D. Shaw. Topical magnetic resonance. Progress in NMR Spectroscopy. 15: 1–47, 1982.

Grimby, G., E. Haggendal, and B. Saltin. Local Xenon 133 clearance from the quadriceps muscle during exercise in man. J. Appl. Physiol. 22: 305–310, 1967.

Hanson, P., A. Claremont, J. Dempsey, and W. Reddan. Determinants and consequences of ventilatory responses to competitive endurance running. J. Appl. Physiol.: Respirat. Environ. Exercise Physiol. 52: 615–623, 1982.

Henriksson, J. and J.S. Reitman. Time course of changes in human skeletal muscle succinate dehydrogenase and cytochrome oxidase activities and maximal oxygen uptake with physical activity and inactivity. Acta Physiol. Scand. 99: 91–97, 1977.

Horstman, D.H., M. Gleser, and J. Delehunt. Effects of altering O_2 delivery on VO_2 of isolated working muscle. Am. J. Physiol. 230: 327–334, 1976.

Horstman, D.H., M. Gleser, D. Wolfe, T. Tyron, and J. Delehunt. Effects of hemoglobin reduction on $\dot{V}_{O_2max}$ and related hemodynamics in exercising dogs. J. Appl. Physiol. 37: 97–100, 1974.

Hughes, R.L., M. Clode, R.H. Edwards, T.J. Goodwin, and N. Jones. Effect of inspired O_2 on cardiopulmonary and metabolic responses to exercise in man. J. Appl. Physiol. 24: 36–347, 1968.

Ivy, J.L., D.L. Costill, and B.D. Maxwell. Skeletal muscle determinants of maximal aerobic power in man. Eur. J. Appl. Physiol. 44: 1–8, 1980.

Ivy, J.L., R.T. Withers, D.J. Van Handel, D.H. Elger, and D.L. Costill. Muscle respiratory capacity and fiber type as determinants of lactate threshold. J. Appl. Physiol.: Respirat. Environ. Exercise Physiol. 48: 523–527, 1980.

Jessup, G.T. Changes in forearm blood flow associated with sustained handgrip performance. Med. Sci. Sports 5: 258–261, 1973.

Jöbsis, F.F. and W.N. Stainsby. Oxidation of NADH during contractions of circulated mammalian skeletal muscle. Resp. Physiol. 4: 292–300, 1968.

Jones, D.A., B. Bigland-Ritchie, and R.H.T. Edwards. Excitation frequency and muscle fatigue: Mechanical responses during voluntary and stimulated contractions. Exp. Neurol. 64: 401–413, 1979.

Kroll, W. Isometric fatigue curves under varied intertrial recuperation periods. Res. Quart. 39: 106–115, 1968.

Lind, A.R. Muscle fatigue and recovery from fatigue induced by sustained contractions. J. Physiol. (London) 127:162–171, 1959.

Lippold, D.C.J., J.W.T. Redfearn, and J. Vuco. The electromyography of fatigue. Ergonomics. 3: 121–131, 1960.

Margaria, R., H.T. Edwards, and D.B. Dill. The possible mechanisms of contracting and paying the oxygen debt and the role of lactic acid in muscular contraction. Am. J. Physiol. 106: 689–715, 1933.

Martin, B.J. and G.M. Gaddis. Exercise after sleep deprivation. Med. Sci. Sports Exercise. 13: 220–223, 1981.

Mathews, D.E., D.M. Bier, M.J. Rennie, R.H.T. Edwards, D. Halliday, D.J. Millard, and G. Clugston. Regulation of leucine metabolism in man: a stable isotope study. Science. 214: 1129–1131, 1981.

Merton, P.A. Voluntary strength and fatigue. J. Physiol. (London) 123: 553–564, 1954.

Meyers, S. and W. Sullivan. Effect of circulatory occlusion on time to muscular fatigue. J. Appl. Physiol. 24:54–59, 1968.

Mitchell, J.H., B.J. Sproule, and C.B. Chapman. Physiological meaning of the maximal oxygen uptake test. J. Clin. Invest. 37: 538, 1958.

Moxham, J., R.H.T. Edwards, M. Aubrer, A. De Troyer, G. Farkas, P.T. Macklem, and C. Roussos. Changes in EMG power spectrum (high-to-low ratio) with force fatigue in humans. J. Appl. Physiol.:Respirat. Environ. Exercise Physiol. 53: 1094–1099, 1982.

Naeije, M. and H. Zorn. Relation between EMG power spectrum shifts and muscle fibre action potential conduction velocity changes during local muscular fatigue in man. Eur. J. Appl. Physiol. 50: 25–33, 1982.

Nakamura, K.D.A. Schoelher, F.J. Winkler, and H.-L. Schmidt. Geographical variations in the carbon isotope composition of hair in contemporary man. Biomedical Mass Spectrometry. 9: 390–394, 1982.

Petrofsky, J.S. Frequency and amplitude analysis of EMG during exercise on the bicycle ergometer. Europ. J. Appl. Physiol. 41: 1–15, 1979.

Pirnay, F., M. Lamy, J. Dujardin, R. Deroanne, and J.M. Petit. Analysis of femoral venous blood during maximum muscular exercise. J. Appl. Physiol. 33: 289–292, 1972.

Pirnay, F., R. Marechal, R. Radermechker, and J.M. Petit. Muscle blood flow during submaximum and maximum exercise on a bicycle ergometer. J. Appl. Physiol. 32: 210–212, 1972.

Reybrouck, T., G.F. Heigenhauser, and J.A. Faulkner. Limitation to maximum oxygen uptake in arm, leg, and combined arm-leg ergometry. J. Appl. Physiol. 38: 774–779, 1975.

Rodbard, S. and M. Farbstein. Improved exercise tolerance during venous congestion. J. Appl. Physiol. 33: 704–710, 1972.

Rodbard, S. and E.B. Pragay. Contraction frequency, blood supply, and muscle pain. J. Appl. Physiol. 24: 142–145, 1968.

Royce, J. Isometric fatigue curves in human muscle with normal and occluded circulation. Res. Quart. 29: 204–211, 1958.

Saltin, B., R.F. Grover, C.G. Blomquist, L.H. Hartley, and R.J. Johnson. Maximal oxygen uptake and cardiac output after 2 weeks at 4300 m. J. Appl. Physiol. 25: 406–409, 1968.

Setchenov, I.M. Zur Frage nach der einwirkung sensitiver reize auf dre muskelarbeit des menschen. In: Selected Works. Moscow: 1935. pp. 246–260.

Secher, N.H., N. Ruberg-Larsen, R. Binkhorst, and F. Bonde-Peterson. Maximal oxygen uptake during arm cranking and combined arm plus leg exercise. J. Appl. Physiol. 36: 515–518, 1974.

Shulman, R.G., T.R. Brown, K. Ugurbil, S. Ogawa, S.M. Cohen, and J.A. den Hollander. Cellular applications of ^{31}P and ^{13}C nuclear magnetic resonance. Science 205: 160–165, 1979.

Simonson, E. (ed.). Physiology of Work Capacity and Fatigue. Springfield, Ill.: Charles C Thomas, 1971.

Stephens, J.A. and A. Taylor. Fatigue of maintained voluntary muscle contraction in man. J. Physiol. (London) 220: 1–18, 1972.

Stull, C.A. and J.T. Kearney. Recovery of muscular endurance following submaximal, isometric exercise. Med. Sci. Sports 10: 109–112, 1978.

Taylor, H.E., E.R. Buskirk, and A. Henschel. Maximal oxygen uptake as an objective measure of cardiorespiratory performance. J. Appl. Physiol. 8: 73–84, 1955.

Waller, A. The sense of effort: An objective study. Brain 14: 179–249, 1881.

LIST OF SYMBOLS AND ABBREVIATIONS

Note: a dash over any symbol indicates a mean value (e.g., $\bar{x}$); a dot above any symbol indicates time derivate (e.g., $\dot{V}$).

RESPIRATORY AND HEMODYNAMIC NOTATIONS

V	gas volume
$\dot{V}$	gas volume/unit time (usually liters/min)
R	ventilatory respiratory exchange ratio (volume CO_2/volume O_2 $= \dot{V}_{CO_2}/\dot{V}_{O_2}$)
RQ	cellular respiratory quotient ($\dot{V}_{CO_2}/\dot{V}_{O_2}$)
I	inspired gas
E	expired gas
A	alveolar gas
F	fractional concentration in dry gas phase
f	respiratory frequency (breath/unit time)
TLC	total lung capacity
VC	vital capacity
FRC	functional residual capacity

Based on *Federation Proceedings* 9:602–605, 1960.

RV	residual volume
T	tidal gas
D	dead space
FEV	forced expiratory volume
$FEV_{1.0}$	forced expiratory volume in 1 sec
MET	multiple of the resting metabolic rate, approximately equal to 3.5 ml O_2/kg body weight/min
MVV	maximal voluntary ventilation (maximum breathing capacity, MBC)
D_L	diffusing capacity of the lungs (ml/min $\times$ mmHg)
P	gas pressure
B or Bar	barometric
STPD	0°C, 760 mmHg, dry
BTPS	body temperature and pressure, saturated with water vapor
ATPD	ambient temperature and pressure, dry
ATPS	ambient temperature and pressure, saturated with water vapor
Q	blood flow or volume
$\dot{Q}$	blood flow/unit time (without other notation, cardiac output; usually liters/min)
SV or V_S	stroke volume
HR or f_H	heart rate (usually beats/min)
BV	blood volume
Hb	hemoglobin concentration (g/100 ml)
HbO_2	oxyhemoglobin
Hct	hematocrit
BP	blood pressure
TPR	total peripheral resistance
C	concentration in blood phase
S	percent saturation of Hb
a	arterial
c	capillary
v	venous

TEMPERATURE NOTATIONS

T	temperature
r	rectal
s	skin
e	esophageal
m	muscle
ty	tympanic

M	metabolic energy yield
C	convective heat exchange
R	radiation heat exchange
E	evaporative heat loss
S	storage of body heat
°C	temperature in degrees centigrade
°F	temperature in degrees Fahrenheit

BODY DIMENSIONS

W	weight
H	height
L	length
LBM	lean body mass
BSA	body surface area

STATISTICAL NOTATIONS

$M, \bar{x}$	arithmetic mean
SD or S.D.	standard deviation
SE or S.E.	standard error of the mean
n	number of observations
r	correlation coefficient
range	smallest and largest observed value
Σ	summation
P	probability
*	denotes a (probably) significant difference; $P < 0.05$

EXAMPLES

V_A	volume of alveolar gas
$\dot{V}_E$	expiratory gas volume/minute
$\dot{V}_{O_2}$	volume of oxygen/minute (oxygen uptake/min)
$\dot{V}_{O_2max}$	maximal volume of O_2 consumed/min
V_T	tidal volume
P_A	alveolar gas pressure
P_B	barometric pressure
$F_{I_{O_2}}$	fractional concentration of O_2 in inspired gas
$P_{A_{O_2}}$	alveolar oxygen pressure

pH_a	arterial pH
Ca_{O_2}	oxygen content in arterial blood
$Ca_{O_2} - C\bar{v}_{O_2}$	difference in oxygen content between arterial and mixed venous blood, often written (a-v)O_2
T_r	rectal temperature
$(a - v)_{O_2}$	arterial venous O_2 difference

CHEMICAL AND PHYSICAL NOTATIONS

ADP	adenosine diphosphate
AMP	adenosine monophosphate
ATP	adenosine triphosphate
Ca^{2+}	calcium ion
Cl^-	chloride ion
CO_2	carbon dioxide
CP	creatine phosphate
e^-	electron
FFA	free fatty acids
H	hydrogen atom
H^+	hydrogen ion
$H^{\cdot}$	hydride ion
HCO_3^-	bicarbonate ion
Hg	mercury
H_2CO_3	carbonic acid
H_2O	water
K^+	potassium ion
kcal	kilocalorie
kcal/min	kilcalories per minute
kg	kilogram
m	meter
meq	milliequivalent
min	minute
ml	milliliter
mm	millimeter
mmol	millimole
Na^+	sodium ion
O_2	oxygen
P	power
PC	phosphocreatine
Pi	inorganic phosphate
PP_i	pyrophosphate
sec	second

ENERGY EXPENDITURE IN OCCUPATIONAL, RECREATIONAL, AND SPORTS ACTIVITIES (KCAL/MIN)

		Body Weight						
	kg	**50**	**53**	**56**	**59**	**62**	**65**	**68**
Activity	**lb**	**110**	**117**	**123**	**130**	**137**	**143**	**150**
Archery		3.3	3.4	3.6	3.8	4.0	4.2	4.4
Badminton		4.9	5.1	5.4	5.7	6.0	6.3	6.6
Bakery, general		1.8	1.9	2.0	2.1	2.2	2.3	2.4
Basketball		6.9	7.3	7.7	8.1	8.6	9.0	9.4
Billiards		2.1	2.2	2.4	2.5	2.6	2.7	2.9
Bookbinding		1.9	2.0	2.1	2.2	2.4	2.5	2.6
Boxing		11.1	11.8	12.4	13.1	13.8	14.4	15.1
Canoeing								
Leisure		2.2	2.3	2.5	2.6	2.7	2.9	3.0
Racing		5.2	5.5	5.8	6.1	6.4	6.7	7.0
Carpentry, general		2.6	2.8	2.9	3.1	3.2	3.4	3.5
Circuit-training		9.3	9.8	10.4	10.9	11.5	12.0	12.6
Climbing hills								
With no load		6.1	6.4	6.8	7.1	7.5	7.9	8.2
With 5-kg load		6.5	6.8	7.2	7.6	8.0	8.4	8.8
With 10-kg load		7.0	7.4	7.8	8.3	8.7	9.1	9.5
With 20-kg load		7.4	7.8	8.2	8.7	9.1	9.6	10.0
Coal mining								
Drilling coal, rock		4.7	5.0	5.3	5.5	5.8	6.1	6.4
Erecting roof supports		4.4	4.7	4.9	5.2	5.5	5.7	6.0
Shoveling coal		5.4	5.7	6.0	6.4	6.7	7.0	7.3
Cricket								
Batting		4.2	4.4	4.6	4.9	5.1	5.4	5.6
Bowling		4.5	4.8	5.0	5.3	5.6	5.9	6.1
Croquet		3.0	3.1	3.3	3.5	3.7	3.8	4.0
Cycling								
Leisure		5.9	6.2	6.6	6.9	7.3	7.6	8.0
Leisure, 5.5 mph		3.2	3.4	3.6	3.8	4.0	4.2	4.4
Leisure, 9.4 mph		5.0	5.3	5.6	5.9	6.2	6.5	6.8
Racing		8.5	9.0	9.5	10.0	10.5	11.0	11.5
Dancing								
Ballroom		2.6	2.7	2.9	3.0	3.2	3.3	3.5
Choreographed, vigorous		8.4	8.9	9.4	9.9	10.4	10.9	11.4
Digging trenches		7.3	7.7	8.1	8.6	9.0	9.4	9.9

| 71 | 74 | 77 | 80 | 83 | 86 | 89 | 92 | 95 | 98 |
157	163	170	176	183	190	196	203	209	216
4.6	4.8	5.0	5.2	5.4	5.6	5.8	6.0	6.2	6.4
6.9	7.2	7.5	7.8	8.1	8.3	8.6	8.9	9.2	9.5
2.5	2.6	2.7	2.8	2.9	3.0	3.1	3.2	3.3	3.4
9.8	10.2	10.6	11.0	11.5	11.9	12.3	12.7	13.1	13.5
3.0	3.1	3.2	3.4	3.5	3.6	3.7	3.9	4.0	4.1
2.7	2.8	2.9	3.0	3.2	3.3	3.4	3.5	3.6	3.7
15.8	16.4	17.1	17.8	18.4	19.1	19.8	20.4	21.1	21.8
3.1	3.3	3.4	3.5	3.7	3.8	3.9	4.0	4.2	4.3
7.3	7.6	7.9	8.2	8.5	8.9	9.2	9.5	9.8	10.1
3.7	3.8	4.0	4.2	4.3	4.5	4.6	4.8	4.9	5.1
13.1	13.7	14.2	14.8	15.4	15.9	16.5	17.0	17.6	18.1
8.6	9.0	9.3	9.7	10.0	10.4	10.8	11.1	11.5	11.9
9.2	9.5	9.9	10.3	10.7	11.1	11.5	11.9	12.3	12.6
9.9	10.4	10.8	11.2	11.6	12.0	12.5	12.9	13.3	13.7
10.4	10.9	11.3	11.8	12.2	12.6	13.1	13.5	14.0	14.4
6.7	7.0	7.2	7.5	7.8	8.1	8.4	8.6	8.9	9.2
6.2	6.5	6.8	7.0	7.3	7.6	7.8	8.1	8.4	8.6
7.7	8.0	8.3	8.6	9.0	9.3	9.6	9.9	10.3	10.6
5.9	6.1	6.4	6.6	6.9	7.1	7.4	7.6	7.9	8.1
6.4	6.7	6.9	7.2	7.5	7.7	8.0	8.3	8.6	8.8
4.2	4.4	4.5	4.7	4.9	5.1	5.3	5.4	5.6	5.8
8.3	8.7	9.0	9.4	9.7	10.1	10.4	10.8	11.1	11.5
4.6	4.8	5.0	5.1	5.3	5.5	5.7	5.9	6.1	6.3
7.1	7.4	7.7	8.0	8.3	8.6	8.9	9.2	9.5	9.8
12.0	12.5	13.0	13.5	14.0	14.5	15.0	15.5	16.1	16.6
3.6	3.8	3.9	4.1	4.2	4.4	4.5	4.7	4.8	5.0
11.9	12.4	12.9	13.4	13.9	14.4	15.0	15.5	16.0	16.5
10.3	10.7	11.2	11.6	12.0	12.5	12.9	13.3	13.8	14.2

					Body Weight			
	kg	50	53	56	59	62	65	68
Activity	lb	110	117	123	130	137	143	150
Eating (sitting)		1.2	1.2	1.3	1.4	1.4	1.5	1.6
Electrical work		2.9	3.1	3.2	3.4	3.6	3.8	3.9
Farming								
Cleaning animal stalls		6.8	7.2	7.6	8.0	8.4	8.8	9.2
Driving harvester		2.0	2.1	2.2	2.4	2.5	2.6	2.7
Driving tractor		1.9	2.0	2.1	2.2	2.3	2.3	2.5
Feeding cattle		4.3	4.5	4.8	5.0	5.3	5.5	5.8
Feeding hens and dogs		3.3	3.4	3.6	3.8	4.0	4.2	4.4
Forking straw bales		6.9	7.3	7.7	8.1	8.6	9.0	9.4
Milking by hand		2.7	2.9	3.0	3.2	3.3	3.5	3.7
Milking by machine		1.2	1.2	1.3	1.4	1.4	1.5	1.6
Shoveling grain		4.3	4.5	4.8	5.0	5.3	5.5	5.8
Field hockey		6.7	7.1	7.5	7.9	8.3	8.7	9.1
Fishing		3.1	3.3	3.5	3.7	3.8	4.0	4.2
Football		6.6	7.0	7.4	7.8	8.2	8.6	9.0
Forestry								
Ax chopping, fast		14.9	15.7	16.6	17.5	18.4	19.3	20.2
Ax chopping, slow		4.3	4.5	4.8	5.0	5.3	5.5	5.8
Barking trees		6.2	6.5	6.9	7.3	7.6	8.0	8.4
Carrying logs		9.3	9.9	10.4	11.0	11.5	12.1	12.6
Felling trees		6.6	7.0	7.4	7.8	8.2	8.6	9.0
Hoeing		4.6	4.8	5.1	5.4	5.6	5.9	6.2
Planting by hand		5.5	5.8	6.1	6.4	6.8	7.1	7.4
Sawing by hand		6.1	6.5	6.8	7.2	7.6	7.9	8.3
Sawing, power		3.8	4.0	4.2	4.4	4.7	4.9	5.1
Stacking firewood		4.4	4.7	4.9	5.2	5.5	5.7	6.0
Trimming trees		6.5	6.8	7.2	7.6	8.0	8.4	8.8
Weeding		3.6	3.8	4.0	4.2	4.5	4.7	4.9
Gardening								
Digging		6.3	6.7	7.1	7.4	7.8	8.2	8.6
Hedging		3.9	4.1	4.3	4.5	4.8	5.0	5.2
Mowing		5.6	5.9	6.3	6.6	6.9	7.3	7.6
Raking		2.7	2.9	3.0	3.2	3.3	3.5	3.7
Golf		4.3	4.5	4.8	5.0	5.3	5.5	5.8
Gymnastics		3.3	3.5	3.7	3.9	4.1	4.3	4.5
Judo		9.8	10.3	10.9	11.5	12.1	12.7	13.3

71	74	77	80	83	86	89	92	95	98
157	163	170	176	183	190	196	203	209	216
1.6	1.7	1.8	1.8	1.9	2.0	2.0	2.1	2.2	2.3
4.1	4.3	4.5	4.6	4.8	5.0	5.2	5.3	5.5	5.7
9.6	10.0	10.4	10.8	11.2	11.6	12.0	12.4	12.8	13.2
2.8	3.0	3.1	3.2	3.3	3.4	3.6	3.7	3.8	3.9
2.6	2.7	2.8	3.0	3.1	3.2	3.3	3.4	3.5	3.6
6.0	6.3	6.5	6.8	7.1	7.3	7.6	7.8	8.1	8.3
4.6	4.8	5.0	5.2	5.4	5.6	5.8	6.0	6.2	6.4
9.8	10.2	10.6	11.0	11.5	11.9	12.3	12.7	13.1	13.5
3.8	4.0	4.2	4.3	4.5	4.6	4.8	5.0	5.1	5.3
1.6	1.7	1.8	1.8	1.9	2.0	2.0	2.1	2.2	2.3
6.0	6.3	6.5	6.8	7.1	7.3	7.6	7.8	8.1	8.3
9.5	9.9	10.3	10.7	11.1	11.5	11.9	12.3	12.7	13.1
4.4	4.6	4.8	5.0	5.1	5.3	5.5	5.7	5.9	6.1
9.4	9.8	10.2	10.6	11.0	11.4	11.7	12.1	12.5	12.9
21.1	22.0	22.9	23.8	24.7	25.5	26.4	27.3	28.2	29.1
6.0	6.3	6.5	6.8	7.1	7.3	7.6	7.8	8.1	8.3
8.7	9.1	9.5	9.8	10.2	10.6	10.9	11.3	11.7	12.1
13.2	13.8	14.3	14.9	15.4	16.0	16.6	17.1	17.7	18.2
9.4	9.8	10.2	10.6	11.0	11.4	11.7	12.1	12.5	12.9
6.5	6.7	7.0	7.3	7.6	7.8	8.1	8.4	8.6	8.9
7.7	8.1	8.4	8.7	9.0	9.4	9.7	10.0	10.4	10.7
8.7	9.0	9.4	9.8	10.1	10.5	10.9	11.2	11.6	12.0
5.3	5.6	5.8	6.0	6.2	6.5	6.7	6.9	7.1	7.4
6.2	6.5	6.8	7.0	7.3	7.6	7.8	8.1	8.4	8.6
9.2	9.5	9.9	10.3	10.7	11.1	11.5	11.9	12.3	12.6
5.1	5.3	5.5	5.8	6.0	6.2	6.4	6.6	6.8	7.1
8.9	9.3	9.7	10.1	10.5	10.8	11.2	11.6	12.0	12.3
5.5	5.7	5.9	6.2	6.4	6.6	6.9	7.1	7.3	7.5
8.0	8.3	8.6	9.0	9.3	9.6	10.0	10.3	10.6	11.0
3.8	4.0	4.2	4.3	4.5	4.6	4.8	5.0	5.1	5.3
6.0	6.3	6.5	6.8	7.1	7.3	7.6	7.8	8.1	8.3
4.7	4.9	5.1	5.3	5.5	5.7	5.9	6.1	6.3	6.5
13.8	14.4	15.0	15.6	16.2	16.8	17.4	17.9	18.5	19.1

Activity	kg lb	50 110	53 117	56 123	59 130	62 137	65 143	68 150
				Body Weight				
Lying at ease		1.1	1.2	1.2	1.3	1.4	1.4	1.5
Machine-tooling								
Machining		2.4	2.5	2.7	2.8	3.0	3.1	3.3
Operating lathe		2.6	2.8	2.9	3.1	3.2	3.4	3.5
Operating punch press		4.4	4.7	4.9	5.2	5.5	5.7	6.0
Tapping and drilling		3.3	3.4	3.6	3.8	4.0	4.2	4.4
Welding		2.6	2.8	2.9	3.1	3.2	3.4	3.5
Working sheet metal		2.4	2.5	2.7	2.8	3.0	3.1	3.3
Marching		7.1	7.5	8.0	8.4	8.8	9.2	9.7
Music playing								
Accordion (sitting)		1.6	1.7	1.8	1.9	2.0	2.1	2.2
Cello (sitting)		2.1	2.2	2.3	2.4	2.5	2.7	2.8
Conducting (standing)		2.0	2.1	2.2	2.3	2.4	2.5	2.7
Drums (sitting)		3.3	3.5	3.7	3.9	4.1	4.3	4.5
Flute (sitting)		1.8	1.9	2.0	2.1	2.2	2.3	2.4
Horn (sitting)		1.5	1.5	1.6	1.7	1.8	1.9	2.0
Organ (sitting)		2.7	2.8	3.0	3.1	3.3	3.4	3.6
Piano (sitting)		2.0	2.1	2.2	2.4	2.5	2.6	2.7
Trumpet (standing)		1.6	1.6	1.7	1.8	1.9	2.0	2.1
Violin (sitting)		2.3	2.4	2.5	2.7	2.8	2.9	3.1
Woodwind (sitting)		1.6	1.7	1.8	1.9	2.0	2.1	2.2
Painting, inside		1.7	1.8	1.9	2.0	2.1	2.2	2.3
Painting, outside		3.9	4.1	4.3	4.5	4.8	5.0	5.2
Planting seedlings		3.5	3.7	3.9	4.1	4.3	4.6	4.8
Plastering		3.9	4.1	4.4	4.6	4.8	5.1	5.3
Printing		1.8	1.9	2.0	2.1	2.2	2.3	2.4
Running, cross-country		8.2	8.6	9.1	9.6	10.1	10.6	11.1
Running, horizontal								
11 min, 30 s per mile		6.8	7.2	7.6	8.0	8.4	8.8	9.2
9 min per mile		9.7	10.2	10.8	11.4	12.0	12.5	13.1
8 min per mile		10.8	11.3	11.9	12.5	13.1	13.6	14.2
7 min per mile		12.2	12.7	13.3	13.9	14.5	15.0	15.6
6 min per mile		13.9	14.4	15.0	15.6	16.2	16.7	17.3
5 min, 30 s per mile		14.5	15.3	16.2	17.1	17.9	18.8	19.7
Scraping paint and sandpapering		3.2	3.3	3.5	3.7	3.9	4.1	4.3
Sitting quietly		1.1	1.1	1.2	1.2	1.3	1.4	1.4

| 71 | 74 | 77 | 80 | 83 | 86 | 89 | 92 | 95 | 98 |
157	163	170	176	183	190	196	203	209	216
1.6	1.6	1.7	1.8	1.8	1.9	2.0	2.0	2.1	2.2
3.4	3.6	3.7	3.8	4.0	4.1	4.3	4.4	4.6	4.7
3.7	3.8	4.0	4.2	4.3	4.5	4.6	4.8	4.9	5.1
6.2	6.5	6.8	7.0	7.3	7.6	7.8	8.1	8.4	8.6
4.6	4.8	5.0	5.2	5.4	5.6	5.8	6.0	6.2	6.4
3.7	3.8	4.0	4.2	4.3	4.5	4.6	4.8	4.9	5.1
3.4	3.6	3.7	3.8	4.0	4.1	4.3	4.4	4.6	4.7
10.1	10.5	10.9	11.4	11.8	12.2	12.6	13.1	13.5	13.9
2.3	2.4	2.5	2.6	2.7	2.8	2.8	2.9	3.0	3.1
2.9	3.0	3.2	3.3	3.4	3.5	3.6	3.8	3.9	4.0
2.8	2.9	3.0	3.1	3.2	3.4	3.5	3.6	3.7	3.8
4.7	4.9	5.1	5.3	5.5	5.7	5.9	6.1	6.3	6.6
2.5	2.6	2.7	2.8	2.9	3.0	3.1	3.2	3.3	3.4
2.1	2.1	2.2	2.3	2.4	2.5	2.6	2.7	2.8	2.8
3.8	3.9	4.1	4.2	4.4	4.6	4.7	4.9	5.0	5.2
2.8	3.0	3.1	3.2	3.3	3.4	3.6	3.7	3.8	3.9
2.2	2.3	2.4	2.5	2.6	2.7	2.8	2.9	2.9	3.0
3.2	3.3	3.5	3.6	3.7	3.9	4.0	4.1	4.3	4.4
2.3	2.4	2.5	2.6	2.7	2.8	2.8	2.9	3.0	3.1
2.4	2.5	2.6	2.7	2.8	2.9	3.0	3.1	3.2	3.3
5.5	5.7	5.9	6.2	6.4	6.6	6.9	7.1	7.3	7.5
5.0	5.2	5.4	5.6	5.8	6.0	6.2	6.4	6.7	6.9
5.5	5.8	6.0	6.2	6.5	6.7	6.9	7.2	7.4	7.6
2.5	2.6	2.7	2.8	2.9	3.0	3.1	3.2	3.3	3.4
11.6	12.1	12.6	13.0	13.5	14.0	14.5	15.0	15.5	16.0
9.6	10.0	10.5	10.9	11.3	11.7	12.1	12.5	12.9	13.3
13.7	14.3	14.9	15.4	16.0	16.6	17.2	17.8	18.3	18.9
14.8	15.4	16.0	16.5	17.1	17.7	18.3	18.9	19.4	20.0
16.2	16.8	17.4	17.9	18.5	19.1	19.7	20.3	20.8	21.4
17.9	18.5	19.1	19.6	20.2	20.8	21.4	22.0	22.5	23.1
20.5	21.4	22.3	23.1	24.0	24.9	25.7	26.6	27.5	28.3
4.5	4.7	4.9	5.0	5.2	5.4	5.6	5.8	6.0	6.2
1.5	1.6	1.6	1.7	1.7	1.8	1.9	1.9	2.0	2.1

Activity	Body Weight						
kg	50	53	56	59	62	65	68
lb	110	117	123	130	137	143	150
Skiing, hard snow							
Level, moderate speed	6.0	6.3	6.7	7.0	7.4	7.7	8.1
Level, walking	7.2	7.6	8.0	8.4	8.9	9.3	9.7
Uphill, maximum speed	13.7	14.5	15.3	16.2	17.0	17.8	18.6
Skiing, soft snow							
Leisure (female)	4.9	5.2	5.5	5.8	6.1	6.4	6.7
Leisure (male)	5.6	5.9	6.2	6.5	6.9	7.2	7.5
Skindiving, as frogman							
Considerable motion	13.8	14.6	15.5	16.3	17.1	17.9	18.8
Moderate motion	10.3	10.9	11.5	12.2	12.8	13.4	14.0
Snowshoeing, soft snow	8.3	8.8	9.3	9.8	10.3	10.8	11.3
Squash	10.6	11.2	11.9	12.5	13.1	13.8	14.4
Standing quietly	1.4	1.4	1.5	1.6	1.7	1.8	1.8
Steel mill, working in							
Fettling	4.5	4.7	5.0	5.3	5.5	5.8	6.1
Forging	5.0	5.3	5.6	5.9	6.2	6.5	6.8
Hand rolling	6.9	7.3	7.7	8.1	8.5	8.9	9.3
Merchant mill rolling	7.3	7.7	8.1	8.6	9.0	9.4	9.9
Removing slag	8.9	9.4	10.0	10.5	11.0	11.6	12.1
Tending furnace	6.3	6.7	7.1	7.4	7.8	8.2	8.6
Tipping molds	4.6	4.9	5.2	5.4	5.7	6.0	6.3
Stock clerking	2.7	2.9	3.0	3.2	3.3	3.5	3.7
Swimming							
Backstroke	8.5	9.0	9.5	10.0	10.5	11.0	11.5
Breast stroke	8.1	8.6	9.1	9.6	10.0	10.5	11.0
Crawl, fast	7.8	8.3	8.7	9.2	9.7	10.1	10.6
Crawl, slow	6.4	6.8	7.2	7.6	7.9	8.3	8.7
Side stroke	6.1	6.5	6.8	7.2	7.6	7.9	8.3
Treading, fast	8.5	9.0	9.5	10.0	10.5	11.1	11.6
Treading, normal	3.1	3.3	3.5	3.7	3.8	4.0	4.2
Table tennis	3.4	3.6	3.8	4.0	4.2	4.4	4.6
Tailoring							
Cutting	2.1	2.2	2.3	2.4	2.5	2.7	2.8
Hand-sewing	1.6	1.7	1.8	1.9	2.0	2.1	2.2
Machine-sewing	2.3	2.4	2.5	2.7	2.8	2.9	3.1
Pressing	3.1	3.3	3.5	3.7	3.8	4.0	4.2
Tennis	5.5	5.8	6.1	6.4	6.8	7.1	7.4

| 71 | 74 | 77 | 80 | 83 | 86 | 89 | 92 | 95 | 98 |
157	163	170	176	183	190	196	203	209	216
8.4	8.8	9.2	9.5	9.9	10.2	10.6	10.9	11.3	11.7
10.2	10.6	11.0	11.4	11.9	12.3	12.7	13.2	13.6	14.0
19.5	20.3	21.1	21.9	22.7	23.6	24.4	25.2	26.0	26.9
7.0	7.3	7.5	7.8	8.1	8.4	8.7	9.0	9.3	9.6
7.9	8.2	8.5	8.9	9.2	9.5	9.9	10.2	10.5	10.9
19.6	20.4	21.3	22.1	22.9	23.7	24.6	25.4	26.2	27.0
14.6	15.2	15.9	16.5	17.1	17.7	18.3	19.0	19.6	20.2
11.8	12.3	12.8	13.3	13.8	14.3	14.8	15.3	15.8	16.3
15.1	15.7	16.3	17.0	17.6	18.2	18.9	19.5	20.1	20.8
1.9	2.0	2.1	2.2	2.2	2.3	2.4	2.5	2.6	2.6
6.3	6.6	6.9	7.1	7.4	7.7	7.9	8.2	8.5	8.7
7.1	7.4	7.7	8.0	8.3	8.6	8.9	9.2	9.5	9.8
9.7	10.1	10.6	11.0	11.4	11.8	12.2	12.6	13.0	13.4
10.3	10.7	11.2	11.6	12.0	12.5	12.9	13.3	13.8	14.2
12.6	13.2	13.7	14.2	14.8	15.3	15.8	16.4	16.9	17.4
8.9	9.3	9.7	10.1	10.5	10.8	11.2	11.6	12.0	12.3
6.5	6.8	7.1	7.4	7.6	7.9	8.2	8.5	8.7	9.0
3.8	4.0	4.2	4.3	4.5	4.6	4.8	5.0	5.1	5.3
12.0	12.5	13.0	13.5	14.0	14.5	15.0	15.5	16.1	16.6
11.5	12.0	12.5	13.0	13.4	13.9	14.4	14.9	15.4	15.9
11.1	11.5	12.0	12.5	12.9	13.4	13.9	14.4	14.8	15.3
9.1	9.5	9.9	10.2	10.6	11.0	11.4	11.8	12.2	12.5
8.7	9.0	9.4	9.8	10.1	10.5	10.9	11.2	11.6	12.0
12.1	12.6	13.1	13.6	14.1	14.6	15.1	15.6	16.2	16.7
4.4	4.6	4.8	5.0	5.1	5.3	5.5	5.7	5.9	6.1
4.8	5.0	5.2	5.4	5.6	5.8	6.1	6.3	6.5	6.7
2.9	3.0	3.2	3.3	3.4	3.5	3.6	3.8	3.9	4.0
2.3	2.4	2.5	2.6	2.7	2.8	2.8	2.9	3.0	3.1
3.2	3.3	3.5	3.6	3.7	3.9	4.0	4.1	4.3	4.4
4.4	4.6	4.8	5.0	5.1	5.3	5.5	5.7	5.9	6.1
7.7	8.1	8.4	8.7	9.0	9.4	9.7	10.0	10.4	10.7

Activity	kg lb	Body Weight						
		50 110	53 117	56 123	59 130	62 137	65 143	68 150
Typing								
Electric		1.4	1.4	1.5	1.6	1.7	1.8	1.8
Manual		1.6	1.6	1.7	1.8	1.9	2.0	2.1
Volleyball		2.5	2.7	2.8	3.0	3.1	3.3	3.4
Walking, comfortable pace								
Asphalt road		4.0	4.2	4.5	4.7	5.0	5.2	5.4
Fields and hillsides		4.1	4.3	4.6	4.8	5.1	5.3	5.6
Grass track		4.1	4.3	4.5	4.8	5.0	5.3	5.5
Plowed field		3.9	4.1	4.3	4.5	4.8	5.0	5.2
Wallpapering		2.4	2.5	2.7	2.8	3.0	3.1	3.3
Watch repairing		1.3	1.3	1.4	1.5	1.6	1.6	1.7
Writing (sitting)		1.5	1.5	1.6	1.7	1.8	1.9	2.0

SOURCES: Bannister, E.W. and S.R. Brown, "The relative energy requirements of physical activity." In: Falls, H.B. (ed), Exercise Physiology, Academic Press, New York, 1968; E.T. Howley and M.E. Glover, "The caloric costs of running and walking one mile for men and women," Med. Sci. Sports 6:235–237, 1974; R. Passmore and J.V.G.A. Durnin, "Human energy expenditure," Physiol. Rev. 35:801–840, 1955.

71	74	77	80	83	86	89	92	95	98
157	163	170	176	183	190	196	203	209	216
1.9	2.0	2.1	2.2	2.2	2.3	2.4	2.5	2.6	2.6
2.2	2.3	2.4	2.5	2.6	2.7	2.8	2.9	2.9	3.0
3.6	3.7	3.9	4.0	4.2	4.3	4.5	4.6	4.8	4.9
5.7	5.9	6.2	6.4	6.6	6.9	7.1	7.4	7.6	7.8
5.8	6.1	6.3	6.6	6.8	7.1	7.3	7.5	7.8	8.0
5.8	6.0	6.2	6.5	6.7	7.0	7.2	7.5	7.7	7.9
5.5	5.7	5.9	6.2	6.4	6.6	6.9	7.1	7.3	7.5
3.4	3.6	3.7	3.8	4.0	4.1	4.3	4.4	4.6	4.7
1.8	1.9	1.9	2.0	2.1	2.2	2.2	2.3	2.4	2.5
2.1	2.1	2.2	2.3	2.4	2.5	2.6	2.7	2.8	2.8

SYMPATHOMIMETIC DRUGS AND THEIR ACTIONS

Receptor	Stimulator	Inhibitor
α	Norepinephrine Phenylephrine Epinephrine (low)	Phentolamine Phenoxybenzamine
$\beta 1$	Isoproterenol Epinephrine Norepinephrine	Atenolol Propranolol Labetalol
$\beta 2$	Isoproterenol Epinephrine	Propranolol Labetalol

SOURCE: Shepherd, J.T. and D.M. Vanhoutle, The Human Cardiovascular System, Raven Press, New York, 1979.

IV

UNITS AND MEASURES

Note: The basic unit of measurement adopted by the System International D'Unites will be denoted as the basic SI unit.

ACCELERATION

Basic SI unit: meter per second squared ($m \cdot s^{-2}$)
1 centimeter per second squared ($cm \cdot s^{-2}$) = 10^2 $m \cdot s^2$
1 kilometer per hour per second ($km \cdot h^{-1} \cdot s^{-1}$ = 2.7×10^{-1} $m \cdot s^{-2}$
Common Anglo-Saxon units of acceleration:
1 foot per second squared ($ft \cdot s^{-2}$) = 3.048×10^{-1} $m \cdot s^{-2}$
1 mile per hour per second ($mile \cdot h^{-1} \cdot s^{-1}$) = 4.470×10^{-1} $m \cdot s^{-2}$

ANGLE

Basic SI unit: radian (rad)
1 degree ($°$) = 1.754×10^{-2} rad
1 minute ($'$) = 2.908×10^{-4} rad
1 second ($''$) = 4.848×10^{-6} rad
1 grade (g) = 1.570×10^{-2} rad

ANGULAR VELOCITY

Basic SI unit: radian per second (rad·s^{-1})
1 degree per minute (° min^{-1}) = 2.908 × 10^4 rads · s^{-1}
1 radian per minute (rad min^{-1} = 1.6 × 10^{-2} rad · s^{-1}
1 degree per second (° s^{-1}) = 1.745 × 10^{-2} rad · s^{-1}

AREA

Basic SI unit: square meter (m^2)
1 square micrometer (m^2) = 10^{-12} m^2
1 square millimeter (mm^2) = 10^{-6} m^2
1 square centimeter (cm^2) = 10^{-4} m^2
1 square decameter (dm^2) = 10^2 m^2
1 square hectometer (hm^2) = 10^4 m^2
1 square kilometer (km^2) = 10^6 m^2
Common Anglo-Saxon units of area:
1 square inch (in.^{2}) = 6.451 × 10^{-4} m^2
1 square foot (ft^2) = 9.290 × 10^{-2} m^2
1 square yard (yd^2) = 8.361 × 10^{-1} m^2
1 acre (^{ac2}) = 4.046 × 10^3 m^2
1 square mile (mi^2) = 2.589 × 10^6 m^2

DENSITY

Basic SI unit: kilogram per cubic meter (kg·m^{-3})
1 microgram per cubic centimeter (μg·cm^{-3}) = 10^{-3} kg·m^{-3}
1 milligram per cubic centimeter (mg·cm^{-3}) = 1 kg·m^{-3}
1 gram per cubic centimeter (g·cm^{-3}) = 10^3 kg·m^{-3}
Common Anglo-Saxon unit of density:
1 pound per cubic foot = 1.601 × 10 kg·m^{-3}

FORCE

Basic SI unit: newton (N)
1 pond (p) = 9.80665 × 10^{-3} N
1 kilopond (kp) = 9.80665 N
1 dyne (dyn) = 10^{-5} N
Common Anglo-Saxon unit of force:
1 foot-pound (ft lb · s^{-2}) = 0.13825495437 N

FREQUENCY

Basic SI unit: hertz (Hz)
1 kilohertz (kHz) = 1000 Hz
Common Anglo-Saxon unit of frequency:
1 cycle per second (c/s) = 1 Hz

LENGTH

Basic SI unit: meter (m):
1 angstrom (Å) = 10^{-10} m
1 nanometer (nm) = 10^{-9} m
1 micrometer (μm) = 10^{-6} m
1 millimeter (mm) = 10^{-3} m
1 centimeter (cm) = 10^{-2} m
1 decimeter (dm) = 10^{-1} m
1 deeameter (dam) = 10 m
1 hectometer (hm) = 10^2 m
1 kilometer (km) = 10^3 m
Common Anglo-Saxon units of length:
1 inch (in.) = 2.540×10^{-2} m
1 foot (ft) = 3.048×10^{-1} m
1 yard (yd) = 9.144×10^{-1} m
1 fathom (fath) = 1.828 m
1 furlong (fur) = 2.011×10^2 m
1 statute mile (mi) = 1.609×10^3

LINEAR VELOCITY

Basic SI unit: meter per second (m $\cdot$ s^{-1})
1 centimeter per sec (cm $\cdot$s^{-1}) = 10^{-2}m $\cdot$s^{-2}
1 kilometer per hour (km $\cdot$h^{-1}) = 2.7×10^{-1} m $\cdot$s^{-1}
Common Anglo-Saxon units of linear velocity:
1 foot per second (ft $\cdot$s^{-1}) = $3.048 \times^{-1}$ m $\cdot$s^{-1}
1 mile per hour (mile $\cdot$h^{-1}) = 4.470×10^{-1} m $\cdot$s^{-1}

MASS

Basic SI unit: kilogram (kg)
1 picogram (pg) = 10^{-15} kg

1 nanogram (ng) = 10^{-12} kg
1 microgram (μg) = 10^{-9} kg
1 milligram (mg) = 10^{-6} kg
1 gram (g) = 10^{-3} kg
1 metric ton (t) = 10^3 kg
Common Anglo-Saxon units of mass:
1 ounce (oz) = 2.834×10^{-2} kg
1 pound (lb) = 4.535×10^{-1} kg
1 short ton (sh tn) = 9.071×10^2 kg

POWER

Basic SI unit: watt (W)
1 joule per second (J·s^{-1}) = 1 W
1 erg per second (erg s^{-1}) = 10^7 W
1 kilopond meter per minute (kpm·min^{-1}) = 0.1635 W
Common Anglo-Saxon units of power:
1 British thermal unit per hour (Btu·h^{-1}) = 2.931×10^{-1} W
1 horsepower (hp) = 7.457×10^2 W

PRESSURE

Basic SI unit: newton per square meter (N·m^{-2})
Common laboratory unit: millimeters of mercury (mmHg)
1 mmHg = 133.322 N·m^{-2}
1 atmosphere (atm) = 101325 N·m^{-2}
1 torr (torr) = 133.322368 N·m^{-2}

TEMPERATURE

Basic SI unit: Kelvin (K)
Common laboratory unit: degrees Celsius (°C). Note: The Celsius scale is subdivided into the same intervals as the Kelvin scale but has its zero point displaced by 273.15 K.
1 degree Celsius (°C) = 1 K
1 degree Celsius (°C) = 9/5 °Fahrenheit
Common Anglo-Saxon units of temperature:
1 degree Fahrenheit (°F) = 5/9 K
1 degree Rankine (°R) = 5/9 K

TIME

Basic SI unit: second (s)
1 minute (min) = 6×10 s
1 hour (hr) = 3.6×10^3 s
1 day (d) = 8.64×10^4 s
1 year (y, 365 d) = 3.1536×10^7 s

VOLUME

Basic SI Unit: cubic meter (m^3)
Common laboratory unit: liter (l)
1 cubic nanometer (nm^3) = 10^{-27} m^3
1 cubic micrometer (μm^3) 10^{-18} m^3
1 cubic millimeter (mm^3) = 10^{-9} m^3 = 1 microliter (μl)
1 cubic centimeter (cc^3) = 10^{-6} m^3 = 1 milliter (ml)
1 cubic decimeter (dm^3) = 10^{-3} m^3 = 1 liter (l)
1 cubic kilometer (km^3) = 10^9 m^3 = 1 hectoliter (hl)
Common Anglo-Saxon units of volume:
1 cubic inch (in.^{3}) = 1.638×10^{-5} m^3
1 cubic foot (ft^3) = 2.831×10^{-2} m^3
1 cubic yard (yd^3) = 7.645×10^{-1} m^3
1 fluid ounce (fl oz) = 2.957×10^{-2} l
1 liquid quart (liq qt) = 9.463×10^{-1} l
1 gallon (gal) = 3.785 l

WORK AND ENERGY

Basic SI unit: joule (J)
1 kilocalorie (kcal) = 4186 J
1 kilopond meter (kp m) = 9.807 J
1 Newton meter (Nm) = 1 J
Common Anglo-Saxon units of power:
1 British thermal unit (Btu) = 1.055×10^3 J
1 horsepower-hour (hph) = 2.685×10^6 J
1 foot pound-force (ft lbf) = 1.356 J

AMERICAN COLLEGE OF SPORTS MEDICINE POSITION STATEMENT ON PREVENTION OF HEAT INJURIES DURING DISTANCE RUNNING

The Purpose of this Position Statement is:

(a) To alert local, national and international sponsors of distance running events of the health hazards of heat injury during distance running, and

(b) To inform said sponsors of injury preventive actions that may reduce the frequency of this type of injury.

The recommendations address only the manner in which distance running sports activities may be conducted to further reduce incidence of heat injury among normal athletes conditioned to participate in distance running. The Recommendations Are Advisory Only.

Recommendations concerning the ingested quantity and content of fluid are merely a partial preventive to heat injury. The physiology of each individual athlete varies; strict compliance with these recommendations and the current rules governing distance running may not reduce the incidence of heat injuries among those so inclined to such injury.

RESEARCH FINDINGS

Based on research findings and current rules governing distance running competition, it is the position of the American College of Sports Medicine that:

SOURCE: Med. Sci. Sports 7(1):vii–viii, 1975. Reprinted by permission of ACSM. Also see Appendix XII.

1. Distance races ($>$ 16 km or 10 miles) should *not* be conducted when the wet bulb temperature—globe temperature (adapted from Minard, D. Prevention of heat casualties in Marine Corps recruits. *Milit. Med.* 126:261, 1961. WB-GT = 0.7 [WBT] + 0.2 [GT] + 0.1 [DBT]) exceeds 28°C (82.4° F).

2. During periods of the year, when the daylight dry bulb temperature often exceeds 27°C (80°F), distance races should be conducted before 9:00 A.M. or after 4:00 P.M.

3. It is the responsibility of the race sponsors to provide fluids which contain small amounts of sugar (less than 2.5 g glucose per 100 ml of water) and electrolytes (less than 10 mEq sodium and 5 mEq potassium per liter of solution).

4. Runners should be encouraged to frequently ingest fluids during competition and to consume 400–500 ml (13–17 oz.) of fluid 10–15 minutes before competition.

5. Rules prohibiting the administration of fluids during the first 10 kilometers (6.2 miles) of a marathon race should be amended to permit fluid ingestion at frequent intervals along the race course. In light of the high sweat rates and body temperatures during distance running in the heat, race sponsors should provide "water stations" at 3–4 kilometer (2–2.5 mile) intervals for all races of 16 kilometers (10 miles) or more.

6. Runners should be instructed in how to recognize the early warning symptoms that precede heat injury. Recognition of symptoms, cessation of running, and proper treatment can prevent heat injury. Early warning symptoms include the following: piloerection on chest and upper arms, chilling, throbbing pressure in the head, unsteadiness, nausea, and dry skin.

7. Race sponsors should make prior arrangements with medical personnel for the care of cases of heat injury. Responsible and informed personnel should supervise each "feeding station." Organizational personnel should reserve the right to stop runners who exhibit clear signs of heat stroke or heat exhaustion.

It is the position of the American College of Sports Medicine that policies established by local, national, and international sponsors of distance running events should adhere to these guidelines. Failure to adhere to these guidelines may jeopardize the health of competitors through heat injury.

The requirements of distance running place great demands on both circulation and body temperature regulation. Numerous studies have reported rectal temperatures in excess of 40.6°C (105°F) after races of 6 to 26.2 miles (9.6 to 41.9 kilometers). Attempting to counterbalance such overheating, runners incur large sweat losses of 0.8 to 1.1 liters/m²/hr. The resulting body water deficit may total 6–10% of the athlete's body weight. Dehydration of these proportions severely limits subsequent sweating, places dangerous demands on circulation, reduces exercise capacity and exposes the runner to the health hazards associated with hyperthermia (heat stroke, heat exhaustion and muscle cramps).

Under moderate thermal conditions [e.g., 65–70°F (18.5–21.3°C)], no cloud cover, relative humidity 49–55%, the risk of overheating is still a serious threat to highly motivated distance runners. Nevertheless, distance races are frequently conducted under more severe conditions than these. The air temperature at the 1967 U.S. Pan American Marathon Trial, for example, was 92–95°F (33.6–35.3°C). Many highly conditioned athletes failed to finish the race and several of the competitors demonstrated overt symptoms of heat stroke (no sweating, shivering and lack of orientation).

The above consequences are compounded by the current popularity of distance running among the middle-aged and aging men and women who may possess significantly less heat tolerance than their younger counterparts. In recent years, races of 10 to 26.2 miles (16 to 41.9 kilometers) have attracted several thousand runners. Since it is likely that distance running enthusiasts will continue to sponsor races under adverse heat conditions, specific steps should be taken to minimize the health threats which accompany such endurance events.

Fluid ingestion during prolonged running (two hours) has been shown to effectively reduce rectal temperature and minimize dehydration. Although most competitors consume fluids during races that exceed 1–1.5 hours, current international distance running rules prohibit the administration of fluids until the runner has completed 10 miles (16 kilometers). Under such limitations, the competitor is certain to accumulate a large body water deficit (−3%) before any fluids would be ingested. To make the problem more complex, most runners are unable to judge the volume of fluids they consume during competition. At the 1968 U.S. Olympic Marathon Trial, it was observed that there were body weight losses of 6.1 kg, with an average total fluid ingestion of only 0.14 to 0.35 liters. It seems obvious that the rules and habits which prohibit fluid administration during distance running preclude any benefits which might be gained from this practice.

Runners who attempt to consume large volumes of sugar solution during competition complain of gastric discomfort (fullness) and an inability to consume fluids after the first few feedings. Generally speaking, most runners drink solutions containing 5–20 grams of sugar per 100 milliliters of water. Although saline is rapidly emptied from the stomach (25 ml/min), the addition of even small amounts of sugar can drastically impair the rate of gastric emptying. During exercise in the heat, carbohydrate supplementation is of secondary importance and the sugar content of the oral feedings should be minimized.

APPENDIX

AMERICAN COLLEGE OF SPORTS MEDICINE POSITION STAND ON WEIGHT LOSS IN WRESTLERS

Despite repeated admonitions by medical, educational and athletic groups, most wrestlers have been inculcated by instruction or accepted tradition to lose weight in order to be certified for a class that is lower than their preseason weight. Studies of weight losses in high school and college wrestlers indicate that from 3–20% of the preseason body weight is lost before certification or competition occurs. Of this weight loss, most of the decrease occurs in the final days or day before the official weigh-in with the youngest and/or lightest members of the team losing the highest percentage of their body weight. Under existing rules and practices, it is not uncommon for an individual to repeat this weight losing process many times during the season because successful wrestlers compete in 15–30 matches/year.

Contrary to existing beliefs, most wrestlers are not "fat" before the season starts. In fact, the fat content of high school and college wrestlers weighing less than 190 pounds has been shown to range from 1.6 to 15.1 percent of their body weight with the majority possessing less than 8%. It is well known and documented that wrestlers lose body weight by a combination of food restriction, fluid deprivation and sweating induced by thermal or exercise procedures. Of these methods, dehydration through sweating appears to be the method most frequently chosen.

SOURCE: Med. Sci. Sports 8(2):xi–xiii, 1976. Reprinted by permission of ACSM.

Careful studies on the nature of the weight being lost show that water, fats and proteins are lost when food restriction and fluid deprivation procedures are followed. Moreover, the proportionality between these constituents will change with continued restriction and deprivation. For example, if food restriction is held constant when the volume of fluid being consumed is decreased, more water will be lost from the tissues of the body than before the fluid restriction occurred. The problem becomes more acute when thermal or exercise dehydration occurs because electrolyte losses will accompany the water losses. Even when 1–5 hours are allowed for purposes of rehydration after the weigh-in, this time interval is insufficient for fluid and electrolyte homeostasis to be completely reestablished.

Since the "making of weight" occurs by combinations of food restriction, fluid deprivation and dehydration, responsible officials should realize that the single or combined effects of these practices are generally associated with (1) a reduction in muscular strength; (2) a decrease in work performance times; (3) lower plasma and blood volumes; (4) a reduction in cardiac functioning during submaximal work conditions which are associated with higher heart rates, smaller stroke volumes, and reduced cardiac outputs; (5) a lower oxygen consumption, especially with food restriction; (6) an impairment of thermoregulatory processes; (7) a decrease in renal blood flow and in the volume of fluid being filtered by the kidney; (8) a depletion of liver glycogen stores; and (9) an increase in the amount of electrolytes being lost from the body.

Since it is possible for these changes to impede normal growth and development, there is little physiological or medical justification for the use of the weight reduction methods currently followed by many wrestlers. These sentiments have been expressed in part within Rule 1, Section 3, Article 1 of the *Official Wrestling Rule Book* published by the National Federation of State High School Associations which states, "The Rules Committee recommends that individual state high school associations develop and utilize an effective weight control program which will discourage severe weight reduction and/or wide variations in weight, because this may be harmful to the competitor. . . ." However, until the National Federation of State High School Associations defines the meaning of the terms "severe" and "wide variations," this rule will be ineffective in reducing the abuses associated with the "making of weight."

Therefore, it is the position of the American College of Sports Medicine that the potential health hazards created by the procedures used to "make weight" by wrestlers can be eliminated if state and national organizations will:

1. Assess the body composition of each wrestler several weeks in advance of the competitive season. Individuals with a fat content less than five percent of their certified body weight should receive medical clearance before being allowed to compete.

2. Emphasize the fact that the daily caloric requirements of wrestlers should be obtained from a balanced diet and determined on the basis of age, body surface area, growth and physical activity levels. The minimal caloric needs of wrestlers in high schools and colleges will range from 1200 to 2400 kcal/day; therefore, it is the responsibility of coaches, school officials, physicians and parents to discourage wrestlers from securing less than their minimal needs without prior medical approval.

3. Discourage the practice of fluid deprivation and dehydration. This can be accomplished by:

(**a**) Educating the coaches and wrestlers on the physiological consequences and medical complications that can occur as a result of these practices.

(**b**) Prohibiting the single or combined use of rubber suits, steam rooms, hot boxes, saunas, laxatives, and diuretics to "make weight."

(**c**) Scheduling weigh-ins just prior to competition.

(**d**) Scheduling more official weigh-ins between team matches.

4. Permit more participants/team to compete in those weight classes (119–145 pounds) which have the highest percentages of wrestlers certified for competition.

5. Standardize regulations concerning the eligibility rules at championship tournaments so that individuals can only participate in those weight classes in which they had the highest frequencies of matches throughout the season.

6. Encourage local and county organizations to systematically collect data on the hydration state of wrestlers and its relationship to growth and development.

AMERICAN COLLEGE OF SPORTS MEDICINE POSITION STATEMENT ON THE USE AND ABUSE OF ANABOLIC-ANDROGENIC STEROIDS IN SPORTS

Based on a comprehensive survey of the world literature and a careful analysis of the claims made for and against the efficacy of anabolic-androgenic steroids in improving human physical performance, it is the position of the American Collete of Sports Medicine that:

1. The administration of anabolic-androgenic steroids to healthy humans below age 50 in medically approved therapeutic doses often does not of itself bring about any significant improvements in strength, aerobic endurance, lean body mass, or body weight.

2. There is no conclusive scientific evidence that extremely large doses of anabolic-androgenic steroids either aid or hinder athletic performance.

3. The prolonged use of oral anabolic-androgenic steroids (C_{17}-alkylated derivatives of testosterone) has resulted in liver disorders in some persons. Some of these disorders are apparently reversible with the cessation of drug usage, but others are not.

4. The administration of anabolic-androgenic steroids to male humans may result in a decrease in testicular size and function and a decrease in sperm production. Although these effects appear to be reversible when small doses of steroids are used for short periods of time, the reversibility of the effects of large doses over extended periods of time is unclear.

SOURCE: Med. Sci. Sports 9(4):xi–xiii, 1977. Reprinted by permission of ACSM.

5. Serious and continuing effort should be made to educate male and female athletes, coaches, physical educators, physicians, trainers, and the general public regarding the inconsistent effects of anabolic-androgenic steroids on improvement of human physical performance and the potential dangers of taking certain forms of these substances, especially in large doses, for prolonged periods.

RESEARCH BACKGROUND FOR THE POSITION STATEMENT

This position stand has been developed from an extensive survey and analysis of the world literature in the fields of medicine, physiology, endocrinology, and physical education. Although the reactions of humans to the use of drugs, including hormones or drugs which simulate the actions of natural hormones, are individual and not entirely predictable, some conclusions can nevertheless be drawn with regard to what desirable and what undesirable effects may be achieved. Accordingly, whereas positive effects of drugs may sometimes arise because persons have been led to expect such changes (''placebo'' effect), repeated experiments of a similar nature often fail to support the initial positive effects and lead to the conclusion that any positive effect that does exist may not be substantial.

1. Administration of testosteronelike synthetic drugs which have anabolic (tissue building) and androgenic (development of male secondary sex characteristics) properties in amounts up to twice those normally prescribed for medical use have been associated with increased strength, lean body mass and/or body weight in some studies but not in others. One study reported an increase in the amount of weight the steroid group could lift compared to controls but found no difference in isometric strength, which suggests a placebo effect in the drug group, a learning effect or possibly a differential drug effect on isotonic compared to isometric strength. An initial report of enhanced aerobic endurance after administration of an anabolic-androgenic steroid has not been confirmed. Because of the lack of adequate control groups in many studies it seems likely that some of the positive effects on strength that have been reported are due to ''placebo'' effects, but a few apparently well-designed studies have also shown beneficial effects of steroid administration on muscular strength and lean body mass. Some of the discrepancies in results may also be due to differences in the type of drug administered, the method of drug administration, the nature of the exercise programs involved, the duration of the experiment, and individual differences in sensitivity to the administered drug. High protein dietary supplements do not insure the effectiveness of the steroids. Because of the many failures to show improved muscular strength, lean body mass, or body weight

after therapeutic doses of anabolic-androgenic steroids it is obvious that for many individuals any benefits are likely to be small and not worth the health risks involved.

2. Testimonial evidence by individual athletes suggests that athletes often use much larger doses of steroids than those ordinarily prescribed by physicians and those evaluated in published research. Because of the health risks involved with the long-term use of high doses and requirements for informed consent it is unlikely that scientifically acceptable evidence will be forthcoming to evaluate the effectiveness of such large doses of drugs on athletic performance.

3. Alterations of normal liver function have been found in as many as 80 percent of one series of 69 patients treated with C_{17}-alkylated testosterone derivatives (oral anabolic-androgenic steroids). Cholestasis has been observed histologically in the livers of persons taking these substances. These changes appear to be benign and reversible. Five reports, document the occurrence of peliosis hepatitis in 17 patients without evidence of significant liver disease who were treated with C_{17}-alkylated androgenic steroids. Seven of these patients died of liver failure. The first case of hepato-cellular carcinoma associated with taking an androgenic-anabolic steroid was reported in 1965. Since then at least 13 other patients taking C_{17}-alkylated androgenic steroids have developed hepato-cellular carcinoma. In some cases dosages as low as 10–15 mg/day taken for only three or four months have caused liver complications.

4. Administration of therapeutic doses of androgenic-anabolic steroids in men often, but not always, reduces the output of testosterone and gonadrotopins and reduces spermatogenesis. Some steroids are less potent than others in causing these effects. Although these effects on the reproductive system appear to be reversible in animals, the long-term results of taking large doses by humans is unknown.

5. Precise information concerning the abuse of anabolic steroids by female athletes is unavailable. Nevertheless, there is no reason to believe females will not be tempted to adopt the use of these medicines. The use of anabolic steroids by females, particularly those who are either prepubertal or have not attained full growth, is especially dangerous. The undesired side effects include masculinization, disruption of normal growth pattern, voice changes, acne, hirsutism, and enlargement of the clitoris. The long-term effects on reproductive function are unknown, but anabolic steroids may be harmful in this area. Their ability to interfere with the menstrual cycle has been well documented.

For these reasons, all concerned with advising, training, coaching, and providing medical care for female athletes should exercise all persuasions available to prevent the use of anabolic steroids by female athletes.

AMERICAN COLLEGE OF SPORTS MEDICINE POSITION STATEMENT ON THE RECOMMENDED QUANTITY AND QUALITY OF EXERCISE FOR DEVELOPING AND MAINTAINING FITNESS IN HEALTHY ADULTS

Increasing numbers of persons are becoming involved in endurance training activities and thus, the need for guidelines for exercise prescription is apparent.

Based on the existing evidence concerning exercise prescription for healthy adults and the need for guidelines, the American College of Sports Medicine makes the following recommendations for the quantity and quality of training for developing and maintaining cardiorespiratory fitness and body composition in the healthy adult:

1. Frequency of training: 3 to 5 days per week.
2. Intensity of training: 60% to 90% of maximum heart rate reserve or, 50% to 85% of maximum oxygen uptake ($\dot{V}_{O_2max}$).

SOURCE: Med. Sci. Sports 10(3): vii–ix, 1978. Reprinted by permission of ACSM.

3. Duration of training: 15 to 60 minutes of continuous aerobic activity. Duration is dependent on the intensity of the activity, thus lower intensity activity should be conducted over a longer period of time. Because of the importance of the "total fitness" effect and the fact that it is more readily attained in longer duration programs, and because of the potential hazards and compliance problems associated with high intensity activity, lower to moderate intensity activity of longer duration is recommended for the nonathletic adult.
4. Mode of activity: Any activity that uses large muscle groups, that can be maintained continuously, and is rhythmical and aerobic in nature (e.g. running-jogging, walking-hiking, swimming, skating, bicycling, rowing, cross-country skiing, rope skipping, and various endurance game activities).

RATIONALE AND RESEARCH BACKGROUND

The questions, "How much exercise is enough and what type of exercise is best for developing and maintaining fitness?", are frequently asked. It is recognized that the term "physical fitness" is composed of a wide variety of variables included in the broad categories of cardiovascular-respiratory fitness, physique and structure, motor function, and many histochemical and biochemical factors. It is also recognized that the adaptive response to training is complex and includes peripheral, central, structural, and functional factors. Although many such variables and their adaptive response to training have been documented, the lack of sufficient in-depth and comparative data relative to frequency, intensity, and duration of training make them inadequate to use as comparative models. Thus, in respect to the above questions, fitness will be limited to changes in $\dot{V}_{O_2max}$, total body mass, fat weight (FW), and lean body weight (LBW) factors.

Exercise prescription is based upon the frequency, intensity, and duration of training, the mode of activity (aerobic in nature [e.g., listed under No. 4 above]), and the initial level of fitness. In evaluating these factors, the following observations have been derived from studies conducted with endurance training programs.

1. Improvement in $\dot{V}_{O_2max}$ is directly related to frequency, intensity, and duration of training. Depending upon the quantity and quality of training, improvement in $\dot{V}_{O_2max}$ ranges from 5 to 25%. Although changes in $\dot{V}_{O_2max}$ greater than 25% have been shown, they are usually associated with large total body mass and FW loss, or a low initial level of fitness. Also, as a result of leg fatigue or a lack of motivation, persons with low initial fitness may have spuriously low initial $\dot{V}_{O_2max}$ values.
2. The amount of improvement in $\dot{V}_{O_2max}$ tends to plateau when frequency of training is increased above 3 days per week. For the non-athlete, there is

not enough information available at this time to speculate on the value of added improvement found in programs that are conducted more than 5 days per week. Participation of less than two days per week does not show an adequate change in $\dot{V}_{O_2max}$.

3. Total body mass and FW are generally reduced with endurance training programs, while LBW remains constant or increases slightly. Programs that are conducted at least 3 days per week, of at least 20 minutes duration and of sufficient intensity and duration to expend approximately 300 kilocalories (kcal) per exercise session are suggested as a threshold level for total body mass and FW loss. An expenditure of 200 kcal per session has also been shown to be useful in weight reduction if the exercise frequency is at least 4 days per week. Programs with less participation generally show little or no change in body composition. Significant increases in $\dot{V}_{O_2max}$ have been shown with 10 to 15 minutes of high intensity training, thus, if total body mass and FW reduction is not a consideration, then short duration, high intensity programs may be recommended for healthy, low risk (cardiovascular disease) persons.

4. The minimal threshold level for improvement in $\dot{V}_{O_2max}$ is approximately 60% of the maximum heart rate reserve (50% of $\dot{V}_{O_2max}$). Maximum heart rate reserve represents the percent difference between resting and maximum heart rate, added to the resting heart rate The technique as described by Karvonen, Kentala, and Mustala,[1] was validated by Davis and Convertino,[2] and represents a heart rate of approximately 130 to 135 beats/minute for young persons. As a result of the aging curve for maximum heart rate, the absolute heart rate value (threshold level) is inversely related to age, and can be as low as 110 to 120 beats/minute for older persons. Initial level of fitness is another important consideration in prescribing exercise. The person with a low fitness level can get a significant training effect with a sustained training heart rate as low as 110 to 120 beats/minute, while persons of higher fitness levels need a higher threshold of stimulation.

5. Intensity and duration of training are interrelated with the total amount of work accomplished being an important factor in improvement in fitness. Although more comprehensive inquiry is necessary, present evidence suggests that when exercise is performed above the minimal threshold of intensity, the total amount of work accomplished is the important factor in fitness development and maintenance. That is, improvement will be similar for activities performed at a lower intensity-longer duration compared to higher intensity-shorter duration if the total energy cost of the activities is equal.

[1] M. Karvonen, K. Kentala, and O. Mustala, The effects of training heart rate: A longitudinal study. Ann. Med. Exptl. Biol. Fenn. 35: 307–315, 1957.

[2] J. A. Davis and V.A. Convertino, A comparison of heart rate methods for predicting endurance training intensity. Med. Sci. Sports 7: 295–298, 1975.

If frequency, intensity, and duration of training are similar (total kcal expenditure), the training result appears to be independent of the mode of aerobic activity. Therefore, a variety of endurance activities (e.g., listed above) may be used to derive the same training effect.

6. In order to maintain the training effect, exercise must be continued on a regular basis. A significant reduction in working capacity occurs after two weeks of detraining with participants returning to near pretraining levels of fitness after 10 weeks to 8 months of detraining. Fifty percent reduction in improvement of cardiorespiratory fitness has been shown after 4 to 12 weeks of detraining. More investigation is necessary to evaluate the rate of increase and decrease of fitness with varying training loads and reduction in training in relation to level of fitness, age, and length of time in training. Also, more information is needed to better identify the minimal level of work necessary to maintain fitness.

7. Endurance activities that require running and jumping generally cause significantly more debilitating injuries to beginning exercisers than other non-weight-bearing activities. One study showed that beginning joggers had increased foot, leg, and knee injuries when training was performed more than 3 days per week and longer than 30 minutes duration per exercise session. Thus, caution should be taken when recommending the type of activity and exercise prescription for the beginning exercise. Also, the increase of orthopedic injuries as related to overuse (marathon training) with chronic jogger-runners is apparent. Thus, there is a need for more inquiry into the effect that different types of activities and the quantity and quality of training has on short-term and long-term participation.

8. Most of the information concerning training described in this position statement has been conducted on men. The lack of information on women is apparent, but the available evidence indicates that women tend to adapt to endurance training in the same manner as men.

9. Age in itself does not appear to be a deterrent to endurance training. Although some earlier studies showed a lower training effect with middle-aged or elderly participants, more recent study shows the relative change in $\dot{V}_{O_2max}$ to be similar to younger age groups. Although more investigation is necessary concerning the rate of improvement in $\dot{V}_{O_2max}$ with age, at present it appears that elderly participants need longer periods of time to adapt to training. Earlier studies showing moderate to no improvement in $\dot{V}_{O_2max}$ were conducted over a short time-span or exercise was conducted at a moderate to low Kcal expenditure, thus making the interpretation of the results difficult.

Although $\dot{V}_{O_2max}$ decreases with age, and total body mass and FW increase with age, evidence suggests that this trend can be altered with endurance training. Also, 5 to 10 year follow-up studies where participants continued their training at a similar level showed maintenance of fitness. A study of older competitive runners showed decreases in $\dot{V}_{O_2max}$ from the fourth to seventh decade of life, but also showed reductions in their training load. More inquiry into the relationship of long-term training (quantity and quality) for both competitors and non-competitors and physiological function with increasing age, is necessary before more definitive statements can be made.

10. An activity such as weight training should not be considered as a means of training for developing $\dot{V}_{O_2max}$, but has significant value for increasing muscular strength and endurance, and LBW. Recent studies evaluating circuit weight training (weight training conducted almost continuously with moderate weights, using 10 to 15 repetitions per exercise session with 15 to 30 seconds rest between bouts of activity) showed little to no improvements in working capacity and $\dot{V}_{O_2max}$.

Despite an abundance of information available concerning the training of the human organism, the lack of standardization of testing protocols and procedures, methodology in relation to training procedures and experimental design, a preciseness in the documentation and reporting of the quantity and quality of training prescribed, make interpretation difficult. Interpretation and comparison of results are also dependent on the initial level of fitness, length of time of the training experiment, and specificity of the testing and training. For example, data from training studies using subjects with varied levels of $\dot{V}_{O_2max}$, total body mass and FW have found changes to occur in relation to their initial values (i.e., the lower the initial $\dot{V}_{O_2max}$ the larger the percent of improvement found, and the higher the FW the greater the reduction. Also, data evaluating trainability with age, comparison of the different magnitudes and quantities of effort, and comparison of the trainability of men and women may have been influenced by the initial fitness levels.

In view of the fact that improvement in the fitness variables discussed in this position statement continue over many months of training, it is reasonable to believe that short-term studies conducted over a few weeks have certain limitations. Middle-aged sedentary and older participants may take several weeks to adapt to the initial rigors of training, and thus need a longer adaptation period to get the full benefit from a program. How long a training experiment should be conducted is difficult to determine, but 15 to 20 weeks may be a good minimum standard. For example, two investigations conducted with middle-aged men who jogged either 2 or 4 days per week found both groups to improve in $\dot{V}_{O_2max}$. Mid-test results of the 16 and 20 week programs showed no difference between groups, while subsequent testing found the 4 day per week group to improve significantly more. In a similar study with young college men, no differences in $\dot{V}_{O_2max}$ were found among groups after 7 and 13 weeks of interval training. These latter findings and those of other investigators point to the limitations in interpreting results from investigations conducted over a short timespan.

In summary, frequency, intensity, and duration of training have been found to be effective stimuli for producing a training effect. In general, the lower the stimuli, the lower the training effect, and the greater the stimuli, the greater the effect. It has also been shown that endurance training less than two days per week, less than 50% of maximum oxygen uptake, and less than 10 minutes per day is inadequate for developing and maintaining fitness for healthy adults.

AMERICAN COLLEGE OF SPORTS MEDICINE POSITION STATEMENT ON THE USE OF ALCOHOL IN SPORTS

Based upon a comprehensive analysis of the available research relative to the effects of alcohol upon human physical performance, it is the position of the American College of Sports Medicine that:

1. The acute ingestion of alcohol can exert a deleterious effect upon a wide variety of psychomotor skills such as reaction time, hand-eye coordination, accuracy, balance, and complex coordination.

2. Acute ingestion of alcohol will not substantially influence metabolic or physiological functions essential to physical performance such as energy metabolism, maximal oxygen consumption ($\dot{V}_{O_2max}$), heart rate, stroke volume, cardiac output, muscle blood flow, arteriovenous oxygen difference, or respiratory dynamics. Alcohol consumption may impair body temperature regulation during prolonged exercise in a cold environment.

3. Acute alcohol ingestion will not improve and may decrease strength, power, local muscular endurance, speed, and cardiovascular endurance.

4. Alcohol is the most abused drug in the United States and is a major contributing factor to accidents and their consequences. Also, it has been documented widely that prolonged excessive alcohol consumption can elicit pathological changes in the liver, heart, brain, and muscle, which can lead to disability and death.

SOURCE: Med. Sci. Sports 14(6): ix–xi, 1982. Reprinted by permission of ACSM.

5. Serious and continuing efforts should be made to educate athletes, coaches, health and physical educators, physicians, trainers, the sports media, and the general public regarding the effects of acute alcohol ingestion upon human physical performance and on the potential acute and chronic problems of excessive alcohol consumption.

RESEARCH BACKGROUND FOR THE POSITION STATEMENT

This position statement is concerned primarily with the effects of acute alcohol ingestion upon physical performance and is based upon a comprehensive review of the pertinent international literature. When interpreting these results, several precautions should be kept in mind. First, there are varying reactions to alcohol ingestion, not only among individuals, but also within an individual depending upon the circumstances. Second, it is virtually impossible to conduct double-blind placebo research with alcohol because subjects can always tell when alcohol has been consumed. Nevertheless the results cited below provide us with some valid general conclusions relative to the effects of alcohol on physical performance. In most of the research studies, a small dose consisted of 1.5–2.0 ounces (45–60 ml) of alcohol, equivalent to a blood alcohol level (BAL) of 0.04–0.05 in the average-size male. A moderate dose was equivalent to 3–4 ounces (90–120 ml), or a BAL of about 0.10. Few studies employed a large dose, with a BAL of 0.15.

1. Athletes may consume alcohol to improve psychological function, but it is psychomotor performance that deteriorates most. A consistent finding is the impairment of information processing. In sports involving rapid reactions to changing stimuli, performance will be affected most adversely. Research has shown that small to moderate amounts of alcohol will impair reaction time, hand-eye coordination, accuracy, balance, and complex coordination or gross motor skills. Thus, while Coopersmith[1] suggests that alcohol may improve self-confidence, the available research reveals a deterioration in psychomotor performance.

2. Many studies have been conducted relative to the effects of acute alcohol ingestion upon metabolic and physiological functions important to physical performance. Alcohol ingestion exerts no beneficial influence relative to energy sources for exercise. Muscle glycogen at rest was significantly lower after alcohol compared to control. However, in exercise at 50% maximal oxygen uptake ($\dot{V}_{O_2max}$), total glycogen depletion in the leg muscles was not

---

[1] S. Coopersmith, The effects of alcohol on reaction to affective stimuli. Q. J. Stud. Alcohol 25: 459–475, 1964.

affected by alcohol. Moreover, Juhlin-Dannfelt et al.[2] have shown that although alcohol does not impair lipolysis or free fatty acid (FFA) utilization during exercise, it may decrease splanchnic glucose output, decrease the potential contribution from liver gluconeogenesis, elicit a greater decline in blood glucose levels leading to hypoglycemia, and decrease the leg muscle uptake of glucose during the latter stages of a 3-h run. Other studies have supported the theory concerning the hypoglycemic effect of alcohol during both moderate and prolonged exhaustive exercise in a cold environment. These studies also noted a significant loss of body heat and a resultant drop in body temperature and suggested alcohol may impair temperature regulation. These changes may impair endurance capacity.

In one study, alcohol has been shown to increase oxygen uptake significantly during submaximal work and simultaneously to decrease mechanical efficiency, but this finding has not been confirmed by others. Alcohol appears to have no effect on maximal or near-maximal $\dot{V}_{O_2}$.

The effects of alcohol on cardiovascular-respiratory parameters associated with oxygen uptake are variable at submaximal exercise intensities and are negligible at maximal levels. Alcohol has been shown by some investigators to increase submaximal exercise heart rate and cardiac output, but these heart rate findings have not been confirmed by others. Alcohol had no effect on stroke volume, pulmonary ventilation, or muscle blood flow at submaximal levels of exercise, but did decrease peripheral vascular resistance. During maximal exercise, alcohol ingestion elicited no significant effect upon heart rate, stroke volume and cardiac output, arteriovenous oxygen difference, mean arterial pressure and peripheral vascular resistance, or peak lactate, but did significantly reduce tidal volume resulting in a lowered pulmonary ventilation.

In summary, alcohol appears to have little or no beneficial effect on the metabolic and physiological responses to exercise. Further, in those studies reporting significant effects, the change appears to be deterimental to performance.

3. The effects of alcohol on tests of fitness components are variable. It has been shown that alcohol ingestion may decrease dynamic muscular strength, isometric grip strength, dynamometer strength, power and ergographic muscular output. Other studies reported no effect of alcohol upon muscular strength. Local muscular endurance was also unaffected by alcohol ingestion. Small doses of alcohol exerted no effect upon bicycle ergometer exercise tasks simulating a 100-m dash or a 1500-m run, but larger doses had a

[2] A. Juhlin-Dannfelt, G. Ahlborg, L. Hagenfeldt, L. Jorfeldt, and P. Felig, Influence of ethanol on splanchnic and skeletal muscle substrate turnover during prolonged exercise in man. Am. J. Physiol. 233: E195–E202, 1977.

deleterious effect. Other research has shown that alcohol has no significant effect upon physical performance capacity, exercise time at maximal levels, or exercise time to exhaustion.

Thus, alcohol ingestion will not improve muscular work capacity and may lead to decreased performance levels.

4. Alcohol is the most abused drug in the United States. There are an estimated 10 million adult problem drinkers and an additional 3.3 million in the 14–17 age range. Alcohol is significantly involved in all types of accidents—motor vehicle, home, industrial, and recreational. Most significantly, half of all traffic fatalities and one-third of all traffic injuries are alcohol related. Although alcohol abuse is associated with pathological conditions such as generalized skeletal myopathy, cardiomyopathy, pharyngeal and esophageal cancer, and brain damage, its most prominent effect is liver damage.

5. Because alcohol has not been shown to help improve physical performance capacity, but may lead to decreased ability in certain events, it is important for all those associated with the conduct of sports to educate athletes against its use in conjunction with athletic contests. Moreover, the other dangers inherent in alcohol abuse mandate that concomitantly we educate our youth to make intelligent choices regarding alcohol consumption. Anstie's rule, or limit,[3] may be used as a reasonable guidline to moderate, safe drinking for adults. In essence, no more than 0.5 ounces of pure alcohol per 23 kg body weight should be consumed in any one day. This would be the equivalent of three bottles of 4.5% beer, three 4-ounce glasses of 14% wine, or three ounces of 50% whiskey for a 68-kg person.

[3] Anstie, F.E. *On the Uses of Health and Disease*. London: Macmillan, 1877. p. 5–6.

AMERICAN COLLEGE OF SPORTS MEDICINE POSITION STATEMENT ON PROPER AND IMPROPER WEIGHT LOSS PROGRAMS

Millions of individuals are involved in weight reduction programs. With the number of undesirable weight loss programs available and a general misconception by many about weight loss, the need for guidelines for proper weight loss programs is apparent.

Based on the existing evidence concerning the effects of weight loss on health status, physiologic processes and body composition parameters, the American College of Sports Medicine makes the following statements and recommendations for weight loss programs.

For the purposes of this position statement, body weight will be represented by two components, fat and fat-free (water, electrolytes, minerals, glycogen stores, muscular tissue, bone, etc.):

1. Prolonged fasting and diet programs that severely restrict caloric intake are scientifically undesirable and can be medically dangerous.

2. Fasting and diet programs that severely restrict caloric intake result in the loss of large amounts of water, electrolytes, minerals, glycogen stores, and other fat-free tissue (including proteins within fat-free tissues), with minimal amounts of fat loss.

3. Mild calorie restriction (500–1000 kcal less than the usual daily intake) re-

sults in a smaller loss of water, electrolytes, minerals, and other fat-free tissue, and is less likely to cause malnutrition.

4. Dynamic exercise of large muscles helps to maintain fat-free tissue, including muscle mass and bone density, and results in losses of body weight. Weight loss resulting from an increase in energy expenditure is primarily in the form of fat weight.

5. A nutritionally sound diet resulting in mild calorie restriction coupled with an endurance exercise program along with behavioral modification of existing eating habits is recommended for weight reduction. The rate of sustained weight loss should not exceed 1 kg (2 lb) per week.

6. To maintain proper weight control and optimal body fat levels, a lifetime commitment to proper eating habits and regular physical activity is required.

RESEARCH BACKGROUND FOR THE POSITION STATEMENT

Each year millions of individuals undertake weight loss programs for a variety of reasons. It is well known that obesity is associated with a number of health-related problems. These problems include impairment of cardiac function due to an increase in the work of the heart and to left ventricular dysfunction; hypertension; diabetes; renal disease; gall bladder disease; respiratory dysfunction; joint diseases and gout; endometrial cancer; abnormal plasma lipid and lipoprotein concentrations; problems in the administration of anesthetics during surgery; and impairment of physical working capacity. As a result, weight reduction is frequently advised by physicians for medical reasons. In addition, there are a vast number of individuals who are on weight reduction programs for aesthetic reasons.

It is estimated that 60–70 million American adults and at least 10 million American teenagers are overfat. Because millions of Americans have adopted unsupervised weight loss programs, it is the opinion of the American College of Sports Medicine that guidelines are needed for safe and effective weight loss programs. This position statement deals with desirable and undesirable weight loss programs. Desirable weight loss programs are defined as those that are nutritionally sound and result in maximal losses in fat weight and minimal losses of fat-free tissue. Undesirable weight loss programs are defined as those that are not nutritionally sound, that result in large losses of fat-free tissue, that pose potential serious medical complications, and that cannot be followed for long-term weight maintenance.

Therefore, a desirable weight loss program is one that:

1. Provides a caloric intake not lower than 1200 kcal $\cdot$ d^{-1} for normal adults in order to get a proper blend of foods to meet nutritional requirements. (Note: this requirement may change for children, older individuals, athletes, etc.).

2. Includes foods acceptable to the dieter from the viewpoints of socio-cultural background, usual habits, taste, cost, and ease in acquisition and preparation.

3. Provides a negative caloric balance (not to exceed $500-1000$ kcal $\cdot d^{-1}$ lower than recommended), resulting in gradual weight loss without metabolic derangements. Maximal weight loss should be 1 kg $\cdot$wk?

4. Includes the use of behavior modification techniques to identify and eliminate dieting habits that contribute to improper nutrition.

5. Includes an endurance exercise program of at least 3 d/wk, 20–30 min in duration, at a minimum intensity of 60% of maximum heart rate (refer to ACSM Position Statement on the Recommended Quantity and Quality of Exercise for Developing and Maintaining Fitness in Healthy Adults, *Med. Sci. Sports* 10:vii, 1978).

6. Provides that the new eating and physical activity habits can be continued for life in order to maintain the achieved lower body weight.

1. Since the early work of Keys et al.[1] and Bloom[2] which indicated that marked reduction in caloric intake or fasting (starvation or semistarvation) rapidly reduced body weight, numerous fasting, modified fasting, and fad diet and weight loss programs have emerged. While these programs promise and generally cause rapid weight loss, they are associated with significant medical risks.

The medical risks associated with these types of diet and weight loss programs are numerous. Blood glucose concentrations have been shown to be markedly reduced in obese subjects who undergo fasting. Further, in obese non-diabetic subjects, fasting may result in impairment of glucose tolerance. Ketonuria begins within a few hours after fasting or low-carbohydrate diets are begun and hyperuricemia is common among subjects who fast to reduce body weight. Fasting also results in high serum uric acid levels with decreased urinary output. Fasting and low-calorie diets also result in urinary nitrogen loss and a significant decrease in fat-free tissue. In comparison to ingestion of a normal diet, fasting substantially elevates urinary excretion of potassium. This, coupled with the aforementioned nitrogen loss, suggests that the potassium loss is due to a loss of lean tissue. Other electrolytes, including sodium, calcium, magnesium, and phosphate have been shown to be elevated in urine during prolonged fasting. Reductions in blood volume and body fluids are also common with fasting and fad diets. This can be associated with weakness and fainting. Congestive heart failure and sudden death have been reported in subjects who fasted or markedly restricted their caloric intake. Myocardial atrophy appears to contribute to sudden death. Sudden death may also occur during

[1] A. Keys, J. Brozek, A. Henshel, O. Mickelson, and H.L. Taylor, The Biology of Human Starvation. Minneapolis: University of Minnesota Press, 1950.

[2] W.L. Bloom, Fasting as an introduction to the treatment of obesity. Metabolism 8: 214–220, 1959.

refeeding. Untreated fasting has also been reported to reduce serum iron binding capacity, resulting in anemia. Liver glycogen levels are depleted with fasting and liver function and gastrointestinal tract abnormalities are associated with fasting. While fasting and calorically restricted diets have been shown to lower serum cholesterol levels, a large portion of the cholesterol reduction is a result of lowered HDL-cholesterol levels. Other risks associated with fasting and low-calorie diets include lactic acidosis, alopecia, hypoalaninemia, edema, anuria, hypotension, elevated serum bilirubin, nausea and vomiting, alterations in thyroxine metabolism, impaired serum triglyceride removal and production, and death.

2. The major objective of any weight reduction program is to lose body fat while maintaining fat-free tissue. The vast majority of research reveals that starvation and low-calorie diets result in large losses of water, electrolytes, and other fat-free tissue. One of the best controlled experiments was conducted from 1944 to 1946 at the Laboratory of Physiological Hygiene at the University of Minnesota. In this study subjects had their base-line caloric intake cut by 45% and body weight and body composition changes were followed for 24 wk. During the first 12 wk of semistarvation, body weight declined by 25.4 lb (11.5 kg) with only an 11.6-lb (5.3-kg) decline in body fat. During the second 12-wk period, body weight declined an additional 9.1 lb (4.1 kg) with only a 6.1-lb (2.8 kg) decrease in body fat. These data clearly demonstrate that fat-free tissue significantly contributes to weight loss from semistarvation. Similar results have been reported by several other investigators. Buskirk et al.[3] reported that the 13.5-kg weight loss in six subjects on a low-calorie mixed diet averaged 76% fat and 24% fat-free tissue. Similarly, Passmore et al.[4] reported results of 78% of weight loss (15.3 kg) as fat and 22% as fat-free tissue in seven women who consumed a 400-kcal·d^{-1} diet for 45 d. Yang and Van Itallie[5] followed weight loss and body composition changes for the first 5 d of a weight loss program involving subjects consuming either an 800-kcal mixed diet, an 800-kcal ketogenic diet, or undergoing starvation. Subjects on the mixed diet lost 1.3 kg of weight (59% fat loss, 3.4% protein loss, 37.6% water loss), subjects on the ketogenic diet lost 2.3 kg of weight (33.2% fat, 3.8% protein, 63.0% water), and subjects on starvation regimens lost 3.8 kg of weight (32.3% fat, 6.5% protein, 61.2% water). Grande and Grande et al.[6] reported similar

[3] E.R. Buskirk, R.H. Thompson, L. Lutwak, and G.D. Whedon, Energy balance of obese patients during weight reduction: Influence of diet restriction and exercise. Ann NY Acad. Sci. 110: 918–940, 1963.

[4] R. Passmore, J.A. Strong, and F.J. Ritchie, The chemical composition of the tissue lost by obese patients on a reducing regimen. Br. J. Nutr. 12: 113–122, 1958.

[5] M. Yang and T.B. Van Itallie, Metabolic responses of obese subjects to starvation and low calorie ketogenic and nonketogenic diets. J. Clin. Invest. 58: 722–730, 1976.

[6] F. Grande, H.L. Taylor, J.T. Anderson, E. Buskirk, and A. Keys, Water exchange in men

findings with a 1000-kcal carbohydrate diet. It was further reported that water restriction combined with 1000-kcal $\cdot$ d^{-1} of carbohydrate resulted in greater water loss and less fat loss.

Recently, there has been some renewed speculation about the efficacy of the very-low-calorie diet (VLCD). Krotkiewski and associates[7] studied the effects on body weight and body composition after 3 wk on the so-called Cambridge diet. Two groups of obese middle-aged women were studied. One group had a VLCD only, while the second group had a VLCD combined with a 55-min/d, 3-d/wk exercise program. The VLCD-only group lost 6.2 kg in 3 wk, of which only 2.6 kg was fat loss, while the VLCD-plus-exercise group lost 6.8 kg in 3 wk with only a 1.9-kg body fat loss. Thus it can be seen that VLCD results in undesirable losses of body fat, and the addition of the normally protective effect of chronic exercise to VLCD does not reduce the catabolism of fat-free tissue. Further, with VLCD, a large reduction (29%) in HDL-cholesterol is seen.

3. Even mild calorie restriction (reduction of 500–1000 kcal $\cdot$ d^{-1} from baseline caloric intake), when used alone as a tool for weight loss, results in the loss of moderate amounts of water and other fat-free tissue. In a study by Goldman et al.,[8] 15 female subjects consumed a low-calorie mixed diet for 7–8 wk. Weight loss during this period averaged 6.43 kg (0.85 kg $\cdot$ wk^{-1}), 88.6% of which was fat. The remaining 11.4% represented water and other fat-free tissue. Zuti and Golding[9] examined the effect of 500 kcal $\cdot$ d^{-1} calorie restriction on body composition changes in adult females. Over a 16-wk period the women lost approximately 5.2 kg; however, 1.1 kg of the weight loss (21%) was due to a loss of water and other fat-free tissue. More recently, Weltman et al.[10] examined the effects of 500 kcal $\cdot$ d^{-1} calorie restriction (from base-line levels) on body composition changes in sedentary middleaged males. Over a 10-wk period subjects lost 5.95 kg, 4.03 kg (68%) of which was fat loss and 1.92 kg (32%) was loss of water and other fat-free tissue. Further, with calorie restriction only, these subjects exhibited a decrease in HDL-cholesterol. In the same study, the two other groups who exercised and/or dieted and exercised

on a restricted water intake and a low calorie carbohydrate diet accompanied by physical work. J. Appl. Physiol. 12: 202–210, 1958.

[7] M. Krotkiewski, L. Toss, P. Bjorntorp, and G. Holm, The effect of a very-low-calorie diet with and without chronic exercise on thryroid and sex hormones, plasma proteins, oxygen uptake, insulin, and c peptide concentrations in obese women. Int. J. Obes. 5: 287–293, 1981.

[8] R.F. Goldman, B. Bullen, and C. Seltzer, Changes in specific gravity and body fat in overweight female adolescents as a result of weight reduction. Ann. NY Acad. Sci. 110: 913–917, 1963.

[9] W.B. Zuti and L.A. Golding, Comparing diet and exercise as weight reduction tools. Phys. Sportsmed. 4(1): 49–53, 1976.

[10] A. Weltman, S. Matter, and B.A. Stamford, Caloric restriction and/or mild exercise: Effects on serum lipds and body composition. Am. J. Clin. Nutr. 33: 1002–1009, 1980.

were able to maintain their HDL-cholesterol levels. Similar results for females have been presented by Thompson et al.[11] It should be noted that the decrease seen in HDL-cholesterol with weight loss may be an acute effect. There are data that indicate that stable weight loss has a beneficial effect on HDL-cholesterol.

Further, an additional problem associated with calorie restriction alone for effective weight loss is the fact that it is associated with a reduction in basal metabolic rate. Apparently exercise combined with calorie restriction can counter this response.

4. There are several studies that indicate that exercise helps maintain fat-free tissue while promoting fat loss. Total body weight and fat weight are generally reduced with endurance training programs while fat-free weight remains constant or increases slightly. Programs conducted at least 3 d/wk, of at least 20-min duration and of sufficient intensity and duration to expend at least 300 kcal per exercise session have been suggested as a threshold level for total body weight and fat weight reduction. Increasing caloric expenditure above 300 kcal per exercise session and increasing the frequency of exercise sessions will enhance fat weight loss while sparing fat-free tissue. Leon et al.[12] had six obese male subjects walk vigorously for 90 min, 5 d/wk for 16 wk. Work output progressed weekly to an energy expenditure of 1000–1200 kcal/session. At the end of 16 wk, subjects averaged 5.7 kg of weight loss with a 5.9-kg loss of fat weight and a 0.2-kg gain in fat-free tissue. Similarly, Zuti and Golding followed the progress of adult women who expended 500 kcal/exercise session 5 d/wk for 16 wk of exercise. At the end of 16 wk the women lost 5.8 kg of fat and gained 0.9 kg of fat-free tissue.

5. Review of the literature cited above strongly indicates that optimal body composition changes occur with a combination of calorie restriction (while on a well-balanced diet) plus exercise. This combination promotes loss of fat weight while sparing fat-free tissue. Data of Zuti and Golding and Weltman et al. support this contention. Calorie restriction of 500 kcal$\cdot$d^{-1} combined with 3–5 d of exercise requiring 300–500 kcal per exercise session results in favorable changes in body composition. Therefore, the optimal rate of weight loss should be between 0.45-1 kg (1–2 lb) per wk. This seems especially relevant in light of the data which indicates that rapid weight loss due to low caloric intake can be associated with sudden death. In order to institute a desirable pattern of calorie restriction plus exercise, behavior modification techniques should be

[11] P.D. Thompson, R.W. Jeffrey, R.R. Wing, and P.D. Wood, Unexpected decrease in plasma high density lipoprotein cholesterol with weight loss. Am. J. Clin. Nutr. 32: 2016–2021, 1979.

[12] A.S. Leon, J. Conrad, D.M. Hunninghake, and R. Serfass. Effects of a vigorous walking program on body composition, and carbohydrate and lipid metabolism of obese young men. Am. J. Clin. Nutr. 32: 1776–1787, 1979.

incorporated to identify and eliminate habits contributing to obesity and/or overfatness.

6. The problem with losing weight is that, although many individuals succeed in doing so, they invariably put the weight on again. The goal of an effective weight loss regimen is not merely to lose weight. Weight control requires a lifelong commitment, an understanding of our eating habits and a willingness to change them. Frequent exercise is necessary, and accomplishment must be reinforced to sustain motivation. Crash dieting and other promised weight loss cures are ineffective.

SOURCE: Reprinted by permission of ACSM.

AMERICAN COLLEGE OF SPORTS MEDICINE OPINION STATEMENT ON THE PARTICIPATION OF THE FEMALE ATHLETE IN LONG-DISTANCE RUNNING

In the Olympic Games and other international contests, female athletes run distances ranging from 100 meters to 3,000 meters, whereas male athletes run distances ranging from 100 meters through 10,000 meters as well as the marathon (42.2 km). The limitation on distance for women's running events has been defended at times on the grounds that long-distance running may be harmful to the health of girls and women.

OPINION STATEMENT

It is the opinion of the American College of Sports Medicine that females should not be denied the opportunity to compete in long-distance running. There exists no conclusive scientific or medical evidence that long-distance running is contraindicated for the healthy, trained female athlete. The American College of Sports Medicine recommends that females be allowed to compete at the national and international level in the same distances in which their male counterparts compete.

SOURCE: Med. Sci. Sports 11(4): ix–xi, 1979. Reprinted by permission of ACSM.

SUPPORTIVE INFORMATION

Studies have shown that females respond in much the same manner as males to systematic exercise training. Cardiorespiratory function is improved as indicated by significant increases in maximal oxygen uptake. At maximal exercise, stroke volume and cardiac output are increased after training. At standardized submaximal exercise intensities after training, cardiac output remains unchanged, heart rate decreases, and stroke volume increases. Also, resting heart rate decreases after training. As is the case for males, relative body fat content is reduced consequent to systematic endurance training.

Long-distance running imposes a significant thermal stress on the participant. Some differences do exist between males and females with regard to thermoregulation during prolonged exercise. However, the differences in thermal stress response are more quantitative than qualitative in nature. For example, women experience lower evaporative heat losses than do men exposed to the same thermal stress and usually have higher skin temperatures and deep body temperatures upon onset of sweating. This may actually be an advantage in reducing body water loss so long as thermal equilibrium can be maintained. In view of current findings, it appears that the earlier studies which indicated that women were less tolerant to exercise in the heat than men were misleading because they failed to consider the women's relatively low level of cardiorespiratory fitness and heat acclimatization. Apparently, cardiorespiratory fitness as measured by maximum oxygen uptake is a most important functional capacity as regards a person's ability to respond adequately to thermal stress. In fact, there has been considerable interest in the seeming cross-adaptation of a life style characterized by physical activity involving regular and prolonged periods of exercise hyperthermia and response to high environmental temperatures. Women trained in long-distance running have been reported to be more tolerant of heat stress than nonathletic women matched for age and body surface area. Thus, it appears that trained female long-distance runners have the capacity to deal with the thermal stress of prolonged exercise as well as the moderate-to-high environmental temperatures and relative humidities that often accompany these events.

The participation of males and females in road races of various distances has increased tremendously during the last decade. This type of competition attracts the entire spectrum of runners with respect to ability—from the elite to the novice. A common feature of virtually all of these races is that a small number of participants develop medical problems (primarily heat injuries) which frequently require hospitalization. One of the first documentations of the medical problems associated with mass participation in this form of athletic competition was by Sutton and co-workers.[1] Twenty-nine of 2,005 entrants in the

[1] J. Sutton, M.J. Coleman, A.P. Millar, L. Lazarus, and P. Russo, The medical problems of mass participation in athletic competition. Med. J. Aust. 2: 127–133, 1972.

1971 Sydney City-to-Surf race collapsed; seven required hospitalization. All of the entrants who collapsed were males, although 4% of the race entrants were females. By 1978 the number of entrants increased approximately 10 fold with females accounting for approximately 30% of the entrants. In the 1978 race only nine entrants were treated for heat injury and again all were males. In a 1978 Canadian road race, in which 1,250 people participated, 15 entrants developed heat injuries—three females and 12 males, representing 1.3% and 1.2% of the total number of female and male entrants, respectively. Thus, females seem to tolerate the physiological stress of road race competition at least as well as males.

Because long-distance running competition sometimes occurs at moderate altitudes, the female's response to an environment where the partial pressure of oxygen is reduced (hypoxia) should be considered. Buskirk[2] noted that, although there is little information about the physiological responses of women to altitude, the proportional reduction in performance at Mexico City during the Pan American and Olympic Games was the same for males and females. Drinkwater et al.[3] found that women mountaineers exposed to hypoxia demonstrated a similar decrement in maximal oxygen uptake as that predicted for men. Hannon et al.[4] have found that females tolerate the effects of altitude better than males because there appears to be both a lower frequency and shorter duration of mountain sickness in women. Furthermore, at altitude women experience less alteration in such variables as resting heart rate, body weight, blood volume, electrocardiograms, and blood chemistries than men. Although one study has reported that women and men experience approximately the same respiratory changes with altitude exposure, another reports that women hyperventilate more than men, thereby increasing the partial pressure of arterial oxygen and decreasing the partial pressure of arterial carbon dioxide. Thus, females tolerate the stress of altitude at least as well as men.

Long-distance running is occasionally associated with various overuse syndromes such as stress fracture, chondromalacia, shinsplints, and tendonitis. Pollock et al.[5] have shown that the incidence of these injuries for males en-

[2] E.R. Buskirk, Work and fatigue in high altitude. In: Simonsen, E. (ed.), Physiology of Work Capacity and Fatigue. Springfield, Ill.: Charles C Thomas. p. 312–324, 1971.

[3] B.L. Drinkwater, L.J. Folinsbee, J.F. Bedi, S.A. Plowman, A.B. Loucks, and S.M. Horvath, Response of women mountaineers to maximal exercise during hypoxia. Avia. Space Environ. Med. 50: 657–662, 1979.

[4] J.P. Hannon, J.L. Shields, and C.W. Harris, A comparative review of certain responses of men and women to high altitude. In: Helfferich, C. (ed.), Proceedings symposia on arctic biology and medicine. VI. The physiology of work in cold and altitude. Fort Wainwright, Alaska: Arctic Aeromedical Laboratory. p. 113–245, 1966; and J.P. Hannon, J.L. Shields, and C.W. Harris, Effects of altitude acclimatization on blood composition of women. J. Appl. Physiol. 26: 540–547, 1969.

[5] M.L. Pollock, The quantification of endurance training programs. In: Wilmore, J.H. (ed.), Exercise and Sport Sciences Reviews. New York: Academic Press, Vol. 1. p. 155–188, 1973.

gaged in a program of jogging was as high as 54% and was related to the frequency, duration, and intensity of the exercise training. Franklin et al.[6] recently reported the injury incidence of 42 sedentary females exposed to a 12-week jogging program. The injury rate for the females appeared to be comparable to that found for males in other studies although, as the investigators indicated, a decisive interpretation of presently available information may be premature because of the limited orthopedic injury data available for women. It has been suggested that the anatomical differences between men's and women's pelvic width and joint laxity may lead to a higher incidence of injuries for women who run. There are no data available, however, to support this suggestion. Whether or not the higher intensity training programs of competitive male and female long-distance runners result in a difference in injury rate between the sexes is not known at this time. It is believed, however, that the incidence of injury due to running surfaces encountered, biomechanics of the back, leg and foot, and to foot apparel.

Of particular concern to female competitors and to the American College of Sports Medicine is evidence which indicates that approximately one-third of the competitive female long-distance runners between the ages of 12 and 45 experience amenorrhea or oligomenorrhea for at least brief periods. This phenomenon appears more frequently in those women with late onset of menarche, who have not experienced pregnancy, or who have taken contraceptive hormones. This same phenomenon also occurs in some competing gymnasts, swimmers, and professional ballerinas as well as sedentary individuals who have experienced some instances of undue stress or severe psychological trauma. Apparently, amenorrhea and oligomenorrhea may be caused by many factors characterized by loss of body weight. Running long distances may lead to decreased serum levels of pituitary gonadotrophic hormones in some women and may directly or indirectly lead to amenorrhea or oligomenorrhea. The role of running and the pathogenesis of these menstrual irregularities remains unknown.

The long-term effects of these types of menstrual irregularities for young girls that have undergone strenuous exercise training are unknown at this time. Eriksson and co-workers have reported, however, that a group of 28 young girl swimmers, who underwent strenuous swim training for 2.5 years, were normal in all respects (e.g., childbearing) 10 years after discontinuing training.

In summary, a review of the literature demonstrates that males and females adapt to exercise training in a similar manner. Female distance runners are characterized by having large maximal oxygen uptakes and low relative body fat content. The challenges of the heat stress of long-distance running or the lower partial pressure of oxygen at altitude seem to be well tolerated by fe-

[6]B.A. Franklin, L. Lussier, and E.R. Buskirk, Injury rates in women joggers. Phys. Sports Med. 7: 105–112, 1979 (Mar.).

males. The limited data available suggest that females, compared to males, have about the same incidence of orthopedic injuries consequent to endurance training. Disruption of the menstrual cycle is a common problem for female athletes. While it is important to recognize this problem and discover its etiology, no evidence exists to indicate that this is harmful to the female reproductive system.

PROPOSED AMERICAN COLLEGE OF SPORTS MEDICINE POSITION STATEMENT ON PREVENTION OF ILLNESS DURING DISTANCE RUNNING[a]

Purpose of the Position Statement:

1. To alert sponsors of distance running events to the potentially serious health hazards during distance running—especially heat illness.
2. To have sponsors identify the environmental heat stress on the race day and communicate this to competitors.
3. To educate competitors regarding heat illness susceptibility.
4. To inform sponsors of preventive actions which may reduce the frequency and severity of this type of illness.

This position statement replaces that of "Prevention of Heat Injury during Distance Running," published by the American College of Sports Medicine in 1975. It is aimed at the general community of joggers and fun runners as well as the elite athlete and the sponsors of such events. Although heat injury is still the most commonly encountered serious problem in North American fun runs

[a]Note: This position statement is now under consideration by the ACSM as an updated replacement for the 1975 position statement on the prevention of heat injuries during distance running (Appendix V). Reprinted by permission of ACSM.

and races, hypothermia also occurs. This is especially so for slow runners in long races such as the marathon, when conducted in cold and/or wet conditions.

As the physiology of each participant varies, strict compliance with the recommendations, while helpful, does not guarantee complete protection from heat illness in susceptible individuals.

POSITION STATEMENT

Based on research findings, it is the position of the American College of Sports Medicine that the following recommendations be employed when conducting distance races or community fun runs. It is ideal to have a medical director who is responsible for the coordination of the preventive and therapeutic aspects related to the fun run and who works closely with the race director.

1. The Race Organization

(a) Races should be organized to avoid the hottest summer months and the hottest part of the day. Organizers should be very cautious of unseasonable hot days in the early spring, particularly in North America, as entrants will almost certainly not be heat acclimatized.

(b) The environmental heat stress on the day should be known and is best measured as WBGT[1]. If WBGT is above 28°C (82°F), the race should be cancelled. If below 28°C, the degrees of heat stress should be conveyed to competitors by the use of color-coded flags at the start of the race to alert them, as follows:

   (i) *A red flag*—high risk: when WBGT is 23–28°C (73–82°F). This signal would indicate that all runners should be aware that heat injury is possible and any person particularly sensitive to heat or humidity should probably not run.

   (ii) *An amber flag*—moderate risk: when WBGT is 18–23°C (65–73°F). It should be remembered that the air temperature and probably also the humidity, and almost certainly the radiant heat at the beginning of the race, will increase during the course of the race if it is conducted in the morning or early afternoon.

   (iii) *A green flag*—low risk: when WBGT is below 18°C (65°F). This in no way guarantees that heat illness will not occur, but indicates only that the risk is low.

(c) All summer events should be scheduled for the early morning, ideally be-

[1] WBGT: Wet Bulb Globe Temperature. See the end of this appendix.

fore 9:00 A.M., or in the evening after 6:00 P.M., to minimize solar radiation.

(d) An adequate supply of fluid should be available before the race and every 2–3 km during the race.

(e) Hoses should be available to cool competitors during the race.

(f) Race officials should be educated as to the warning signs of an impending collapse. They should wear an identifiable arm band or badge and be empowered to stop runners who appear to be in difficulty.

(g) Adequate traffic control is necessary.

(h) There should be a ready source of communications from various points on the course to a central organizing point, to meet emergencies.

2. Medical Support

(a) *Medical organization*

Race organizers should alert local hospitals and ambulance services to the event and should take prior arrangements with medical personnel for the care of casualties, especially those suffering from heat injury. The mere fact that an entrant signs a waiver in no way absolves the organizers of the moral and/or legal responsibility.

(b) *Medical facilities*

(i) These should be available at the race site.

(ii) Staffed with personnel capable of instituting immediate and full-scale resuscitation measures. Apart from the routine resuscitation equipment, ice packs and fans for cooling are required.

(iii) Persons trained in first aid should be stationed along the course with the right to stop runners who exhibit signs of impending heat stroke or other abnormalities.

(iv) One or more ambulances or vans with accompanying medical personnel should follow the competitors at intervals.

(v) Although the emphasis has been on the management of hyperthermia, on cold, wet, and windy days athletes will be cold and require "space blankets," blankets, and warm drinks to prevent hypothermia. Especially vulnerable are the slower athletes who, when lightly clad, will lose heat faster than their rate of metabolic heat production.

3. Competitor Education

The education of fun runners has greatly increased in recent years, due largely to the lay person runners' magazines. Distributing sample runners' guidelines at the time of registration, if pre-registration occurs, and also holding clinics before runs are valuable.

The following persons are particularly prone to heat illness: the obese, unfit, dehydrated, those unacclimatized to the heat, the very young and the old, those with a previous history of heat stroke, and anyone who runs while ill. Based on the above information, all competitors should be advised of the following.

(a) Adequate training and fitness are important for full enjoyment of the run and also to prevent heat stroke.

(b) Prior training in the heat will produce heat acclimatization and also reduce the risk of heat injury. It is wise to do as much training as possible at the time of day at which the race will be held.

(c) Fluid consumption before and during the race will also reduce the risk of heat injury, particularly in the longer runs such as the marathon.

(d) Splashing with water or running under available hoses during a race will make runners more comfortable, but is unlikely to reduce the risk of heat injury.

(e) Illness prior to or at the time of the event should preclude competition. This applies to any febrile illness or gastroenteritis.

(f) Competitors should be advised of the early symptoms of heat injury which include excessive sweating, headache, nausea, dizziness, and any gradual impairment of consciousness.

(g) Competitors should be advised to choose a comfortable speed and not to run faster than they have when training.

(h) Competitors are advised to run with a partner, each being responsible for the other's well-being.

BACKGROUND FOR POSITION STATEMENT

There has been an exponential rise in the number of fun runs and races in recent years and, as would be expected, a similar increase in the number of running-related injuries. Minor injuries such as bruises, blisters, and muscular-skeletal injuries are most common. Myocardial infarction or cardiac arrest is, fortunately, very rare and almost always occurs in patients with symptomatic heart disease. Hypoglycemia may occasionally be seen in normal runners and has been observed following the marathon and also in shorter fun runs.

The most serious injuries in fun runs and races are related to problems of thermoregulation. In the shorter races, 6 km (10 miles) or less, hyperthermia with the attendant problems of heat exhaustion, heat syncope and heat stroke dominate, even on relatively cool days. In longer races, heat problems are common on hot days, but on moderate to cold days, hypothermia is more common in the fun runner.

Thermoregulation and Hyperthemia

It is surprising that fun runners may experience both hyperthermia and hypothermia, sometimes in the same race. While the adequately clothed runner is capable of withstanding great environmental temperature extremes, the scantily clad runner is totally at the mercy of the environment. This can be a problem in warm and hot weather when the rate of heat production is too great, but also in cold weather when heat production is too small. Body core temperature increases with exercise and is the dynamic balance between those processes producing heat and those responsible for heat loss. This is embodied in the equation of Winslow et al.[2]:

$$S = M \pm R \pm C_d \pm C_v \rightarrow E$$

where

S = amount of stored heat
M = metabolic heat production
R = heat gained or lost by radiation
C_d = conductive heat gained or lost
C_v = convective heat gained or lost
E = evaporative heat loss

The most important process for losing heat during exercise is the evaporation of sweat. When humidity rises, this avenue of heat loss is reduced.

Numerous studies have reported rectal temperatures above 40.6°C after races and fun runs and even as high as 42–43°C has been seen in fun run participants who have collapsed. In long races, sweat loss can be significant and result in a total body water deficit of 6–10% of body weight. Such dehydration will subsequently reduce sweating and predispose the runner to hyperthermia, heat stroke, heat exhaustion and muscle cramps.

Fluid ingestion before and during prolonged running will minimize dehydration (and control the rate of increase in rectal temperature). However, in the short fun runs of less than 10 km, hyperthermia may occur in the absence of significant dehydration. Runners should avoid consuming large quantities of sugar solution during runs as this may result in gastric discomfort and a decrease in gastric emptying.

Thermoregulation and Hypothermia

The previous equation of Winslow et al. demonstrates that heat can be lost readily from the body when the rate of heat production is outstripped by that of heat loss. Even on moderately cool days, if a runner is slow and/or if weather

[2] C.E.A. Winslow, L.P. Herrington, and A.P. Gagge, Physiological reactions of the human body to various atmospheric humidities. Am. J. Physiol. 120:288–299, 1937.

conditions become cooler en route, hypothermia may ensue. Several deaths have been reported from hypothermia during fun runs in mountain environments. When cold and wet, runners become inefficient at low speeds and rectal temperature will fall. As the pace slows and heat production drops, body core temperature will plummet. Hypothermia is now common in inexperienced marathon runners who frequently run the second half of the race much more slowly than the first half. Such runners may be able to maintain core temperature initially, but with the slow pace of the second half on cool or wet days, hypothermia will follow.

MEASUREMENT OF ENVIRONMENTAL HEAT STRESS

Ambient temperature is only one component of environmental heat stress; others are humidity, air movement, and radiant heat. Therefore, measurement of ambient temperature, dry bulb alone, is inadequate. The most useful and widely applied approach is Wet Bulb Globe Temperature (WBGT).

$$WBGT = 0.7 \times T_{wb} + 0.2 T_g + 0.1 T_{db}$$

where T_{wb} = temperature, wet bulb thermometer
T_g = temperature, black globe thermometer
T_{db} = temperature, dry bulb thermometer

The importance of wet bulb temperature can be readily appreciated as it accounts for 70% of the index, whereas dry bulb temperature accounts for only 10%. A simple portable heat stress monitor which gives direct WBGT in degrees Celsius or degrees Fahrenheit to monitor conditions during fun runs has proven useful.

Simple instruments available to measure WBGT include the "Heat Stress Monitor" from Reuter-Stokes, Cambridge, Ontario and "Botsball" from Howard Engineering Co., Bethlehem, Pa.

ROAD RACE CHECKLIST

Medical Personnel

1. Have aid personnel available if race is 6.2 miles (10 km) or longer, run in warm weather, or exceeds 500 participants.
2. Recruit back-up from existing emergency medical services (police, fire rescue, emergency medical service).
3. Notify local hospitals of time and place of road race.

Aid Stations

1. Arrange major aid station at finish point cordoned off from public access.

2. Equip major aid station with the following supplies:

Tent

Cots

Bath towels

Water in large containers

Ice in bag or ice chest

Hose with spray nozzle

Tables for medical supplies and equipment

Stethoscopes

Blood pressure cuffs

Rectal thermometers or meters (range up to 43°C–110°F)

Dressings

Ace bandages

Splints

Skin disinfectants

Intravenous fluids (supervision by physician required)

3. Position aid stations along route at 4 km (2.5 mile) intervals for races over 10 km at half way point for shorter races.

4. Stock each aid station with enough fluid (cool water is the optimum) for each runner to have 10–12 ounces (300–360 ml), at each 2.5 mile aid station. A margin of 25% extra cups should be available to account for spillage and double usage.

Communications/Surveillance

1. Set up communication between medical personnel and major aid station.

2. Arrange for radio equipped car or van to follow race course and provide radio contact with director.

Instructions to Runners

1. Appraise race participants of potential medical problems in advance of race so precautions may be followed.

2. Advise race director to announce the following information by loudspeaker immediately before the race:

Current temperature

Humidity

Predicted maximum (or minimum) temperature level

Location of aid stations and type of fluid available

Reinforcement of warm weather or cold weather self-care

3. Advise race participants to print their name, address, and any medical problems on back of registration number.

MEDICAL STATIONS

General Guidelines
Staff for large races

1. Physician, podiatrist, nurse, or EMT—team of 3 per 1,000 runners. Double or treble this number at the finish area.

2. Ambulance—1 per 3,000 runners at finish area; one cruising vehicle.

3. One (1) physician to act as triage officer at finish.

Water Estimate 1 quart or about 1 liter (0.26 gallon) per runner per 10 miles (16 km) (or roughly 60–90 minutes running time, and depending on number of stations)

For 10 kilometers, the above rule is still recommended.

Cups: = Number of entrants × number of stations + 25% additional per station.

        2 × number of entrants extra at finish area.

        Double total if course out-and-back.

TABLE 1
Equipment Needed at Aid Stations (per 1,000 runners)

Number	Item
	Ice in small plastic bags
5	Stretchers (10 at 7.5 miles and beyond)
5	Blankets (10 at 7.5 miles and beyond)
6 each	6 in. and 4 in. Ace wraps
½ case	4 in. by 4 in. gauze pads
½ case	1½ in. tape
½ case	Surgical soap
	Small instrument kits
	Bandaids
	Moleskin
½ case	Petroleum jelly
2 each	Inflatable arm and leg splints
	Athletic trainer's kit

TABLE 2
Equipment Needed at Field Hospital (per 1,000 runners)

Number	Item
10	Stretchers
4	Sawhorses
10	Blankets
10	Intravenous setups
2 each	Inflatable arm and leg splints
2 cases	Tape (1½ in.)
2 cases each	Ace wraps (6 in., 4 in., and 2 in.)
2 cases	Sheet wadding Underwrap
2 cases	4 in. by 4 in. gauze pads Bandaids Moleskin
½ case	Surgical soap
2	Oxygen tanks with regulators and masks
2	ECG monitors with defibrillators
	Ice in small plastic bags
	Small instrument kit

Adapted from B.H. Noble, and D. Bachman, Medical aspects of distance race planning. Phys. Sports Med. 7:78–84, June 1979.

INDEX